PLUTARCH'S SCIENCE OF NATURAL PROBLEMS
A STUDY WITH COMMENTARY ON *QUAESTIONES NATURALES*

PLUTARCHEA HYPOMNEMATA

Editorial Board

Jan Opsomer (KU Leuven)
Geert Roskam (KU Leuven)
Frances Titchener (Utah State University, Logan)
Luc Van der Stockt (KU Leuven)

Advisory Board

F. Alesse (ILIESI-CNR, Roma)
M. Beck (University of South Carolina, Columbia)
J. Beneker (University of Wisconsin, Madison)
H.G. Ingenkamp (Universität Bonn)
A.G. Nikolaidis (University of Crete, Rethymno)
Chr. Pelling (Christ Church, Oxford)
A. Pérez Jiménez (Universidad de Málaga)
Th. Schmidt (Université de Fribourg)
P.A. Stadter (University of North Carolina, Chapel Hill)

PLUTARCH'S SCIENCE OF NATURAL PROBLEMS

A STUDY WITH COMMENTARY ON *QUAESTIONES NATURALES*

By

MICHIEL MEEUSEN

Leuven University Press

Published with support of the Universitaire Stichting van België

© 2016 by Leuven University Press / Presses Universitaires de Louvain / Universitaire Pers Leuven
Minderbroedersstraat 4, B-3000 Leuven (Belgium)

All rights reserved. Except in those cases expressly determined by law, no part of this publication may be multiplied, saved in an automated datafile or made public in any way whatsoever without the express prior written consent of the publishers.

ISBN 978 94 6270 084 0
D/2016/1869/38
NUR 735-635

Cover design: Joke Klaassen

ἐγὼ νέος ὢν θαυμαστῶς ὡς ἐπεθύμησα ταύτης τῆς σοφίας ἣν δὴ καλοῦσι περὶ φύσεως ἱστορίαν· ὑπερήφανος γάρ μοι ἐδόκει εἶναι, εἰδέναι τὰς αἰτίας ἑκάστου, διὰ τί γίγνεται ἕκαστον καὶ διὰ τί ἀπόλλυται καὶ διὰ τί ἔστι.

When I was young, I was tremendously eager for the kind of wisdom which they call investigation of nature. I thought it was a glorious thing to know the causes of everything, why each thing comes into being and why it perishes and why it exists.

Pl., *Phd.* 96a

Contents

Acknowledgements   11

Prologue

Plutarch and the history of science: the case of *Quaestiones naturales*   15
1. Plato, Plutarch and scientific infancy   18
2. Date and chronology of *Quaestiones naturales*: a 'life's work'?   24
3. The value of Plutarch's natural problems   30
4. Classical philology and the petrification of science   46
5. *Status quaestionis*   51
6. Note on translations and abbreviations   53

Introduction

1. Problems, problems, problems (and Aristotelian precedents)   61
   1.1. *Quaestiones naturales* and the Aristotelian genre and tradition of natural problems   61
      1. Preliminary remarks on Plutarch's *Naturwissenschaft*   61
      2. *Quaestiones naturales*: the work of a *Plutarchus Aristotelicus*?   67
      3. The genre of problems and the Aristotelian tradition of natural problems   75
      4. Internal organisation of Plutarch's natural problems (microstructure)   84
      5. Coherent reading in *Quaestiones naturales* and *convivales* (macrostructure)   92
      6. The title and its programmatic value   102
   1.2. Problems related to Plutarch's scientific discourse   110
      1. Trifles unworthy of Plutarch? Some remarks on authenticity   110
      2. The rhetoric of scientific discourse according to Plutarch   112
      3. The problem of style   117
      4. The problem of morality   119

5. A 'generic' solution	127
6. Conclusion and new questions	129

2. The position of *Quaestiones naturales* in the *corpus Plutarcheum* — 131
   2.1. Scientific traits in the *corpus Plutarcheum* — 131
     1. Intellectual and literary interest of natural phenomena — 132
     2. Cluster analysis in *Quaestiones naturales* — 138
     3. Scientific digressions in the *Vitae* — 141
     4. Indirect references to *Quaestiones naturales* — 148
   2.2. A comparative study of *Quaestiones naturales* and *Quaestiones convivales* — 150
     1. The level of *elocutio* — 151
     2. The level of *dispositio* — 156
     3. The level of *inventio* — 159
   2.3. Hypomnematic text genetics of *Quaestiones naturales* and *Quaestiones convivales* — 161
     1. Historicity and fiction in *Quaestiones convivales* — 162
     2. Problems and personal notes — 165
     3. Zetetic autonomy in *Quaestiones naturales* — 173
   2.4. Opening up Plutarch's zetetic archive — 177
     1. The issue of publication: problems as functional literature — 177
     2. Classification and overlap — 182
     3. Conclusion and new questions — 184

3. *Quaestiones naturales* and zetetic παιδεία — 187
   3.1. *Sitz im Leben*: readership and educational context — 187
     1. Natural problems and philosophical σχολή — 188
     2. Plutarch's academy — 190
     3. Digestive discussions and problematic promenades — 195
     4. *Quaestiones naturales* as school text: technicality and complexity — 204
     5. The dialogue between author and reader: vivacity and historicity — 212
   3.2. *Quaestiones naturales* as a preamble to metaphysics — 219
     1. Natural problems as a means of exercising the mind — 219
     2. Natural problems as a means of easing the mind — 225
     3. Conclusion and new questions — 232

4. Plutarch's Platonic world view: the aetiological design of *Quaestiones naturales* and its scientific context — 235
   4.1. Science and its foes? The ancient scientific value of

	Quaestiones naturales	235
4.1.1.	Saving popular beliefs: the wonders and paradoxes of nature	235
	1. Natural problems and the fabric of strangeness	237
	2. Democritus and the cucumber	244
	3. Plutarch's popular beliefs: anti-Aristotelian and anti-Stoic dynamics	248
4.1.2.	Plutarch's dualistic causality: rationalising the divine and the use of myth and poetry	258
	1. Plato's scientific revolution	258
	2. Science, religion and mythology	264
	3. Science and poetry	274
4.2.	Constructing scientific authority: between continuity, ingenuity and innovation	278
4.2.1.	Character and use of the scientific tradition	279
	1. Quotations from scientific prose authors	279
	2. Problematisation of scientific knowledge	288
4.2.2.	Scientific innovation and performance	291
	1. A note on the sociology of knowledge and παιδεία	291
	2. The pragmatics of Plutarch's scientific ingenuity and creativity	293
4.3.	Plutarch's scientific methodology: a rough guide to explaining natural phenomena	299
4.3.1.	Material principles and natural processes	299
	1. Material principles	301
	2. Natural processes	306
4.3.2.	Towards the limits of natural science	312
	1. A 'sceptical' Plutarch: ἐμπειρία, ἐποχή and εὐλάβεια	313
	2. Truth and probability in *Quaestiones naturales*	321
	3. Sense perception and the issue of autopsy in *Quaestiones naturales*	328
4.3.3.	Logical-rhetorical dynamics	339
	1. Contradiction, non-contradiction and aetiological freedom	340
	2. Aetiological comprehensiveness and pluricausality	345
	3. Aetiological subtlety and sophistication	349
4.3.4.	Uniformity and technicality of the scientific terminology	354
	1. Let's talk science: the birth and use of technical vocabulary	355

    2. Big words? High-tech vs. low-tech vocabulary   359
    3. Conclusion: Plutarch, Plato and Aristotle
       (again)   362

Commentary

0. Approach and structure   367
1. Salt and water (*Q.N.* 1–13)   368
2. Wheat and barley (*Q.N.* 14–16)   420
3. Sea animals and fishing (*Q.N.* 17–19)   425
4. Land animals and hunting (*Q.N.* 20–28)   437
5. Viniculture (*Q.N.* 30–31)   464
6. Longolius (*Q.N.* 32–39)   469
7. Psellus (*Q.N.* 40–41)   487

Synopsis   492

Bibliography   495

Index Locorum   531

Acknowledgements

The volume at hand is a revised version of my PhD dissertation completed under the supervision of Luc Van der Stockt (KU Leuven, 2013). I cannot thank him enough for the intellectual support during my years as a doctoral student and after, and for the numerous opportunities he made possible for me in Academia.

It is a pleasure to extend thanks to Geert Roskam, co-supervisor of the project in Leuven, who was and continues to be always happy to offer help and good advice whenever asked. I am much indebted also to Katerina Oikonomopoulou, Jan Opsomer, Françoise Frazier and Toon Van Hal, all of whom read an earlier draft and offered many pertinent suggestions. Special thanks are due to Sean Winkler for his corrections of the English. Any remaining inaccuracies are my own.

I would like to thank my KU Leuven colleagues, both from the Literary Studies Department and from the Institute of Classical Studies. I also gratefully acknowledge the financial support of the KU Leuven Research Council, the Research Foundation – Flanders (FWO) and the University Foundation for making this publication possible.

Finally, my heartfelt gratitude goes to my parents, my brothers, and partner for their love and caring, unsayable in words.

PROLOGUE

Plutarch and the history of science: the case of *Quaestiones naturales*

Plutarch was a man of many talents as his life and writings show. Between being a political delegate and representative of his small hometown of Chaeronea, a philosophy tutor specialising in the teachings of Plato, and in his final years a priest worshipping Apollo at the oracle of Delphi, he was a full time intellectual and a true paragon of ancient learning, who found a great joy in collecting and critically transmitting many different forms of knowledge that personally caught his attention (history, literature, philosophy, science etc.). Among many other branches of ancient learning – the code word here is πολυμάθεια –, the Chaeronean was very interested in the natural world around him in terms mainly of its underlying material principles, physical processes and its providential ordering. If Plutarch's so-called natural scientific writings can teach us one thing about his perception of physical reality, it is the fact that they are based on a very different outlook on the world than is generally promoted by scientists today. Plutarch lived in the same physical world as we do, but he saw it in a very different way and from a very different perspective. In line with his Platonic philosophy and the corresponding division between the sensible and intelligible realms in the cosmos, he ascribed a divine providence to the world, which can partly explain his interest in more fanciful beliefs regarding nature and natural phenomena, as this study will show.

In his dual role as a *homo philosophicus* and a *homo religiosus*, Plutarch did not draw a clear distinction between, what people today would call, natural science, on the one hand, and religion and mythology, on the other – that is, the traditionally ill-conceived distinction between 'reason' and 'myth'. In fact, the opposite is true, as is clear, for instance, from his *De facie*. In this work, Plutarch concludes an astrophysical dialogue about the substance and nature of the moon with a mythological account of the moon's purpose in the universe, explaining its importance for the life-cycle of human souls. This dualistic approach is not at all new to contemporary Plutarch scholars, but the claim that the same approach is also subtly present in Plutarch's discussions of more particular scientific topics as treated in *Quaestiones naturales* – a collection of 41[1] natural

---

[1] This number (41) includes the additional problems from Gybertus Longolius' 1542 Latin translation (*Q.N.* 32–39) and from Michael Psellus' *De omnifaria doctrina* (*Q.N.* 40–41 = §§ 170 and 188 Westerink). There is some controversy about the authenticity of

problems modelled after the Ps.-Aristotelian *Natural problems* (from here on simply *Problems*) – has not yet been made, or has even been doubted. One of the goals of this study will be to resolve this issue, and to show that Plutarch's natural problems form an integral part of his wider natural philosophical project, fully consistent with the method and conceptual framework of his other scientific writings, albeit perhaps in a less obvious manner.

Plutarch discusses natural problems throughout his entire oeuvre. In *Quaestiones convivales* he cross-fertilises the natural problem format with the literary genre of the *symposium*, and in the *Vitae* he sporadically incorporates natural scientific digressions (παρεκβάσεις) in his biographical narratives [see 2.1.3.]. Plutarch uses the problem format in its traditional form only in *Quaestiones naturales*, where he treats such problems in an autonomous fashion. By straddling a large variety of questions (and answers) related to ancient Greek zoology, botany, meteorology and their respective subdisciplines, the collection is firmly rooted in ancient Greek physical theory[2], especially as conceptualised by the Peripatetics [see 1.1.3.]. A few examples of particular – and at times very peculiar – problems Plutarch tries to solve are: 'Why does seawater not provide nourishment to trees?' (*Q.N.* 1), 'Why do the tears of boars taste sweet, while those of deer taste salty and ordinary?' (*Q.N.* 20), 'Why does a vine wilt if it is sprinkled with wine, and especially with wine made from its own grapes?' (*Q.N.* 31), 'Why is water that is drawn from wells less nutritious than water that flows from a spring or falls from the sky?' (*Q.N.* 33), 'Why are bees quicker to sting people who have just committed adultery?' (*Q.N.* 36).

Due to Plutarch's primary focus on the natural causes of such phenomena and not *also* on their higher, divine motivation, scholars have argued that the place of *Quaestiones naturales* among Plutarch's other natural philosophical writings is puzzling. However, as we will see, these scholars have often neglected the deeper philosophical-religious motivations and mythological references that discreetly accompany the collection's scientific discourse [4.1.2.2.]. The few attempts that have been made to evaluate the work's scientific character – mainly in terms of its physical-aetiological approach and referential and impersonal style – were mostly

---

the two chapters in Psellus' text, which may contain the remains of two lost *Quaestiones naturales*. The least that can be said is that there is a Plutarchan core to these two chapters. The authenticity of Longolius' additional chapters is beyond debate. See *ad loc.* in the commentary for further detail and literature.

[2] Cf. K. Ziegler, 1951, col. 857: "'physische', d.h. nach unserer heutigen Ausdrucksweise zumeist physiologisch-biologisch-medizinische Fragen". Cf. also R. Flacelière, J. Irigoin, J. Sirinelli and A. Philippon, 1987, p. lxxxii: "On voit qu'il ne s'agit pas là uniquement de "physique" proprement dite, mais aussi de biologie et de plusieurs autres matières."

biased by modern hindsight, at the risk of neglecting the broader scientific and socio-cultural context from which the text originates. In other cases that emphasise the collection's strange and exotic character, scholars have tried to cover up their interpretative misguidance by conveniently, though often silently, siding with that lovely profundity – or scholarly fig leaf rather? – that the past is a 'foreign country', that is, a country with natural laws and scientific conventions of its own, impenetrable to modern understanding. As a result, a proper attempt to make the collection more comprehensible for the modern reader was left to others, and rightly so.

Owing to a growing awareness of the particular and, in many cases, very different intellectual-philosophical and socio-cultural background of ancient scientific texts, there are several methodological tools available now for interpreting the scientific set-up of this type of literature in conjunction with its historical context. This endeavour forms one of the basic objectives for the study at hand. Indeed, only since relatively recently scholars have started to take a more positive stance towards Plutarch's natural science, but even so *Quaestiones naturales* has continued to lag behind. This general reappraisal can be linked with the wider scholarly tendency in the contemporary history of science to draw a realistic and detailed picture of ancient scientific literature without in any way idealising it. Thus, it is expected that an attempt to set the game straight for *Quaestiones naturales* will certainly be of interest both to Plutarchists and to historians of (ancient) science. The principal aim of this prologue, therefore, is to establish a broad conceptual and methodological framework within which we can approach Plutarch's natural problems in a suitable fashion, that is, in light both of contemporary Plutarch scholarship and of the history of (ancient) science. Historians of science will be familiar with many of the points raised here, but this may not be the case for scholars working in the field of Plutarch studies. In either case, examining Plutarch's role in the history of science may offer something new to both types of readers.

The prologue at hand aims to explore how the scientific value of *Quaestiones naturales* can properly be assessed and what is its place in its contemporary scientific context. Through outlining a *status quaestionis* of the research that has already been conducted on this text, I will try to show that it is only fair to study Plutarch's scientific endeavours at face value, that is from an ancient rather than from a modern perspective. As such, the goal of this study will be of a mainly historical-antiquarian kind. It should be noted, though, that the principle of charity, which this assessment will be established on – by holding that an author's tenets and convictions be valued on the basis of his own intellectual standards and that of his time –, should not, of course, exclude a diachronic evaluation[3] (see further).

---

[3] See J. Opsomer, 2014, p. 91: "The principle of charity, as I understand it, […]

A first important terminological question will, then, be whether the concept of 'ancient science' is actually legitimate, and, if so, why one would want to apply it here. The overarching question is whether Plutarch's scientific programme is really as 'immature', if not straightway unscientific, as it may seem to some (modern readers). This question may sound naïve, but its apologetic significance will soon become clear. In other words, what grounds do we have to take the more marginal(ised) aspects of ancient science – including Plutarch's natural problems – seriously, and why would we even care to do so? In order to provide a convincing answer to these questions, we should first take a look at the scientific programme of Plutarch's philosophical hero, Plato.

1. *Plato, Plutarch and scientific infancy*

As the attentive reader may have noticed, the quotation from Plato's *Phaedo* in the epigraph to this study is a rather misleading excerpt from the original Platonic dialogue. That is, it stands somewhat bare and decontextualised from the original Platonic text. This is deliberate, and it will become clear why, once we have considered the passage in greater depth.

In *Phd.* 95e–99d, Plato incorporates an important intermezzo, where Socrates is conversing with Cebes in an autobiographical mode about his own intellectual *Werdegang*. As a youngster (νέος), Socrates was very enthusiastic about natural science (περὶ φύσεως ἱστορία), because it seemed to acknowledge for the existence, coming to be and perishing of everything in the world by means of a suitable causal approach. Unfortunately, we can only guess at the real, historical extent of Socrates' interests in natural scientific matters[4]. From this passage in the *Phaedo*, however, we learn

---

demands that we assume, at least for the sake of a rational reconstruction of his views, that Plutarch advocated his views because he was convinced of their truth, and condemned incompatible views because he believed them to be false. (One could of course argue that his real reasons for believing certain views were opaque to him. Plutarch's psychological motives, however, are not accessible to us.)"

[4] It remains to be seen, after all, how much truth and how much slandering there is precisely in Meletus' attribution of Anaxagoras' physical theories to Socrates at his trial (viz. that the sun is a stone and the moon, earth), or to what precise historical extent Aristophanes' portrait of him as a mad scientist is a caricature or not. Pl., *Apo.* 26d (= Anaxag., DK59A35): τὸν μὲν ἥλιον λίθον φησὶν εἶναι, τὴν δὲ σελήνην γῆν. Cf. also *Apo.* 19b for the accusation of Socrates' excessive interest in the study of 'what is beneath the earth and in the heavens' (Σωκράτης ἀδικεῖ καὶ περιεργάζεται ζητῶν τά τε ὑπὸ γῆς καὶ οὐράνια). The slandering is very clear in Aristophanes, (e.g., *Nub.* 174: ἥσθην γαλεώτῃ καταχέσαντι Σωκράτους). But Plutarch is also very clear on the matter in *Nic.* 23, 3: Σωκράτης, οὐδὲν αὐτῷ τῶν γε τοιούτων προσῆκον, ὅμως ἀπώλετο διὰ φιλοσοφίαν. Cf. also Pl., *Phdr.* 230d: φιλομαθὴς γάρ εἰμι· τὰ μὲν οὖν χωρία καὶ τὰ δένδρα οὐδέν μ' ἐθέλει διδάσκειν, οἱ δ' ἐν τῷ ἄστει ἄνθρωποι.

that Socrates became interested in Anaxagoras' theory of an all-embracing νοῦς when he heard someone – perhaps Archelaus, Socrates' supposed teacher – reading from Anaxagoras' book. Socrates expected that this νοῦς would arrange everything 'in such a way as it is best for it to be' (ταύτῃ ὅπῃ ἂν βέλτιστα ἔχῃ). Anaxagoras' theory seemed very promising at first in this regard, but eventually – and quite ironically so, considering the kind words Socrates first had for Anaxagoras – it failed to meet Socrates' initial expectations. This disappointment is due to 'Mr. Mind's' (cf. *Per*. 4, 6) main focus on natural causes, a rather narrow approach in Socrates' opinion. Socrates gives the following absurd example to disprove Anaxagoras: one could say that it is due to certain positions and movements of his muscles that he sits there (in prison) with his legs bent, but when it comes to those muscles, he could just as easily have set course for Megara or Boeotia to escape the impending death penalty (the allusion is to the *Crito*). Therefore, the main cause for Socrates' stay in prison is his personal choice to accept the judges' verdict and not go into exile. In this sense, his muscles are only the *means* by which he can sit or run away, but to call them the *real* cause of his action is, for Socrates, most absurd (ἀλλ' αἴτια μὲν τὰ τοιαῦτα καλεῖν λίαν ἄτοπον).

It is interesting – at least for the sake of the argument – that natural science and a person's interest in natural causality is depicted in the *Phaedo* passage as a puerile practice for immature intellects. Socrates is young, so his interest in natural phenomena could be pardoned as a youthful sin. Even as a youth, however, Socrates frowned at the flaws in natural science, mainly because he did not see it as shedding light on the *real* causes (τὰς ὡς ἀληθῶς αἰτίας). For Socrates, explaining natural phenomena, such as the working of the muscles, in a purely physical-aetiological way is a 'childish' and truly 'infantile' procedure[5]. It is an oversimplified manner of speaking (πολλὴ ἂν καὶ μακρὰ ῥᾳθυμία εἴη τοῦ λόγου) and shows an inability to make proper distinctions (τὸ γὰρ μὴ διελέσθαι οἷόν τ' εἶναι). People who equate the natural cause with the real cause are only groping in the dark and use the wrong word (ὃ δή μοι φαίνονται ψηλαφῶντες οἱ πολλοὶ ὥσπερ ἐν σκότει, ἀλλοτρίῳ ὀνόματι προσχρώμενοι).

Scholars have argued that Plato is most likely projecting his own philosophical *Werdegang* on that of Socrates in this passage. Thus, one may wonder why Plato himself, in spite of his well-known disdain for experimental science (cf., e.g., *Tht*. 162e), still had an interest in biology and physical theory, as is clear from the *Timaeus*[6]. In this late work, Plato

---

[5] For more detail on the ancient belief that philosophical education can commence only after *infantia*, when reason sets in and logical thought starts to develop, see C. Laes, 2011, p. 84 (cf. Aët., *Plac*. 4, 11 = Ps.-Plut. 900BD and Sen, *Ep*. 118, 14).

[6] See H. Görgemanns, 1999 (with p. 75 for the theory of Plato's autobiographical writing in the *Phaedo* passage).

clearly demonstrates that the contemplation of natural causes does hold great interest for a true and mature philosopher, but at the same time he makes it very clear that the study of natural phenomena should remain completely subjugated to the contemplation of a higher, intelligible cause (i.e. the idea of the beautiful and the good, the divine demiurge etc.). This intelligible and divine principle is the corner-stone of Plato's philosophical doctrine.

The point I am trying to make is different, though. To stay with Plato's *Timaeus* for a moment, it would be an understatement that historians of science have not always been very cheerful about its contents. Clagett, for instance, notes that "[e]ven the most apologetic Platonist will not stand behind Plato's *Timaeus* as a work of high scientific caliber, although it is true that some of the ideas suggested therein were not without their influence on Aristotle and later authors"[7]. It remains to be seen, of course, whether the significance of the *Timaeus* for the history of science is, as Clagett here suggests, only extrinsic (viz. to be valued by its influence on later authors) and not also intrinsic (viz. to be valued in the text itself qua ancient scientific product). Plutarch, for one, can certainly be counted among these later authors: he is, indeed, an intellectual heir of Plato's science, for whom the *Timaeus* served as some kind of a scientific manifesto and a methodological guide to be followed when contemplating the natural world [see 4.3.2.]. From Clagett's perspective, though, this is not exactly a cause for celebration.

In fact, the problem remains, if it does not become even worse in the case of Plutarch, whose scientific project not only stems from that of Plato, but also shows a peculiar inclination to accept popular beliefs regarding nature without great concern for their reliability [see 4.1.1.]. It is not my goal in this preface to concentrate on Plutarch's own Platonically inspired focus on higher causes (this will be fleshed out later, when dealing with his dualistic view on causality [see 4.1.2.]) but to take a step downwards on the causal ladder and shed a few preliminary thoughts on the scientific value of his natural problems and their attempt to formulate plausible physical explanations for often rather peculiar natural phenomena. This question is particularly relevant in light of the physical aetiologies Plutarch provides in *Quaestiones naturales*, or in Greek: Αἰτίαι φυσικαί [see 1.1.6.].

---

[7] M. Clagett, 1955, p. 64 (cf. also, e.g., B. Farrington, 1961, p. 120: "from the scientific point of view the *Timaeus* is an aberration"). A review of Clagett's work is found in J.T. Vallance, 1990 (esp. pp. 717 and 719 on the *Timaeus*), whose observations are of general interest for the arguments in this prologue. For a more positive reappraisal of Plato's science, see, e.g., G.E.R. Lloyd, 1968. See also J.P. Anton, 1980, A.F. Ashbaugh, 1988, A. Gregory, 2000, D.J. Zeyl, 2000, pp. xiii–xv, T.K. Johansen, 2004, R.D. Mohr and B. Sattler, 2010 and S. Broadie, 2012.

Indeed, one may wonder how scientific the answer can be to a question as 'Why are bees quicker to sting people who have just committed adultery?', which is *Q.N.* 36 in the collection. Two basic reactions to such questions can be distinguished in modern scholarship: either eager justification of the actual occurrence of such a phenomenon in nature or outright dismissal of the problem as a bizarre, if not completely fictitious, invention. As it turns out, however, both reactions are often equally prejudiced in principle. To take the example of the bee-problem just quoted, Sandbach tried to save the phenomenon by noting that "[t]he belief that bees readily attack those who carry the odour of sexual intercourse may be true, since they appear to be provoked by other body odours."[8] Flashar, by contrast, is less enthusiastic about this problem, which he claims to be "ausgesprochen gesucht und naturwissenschaftlich unbegründet"[9]. Arguably, however, Sandbach is asking the wrong question while Flashar is using the wrong standard. When it comes to the scientific appeal of natural *mirabilia*, and of those recorded in Plutarch's natural problems more specifically, it seems only fair – and certainly much more pertinent – to evaluate these problems according to contemporary (i.e. ancient) natural philosophical standards, as opposed to Flashar's modern standards. Seeing that Plutarch makes a considerable attempt to explain these and similar phenomena in a plausible, physical way, it is only reasonable to ask why he takes such curiosa seriously. The main point of interest for us, however, should not be to find out whether Plutarch got it right or wrong (which is Sandbach's main concern): there is no use to testing the bee-problem – by any means! – in order to assess Plutarch's natural scientific inquisitiveness. What will mainly concern us in this study, then, is Plutarch's actual intention and underlying motive to account for such problems in a serious way, and what constitutes proper scientific conduct for him.

There is no denying to the fact that some of these natural questions appear to be quite playful, but the modern reader may consider Plutarch's answers to these questions even more perplexing. One might find that the physical explanations Plutarch provides contain several argumentative flaws. These can occasionally be attributed to the author's untended and careless writing, but it would certainly be too easy to ascribe each aetiological flaw solely to Plutarch's sloppy authorship[10]. What is probably most alarming for the modern reader is Plutarch's mainly theoretical approach to natural phenomena [see 4.3.2.]. The physical aetiologies are not infrequently based on what, for many modern readers, would seem to

---

[8] F.H. Sandbach, 1965, p. 219, n. c.

[9] H. Flashar, 1962, p. 370. Cf. also G. Nuzzo, 1991, p. 410.

[10] Cf., e.g., P. Donini, 2011, p. 20 (with n. 25) for the point that some errors in *De facie* could, in fact, already have been avoided in Plutarch's time.

be unsound assumptions, if not absurd sophisms, about nature, many of which provoke the author's argumentative ingenuity and rhetorical talent for vindicating the untenable. This rhetorical ingenuity and aetiological sophistication on Plutarch's behalf has often been cast in a bad light by modern critics, but it should be noted that it is, in fact, an essential feature of many scientific writings from the Greco-Roman era and also of Plutarch's scientific programme in particular (think, for instance, of Galen's rhetorical debunking of rivals and predecessors). Especially the natural problems of Plutarch's *Quaestiones convivales* have been severely criticised in this way[11], but the same criticism applies to the problems collected in *Quaestiones naturales* just as well[12].

The physical explanations Plutarch provides often seem to end up in idiosyncratic associations (e.g., *Q.N.* 16, 915F: the attribution of hot and cold properties to wheat and barley, respectively), absurd value patterns (e.g., *Q.N.* 17, 915F: the strength of hair is gender-related), tentative speculations (e.g., *Q.N.* 3, 912E: salt literally contains sharpness, δριμύτης, because of its taste), and plain contradictions (e.g., *Q.N.* 2, 912BC: rainwater is suggested to be flavourless, but a little further Plutarch

---

[11] See, e.g., Z. Abramowiczówna, 1962, pp. 82–83, F. Fuhrmann, 1972, pp. xxiii–xxiv ("Ainsi présentées, ces recherches [scientifiques] rappellent la déclamation rhétorique qui ne constitue, elle aussi, qu'une imitation d'impression et de sentiments véritables. [...] ce ne sont que des exercices de l'esprit"), E. Teixeira, 1992, p. 219 ("Mais les explications données par Plutarque sont le plus souvent assez fantaisistes, semble-t-il."). For a more positive evaluation, see, e.g., F. Frazier and J. Sirinelli, 1996, p. 206 ("Ce goût du paradoxal et de l'étrange peut, à première vue, sembler un tribut payé à la rhétorique d'époque, mais il correspond aussi à une conception de la philosophie comme réflexion suscitée par l'étonnement.") and J. König, 2007, p. 51 (quoted n. 71).

[12] The same is true, moreover, for the Ps.-Aristotelian *Problems*, after which Plutarch's natural problems are modelled. In Antiquity, however, Aristotle's *Problems* were praised as being 'most delightful and filled with choice knowledge of all kinds' (Gell., *NA* 19, 4: *lepidissimi et elegantiarum omnigenus referti*). Of course, times have changed considerably since; as is well known, Aristotle's causal model of scientific research did not survive the Middle Ages. It will not come as a surprise, therefore, that modern readers have not always been positive about the scientific value of Ps.-Aristotle's *Problems*. For a compilation of modern disapproval (but sometimes also appreciation), see H. Flashar, 1962, pp. 377–378. A. Schopenhauer, 1976 (1851), pp. 478–479, for instance, was very pessimistic – how could he not be? – in his evaluation: "Wer wissen will, wie unglaublich weit die *Unwissenheit* der Alten in der Physik und Physiologie ging, lese die 'Problemata' *Aristotelis*: sie sind ein wahres specimen ignorantiae veterum. Zwar sind die *Probleme* meistens richtig und zum Teil fein aufgefaßt: aber die Lösungen sind größtenteils erbärmlich, weil er keine anderen Elemente der Erklärung kennt als nur immer τὸ θερμὸν καὶ ψυχρόν, τὸ ξηρὸν καὶ ὑγρόν." R. Mayhew, 2011a has recently produced a new English translation of Ps.-Aristotle's *Problems*, and the papers collected in B. Centrone, 2011 and R. Mayhew 2015a also mark a renewed scholarly interest in this peculiar branch of scientific literature. For its reception in the Middle Ages, see esp. P. De Leemans and M. Goyens, 2006.

asserts that it nevertheless contains a sweet constituent; *Q.N.* 3, 912DE: salt can cause both an increase and a reduction in bulk). Furthermore, the aetiologies are at some points infested with *non sequitur* inferences, because Plutarch seems to too easily accept what he hypothesises to be true (e.g., *Q.N.* 18, 916B: the fact that the cephalopod has no body armature and is composed entirely out of soft flesh is not necessarily a cogent reason for it also being more sensitive to cold and disturbance in the sea). One may wonder, however, how fair such an evaluation of Plutarch's science really is, since many of these theories were commonly accepted by ancient natural philosophers.

Bearing in mind the passage from Plato's *Phaedo*, a first – somewhat introductory – point of interest for us will be whether Plutarch's *Quaestiones naturales* is perhaps the work of a young author, that is a youthful lapse, representative of Plutarch's juvenile enthusiasm for things natural. This is not a futile question, let alone an easy one to settle. In order to illustrate this, a link can be drawn with Plutarch's works on animal psychology (*De sollertia animalium, Bruta animalia ratione uti, De esu carnium* [see 1.1.1.]). Many scholars have assumed the hand of a young author in these writings, mainly for reasons of their ostensibly playful and juvenile contents and rhetorical style. Such a biographical reading, traditionally based on a text's style and contents, is generally considered somewhat trite today, though[13]. Indeed, one may object that the link between zoology and ethics, as is present in these writings (on animal psychology), was not without precedent in ancient literature, and that the philosophical (esp. anti-Stoic and Pythagorean) overtones present therein can just as well point at a more mature authorship[14]. Of course, *Quaestiones naturales*

---

[13] *Pace*, e.g., F. Krauss, 1912, pp. 80–83 and H. Cherniss and W.C. Helmbold, 1957, pp. 314, 490, 537. K. Ziegler, 1951, col. 732 remains hesitant when he says that Plutarch's works on animal psychology are "der rhetorischen Jugendperiode entweder noch zugehörig oder ihr doch nicht fernstehend". The belief that a highly rhetorical and literary discourse can only be ascribed to an author's youthful character, whereas a simple, unadorned style is indicative of a more settled and mature authorship is generally rejected today. Cf. C. Pelling, 2011, p. 211, n. 14 (with further references), who objects to "the crude interference that declamatory style is a mark of immaturity". The same point was made regarding the date of *De sollertia animalium* by J. Bouffartigue, 2012, p. xxi. Cf. *De soll. an.* 959C: καὶ γὰρ ἐκεῖνος [sc. Plutarch?] ἔδοξέ μοι τὸ ῥητορικὸν ἐγεῖραι διὰ χρόνου, χαριζόμενος καὶ συνεαρίζων τοῖς μειρακίοις. Cf. also T. Schmitz, 2014, pp. 32–33.

[14] For the anti-Stoic tendencies in these works, see D. Babut, 1969, pp. 61ff., who assigns them "[à] l'époque de sa pleine maturité". Cf. also J. Bouffartigue, 2012, pp. xx–xxi. For their Pythagorean (and Orphic) tendencies, see D. Tsekourakis, 1986 (who, however, adheres to the youth theory on p. 127, n. 3). For the natural philosophical value of Plutarch's writings on animal psychology more generally, see R. French, 1994, pp. 178–184 (esp. p. 182 for their Academic leanings) and S.T. Newmyer, 2006 and 2014. Moreover, for the anti-Epicurean tendencies in *De amore prolis*, see G. Roskam, 2011a, pp. 200–201.

does not belong to the category of Plutarch's writings on animal psychology, but if one considers the rhetorical dynamics and often peculiar contents of the problems collected there, the situation is worth considering more closely.

2. Date and chronology of Quaestiones naturales: *a 'life's work'?*

As stated, a first hurdle that cannot simply be avoided in this study – and that should best be cleared early on – is that of the date and chronology of *Quaestiones naturales*. Most scholars agree that this collection does not belong among Plutarch's early works but should be dated to a much later period of his literary career. There are no points of reference in the text to conclusively date the collection, though. The precise *floruit* of Laetus (presumably the Platonic philosopher Ofellius Laetus, quoted in *Q.N.* 2, 911F and 6, 913E), remains uncertain, and the same is true for Dionysius (ὁ ὑδραγωγός, quoted in *Q.N.* 9, 914B) [see 4.2.1.1., nn. 114–115]. It is generally accepted that the collection was composed more or less contemporaneously with *Quaestiones convivales*, around 100–110 AD[15] (a theory that was first formulated by Sandbach)[16], but this date is, in my opinion, uncertain for *Quaestiones naturales*[17]. Scholars have

---

[15] For the late date of *Quaestiones convivales*, see K. Ziegler, 1951, cols. 713 and 888: "mindestens ins 1. Jahrzehnt des 2. Jhdts., und zwar eher an sein Ende als an seinen Anfang". C.P. Jones, 1966a, pp. 72–73 dates *Quaestiones convivales* after 99 and before 116 AD. Cf. also F. Fuhrmann, 1972, p. xxvi ("les *Propos de Table* ont été écrits probablement au cours de la 2ᵉ décennie du IIᵉ siècle, et [...] ils représentent une des dernières œuvres de Plutarque") and J. Sirinelli, 2000, p. 370 ("entre 107 et 115 et plus vraisemblablement entre 107 et 110") and p. 380 ("Il [sc. Plutarque] touche ou vient de toucher à la soixantaine.").

[16] F.H. Sandbach, 1965, p. 138. According to G.W.M. Harrison, 2000b, pp. 242–243 (cf. also 2000a, p. 197) both collections were composed more or less simultaneously, and, indeed, "[t]he *Quaestiones* in general are written in the second half of Plutarch's career". L. Senzasono, 2006, pp. 46–48 situates *Quaestiones naturales* at the end of the first decennium of the second century, a little earlier than *Quaestiones convivales* and probably after the *Vitae* (p. 25, n. 36; on the contemporaneous composition of *Quaestiones convivales* and the *Vitae*, see, however, C. Pelling, 2011, pp. 207–208). Problematic in Senzasono's argument is the idea that the unadorned style of *Quaestiones naturales* [see 1.2.3.] would imply a late date, while the same late date is also presupposed for the literary vivacity of *Quaestiones convivales* and the *Vitae*, i.e. writings that are highly embellished from a stylistic perspective. According to F. Fuhrmann, 1964, pp. 19, 22 and 77, *Quaestiones naturales* was composed after 100 AD, because it contains only one literary image (viz. in *Q.N.* 29, 919B: "Les météores éclatent comme des bulles"), but this is not the only case of imagery, as we will see later [1.2.3.], and, even so, this stylistic argument is not really convincing (see n. 13; it is certainly outmoded, see already F. Krauss, 1912 and J. Kowalski, 1918).

[17] In fact, only a small number of Plutarch's works allow for determining an absolute date. A study that cannot remain unmentioned here is C.P. Jones, 1966a (but for a critical

made great efforts in dating some of the sympotic discussions recorded in *Quaestiones convivales* by basing their argument on the historical contexts and the prosopography of the attending symposiasts[18]. It remains to be seen, though, whether in those cases where a precise date can be deduced (assuming that we can at least accept a certain aspect of historicity for *Quaestiones convivales* [see 2.3.1.]) the same date should *necessarily* be accepted for the parallel passages in our collection. For example, the specification ἕν γε τῷ παρόντι κτλ. ('at least in the present discussion') in *Quaest. conv.* 664D is probably an implicit allusion to the parallel account about the generative property of 'lightning water' in *Q.N.* 4, 912F–913A (I will discuss this parallel later [see 3.1.4.]), but the sympotic discussion at issue cannot be precisely dated, since it is set at a random dinner in Elis and is hosted by the otherwise unknown Agemachus. My point is, however, that even if we had been able to date this sympotic discussion precisely, this would not necessarily have implied that the passage from *Quaestiones naturales* is contemporaneous with the one from *Quaestiones convivales*, since an earlier or later composition is at least equally plausible. This is actually true for each and every parallel between the two works (and these are numerous: cf., e.g., *Q.N.* 1, 911CF and *Quaest. conv.* 627AD; *Q.N.* 3, 912F and *Quaest. conv.* 685D etc. [see 2.2.3.]). Therefore, the contemporaneous composition of *Quaestiones naturales* and *Quaestiones convivales* is not a given fact. I am, in any case, inclined to be open-minded in this matter[19].

It is important to draw a clear distinction between the date of a text's publication and its period of composition. This is not irrelevant, especially if one considers that some scholars argue in favour of a posthumous publication of *Quaestiones naturales*[20] (I will come back to the issue of

---

review, see R. Flacelière, J. Irigoin, J. Sirinelli and A. Philippon, 1987, pp. x–xi). See also G. Hein, 1916, C. Stoltz, 1929, K. Ziegler, 1951, cols. 708–719 and J. Sirinelli, 2000, pp. 476–483.

[18] See esp. the commentary of S.-T. Teodorsson, 1989, 1990a, 1996. The main prosopographical study of Plutarch's oeuvre is B. Puech, 1992.

[19] Only in a few exceptional cases one can attempt to reconstruct the approximate chronological order of the text based on specific clusters of parallel passages in the *corpus Plutarcheum*. However, this attempt will still remain conjectural. This is the case, most notably, with *Q.N.* 19 (on the octopus' metachrosis): see F.H. Sandbach, 1965, p. 137 and M. Meeusen, 2012a, pp. 250–252. It is perhaps best in such cases not to claim that Plutarch worked on several works at the same time, since this is at the risk of coming to the absurd conclusion that Plutarch wrote his entire oeuvre all at once. Indeed, Plutarch's use and reuse of personal notes (ὑπομνήματα) at different occasions offers a plausible way out [see 2.1.2.].

[20] See F. Fuhrmann, 1964, p. 19: "Il est évident que ces ébauches [dont il est malaisé de déterminer le but exact aux yeux de leur auteur] n'ont pas été publiées par Plutarque, mais après sa mort par les membres de sa famille, ses amis ou familiers.

publication later [see 2.4.1.]). Whereas nothing is known about the date of publication of *Quaestiones naturales*, we do have some indications of its period of composition. I am inclined to accept that Plutarch may very well have worked on his natural problems over several periods throughout his life, perhaps from his years as an Athenian student onwards, adding something new or omitting (or at least reorganising) older material from time to time. These editorial interventions in the text probably involved the addition of new (sets of) problems and new answers to older problems. This is more or less the idea behind Harrison's theory of a long and intermittent composition of *Quaestiones naturales*[21]. The central point of Harrison's

---

De cette sorte sont les *Aetia romana*, les *Aetia graeca*, les *Aetia physica*." See also K. Ziegler, 1951, col. 857: "man [kann] zweifeln [...], ob sie von P. selbst oder erst aus seinem Nachlaß herausgegeben worden sind. [...] Die Möglichkeit der Herausgabe durch P. selbst muß man jedenfalls im Hinblick auf die ganz gleich gearteten Αἴτια Ῥωμαϊκά und Ἑλληνικά zugestehen, für die die Selbstzitate die Veröffentlichung durch P. selbst sicher stellen." See already R.W. Emerson, 1891, pp. 309–310 (quoted below, n. 32).

[21] See G.W.M. Harrison, 2000b, pp. 239–243 (cf. also 2000a, pp. 197–198). According to L. Van der Stockt, 2011, p. 450: "Harrison's hypothesis of the genesis of the composition (through 'intermittent composition', as he calls it) is most interesting, but [...] needs to be tested against the hypothesis of genesis through hypomnemata [...]." I will deal with this "genesis through hypomnemata" of *Quaestiones naturales* later [see 2.3.2.]. See already A. Gudeman, 1927, col. 2523, who argues that collections of problems (in general) are "im Laufe der Zeit aufgehäufte Kollektaneen". The same point was made more specifically for Plutarch's *Quaestiones Romanae* and *Graecae* by W.R. Halliday, 1924, p. 13: "Plutarch has put together [these collections], perhaps over a fairly long period, from his miscellaneous reading". Halliday's chronological vagueness puts the stricter theories about the publication date of Plutarch's collections of problems (c.q. *Quaestiones Romanae* and *Graecae*) into perspective. According to J.-M. Pailler, 1998, p. 77, for instance, "les matériaux des *Questions* ont été réunis en même temps que ceux des *Vies*", but even this is perhaps a bit too restrictive (cf. also G. Hein, 1916, p. 11: "quaestiones Romanas [...] conscripsit cum vitas parallelas componeret; quam ob rem illae in posterioribus Plutarchi scriptis numerandae sunt"; K. Ziegler, 1951, col. 862). It seems unlikely that Plutarch read and, by implication, made extractions from historical texts *only* while he was working on the *Vitae* – though, of course, the production may very well have increased at that time. It is not impossible that those specific passages in the *Quaestiones Romanae* which Plutarch refers to in *Cam*. 19, 8 and *Rom*. 15, 7 [see 2.1.4.] were composed shortly before their correlates in the *Vitae* (and are therefore late), but this is not at all certain and neither does it necessarily hold true for the collection as a whole. Nevertheless, J. Boulogne, 1992, p. 4687 (with n. 30) dates *Quaestiones Romanae* to around the end of the first century AD: "elles profitent de la pleine maturité intellectuelle de Plutarque, qui a, alors, atteint la cinquantaine". The allusion to Domitianus in *Quaest. Rom*. 276E would serve as a *terminus post quem*, pushing the date of the collection's composition after 96 AD. One should acknowledge, though, that this date perhaps only applies to that specific chapter. *Pace* also K. Ziegler, 1951, col. 860, but with more nuance in col. 712: "die *Stelle*

theory is that the collection was probably not put together in one and the same breath but at several distinct moments:

> "like someone who solves cross word or jig saw puzzles over a number of days, Plutarch picked up and put down and came back to a series of questions that had begun to excite his curiosity [...]. While engaged on other projects, as he had further thoughts, Plutarch made additions to each of the *quaestiones* just as trains may add on cars at various stops but always in a determined sequence."[22]

The possibility of smaller or larger chronological intervals, during which Plutarch let the material sink in for a while in order to revisit it afterwards, is very difficult, if not impossible, to determine within the text[23]. It is not unlikely that the thematic clusters of problems that are found throughout the collection hint at such an intermittent composition (see the scheme in the introduction to the commentary), but, even then, it is a hopeless task to prove exactly where and when Plutarch put his pen down or picked it up again[24]. On the assumption that such thematic clustering implies

---

[scheint] zwar nicht sicher, aber doch sehr wahrscheinlich nach seinem [sc. Domitians] Tode geschrieben" (my italics). See also H.J. Rose, 1924, pp. 47–48 and C.P. Jones, 1966a, p. 73 (after ca. 105 AD).

[22] G.W.M. Harrison, 2000b, p. 240. A similar notion of intermittent composition was entertained for *Quaestiones Platonicae* by H. Cherniss, 1976, pp. 4–5 (cf. J. Opsomer, 1994a, pp. 4–5): "The ten ζητήματα may not all have been written at one time and for a single work. It is at least as likely that at some time Plutarch put together ten separate notes on Platonic passages that he had written at different times and had found no suitable occasion to incorporate into his other compositions. If this is so, any indication of the relative chronology of one of the ten would not necessarily be pertinent to that of the others." Similarly, for the intermittent composition of Plutarch's *Apophthegmata Laconica*, see F. Fuhrmann, 1988, p. 135.

[23] See, e.g., the transition between *Q.N.* 23–24, and more specifically the ghost-reference in the opening phrase in *Q.N.* 24, 917F: Ἡ διὰ τὴν εἰρημένην αἰτίαν; (see the commentary *ad loc.* for further detail). Cf. also G.W.M. Harrison, 2000b, p. 241.

[24] According to G.W.M. Harrison, 2000b, p. 242 (with a synopsis in 2000a, pp. 197–198) there is a substantial chronological rupture between the composition of *Q.N.* 1–18 and 19–31, but the stylistic arguments he adduces (viz. the use of the present tense and the incorporation of literary quotations) are not very convincing, in my opinion. Harrison bases his theory on the chronology put forward by F.H. Sandbach, 1965, pp. 136–137. Sandbach does not, however, postulate a chronological rupture between *Q.N.* 1–18 and 19–31 as such, but focuses only on the chronology of *Q.N.* 19 and its parallel passages [see the scheme in 2.1.2.]. Considering the thematic clustering of problems on fishing and sea animals in *Q.N.* 13, 17–19 and the parallel passages – presumably based on the same or similar hypomnematic material – in a relatively small section of *De sollertia animalium* (viz. 976E–977A, 978EF, 979B), it seems rather unlikely that there is a major chronological rupture in composition between *Q.N.* 18 and 19. The inclusion of the cluster of problems

synchronous composition, it is not impossible that the problems in each of these clusters were composed more or less contemporaneously. After all, the solution to one problem may trigger the formulation of a new one that is closely related to it. But it is just as likely that some problems were added later on to these clusters, or that the clustering itself is perhaps the result of an editorial intervention. There is no need to go into the details of this theory; it is simply the general idea behind the unspecifiable chronology of *Quaestiones naturales* that matters here.

The idea that Plutarch perhaps worked on his natural problems from his early years as an Athenian student onwards can be corroborated by the fact that, at several points in the discussions described in *Quaestiones convivales*, Plutarch stages his own literary *alter ego* as a young symposiast and interlocutor[25]. The narrated time (*erzählte Zeit, temps de l'histoire*) of *Quaestiones convivales* goes back to the time when Plutarch was an Athenian student under the tutorship of Ammonius, with whom he joins in discussion on several occasions. This is the case in the first two talks of Book three (*Quaest. conv.* 645D–649F), wherein two natural problems are discussed (concerning the natural properties of flower garlands and ivy) and also in Book nine, which is set at the festival of the Muses that probably took place at Ammonius' private Academy in Athens. Ammonius there (in Book nine) appears on the scene several times, and as the titles of the lost talks indicate (especially talks ten through twelve), some of the discussions originally dealt with particular natural problems (concerning solar and lunar eclipses, the body's permanent state of flux and the number of the stars respectively). Notably, Ammonius is also an interlocutor in the discussion held in *De E*, where Plutarch again appears as his young student (the topic of the body being in flux is treated in *De E* 392AE). Jones has dated these talks in *Quaestiones convivales* and in *De E* to around 66–67 AD (when Nero was visiting Greece: *De E* 385B), pointing out that Plutarch counts himself among the νέοι (*Quaest. conv.* 649A, 646A, cf. also *De E* 391E) and that his brother Lamprias (who is also present in *De E*) is still a παῖς (*Quaest. conv.* 747B)[26]. If there is any historicity to

---

regarding the natural properties of wheat and barley (*Q.N.* 14–16) can be explained, then, in light of structural *variatio* [see 1.1.5.]. For further detail, see the commentary *ad loc.* and M. Meeusen, 2012a, p. 252, n. 75.

[25] For Plutarch's self-presentation as a young symposiast in *Quaestiones convivales*, see F. Klotz, 2007, p. 655 and J. König, 2011, p. 179. Notably, P. Louis, 1991, pp. xxviii–xxix argues, in a similar fashion, that Aristotle probably started composing his natural problems shortly after his entry in Plato's Academy. Cf. also C. Jacob, 2004, p. 44.

[26] C.P. Jones, 1966b, pp. 206–207 (in my opinion, S.-T. Teodorsson, 1989, pp. 289–290 has no convincing argument against Jones' early date; and for Book nine he argues that "[i]t is quite possible that considerable parts of the contents are student reminiscences of Plutarch": *id.*, 1996, p. 299). Talk three of Book eight (*Quaest. conv.* 720C–722F: 'Why

these passages [see 2.3.1.], this would imply that Plutarch was engaged in solving natural problems – and other kinds of problems just as well – from a relatively early age up until his more settled years as a mature philosopher. Thus, he had a strong grasp of the genre of (natural) problems by the time he started writing *Quaestiones convivales* (which he did at the behest of Sossius Senecio, cf. *Quaest. conv.* 612E: ᾠήθης τε δεῖν ἡμᾶς κτλ.). So he probably already had some material on hand, and had probably already composed some or most parts of *Quaestiones naturales* by that time.

In conclusion, if it is true that Plutarch worked on his collection of *Quaestiones naturales* during several stadia of his life, there may very well have been an increase and overlap in productivity in the period that he composed *Quaestiones convivales*, hence their mutual influence. This is no reason, however, to restrict the chronology to that period only (nor to take *Quaestiones naturales* as a set of preparatory drafts *for* the composition of *Quaestiones convivales* [see 2.3.2.]). If we bear in mind Harrison's theory of intermittent composition, it is not unlikely that Plutarch's *Quaestiones naturales* (and his other collections of *quaestiones* just as well) were perhaps a 'life's work' in a chronological sense, that is, a lifelong project, representative of his relentless interest in the natural world around him. In that case, it is only likely that Plutarch sporadically made additions and adjustments to the collection after reading something noteworthy or after a discussion with his colleagues. The bottom line is that there are no certainties about the exact date and chronology of *Quaestiones naturales*, but everything seems to indicate that Plutarch started discussing, and presumably also composing, natural problems from a relatively early age onwards up to his more settled years as a mature philosopher.

What we learn from this, is that we cannot effectively vindicate the value of Plutarch's *Quaestiones naturales* as an ancient scientific product on the mere and most uncertain basis of the text's chronology. After all, it may still seem to serve a juvenile, pseudo-scientific, if not unscientific purpose. Let us, therefore, return to our initial problem.

---

is the night more sonorous than the day?') should probably be dated later, supposedly in 81 AD, because at that occasion Ammonius participates in the discussion with his (presumably) adult son (cf. S.-T. Teodorsson, 1996, p. 181). The difference in date is not necessarily complicated by the fact that Plutarch notes that Ammonius is in office as Athenian στρατηγός both in *Quaest. conv.* 720C (Book eight, talk three) and in *Quaest. conv.* 736D (Book nine), because Ammonius held this office three times throughout his life. To complete the list: Ammonius is also an interlocutor in *De defectu oraculorum* (which Jones dates "in the 70's or early 80's"), and he is also mentioned in *De ad. et am.* 70E and *Them.* 32, 6. For further reading on Ammonius, see K. Ziegler, 1951, cols. 651–653, P. Donini, 1986b and J. Opsomer, 2009 (esp. his conclusion at p. 177 for Ammonius' interest in Peripatetic science).

3. The value of Plutarch's natural problems

If introductory companions to Plutarch's life and work or to ancient Greek literature and science more generally do not simply gloss over the Chaeronean's natural scientific achievements, they are not always very constructive in their value judgements[27]. The general approach of such studies is often underpinned by a certain feeling of astonishment that an author who is mainly known as a biographer and a moralist ventured to take a few humble steps in the field of natural science. Especially the humble character of Plutarch's science is underlined, in such cases, by depicting it as an often absurd, if not completely insignificant specimen of ancient thinking, or at least as the work of an amateur: that is, a trivial pursuit of inferior knowledge.

If a cursory light is shed on *Quaestiones naturales* (the work is mostly ignored, though), this often results in discrediting its scientific appeal. The collection's content is depicted as obscure, trivial, and only marginally scientific at best. This view is typical especially of 19th and early 20th century scholarship (but traces can still be found in more recent literature)[28].

---

[27] R.H. Barrow, 1967, p. 117, for instance, notes in passing that "[t]here are scientific works like the *Natural Questions*, or *On the face on the moon*. [...] But after making acquaintance with some of the better dialogues or essays a reader must not expect too much from some of those that remain unnoticed here. [...] They all contain much of interest, but Plutarch should not be judged by them." Neither *Quaestiones naturales* nor *Quaestiones convivales* are mentioned among Plutarch's natural philosophical writings by G.E. Karamanolis, 2010. Moreover, A. and M. Croiset, 1899, p. 511 (with n. 1) seem to underestimate the general literary value of several of Plutarch's scientific writings: "Mais ces traités [sc. écrits relatifs aux sciences naturelles] n'ont pas un rapport assez direct à l'histoire littéraire, pour qu'il soit à propos de les étudier ici."

[28] Similarly, regarding the scientific value of the natural problems in *Quaestiones convivales*, see, e.g., R. Flacelière, J. Irigoin, J. Sirinelli and A. Philippon, 1987, p. lxxxiii: "Il faut reconnaître que beaucoup de ces questions nous paraissent aujourd'hui futiles, quelquefois absurdes, et que les explications proposées sont souvent peu convaincantes. [...] La curiosité de Plutarque, extraordinairement vaste, avait un goût marqué pour l'insolite et le paradoxal, et les arguments d'Aristote, qu'il cite si fréquemment, peuvent aujourd'hui faire sourire. *Quandoque bonus dormitat ... Plutarchus*." Regarding *De facie* and *Quaestiones naturales*, G. Guidorizzi, 2000, p. 559 speaks (without further specification) of Plutarch's "interesse [...] in un certo senso 'amatoriale' [...] per la scienza" (for Plutarch's ostensible dilettantism in *Quaestiones naturales*, see also V. Ramón Palerm, 2005, p. 398). According to P. Levi, 1985, p. 477, "[i]t would be a mistake to value Plutarch only for his *Questions* and *Table Talks* and his infinite fund of gossip." Cf. also p. 479: "Among the stranger themes that attracted Plutarch now and then – relics, I suppose, of sophistic playfulness about science – the *Man in the moon* [sic] (*De Facie in Orbe Lunae*) gives the most pleasure." A.M. Battegazzore, 1992, p. 32 even connects the lack of specialisation in Plutarch's science with his 'humanism' (cf. also M. Vamvouri Ruffy, 2012, p. 75).

Croiset, for instance, resentfully considered the collection "un ouvrage sans valeur"[29], which aggravates Doehner's label of "quisquilias"[30], trifles, unworthy of Plutarch's authorship [see 1.2.1.]. Typical in this regard is also Huit's remark:

> "On éprouve quelque surprise à voir nommer dans une étude sur la philosophie de la nature le célèbre historien de Chéronée, plus connu évidemment comme biographe que comme physicien. Mais ouvrons celui de ses ouvrages qui porte le titre de *Causes naturelles*. Il contient la réponse (en général, il faut l'avouer, *aussi peu scientifique que possible*) à trente-neuf questions, les unes assez curieuses [...] les autres singulières [...] d'autres enfin absurdes [...]."[31]

To claim that it really was Plutarch's goal to be as unscientific as possible in this "compilation des plus médiocres" and that its bizarre content is necessarily an indication of its lack of proper science is not the most unbiased position to take. To give another example, Emerson also underlines the curious character, rather than scientific merit, of Plutarch's natural problems:

> "Except as historical curiosities, little can be said in behalf of the scientific value of the Opinions of the Philosophers, the Questions and the Symposiacs. They are, for the most part, very crude opinions; many of them so puerile that one would believe that Plutarch in his haste adopted the notes of his younger auditors, some of them jocosely misreporting the dogma of the professor, who laid them aside as *memoranda* for future revision, which he never gave, and they were posthumously published."[32]

In what follows, Emerson adds – on the positive side – that there are some occasional "hints of superior science". These are "statements that are predictions of facts established in modern science" that can be *culled* (to use his wording) from Plutarch's texts (he mentions "[t]he explanation of the rainbow, of the floods of the Nile, and of the remora"), but in general, Plutarch's "Natural History is that of a lover and poet, and not

---

[29] A. and M. Croiset, 1899, p. 511, n. 1.

[30] T. Doehner, 1858, p. 14.

[31] C. Huit, 1901, pp. 479–480 (my italics). One may find it odd, moreover, that Huit considers *De E* 386E (regarding the Delian problem) an actual introduction to *Quaestiones naturales*.

[32] R.W. Emerson, 1891, pp. 309–310 (part of the account also quoted by F. Klotz and K. Oikonomopoulou, 2011, p. 31). For a more recent evaluation of *Quaestiones convivales* as treating "questioni minime e marginali", see A.M. Scarcella, 1998, p. 133.

of a physicist". The message is clear: Plutarch's natural problems are strange and not all that valuable in themselves, even if they contain some exceptional glints of scientific ingenuity. These exceptional glints are valuable in light of certain achievements in modern science, which they are even believed to predict, but the rest remains obscure and should be bracketed, if only to save the author's reputation. There are good reasons to discard the underlying logic of this ambiguous compliment.

Most glaringly, It goes without saying, that modern scientists are no longer concerned with the problems raised in *Quaestiones naturales*, or that they would at least frown at Plutarch's solutions. Take, for example, *Q.N.* 10, where Plutarch deals with the usefulness of adding salt (seawater) or baked gypsum to wine. He gives two alternative explanations: 1) the heat of these substances is an aid against the chilling of the wine, and 2) their earthy constituents help against unpleasant odours, putrefaction, or turbidity in the wine. A modern explanation of the chemical reactions that occur in this process is provided by Sandbach in his commentary to the passage:

> "Sea-water is perhaps no longer used in Greek wine-making. It would slightly increase acidity, since chlorine ions, produced by hydrolysis of sodium chloride, decrease the pH value. This increased acidity might improve the wine by inhibiting the growth of micro-organisms that cause cloudiness and instability. The use of gypsum, baked or unbaked, which is still practiced in some places, has the same good effect by a different means: added to the unfermented juice, the gypsum (calcium sulphate, hydrated when unbaked) reacts with potassium hydrogen tartrate contained in the juice and stalks to produce calcium tartrate, potassium sulphate, and tartaric acid: the last, being soluble in alcohol, is not precipitated (unlike the insoluble tartrate), but remains in the wine and increases its acidity. Calcium sulphate also has clarifying properties, since it causes colloidal suspended matter to settle out. Plutarch therefore *correctly states the effects* of adding sea-water and gypsum in wine-making, although *he had no means of knowing how they are produced.*"[33]

Presumably, Sandbach had the best of intentions by providing such information, but the conclusion he draws is rather partial. Of course, Plutarch did not *know*, or more precisely, "had no *means of knowing*" these effects the way modern scientists do. Today sulphites are added to the wine, because we *know*, so to say, that they eliminate unwanted bacteria and yeasts in the wine, and that they slow down the oxidation process, thus extending the wine's shelf life and increasing its tolerance

---

[33] F.H. Sandbach, 1965, pp. 174–175, n. a (my italics).

for changing temperatures. But the very existence of bacteria was unknown in Antiquity, and Plutarch did not have Mendeleev's periodic table at his disposal to explain such chemical reactions in, what for Sandbach would probably be, a more accurate way. Nor was he privy to an outlook of the physical world as a man like Lavoisier or, for that matter, any radical atomist was – Plato's geometric atomism not included.

The same conclusion can be drawn for *Q.N.* 4, where Plutarch examines why rainwater that is accompanied by thunder and lightning is said to be more fertilising for the growth of seeds. This is considered a proven fact today, but, for obvious reasons, no explanation in Plutarch's tripartite aetiology comes close to a modern chemical explanation of this phenomenon. I am told that lightning electrifies moist air in the atmosphere so that nitric acid (HNO_3) is formed, which stimulates the growth of plants (it is used today primarily to manufacture fertilisers). The electric currents of lightning flashes trigger a reaction between atmospheric nitrogen dioxide (NO_2) in the presence of water vapour (H_2O). Perhaps the following chemical equation can make things clearer: $3\,NO_2 + H_2O \rightarrow 2\,HNO_3 + NO$. This is what Sandbach (and probably also many modern scientists with him) would consider to be the "true reason"[34]. Plutarch, however, explains this phenomenon by referring to 1) the admixture of air and breath to the rain, 2) the rainwater's concocted nature and 3) the chilling effect of spring rains. Strictly speaking, these explanations must be false, then.

Several other such examples could be added, but this is not the right place to deal with each and every one of them. After all, one may probably doubt about the scientific character of many other natural phenomena that Plutarch tries to explain, but which are less likely to be considered proven fact today. What are we to make, for instance, of the natural phenomenon at issue in *Q.N.* 32, where Plutarch wonders why the palm tree alone among all trees rises against a weight imposed upon it? In what sense exactly does the palm rise against an imposed weight? Is it even plausible that this phenomenon really occurs in nature? Apparently, this last question – relating to matters of empirical verification – did not really matter for Plutarch to address the phenomenon in the same aetiological way as he has been doing all along. As we will see later on, he had good epistemological motives for doing so [see 4.1.1. and 4.3.2.]. Laying bare these methodological dynamics will make Plutarch's scientific project more comprehensible to the modern reader.

---

[34] F.H. Sandbach, 1965, p. 161, n. d: "The true reason is that rain of a thunderstorm contains nitric acid, formed by the passage of electric currents through the air in the presence of water: the nitrogen is a rapid stimulant of plant growth." Cf. also S.-T. Teodorsson, 1990a, p. 50.

It still remains to be seen what we actually mean by 'science', then, and how the present study aims to contribute to its history. The fact that the majority of Plutarch's natural problems are, in many ways, obsolete, worn-out, and truly ancient, explains why they are generally abandoned by modern physicists today[35]. This does not imply, though, that they are *un*-scientific in principle. After all, if these ancient beliefs are to be deemed unscientific myths, they are subject to scientific inquiry for much the same reasons as those which lead to scientific knowledge today[36]. If they are to be considered representative of genuine science, on the other hand, we see that ancient science contains a large amount of convictions that are, in many ways, incommensurable with today's perception of it[37]. How to resolve this tension?

---

[35] One of the exceptions would be "die auch heute noch ungelöste Aporie von dem Verschwinden des Salzes beim Durchgang des Meerwassers durch die Pflanzen", as reported in *Q.N.* 5, 913C (H. Diels, 1905, p. 312). Plutarch explains that the pores, on account of their narrowness, filter the earthy and large particles of the salt, but "diese Tatsache ist, wie die moderne Forschung festgestellt hat, in dieser allgemeinen Fassung irrig" (p. 314). Cf. also C.F. Schnitzer, 1860, p. 2709, n. 1: "Daß zum Theil die Stellung der Fragen, noch mehr aber ihre Lösung dem jetzigen Stande der Wissenschaft nicht entspricht, wird der Leser begreiflich finden; indeß ist es, abgesehen von dem historischen Interesse das diese Mittheilungen aus der Physik der Alten haben, doch bemerkenswerth wie nahe manche der gegebenen oder versuchten Erklärungen an die richtige hinstreift."

[36] More universal scientific features would include the fact that Plutarch's natural science originates from a genuine wonder for the 'natural spectacle' and pursues a serious and detailed physical explanation for specific natural phenomena. To this end, Plutarch employs a scientific procedure that takes into account the intellectual tradition but, at the same time, aims to advance traditional theories by looking for innovative viewpoints [see 4.2.]. He also employs a standardised set of scientific terminologies to describe the physical processes that occur in nature in a more or less uniform way, and provides explanations that receive the necessary circumspection and prudence from a logical and epistemological perspective [see 4.3.].

[37] In comparison to modern scientific practice, Plutarch's approach might seem to be too theoretical (there is no interest for personal observation or experimentation) and inexact (it estimates relative qualities rather than measuring absolute quantities). Modern science, by contrast, formulates its claims in terms of universal laws and preferably in a mathematical fashion. It underpins its claims with repeatable experimentation, has a strong link with technological application and often involves an 'unnatural' manipulation of nature itself (e.g., in laboratories). It also claims to hold to 'objectivity', and – most notably – it is generally considered the counterpart of any religiously inspired discourse that is based on any 'subjective' acceptance of dogma or certain belief systems. Plutarch, however, also incorporates mythological and poetical material into his physical aetiologies, and his general outlook on the world is based on a dualistic view on causality, according to which natural phenomena are subjugated to a higher, intelligible cause and divine principle [see 4.1.]. These features are clearly incommensurable with modern science. Cf. also D.C. Lindberg, 1992, pp. 1–2 and T. Barton, 1994b, p. xii.

Lack of appreciation of the scientific appeal of Plutarch's natural problems mainly arose – and, in many cases, continues to arise – from a basic unfamiliarity not only with the ancient scientific paradigm to which they adhere (viz. as furnished by Ps.-Aristotle's *Problems* and its causal model of scientific inquiry) and from an inferior understanding of Plutarch's more general natural philosophical project (viz. as inspired by Plato's *Timaeus*), but also, and often primarily so, from an unfounded belief in the teleological nature of the history of scientific thought. Such a bias, which is representative of more traditional studies in the history of science, is characterised by the acceptance of a progressive and evolutionary development of scientific knowledge, where it is assumed that scientific truth becomes gradually more unveiled over time. This model draws a linear, but inherently distorted, picture of the development of science that is principally unhistorical from the very outset. Only those ancient theories that were considered relevant for contemporary scientific research or were proven to be valid by modern scientists were included in the historical framework under the label of genuine science (as is the case in Emerson's notion of Plutarch's "predictions" above)[38]. This approach is obviously biased by modern prejudice, which sees the results attained by contemporary science as the culmination of ages of continuous research and scientific progress. In this approach, the category of preconceived scientific 'correctness' serves as a historical measure in the evaluation of any scientific theory, without acknowledging that the category in itself is not necessarily a universal or transcultural given (see further). The great ancient Greeks were, thus, presented as having made breath-taking and even 'miraculous' advances in the field of science, for which they should be held in honour, but in the long run their discoveries remain rather immature[39].

In this regard, Barton speaks of a 'genetic history' of science, according to which the teleological approach entails some kind of a historical eugenics of scientific ideas, where their individuality in the succession of

---

[38] Exemplary is the remark in R. Flacelière, J. Irigoin, J. Sirinelli and A. Philippon, 1987, p. lxxxi: "[I]l faut bien avouer que les ouvrages ou les passages isolés où Plutarque traite de physique sont généralement décevants, sans doute parce que cette science a fait […] de tels progrès que beaucoup de théories antiques […] nous paraissent aujourd'hui irrémédiablement périmées."

[39] Cf., e.g., W.H.S. Jones, 1931, p. xxii: "The hypotheses of early Greek thought are mere guesses, brilliant guesses no doubt, but related to the facts of experience only in the most casual way." Cf. also, e.g., P. Raingeard, 1935, p. viii: "Ensuite le *De facie* nous est un témoin précieux de l'état de la Science vers la fin du premier siècle. Si certaines ignorances nous font sourire […], nous éprouvons par contre quelque stupeur à découvrir l'antiquité de theories relativement modernes à qui il ne manqua pour figurer plus tôt parmi les acquisitions de l'esprit humain que de triompher des opinions régnantes."

particular world systems is neglected[40]. It goes without saying that such an approach was bound to remain restrictive, resulting in a fragmentation, decontextualisation and, eventually, isolation of specific scientific theories by stripping them from the underlying world view that gave birth to them in the first place (think of Emerson's *culling* of the "hints of superior science" from Plutarch's text). As an alternative to this historical model (that is commonly referred to as 'whiggish' history)[41], contemporary history of science takes a more relativistic and contextual stance[42]. Notably, in Antiquity there was no term for denoting what we today call 'natural science' (perhaps the notion of 'natural history', φυσικὴ ἱστορία, understood as the universal study of things natural, comes closest). The concept of ἐπιστήμη, that is, the Latin *scientia*, primarily refers to 'scientific knowledge' (as opposed to δόξα, τέχνη, ἄνοια etc.), and it does not refer to the actual pragmatics behind scientific thinking as such (which is not per se an infallible practice, of course, science being a human enterprise)[43]. In accordance with Plutarch's Platonism, for instance, no natural science could amount to the level of genuine 'science' in the philosophical-epistemic sense

---

[40] T. Barton, 1994b, pp. xiv–xvi. Barton's introduction in 1994b, pp. x–xxiii is relevant for the points made in this prologue. See also her more extensive (and at some points more abstract) introduction in 1994a, pp. 1–25.

[41] For further objections to the methods of 'whig' history, see A. Cunningham, 1988, p. 387 and T. Barton, 1994b, pp. xiv–xvi. The term was coined by H. Butterfield, 1931.

[42] E.g., for a description of Plutarch's natural science, as what stands in relation to *ancient* theories that explain natural phenomena, cf. J. Boulogne, 2008, p. 734 (who uses this description in demarcating the scientific digressions in the *Vitae* [see 2.1.3.]). Compare also, for instance, the definition of ancient science by P.J. van der Eijk, 1997, p. 77, n. 1: "As for 'scientific', this is used […] in a non-sophisticated, non-evaluative manner to refer to any serious attempt at studying and understanding the nature of things (or part thereof), and – as far as texts are concerned – to any verbal expression intended to communicate about this with an audience […]. Of course I do not wish to claim that the texts in question meet the criteria of what would count as 'scientific' in any modern sense of the word; and I am aware that there is no clear distinction between 'science' and 'philosophy' in Antiquity […]." Similarly, regarding early Greek and Chinese scientific practices G.E.R. Lloyd, 1999, p. 314 (also partly in 1996, p. 227) states that: "the history of early investigations in ancient civilizations is the history of the acquisition of a potential for cognitive development, not just with respect to what was believed, but also with regard to the ways of getting to believe it. Where Greece and China are concerned, to go no further afield, history shows both that the ways of acquiring that potential differed and indeed that the potential acquired did. Not that, in either case, the new potential corresponded closely to the expectations that might be generated by naive retrospection from the eventual emergence of modern science." In this sense, Plutarch's natural scientific programme (including his natural problems) certainly testifies to a certain scientific potential, how essentially different (or not) it may be from that of modern science.

[43] Cf. LSJ, s.v. ii, 2.

of the word[44], as it always involves a basic uncertainty towards sensible objects and the kind of knowledge that they yield [see 4.3.2.]. In this sense, *scientia* was not practiced but rather strived after in Antiquity.

As is well known, moreover, in Antiquity the study of natural phenomena was generally integrated in a wider philosophical programme, so that it was not considered an independent branch of research that was conducted for its own sake (as is the case rather with modern science). Also the ancient concept of 'physiology' (φυσιολογία), which is often used interchangeably with 'natural history', cannot be separated from a wider natural philosophical framework, so that it is not an independent science. There is no denying to the fact that there is a certain convenience in speaking about 'physiology', 'natural history' or 'natural philosophy', for the simple fact that the ancient thinkers themselves would have given it that name, and this is basically what it *is* from an ancient perspective. This does not necessarily imply, however, that the term 'science' is wrong, or that ancient natural philosophy cannot be considered 'scientific'. In any case, the fact that ancient Greek and Roman thinkers did not, or could not, call their natural philosophy to be 'scientific' does not imply that it is, therefore, fundamentally unscientific.

For many centuries 'a science' in English denoted anything that was taught in the schools (such as grammar, logic, theology etc.). The word 'scientist' came into use not long before the middle of the nineteenth century, when the concept of 'science' began to take on its modern meaning. In this period, the 'science of natural philosophy' began to predominate the other 'sciences' and eventually acquired the monopoly of the term 'science'. The practitioners of that 'science' were called 'scientists' (and it is in this sense that 'science' existed before 'scientists'), but by that time, the 'science' itself had become very different from that of the ancients[45]. So, to reformulate the problem: why do we even care about 'science' in Plutarch's natural philosophical writings in the first place or in his *Quaestiones naturales* more specifically? Why not simply use his own terminology, thus avoiding the risk of any terminological ambiguity or of an anachronistic misconception of what Plutarch himself believed he was doing?

I believe that the connotations connected with such a reticence are undesirable. Not calling Plutarch's natural philosophy 'scientific' might raise the false impression 1) that it is only of *pseudo*-scientific or *para*-scientific significance[46], or 2) that it has no scientific value at all, or

---

[44] Cf. G.E.R. Lloyd, 1968, p. 92.

[45] I here rely (in part) on T. Barton, 1994b, pp. xi–xii. Cf. also A. Cunningham, 1988, p. 380 and R.W. Sharples, 2005, pp. 1–3.

[46] According to G. Nuzzo, 1991, p. 410 (repeated in G. D'Ippolito and G. Nuzzo, 2012, p. 58), *Quaestiones naturales* is nothing more than "un singolare 'zibaldone' di curiosità

in any case less so in comparison to superior (viz. modern) scientific achievements. This is exactly what I intend to avoid. 1) As regards the first point, Barton has correctly pointed out that "[p]seudoscience is a label that implies a deliberate falsehood on the part of its practitioners"[47], but this does not apply to Plutarch, who himself took his natural philosophical programme seriously. In any case, there is no reason to assume that he did not (see n. 3 above). Even in the 'spoudogelastic' context of Plutarch's symposia we see that serious efforts are made to provide plausible explanations for natural problems. This is at least true within the conceptual scope of ancient physical theory. 2) Therefore, in light of the second point, it is only reasonable to evaluate Plutarch's natural problems according to the parameters of ancient rather than modern scientific thinking, even if there are certain points of convergence[48]. In the end, there are also obvious divergences between the two, meaning that it is best not to assess them in comparative terms (see nn. 36–37 above). It goes without saying, therefore, that when reference is made to ancient 'science' in the present study, this must be understood in light of Plutarch's theories and concepts about the natural world and those of his philosophical role models (unless otherwise suggested). As such, we accept that the distinction in terminology between actor's terms ('natural philosophy') and analytical terms ('natural science') is a matter of formality rather than semantics[49].

---

pseudo- o parascientifiche, che non un' opera di impianto organico ed unitario, tanto che qualche studioso fu da ciò indotto a negarne la paternità plutarchea" (I will come back later to Plutarch's authorship of *Quaestiones naturales* [see 1.2.1.]). The same label of pseudo-science was used by R. Flacelière, J. Irigoin, J. Sirinelli and A. Philippon, 1987, p. lxxxii and by R. Caballero Sanchez, 1992, p. 91. Z. Abramowiczówna, 1962, p. 82 speaks of the "halbwissenschaftliche Atmosphäre" in *Quaestiones convivales*.

[47] T. Barton, 1994a, p. 15 (see also pp. 16–17 on the inappropriate normativity of the concept of 'pseudo-science').

[48] E.g., according to A.M. Battegazzore, 1992, p. 48, Plutarch's attitude in scientific matters presupposes "un modello di ricerca vicino a quello di un 'sapiente rinascimentale' dilettante di tutto e di tutto curioso". However, linking Plutarch with a Renaissance model of science probably produces more contextual problems than it solves. See also L. Inglese, 1996, p. 151 (regarding Battegazzore's remark): "Questo non può comportare, nel giudizio su Plutarco, l'adozione di parametri 'moderni' di scientificità della ricerca."

[49] There is much scholarly debate on this topic. A. Cunningham, 1988 prefers using the actor's terms (such as 'natural philosophy') over analytic terms and categories (such as 'science') in order to avoid the anti-historical fallacy of imposing present thinking upon the past (pp. 378–385). W.H. Stahl, 1962, pp. 3–14 also doubts the appropriateness of calling Roman science 'scientific' (or even 'Roman', for that matter). Some scholars would, indeed, prefer the actor's terminology (see also, e.g., T. Barton, 1994b, pp. xiv, xxi, xxiii, n. 13), whereas others do not (see, e.g., G.E.R. Lloyd, 1983, pp. 210–217, who vindicates the appellation of 'science' for ancient Greek medicine and biology). Most

Indeed, Plutarch's subordination of physics to prime philosophy is not an adequate reason to reject it as unscientific. This is significant if for no other reason than that there is a discussion among Plutarchists about the scientificity of the Chaeronean's natural philosophical works, although it is not very widespread, let alone profound. The ill-considered 19th and early 20th century claims quoted above are clear examples of this. A more recent account is found in Donini's 1994 contribution on Plutarch and the Platonic renaissance in *Lo spazio letterario della Grecia antica*, where the author – in a footnote – tends to adhere to this traditional view[50]. Donini concludes: "A mio giudizio, non c'è nessuna opera di Plutarco che meriti di essere definita semplicemente 'scientifica'."[51] This is intended as a frontal attack on Ziegler's designation of the category of Plutarch's "scritti di fisica e di scienza naturale" ("naturwissenschaftliche Schriften" [see 1.1.1.]), but Donini in his protest employs the concept of 'science' mainly – and problematically so – from a modern perspective (in this regard, his approach does not seem to be all that different from Ziegler's). Donini demarcates Plutarch's 'science' as the attempt to explain natural phenomena solely in terms of natural causes. In my opinion, he thus too strictly divorces the Chaeronean's physical theory from its philosophical implications. It is absolutely true that Plutarch distinguishes between the physical causes of natural phenomena and their higher, divine motivation, but the distinction is not strict, since he subordinates the former category to the latter [see 4.1.2.]. As such, both categories are not, and should not be, disconnected. As Donini himself has shown in his pioneering work on Plutarch's view on causality, both modes of explanation actually go hand in hand (viz. physical and meta-physical). In the account at hand, then, Donini's restriction of "scienza" to natural causality alone might be problematic – in opposition to other, more nuanced accounts of his on the topic (see n. 54 below). In what follows, Donini successively sheds his

---

notably, according to E. Grant, 2007, p. 319, the difference between actor's terms and analytic terms is superficial, because actor's terms are "mere names, or labels". However, the application of "mere" is perhaps somewhat too dismissive (as E. Lao argues in her 2010 BMCR review of Grant's work). It should be noted, moreover, that a broadening of the semantic field of 'science' may risk the concept becoming meaningless, because it enables the modern historian to identify the category of 'science', or whatever it was called in the past, as always having existed. The correct procedure is probably somewhere in between, viz. in applying the concept of 'science' in historical contexts only with the necessary conceptual circumspection and qualifications (which is my aim in this study).

[50] P. Donini, 1994a, p. 48, n. 32 (the quotations that follow are drawn from this passage).

[51] Compare the use of inverted comma's by I. Gallo, 1998, p. 3527 (= 1999, p. 64) in his description of *Quaestiones naturales* as "une raccolta di quesiti 'scientifici' con le relative risposte".

light on Plutarch's *De facie*, *De primo frigido* and *Quaestiones naturales*. As to the *De facie*, first of all, he seems to overdo things when he writes that the work:

> "contiene certamente molta buona scienza astronomica e d'altro genere, ma non è affatto un trattato scientifico: discute un importante problema filosofico, al quale è totalmente subordinato il contenuto che i moderni [!] chiamano scientifico."

Further on, however, Donini is more at ease to underline the conceptual unity of *De facie*, pointing out that the closing myth:

> "vuolo ridurre, non distruggere il credito delle scienze; è lí precisamente a dirci che il discorso delle scienze fisiche e matematiche, se non è teologicamente fondato, sarà sempre inadeguato."[52]

When Donini, in what follows, claims that Plutarch in *De primo frigido* – which he interprets "a rovescio dalla maggior parte dei moderni" – actually discusses an epistemological problem, he restricts the entire treatise to the ἐποχή statement in the very last paragraph (in combination with the eighth [see 4.3.2.1.]). This seems to entirely overlook Plutarch's introductory discussion about whether heat and cold are principles of their own and also the lengthy doxography that follows on the principle of cold[53]. However, I see no reason to assume that both types of discourse (viz. physical and epistemological) were seen as completely distinct by Plutarch himself.

With respect to *Quaestiones naturales*, lastly, Donini remains pensive in making his judgements:

> "A conti fatti, ho l'impressione che resti disponibile per questa classe [sc. scritti di fisica e di scienza naturale] forse (forse) la sola raccolta delle *Cause naturali*, che potrebbe essere soltanto un insieme di appunti da elaborare in altre opere di tutt'altro genere."

---

[52] P. Donini, 1994a, p. 56. Cf. also Donini's conclusion in 1984, p. 374: "[M]i chiedo però se abbiamo con ciò stesso il diritto di dire che l'*historia* di Plutarco non era autentica curiosità scientifica. Era, quanto meno, la base adeguata per una scienza come quella allora possibile, così solidale con quel platonismo, con quei demoni, con quell'irrazionalità." Cf. also *id.*, 1988 and 2011, p. 21, n. 26.

[53] Other scholars have interpreted the ancient scientific character of *De primo frigido* in terms of its allegedly playful and theoretical approach. See O. Longo, 1992, p. 229: "*Lusus* di letterato? Vera e propria discussione scientifica? Un po' dell'uno e un po' dell'altro, e in ogni caso una fisica fatta a tavolino e sui libri, assai più che mettendo il naso fuori dalla finestra."

Donini's hesitation is based on a conviction that the preceding objections (regarding the dualistic and epistemological aspects of Plutarch's science) do not, at first, seem to apply to this collection. However, as already noted at the beginning of the prologue, one of the goals of this study will be to show that the reverse is in fact true, by demonstrating that *Quaestiones naturales* is fundamentally in line, despite its main focus on physical causality, with Plutarch's general scientific project, including its dualistic and epistemological backdrop [see 4.1.–3.].[54]

Another point that needs to be stressed in evaluating the scientificitiy of Plutarch natural philosophical writings is the contingent nature of Plutarch's truth claims. For instance, regarding the claim in *Quaest. conv.* 641C that there is empirical proof of the fact that a magnet rubbed with garlic loses its attractive powers, Lehoux rightly points out that:

> "Any given way of framing questions of truth and falsity about the natural world is bound up in networks of relations and requires a background of standards, concepts, methods, tools, and objects against which truth and falsity can be judged."[55]

As such, the ultimate scientific truth – whatever this may be – will only be a circumstantial aspect for properly evaluating the scientific character of Plutarch's theories about the natural world[56]. This boils down to the idea that science is – at least in a historical sense – not all that concerned with 'truth' per se.

---

[54] Donini's judgement on the scientific character of Plutarch's "scritti di fisica e di scienza naturale" (c.q. *De facie*) is more nuanced in 1992, pp. 106–107: "Ma, dato che noi moderni siamo abituati a pensare che sia 'scientifica' quella spiegazione che del fenomeno in discussione dice veramente tutto l'essenziale e ne chiarisce i fondamenti ultimi, è evidente che parlando della normalità dei casi di cui si occupava Plutarco non dovremmo definire 'scientifica' altra spiegazione che quella che dica entrambi gli ordini delle cause e prima di tutte, anzi, quelle 'divine': dove infatti siano coinvolte entrambi le classi, la spiegazione che si limitasse a far rilevare le sole cause materiali e strumentali sarebbe sì ancora 'fisica', ma non certamente 'scientifica'. Volendo proprio usare questo termine in riferimento al nostro autore, bisognerà adattarsi ad accettare che sia 'scientifico' l'argomento che è anche e prima di tutto teologico." Donini draws a similar conclusion for *De primo frigido* in 1986a, p. 211: "il problema del *de primo frigido* insomma non è esclusivamente fisico, ma in ultima istanza rinvia agli agenti metafisici e divini dell'ordinamento del cosmo". See also J. Opsomer, 1998, p. 214.

[55] D. Lehoux, 2007, p. 448, see also 2003, pp. 339–340 (cf. also, e.g., T. Barton, 1994a, p. 4). For further discussion of Lehoux' account, see M. Meeusen, 2014, 337–338.

[56] Contrast the concern of P. Donini, 1988, pp. 126–127 regarding *De facie*: "Physics, astronomy, and geometrical optics are here used to explain the nature of the moon and its spots, and the explanation proposed is the closest to scientific truth that we know from Antiquity."

In order to clarify this, an analogy can be drawn with the revolutionary finds physicists working at CERN near Geneva, home of the Large Hadron Collider, have done in the last years. Besides from further specifying our knowledge about the basic structure of our universe (think of the discovery of the Higgs-boson or 'God particle'), these scientists have recently found clues that there may be a deeper kind of physics, a dark sector that we have not been able to reach yet, and that may even be unreachable to us. It is this deeper kind of physics that, so it is often metaphorically described, shines through the cracks of the Standard Model, which up to now served – and continues to serve – as the basic fundament of particle physics. It is not so much that the Standard Model would be wrong, of course, but one may still wonder how scientific it really is, considering the fundamental and thus far unanswered problems these discoveries raise about the existence of the universe (not to mention the potential existence of other universes). What is relevant for the history of science, then, is not so much *whether* any truth comes out of smashing atoms together, but *what kind* of truth people make of it, *how* it is reached, and *why* such research is conducted to begin with. What the study at hand aims to do, then, is to put the kind of natural problems Plutarch sought to explain in a contextual perspective, to find out what ultimate goals he had in mind in doing so, and which conceptual limits he faced. To this end, we can learn a great deal from the kind of answers he provides and from the general methodology and theoretical-conceptual framework that he employs.

Apart from revealing these intellectual mechanisms that underlie a person's or a society's world view[57], contemporary studies of ancient scientific literature often also bring into consideration how specific socio-cultural factors play along in the authorisation, validation and dissemination of scientific knowledge in particular societal contexts[58]. A study of these features is relevant to us, as it provides a valuable perspective on how the natural world was perceived of in Antiquity and how this view became entrenched in a real-life community and

---

[57] For the concept of 'world view' as a discursive category in ancient scientific texts, cf. H. Flashar, 1962 (e.g., pp. 318 and 331), who uses its German equivalent (*Weltbild*) to designate the general explanatory scheme that can be deduced from the problems and explanations recorded in Ps.-Aristotle's *Problems*. Cf. also, e.g., the title in D. Lehoux, 2012.

[58] This socio-cultural contextualisation is, in fact, very central to contemporary studies in the field of the history of science: cf., e.g., T.E. Rihll, 1999, p. iv. Cf. also J.T. Vallance, 1990, pp. 716–717: "First, it is now generally agreed that we should not – indeed, cannot – separate ancient science from its philosophical background. Ethical, metaphysical, and scientific motivations must be understood side by side. Second, some idea of the social and cultural context in which scientific ideas developed is now widely seen as essential to any generally useful appreciation."

civilisation – the Greco-Roman civilisation, which had a seminal influence on our own. This has reflective value for our contemporary outlook on the natural world and for the place that is allotted to it in our own modern society, where science has become omnipresent.

Of course, world views can shift over time, and can be different from person to person or from society to society. Therefore, it seems useful to study scientific concepts and theories on their own terms and in view of the social and intellectual contexts from which they originate. This is especially to be understood as a caveat to the teleological approach as described above (i.e. the 'genetic history'). It does not imply that a diachronic approach should be avoided at all costs. This approach has its use, for instance, when examining the scientific or philosophical importance of a person's outlook on the world in view of a scientific or philosophical historiography[59]. From this perspective, the value of Plutarch's *Quaestiones naturales* for the history of science is reflected in the work's reception and transmission by later authors and can be examined by studying how they picked it up and adopted it to suit their specific authorial needs. In what follows, I will not discuss this in full detail but will only highlight a number of cases that were of importance for the text's later history.

The 11th century Byzantine polymath and 'Chief of Philosophers' Michael Psellus is especially worth mentioning here, as he incorporated and adapted several of Plutarch's natural problems – both from *Quaestiones naturales* and *convivales* – in his encyclopaedic *De omnifaria doctrina* (two additional chapters, viz. *Q.N.* 40–41, derive from this work: see the commentary *ad loc.*). In global, Psellus' work nicely illustrates how Pagan knowledge, thus including Plutarch's natural problems, was hesitantly accepted by the author and which intellectual restrictions were imposed on it by the religious (c.q. Christian-Orthodox) establishment of his time. Notably, Psellus did not lable his excerpts as being drawn from Plutarch (thus, perhaps, implicitly rejecting the Chaeronean's scientific authority?). On the contrary, it seems that Psellus, through Plutarch's lens, looked at, and approved of, *Aristotle's* scientific authority by quoting the *Problems* via Plutarch. The merit of Psellus' *De omnifaria doctrina* (or at least the work's first redaction) lies in its attempt to create a genuine Christian cosmology, which is firmly based on ancient authority. Importantly, Psellus addressed his work to the Byzantine emperor, God's regent on earth. The relationship between such encyclopaedic knowledge and

---

[59] E.g., for the influence of Plutarch's *De facie* on Kepler, see H. Görgemanns, 1970, pp. 157–161. Kepler found the text so important that he edited it with his own Latin translation and astronomical commentary. In what follows, I repeat some of the insights gained in previous publications on the reception of Plutarch's natural problems (for a short overview, see M. Meeusen, forthcoming b).

imperial power is not of disinterest, as it provides a better understanding of what highly placed Byzantine figures were expected to know, or, at the very least, to have read[60].

A similar case of religious adoption of Plutarch's natural problems is found in the *Diálogos Familiares de la Agricultura Cristiana* (1589) by the Spanish humanist and Jesuit Juan de Pineda. In this work, the author relies heavily on Plutarch's authority (amongst that of other Pagan authors) and, at points, incorporates several passages from *Quaestiones naturales* in his Christian discourse. As Ramón Palerm has shown, the author of this work "through an ongoing confrontation of the Christian and Pagan worlds, struggles to win for the Christian cause the content of the ancient traditions, to which he gives an obvious moral sense in a didactic-doctrinal tone"[61]. As such, the cases of Psellus and de Pineda show how later, Christian authors – both in the Orthodox East and in the Reforming West (de Pineda speaks with little respect of the Spanish Inquisition) – used Plutarch's natural problems as a basis for their own inquiries, not so much by addressing them anew, but by exploiting them as a *Fundgrube* of exotic materials to be assimilated into the context of a new (c.q. Christian) world system.

The situation is different in other cases, though. The scholarly interest in Plutarch's *Quaestiones naturales* in the Humanist era is reflected mainly in the production of new editions and Latin translations (mostly in collective volumes with other works from the *corpus Plutarcheum*). The 1542 Latin translation by the Dutch Protestant scholar, professor and doctor Gybertus Longolius deserves specific mention here. In this Latin version, the Aldine problems (*Q.N.* 1–31) are followed by eight additional problems (*Q.N.* 32–39) that were extracted, so Longolius indicates in a marginal note, from a Milanese manuscript. Unfortunately, this manuscript has been lost ever since, and the Greek text is still missing today. Considering the numerous parallels in Plutarch's other works and the same general style and method of explanation, it is beyond doubt that these additional problems are authentic (see the commentary *ad loc.*)[62].

Another Latin translation of the Aldine problems was produced by Pedro Juan Núñez in 1554. Interestingly, this translation served as an appendix to Theodor Gaza's version of Ps.-Aristotle's and Ps.-Alexander's

---

[60] For further discussion of the reception of Plutarch's natural problems in Psellus' *De omnifaria doctrina*, see M. Meeusen, 2012b (*Quaestiones naturales*) and forthcoming c (*Quaestiones convivales*).

[61] V. Ramón Palerm, 2011, p. 621 (see pp. 629–632 for an analysis of the *Quaestiones naturales* material).

[62] For further detail on Longolius' translation, see A. Morales Ortiz, 1999 and M. Meeusen, forthcoming a. See also R. Flacelière, J. Irigoin, J. Sirinelli and A. Philippon, 1987, p. 283.

*Problems*⁶³. This is important, as it seems to imply that Plutarch was seen as continuous with a unified and long-lasting scientific tradition that was initiated by Aristotle and his Peripatetic successors (see also Psellus' case above). The practice of solving Aristotelian natural problems lasted well until the Middle Ages and the Renaissance, when new collections of problems made their appearance and older ones were constantly copied, translated and commented upon [see 1.1.3., n. 78]. As such, Plutarch's *Quaestiones naturales* were seen as a genuine contribution to the Aristotelian study and understanding of natural phenomena and to the development of a scientific method for approaching them. One of the goals of this study will be to nuance this view, and to show that Plutarch's natural problems are not the product of the author's Aristotelian aspirations – despite the fact that history clearly suggests otherwise [see 1.1.2. and 4.3.4.3.].

In conclusion, the study at hand takes inspiration from the plea often heard in recent scholarship to study ancient scientific texts impartial to considerations of quality or centrality. Rihll, for instance, has argued:

> "The primary sources for any period consist not just of the well-known and well-ploughed texts: there are a lot of grossly underutilized 'scientific' texts too, which cast a different and sometimes brilliant light on ancient society. This incidental information is also important to the historian of science, for the society in which the science was created shapes the science itself, and neither can be properly understood in isolation from the other."⁶⁴

By taking these contextual facets into consideration, the study at hand can be seen as a contribution to broadening the traditional image of ancient science and, by implication, of the history of science in general⁶⁵. Even though we are dealing with a rather obscure, non-canonical source in the case of *Quaestiones naturales*, this should not complicate our efforts to

---

⁶³ The work was printed in 1554 in Valencia by Joan Mei from Flanders as *Problematum Aristotelis sectiones duae de quadraginta. Problematum Alexandri Aphrodisiei libri duo Theodoro Gaza interprete ad haec Eruditissima problemata Plutarchi. Extant apud Borbonium bibliopolam. Valentiae, Typis Ioannis Mey, Flandri.* (Raya) 1554. See A. Morales Ortiz, 1998 and 2000, p. 90. For further detail on the Latin translations of Plutarch's *Moralia* more generally dating from the 13th to the 16th century, see F. Becchi, 2009.

⁶⁴ T.E. Rihll, 1999, p. 7.

⁶⁵ I, thus, accept Rihll's challenge (1999, p. xii): "there is a role and a need for 'ordinary' historians in the study of the history of Greek science, which is a land of opportunity for adventurous scholars."

pull it from the margins of the *corpus Plutarcheum*[66]. In this light, Van der Stockt has recently claimed that "it are precisely the more 'irrational' and 'absurd' beliefs and practices that are most fascinating"[67]. His plea for a profound study of Plutarch's *Quaestiones naturales* more in specific is worth quoting in full:

> "[I]t may have become clear that Plutarch's *Causes of natural phenomena* is in dire need of an interpretation that does justice to its peculiar nature. Provided that this research is conducted with philological tact as well as through a contextualising approach, the results are likely to shed light on the worldview of Plutarch as well as on the practice of authentic ancient 'science'."[68]

The phrase "philological tact", especially in conjunction with "a contextualising approach", is well-put from a methodological perspective but may require some further specification, which the following section will provide.

4. Classical philology and the petrification of science

Historians of ancient science generally hold that there was a rapid decline of scientific creativity after its 'Golden Age'[69] in the Hellenistic era, an era when several scientific disciplines, such as medicine, biology, alchemy, mathematics, geography, astronomy and mechanics, flourished more than ever before or ever after (esp. the 3rd and 2nd centuries BC). The Imperial Era, by contrast, is presented as a period of consolidation and transmission of received knowledge. The *Zeitgeist* of this era finds its incarnation in the figure of Pliny the Elder, author of the encyclopaedic *Naturalis historia*. Traditional Plutarchists have not resisted this view of a scientific decline in Plutarch's days, yet some nuancing is in place, at least in Plutarch's case.

In light of the period in which Plutarch lived, someone like Sambursky favoured the idea of a "petrification of science" by arguing that:

> "the first century A.D. marks the beginning of the work of compilers and interpreters which went on for more than four hundred years and

---

[66] This idea is inspired by the new historicist perspective that ascribes historical value to each historical product. For problems of quality and canon in Greek literature, see T. Whitmarsh, 2004, pp. 8–10.

[67] L. Van der Stockt, 2011, p. 447. A similar conclusion was made regarding ancient paradoxes and puzzles by G.E.R. Lloyd, 2004, pp. 5–7.

[68] L. Van der Stockt, 2011, p. 450.

[69] For this designation, see, e.g., G. Sarton, 1970 and S. Sambursky, 1963, p. 204.

which is the mirror wherein we see a large part of ancient Greek science."[70]

Regarding the natural problems discussed in Plutarch's *Quaestiones convivales*, Fuhrmann bade the reader, in a rather condescending fashion, to not be too hard on Plutarch, given that his time was afflicted with "un affaiblissement général de l'esprit scientifique"[71]. Similar conclusions have been

---

[70] S. Sambursky, 1963, pp. 204 and 242. Cf. also the references in P. Culham, 1992, p. 197, n. 30 more generally. Regarding the Greeks of the 2nd century AD, B.A. Van Groningen, 1965, p. 56 draws the following conclusion (with what seems to be an unhealthy sense for exaggeration): "There is no real activity; nobody sets out on an exploration; everybody walks on trodden paths. Why? Because they themselves are weak, unable to display psychic energy. They are tired; they sit down comfortably in well-known surroundings, and they are waiting, waiting for something they will not find, because it is not really looked for. [...] The Greek literature of the second century is the work of a powerless community, which, on the other hand, overstrains its faculties in unhealthy exaggerations. It is a neglected one in a neglected century, and, generally speaking, it deserves this neglect." The view that all post-Classical societies are of secondary importance, non-original and non-authentic is generally abandoned in contemporary scholarship (see, e.g., T. Whitmarsh, 2001, p. 28).

[71] F. Fuhrmann, 1972, p. xxiii. Regarding Plutarch's method of solving natural problems, more in particular, Fuhrmann draws the following conclusion (p. xxiv): "Pour les questions d'ordre scientifique, il est particulièrement grave de ne pas discuter les problèmes dans leur fond. Or c'est ce qui se passe ici. Au lieu de chercher les causes véritables des phénomènes, Plutarque se contente en général de la vraisemblance, en citant plusieurs théories qui s'y rapportent, ou en rappelant ce que divers auteurs en ont dit. Les différentes opinions se succèdent ainsi sans aucune analyse et le plus souvent sans solution, comme si ceux qui sont chargés de les défendre s'amusaient avec elles. [...] Quand par hasard Plutarque veut faire preuve d'esprit critique, il n'y réussit guère". This view is accepted by E. Teixeira, 1992, p. 221, who concludes, however, that: "Par l'intermédiaire de renseignements scientifiques, entre autres, Plutarque a, à n'en point douter, le mérite de contribuer pour une part important à mieux nous faire connaître la pensée et la culture grecques." In a similar fashion, R.H. Barrow, 1967, p. 22 claimed that "[t]he importance of the *Table Talks* rests not on their subject-matter, which [...] is often trivial, but upon the picture which they give of the society in which Plutarch moved, the texture of social life, and the ease and frequency of movement of people from place to place". For the inferior scientific *Zeitgeist* of Plutarch's time, see also K. Ziegler, 1951, col. 889: "Nicht verhehlen darf man sich, daß die Behandlung, besonders der naturwissenschaftlichen Probleme, oft recht oberflächlich und spielerisch ist; ein Vorwurf, der freilich nicht unserm P. zur Last fällt, sondern die seiner Zeit eigentümliche Erschlaffung des ernsthaften wissenschaftlichen Geistes kennzeichnet." Cf. also R. Flacelière, J. Irigoin, J. Sirinelli and A. Philippon, 1987, p. lxxxiii (quoted n. 28). For opposition against the idea of a contemporary "déclin du rationalisme", see, however, *ibid.*, p. lxxi. See also J. König, 2007, p. 51, who criticises Führmann's remark as "an assumption which exemplifies a common failure to understand the rhetorical idiom of so much ancient scientific writing". It was only since relatively

reached for Pliny the Elder's *Naturalis historia*[72]. Yet, as Stahl rightly nuances, "Plutarch demonstrates greater aptitude for assimilating and reporting scientific information than Pliny, but he is no less credulous and uncritical of quaint and incongruous data."[73] In light of Plutarch's allegedly derivative and compilatory authorship, Jeanneret even claimed that the Chaeronean (again in his *Quaestiones convivales*) could be categorised as an author who, just like Athenaeus or Macrobius, aimed for a variegated, though bloodless accumulation of knowledge, without any sense of critical evaluation. The lack of scientific creativity and ingenuity thus becomes connected with a faceless authorship that is characterised by the absence of any serious intellectual talent and ambition for personal creativity:

> "What is often said of Plutarch is also true of the others: they are basically eclectic. They neither judge nor criticize, but rather put things on show [...]. The author melts into an anonymous collector and mediator; he lets the books, of which he is a mere interpreter, speak for themselves."[74]

The idea that Plutarch should be ranked among other late compilers and interpreters risks grossly oversimplifying the real accomplishments of his work. Throughout his scientific writings, to go no further afield, Plutarch unambiguously aims to formulate inventive and innovative explanations for the phenomena he studies [see 4.2.2.2.]. In fact, he often explicitly marks his own, personal views (e.g., via his literary *alter ego* in *Quaestiones convivales*), thus emphasising that he does not blindly rely on received knowledge. Of course, the aspect of Plutarch's scientific innovativeness should be placed *within* the ancient scientific paradigm of his time: Plutarch was no scientific revolutionary. He did not leave the confines of the 'normal science' of his day. But even so, this did not stop him from contriving explanations that were certainly original *within* the scientific paradigm at that time[75].

---

recently that scholars have started to recognise and revalue the argumentative style of *Quaestiones convivales*. See, e.g., F. Frazier and J. Sirinelli, 1996, pp. 177–207, J. König, 2008, p. 88, n. 11 and the contributions in F. Klotz and K. Oikonomopoulou, 2011.

[72] See V. Naas, 2011, pp. 61, and esp. 66–67.

[73] W.H. Stahl, 1962, p. 133.

[74] M. Jeanneret, 1991, p. 167. For the alleged facelessness of Plutarch's authorship, see also R.H. Barrow, 1967, p. 15: "he [Plutarch] was a tantalisingly modest man and he effaces everything personal from his narrative". The same passage from Jeanneret was quoted and criticised by J. König, 2011, p. 189 (cf. also *id.*, 2007, pp. 51–52 and J.C. Relihan, 1992, p. 218).

[75] Cf. R. Flacelière, J. Irigoin, J. Sirinelli and A. Philippon, 1987, p. lv: "Pour les sciences physiques et biologiques, par exemple, encore qu' il émette parfois des opinions qui semblent bien lui être personnelles, il suit d' ordinaire Aristote et l' école péripatéti-

Only since relatively recently, scholars have started to recognise that Plutarch's scientific project, and his science of natural problems more in specific, is not just an impersonal echolalia of past authorities. But even so, most of the existing scholarship remains restricted to matters of source criticism. Of course, traditional *Quellenforschung* has great value for properly understanding the composition of a text on the basis of its specific sources and, more broadly, of the tradition in which it is anchored (this will prove a particularly useful approach in the commentary), but it also has a number of interpretive deficits. By largely restricting its scope to the analysis of the sources of a text, such an approach tends to downplay, isolate, and eventually even exclude the author from his own text[76]. The author is, in fact, reduced – often in a highly speculative fashion – to the sources that he is claimed to rely on, while his personal adaptations of the tradition are generally ignored[77]. If there is any place left for the author's personal contributions, these are pushed to the margin of the text, if for no other reason than that scholars are embarrassed not to have found any matching *Quellen* for them[78].

On the other hand, even if Plutarch clearly pursues argumentative creativity and originality in his scientific writings, traditional authorities still play an important role in his arguments. As such, his innovativeness is perhaps not as 'adventurous' as modern critics may have expected[79].

---

cienne. [...] Mais il ne semble pas avoir fait progresser lui-même les connaissances scientifiques de son temps. Il fut un 'honnête homme' au sens du xviie siècle [*cave*: see n. 48], un véritable érudit, très informé et très éclairé. Il réalisa parfaitement en lui l'idéal du πεπαιδευμένος ἀνήρ."

[76] See the introduction "On the impoliteness of *Quellenforschung*" in L. Van der Stockt, 2004, pp. 331–333. See also J. Mansfeld, 1999, pp. 13–16. For similar criticism of the presentation of Plutarch as "un simple compilateur, tout juste capable de recopier des œuvres antérieures en les démarquant", see also R. Flacelière, J. Irigoin, J. Sirinelli and A. Philippon, 1987, p. ccii (with further references). See also I. Gallo *apud* A.M. Battegazzore, 1992, p. 37.

[77] E.g., regarding the *Quellenforschung* tradition on Plutarch's writings on animal psychology, see S.T. Newmyer, 1992, p. 41: "Scholars eager to detect the sources of Plutarch's arguments have paid relatively little attention to the points which he makes or to how he defends his positions." On Plutarch's adaptation of his source-material in the *Vitae*, cf. also C. Pelling, 1980.

[78] See, e.g., K. Giesen, 1901, p. 448 (with the quote from F. Leo, 1864). Such an approach would be sound if the text were simply a derivate of traditional sources, but things are obviously not always so simple.

[79] See, e.g., R.H. Barrow, 1967, p. 74: "It is not too much to say that Plutarch had at his command Hellenic and Hellenistic thought and literature. His mind was not adventurous; it did not use its accumulated knowledge as a springboard to make a leap; it may have lacked imagination." See also Barrow's strong words on p. 77: "Plutarch was a teacher, not a constructive thinker; he created no new system. Few teachers do; they may have

Even still, to claim that his scientific programme is representative of, what Sambursky called, the "petrification of science in the period of commentators and scholastics"[80] is a one-dimensional presentation of the facts, as this study will show [see 4.2.2.2.].

An alternative way to examine how Plutarch deals with received knowledge and how he constructs his own scientific authority is by studying his 'authorial voice' – understood as an analytical concept that has recently been introduced by scholars of ancient scientific and technical writing[81]. This 'authorial voice' serves as an important discursive category in the text that resonates, for instance, in the explicit evaluation of the theories of predecessors but is also situated at a structural level, viz. in the development and ranking of the different arguments, and can also be seen in his method of quoting and adapting the available source material[82]. The concept of the author's voice proves to be a worthy analytical tool in examining the discursive construction (and deconstruction) of scientific authority and, more particularly, in detecting the author's scientific creativity that manifests itself in this process. The question of scientific authority is highly relevant in examining Plutarch's

---

at the back of their minds guiding principles, but they do not readily build theoretical constructions but rather make practical application. They are bound to be opportunists, availing themselves of the openings furnished by their pupils, responding to questions in answers adapted to the intelligence and experience of the questioner and then abandoned till some later time." This is in keeping with what he says earlier about *Quaestiones convivales* (p. 27): "table talks are interesting for their incidental matter, seldom for the original question or its answer". For the question of Plutarch's originality in the field of ancient science, see also A.M. Battegazzore, 1992, p. 32 and pp. 35–36. *Pace* J. Sirinelli, 2000, p. 359: "Les explications qu'il nous donne sur la patine des statues des navarques lacédémoniens à Delphes [= *De Pyth. or*. 395A–396C; see 2.1.1.] sont un peu naïves et ses traits *Sur le froid primitif* et *Sur les causes physiques* sont des recueils d'opinions glanées au gré de ses lectures. Le résultat n'est pas plus brillant que les considérations qui jalonnent les *Propos de table*. Ce sont des 'curiosités' dont, avec ses convives, il cherche l'explication, et non des démarches scientifiques. En résumé, on peut difficilement présenter Plutarque comme un savant, même à l'aune de ce temps." For further criticism of the general scholarly contempt of Plutarch's originality in the field of philosophy, see J. Opsomer, 1994a, pp. 17–19 (with further references).

[80] S. Sambursky, 1963, p. 242.

[81] See the contributions in L. Taub and A. Doody, 2009 and M. Asper, 2013.

[82] To employ the alternative terminology of the classical hermeneutical schema set out by W. Babilas, 1961, the 'authorial will' (*voluntas auctoris*) has to be taken into consideration at any point in the text. In light of the philological method employed by traditional *Quellenforschung*, Babilas has warned that traditional *materia* is only seldom simply copied by the author. It is not only commented on and criticised, but also often adapted according to the particular needs of the author. The reconciliation of the *materia* to the new context occurs at three distinct levels, viz. of content, arrangement and phrasing (*inventio, dispositio, elocutio*).

position towards the scientific tradition, and more precisely how he tries to inscribe himself within it (or opposes it). Also the use and avoidance of a self-referential *ego* and alternative personal forms are relevant to us, as they can tell us something about the underlying sociology of the text[83]. In particular, the addresses to the reader can inform us about the envisaged knowledge transfer from author to reader (and, thus, the text's intended reading). Analysing these discursive features in Plutarch's *Quaestiones naturales* will yield a much richer interpretation of the text than has been offered thus far. Let us, therefore, take a closer look at the existing scholarship and how this study aims to contribute to it.

5. Status quaestionis

Scholars have only relatively recently started to reappraise the actual appeal of Plutarch's scientific writings. In 1976 Flacelière could still write that "'Plutarque et la science de son temps' est un sujet qui a été à peine effleuré jusqu'ici, et qui mériterait, à mon avis, des recherches approfondies"[84]. However, there has been considerable progress since that time. Scholarly attention mostly goes to the more literary and essayistic treatises like *De facie*, *De primo frigido*, *Quaestiones convivales* and Plutarch's writings on animal psychology, whereas *Quaestiones naturales* often remains undiscussed[85]. It may well be the case that these other treatises offer a more efficient introduction into the author's scientific thought than the one at issue here, but then again *Quaestiones naturales* originates from the same 'genius', and, if we may assume that the much praised unity of Plutarch's works also applies to this work[86], there are clear

---

[83] On Plutarch's self-presentation in *Quaestiones convivales*, see F. Klotz, 2007 and J. König, 2011.

[84] R. Flacelière, 1976, p. 195. Several editions and commentaries of Plutarch's scientific writings had already appeared by that time, but no further study of these texts had been conducted – with the important exception of H. Görgemanns, 1970. For a general introduction into 'Plutarch and the sciences', see R. Flacelière, J. Irigoin, J. Sirinelli and A. Philippon, 1987, pp. lxix–lxxxvii, J. Sirinelli, 2000, pp. 355–366 and the contributions in I. Gallo, 1992; see also recently M. Meeusen and L. Van der Stockt, 2015. Scholars have devoted much attention to the presence and tenets of several separate scientific disciplines and fields of technical knowledge throughout Plutarch's works, such as medicine (e.g., J. Boulogne, 1996; M. Vamvouri Ruffy, 2012), astronomy (A. Pérez Jiménez, 1992), mathematics (R. Seide, 1981; M. Isnardi Parente, 1992), music (J.P.H.M. Smits, 1970) and linguistics (O. Göldi, 1920).

[85] Characteristic of the general scholarly neglect of *Quaestiones naturales* is the fact that in the proceedings of the 1991 conference on "Plutarco e le scienze" (organised by the Italian section of the *International* Plutarch Society), there is only one reference to this work (in a footnote): see O. Longo, 1992, p. 230, n. 4. The same observation was made by L. Van der Stockt, 2011, p. 449.

[86] See J. Barthelmess, 1986, pp. 62–64 and the contributions in A.G. Nikolaidis, 2008.

indications that it is in several respects (including contents, method and composition) fundamentally in line with his other writings. In light of this understanding, Harrison even wrote – in a lyrical vein – that Plutarch's problems "offer endless delight to the literary critic since they are so well written and so deeply infused with the warmth of Plutarch's personality". Regarding *Quaestiones naturales*, he asserts that they "offer a coherent and manageable collection for an investigation of Plutarch's style of composition and literary techniques within all of the *quaestiones*"[87]. One of the goals of this study will be to prove Harrison right, while also showing that there is even more to the collection than matters of style, composition and literary technique alone.

The situation has improved for *Quaestiones naturales* over the past few years. We have several thematic studies which examine specific aspects of the collection in more detail, viz. specific textual problems[88], its textual history[89], the relationship between Plutarch and Aristotelian science[90], Plutarch's use of scientific terminology[91], the collection's literary value[92] as well as its 'encyclopaedic' appeal[93]. A programmatic study of *Quaestiones naturales* with specific attention to the cluster analysis of parallel passages was conducted by Van der Stockt[94] [see 2.1.2.]. Until recently, a systematic study of the collection, which includes both an analytical and a descriptive approach, remained a scholarly desideratum. The 2006 edition, with an Italian introduction, translation and commentary by Senzasono was intended to fill in this lacuna (in the *Corpus Plutarchi Moralium* series). Senzasono's work offers a useful contribution especially as a scholarly instrument for consultation of the text, its translation and specific lemma's, but the general approach in the introduction is somewhat disappointing in some regards[95]. Senzasono's main focus is on the text and its translation, and this is certainly where the main value of his work rests. In his introduction, he is not, however, concerned with the actual place of the collection in its wider socio-cultural and intellectual-philosophical context, especially its educational goals. Senzasono does not

---

[87] G.W.M. Harrison, 2000b, p. 237.
[88] S.-T. Teodorsson, 1990b, V. Ramón Palerm, 2005, M. Meeusen, 2015a and 2015b.
[89] A. Morales Ortiz, 1999, M. Meeusen, 2012b and forthcoming a, b, c.
[90] S.-T. Teodorsson, 1999a, K. Oikonomopoulou, 2011, M. Meeusen, 2011 and 2016.
[91] L. Senzasono, 1999, J. Opsomer, 1999, M. Meeusen, 2013b, L. Van der Stockt, 2013.
[92] G.W.M. Harrison, 2000a and 2000b.
[93] K. Oikonomopoulou, 2013a.
[94] L. Van der Stockt, 2011. For a preliminary study of Plutarch's science of natural problems more generally, see M. Meeusen, 2014 and 2015b.
[95] See also the criticism by L. Van der Stockt, 2011, p. 449. The notes in Senzasono's commentary are often very extensive, but they do not always clarify the actual logic behind Plutarch's arguments and they are sometimes dizzyingly off track (moreover, for a scholarly instrument the "indici" at the end of the book are rather meagre).

discuss Plutarch's dualistic view on causality, a remarkable interpretative inadvertence considering the collection's main aetiological concern. In general, Senzasono fails to provide a coherent and comprehensive account of the text's place among Plutarch's other (natural scientific) writings, its social and intellectual *Sitz im Leben* and its scientific methodology.

The more dated, though still recommendable, 1965 Loeb edition by Sandbach is by no means adumbrated by this recently renewed scholarly interest in *Quaestiones naturales*. The relatively circumstantial introduction and clear notes have been a welcome source of inspiration for this study. Sandbach[96] is obliged to Hubert's 1960 Teubner edition for the collection of parallel passages, from which he admits to "have drawn heavily" in his own clarifying notes. For my part, I owe them both a serious debt of gratitude in this regard[97]. *Quaestiones naturales* has been translated into several modern languages[98], but nowhere do we find a comprehensive and monographic study of the collection as a whole, accompanied by a thorough lemmatic commentary. Therefore, the present study will attempt to restore this often marginalised and undervalued Plutarchan work by rehabilitating its significance to contemporary scholarship[99]. This will be done in two ways. The first part of this study contains four chapters, which successively discuss 1) the collection's relation to the Aristotelian genre and tradition of natural problems and its sub-literary and a-moralistic discourse, 2) its relationship with other works in the *corpus Plutarcheum* (esp. *Quaestiones convivales*) and its alleged hypomnematic nature, 3) its educational and intellectual-philosophical value as a propaedeutic school text, and finally, 4) its aetiological design and scientific method. The first part will provide the preliminaries required for an informed reading and proper understanding of the text itself, which is presented in the form of a commentary in part two (see the introduction *ad loc.*).

6. *Note on translations and abbreviations*

Translations are borrowed from the Loeb Classical Library (with sporadic adaptations). Commonly used abbreviations are as follows.

---

[96] F.H. Sandbach, 1965, p. 147.

[97] The collection has not yet been edited in the *Collection des Universités de France* (Budé) but will be soon by Filippomaria Pontani and myself.

[98] See, e.g., D. Ricard, 1844 (French), C.F. Schnitzer, 1860 (German), V. Bétolaud, 1870 (French), W.W. Goodwin, 1878 (English), F.H. Sandbach, 1965 (English), V. Ramón Palerm and J. Bergua Cavero, 2002 (Spanish), G. Janssen, 2004 (Dutch), L. Senzasono, 2006 (Italian).

[99] In much the same way as J. Boulogne, 1992, p. 4683 tried to revalue *Quaestiones Romanae* by saving it from a classification among Plutarch's "écrits secondaires, négligeables ou à jeter aux oubliettes".

General abbreviations

DK	Diels, H. and Kranz, W., *Die Fragmente der Vorsokratiker*, Dublin – Zurich, 1966–1967.
D.L.	Diogenes Laertius, *Vitae philosophorum*.
FGrHist	Jacobi, F., *Die Fragmente der griechischen Historiker*, Leiden, 1957–1969.
L&S	Lewis, C.T. and Short, C., *A Latin Dictionary*, Oxford, 1969.
LSJ	Liddell, H.G. and Scott, R., *A Greek-English Lexicon*, 9th ed. (rev. Jones, H.S.; with a revised supplement), Oxford, 1996.
OED	Trumble, W. and Brown, L., *The Oxford English Dictionary*, 5th ed., Oxford, 2002.
RE	Wissowa, G. *et al.* (eds.), *Paulys Realencyclopädie der classischen Altertumswissenschaft*, München – Stuttgart, 1893–1980.
SVF	von Arnim, H., *Stoicorum Veterum Fragmenta*, Leipzig, 1903–1924.
TGF	Snell, B., Kannicht R. and Radt, S., *Tragicorum Graecorum Fragmenta*, Göttingen, 1977–2004.
FHSG	Fortenbaugh, W.W., Huby, P., Sharples, R.W. and Gutas, D., *Theophrastus of Eresus. Sources for his Life, Writings, Thought and Influence*, Leiden, 1992–…

Plutarch's works

Moralia

De liberis educandis	*De lib. educ.*
De audiendis poetis (*Quomodo adolescens poetas audire debeat*)	*De aud. poet.*
De audiendo (*De recta ratione audiendi*)	*De aud.*
De adulatore et amico (*Quomodo adulator ab amico internoscatur*)	*De ad. et am.*
De profectibus in virtute (*Quomodo quis suos in virtute sentiat profectus*)	*De prof. in virt.*
De capienda ex inimicis utilitate	*De cap. ex inim.*
De amicorum multitudine	*De am. mult.*
De fortuna	*De fortuna*
De virtute et vitio	*De virt. et vit.*
Consolatio ad Apollonium	*Cons. ad Apoll.*
De tuenda sanitate praecepta	*De tuenda*
Coniugalia praecepta	*Coni. praec.*
Septem sapientium convivium	*Sept. sap. conv.*
De superstitione	*De sup.*
Regum et imperatorum apophthegmata	*Reg. et imp. apophth.*

Apophthegmata Laconica – Instituta Laconica – Lacaenarum apophthegmata	*Apophth. Lac.*
Mulierum virtutes	*Mul. virt.*
Quaestiones Romanae	*Quaest. Rom.*
Quaestiones Graecae	*Quaest. Graec.*
Parallela Graeca et Romana	*Parall. Graec. et Rom.*
De fortuna Romanorum	*De fort. Rom.*
De Alexandri Magni fortuna aut virtute	*De Al. Magn. fort.*
Bellone an pace clariores fuerint Athenienses (*De gloria Atheniensium*)	*Bellone an pace*
De Iside et Osiride	*De Is. et Os.*
De E apud Delphos	*De E*
De Pythiae oraculis	*De Pyth. or.*
De defectu oraculorum	*De def. or.*
An virtus doceri possit	*An virt. doc.*
De virtute morali	*De virt. mor.*
De cohibenda ira	*De coh. ira*
De tranquillitate animi	*De tranq. an.*
De fraterno amore	*De frat. am.*
De amore prolis	*De am. prol.*
An vitiositas ad infelicitatem sufficiat	*An vitiositas*
Animine an corporis affectiones sint peiores	*Animine an corp.*
De garrulitate	*De gar.*
De curiositate	*De cur.*
De cupiditate divitiarum	*De cup. div.*
De vitioso pudore	*De vit. pud.*
De invidia et odio	*De inv. et od.*
De se ipsum citra invidiam laudando (*De laude ipsius*)	*De se ipsum laud.*
De sera numinis vindicta	*De sera num.*
De fato	*De fato*
De genio Socratis (*De Socratis daemonio*)	*De genio Socr.*
De exilio	*De exilio*
Consolatio ad uxorem	*Cons. ad ux.*
Quaestiones convivales	*Quaest. conv.*
Amatorius	*Amatorius*
Amatoriae narrationes	*Am. narr.*
Maxime cum principibus philosopho esse disserendum (*Maxime cum principibus philosophandum esse*)	*Maxime cum principibus*
Ad principem ineruditum	*Ad princ. iner.*
An seni respublica gerenda sit	*An seni*
Praecepta gerendae reipublicae	*Praec. ger. reip.*

De unius in republica dominatione, populari statu, et paucorum imperio	*De unius*
De vitando aere alieno	*De vit. aer.*
Decem oratorum vitae	*Dec. or. vit.*
Comparationis Aristophanis et Menandri epitome	*Comp. Ar. et Men.*
De Herodoti malignitate	*De Her. mal.*
Placita philosophorum	*Plac.*
Quaestiones naturales	*Q.N.*
De facie quae in orbe lunae apparet	*De facie*
De primo frigido	*De prim. frig.*
Aqua an ignis sit utilior	*Aqua an ignis*
De sollertia animalium (*Terrestriane an aquatilia animalia sint callidiora*)	*De soll. an.*
Gryllus (*Bruta animalia ratione uti*)	*Gryllus*
De esu carnium	*De esu*
Quaestiones Platonicae	*Quaest. Plat.*
De animae procreatione in Timaeo	*De an. procr.*
De Stoicorum repugnantiis	*De Stoic. rep.*
Stoicos absurdiora poetis dicere	*Stoic. absurd. poet.*
De communibus notitiis adversus Stoicos	*De comm. not.*
Non posse suaviter vivi secundum Epicurum	*Non posse*
Adversus Colotem	*Adv. Col.*
De latenter vivendo (*An recte dicendum sit latenter esse vivendum*)	*De lat. viv.*
De vita et poesi Homeri	*De vit. et po. Hom.*
Parsne an facultas animi sit vita passiva	*Pars an fac.*
De musica	*De mus.*
Fragments	fr. Sandbach
Lamprias catalogue	Lampr. cat.

Vitae

Theseus	*Thes.*
Romulus	*Rom.*
Comparatio Thesei et Romuli	*Comp. Thes. et Rom.*
Lycurgus	*Lyc.*
Numa	*Num.*
Comparatio Lycurgi et Numae	*Comp. Lyc. et Num.*
Solon	*Sol.*
Publicola	*Publ.*
Comparatio Solonis et Publicolae	*Comp. Sol. et Publ.*
Themistocles	*Them.*
Camillus	*Cam.*

Aristides	*Arist.*
Cato Maior	*Ca. Ma.*
Comparatio Aristidis et Catonis	*Comp. Arist. et Ca. Ma.*
Cimon	*Cim.*
Lucullus	*Luc.*
Comparatio Cimonis et Luculli	*Comp. Cim. et Luc.*
Pericles	*Per.*
Fabius Maximus	*Fab.*
Comparatio Periclis et Fabii Maximi	*Comp. Per. et Fab.*
Nicias	*Nic.*
Crassus	*Crass.*
Comparatio Niciae et Crassi	*Comp. Nic. et Crass.*
Alcibiades	*Alc.*
Marcius Coriolanus	*Cor.*
Comparatio Alcibiadis et Marcii Coriolani	*Comp. Alc. et Cor.*
Lysander	*Lys.*
Sulla	*Sull.*
Comparatio Lysandri et Sullae	*Comp. Lys. et Sull.*
Agesilaus	*Ages.*
Pompeius	*Pomp.*
Comparatio Agesilai et Pompeii	*Comp. Ages. et Pomp.*
Pelopidas	*Pel.*
Marcellus	*Marc.*
Comparatio Pelopidae et Marcelli	*Comp. Pel. et Marc.*
Dion	*Dion*
Brutus	*Brut.*
Comparatio Dionis et Bruti	*Comp. Dion. et Brut.*
Timoleon	*Timol.*
Aemilius Paulus	*Aem. Paul.*
Comparatio Timoleontis et Aemilii Pauli	*Comp. Tim. et Aem.*
Demosthenes	*Dem.*
Cicero	*Cic.*
Comparatio Demosthenis et Ciceronis	*Comp. Dem. et Cic.*
Alexander	*Alex.*
Caesar	*Caes.*
Sertorius	*Sert.*
Eumenes	*Eum.*
Comparatio Sertorii et Eumenis	*Comp. Sert. et Eum.*
Phocion	*Phoc.*
Cato Minor	*Ca. Mi.*
Demetrius	*Demetr.*
Antonius	*Ant.*
Comparatio Demetrii et Antonii	*Comp. Demetr. et Ant.*

Pyrrhus	*Pyrrh.*
Caius Marius	*Mar.*
Agis	*Agis*
Cleomenes	*Cleom.*
Tiberius Gracchus	*TG*
Caius Gracchus	*CG*
Comparatio Agidis et Cleomenis cum Tiberio et Caio Graccho	*Comp. Ag., Cleom. et Gracch.*
Philopoemen	*Phil.*
Titus Flamininus	*Flam.*
Comparatio Philopoemenis et Titi Flaminini	*Comp. Phil. et Flam.*
Aratus	*Arat.*
Artaxerxes	*Art.*
Galba	*Galba*
Otho	*Oth.*

INTRODUCTION

I
Problems, problems, problems (and Aristotelian precedents)

1.1. Quaestiones naturales *and the Aristotelian genre and tradition of natural problems*

1. Preliminary remarks on Plutarch's *Naturwissenschaft*

Plutarch's collection of Αἰτίαι φυσικαί – which is nr. 218 in the Lamprias catalogue, nr. 50 in the Planudean order, nr. 67 in the 1509 Aldine edition and nr. 59 in the 1572 Stephanus edition of the *Moralia* (911C–919E in the traditional pagination) – is commonly referred to by its Latin title, *Quaestiones naturales*[1]. According to Harrison, the Φυσικὴ ἐπιτομή and the Μελετῶν φυσικῶν καὶ πανηγυρικῶν, both of which are listed in the Lamprias catalogue (as nrs. 183 and 200a respectively), "would surely have contained much comparative information"[2], but since these works are no longer extant this remains uncertain. One cannot even be sure that these entries were authentic to begin with[3]. The Φυσικὴ ἐπιτομή, first of all, bears a very similar title to Book 10 of Ps.-Aristotle's *Problems*, the Ἐπιτομὴ φυσικῶν, which is the longest in the collection and draws heavily on Aristotle's zoological and biological writings (it often specifically deals with copulation, generation and with the number and nature of offspring in animals: cf. *Q.N.* 21 and 38)[4]. Due to the generality of this type of

---

[1] In the traditional order of Plutarch's *Moralia* in the Stephanus edition, *Quaestiones naturales* precedes *De facie* (920B–945E), a work of astrophysical interest which deals with the nature of the moon and its role in the universe. It succeeds Ps.-Plutarch's (Aëtius') *Placita philosophorum* (874D–911C), which is a doxography in five Books on physical and natural philosophical matters (see n. 5). Among the 123 *Placita* collected there, 54 are also composed in question-and-answer form (A. Gudeman, 1927, col. 2525).

[2] G.W.M. Harrison, 2000b, p. 237, n. 2.

[3] For the presence of spuria in the Lamprias catalogue, see F.H. Sandbach, 1969, p. 6. See also J. Irigoin, 1986 more generally.

[4] According to R. Mayhew, 2011a, p. 280 (with n. 3) it is not improbable that this Book originates from (or is based on) an epitome by Theophrastus: perhaps the Φυσικῶν ἐπιτομῆς (D.L. 5, 48; cf. also 9, 21) or Περὶ φυσικῶν ἐπιτομῆς (D.L. 5, 46). This remains uncertain, because the first title (Φυσικῶν ἐπιτομῆς) may also be an epitome of a doxography of ancient natural philosophers (in any case, it is listed after Φυσικῶν δοξῶν in 16 Books in D.L. 5, 48, and D.L. 9, 21 mentions Theophrastus' Ἐπιτομὴ Ἀναξιμάνδρου), and the second title (Περὶ φυσικῶν ἐπιτομῆς) was perhaps an introduction to the former (in this sense, it is a work *about*, περί, the Φυσικῶν ἐπιτομῆς), or – more likely – an epitome of the Περὶ φυσικῶν in

title, though, nothing can be said with any certainty about the original form or content of the entry in the Lamprias catalogue[5]. The same is true for the Μελετῶν φυσικῶν καὶ πανηγυρικῶν. Boulogne classified this work as a collection of Plutarch's philosophical school manuals, calling it "séminaires et conférences ouvertes au grand public sur la nature"[6]. The term πανηγυρικῶν is particularly intriguing, since it may imply that we are dealing with declamations held during festive occasions or public festivals. One may draw a link with the sympotic discussions of natural problems in *Quaestiones convivales*. Again, though, any further specification in this regard is speculative and inconclusive.

Ziegler classified *Quaestiones naturales* among Plutarch's "naturwissenschaftliche Schriften"[7], a category that also includes *De facie*, *De primo frigido*, the spurious (?)[8] *Aqua an ignis sit utilior* and a large part –

---

eighteen Books (or Περὶ φύσεως in three Books) listed precedingly in D.L. 5, 46. For these and similar titles among Theophrastus' writings, see 137 FHSG (esp. nrs. 1a, 3–7).

[5] For that matter, the Φυσικὴ ἐπιτομή could just as well have been a synopsis, e.g., of Aristotle's *Physica* or of Epicurus' physical writings. Indeed, the latter introduces his *Epistula ad Herodotum* (which concerns matters περὶ φύσεως) as being an ἐπιτομή (35, 2–5; cf. also D.L. 10, 27: Ἐπιτομὴ τῶν πρὸς τοὺς φυσικούς). In addition, D.L. 1, 10 mentions Manetho's Τῶν φυσικῶν ἐπιτομή concerning Egyptian theology and cosmology (FGrHist 609, 17). This Manetho (from Sebennytos in Egypt) is quoted several times in Plutarch's *De Iside et Osiride* (see C. Froidefond, 1988, p. 61; FGrHist 609, 19–22). By contrast, H. Diels, 1879, p. 27 has argued that the Φυσικὴ ἐπιτομή from the Lamprias catalogue (nr. 183) along with the five volumes Περὶ τῶν ἀρεσκόντων φιλοσόφοις φυσικῆς ἐπιτομῆς βιβλία (nr. 61) and Φυσικῶν ἀρεσκόντων (nr. 196) are actually three versions of the same Ps.-Plutarchan *Placita* (see n. 1). For further detail on the complex tradition of Ps.-Plutarch's/Aëtius' doxography, see J. Mansfeld and D.T. Runia, 1997.

[6] J. Boulogne, 2003, p. 37, n. 98. F.H. Sandbach, 1969, p. 27 translates the entry as "A Collection of Scientific Lectures and Public Addresses". Wyttenbach separated it from nr. 200: Περὶ ἡμερῶν.

[7] K. Ziegler, 1951, cols. 637, 706 and 851–858. This category also includes a number of other works that are listed in the Lamprias catalogue but which are now lost: Φυσικὴ ἐπιτομή (nr. 183), Περὶ προβλημάτων (nr. 193; see n. 90), Φυσικῶν ἀρεσκόντων (nr. 196) and Περὶ σεισμῶν (nr. 212). Perhaps Περὶ κομητῶν (nr. 99) can be added. G. Nuzzo, 1991, p. 410 would also add *De tuenda sanitate praecepta* (cf. also W. von Christ, 1959, p. 512, who uses the category of "naturwissenschaftlichen Fragen"). For protest against the Zieglerian designation ("scritti di fisica e di scienza naturale"), see P. Donini, 1994a, p. 48, n. 32 [see the prologue, n. 50]. Cf. also G. D'Ippolito and G. Nuzzo, 2012, pp. 56–59.

[8] This treatise was considered "a miserable sophistical exercise" by F.H. Sandbach, 1939, p. 201 (cf. also pp. 198–202 more generally), and it was claimed to be spurious already by J. Kowalski, 1918, pp. 258–262 (cf. also H. Cherniss and W.C. Helmbold, 1957, pp. 288–289). It remains to be seen, however, if this work (disregarding whether we have it in its final version or not) is as naïve as Sandbach takes it to be. In any case, the format of the 'contradictory discussion' (ἐπιχείρησις εἰς ἑκάτερον, *disputatio in utramque partem*), where topics are argued from both sides, is employed more often in Plutarch's writings

roughly one third – of the table talks in *Quaestiones convivales* that are concerned with very similar natural problems. Ziegler was well aware, however, that there may be certain problems in categorising these works in this way. *De facie*, so he notes, deals with topics relating to ancient lunar science in the form of a dialogue, which culminates in the domain of religion by the integration of an eschatological myth at the end of the work. Therefore, in spite of the fact that a major part of the work relates to what Ziegler calls Plutarch's "Naturwissenschaft"[9], the dialogue – or at least part of it – just as easily fits among his "theologische Schriften" (including *De superstitione*, *De Iside et Osiride*, the Pythian dialogues, *De genio Socratis* and *De sera numinis vindicta*). What is important here is that natural science clearly did not exclude traditional religion and mythology for Plutarch (I will come back to this later [see 4.1.2.]). Therefore, the modern attempt to dissect the Chaeronean's work into different categories probably tells us more about our own departmentalised view on science and other fields of knowledge than about Plutarch's take on the matter.

Similarly, regarding *De primo frigido* – where Plutarch demonstrates that cold has a principle of its own, which he tries to identify in a lengthy doxography – scholars have argued that Ziegler's classification of this essay among the "naturwissenschaftliche Schriften", as distinct from the "wissenschaftlich-philosophische Schriften"[10] (including *Quaestiones Platonicae*, *De animae procreatione in Timaeo*, *De Stoicorum repugnantiis*, *Adversus Colotem* etc.), is ambiguous[11]. Indeed, there may again be certain difficulties in categorising this text. If one considers the epistemological implications of some of its chapters (esp. the last in combination with the eighth [see 4.3.2.1.]), this work can – again at least in part – be just as easily grouped in with the "wissenschaftlich-philosophische Schriften". However, if one considers the technical-philosophical nature of these writings, such a recategorisation would clearly result in an overly restrictive interpretation of the text. In any case, the classification of *De primo frigido* among the "naturwissenschaftliche Schriften" is not unsound since the

---

[see 4.3.3.1.], and the essay's general style of argumentation is no more 'sophistical' than that of an average Plutarchan natural problem. For the description of this treatise as a rhetorical *tour de force*, see R. Flacelière, J. Irigoin, J. Sirinelli and A. Philippon, 1987, pp. ccvi–ccvii. The athetesis, therefore, seems to be based on doubtful grounds (cf. also C. Hubert, 1959, p. 1). For further discussion, see the introduction in G. D'Ippolito and G. Nuzzo, 2012, pp. 177–198.

[9] K. Ziegler, 1951, col. 851 (for Plutarch's writings on theology, see cols. 637, 705, 825–851). This feature of thematic overlap relates to the "Kreuzung der Gattungen" discussed as a typical feature of Plutarch's *Moralia* by I. Gallo and C. Moreschini, 2000, pp. 14–15. Cf. also G. D'Ippolito and G. Nuzzo, 2012, pp. 56–59.

[10] K. Ziegler, 1951, cols. 636–637, 704–705, 744–768.

[11] See J. Opsomer, 1998, p. 214. Cf. already P. Donini, 1986a, p. 211 and 1994a, p. 48, n. 32 [quoted in the prologue, n. 50].

work spends much, if not most, of its attention to specific natural phenomena and physical processes related to cold and heat. Plutarch explicitly notes that he is deliberately taking a step downwards on the ladder of epistemology, rather than upwards, turning most of his attention to the domain of sensory observation and plausible argumentation[12]. In fact, the more abstract discussion of what can be *known* about these issues remains rather peripheral (it introduces and concludes the doxographical part). What is central, in the end, is the principle of cold.

Third, Ziegler also distinguished the category of the "naturwissenschaftliche Schriften" from the "tierpsychologische Schriften"[13] (including *De sollertia animalium*, *Bruta animalia ratione uti*, *De esu carnium*), insofar that the latter primarily deal with the topic of animal behaviour, morality and intelligence, rather than animal biology and physiology in a strict naturalist sense (as known from Aristotle's zoological writings)[14]. Indeed, the ethical fundament of these writings is generally absent in Plutarch's "naturwissenschaftliche Schriften". However, even if the "naturwissenschaftliche Schriften" are far less concerned with matters pertaining to animal psychology, the distinction is not based on the traditional division between animate and inanimate nature (ἔμψυχα – ἄψυχα). The eschatological myth at the end of *De facie* nicely illustrates this, speculating as it does about the purpose of the moon in the universe, more specifically by explaining its importance for the life-cycle of human souls (940F–945D). Moreover, specific psychic and psychological phenomena are also discussed, sometimes at considerable length, throughout Plutarch's other natural scientific works (see, e.g., *Quaest. conv.* 654E, *De prim. frig.* 946C, *Aqua an ignis* 958E). In *Quaestiones naturales*, we find one explicit reference to a psychic process, viz. the psycho-somatic effects of fear caused at sea (*Q.N.* II, 914F: ἡ ψυχὴ σάλον ἔχουσα καὶ θορυβουμένη συγκινεῖ καὶ ἀναπίμπλησι τὸ σῶμα τῆς ταραχῆς)[15]. Plutarch also deals with

---

[12] See *De prim. frig.* 948D: Οὐ μὴν ἀλλὰ καὶ τὰ αἰσθητὰ ταυτὶ προανακινῆσαι βέλτιόν ἐστιν, ἐν οἷς Ἐμπεδοκλῆς τε καὶ Στράτων καὶ οἱ Στωικοὶ τὰς οὐσίας τίθενται τῶν δυνάμεων. For this deliberate epistemological step downwards, cf. also P. Donini, 1986a, p. 211 and J. Opsomer, 1998, p. 216, n. 11. Moreover, the category of the divine and the intelligible is still present in *De primo frigido*, though less central to the main argument [see 4.1.2.2.].

[13] K. Ziegler, 1951, cols. 636, 706, 732–744.

[14] Cf. R. Flacelière, J. Irigoin, J. Sirinelli and A. Philippon, 1987, p. lxxxiv: "Quand il s'agit des bêtes, que Plutarque semble avoir aimées, il s'intéresse à leur 'psychologie' plus souvent qu'à leur physiologie." See also S.T. Newmyer, 2014, p. 223: "It has become clear to students of Plutarch's animal treatises that to dismiss him as a failed Aristotle is to misunderstand the intention of his animal philosophy, since he approaches the question of what constitutes animality and what distinguishes it from humanity not as a biologist but as a moralist." Cf. also M. Vegetti, 1979, T. Barton, 1994a, p. 126 and R. French, 1994, pp. 178–184.

[15] Cf. L. Senzasono, 2006, p. 185, n. 73: "Del resto in esso [opuscolo] l'elemento

psychological-pathological topics, for instance, in discussing the physiological mechanisms behind hate, courage and their opposites[16]. These are, in fact, the same ethical categories that structure Plutarch's arguments in the "tierpsychologische Schriften", yet they lack any obvious moralising agenda here [see 1.2.4.]. In *Q.N.* 20, for instance, Plutarch does not explain the bravery of wild boars, as opposed to the cowardice of deer, in an ethical but a physical fashion (viz. by associating the difference in their character with their hot and cold bodily constitutions respectively). Chapters like *Q.N.* 26 and 37 are also particularly relevant in light of animal psychology, since they deal with animal intelligence – though specifically the lack of it. In *Q.N.* 37, Plutarch argues that dogs cannot understand anything by means of their intellect and have no memory, since these virtues are set aside for human beings (*An quia neque cogitatione comprehendere quicquam nec reminisci (quibus solus homo virtutibus valet) potest?*). Similarly, in *Q.N.* 26, Plutarch wonders why animals seek and pursue substances that have remedying properties when they are ill, and often restore themselves to health by using them. He points out that these animals have no previous experience or have never tried these remedies before (τούτων δ' οὔτε πεῖρα οὔτε περίπτωσις γέγονεν αὐτοῖς). This means that they do not act on the basis of knowledge or insight, so that there must be another, more physical, reason for it. We are very remote from the arguments in the "tierpsychologische Schriften", where Plutarch, in an overt anti-Stoic fashion, often emphasises the rational abilities of animals in combination with their moral capacities. Interestingly, the irrational nature of animals is also a common topic in Ps.-Aristotle's *Problems*, which served as Plutarch's model in his own natural problems[17]. One may presume, then, that the conceptual differences and inconsistencies in Plutarch's writings on animals are genre-related [see 1.2.5.].

As these cases show, any attempt to strictly categorise Plutarch's natural scientific works runs the risk of simply being artificial or ambiguous. I will further specify this point for *Quaestiones naturales* later on in view of the thematic overlaps with Plutarch's other collections of *quaestiones* [see 2.4.2.]. I do not intend to do away with Ziegler's categories altogether,

---

psicologico è assai raro, sia come oggetto di problematica, sia come strumento di spiegazione dei fenomeni."

[16] Cf. *Q.N.* 11, 914E (τῶν δὲ παθῶν ὁ φόβος), 19, 916B (δειλόν ἐστι φύσει ζῷον), 20, 917A (διὰ τὸν θυμὸν ἐκβάλλει τὸ δάκρυον), 21, 917B (μετὰ φόβου), *Q.N.* 35 (*detestantur*), 36 (*odium vehementius*), 37 (*odit*).

[17] Cf., e.g., Ps.-Arist., *Probl.* 887a11 (ἄνευ λόγου), Ps.-Arist./Alex. Aphr., *Suppl. probl.* 2, 51 (ἀλόγων ζῴων). Moreover, Book 27 of Ps.-Aristotle's *Problems* is specifically devoted to the physical aspects of fear and courage (ὅσα περὶ φόβον καὶ ἀνδρείαν), and especially to the physiological symptoms of shivering, pallor, urination etc. [see 1.2.5.]. See L.M. Castelli, 2011.

though. After all, the "naturwissenschafliche Schriften" do, in several regards, distinguish themselves from Plutarch's treatises on theology, philosophy and animal psychology[18]. As opposed to the "tierpsychologische Schriften", the "naturwissenschafliche" do not serve an overt ethical agenda, while the strictly technical-philosophical and at times dogmatic features of the "wissenschaftlich-philosophische Schriften", as well as the overtly religious-mythological approach of the "theologische Schriften", makes way for a more central focus on the tentativeness of plausible natural explanations in the "naturwissenschafliche Schriften".

If we now turn to the internal organisation of these "naturwissenschafliche Schriften", an essential observation regards their basic structure and set-up. Scholars have argued that the question-and-answer format of the genre of natural problems, as found in *Quaestiones naturales*, offers an epistemological matrix that lies at the basis of Plutarch's other scientific writings as well[19]. Each of these works is in principle concerned with supplying multiple explanations to one main problem. From a text-genetic perspective, one could even argue that these treatises are composed on the basis of actual problems, where several answers are provided to one central *quaestio*: as such, *De primo frigido* deals with the question 'Which, if any, is the active principle or substance of cold?' (cf. 945F), *Aqua an ignis sit utilior* investigates 'Whether fire or water is more useful?', while *De facie* asks 'What is the substance and purpose of the moon in the cosmos?'[20]. Thus, it can be argued that the problem format of *Quaestiones naturales* (which I will analyse in detail later on [see 1.1.4.]) provides a structural mould for the orderly organisation of Plutarch's scientific writings, the development of which can vary in length depending on the degree of argumentative elaboration[21].

---

[18] In addition, the distinction between the treatises on theology, philosophy and animal psychology may not be that strict either. For instance, as D. Babut, 1969, p. 61 shows, *De sollertia animalium* belongs to Plutarch's anti-Stoic works (cf. also J. Mossman and F. Titchener, 2011, p. 278), while G. Roskam, 2011a, pp. 200–201, proposes to classify *De amore prolis* with Plutarch's anti-Epicurean works. For animals serving as instruments of the divine, on the other hand, cf., e.g., *De soll. an.* 975B and *De Is. et Os.* 382AB [quoted 1.2.4.].

[19] See J. Boulogne, 2005b, p. 198, n. 6: "Il s'agit véritablement d'une matrice épistémologique qui semble être la base de l'écriture scientifique chez Plutarque, et qui, selon l'ampleur du développement rhétorique, génère des compositions plus ou moins longues."

[20] See also L. Senzasono, 2006, p. 9. According to H. Görgemanns, 1968, p. 11 (cf. also 1970, pp. 90–120), the central cosmological section in *De facie* (§§ 6–15, 922F–928D) was, in any case, composed on the basis of an earlier draft (perhaps a set of problems?).

[21] That the format of the problem genre had its compositional advantages, is also clear from several other Plutarchan treatises, such as *De sollertia animalium*: 'Whether land or sea animals are more clever?', *De E apud Delphos*: 'What is the meaning of the enigmatic

However, as opposed to treatises like *De facie* or *De primo frigido*, Plutarch's natural problems, as treated in *Quaestiones naturales* and *convivales*, do not explicitly portray a systematic or comprehensive vision of the object they study (i.e. the realm of natural phenomena). Instead, they deal with a range of particular questions that each concern one specific problematic issue, viz. one enigmatic natural phenomenon and its underlying causes. Plutarch clearly borrows this fragmentary and piecemeal approach from the Ps.-Aristotelian *Problems*, where over nine hundred such chapters are dealt with in 38 thematic Books (making it the third largest work in the *corpus Aristotelicum*). Therefore, Teodorsson claimed that Plutarch's "*quaestiones* are truly Peripatetic in character"[22]. To be sure, Plutarch in solving natural problems employs a wide range of scientific concepts and theories that have a specific Aristotelian imprint (or more generally Peripatetic). In fact, *Quaestiones naturales* as a whole may give the impression that it was written by a Peripatetic author. However, it is not clear from this that Plutarch actually intended to compose this work with a fundamentally Aristotelian predisposition and purpose. This is, in any case, problematic in light of Plutarch's well known allegiance to Plato and Academic philosophy. In what follows, I will sketch the problem more in detail and suggest a number of explanations (a conclusive answer will only be given at the very end of chapter four [see 4.3.4.3.]).

2. *Quaestiones naturales*: the work of a *Plutarchus Aristotelicus*?

It goes without saying that Aristotle counted as a pre-eminent authority in the field of natural philosophy in Antiquity and that Plutarch, therefore, repeatedly quotes him in his writings on natural science (and also his acolytes, esp. Theophrastus)[23]. Yet, the question as to whether Plutarch

---

E at Delphi?', and *De defectu oraculorum*: 'Why have many oracles in Greece seized to function?' (cf. 411EF). For the link between Plutarch's philosophical dialogues and the genre of problems, see also J. Opsomer, 2005, pp. 198–199 and 2010, p. 115.

[22] S.-T. Teodorsson, 1999a, pp. 665–666. See also K. Oikonomopoulou, 2013a, p. 152: "Thus the *QN* as a whole forges strong links with an ideal of encyclopaedic comprehensiveness that is specifically Peripatetic, and pays homage to the Peripatetic encyclopaedic achievement." For the Ps.-Aristotelian *Problems* as a model for Plutarch's natural problems, see, e.g., H. Flashar, 1962, p. 370: "So kann man im ganzen sagen: es ist sehr wahrscheinlich, daß Plutarch zur Anlage dieser Sammlung angeregt ist durch die peripatetische Problemata-Literatur." See also F.H. Sandbach, 1965, p. 134, L. Senzasono, 2006, p. 7, G. Roskam, 2011b, pp. 45–46, K. Oikonomopoulou, 2011, M. Meeusen, 2011 and 2016.

[23] For Plutarch's quotations from Aristotle's physical and biological writings, see G. Roskam, 2011b, pp. 45–46. For a systematic overview of Plutarch's quotations from Theophrastus, see J. Boulogne, 2005c. Strato of Lampsacus, the third in line as head of the Lyceum, is also an important Peripatetic authority for Plutarch (quoted as a φυσικός in *De*

actually tried to attain Peripatetic philosophership in his natural problems more in specific requires closer examination. This is, indeed, a relevant question in light of the reception of Aristotelian science, or at least a very specific part of it (c.q. the Ps.-Aristotelian *Problems*), in the Imperial Era[24]. Some scholars would actually claim that Plutarch, in conformity with other middle- and neo-Platonists, intended to ally Platonism with Aristotelianism in his scientific writings[25]. This would be of great interest for the history of Greek science in the Imperial Era, but one should be cautious in attributing Plutarch this intent, bearing in mind his primary philosophical allegiance to Plato and the Academic tradition.

The debate about the intellectual relationship between Plutarch and Aristotle, of course, is not anything new. It has been a particularly potent topic in the last few decades, giving rise to a variety of – often times irreconcilable – claims[26]. This discussion cannot simply be ignored here, but neither is this the place to deal with it in full detail, so I will only

---

tranq. an. 472E and *De soll. an.* 961A; cf. also *De prim. frig.* 948CD, *De Stoic. rep.* 1045F, *Adv. Col.* 1115B, fr. 216g Sandbach).

[24] The relevance of this problem was already stressed by W. Capelle, 1910, p. 328: "Doch bedürfen diese Dinge, überhaupt die Quellen der Plutarchischen Symposiaka [*Quaestiones naturales* can be added] sowie die Art ihrer Benutzung durch Plutarch, einer größeren Untersuchung, die für die Geschichte der nacharistotelischen Physiologie, zumal wenn man die übrigen physikalischen Schriften Plutarchs heranzieht, von besonderer Bedeutung sein wird."

[25] See, e.g., J. Sirinelli, 2000, pp. 361–362: "Il y a des pans entiers des connaissances de Plutarque qui proviennent d'Aristote et de son école, notamment tout ce qui concerne ce que nous appellerions les sciences naturelles. Certes, c'est surtout dans ce courant de pensée que s'est développée la connaissance systématique de la nature et il n'y a rien d'étonnant à cela. Ce qui doit être noté est l'aisance avec laquelle Plutarque allie ses deux philosophies, sans les faire interférer et sans problèmes, chacune pour ce qu'elle lui offre, Platon apportant l'essentiel, c'est-à-dire la connaissance de la divinité du Monde et de l'Esprit, Aristote celle de la nature, de la vie et très probablement de la logique. Ce n'est pas une position aberrante et isolée. Au contraire, chez Plutarque se manifeste de façon claire et pratique cette complémentarité entre l'Académie et le Lycée qui va devenir institutionnelle dorénavant dans le platonisme." For the view that Plutarch employs Aristotelian theories in his scientific writings in order to 'save' Plato, see I. Rodríguez Alfageme, 1999b, pp. 624–625. For the concept of 'harmonisation' between both strands of philosophy, see also P. Donini, 1992, p. 108 ("armonizzarne"). As to Plutarch's general aetiological project and its link with Aristotle, see C. Darbo-Peschanski, 1998, p. 21: "Plutarque médio-platonicien, héritier d'un mixte de philosophie académique et péripatéticienne, est, dans les Αἴτια, largement influencé par Aristote. Cette influence lui fournit le cadre dans lequel inscrire les préoccupations érudites qu'il partage avec bien des Grecs d'époque romaine".

[26] Seminal studies are those of G. Verbeke, 1960, F.H. Sandbach, 1982, P. Donini, 1986a, 1988, 1999, D. Babut, 1996, F. Becchi, 1975, 1978, 1999, 2014, G.E. Karamanolis, 2006, pp. 85–126 and G. Roskam, 2008/9, 2011b. See also more generally the contributions in A. Pérez Jiménez, J. Garciá López and R.M. Aguilar, 1999.

sketch the debate along very general lines and with a specific focus on Plutarch's natural scientific writings.

An interesting starting-point is found in Plutarch's argument in *De primo frigido*. The fact that the frequency of explicit (i.e. *nominatim*) quotations from the Stagirite is rather low in this work for some scholars marks a clear contrast with the work's generally 'Aristotelian' style of discourse. Regarding the recurrent allusion to specific natural scientific theories and concepts in this treatise, Opsomer (following Glucker) speaks of Plutarch's use of "a mock-Aristotelian style", that is, an "outward appearance of Aristotelianism [in honour of his youthful friend and/or pupil, Favorinus, which] should not be taken seriously"[27]. It is only at the very end of the treatise, then, that Plutarch throws off his Peripatetic mask and shows his true, Academic face, by promoting ἐποχή as a superior philosophical attitude in natural scientific matters (*De prim. frig.* 955C [quoted 4.3.2.1.]). This view, however, is, in my opinion, open to debate. At least some nuance seems in place.

First of all, Plutarch is nowhere explicitly making fun of Aristotle or his style of discourse in *De primo frigido*. In fact, the situation seems more or less similar as in Plutarch's writings on animal psychology, where Aristotelian science (c.q. biology) is also instrumentalised in a new philosophical (c.q. moralising) framework (see n. 14). In the case of *De primo frigido*, then, the framework is provided by Academic ἐποχή. If there is anything humoristic about this procedure, it is, indeed, very subtle. Second, to denote anything natural scientific as 'Aristotelian' would seem a gross oversimplification. Notably, *De primo frigido* contains some overt anti-Aristotelian and anti-Peripatetic traits, such as Plutarch's dismissal of the theory, elaborated in the first part of the treatise, that cold is a privation (στέρησις) of heat. This theory has clear parallels in Aristotle but is rejected at length here (*De prim. frig.* 945F–948A; see the parallel in *Q.N.* 29, 919A with the commentary *ad loc.*). Plutarch also openly rejects Strato's theory identifying water as the principle of cold (*De prim. frig.* 948CD). In light of the infrequent quotes from Aristotle in *De primo frigido*, Helmbold concludes that "[n]o doubt it is in virtue of Favorinus' youth that his idol is treated so lightly", adding in the same breath that "Plutarch [...] became much more favourable to Peripatetics later in

---

[27] J. Opsomer, 1998, p. 219; see J. Glucker, 1978, pp. 286–290. For Favorinus' admiration of Aristotle and his adherence to Peripatetic philosophy, see *Quaest. conv.* 734F: δαιμονιώτατος Ἀριστοτέλους ἐραστής ἐστι καὶ τῷ Περιπάτῳ νέμει μερίδα τοῦ πιθανοῦ πλείστην. See also H. Cherniss and W.C. Helmbold, 1957, p. 228, P. Donini, 1986a, p. 223, n. 28, L. Holford-Strevens, 1997, p. 204 (who rightly notes that "assigning the greatest share of the πιθανόν to the Peripatetics is not incompatible with Academic scepticism"; cf. *De prim. frig.* 949F and 955C).

his life (*e.g.* in the *Life of Alexander*)"[28]. The latter part, based on the chronology of Plutarch's life and writings, is not further motivated, and it can, indeed, be contested on account of the fact that it is not unlikely that Plutarch was interested in discussing Aristotelian-style scientific topics from a relatively young age onwards [see the prologue]. In Plutarch's natural problems, Aristotle and his followers – with Theophrastus as the first in line – are regarded as precious authorities in the development of the arguments: *Quaestiones convivales* shows that personal acquaintance with their natural scientific writings is not at all a thing to scorn but rather to display, and in *Quaestiones naturales* Plutarch quotes Aristotle and Theophrastus by name at numerous occasions, while remaining critical in his evaluations [see 4.2.1.1. and 4.2.2.2.]. So if Plutarch had any interest in Aristotle and the Peripatetics, it is probably for the best not to conceive of it in terms of strict chronological phases in the Chaeronean's philosophical career, but to draw a more general model of intellectual allegiance – or the lack of it.

The most recent study (I know of) to provide a global account of Plutarch's reliance on Aristotelian knowledge is that of Roskam (2011b). Roskam collects and analyses the explicit (i.e. *nominatim*) quotations from Aristotle throughout the entire *corpus Plutarcheum*, thus employing a positive method that is rigorous but safe. Roskam expresses considerable opposition to a number of Becchi's and Karamanolis' views, both of whom have defended Plutarch's Aristotelianism on the basis of more lenient readings of the available material[29]. There is no need to go into the details of these studies here; what matters most for us is Roskam's conclusion:

"[I]f Plutarch was prepared to embrace him [sc. Aristotle] [...] he was rather pursuing a marriage of convenience than acting out of true love. His true love was Timoxena in everyday life, and Plato in philosophy."[30]

It is beyond doubt that Aristotle was not Plutarch's "true love". This does not necessarily imply, however, that Aristotle did not hold a privileged rank in Plutarch's thought. Donini, for instance, has an open mind on this matter, arguing that while Aristotle did not belong to the Platonic or Academic tradition for Plutarch, one can, nevertheless, assume a certain sense of philosophical evolution, according to which, in Plutarch's mind,

---

[28] H. Cherniss and W.C. Helmbold, 1957, p. 228. Helmbold finds it "odd that of the three quotations from Aristotle one is a rebuke (950B), one is apparently a partial miscalculation (948A [...]), while the third is of no importance".

[29] Viz. as based, respectively, on Plutarch's ethical (F. Becchi, 1975, p. 179 and 1978, p. 264) and polemical writings (G.E. Karamanolis, 2006, pp. 92–100).

[30] G. Roskam, 2011b, p. 61.

Aristotle was Plato's closest pupil not only chronologically but also intellectually[31]. Although Aristotle is dealt a great deal of criticism by Plutarch[32], the Chaeronean also has kind words for him, which cannot be said of representatives of other philosophical schools (esp. the Stoics and Epicureans). Aristotle was a 'prominent'[33] and 'famous'[34] thinker for Plutarch, and is even called 'the most famous and learned of the philosophers'[35]. Sporadically[36], Plutarch even mentions Aristotle along with his own favourite Plato – whom he considered a 'divine'[37] philosopher 'pre-eminent in reputation and in influence'[38] –, presenting both thinkers as 'the best authorities in the field of philosophy'[39].

In a certain way, then, Roskam's and Donini's positions can be reconciled: Plutarch's explicit quotations in no way hint at a decisive adherence to Aristotelian philosophy but in more general utterances, Plutarch does not hide his appreciation of Aristotle and the man's philosophy, if for no other reason than that he was Plato's closest pupil. It remains to be seen, of course, to what extent Plutarch in some cases affirms Aristotle in order to support his own philosophical convictions with the authority of his predecessor (without any further implicit value judgement), whereas in other cases, where he openly criticises Aristotle, he does this in order to prove Aristotle wrong or to demonstrate his own

---

[31] P. Donini, 1986a, pp. 220–221 (see also 1988, pp. 139–140 and 144): "Ma [...] non lo considerasse affatto irrimediabilmente lontano dal platonismo: soprattutto nel confronto con le altre scuole filosofiche l'aristotelismo rivelava a suo giudizio qualche persistente e profonda affinità con il pensiero di Platone in alcune delle questioni di maggior peso nella filosofia. [...] Se infatti Aristotele era, secondo Plutarco, colui che, senza essere precisamente platonico, o academico, rimaneva tuttavia pur sempre di tutti i filosofi non platonici il più vicino al pensiero di Platone." For Plutarch's notion of a close alliance between the philosophy of Aristotle and Plato, cf. also P.H. De Lacy, 1953, p. 79. It should be noted that this view is *not* as such rejected by G. Roskam, 2008/9, p. 25, who argues that the Peripatetic tradition, indeed, "remains fairly close to Platonism [for Plutarch], so that it can often function as Plutarch's privileged ally in his attacks against other schools". *Pace* G.E. Karamanolis, 2006, p. 115, who argues that "Plutarch considers Aristotle part of the Platonist tradition".

[32] D. Babut, 1996, pp. 23–22 points out that "l'auteur des *Moralia* n'hésite pas, à l'occasion, à critiquer nommément le fondateur de l'école péripatéticienne, *ce qu'il ne fait jamais*, il faut le souligner, *quand il s'agit de Platon*" (Babut's italics).

[33] *Non posse* 1086E: ἐπιφανής.

[34] *Adv. Col.* 1124C: ἐλλόγιμος.

[35] *Alex.* 7, 2: τῶν φιλοσόφων ὁ ἐνδοξότατος καὶ λογιώτατος.

[36] Viz. in *De aud. poet.* 26B, *De Is. et Os.* 382D, *De E* 389F, *De Stoic. rep.* 1040A–1041B and esp. *De Is. et Os.* 375C and *Quaest. Plat.* 1006D.

[37] *De cap. ex inim.* 90C, *Per.* 8, 1: θεῖος.

[38] *Quaest. conv.* 700B: φιλόσοφος δόξῃ τε καὶ δυνάμει πρῶτος.

[39] *De Stoic. rep.* 1041A: τῶν ἀρίστων φιλοσόφων.

critical skills, thus trying to pass himself as even more philosophically pre-eminent than the Stagirite – a philosophical heavyweight himself.

The collection of *Quaestiones naturales* offers a manageable text for taking a closer look at Plutarch's reception and evaluation of Aristotle's natural scientific theories (as formulated primarily in the *Problems*). In some cases, Plutarch criticises specific Aristotelian accounts (*Q.N.* 2, 911F, 12, 914F), while on other occasions he gives a more positive evaluation (*Q.N.* 1, 911E, 21, 917D), even praising him for his excellent insights into natural science (*Q.N.* 40; cf. also *Quaest. conv.* 656C). It would be unwise to make generalisations on the basis of these few accounts (which I will analyse in further detail later on [see 4.2.1.1.]), unless, perhaps, by pointing out that there is not much consistency among them. Roskam is probably right, therefore, that "Plutarch's evaluation of Aristotle is always *ad hoc* and does not reflect a thoroughly considered general attitude towards the Stagirite"[40]. Nevertheless, Aristotle's authority holds a privileged rank in Plutarch's natural problems, which is why Roskam correctly adds that "Plutarch appears to consider Aristotle [...] a kind of 'secretary of nature' [...]. For Plutarch, Aristotle is primarily a *Fundgrube* of erudition."[41] What we learn from this is that Plutarch's intellectual marriage with Aristotle is not borne out of "true love" but of utility.

Of course, natural science is only one specific part of Aristotle's much wider philosophical project, and Aristotle himself is only one – seminal – link in the tradition of ancient Greek science more generally[42]. It is perhaps not so remarkable, therefore, that Plutarch recurrently invokes Aristotle's authority in his natural problems, while also quoting a large number of other authoritative φυσικοί. Notably, Empedocles is the most frequently quoted authority in *Quaestiones naturales* with a total of seven quotations (while Aristotle and Homer share a second place with five quotations each [see 4.1.2.3. and 4.2.1.1.]), and there are also numerous references and allusions to his physical theories (e.g., of emanations)[43]. It would

---

[40] G. Roskam, 2011b, p. 47.

[41] G. Roskam, 2011b, p. 48.

[42] G. Roskam, 2011b, p. 47 is probably right, therefore, that "Aristotle is a rather small aspect of Plutarch's impressive παιδεία". See also J.P. Herhsbell, 1971, p. 181: "Although he [sc. Plutarch] was educated by Ammonius in the Academy, and considered Plato the greatest of philosophers, his own thought was somewhat eclectic. He was, for example, open to the influence of the Peripatetics and in some details to the Stoics. Although he polemicized against their principles, he rejected absolutely only the Epicurean system [...]. The influence of the Peripatetics is clear in *Quaest. nat.* where Plutarch discusses many of the problems propounded by Aristotle and Theophrastus, and where he also uses language employed by the Peripatetics." For Ammonius' interest in Peripatetic science, see the conclusion in J. Opsomer, 2009, p. 177.

[43] According to J.P. Hershbell, 1971, pp. 172–173 it is not implausible that some of Plutarch's citations of Empedocles derive from the works of Aristotle and Theophrastus,

seem a bit impetuous to conclude that the Chaeronean was, therefore, an Empedoclean scientist. In any case, Empedocles' elemental and emanation theory had become common currency in ancient scientific thinking already by the time of Plato[44].

Nevertheless, there is an obvious connective thread throughout Plutarch's natural problems that strikes the reader as being highly Aristotelian and obviously in line with the Stagirite's causal model of scientific research. Even if these natural problems portray a peculiar aspect of Plutarch's versatile inquisitive interests in the natural world, we do not necessarily have to rely on the *nominatim* quotations from Aristotle alone (or of Theophrastus, or other φυσικοί) in order to grasp the collection's Aristotelian character. One can simply take into consideration the obvious, but often implicit, parallels in Aristotle's or Theophrastus' writings and the general scientific discourse and style of Plutarch's physical aetiologies: these are teeming with scientific terminology and physical theories that are commonly attested to in the works of the Peripatetics, and most obviously in Ps.-Aristotle's *Problems*[45] [see 4.3.4.]. So even though Plutarch presents himself as a faithful Platonist, it might, thus, seem as though he did not see anything wrong with openly flirting with what he himself, and his contemporaries, would probably consider to be Peripatetic reasoning. One might find it remarkable, in any case, that as long as this Peripatetic affair lasts, Plato and Platonic thought seems to be pushed to the margins.

When it comes to statistics, Plato is quoted only three times in *Quaestiones naturales* (on a par with Theophrastus [see 4.2.1.1.]). This could imply that Plutarch's thoughts in this collection were not, or at least far less, with his "true love" in philosophy. But insofar that the number of explicit quotations does not necessarily provide conclusive proof of this, it should not be taken for granted that *Quaestiones naturales* exhibits Plutarch's Peripatetic aspirations in the absence of a potentially underlying

---

although the large quantity of quotations may also suggest that he knew the complete poems, or worked with one or more collections of fragments [see 4.1.2.3., n. 92]. See also D. Babut, 1976, p. 143 (with n. 22).

[44] I here allude to the conclusion that was drawn for Empedocles' alleged influence on Plato's *Timaeus* by J.P. Hershbell, 1974, pp. 165: "No doubt it can be presumed that Empedocles' views became common currency in Antiquity and that Plato was familiar with them [...], but from this it does not follow that Empedocles' influence on the *Timaeus* was as great as Taylor claimed." See A.E. Taylor, 1928, p. 11.

[45] Cf. G. Roskam, 2011b, p. 39: "In his *Quaestiones naturales* [...] Aristotle is mentioned only four times (911E, 912A, 914F and 917C [excluding Psellus' *Q.N.* 40]), although Plutarch elsewhere too makes use of Aristotelian (and Peripatetic) material. Even the genre itself of the work can obviously be traced back to the Ps.-Aristotelian Προβλήματα literature." G.W.M. Harrison, 2000b, p. 239 argues that "[o]f the thirty-one *quaestiones* preserved in the Planudean tradition, only nine do not have recognisable references to passages within surviving works of Aristotle and Theophrastus".

Platonic agenda – a view that has, in any case, been suggested for the Peripatetic tendencies in the natural problems of *Quaestiones convivales*[46]. Oikonomopoulou saves the day when she notes that "Peripateticism is a key intellectual strand within Plutarch's *Table Talk*, second only to the position Platonism holds in its intellectual universe"[47]. Exemplary is the way in which Plutarch in *Quaest. conv.* 700B tries to vindicate Plato's contested view that drink passes through the lungs. This passage (which I will discuss in detail later [see 4.3.2.1.]) clearly illustrates Plutarch's high opinion of Plato, Platonic doctrine and Platonic epistemology (c.q. his sceptical attitude towards natural phenomena and observational data). One may still wonder, though, in which precise sense Peripateticism is "second" to Platonism for Plutarch. 1) Is Plutarch, perhaps, deliberately assuming a Peripatetic *persona* in his natural problems in order to demonstrate his all-round philosophical acumen and education, without wanting to be counted among Aristotle's ranks? 2) And/or is he perhaps trying to get a few steps closer to his "true love" in philosophy (Plato) by approaching the thought of his closest peer (Aristotle)?

1) As to the first option, Kechagia has recently made an interesting suggestion[48]. She argues that while the Peripatetic character of the natural problems in *Quaestiones convivales* is an identifying feature of the symposiasts' explanations (including those of Plutarch's literary *alter ego*), this most likely does not provide a strict indication of Plutarch's philosophical allegiance. After all, the symposiasts that Plutarch stages in his dialogues often adhere to different philosophical schools but they still share the very same interests and knowledge of the, in that case, more 'generic' Peripatetic tradition. This would imply that the theories and concepts of the Aristotelian tradition of natural problems (much like Empedocles' physical theories, as we saw) had become common currency by the time of Plutarch [see also 4.3.4.3.]. This argument is not necessarily incompatible with the second.

2) Plutarch most likely conducted this kind of physical-aetiological research with a basically Platonic motive in mind, through which explaining natural phenomena serves as a preamble to more metaphysical contemplations. I will elaborate this view in full detail later on, but it can already be said that the inquisitive method Plutarch employs in explaining particular phenomena – as provided by the interrogative structure and anti-dogmatic approach of the natural problem genre – is, as far as the

---

[46] For the Peripatetic influence in *Quaestiones convivales*, see S.-T. Teodorsson, 1999a, K. Oikonomopoulou, 2011 and M. Meeusen, 2016. As to Plutarch's Platonism and "peut-être le côté aristotélicien de sa formation" in *Quaestiones convivales*, cf. J. Sirinelli, 2000, p. 387. See also Z. Abramowiczówna, 1962, p. 88.

[47] K. Oikonomopoulou, 2011, p. 105. Cf. also R. Lopes, 2009, p. 419.

[48] E. Kechagia, 2011a, p. 98.

Chaeronean is concerned, informed by the author's Platonic-Academic convictions, according to which one cannot reach definite certainty in the study of nature, so that one should hold to plausible arguments (τὸ πιθανόν) and postpone final judgement (ἐποχή) [see 4.3.2.]. Moreover, in line with Plutarch's Platonic dualism, the practice of solving enigmatic natural problems is a useful exercise in looking for natural explanations for wonder-inducing phenomena [see 4.1.2.]. As such, physical aetiology trains the philosopher's mind and opens up the possibility for attaining a more stable stance towards natural 'miracles', and hence towards the working of divine providence in the world. This, in turn, enables a more intellectual devotion to the divine (εὐσέβεια) that does not succumb to the irrational reflex of superstition or, worse even, atheism. As such, so I will elaborate later on, Plutarch's science of natural problems is not just the product of a πεπαιδευμένος' scholarly pastime, but a lighter version of genuine philosophy [see 3.2.2.].

Clearly, this is not the right place to elaborate these theories in any detail. At this point, it is better to start from the beginning and situate Plutarch's *Quaestiones naturales* in the wider context of the Aristotelian tradition of natural problems – to which it, if not from a strictly intellectual, at least from a genre perspective, adheres.

3. The genre of problems and the Aristotelian tradition of natural problems

The genre of problems (προβλήματα) has a long history that starts with early Homeric scholarship. The study of the works of the Poet raised numerous problems regarding literary quality, consistency, interpretation etc. Ancient scholars employed several approaches in order to solve these problems, ranging from purely philological (e.g., regarding problems of grammar, style and prosody)[49] to allegorical (e.g., by explaining the actions of the Homeric Gods in terms of cosmic principles) and moralising-apologetic (using Homer's texts as a platform for moral education by demining – i.e. providing an 'apology' for – passages that were considered morally objectionable: cf. *De aud. poet.* 28E). Problems related to Homer's writings were still very popular in Plutarch's time – indeed, Plutarch himself composed a collection of Ὁμηρικαὶ μελέται (from which only frs. 122–127 Sandbach remain)[50].

---

[49] A. Gudeman, 1927, col. 2512 even speaks of the "Geburtstunde der philologische Wissenschaft". See also W.J. Verdenius, 1966 and W. Bühler, 1977 more generally.

[50] Notably, Aristotle also composed a series of questions related to Homer's writings (see G.L. Huxley, 1979, S. Halliwell, 1989 and R. Mayhew, 2015c). We also have a collection of *Quaestiones Homericae* (also known as the *Allegoriae*) by Heraclitus, the Stoic grammarian and rhetorician from the 1st century AD (see D.A. Russell, 2003 and D.A. Russell and D. Konstan, 2005).

Relatively soon, the genre expanded to specific passages in other texts as well, and even to particular statements, opinions and realia of other sorts (e.g., cultural, historical, scientific etc.). Along with the art of philology, the genre of problems flourished during the Hellenistic era, particularly in the Museum of Alexandria[51]. The custom of solving problems was still prevalent during Plutarch's time, even amongst the most highly-placed authorities. For instance, it is said of Emperor Hadrian – arguably the most Hellenophilic of them all – that he posed a number of questions to the scholars at the Museum and also provided responses to those addressed to himself[52]. This clearly reveals the popularity of the question-and-answer genre and its established reputation in the highest socio-political circles at the time.

Callimachus' famous Αἴτια was an exponent of this Hellenistic tradition. In this poem, the author used elegiac form and metre to explain obscure cultural phenomena (such as cults and temple ceremonies, origin stories of cities, odd local institutions and habits throughout the Hellenic world)[53]. Plutarch does not cite Callimachus' Αἴτια very frequently; yet, he is well acquainted with such aetiological literature more generally[54]. Indeed, Plutarch's own cultural and antiquarian enthusiasm is clearly exhibited in his *Quaestiones Romanae* and *Graecae*, albeit in prose form.

As to the genre of problems, the Greek term πρόβλημα implies in its most basic meaning "anything thrown forward or projecting"; it most likely originates from military tactics, where it refers to "anything put before one as a defence", that is, an entrenchment, a barricade or barrier, a bulwark, or an obstacle set up to guard oneself against an adversary[55]. The metaphorical value of this concept in light of the literary-intellectual genre of problems

---

[51] Cf. Porph., *Quaest. Hom. ad Il.* 9, 682: ἐν τῷ Μουσείῳ τῷ κατὰ Ἀλεξάνδρειαν νόμος ἦν προβάλλεσθαι ζητήματα καὶ τὰς γινομένας λύσεις ἀναγράφεσθαι. On the practice of solving problems in the Museum, see W.J. Slater, 1982, pp. 346–349. An allusion to this Museum-like setting is perhaps found in Book 9 of Plutarch's *Quaestiones convivales*, which is set ἐν τοῖς Μουσείοις in Athens (cf. 736C, 737D, 748D) and were several Homeric problems are discussed. Cf. Also, e.g., *Sept. sap. conv.* 153E–154A.

[52] See Spart. Ael., *Hadr.* 20, 2: *apud Alexandriam in Museo multas quaestiones professoribus proposuit et propositas ipse dissolvit*.

[53] For further detail on Callimachus' Αἴτια, see M. Asper, 2004 and A. Harder, 2012.

[54] For a collection of Callimachus passages in Plutarch, see E. Magnelli, 2005, pp. 218–220. Callimachus' Αἴτια are only mentioned in *Parall. Graec. et Rom.* 315CD (ὡς Καλλίμαχος ἐν δευτέρῳ Αἰτίων). For Plutarch's references to other aetiological authors, see *Amatorius* 761B (ὡς ἐν τοῖς Αἰτίοις Διονύσιος ὁ ποιητὴς ἱστόρησε) and *Rom.* 21, 8 (Βούτας δέ τις, αἰτίας μυθώδεις ἐν ἐλεγείοις περὶ τῶν Ῥωμαϊκῶν ἀναγράφων). For aetiology in ancient literature more generally, see A. Harder, 2012, pp. 24–27 (esp. p. 26) and the contributions in M. Chassignet, 2008.

[55] LSJ, s.v. i and ii. Cf., e.g., *Comp. Arist. et Ca. Ma.* 2, 4, where Cato Maior's eloquence is called a πρόβλημα τοῦ βίου καὶ δραστήριον ὄργανον.

is relatively plastic: in order to reach the truth that lies within or behind the problematic stronghold or obstacle, the difficulties that surround it like walls or defend it like weapons are assailed by means of a suitable solution. The verbal pun is clear in a passage in Plato, where sophists are said to entrench themselves behind their 'problems' (προβλημάτων). In order to catch and expose these men, so Plato writes, the adversary must first 'fight through' (διαμάχεσθαι), that is, solve, the 'problems' they put forward[56].

Traces of the genre of problems can already be found in the Hippocratic writings[57]. According to the tradition, however, Democritus was the first to compose an actual collection of problems, viz. the Χερνικὰ (or Χειρόκμητα?) προβλήματα (D.L. 9, 49 = DK68A33 and DK68B299h). Unfortunately, this collection is no longer extant[58]. Some scholars even saw Democritus as a precursor to Peripatetic natural science[59], but Aristotle was, without a doubt, the most preeminent philosopher to attach his name to the tradition of natural problems with such authority. As we just saw, before Aristotle, Plato also used the word πρόβλημα in his philosophical dialogues (and derived from it προβάλλειν: 'to propose a problem')[60]. Therefore, scholars have come to consider the Socratic elenchus – that is, the method of cross-examination by question and answer – an important precursor of the Aristotelian concept of πρόβλημα[61]. Aristotle, however, incorporated the concept of πρόβλημα into his analytical system by attributing it with its own dialectical designation and by underlining its use in the context of scientific inquiry[62].

---

[56] Pl., *Soph.* 261a: ἔοικεν ἀληθὲς εἶναι τὸ περὶ τὸν σοφιστὴν κατ' ἀρχὰς λεχθέν, ὅτι δυσθήρευτον εἴη τὸ γένος. φαίνεται γὰρ οὖν προβλημάτων γέμειν, ὧν ἐπειδάν τι προβάλῃ, τοῦτο πρότερον ἀναγκαῖον διαμάχεσθαι πρὶν ἐπ' αὐτὸν ἐκεῖνον ἀφικέσθαι. Cf. H. Flashar, 1962, p. 297 and A. Blair, 1999, p. 172.

[57] See, e.g., Hipp., *De diaet. in morb. ac.* 7 (3, 5–14 Littré) and *Epid.* 6, 2, 5 (5, 278–280 Littré). Cf. H. Diller, 1934 and H. Flashar, 1962, pp. 298–299. See also n. 74 below.

[58] For further discussion of the title of this work (Χερνικά is a hapax and Χειρόκμητα uncertain; cf. DK68B300), see H. Flashar, 1962, pp. 302–303, F. Krafft, 1969.

[59] See O. Regenbogen, 1931, p. 349. For a study of Democritus' model of causal research, see P.-M. Morel, 1996. See also esp. S. Menn, 2015 for the link between Democritus' aetiological project and the Ps.-Aristotelian *Problems*.

[60] Cf., e.g., Pl., *Rep.* 530b, 531c, *Tht.* 180c, *Soph.* 245b, 261a, *Pol.* 285d.

[61] For the link between Socratic elenchus and the Aristotelian concept of πρόβλημα, see J.G. Lennox, 2001, p. 72 and 2015, p. 36 (on Socratic elenchus itself, see G. Vlastos, 1983). However, as opposed to the style of Plato's dialogues, Ps.-Aristotle's *Problems* have a more monologic nature, see P. Louis, 1991, pp. xxi–xxii: "Au contraire de la dialectique platonicienne qui met en présence deux interlocuteurs, l' un qui questionne et l' autre qui répond, le problème aristotélicien ne met en scène qu' un seul personnage, l' auteur qui formule la question et qui suggère lui-même une ou plusieurs réponses. Il appartient ensuite au lecteur de se faire une opinion."

[62] Cf. LSJ, s.v. iv, 3.

There are several passages in Aristotle's writings that are seminal for an analysis of his notion of πρόβλημα[63]. According to the definition in the *Topics*, a πρόβλημα is a dialectical subject of investigation (διαλεκτικὸν θεώρημα) about which people either have no opinion or one that differs among intellectuals and/or the people[64]. Aristotle notes that problems involve conflicting arguments, provoking doubt about *whether* (πότερον) something is so or not by the fact that there are plausible arguments to support both sides. They can also concern important topics about which we have no arguments at all, since we think it is difficult to give a proper reason (τὸ διὰ τί): e.g., *whether* the universe is eternal or not[65]. In *Top.* 105b19–21, Aristotle makes a distinction between three kinds of problems: ethical, natural and logical. For the second category, Ps.-Aristotle's *Problems* come to mind, considering their mainly natural scientific content. In this work, often more than one explanation is given to a problem, and the explanations are mostly formulated in an interrogative and anti-dogmatic fashion, leaving place for further debate. This means that multiple plausible explanations can be formulated in order to solve the problem[66].

Strictly speaking, however, the natural problems collected in Ps.-Aristotle's *Problems* are no dialectical problems, because they do not investigate *whether* (πότερον) or not a thing is so, but *why* (διὰ τί) this is the case (see n. 90). As such, it is assumed that the subject of investigation, that is, the problematic natural phenomenon at issue, is a positive fact, the legitimacy of which is not subject to debate. Arguably, the real dialectical problems are the tentative explanations provided in the aetiologies, since they are formulated interrogatively (expecting acceptance or rejection) and, thus, serve as starting-points for further discussion. As we shall see later on, the causal inquiry marked by διὰ τί, as in the case of Ps.-Aristotle's *Problems*, ties in more closely with the framework of Aristotle's

---

[63] Esp. Arist., *Top.* 104b1–28. Aristotle distinguishes a πρόβλημα from a proposition (πρότασις) and a thesis (θέσις). Cf. Arist., *Top.* 101b28–37, *APr.* 24a16–17 and *APo.* 72a8. On this distinction, see P. Moraux, 1951, p. 71 and P. Louis, 1991, pp. xx–xxi. For further study of Aristotle's types of questions concerned with problems, see J.G. Lennox, 2001 (esp. pp. 77 and 87 with reference to the Ps.-Aristotelian *Problems*), 2015 and J. Mansfeld, 2010, pp. 41–49 (esp. p. 43, n. 34 with reference to the *Problems*).

[64] Arist., *Top.* 104b3–5: περὶ οὗ ἢ οὐδετέρως δοξάζουσιν ἢ ἐναντίως οἱ πολλοὶ τοῖς σοφοῖς ἢ οἱ σοφοὶ τοῖς πολλοῖς ἢ ἑκάτεροι αὐτοὶ ἑαυτοῖς.

[65] Arist., *Top.* 104b13–18: ἔστι δὲ προβλήματα καὶ ὧν ἐναντίοι εἰσὶ συλλογισμοί (ἀπορίαν γὰρ ἔχει πότερον οὕτως ἔχει ἢ οὐχ οὕτως, διὰ τὸ περὶ ἀμφοτέρων εἶναι λόγους πιθανούς), καὶ περὶ ὧν λόγον μὴ ἔχομεν, ὄντων μεγάλων, χαλεπὸν οἰόμενοι εἶναι τὸ διὰ τί ἀποδοῦναι, οἷον πότερον ὁ κόσμος ἀΐδιος ἢ οὔ.

[66] Notably, in *Top.* 105b13–19 Aristotle underlines the use of drawing up thematic lists of doxographical material for the formulation of propositions and problems (cf. also *Met.* 995b2–4: βέλτιον ἀνάγκη ἔχειν πρὸς τὸ κρῖναι τὸν ὥσπερ ἀντιδίκων καὶ τῶν ἀμφισβητούντων λόγων ἀκηκοότα πάντων). See J. Mansfeld, 2010, pp. 45–46.

natural scientific (rather than strictly dialectical) method of investigation as expounded in the *Posterior analytics* and put to action in the Stagirite's writings on natural science (e.g., *Parts of animals*, *Meteorology* etc.). In these writings, the examination of the διὰ τί of a specific natural phenomenon introduces a second phase in the scientific inquiry, after having determined the actual, empirical veracity, that is, the ὅτι, of that phenomenon (see *APo*. 89b24–35 – I will discuss this passage later in view of Plutarch's own approach of natural problems [see 4.1.1.3.]).

There can be no doubt that the Ps.-Aristotelian *Problems* served as a model for Plutarch's natural problems. This is true at least for the kind of problems dealt with and for their overall structure and organisation. Scholars have pointed out that it is unlikely, however, that Plutarch relied on the same collection as we have today. The collection of Ps.-Aristotelian *Problems* that came down to us (in 38 Books) is the result of a complex text-genetic process, being the product of centuries of textual accumulation, contamination and reorganisation. It is generally accepted that Aristotle initiated the work by authoring an unknown number of chapters in it, but most of the content should be ascribed to his acolytes in the Lyceum. Indeed, Aristotle's students probably continued the Stagirite's original collection by adding new problems and revising older ones[67]. Therefore, scholars agree that only parts of the *Problems*, in its current form, are authentic, but it is impossible to determine with absolute certainty which are and which not. Notably, the question of authenticity does not, however, seem to have been of concern to Plutarch and his peers[68] – and, indeed, the *Problems* do not necessarily have to be authentic to be considered 'Aristotelian'.

The formal edition of Aristotle's works by Andronicus of Rhodes in the 1st century BC introduced an important phase in the textual history of the *Problems*. It was probably Andronicus' edition that was used by Plutarch (and a wide range of other authors, such as Cicero, Strabo, Seneca, Pliny, Gellius, Apuleius, Galen and Athenaeus), but this is not the same collection of *Problems* as the one we have today[69]. The final

---

[67] For a detailed study of this compositional process, see the introductions in H. Flashar, 1962 and P. Louis, 1991, pp. xxiii–xxxv. See also R. Mayhew, 2011a, pp. xvii–xxi. Among Theophrastus' works, for instance, D.L. lists one Book Περὶ τῶν προβλημάτων φυσικῶν (5, 48; 49), a collection of Προβλήματα πολιτικά, φυσικά, ἐρωτικά, ἠθικά (5, 47), and also a Προβλημάτων συναγωγή in one (5, 48) and five (5, 45) Books (= 137, 26 FHSG).

[68] See, e.g., K. Ziegler, 1951, col. 922, F. Fuhrmann, 1972, p. xxi, P. Louis, 1991, p. xvi, K. Oikonomopoulou, 2011, p. 106, M. Meeusen, 2016.

[69] See H. Flashar, 1962, pp. 312–314 (and 369–370) and P. Louis, 1991, pp. xxxi–xxxiii. Regarding the passage in *Quaest. conv.* 734CD [quoted 3.2.1.], where Florus is reading and discussing a copy of the Προβλήματα Ἀριστοτέλους φυσικά, Flashar remarks (p. 313): "Hier hat man den Eindruck, daß es sich um ein bestimmtes Werk 'Problemata' handelt, nicht

redaction of the Ps.-Aristotelian *Problems* probably took place at the end of the 2nd century AD[70]. Andronicus' new edition is generally considered a catalyst for the revival of the genre of natural problems in the early Greco-Roman Empire (esp. in the 1st and 2nd centuries AD), a period that also witnessed the composition of several new collections of natural and medical problems by various authors. In his monumental study of the *Problems*, Flashar classified Plutarch's natural problems among these new collections, next to 1) the *Supplementary problems* in three Books (formerly known as the *Problemata inedita*), variously attributed to Aristotle and Alexander of Aphrodisias (but considered spurious today)[71], 2) the two Books of *Medical puzzles and natural problems* ascribed to Alexander of Aphrodisias (but probably spurious)[72], and 3) the *Medical difficulties and natural problems* of Cassius (Felix?), surnamed the Iatrosophist[73]. These collections demonstrate how the Aristotelian genre of natural problems became embedded in the medical tradition in the Imperial Era. As we saw, medical problems were already raised in the Hippocratic writings.[74] In fact, the Ps.-Aristotelian *Problems* themselves contain a section specifically devoted to medical issues (viz. the very first

---

um verschiedene Schriften dieses Genus, die untereinander nur locker oder überhaupt nicht verbunden wären."

[70] See P. Louis, 1991, pp. xxxiii–xxxvi. For the complex date of Ps.-Aristotle's *Problems*, see also H. Flashar, 1962, pp. 356–358.

[71] This collection was first edited as *Problemata inedita* by U.C. Bussemaker in 1857 (pp. 291–334) and has recently been re-edited under the heading of *Supplementa problematorum* by S. Kapetanaki and R.W. Sharples, 2006. For the issue of authenticity, see n. 72 below.

[72] This work was first edited by J.L. Ideler, 1841, pp. 3–80 (not to be confused with Alexander's three Books of φυσικαὶ σχολικαὶ ἀπορίαι καὶ λύσεις, which are often cited in modern literature as Alexander's *Quaestiones*). In 1859, Usener annexed the two first Books of Bussemaker's *Problemata inedita* to Ideler's edition, by heading them *Alexandri Aphrodisiensis quae feruntur problematorum libri 3 et 4*. Thus, he believed that all four Books formed a collection of medical puzzles and physical problems circulating under the name of Alexander of Aphrodisias, but, as S. Kapetanaki and R.W. Sharples, 2006, pp. 16 and 27 have pointed out, the association of Bussemaker's first two Books (= Usener's Books three and four) with Alexander is no stronger than their connection with Aristotle. Ideler's two Books are also generally considered spurious. It is notified by Kapetanaki and Sharples (p. 1, n. 1) that Carl-Gustaf Lindqvist of Gothenburg university is preparing a new edition of this text, the forthcoming of which is still eagerly awaited. For an attempt to outline the complex bibliographical details on Ps.-Alexander's collection, see R.W. Sharples, 1987, pp. 1198–1199.

[73] This collection was first edited by J.L. Ideler, 1841, pp. 144–167, and has recently been re-edited by A. Garzya and R. Masullo, 2004.

[74] See n. 57 above. There is also a Byzantine collection of Hippocratic problems, on which, see A. Guardasole, 2007.

Book: ὅσα ἰατρικά)⁷⁵. Moreover, medical question-and-answer literature was common in the Greco-Roman period. It can be found, for instance, in Soranus' *Gynaecia*, in Ps.-Soranus' *Quaestiones medicinales*⁷⁶ and in a number of medical catechisms written on papyrus⁷⁷. Plutarch's natural problems incorporate much medical material (e.g., by quoting renowned doctors [see 4.2.1.1., nn. 110–111]), but they are more generally naturalist in kind, as is its Aristotelian model.

The longstanding popularity of the genre of natural problems is confirmed by the fact that it flourished well beyond the chronological and geographical boundaries of Antiquity and the Occident. Collections of natural problems were still composed during the Middle Ages and the Renaissance, and the Ps.-Aristotelian *Problems* were transmitted in a number of different languages such as Syriac, Arabic, Hebrew, Latin and the vernacular⁷⁸. Perhaps the most important aspect of this brief outline of the history of the *Problems* is the idea that we are dealing with a vast corpus of ancient scientific knowledge that is open to textual evolution, reorganisation and accumulation and is deeply rooted in Aristotle's causal project of scientific research⁷⁹. An important feature of Plutarch's natural problems, then, is the fact that they – at least from a genre perspective – contribute to the Aristotelian tradition, to which they add new problems and from which they revise older ones by looking for new solutions⁸⁰.

---

⁷⁵ Much of the medical content in Ps.-Aristotle's *Problems* (not only in the first Book) is, indeed, Hippocratic in nature. Fur further detail, see F. Poschenrieder, 1887, pp. 38–66, H. Flashar, 1962, pp. 338–340, J. Bertier, 1989, J. Jouanna, 1996, C. Jacob, 2004, pp. 44–45, A. Ulacco, 2011, R. Mayhew, 2015b, K. Oikonomopoulou, 2015, O. Thomas 2015. The *Problems* and their inquisitive method were well-known to Galen. See, e.g., *SMT* 11, 474 Kühn (= Arist. fr. 223 Rose): Ἀριστοτέλης καὶ Θεόφραστος, ἕτεροί τέ τινες ἄνδρες φιλόσοφοι, τὰ τοιαῦτα τῶν προβλημάτων ἐν τοῖς φυσικοῖς ζητήμασιν προβάλλουσί τε καὶ λύουσι.

⁷⁶ First edited by V. Rose, 1870, pp. 161–240. A new edition of this text is being prepared by Klaus-Dietrich Fischer of Mainz university (for further detail, see K.-D. Fischer, 1998).

⁷⁷ These medical catechisms are doctrinal manuals organised according to a clear structural pattern, viz. by a sequence of definitions, causes, signs, characteristic features and therapies of diseases. For further literature, see D. Leith, 2009. For the question-and-answer format in medical literature more generally, see A.M. Ieraci Bio, 1995 (esp. pp. 191–192).

⁷⁸ Particularly useful here are H. Flashar, 1962, pp. 370–382, A. Blair, 1999 and the contributions in P. De Leemans and M. Goyens, 2006 (with a selected bibliography at pp. 295–317). See also esp. B. Lawn, 1963 and L. Filius, 1999.

⁷⁹ Cf., e.g., F.H. Sandbach, 1982, p. 225 and C. Jacob, 2004, p. 43. For the 'encyclopaedic' nature of problem literature (with a central focus on Plutarch's collections of *quaestiones*), see K. Oikonomopoulou, 2013a.

⁸⁰ Cf. F.H. Sandbach, 1965, p. 138: "As things are, we can say no more than that *Quaestiones Naturales* seem to be a compound, in unknown proportions, of traditional and newly adduced solutions."

Notably, in several of his natural problems, Plutarch emphatically deals with phenomena that remained unexplained or were, in his opinion, explained inadequately by Aristotle. In *Quaest. conv.* 650A, for instance, (concerning the problem of why old men especially are susceptible to drunkenness and women least) Florus works out an explanation (αἰτία) that was not elaborated upon (οὐκ ἐξειργάσατο) by Aristotle himself in his Περὶ μέθης (the allusion may be to Book three of the *Problems*: ὅσα περὶ οἰνοποσίαν καὶ μέθην). Again in *Quaest. conv.* 690F, Plutarch says that the natural phenomenon at issue (according to which pebbles or lumps of metal seem to cool and temper the water in which they are thrown) is recorded by Aristotle in his *Problems* but is not explained there, so that it is up to the symposiasts to try to solve it themselves (αὐτὸ τοῦτ' ἔφη μόνον ἐν προβλήμασιν εἴρηκε τὸ γινόμενον· εἰς δὲ τὴν αἰτίαν ἐπιχειρήσομεν ἡμεῖς). As we will see later on, similar passages are found in *Quaestiones naturales* where Aristotle's (or more broadly Peripatetic) theories often serve as a point of departure for Plutarch's physical aetiology. Arguably, then, Plutarch tried to fill in a number of gaps in the Aristotelian tradition of natural problems, and, by extension, in the contemporary scientific paradigm of his time – at several occasions he is, in any case, proud enough to have added something of his own to this tradition[81] [see 4.2.2.2.].

Plutarch was well acquainted with the Aristotelian natural problem tradition (from which he not only draws in his natural scientific works, for that matter)[82]. Apart from the explicit (i.e. *nominatim*) quotations, Plutarch probably also alludes to the *Problems* a number of times without mentioning Aristotle's name[83]. In these cases, exact source passages cannot be conclusively determined. Indeed, it may well be that in these allusive accounts Plutarch was generally inspired by what Aristotle wrote, so that a clear correspondence is not only unlikely to be found but also

---

[81] For the idea that Plutarch in his natural problems progresses the science of his day one step at a time, see M. Meeusen, 2016. The same counts, *mutatis mutandis*, for *Quaestiones Platonicae*. In *Quaest. Plat.* 7, 1004E, Plutarch notes that Plato τὴν καθ' ἕκαστον ἐξεργασίαν ἡμῖν ἀφῆκε, implying that it is his and his students' task to solve a problem left unanswered by Plato.

[82] E.g., in *De coh. ira* 458F–459A, Plutarch relies on Ps.-Arist., *Probl.* 875a34–35 (about Satyrus of Clazomenae stopping up his ears with wax in order not to hear the insults of his opponents).

[83] See W. Capelle, 1910, pp. 329–330, n. 2: "Unzweifelhaft sind aber solche 'aristotelischen' Problemata noch an ungezählten Stellen von Plutarch, Gellius, Galen u.a. benutzt, wo der Name Aristoteles nicht genannt bzw. die Quelle verschwiegen wird." Cf. also H. Flashar, 1962, pp. 308 and 312–313. For lists of Plutarch's references and allusions to the *Problems*, see W.C. Helmbold and E.N. O'Neil, 1959, pp. 9–10, H. Flashar, 1962, pp. 369–370, F.H. Sandbach, 1982, pp. 223–225, P. Louis, 1991, pp. xvi and xxxi–xxxii, E.N. O'Neil, 2004, pp. 82–83 and G. Roskam, 2011, pp. 45–46.

unnecessary to be looked for – the *Problems*, thus, serving as a more abstract and flexible intertext (this is, indeed, much in line with the argumentative creativity that is so central to the genre [see 4.2.2.2.]). As to the explicit quotations, on the other hand, some are probably derived from the *Problems* but cannot be traced in the extant collection (or in other Aristotelian works), so that we are presumably dealing with excerpts from problems that are now lost – of course, this also remains hypothetical[84]. On other occasions, Plutarch's text – either in the explicit quotations or in the implicit allusions that are not labelled with Aristotle's name – contains significant parallels with Ps.-Aristotle's/Alexander's *Supplementary problems* and Ps.-Alexander's *Medical puzzles and natural problems* rather than with Ps.-Aristotle's extant *Problems*[85]. This has led scholars to assume that there must have been a common source, viz. an earlier version of the *Problems* that is now lost – presumably Andronicus' edition (see above)[86]. There are indications that Plutarch, indeed, drew on a version of Peripatetic problems that was more extensive than the one we have today. But how much more extensive this collection really was cannot be determined[87]. Bottom-line is that the Aristotelian genre and tradition of natural problems, with its typical content, style and organisation, had a major influence on Plutarch's own composition of natural problems[88].

---

[84] See H. Flashar, 1962, p. 313. For the remains from the lost *Problems*, see frs. 209–245 Rose.

[85] See H. Flashar, 1962, pp. 360 (with n. 1), 367, 369 and S. Kapetanaki and R.W. Sharples, 2006, p. 12. Cf., e.g., *De Pyth. or.* 395F and Ps.-Arist./Alex. Aphr., *Suppl. probl.* 3, 17; *Quaest. conv.* 689E–690B and Ps.-Arist./Alex. Aphr., *Probl. ined.* 3, 51 (not recorded in the *Supplementary problems*; see S. Kapetanaki and R.W. Sharples, 2006, p. 7, n. 47); *Q.N.* 12, 914F and Ps.-Arist./Alex. Aphr., *Suppl. probl.* 3, 29 and 47; *Q.N.* 21, 917BD and Ps.-Arist./Alex. Aphr., *Suppl. probl.* 2, 144, 145, 155 (= *Probl. ined.* 2, 141, 142, 152 respectively); *De facie* 932BC and Ps.-Alex. Aphr., *Probl.* 2, 46 (J.L. Ideler, 1841, p. 65, 18–32) etc.

[86] See H. Flashar, 1962, p. 369 and P. Louis, 1991, p. xxxii. For a list of parallels between Ps.-Aristotle's *Problems* and the *Problemata inedita/Supplementa problematorum*, see U.C. Bussemaker, 1857, p. x and S. Kapetanaki and R.W. Sharples, 2006, p. 4 (see also the index, at p. 283). H. Flashar, 1962, pp. 364–365 counts 31 parallel problems between Ps.-Aristotle's *Problems* and those ascribed to Alexander of Aphrodisias, and 16 in the collection of Cassius the Iatrosophist (p. 368). See also A. Garzya and R. Masullo, 2004, p. 13.

[87] The lists of Aristotle's works mention a collection of *Problemata* (*physica*) in 70 Books (to which the name of the otherwise unknown Eucaerus is sometimes linked). The ancient evidence for this 70 Book version is provided by *Vita Marciana* 427, 8 R³, index Hesychii nr. 168 and Elias' commentary on Aristotle's *Categories* 114, 12–13. See P. Moraux, 1951, pp. 280–281, H. Flashar, 1962, pp. 312–314, P. Louis, 1991, pp. xxxi–xxxiii.

[88] For the conceptual distinction between sources and traditions, see J. Mansfeld, 1999, pp. 29–30.

In what follows, I will further explore the 'problematic' organisation of *Quaestiones naturales*, first at micro- and then at macrostructural level.

4. Internal organisation of Plutarch's natural problems (microstructure)

There can be no decisive answer as to whether Plutarch was familiar with Aristotle's definition of πρόβλημα and its systematic integration in Aristotelian logic, as outlined above[89]. The Lamprias catalogue does mention a work entitled Περὶ προβλημάτων (nr. 193), which probably provided a great deal of useful information about the proper phrasing, structuring, method and general purpose of the genre of problems (not only natural problems for that matter), but the text is now lost[90]. Perhaps this work could also have shed light on the author's own appreciation of the genre. Unfortunately, though, we can only speculate about its original content. Nevertheless, we can still learn a great deal from Plutarch's collections of problems that are preserved.

The genre of problems covers a considerable share of Plutarch's entire literary production. The Lamprias catalogue lists a significant number of collections of Αἰτίαι among Plutarch's writings, several of which are still extant today, while others are now lost or partially preserved in fragmentary form[91]. Among these collections, *Quaestiones naturales*

---

[89] C. Darbo-Peschanski, 1998, p. 22 remains sceptical, but for the possibility of Plutarch's acquaintance with Aristotle's *Topics*, see F.H. Sandbach, 1982, pp. 212–213 (and 230).

[90] It is unknown whether this work was authentic. Notably, D.L. 5, 23 mentions the same title (viz. Περὶ προβλημάτων) among Aristotle's works. See P. Moraux, 1951, p. 88 and P. Louis, 1991, p. xx, n. 50. One fragment remains from Aristotle's Περὶ προβλημάτων, which concerns the conceptual distinction between natural and dialectical problems, based on their different form of inquiry (fr. 112 Rose = Alex. Aphr., *In Ar. Top.* 62, 30–63, 19). In short, the fragment explains that Aristotle defined a dialectical problem as a question concerning alternatives, where a positive or a negative answer is expected ('*Whether* a thing is so, or not?'). A natural problem, by contrast, investigates the cause or nature of a natural phenomenon ('*Why* is this so?', '*What* is this?'), so that another type of answer is expected (viz. an explanation or a definition). Strictly speaking, then, the natural problems collected in Ps.-Aristotle's *Problems* are no dialectical problems. Cf. J.G. Lennox, 2001, pp. 77–78 and J. Mansfeld, 2010, p. 43, n. 34. Aristotle also deals with problems and their solutions in *Poet.* 1460b6ff. (Περὶ προβλημάτων καὶ λύσεων), but this concerns specific textual problems (viz. difficult passages or expressions in a text). D.L. 5, 48 and 49 also mentions one book Περὶ τῶν προβλημάτων φυσικῶν among Theophrastus' writings (26a FHSG).

[91] The Lamprias catalogue mentions Αἰτίαι τῶν Ἀράτου Διοσημιῶν (nr. 119 = frs. 13–20 Sandbach), Αἰτίαι Ῥωμαϊκαί (nr. 138 = *Quaestiones Romanae*), Αἰτίαι βαρβαρικαί (nr. 139; see T.S. Schmidt, 2008), Αἰτίαι τῶν περιφερομένων Στωικῶν (nr. 149; ἱστοριῶν? Sandbach), Αἰτίαι καὶ τόποι (nr. 160), Αἰτίαι ἀλλαγῶν (nr. 161), Αἰτίαι Ἑλλήνων (nr. 166 = *Quaestiones Graecae*), Αἰτίαι γυναικῶν (nr. 167 = nr. 126 (Γυναικῶν ἀρεταί = *Mulierum virtutes*)? Nachstädt), and

especially exhibits a remarkable degree of similarity to Ps.-Aristotle's model. This is clear especially at a structural level. The internal format of the genre of natural problem is based on a typical 'erotapocritical' scheme, where questions are raised and answered in a tight repetitive structure. The question (*quaestio*) remains fairly short in most cases[92] and is normally introduced with διὰ τί[93]. The question is always 'Why?' and not 'What is this?' or 'Is it true that?', because it is not the author's intention to define or verify the actual existence of the natural phenomenon at hand. Rather, he is interested in finding their physical origin by formulating plausible explanations (*causae*) for them[94]. A collection like *Quaestiones Graecae*, by contrast, is more concerned with defining obscure Greek cultural phenomena, in the manner of an encyclopaedia or a dictionary, than with providing their aetiology or origin: therefore, 'why'-questions are far less frequent there[95].

For nearly every natural problem, Plutarch provides a range of different answers (two and often more), some of which contain received knowledge, while others contain his own innovative contributions. These solutions are mostly formulated in an interrogative, anti-dogmatic fashion, implying that any criticism and new solutions can always be added[96]. The disjunctive

---

eventually Αἰτίαι φυσικαί (nr. 218 = *Quaestiones naturales*). *Quaestiones convivales* is not recorded in the Lamprias catalogue (perhaps to be identified with nr. 125: Ἀπομνημονεύματα). The Πλατωνικὰ ζητήματα (nr. 136) and Ἀποριῶν λύσεις (nr. 170) can be added. See also A. Gudeman, 1927, cols. 2525–2527 and G.W.M. Harrison, 2000a, p. 195 (who notes that "[b]ecause of their placement just before Περὶ ταυτολογίας, both the Αἰτίαι καὶ τόποι and the Αἰτίαι ἀλλαγῶν would seem to be concerned with rhetoric and style, and, therefore, the meaning of αἰτία might be quite different [sc. from that of 'explanations', viz. 'invective'; see LSJ, s.v. i, 2]").

[92] Sometimes an intermediate part is incorporated to illustrate or specify a particular point in the *quaestio*. See *Q.N.* 5, 913B, 21, 917B, 26, 918BC, 34. For intermediate pieces in the *quaestiones* in Ps.-Aristotle's *Problems*, see H. Flashar, 1962, pp. 342–343.

[93] Only a few variant phrases are used, viz. τίς ἡ αἰτία δι' ἥν (*Q.N.* 29 and 40) and διὰ τίν' αἰτίαν (*Q.N.* 20), which have the same basic meaning. R. Mayhew, 2011a, p. xiii, n. 1 has calculated that 98% of the Ps.-Aristotelian *Problems* (counting 903 in total) also begins with διὰ τί.

[94] Cf. *Quaest. conv.* 700C: οὐχ ὅστις εἴη [...] ἀλλὰ περὶ αὐτῆς διηπορεῖτο τῆς αἰτίας καθ' ἥν κτλ.

[95] In *Quaestiones Graecae* most of the questions are introduced with τί, τίς or τίνες (cf. *Quaest. Graec.* 1–25, 29, 30, 32, 34, 38, 40, 44). The introduction with διὰ τί is less frequent but not absent (see *Quaest. Graec.* 9b, 31, 36, 37, 39, 45–51, 53, 55, 58 and K. Oikonomopoulou, 2013b); there are also alternative formulations, see the table in P. Payen, 1998b, p. 41 (cf. also A. Carrano, 2007, p. 7). Moreover, each problem in this collection often receives only one clear-cut solution (or better: definition) rather than a number of successive explanations, as is the case rather in *Quaestiones Romanae*. See R. Preston, 2001, p. 96, J. Boulogne, 2002, pp. 179–180 and P. Payen, 2014, pp. 246–247.

[96] The only (and therefore unintentional?) exceptions are *Q.N.* 20, 29, 35 (first *causa*)

sequence of multiple interrogative solutions to one and the same question is ordered in accordance with the standard problem arrangement: the *causae* are commonly phrased as a compound question introduced with πότερον (…); ἤ (…); ἤ (…);[97], and as the aetiology advances, the degree of plausibility often increases. A solution most often departs from a general argument or observation, which is phrased interrogatively, and is further clarified and justified in a dogmatic fashion in the elaborative explanations, deductions, illustrations, etc., that follow – these I call the 'sub-arguments'. They are commonly introduced with a wide array of conjunctions, particles etc. (such as διό, γάρ, ὅθεν, μέν, δέ), by which the author maintains an essential and coherent structure in the development of the explanations[98].

---

and 38. *Q.N.* 40–41 can be neglected, since they are reformulations by Psellus (see M. Meeusen, 2012b).

[97] For Aristotle's antithetic use of πότερον, see *Met.* 1055b32ff.: τὸ πότερον ἀεὶ ἐν ἀντιθέσει λέγομεν. As a rule, in cases where only one solution is given, the editors of *Quaestiones naturales* decided not to use the disjunctive ἤ (which would then have a modest affirmative, rather than interrogative nuance: cf. H. Bonitz, 1870, pp. 312b57–313a18) but the interrogative ἤ (see J.D. Denniston, 1966, p. 283). Among the first 31 problems (i.e. those containing the original Greek text), 12 are without the opening πότερον *causa*, nine of which have only one solution: *Q.N.* 8, 9, 11, 14, 15, 18, 22, 24, 30 (not so in *Q.N.* 20, 21, 29, where the regular problem scheme is not followed; remarkably enough, *Q.N.* 23 has only the πότερον *causa*: see the commentary *ad loc.*). See also G.W.M. Harrison, 2000b, p. 240. Regarding such solitary solutions (in *Quaestiones Romanae*) J. Boulogne, 1992, p. 4688 (cf. also 2002, p. 94) rightly observes that "l' unité numérique ne signifie pas unicité. Plutarque isole une cause, mais il laisse entendre qu' elle n' épuisse pas la causalité. L' investigation n' est pas fermée, il est possible de toujours l' enrichir. C' est ce à quoi invite la structure ouverte de ces interrogations qui énoncent une seule réponse." The same idea applies to Plutarch's natural problems (see below).

[98] In addition, Plutarch uses several standard phrases in order to indicate a specific aetiological direction to the reader. These phrases can point, for instance, to a logical consequence (e.g., *Q.N.* 5, 913D: οὕτω δὲ τούτων ἐχόντων), something obvious (δῆλον: *Q.N.* 7, 914A, 26, 918D, 29, 919A), something necessary (δεῖ, δεῖται: *Q.N.* 15, 915D, 25, 918B; ἀνάγκη: *Q.N.* 9, 914D; ἀναγκαιότερα: *Q.N.* 4, 913A; θετέον: *Q.N.* 5, 913D), something credible (*Q.N.* 34: *credendum est*), an indication (*Q.N.* 2, 912B: κατηγοροῦσιν), a sign (τεκμήριον, σημεῖον: *Q.N.* 19, 916F, 30, 919C), some evidence (μαρτύρια, μαρτυρεῖ: *Q.N.* 2, 912C, 6, 913F, 8, 914B, 15, 915D), an inductive proof (*Q.N.* 1, 911D: ἄλλοις τε πολλοῖς ἀποδείκνυται, *Q.N.* 7, 914A: ὡς ἔστιν […] καταμαθεῖν) etc. These phrases clearly reflect the elementary logic of the argumentation in its most basic and transparent form. According to L. Senzasono, 2006, p. 9, this structural grid of logical connections aims at an essential and unembellished representation of the natural phenomena. As such, it would testify to the collection's scientific style: "Questi nessi collegano rilievi di fenomeni naturali in una struttura essenziale, priva di qualsiasi abbellimento, come s'addice a un'esposizione che intenda essere scientifica, nel mondo Greco come in altre epoche fino ad oggi." However, as we will see later, the embellishment-free discourse and lack of literary bravura is not an inherent feature of Plutarch's scientific discourse, nor of ancient scientific texts in general [see 1.2.5.].

In short, the problem arrangement develops in three subsequent echelons, three layers of encapsulation, viz. question – argument – sub-argument[99]. The questions and arguments are mostly formulated interrogatively, the sub-arguments assertorically.

This repetitive tripartite structure permeates the entire collection and gives the problems some kind of a 'Chinese-box-effect'. The aetiological movement is upwards, viz. from the particular and atomic sub-arguments to the argument and hence further on to the *quaestio* itself, which is the *explanandum*, and serves as the heading of the chapter. This means that the smallest element in the aetiological hierarchy (the sub-argument), is in support of the higher strata and can, therefore, be considered the actual fundament of the problem construct. The concrete phrasing of the problems develops the other way round, from top to bottom: the *quaestio* comes first and is explained over several *causae* into its smallest details. The opposition between the aetiological movement (upwards) and the actual phrasing of the problem (downwards) does not remain without further interest for the actual writing process and, more in specific, for Plutarch's sophisticated use of sources (to which I will come back later [see 4.2.1.2.]).

It is not my intention to deal with every single deviation and particularity in the phrasing of the *quaestiones* and *causae* here, but special attention should be directed to the following more general observations (these points will also be relevant when dealing with Plutarch's scientific methodology later on [see 4.3.]).

1) The general interrogative and anti-dogmatic formulation of the explanations implies that the aetiological structure remains open, so that the problems do not receive final closure. In this sense, the questions are not fundamentally 'solved', and the author leaves it to the reader to make up his own mind (I will come back later to this in light of the educational interests of the collection [see 3.2.1.]). Another consequence of the interrogative formulation of the aetiologies is that the solutions do not always necessarily exclude one another, unless, of course, in those cases where an explicit disjunction is made (e.g. *Q.N.* 2, 911F: ἢ τοῦτο μὲν οὐκ ἀληθές, 27, 918E: ἢ τοὐναντίον). In many cases, however, the introductory ἢ can actually indicate a sense of argumentative complementarity, rather than a disjunction, between separate *causae*, each solution discussing a specific facet of the complex problem at hand[100]. Such a notion of inclusivity can also be articulated more periphrastically with a phrase like ἢ δεῖ μὴ μόνον κτλ. (*Q.N.* 25, 918B; or ἅμα συνημμένον: *Q.N.* 21, 917C).

---

[99] See also L. Senzasono, 2006, pp. 34–41 on the mutual relations between the causes in *Quaestiones naturales* (viz. their "griglia strutturale di relazione", p. 37).

[100] J. Boulogne, 1994, p. 128 speaks of a "disjonction inclusive", and he even claims (1992, p. 4690) that ἢ always has inclusive value (in *Quaestiones Romanae*).

2) The solutions are primarily structured around the notion of argumentative plausibility (τὸ πιθανόν) rather than certainty (τὸ ἀληθές), which reflects a fundamental uncertainty on the author's side about what is said (the epistemic range of these concepts will be examined later [see 4.3.2.2.]). Even if each specific explanation often approaches the problem from another angle and thus has no lesser or greater claim to τὸ πιθανόν than another, this is not always the case. Some solutions are explicitly suggested to be more plausible than others (e.g., *Q.N.* 39: *an probabilius est?*), and in some cases, a previous explanation is suggested to be incorrect but not necessarily implausible (e.g., *Q.N.* 2, 912A: ἢ καὶ τοῦτο πιθανὸν μᾶλλον ἢ ἀληθές ἐστι;). The fact, however, that Plutarch also records these less plausible solutions seems to testify to his urge for aetiological exhaustivity, each solution attributing at least a certain aspect of plausibility to the aetiology. At the same, time this procedure may also have a heuristical motivation, as being a useful argumentative strategy in the search and development of ever new and increasingly plausible explanations (*reculer pour mieux sauter*)[101].

3) Even though Plutarch prefers to suggest rather than to assert in his physical aetiologies, and thus takes no responsibility in formulating a definite solution, he still has several ways of expressing his own personal preference for one explanation over another. He does this by either explicitly evaluating a certain explanation by venting his criticism or by showing his appreciation for it, mostly assessing the theory's plausibility. This is the case, for instance, in *Q.N.* 2, 911F–912A, where Laetus' explanation is suggested to be 'untrue' (οὐκ ἀληθές, ranked in the first *causa*), whereupon Plutarch wonders whether Aristotle's solution is 'true' (ἀληθές, ranked in the second *causa*). He concludes, however, that it is 'plausible rather than true' (πιθανὸν μᾶλλον ἢ ἀληθές). There are also more subtle ways of doing this. When a *causa* is introduced with the phrase 'or rather' (ἢ μᾶλλον: e.g., *Q.N.* 3, 912E, 17, 916A), one can presume that the solution at issue has more legitimacy towards τὸ πιθανόν than the preceding one. However, this phrase can also point to a simple, unqualified transition to another explanation. In that case, it indicates that we are simply dealing with an alternative solution that is at least equally plausible. Furthermore, the positioning of the explanations themselves often seem to imply a certain ranking by the author. In a number of cases, the explanations are not enumerated in a thoughtless order but more or less hierarchically, according to the principle of increasing plausibility[102]. As scholars have

---

[101] The same conclusion was reached (for *Quaestiones Romanae*) by J. Boulogne, 1992, p. 4689: "En définitive, l'inventaire favorise la découverte en provoquant le surgissement des idées auxquelles on n'a pas pensé, et plus il est étendu, moins la solution conservée court le risque d'être infirmée par ce qui n'est pas venu d'emblée à l'esprit."

[102] For this progressive structure in Plutarch's aetiologies more generally, see also, e.g.,

argued, the first solution (i.e. the πότερον *causa*) is often reserved for *communis opinio*, thus representing a first step in the direction of a correct explanation, whereas the last solution contains most progress in that direction¹⁰³. In this sense, it can be presumed that Plutarch's evaluation of past authorities is often implicit in the ranking of the *causae* in which they are quoted (this ranking will be of interest later in the analysis of Plutarch's quotations from traditional authorities [see 4.1.2.3. and 4.2.1.1.]). This does not necessarily imply, however, that the last position is also reserved for an absolutely correct solution. At most, the last position will be an indication that the explanation expresses the author's preference¹⁰⁴, but it is the *causa palmaris* only in a relative, not an absolute, sense. After all, the interrogative structure and absence of closure implies that the addition of other explanations is still possible and that the aetiological potential has not necessarily been exhausted. Also note that the same aetiological mindset is present, for instance, in *Quaestiones convivales*. There, Plutarch often has the courtesy to let his fellow symposiasts speak before his own sympotic character¹⁰⁵. In this way, he can (via his sympotic *alter ego*) speak last and, thus, most authoritatively¹⁰⁶. There is, indeed, a certain tendency towards authorial self-promotion in *Quaestiones convivales*, but as König observes, some nuancing is in place: "Even where Plutarch does speak last, or take some other prominent role in discussion, he

---

J. Boulogne, 1992, pp. 4694–4696 and P. Donini, 1992, p. 111. The same gradual progression from less to more probable explanations has also been discussed for other works, such as *De E* and *De Iside et Osiride* (see P.R. Hardie, 1992, p. 4755 and G. Roskam, 2011c, p. 425).

¹⁰³ For the place of *communis opinio* in *Quaestiones naturales*, see G.W.M. Harrison, 2000b, pp. 238–239.

¹⁰⁴ See already C. Kahle, 1912, pp. 63–64: "Si sententiae complures quae magni momenti sint, inter se pugnant, omnes deinceps perpetuis orationibus proferri facit, sed nulla refutatur et quae sit vera, positione fere sola quia postremo loco describitur, significatur." This technique is also used in the *Vitae* (cf., e.g., *Rom.* 3, 1: Τοῦ δὲ πίστιν ἔχοντος λόγου μάλιστα καὶ πλείστους μάρτυρας κτλ.). It was also discussed, e.g., for *Quaestiones convivales* by J.C. Relihan, 1992, pp. 232 and G. Roskam, 2011c, p. 425, for *Quaestiones Platonicae* by J. Opsomer, 1994a, p. 12 (with n. 32), 1996a, p. 83 (with n. 42), and 2010 (with n. 3), and for *Quaestiones Graecae*, *Romanae* and *naturales* by D.A. Russell, 1973, p. 45. For the idea that Plutarch endorses his own thought at the end of the dialogue in *De E*, see F.E. Brenk, 2005, p. 29 (with n. 10). The same technique has also been discussed for *De genio Socratis* and for *De facie* by P. Donini, 2009, p. 202 and 2011, p. 86 respectively.

¹⁰⁵ This is the case in *Quaest. conv.* 627EF, 635CD, 673D–674C, 674E–675D, 677E–678B, 690DE, 690F–691C, 691D–692A. Cf. J. König, 2007, p. 51.

¹⁰⁶ After all, as G. Roskam, 2010, p. 47 has noted, Plutarch is the one who is holding the pen. On Plutarch's authorial self-presentation in *Quaestiones convivales* as a complex mean between self-promotion and self-effacement, see J. König, 2011. See also F. Klotz, 2007.

sometimes stresses his own reluctance, going out of his way to avoid the impression of grandstanding."[107] Indeed, *Quaestiones convivales* is not an ego-document, or, at least, it is not Plutarch's intention to let his own authority prevail over that of his companions at any cost. In view of the relative plausibility of the arguments, Plutarch may have had good epistemological reasons for this, besides from purely sociological ones [see 4.3.2.].

4) Sandbach argues that the use of semi-synonymous pairs in the *causae* may indicate (or "provide a clue") that we are dealing with Plutarch's personal contributions to the problems[108]. The use of such semi-synonyms is, indeed, a typical feature of Plutarch's general style, but Sandbach's theory is not, therefore, necessarily correct. The most problematic point is that these pairs are also found in the quotations from traditional authorities (see, e.g., *Q.N.* 2, 912A: τὸ δὲ τοῦ Ἀριστοτέλους ἀληθές, ὅτι πρόσφατόν ἐστι καὶ νέον ὕδωρ τὸ ὑόμενον ἕωλον δὲ καὶ παλαιὸν τὸ λιμναῖον;). But then again Plutarch more often paraphrases quotes by rendering them in his own words rather than by presenting them κατὰ λέξιν[109] [see 4.2.1.1., n. 116]. Sandbach also argues that Plutarch explicitly marks his personal and innovative contributions to the problems with introductory imperatives like σκόπει δὲ μή (*Q.N.* 3, 912E), σκόπει δὴ μή (*Q.N.* 12, 915A), σκόπει δή (*Q.N.* 19, 916C)[110]. Many scholars have shared Sandbach's opinion, supposing that these imperatives mark Plutarch's own contributions and/or highlight the greater reliability of the solutions introduced by them[111]. In

---

[107] J. König, 2011, p. 195.

[108] F.H. Sandbach, 1965, p. 136. Sandbach refers to *Q.N.* 2, 6, 10, 13, 16, 19, 21, 23, 24, 26 (he omits *Q.N.* 3, which is misprinted as *Q.N.* 2 on p. 135).

[109] For Plutarch's general use of semi-synonymous pairs, see U. von Wilamowitz, 1902, p. 203, B.P. Hillyard, 1981, p. xxiii, T.S. Schmidt, 1999, pp. 15–26, S.-T. Teodorsson, 2000b, L. Senzasono, 2006, pp. 15–17. Plutarch's use of such semi-synonyms in *Quaestiones naturales* has been analysed by G.W.M. Harrison, 2000b, pp. 246–249, who also puts Sandbach's theory into perspective. The following reservation was made by Sandbach himself (1965, p. 136): "The hypothesis that passages marked by these semi-synonymous pairs, many of which are drawn from a richer vocabulary than that of the bulk of the work, may be original, implies no claim of absolute priority for Plutarch, but only that he was not here abbreviating or copying a text before him, but drawing on the resources of his well-stocked mind and memory."

[110] F.H. Sandbach, 1965, p. 135. These and similar imperatives (σκόπει μή, ὅρα μή, ὅρα δὲ/δὴ μή) occur frequently in Plutarch's other collections of problems (except for the *Quaestiones Graecae*), as well as in his other writings. Notably, they are also frequent in Plato, whose style of writing Plutarch perhaps imitates (cf., e.g., *Apo.* 27a, *Phd.* 74a, *Tht.* 162e). See also L. Senzasono, 2006, pp. 23–24.

[111] Cf. F. Leo, 1864, p. 6, H.J. Rose, 1924, p. 49, J. Boulogne, 1992, p. 4688 ("plus d'intérêt que les autres"), p. 4695 ("plus favorablement"), 2002, p. 93 ("la solution préférée"), J. Opsomer, 1994a, p. 11 and 1996a, p. 77, G.W.M Harrison, 2000b,

De prim. frig. 952C, for instance, Plutarch opens his (presumably) personal explanation about the essence of coldness with the phrase σκόπει δή. His personal preference may, indeed, go to this theory, which is placed last, but at the same time the treatise's anti-dogmatic conclusion emphatically marks that it is perhaps 'neither less nor much more plausible' than other theories (*De prim. frig.* 955C: κἄν μήτε λείπηται τῇ πιθανότητι μήθ' ὑπερέχῃ πολύ κτλ.) [see 4.3.2.1.]. In the case of *Quaestiones naturales*, it is not always clear whether such imperatives mark Plutarch's personal theory or the one he prefers. For instance, the phrase σκόπει δέ μή in the last *causa* in *Q.N.* 3, 912E (about the aphrodisiac effects of salt on bitches and mice) might suggest that we are dealing with Plutarch's personal contribution. However, the same theory is formulated not by Plutarch, but by Philinus in *Quaest. conv.* 685DE. To give another example, in *De Pyth. or.* 395F an explanation of Aristotle is introduced with σκόπει δ'. These theories may, of course, carry Plutarch's preference, but they are not, therefore, necessarily *his*. Remarkably enough, in some cases a new solution even follows after the *causa* that contains an imperative (as is the case in *Q.N.* 12, 915A). This can cast considerable doubt on the common assumption that these imperatives necessarily contain Plutarch's preferred contributions. The least that can be said about these imperatives, then, is that they shift the interrogative style of the *causa* to the affirmative – or better, to the imperative –, and, thus, draw the reader's attention to an important point that the author wishes to make ('Consider this!'; 'Attention please!'). As such, I believe the actual use of these imperatives should be explained in light of the fact that they emphatically address the intended reader (in the second person) to take something important into consideration. Plutarch, thus, makes the reading process more engaging and provokes a direct response to what is read[112].

---

p. 243. L. Senzasono, 2006, p. 24 is probably right in pointing out that these imperatives "presentano la verità come una conquista de compiere, non come qualcosa di acquisito. [...] Si tratta, in Plutarco come in Platone [see n. 110 above], d'una concezione del sapere come ricerca ed esame, prima che come risultato."

[112] *Pace* G.W.M. Harrison, 2000b, p. 241, who argues that the imperative in *Q.N.* 12, 915A is an indication of intermittent composition [see the prologue, n. 21]. He adds that "Michael Psellus was so bothered by the third *causa* to *Quaestio* 12 that he re-wrote it. The problem may not be so much the content or syntax as its mere presence. It would seem quite natural that an investigator would set up the rival theories first before concluding his own." In my opinion, however, it is highly unlikely that Psellus actually 'knew' that Plutarch introduced his own explanations with these imperatives: this is still not a matter of course today, as I have tried to show. Rather, whereas the *quaestio* from *Q.N.* 12 first mentions the καταφάνεια and then the γαλήνη caused by oil on seawater, Plutarch does the reverse in his aetiology, while Psellus' reorganisation of the explanations keeps more to the sequence of the *quaestio*. For Psellus' adaptations more generally of Plutarch's *Quaestiones naturales* in his *De omnifaria doctrina*, see M. Meeusen, 2012b.

It goes to show, conclusively, that from a methodological perspective, the problem format provides several structural advantages for the orderly and efficient presentation of enigmatic topics. It allows the author not only to formulate a very specific problem on its own terms, and, thus, to concentrate his full attention to it, but also to solve it in an organised fashion. It also enables the author both to summarise and, if appropriate, to criticise the scientific tradition and doxography at issue and to add his personal observations and remarks to it. Moreover, the structural open-endedness allows for further elaboration and review, either by the author himself or by his reader[113]. Even if, in the case of *Quaestiones naturales*, there are strong deviations in length, number of *causae*, and degree of elaboration among the problem chapters, the basic structure remains almost identical at all times. Indeed, one problem can be (much) more elaborate than another. For instance, while *Q.N.* 19 is more like a short essay, *Q.N.* 17 is nothing more than a gloss[114]. As such, the problem format guarantees a neat and transparent exposition of an often chaotic assortment of arguments and view-points leaving space for further insights, all of which is presented in a compendious and highly organised fashion, under the leading principle of plausibility (τὸ πιθανόν). Now that we have dealt with the internal arrangement of *Quaestiones naturales*, let us turn to its external macrostructure, viz. the collection's general organisation in terms of structural order and disorder of the problem chapters, and the possible impact thereof on the intended reading process.

5. Coherent reading in *Quaestiones naturales* and *convivales* (macrostructure)

As noted earlier on, Plutarch's natural problems do not convey a systematic or comprehensive vision of the realm of nature and its phenomena. There is no intention to capture the entire world, or entire facets of it, in a consistent and monographic study, as is rather the case in a work like Aristotle's *Meteorology* or *Physics*. When it comes to the macrostructural arrangement of the successive problem chapters in *Quaestiones naturales*, it is obvious that the collection is not ordered in an orderly but rather in a disorganised, haphazard fashion. Unlike the Ps.-Aristotelian *Problems*, Plutarch's *Quaestiones naturales* has not been rubricated according to

---

[113] For a similar appreciation of the genre of problems, see, e.g., R.H. Barrow, 1967, p. 67: "This method of writing has advantages; it is compendious; the point at issue is clearly defined and shortly stated; the rival solutions are put forward tentatively; there is no dogmatic answer; the writer takes no responsibility and the reader is left to make up his own mind."

[114] A similar distinction was made by H. Dörrie, 1959, p. 2: "Damit steht ζήτημα, was den Umfang anlangt, zwischen der einfachen Worterklärung (Glosse) und dem monographischen Exkurs".

thematic categories that would follow the divisions in nature or the specialised sub-disciplines in science based upon them[115]. On the contrary, Plutarch's arrangement of the problems is considerably chaotic and ordered in a rather spontaneous, organic way, if there is, indeed, any intentional arrangement to the work at all. As we will see, some problem chapters seem to follow specific meandering lines and thematic sequences, so that they can be grouped in together, but this process remains implicit and the problem clusters are eventually interrupted by other (sets of) problems. In what follows, I will deal with these different structural dynamics in greater detail, starting with the issue of structural coherence and later dealing with that of structural variation. I will also focus on the actual reading that seems to be favoured by these dynamics, which will be further substantiated in chapter three [see 3.2.1.].

As noted, there is a certain principle of coherence and structural sophistication in Plutarch's *Quaestiones naturales* that binds many (but not all) of the problem chapters together. Regarding this clustering dynamic, Oikonomopoulou rightly speaks of "an incipient classificatory scheme"[116]. Several such problem clusters can be detected throughout the collection (see the introduction to the commentary for a schematic representation of these clusters). This is the case most obviously in *Q.N.* 1–13, where Plutarch deals with problems related to salt and water, and where the recurrent opposition between salty and sweet water is at issue[117]. Other such clusters are found in *Q.N.* 14–16 (on wheat and barley), 17–19 (on sea animals and fishing), 20–28 (on land animals and hunting), 30–31 (on viniculture) and 35–36 (on apiculture). These associative connections between successive problem chapters guarantee a certain aspect of unity throughout the collection.

Notably, the same structuring process is present in Plutarch's other collections of *quaestiones*, as is the case most notably in *Quaestiones Romanae* (there is, for instance, an obvious "fil thématique" concerning

---

[115] Cf. J. Boulogne, 1992, p. 4684: "Rien de tel [sc. la formation de chapitres unifiés autour d'un contenu commun] dans les 'Questions d'histoire naturelle' [...] où n'est effectué aucun classement systématique." This is not, of course, to object to the claim of E.S. Forster, 1928, p. 165 that "the compiler [of Ps.-Aristotle's *Problems*] seems to have had as his object the collection of as many problems as possible without being greatly at pains to harmonize them into a consistent and logical whole."

[116] K. Oikonomopoulou, 2013a, p. 152. Regarding the thematic rubrications in Ps.-Aristotle's *Problems*, Oikonomopoulou rightly notes: "If [...] the re-organisation of this text into thematic units took place at the end of the 2nd century CE, the thematic clusters offered by the *QN* might be taken as a hint that thematic versions circulated as early as Plutarch's time".

[117] The link with Book 23 of Ps.-Aristotle's *Problems* is clear (ὅσα περὶ τὸ ἁλμυρὸν ὕδωρ καὶ θάλατταν). Cf. H. Flashar, 1962, pp. 315 and 649.

Roman nuptial rites)[118]. This process seems to be more continuous in *Quaestiones naturales*, though, but it still remains very implicit[119]. As an effect of this subtle structuring process, the separate problems are often linked to each other by means of specific thematic and verbal 'synapses'. One of the ways of tying problems together can be seen in the insertion of parallel argumentations (take, for instance, the allusion to the octopus' flesh and skin in *Q.N.* 18-19, or the idea that bees hate what is harmful to them in *Q.N.* 35-36). At other places, the overarching principle of a coherent arrangement is more tangible. This is especially the case in those problems where we can find an explicit reference to the immediately preceding problem (see *Q.N.* 16, 915E: ὡς εἰρήκαμεν, and *Q.N.* 24, 917F: διὰ τὴν εἰρημένην αἰτίαν). These connective phrases signal and even stimulate a coherent, linear reading process[120].

Finding my inspiration in recent scholarship on Catullus' *Carmina*, I will refer to this principle of structural clustering with the concept of *concatenatio*. The concept of a *concatenatio Plutarchea* will facilitate a more or less coherent approach to the problem chapters in the commentary. The notion of *concatenatio* was coined by Claes, who describes it as "[a] way to counteract the disintegrating force of the technique [of variation]" and as a "principle [that] interlinks consecutive poems by repeating themes and phrases"[121]. The connection between problems and poems is not, of course, unproblematic, since these are two completely different genres. Yet, this distinction is not necessarily strict, at least when it comes to their shared scholarly origins and piecemeal approach (Callimachus' Αἴτια, for instance, utilises poetry as a discursive medium for the aetiologies it collects)[122]. Indeed, the search for subtle structural unity is a typical

---

[118] See *Quaest. Rom.* 1-2, 29-31, 65, 85-87, 105, 108; discussed by J. Boulogne, 1998, p. 32: "Cette dispersion thématique a pour effet d' unifier l' ensemble du texte comme un fil de chaîne, un mode de composition correspondant exactement à l' importance qu' accorde Plutarque au couple dans la cohésion du tissu social romain." Similarly, for the notion of "paquet thématique", cf. L. Démarais, 2005, p. 168, n. 32. See also H.J. Rose, 1924, pp. 50-51, J. Boulogne, 1994, pp. 87-88, 2002, pp. 99-100 and J. Scheid, 2012, pp. 11-12.

[119] In fact, J. Schellens, 1864, pp. 18-19 has rightly underlined: "Naturalium denique quaestionum ordinem minus quam romanarum esse turbatum, tibi epigrammata quaestionum celeriter perlegenti patebit."

[120] Cf. L. Van der Stockt, 2011, p. 453.

[121] P. Claes, 2002, p. 27 (see also pp. 51-55 for *concatenatio* in classical poetry more generally). With this concept, Claes aims to demonstrate that Catullus calculatedly structured his *Carmina* as a collection of poems that fit like links in a coherent chain. He distinguishes between thematic and lexical *concatenatio* (with further sub-divisions). A similar structural principle has also been discussed for Martial's epigrams by N. Holzberg, 2002, pp. 37-39, who introduces the concept of "paradeepigramme", and by W. Fitzgerald, 2007, pp. 106-138, who uses the term "juxtaposition".

[122] In regards to the structural set-up of Callimachus' Αἴτια, the famous introductory

feature of Hellenistic poetics that can be linked to the highly specialised activities of Hellenistic scholars in the Museum[123]. Regarding Plutarch's personal decision to employ the problem format in order to address specific case studies, rather than an all-encompassing discourse, Sirinelli rightly observes that:

> "Plutarque est véritablement un intellectuel au sens où nous l'entendons aujourd'hui, c'est-à-dire un homme qui réfléchit sur les problèmes qu'il rencontre ou qui lui sont soumis. C'est sans doute cette image-là qu'il faut garder à l'esprit plutôt que celle d'un érudit parcourant méthodiquement tous les domains du savoir."[124]

The piecemeal and (largely) unsystematic approach of the problem chapters explains why the concatenative line is disrupted at certain points in *Quaestiones naturales*. While *Q.N.* 1–13, for instance, deal with salt and water, Plutarch's attention in *Q.N.* 14–16 abruptly shifts towards agricultural problems regarding wheat and barley. In addition, some topics that belong together thematically are broken apart, as is the case with the problems concerning wine and drinking (*Q.N.* 10, 27, 30–31). From these instances, we learn that there is yet another ordering – or better, disordering – principle that disturbs the process of *concatenatio*, viz. the principle of *variatio* (ποικιλία, 'intricacy'), which is yet another compositional principle that is key to Hellenistic poetics.

Due to the fact that the structural unity of *Quaestiones naturales* remains rather implicit and is disrupted at several points, we are dealing with a digressive organisation of the problem chapters that is, perhaps, meant to encompass, or at least give an idea of, the plurality of disordered natural phenomena and the complex causality that they each involve. By the implicit interconnectedness of several natural problems, *Quaestiones naturales* may, thus, hint at a 'coherent diversity' in the contingent world

---

verses (1–5) of the prologue are particularly relevant. In this passage, Callimachus defends his own literary project against the criticism of the Τελχῖνες for not having accomplished a *continuous* song (an ἄεισμα διηνεκές) of thousands of lines on heroes and lords. To the contrary, he composed a small epic (an ἔπος τυτθόν), which echoes Callimachus' μέγα βιβλίον μέγα κακόν (fr. 465 Pfeiffer = 511 Asper). On Callimachus' "tendency to brevity" (i.e. the *Hang zum kleinen*) as a typically Hellenistic feature in his Αἴτια, see A. Harder, 2012, p. 38 (see also Harder's commentary *ad loc.*).

[123] See n. 51. Cf. S. Saïd, M. Trédé and A. Le Boulluec, 1997, p. 434: "la curiosité de Plutarque est universelle. A la manière d'un Callimaque et des érudits hellénistiques, il s'interroge sur les origines de tel rituel ou de telle coutume étrange chez les Grecs comme chez les Romains. Comme Aristote, il s'intéresse à des questions de physique (les *Questions naturelles*) ou de zoologie (*Sur l'intelligence des animaux*)." For the connection between Aristotle and Hellenistic scholarship, see N.J. Richardson, 1994.

[124] J. Sirinelli, 2000, p. 364.

of natural phenomena and the physical processes and mechanisms that underlie them. Scholars have also detected this principle of a 'coherent diversity' in Plutarch's *Quaestiones Romanae* and *Graecae*, where the reader finds a pluralistic, but at the same time generally coherent, picture of Roman and Greek culture. In these collections, too, unity is, in a certain sense, procured *through* diversity, in that the text as whole reflects a general cultural identity – viz. the categories of Greekness or Romanness – *through* a pluralism of particular cultural manifestations that build up this identity[125].

Recent scholarship has considered Plutarch's collections of *quaestiones* more generally as part and parcel of ancient miscellaneous literature[126]. Specific attention went to the organisation and reading of Plutarch's *Quaestiones convivales* as representative of this literary branch[127]. As we will see, the results of these studies are not irrelevant in light of the literary techniques of *concatenatio* and *variatio* in *Quaestiones naturales* either (or in Plutarch's other collections of *quaestiones*)[128]. An interesting point of departure is Harrison's theory of a fragmented, piecemeal reading of *Quaestiones convivales*. According to this theory, the work's miscellaneous content and organisation allows the reader to decide for himself which specific problem chapters he would like to read and which ones he would like to leave aside:

> "Beyond expanding its scope so that it could encompass all the different genres of *quaestiones*, Plutarch brought an episodic structure to the symposium, which allowed the reader to take up and put down his convivial reminiscences at will and browse through them rather like a collection of poems or fables instead of a work whose argument had to be followed sequentially."[129]

---

[125] See J. Boulogne, 1992, pp. 4698–4707 (who for *Quaestiones Romanae* distinguishes between "une étiologie politique, grécisante et anthropologique") and P. Payen, 1998b, pp. 49–54 (who argues for a coherent cultural landscape in both collections, based on geographical markers in the text). See also R. Preston, 2001, J. Boulogne, 2002, pp. 99–100 and 183–184 and K. Oikonomopoulou, 2013a.

[126] On miscellaneous literature from the Imperial Era more generally, see the introductions in J. König and T. Whitmarsh, 2007 and F. Klotz and K. Oikonomopoulou, 2011, pp. 22–24. See also T. Morgan, 2011, esp. pp. 70–73 for Plutarch's miscellanies in specific (see also pp. 49–54 for problems with the demarcation of the genre of ancient miscellanies).

[127] See esp. J. König, 2007, pp. 62–67 and the contributions in F. Klotz and K. Oikonomopoulou, 2011.

[128] Note, for instance, that among other typical titles of miscellaneous writings Gell., *NA Praef.* 9 specifically mentions *Epistolarum Quaestionum aut Confusarum*.

[129] G.W.M. Harrison, 2000a, p. 197. A similar piecemeal reading of *Quaestiones convivales* was supported by S. Goldhill, 2009, p. 109 (criticised by K. Oikonomopoulou, 2011, p. 125, with n. 59). Regarding the 'episodic' structure of *Quaestiones convivales*, see

One could argue in favour of Harrison's theory that the information furnished by the problem headings (the *quaestiones*) of the specific sympotic discussions are very useful in informing the reader as to whether or not a specific passage is of any interest for him. However, these headings do not always entirely – let alone clearly – cover the expanse of the problem(s) dealt with (if they are authentic to begin with: e.g., *Quaest. conv.* 623D, 738C, 741D). Furthermore, a table of contents at the beginning of the work listing the titles or a general description of the problem chapters would certainly have made the piecemeal reading more efficient[130]. However, there is none. More concrete criticism was formulated by Titchener, who objects that Harrison's theory "depends on the physical existence of the *Table talk* in a particular physical format, something that cannot be assumed"[131]. It is, indeed, uncertain as to whether Plutarch sent his nine βιβλία of *Quaestiones convivales* to Senecio in the form of 'books', rather than 'bookrolls' – a medium that would obviously hamper a convenient browsing through the text. Finally, and most importantly perhaps, one could object that the implied reader – the all-round πεπαιδευμένος – would be interested in *each* of these subjects without distinction, so that a linear reading of the miscellaneous problems is not that problematic to begin with.

Bearing in mind Harrison's notion of an "episodic structure" in *Quaestiones convivales*, it seems that the piecemeal structure of the problem chapters and their lack of a clear overall ordering produce a certain literary effect by directing the reader's focus toward the private character of the sympotic discussions and scenes, which are time and again re-set, so that the reader finds himself invited (by means of a *quaestio*) to join at an ever new literary dinner table, not just as a passive witness but as an active participant in an ever new discussion (I will further ruminate on this theory later, when dealing with the text's educational goals [see 3.2.1.]). It is perhaps not unlikely that the structural aspects of *concatenatio* and *variatio* in *Quaestiones convivales* somehow resemble the conditions of

---

already A. Gudeman, 1927, col. 2526: "Auch verzichtet der Verfasser [...] fast gänzlich auf die übliche szenische Einkleidung und reiht die einzelnen Lösungen gleichsam wie Perlen an einer Schnur aneinander." Cf. also L. Senzasono, 2006, p. 8 for the "carattere frammentario, o almeno episodico" of Ps.-Aristotle's *Problems*.

[130] Such tables of contents can be found, for instance, at the beginning of each Book of Ps.-Plutarch's/Aëtius' *Placita*. Gellius also includes a table of contents after the preface of his miscellaneous *Noctes Atticae*, where he explicitly underlines its use for the reader to look up specific passages (*Praef.* 25: *ut iam statim declaretur quid quo in libro quaeri invenirique possit*). See E. Gunderson, 2009, pp. 45–47. There is also a table of contents, for instance, in Pliny's *Naturalis historia*, but it is not a very efficient one, since it often misleads the reader when trying to track material within the work. See T. Murphy, 2004, p. 32.

[131] F. Titchener, 2011, p. 47 (cf. also 2009, p. 397).

intellectual dialogue in real life, albeit, of course, in a dramatised form [see 2.3.1.]. It is a logical consequence of the natural development of sympotic discussions that the solution of one problem provokes the formulation of a new problem that can, but need not be, closely connected. The first three problems of Book six, for instance, concern hunger and thirst, and the following three, cold water and snow. Sometimes, the symposiasts also deal with a wide variety of themes at one occasion, as is the case, most notably, in Book nine as a whole, where Plutarch gives a report on the conversations held at Athens during the so-called festival of the Muses.

From the perspective of Plutarch's literary technique, it is not, of course, unlikely that the arrangement of *Quaestiones convivales* emerges spontaneously from the author's associative memory and use of personal notes. A seminal account of the miscellaneous organisation of *Quaestiones convivales* is found in the preface to the second Book, where Plutarch says that he simply jotted down the conversations 'without any systematic order, as each came to mind' (*Quaest. conv.* 629D: σποράδην δ' ἀναγέγραπται καὶ οὐ διακεκριμένως ἀλλ' ὡς ἕκαστον εἰς μνήμην ἦλθεν). Scholars agree that this passage is very relevant for the work's underlying writing process and method of composition, including the structuring principles that guide it, but there is debate about its precise meaning. In light of this passage, some scholars have argued that the distribution of the questions and the transitions from one subject to another seem to be entirely capricious[132]. There is obviously no prepared order of subjects. Yet, regarding the first part of Plutarch's statement ('without any systematic order'), there may be more to what Plutarch actually says, as König has influentially argued in his recent study of the complex dynamics of coherence and fragmentation in the work:

> "Many ancient miscellanists [...] gesture towards thematic order, drawing us into a search for patterns while also at the same time disrupting and frustrating that search. On that argument, the claim that many miscellanists make, that they are composing at random, turns out, at least in some cases, to be a matter of convention, a miscellanistic *pose* which can hide careful structuring beneath it [...]."[133]

In what follows, König argues that the reader is encouraged to navigate through the separate problems with close attention to several thematic and verbal reminiscences. This makes the reading process both more intriguing as well as more challenging (I will deal with his theory of an 'active reading' process later on [see 3.2.1.]).

---

[132] S.-T. Teodorsson, 1989, p. 169 considers it "a literary device to announce unconnected disposition".

[133] J. König, 2007, p. 44 (my italics).

In regards to the second part of Plutarch's statement ('as each thing came to my mind'), König argues:

> "[This] looks, on closer inspection, not like a statement of the work's randomness, but rather like an attempt to equate the ordering of the work with the retrospective patterning which memory inevitably imposes."[134]

In my opinion, this is as ingenious as it may sound casuistic to some[135]. But the intriguing thing about it is that it enables us to kill two troublesome birds with one stone. That is, it can explain both principles of *variatio* and *concatenatio* on the basis of the author's memory and recollection, and the vagaries that they involve.

To return to the first part of Plutarch's statement now, König is absolutely right that other miscellanistic authors (like Gellius, Pamphila, Aelian and Athenaeus)[136] also emphasise the artless and haphazard organisation of their writings, so that we can rightfully speak of a genuine miscellanistic τόπος. In my opinion, this does not tell us much insofar that a τόπος is not necessarily a ψεῦδος. Plutarch may very well be playing along with this miscellanistic convention (perhaps even intentionally), but even so, it would be wrong, in my opinion, to claim that he is 'posing' (and this may also count for those other miscellanistic authors). Plutarch is no 'poser' – or otherwise, he is extremely good at it. Indeed, with the emphasis on the adjuncts σποράδην and οὐ διακεκριμένως, he primarily

---

[134] J. König, 2007, p. 62.

[135] The germ of König's theory is present already in J.P. Small, 1997, p. 180: "We might have discounted Plutarch's 'random' order of topics, if it were not for the Preface of Gellius. Instead, his explanation told the ancient reader that what was to follow fitted a well-known genre of collections of discussions on diverting topics."

[136] See, e.g., Gell., *NA Praef.* 2 (*Usi autem sumus ordine rerum fortuito, quem antea in excerpendo feceramus*), Pamphila in Phot., *Bibl.* Cod. 175, 119b (οὕτως εἰκῇ καὶ ὡς ἕκαστον ἐπῆλθεν ἀναγράψαι, ὡς οὐχὶ χαλεπὸν ἔχουσα, φησί, τὸ κατ' εἶδος αὐτὰ διελεῖν, ἐπιτερπέστερον δὲ καὶ χαριέστερον τὸ ἀναμεμιγμένον καὶ τὴν ποικιλίαν τοῦ μονοειδοῦς νομίζουσα), Ael., *NA Epil.* 43–46 (οἱονεὶ λειμῶνά τινα ἢ στέφανον ὡραῖον ἐκ τῆς πολυχροίας, ὡς ἀνθεσφόρων τῶν ζῴων τῶν πολλῶν, ᾠήθην δεῖν τήνδε ὑφᾶναί τε καὶ διαπλέξαι τὴν συγγραφήν). For further references, see J. König, 2007, p. 44, n. 3. As to the associative style of arrangement in Plutarch's *Quaestiones convivales*, Athenaeus' *Deipnosophistae* and Gellius' *Noctes Atticae*, see also K. Oikonomopoulou, 2013a, pp. 148–149: "In texts like Plutarch's *QC* and Athenaeus' *Deipnosophistae*, this is meant to reflect the associative twists and turns of sympotic conversation. In texts like Gellius' *Attic Nights*, it is meant to reflect the author's own associative leaps at the moment of composition. The common denominator is the link drawn between the order of the textual product and a cognitive process (such as memory, or mental association) or work method (such as reading, excerpting) from which it emerged." See also J.P. Small, 1997, pp. 179–181.

intends to capture the text's obvious lack of structural organisation, rather than to proclaim his own alleged pseudo-nonchalant writing technique[137]. As a matter of fact, there is no obvious general structure discernible in the succession of the questions that might indicate a clear-cut organisation of the problems, even if it would be intentionally complex or subtle. Nowhere does Plutarch explicitly formulate that intention and perhaps least of all in *Quaest. conv.* 629D.

To reject the question of an overt plan does not, however, eliminate the possibility of an intelligent design to the collection as a whole. Even though there is no obvious organisation of the problems, a more concealed design is, indeed, palpable and can be read somewhere in between the lines of the discussions. In this light, Titchener is convinced that:

> "Plutarch goes to so much trouble to describe the well-made dinner party as something that has little obvious, but much concealed structure that it is counterintuitive to assume that there is NOT a similar structure to this work"[138].

Let it be clear, therefore, that König is absolutely right – and this is, in my opinion, the most important and convincing point of his argument – that the implied reader can detect several repeating themes, terminologies, theories and conversational turns throughout the work. Thus, there is some kind of a subtle explanatory scheme that overarches the text's chaotic surface and gives it a certain sense of unity[139]. This overarching explanatory scheme will be important in determining the educational value of Plutarch's collections of *quaestiones* later on [see 3.2.1.].

What matters here is that there is no tangible structure to miscellaneous texts like Plutarch's *Quaestiones convivales*. One can presume that this is deliberate, in the sense that Plutarch's πολυμάθεια project not only requires a playfully digressive structure but also generates it. In view of the literary aesthetic of *variatio*, the associative continuum of the concatenated problem chapters breaks off at certain points only to be re-initialised later. This certainly complicates a linear reading, but it does not make such a reading impossible. Perhaps it is better, then, to speak of a staccato reading process (as an alternative for Harrison's fragmentary, piecemeal

---

[137] Cf. also, e.g., *Mul. virt.* 253E: τὰς δὲ καθ᾽ ἑκάστην ἀρετάς, ὅπως ἂν ἐπίῃ, σποράδην ἀναγράψομεν. See S.-T. Teodorsson, 1989, p. 169. Cf. also H. Bolkestein, 1946, p. 36 and F. Fuhrmann, 1972, p. xi. A little earlier in *Quaest. conv.* 629D, Plutarch refers to the discussions from the first Book as μεμιγμένα δείγματα. Incidentally, at the very beginning of Book one, in *Quaest. conv.* 612E, Plutarch notes that the discussions themselves took place σποράδην (viz. πολλάκις ἔν τε Ῥώμῃ μεθ᾽ ὑμῶν καὶ παρ᾽ ἡμῖν ἐν τῇ Ἑλλάδι).

[138] F. Titchener, 2009, p. 396.

[139] J. König, 2007, p. 61. See also E. Kechagia, 2011a, pp. 97–99.

reading process). Read in this way, Plutarch's variegated writing strategy guarantees a certain degree of structural unity and disunity, offering the reader an opportunity to focus, divert, and refocus his attention from time to time. After all, the intended reader of such miscellanies – that is, the all-round πεπαιδευμένος – was interested in the broad field of ancient learning. In *Quaestiones convivales*, one can now read something on literature, then something on science, then something on music, history, etc. It would, indeed, go against the basic prerequisites of the miscellaneous genre to expect a ready-made and fully transparent ordering of the material, and Plutarch very well knew this.

If we now transpose these ideas onto the miscellaneous macrostructure of *Quaestiones naturales* (while trying to avoid too much repetition), we can easily see that there is greater thematic unity in this collection than in Plutarch's *Quaestiones convivales*, as it exclusively deals with natural problems and their explanations. Even so, there is no clear-cut macrostructure to its contents either. Plutarch does not provide a systematic survey or catalogue of nature, but a collection of very diverse natural problems that are only loosely connected with each other through specific connective phrases or repeated themes and concepts[140]. Despite the collection's superficial disunity, it seems that the theory of a linear, staccato reading, as has just been described for *Quaestiones convivales*, also applies to *Quaestiones naturales*. Several arguments can be adduced in favour of this type of reading (which I summarise from Morgan's article on Plutarch and the miscellany)[141]. 1) The lack of an obvious structure complicates looking up any specific passage, and the same goes for the lack of a table of contents[142]. 2) Collections of problems were used in education [see 3.1.] and, by implication, were read *in extenso*. What was useless and could be neglected was simply left out in the first place. 3) The thematic sequences (*concatenatio*) are intended by the author, and they would have no use, unless the author did not expect the reader to

---

[140] J. Boulogne, 1992, p. 4699, n. 111 draws the same conclusion for *Quaestiones Romanae*: "l'ouvrage n'a pas été conçu pour être consulté à la manière d'une espèce de catalogue, qu'on peut ouvrir à n'importe quelle page. Il a été rédigé de façon liée, afin d'être parcouru dans son déroulement linéaire". When G. Nuzzo, 1991, p. 410 regarding *Quaestiones naturales* speaks of the "rigida cadenza catalogica scandita dall'*incipit* διὰ τί", this should not, of course, be understood as a reference to the overall structure of the collection but to the typical introduction of the *quaestiones* themselves. Nuzzo refers to Ps.-Aristotle's model and draws a comparison with the structure of Hesiod's ἢ οἵα (i.e. the *Catalogue of women*). Cf. also E. Kechagia, 2011a, p. 99 (regarding *Quaestiones convivales*).

[141] T. Morgan, 2011, pp. 69–70.

[142] Longolius' Latin translation does contain an *Index Problematum de causis naturalibus Plutarchi*, but this is not original, and the book-format may explain why it is included (cf. also the index in L. Senzasono, 2006, pp. 53–55).

notice them. 4) Bridge passages between two sections would have no use if the sections were not read consecutively. As we saw, such phrases as ὡς εἰρήκαμεν (*Q.N.* 16, 915E) and διὰ τὴν εἰρημένην αἰτίαν (*Q.N.* 24, 917F) draw an explicit link with the previous problem chapter and, as such, mark and stimulate a linear reading[143]. 5) An argument that can be added here, but that does not apply to *Quaestiones naturales* (though it does to the prefaces of *Quaestiones convivales*), is that the author expects the reader to have read the previous section before reading the following one by the use of introductions and digressions.

In conclusion, the fact *that* Plutarch's collections of *quaestiones* promote a linear reading does not explain *why* this exactly is the case. However, this is a point that I will discuss later on, when dealing with the 'active reading' process promoted by such works [see 3.2.1.]. If it is true, moreover, that Plutarch expected the reader of *Quaestiones naturales* to go through the collection in a systematic way and not by browsing through it at random, it seems useful to provide an interpretation of the very first words of the text as provided by the title. As is the case with most texts, there lies essential programmatic value in their title, since it creates a broad horizon of expectation on the side of the intended reader. Getting the title correct, then, will not just be a matter of scholarly pedantry but will allow a better understanding of the author's intellectual project and intentions with this work.

6. The title and its programmatic value

As noted earlier on, the Lamprias catalogue lists a significant number of collections of Αἰτίαι among Plutarch's writings, several of which we still have today, whereas others are now lost or partially preserved in fragmentary form (see n. 91). Among these collections only *Quaestiones Romanae* (nr. 138), *Quaestiones Graecae* (nr. 166) and *Quaestiones naturales* (nr. 218) have been preserved at any considerable length (*Quaestiones convivales* is not listed and *Quaestiones Platonicae* belongs to the genre of ζητήματα; see further). The original title of *Quaestiones Romanae* and *Graecae* is not obvious from the manuscripts[144], but on the basis of Plutarch's self-reference in *Rom.* 15, 7 (ἐν τοῖς Αἰτίοις) and in analogy with Callimachus' famous Αἴτια scholars have inferred Αἴτια Ῥωμαϊκά and Ἑλληνικά (which seems reasonable), and hence also conjectured Αἴτια φυσικά – all *neutral* forms, in opposition to the feminine forms of the Lamprias catalogue. The latter title (Αἴτια φυσικά) was in general use after the publication of Bernardakis' 1893 Teubner edition

---

[143] Cf. also L. Van der Stockt, 2011, p. 453. For consecutive reading in Plutarch's collections of *quaestiones* more generally, see K. Oikonomopoulou, 2013a, pp. 147–152.

[144] See J. Boulogne, 2002, pp. 91 and 179.

of *Quaestiones naturales*[145], until Sandbach in his 1965 Loeb edition rejected it as a false conjecture and preferred the feminine to the neutral form: Αἰτίαι φυσικαί[146]. Sandbach's choice seems valid, since it is based on the reading of (each of) the manuscripts, but, unfortunately, Plutarch does never explicitly refer to *Quaestiones naturales* in his other writings (or at least not by the collection's title [see 2.1.4.]). As such, except from the manuscript reading, there is no firm evidence for the title's original wording. Moreover, the evidence that is furnished by the Lamprias catalogue (which has the feminine form) has been contested by Boulogne, who pointed at the catalogue's established unreliability (see n. 3). Boulogne's guardedness may, indeed, be justifiable as far as the *Lamprias catalogue* is concerned, but the reading of the manuscripts still provides a sufficient argument for Sandbach's correction[147]. A further indication is furnished by the fact that the female noun αἰτία recurs more often in the collection (than the neutral form αἴτιον) and is even found at the very beginning of the opening chapter (*Q.N.* 1, 911C: Πότερον δι' ἥν αἰτίαν)[148]. So even in the – uncertain – case that the Greek title of our collection would be apocryphal (was it perhaps attributed to it posthumously on the basis of the opening in *Q.N.* 1? [see prologue, n. 20]), Αἰτίαι φυσικαί is closest to the terminology Plutarch uses throughout the

---

[145] Bernardakis' conjecture was adopted by C. Hubert in his 1960 Teubner edition. See also, e.g., H.J. Rose, 1924, p. 49 and J. Boulogne, 1992, pp. 4683–4684 with n. 11. P. Payen, 2014, pp. 243–244 sketches the situation in a more confused way.

[146] F.H. Sandbach, 1965, p. 133 (cf. also W. von Christ, 1959, p. 512, n. 2). A. Gudeman, 1927, col. 2525 before Sandbach already preferred Αἰτίαι φυσικαί. Perhaps ἐν τοῖς Αἰτίοις in *Rom.* 15, 7 should be emended in ἐν ταῖς Αἰτίαις, but this is conjecture [see 2.4.1., n. 136].

[147] J. Boulogne, 1992, p. 4684, n. 11. Sandbach also pointed out that there is nothing fundamentally objectionable in the title Αἰτίαι φυσικαί, a name which is, in any case, "known to have been used by his Greek-speaking readers" (p. 133). Notably, the feminine form (Αἰτίαι φυσικαί) is also found as the title of manuscripts Hieros. gr. 108 and Laur. vii 35 of the first redaction of Psellus' *De omnifaria doctrina*, of which §§ 92–107 are extracted from Plutarch's *Quaestiones naturales*: see L.G. Westerink, 1948, p. 3. For further detail on Psellus' incorporation of *Quaestiones naturales* material in his *De omnifaria doctrina*, see M. Meeusen, 2012b.

[148] See also *Q.N.* 4, 913A (διὰ τὴν εἰρημένην αἰτίαν), 19, 916C (τὰ δὲ κύρια τῆς αἰτίας), 916F (τεκμήριον δὲ τῆς αἰτίας μέγα), 20, 916F (Διὰ τίν' αἰτίαν), 917A (Αἰτία δέ), 24, 917F (Ἡ διὰ τὴν εἰρημένην αἰτίαν), 27, 918F (αὕτη δ' ἐστὶν αἰτία), 29, 919A (Τίς ἡ αἰτία [...] θερμότης αἰτία), 33 (*in causa est*), 35 (*in causa est* [...] *qua de causa*), 40 (Τίς ἡ αἰτία [...] ταύτην τὴν αἰτίαν). The neutral form is used only once, albeit in a more abstract context (*Q.N.* 29, 919B: πλειόνων αἴτιον φαίνεται τὸ μὴ ὂν τοῦ ὄντος). The verb αἰτιάσασθαι occurs twice, viz. in *Q.N.* 2, 912C and 12, 915A. Cf. F.H. Sandbach, 1965, p. 134 (who mentions only *Q.N.* 20, 916F, 29, 919A and 40). For Plutarch's mentioning of αἰτίαι φυσικαί, cf. also, e.g., *De def. or.* 424B, 435F, 436D. The observation of G.W.M. Harrison, 2000b, p. 244 that the term αἰτία occurs only once in the collection is inaccurate (viz. in *Q.N.* 20, 917A; his assertion that this is the only *causa* that is formulated as a dogmatic statement is also incorrect: see n. 96).

work, so that it at least provides an adequate description of the collection, in terms of its causal approach (see further).

As to the term Αἰτίαι in the title, Harrison believes that "Plutarch's use of the term seems to be by metonymy for αἰτιολογία or αἰτιολογέω [...], which would seem in turn to imply that Plutarch looked to one or another of Epicurus' lost works for inspiration in modifying the genre."[149] The first part of Harrison's assertion may be correct (though αἰτιολογία is nowhere to be found in Plutarch and αἰτιολογεῖν only once, viz. in *Quaest. conv.* 689B), but the second part (about the influence of Epicurus) seems rather unlikely. In any case, D.L. (10, 26–28) does not list a collection of Αἰτίαι among Epicurus' works (although there is a collection of Διαπορίαι and a Συμπόσιον, which were both known to Plutarch: see Epic., frs. 18–21 and 57–65 Usener respectively). Moreover, the argumentative strategy of providing several plausible explanations for one and the same natural phenomenon was also an important feature of Epicurus' scientific method, albeit for motives completely different than Plutarch's (as we will see later [see 4.3.3.2.]).

It seems more suitable, I believe, to refer to Democritus in this regard, who is cited in the very first αἰτία in *Q.N.* 1, 911CD, and who, as we saw earlier, was the first to compose an actual collection of problems (see n. 58). Democritus' aetiological interests are captured in his reported saying that he would rather prefer to find a single aetiology (αἰτιολογίαν) than to reign over the Persian empire (DK68B118). In fact, we know from D.L. 9, 47 that Democritus composed eight thematic sets of natural scientific Αἰτίαι (*sic*), which are mentioned under the ἀσύντακτα in Thrasyllus' catalogue of Democritus' works. These sections concern Αἰτίαι on heaven, air, earth, fire, sounds, seeds, plants, fruits, animals (in three Books) and miscellanea. As we saw earlier, scholars have even considered Democritus a precursor to Peripatetic natural science, drawing a close parallel between his and Aristotle's aetiological project (see n. 59). Diels claimed that many *Democritea* were actually integrated in Ps.-Aristotle's *Problems*, so that Democritus' indirect influence on Plutarch's natural problems is not unlikely, although it is not always clearly traceable either[150]. Remnants of Democritus' theories about animals, plants and water can be traced, for instance, in *Q.N.* 1, 911CD (regarding seawater being undrinkable and bad for humans but nourishing for fish) and 5, 913CD (regarding the filtration

---

[149] G.W.M. Harrison, 2000a, p. 195, n. 6.

[150] H. Diels, 1905, p. 316. Notably, D.L. 5, 26 lists two Books of Προβλήματα ἐκ τῶν Δημοκρίτου among Aristotle's works, and the index Hesychii contains a similar title (nr. 116: Προβλημάτων Δημοκριτείων Β'). See P. Moraux, 1951, pp. 120–121. If the link with the Ps.-Aristotelian *Problems* is real, it is not impossible that the thematic rubrications in this work was inspired by the thematic categories in Democritus' Αἰτίαι. This remains uncertain, though.

of seawater) (see the commentary *ad loc.* for more detail). In addition, Democritus' causal research is also relevant for interpreting Plutarch's own approach to natural problems, but this will be treated later on [see 4.1.1.2.].

To come back to the wording in the Greek title, there is no essential distinction in semantics between αἰτίαι and αἴτια, but Plutarch's word choice is not, therefore, insignificant. As Grandjean recently observed, "[c]ontrairement à *Aitia*, les *Aitiai* ne renvoient pas à une tradition littéraire. C'est un terme emprunté à la physique, à la métaphysique et au droit. Un tel titre laisse présager une réflexion philosophique plutôt qu'un récit ou une explication littéraire."[151] The Latin title, *Quaestiones naturales*, by contrast, which was assigned to the collection since the very first editions of Xylander etc., is clear but not really apposite, because the concept of *quaestiones* is too generic to map the subtle distinctions between the Greek concepts of αἰτίαι, ζητήματα, προβλήματα etc. Two further remarks should be made in this regard.

First of all, the Latin title obfuscates the distinction between the concepts of αἰτίαι and ζητήματα. Plutarch probably composed more collections of αἰτίαι than ζητήματα (see n. 91). As scholars have pointed out regarding Plutarch's *Quaestiones Platonicae* (in Greek: Πλατωνικὰ ζητήματα), the Chaeronean's use of ζητήματα often has an exegetical connotation, as it is mostly concerned with the elucidation and interpretation of particular enigmatic passages in a given philosophical or poetic text[152]. By contrast, Plutarch's collections of αἰτίαι mostly treat more general intellectual topics, such as the origins of specific cultural traditions or the causes of natural phenomena, so that the strict connection with a text is absent[153]. This is not, of course, to deny that Plutarch often relies on written sources in his collections of αἰτίαι. Moreover, there is not always much consistency in Plutarch's own wording, so that at least in some cases the nuances in semantics seem artificial[154]. In addition, both types of inquiry do not

---

[151] T. Grandjean, 2008, p. 147, n. 2.

[152] Cf. *Quaest. Plat.* 1006F: τοῦτο μὲν οὖν τοιαύτην ἔχει τὴν ἐξήγησιν. Cf. also *De tranq. an.* 464F: περὶ τῶν ἐν Τιμαίῳ δεομένων ἐπιμελεστέρας ἐξηγήσεως. See H. Dörrie, 1959, p. 2, J. Opsomer, 1994a, p. 10, 1996a, p. 72, 2010, p. 93 and J. Boulogne, 1992, p. 4684, with n. 15: "Il [sc. Plutarque] leur [sc. les αἰτίαι] substitue le substantif ζητήματα quand il s'agit de difficultés soulevées par les assertions de poètes ou de philosophes." See also G.W.M. Harrison, 2000a, p. 195 and L. Van der Stockt, 2000b, p. 96.

[153] The exegetical nature of the ζητήματα, which requires that the passage under discussion be reissued and explained more or less κατὰ λέξιν, can perhaps explain why Plutarch's αἰτίαι are generally more restricted in length. Plutarch's αἰτίαι are, in any case, significantly shorter than an average ζήτημα. *Q.N.* 1, 2, 5, 19, 21 and 26, for instance, are relatively lengthy, but even so, each of them is only about half as long as an average *Quaestio Platonica*.

[154] Regarding Plutarch's terminology in *Quaestiones convivales*, G.W.M. Harrison,

strictly exclude one another: for instance, the eight fragments that remain from Plutarch's Αἰτίαι τῶν Ἀράτου Διοσημιῶν combine an aetiological and an exegetical approach (= frs. 13–20 Sandbach; cf. also the Aratus quote in *Q.N.* 2, 912D). Importantly, the more scholarly feature of ζητήματα does not devalue their philosophical interest for Plutarch. This is true at least for *Quaestiones Platonicae*, because, for Plutarch, a correct understanding of Plato's texts would enable him to grasp the philosophical truth that they contained[155].

The second point concerns the discrepancy in meaning between the concepts of αἰτίαι and προβλήματα. Notably, the Greek title of *Quaestiones naturales* does not mention the word προβλήματα. Even if the 'problematic' form, style and content of Plutarch's natural problems and those collected in Ps.-Aristotle's *Problems* (in Greek: Προβλήματα φυσικά) are very similar, the variable terminologies in the titles may suggest a subtle difference in purpose. With the term αἰτίαι more emphasis is, then, put on the explanations – the *quaerenda* – of the natural phenomena under scrutiny, while the term προβλήματα rather stresses the enigmatic character of these phenomena themselves – as presented in the *quaestiones*[156]. As we will see later on, the distinction between *quaerenda* and *quaestiones* is very relevant in light of the scientific methodology Plutarch employs in dealing with problematic natural phenomena. In short, Plutarch's main concern is with the διὰ τί, that is, with the natural causes and explanations of the phenomena, rather than with the ὅτι, that is, the aspect of empirical verification [see 4.1.1.]. In other words, Plutarch is not so much interested in the actual physical reality of the problematic natural phenomena he studies (outside of reported ἱστορία) but with their physical aetiology. Therefore, I agree with Boulogne when he prefers to speak of "*Étiologies* plutôt que [...] *Questions*"[157]. Regarding Plutarch's choice of words in the titles of his aetiological collections, Boulogne affirms that:

> "les mots ne sont pas employés indifféremment, comme s'ils étaient synonymes. [...] Il semblerait, partant, que Plutarque choisisse le vocable αἴτια, et peut-être aussi celui d' αἰτίαι [...], afin de placer l'accent sur la détermination de l'origine des faits observés, plutôt que sur leur caractère intrigant, et par là, de signaler une étude

---

2000a, p. 196 rightly concludes: "The terms ζήτημα and πρόβλημα would appear to be interchangeable in this work since no pattern is detectable".

[155] See J. Opsomer, 1996a, p. 74.

[156] The distinction in semantics, again, may not be very strict: cf., e.g., *De tuenda* 133E (φυσικὰ προβλήματα), *De fato* 568F (ζητήματα φυσικά).

[157] J. Boulogne, 1998, p. 31. F.C. Babbitt, 1936a, p. 2 speaks of "The Reasons Why". G.W.M. Harrison, 2000a, p. 195 prefers "*Explanations*, that is, clarifying information".

particulièrement approfondie des causes sur des objets séparés et, pour ainsi dire, isolés les uns des autres, comme s'il voulait limiter le plus étroitement possible le champ de son examen et rendre, de la sorte, son investigation plus pointue, sans que vienne s'ajouter d'autre préoccupation."[158]

In this sense, the Greek title of Αἰτίαι φυσικαί indicates that the collection is meant to be a set of profound physical aetiologies, that is, a thorough study of the natural causes of the particular enigmatic phenomena that drew the author's attention. I am not so sure, however, whether Ps.-Aristotle's model lacks this aetiological profundity, as Boulogne claims (subsequently to the given quote), or, *a fortiori*, that the aetiological openness of the problems there has become a purpose in itself so that the *Problems* remain immanently 'problematic', in opposition to Plutarch's natural problems, where the explanations would be more 'conclusive'[159]. It is, indeed, true that the average number of explanations for each problem is higher in Plutarch's *Quaestiones naturales* than in Ps.-Aristotle's *Problems*, where often only one explanation is given for a problem, but even so, the natural problems remain fundamentally unsolved also in Plutarch's case, since they are formulated anti-dogmatically and, thus, invite for further research. The only thing that is fundamentally different is the epistemological basis on which this research was grounded, Plutarch postponing final judgement, since he, in opposition to Aristotle, refused to put much confidence in knowledge derived from sensory data [see 4.3.2.1.]. By contrast, Aristotle's avoidance of argumentative conclusiveness was more practically motivated, aiming to foster further research in the Lyceum

---

[158] J. Boulogne, 1992, p. 4684.

[159] Regarding the difference between Plutarch's αἴτια/αἰτίαι and Ps.-Aristotle's *Problems*, J. Boulogne, 1992, p. 4684 (see also C. Darbo-Peschanski, 1998, p. 22) argues that "l'exploration de la causalité y [sc. dans les 'Questions d'histoire naturelle' et les Q.R. de Plutarque] est poussée bien plus loin et l'on n'éprouve pas l'impression d'être en présence d'un répertoire d'énigmes ou de mystères, pour l'éclaircissement desquels est chaque fois consignée une ébauche de solution, comme pour mémoire, dans l'attente d'enquêtes ultérieures". It should be noted, however, that the interrogative structure of the explanations is a common feature of both Plutarch's αἴτια/αἰτίαι and Ps.-Aristotle's προβλήματα, so that it is problematic to maintain that the genre of προβλήματα is more superficial or open from an aetiological perspective than that of αἴτια/αἰτίαι in this regard. *Pace* also J. Boulogne, 1994, p. 76 (regarding *Quaestiones Romanae*): "Toutefois, loin de se contenter de recenser les sujets énigmatiques et de suggérer des solutions possibles qui restent à explorer, il prétend fournir des réponses et apporter le résultat de la recherché, *au lieu d'inviter à l'investigation*" (my italics). As a matter of fact, *both* Plutarch's and Ps.-Aristotle's natural problems invite for further investigation, as they both follow an anti-dogmatic approach. What is different is the epistemological basis for this approach (see below).

context. In the end, Aristotle put much more trust in the feasibility of natural scientific research than Plutarch and with him any Platonist ever did.

The bottom line is that, as the Greek title suggests, Plutarch's main attention clearly goes to the causes and not so much to the problems themselves in *Quaestiones naturales*. As such, the problematic natural phenomena formulated in the *quaestiones* are considered as given facts that do not so much require verification but rather explanation. In this sense, the Greek title captures the scope of the collection rather well, but the same cannot be said of the unlucky Latin rephrasing: *Causae naturales* would be a more accurate rendering than *Quaestiones naturales*[160].

In order to now make Plutarch's primary focus on the physical causes in his natural problems more concrete, a short comparison with Seneca's *Naturales quaestiones* is in place. First of all, there is no reason to assume that Plutarch had actually read Seneca's work or that he was even acquainted with its existence. Apart from a few inevitable parallels in subject matter[161], Seneca's work is completely different in several regards from Plutarch's, so that the significance of their resembling *Latin* title should not be exaggerated. This is true not only from the perspective of 1) structural arrangement but also 2) of their scientific purpose[162].

1) Seneca's *Naturales quaestiones* does not belong to the tradition of Aristotelian natural problems from a formal perspective. Hine rightly suggests that "[i]f S. himself chose the present title of the *NQ*, it was not because he was adhering to that genre [sc. of the *problemata* and *zetemata*] as we know it. A possible reason is that such a title allowed him to treat a miscellany of such subjects as took his fancy, without committing him to a systematic and comprehensive treatise, such as a title like *De rerum natura* might have suggested."[163] In this work, Seneca does, however, name two authorities that are of potential relevance for the tradition of natural problems: viz. Asclepiodotus and Papirius Fabianus. Asclepiodotus lived in the 1st century BC. He was one of Posidonius'

---

[160] The same point was made by H.M. Hine, 1984, p. 29. Sandbach's *Causes of Natural Phenomena* is periphrastic but closer to the Greek original. Longolius translates as *Causarum naturalium liber unus*. Flacelière speaks of "*Causes physiques* [...] et peut-être ferait-on mieux de parler de *Questions de physique*" (in R. Flacelière, J. Irigoin, J. Sirinelli and A. Philippon, 1987, pp. lxxxi–lxxxii). K. Ziegler, 1951, col. 857 has *Quaestiones physicae*. And so the list can go on (see F. Tanga, 2015, pp. 114–115, n. 21).

[161] Cf., e.g., *Q.N.* 5, 913C and *NQ* 3, 5; *Q.N.* 6, 913E and *NQ* 3, 25, 11; *Q.N.* 13, 915B and *NQ* 6, 13, 2; *Q.N.* 29, 919B and *NQ* 7, 1–4. There are also obvious divergences, though: cf., e.g., *Q.N.* 4, 913A vs. *NQ* 2, 57, 2; *Q.N.* 40 vs. *NQ* 2, 31, 2.

[162] See H.M. Hine, 1984, pp. 28–29 (esp. at p. 29 on Plutarch's *Quaestiones naturales*).

[163] H.M. Hine, 1984, pp. 28–29. See also A. Blair, 1999, p. 192, n. 14: "[t]he most famous *Natural Questions*, by Seneca [...], are never called *problemata* [in Renaissance editions] and take a somewhat different form, with longer discursive answers within a dialogue".

students and composed a collection of *Quaestionum naturalium causae* (?). He is quoted several times by Seneca. The title of his work is mentioned in *NQ* 6, 17, 3 – though possibly in a corrupt gloss¹⁶⁴. Papirius Fabianus, on the other hand, lived at the beginning of our era and composed a collection of *Causae naturales*. This work may have had more in common, from a formal perspective at least, with Plutarch's *Quaestiones naturales* than with Seneca's *Naturales quaestiones*. In the end, Fabianus is mentioned only once by Seneca (*NQ* 3, 27, 3), so that it is unlikely that his work served as Seneca's model¹⁶⁵. There is a clear link between Seneca's *Naturales quaestiones* and Aristotle's natural science, but an important intermediate source of inspiration for Seneca may have been Posidonius (perhaps precisely via Asclepiodotus)¹⁶⁶, whom Strabo denounces as a would-be Peripatetic in light of his interests in causal research¹⁶⁷.

2) Seneca presents his *Naturales quaestiones* as a thoroughly wrought piece of literature in dialogue form, which, as Hine has argued, meets the stylistic level of Plutarch's *Quaestiones convivales*, rather than *Quaestiones naturales*¹⁶⁸. Of course, the comparison does not hold up to further analysis: in Seneca's work, there is obviously not much dialogue, where several characters are put on scene. Seneca only addresses the dedicatee, Lucilius, directly, and at times an imaginary interlocutor falls in, but even so, a monologic discourse clearly prevails¹⁶⁹. The question-and-answer format, which is so typical of the structural organisation of the genre of problems, yields to a more essayistic and prosaic discourse in Seneca. Seneca only formulates questions on occasion, but the explanations are phrased dogmatically and there is not much place for an enumeration of several plausible explanations¹⁷⁰. Notably, in *Ben.* 7, 1, 5, Seneca actually agrees with the Cynic Demetrius in rejecting the practice of solving natural problems (e.g., concerning the ocean tides and optical illusions), since they provide only useless knowledge¹⁷¹. This

---

¹⁶⁴ See H.M. Hine, 1984, pp. 24–25. See also K.K. Müller, 1896 and P.T. Keyser and G.L. Irby-Massie, 2008b, p. 172.

¹⁶⁵ Cf. H.M. Hine, 1984, p. 29. Pliny, by contrast, names him several times and calls him 'most experienced in natural scientific affairs' (*NH* 36, 125: *rerum naturae peritissimus*). See W. Kroll, 1949 and A. Zucker, 2008, pp. 610–611.

¹⁶⁶ See P. Oltramare, 1961, p. xvii. For Seneca's reliance in *Naturales quaestiones* on Greek science, see P. Parroni, 2002, pp. xxii–xxvi.

¹⁶⁷ Strabo, *Geogr.* 2, 3, 8: πολὺ γάρ ἐστι τὸ αἰτιολογικὸν παρ' αὐτῷ καὶ τὸ Ἀριστοτελίζον, ὅπερ ἐκκλίνουσιν οἱ ἡμέτεροι διὰ τὴν ἐπίκρυψιν τῶν αἰτιῶν. For the place of natural problems in Stoic scientific thinking more generally, cf. D.L. 7, 132–133.

¹⁶⁸ H.M. Hine, 1984, p. 28.

¹⁶⁹ F.H. Sandbach, 1965, p. 134 is right in saying that "[u]nlike Seneca's *Naturales quaestiones*, it [sc. Plutarch's *Quaestiones naturales*] is not a literary work."

¹⁷⁰ See, e.g., *NQ* 2, 58; 3, 11; 16; 20.

¹⁷¹ Seneca also includes natural problems among the 'leisurely delights' in *Ben.* 7, 1, 7

criticism should be put in perspective by the fact that the right study of natural phenomena has significant *moral* implications for Seneca, in that it aims to provide, what Williams has recently called, a "cosmic viewpoint" for his ethical philosophy[172]. As opposed to what Plutarch does in his natural problems, Seneca frequently incorporates a strand of ethical paraenesis in his scientific discourse, so that the text is lifted to a moralising echelon. By issuing imperatives of traditional social *mores*, there is a direct relation between the science of nature and that of life in Seneca's *Naturales quaestiones*[173]. This *trait-d'union* is – quite notably, considering his otherwise well-known moralising inclinations – absent in Plutarch's natural problems, as is the feature of literary stylistics.

In the following section, I will, therefore, draw further attention to the central focus in Plutarch's *Quaestiones naturales* on the physical causes of natural phenomena. I will do this specifically by analysing its 'problematic' discourse at two levels, viz. by examining its sub-literary style and its lack of moralising dynamics. I conclude here that it seems imprudent to include Seneca's *Naturales quaestiones* in the wider tradition of Aristotelian natural problem literature, and that it has only little in common with Plutarch's alleged Greek counterpart – except perhaps from its Latin title, but even this is not so unproblematic, as we saw.

1.2. Problems related to Plutarch's scientific discourse

1. Trifles unworthy of Plutarch? Some remarks on authenticity

In contemporary scholarship, Plutarch's *Quaestiones naturales* has come to be accepted as authentic without reservation, but this has not always been the case. Among 19th century scholars in particular, the collection's spuriousness was commonly accepted. This changed, however, around 1900. Doehner, for instance, severely rejected the work's authenticity by disparaging it as "miseras Plutarchi imitatorum quisquilias"[174], and

---

(*oblectamenta otii*). In fact, he considers the genre of problems more generally as 'useless furniture of learning' (*Ep.* 88, 36: *supervacua litterarum supellectile*). Cf. also *Brev. vit.* 13, 1–3. See W. Bühler, 1977, p. 44 and M. Beagon, 2011, p. 73.

[172] G.D. Williams, 2012. Cf., e.g., *NQ* 3, *Praef.* 18. Notably, Chrysippus composed Φυσικὰ ζητήματα (see SVF 3, p. 205, lx), but he also held that 'physical speculation should be undertaken for no other purpose than for the distinction of good and evil' (*De Stoic. rep.* 1035D = SVF 3, p. 17, fr. 68: οὐδ' ἄλλου τινὸς ἕνεκεν τῆς φυσικῆς θεωρίας παραληπτῆς οὔσης ἢ πρὸς τὴν περὶ ἀγαθῶν ἢ κακῶν διάστασιν).

[173] See esp. *NQ* 6, 32, 1: *alterum sine altero non fit*. G. Stahl, 1964, p. 426 distinguishes three thematic categories in Seneca's *Naturales quaestiones*: "exakt-wissenschaftliche Naturuntersuchungen, *mirabilia*-Geschichten und natur- bzw. moralphilosophische Paränese". Only the first two categories are also present in Plutarch's natural problems.

[174] T. Doehner, 1858, p. 14 (cf. also 1864, p. 61: "Plutarchum, vel quicunque quaestionum naturalium auctor est").

scholars like Volkmann[175] followed suit. Weiss claimed in a similar vein that "quaestiones naturales non a Plutarcho profectas esse inter omnes constat"[176]. Croiset's evaluation was also negative: "Les *Questions naturelles* sont un ouvrage sans valeur, qui ne peut être attribué à Plutarque."[177] The authenticity of the collection was vindicated, however, on linguistic and stylistic grounds by Weissenberger[178]. Diels[179] rejected Doehner's false evaluation of the unpretentious "Schriftchen", while Hartman[180] could find no reason to reject its authenticity. Ziegler[181] was also convinced of its authenticity, as was Hubert[182]. Eventually, Sandbach[183] did not even mention the problem of authenticity, and, as Senzasono pointed out[184], this problem has simply been superseded today.

It seems that the initial doubt about the work's authenticity was fed by the scholarly prejudice that Plutarch is first and foremost a moralist and eloquent story-teller, author of popular-philosophical speeches and of dramatised dialogues. Still today, the Chaeronean is seen as a flamboyant litterateur and lecturer with a balanced enthusiasm for moral instruction and stylistic embellishment. Clearly, this image does not apply to his more technical writings and to his collections of *quaestiones* more in specific (with the exception of *Quaestiones convivales*), where the text's aesthetic is restrained by a rather simple and rudimentary style of discourse. We will see that there are some glints of literary *ornatus* in *Quaestiones naturales*, though, but even so Plutarch is clearly preoccupied with the discursive concerns of brevity and clarity, thus making it absolutely clear that his main concern is with the argument rather than with the form[185].

---

[175] R. Volkmann, 1869, p. 188 (in footnote).

[176] D. Weiss, 1888, p. 18.

[177] A. and M. Croiset, 1899, p. 511, n. 1.

[178] B. Weissenberger, 1895, pp. 81–82.

[179] H. Diels, 1905, p. 315, n. 1.

[180] J.J. Hartman, 1916, p. 556: "*Aetia* vero *Physica* cur Plutarcho abiudicemus nullam causam video. Haud pauca in eo libro admodum lepida leguntur quaeque Plutarcho cordi fuisse minime mirum sit [...]."

[181] K. Ziegler, 1951, col. 857: "Keinesfalls besteht ein Grund, die Schrift [...] dem P. abzusprechen".

[182] C. Hubert, 1960, p. v: "quin genuinae sint dubitare non licet".

[183] F.H. Sandbach, 1965, p. 133.

[184] L. Senzasono, 2006, p. 44.

[185] As Plutarch notes himself, philosophical discourse is concerned with 'the lucid and the instructive' (*De Pyth. or.* 406E: τὸ σαφὲς καὶ διδασκαλικόν). In *De aud.* 42C, he warns his students that they should first and foremost focus on the content of what is said during philosophical lectures rather than on the form (καὶ γὰρ εἰ τοῖς λέγουσι προσήκει μὴ παντάπασιν ἡδονὴν ἐχούσης καὶ πιθανότητα λέξεως παραμελεῖν, ἐλάχιστα τούτου φροντιστέον τῷ νέῳ, τό γε πρῶτον).

Even if Plutarch's literary and moralising flair are not exactly on display in *Quaestiones naturales*, the text's authenticity can be confirmed on a number of grounds. First of all, several typical features of Plutarch's basic style are clearly evident in the collection. I have already mentioned the use of semi-synonymous pairs (see n. 109 above); to this the general avoidance of hiatus and (less evidently) the use and avoidance of certain rhythms at sentence endings can be added[186]. Furthermore, a large amount of parallel passages can be found throughout Plutarch's other works. The similarities between these parallels are so prominent that they cannot be treated as coincidental. The fact, moreover, that the collection is recorded in the Lamprias catalogue could further support the text's authenticity but this does not serve as any definitive proof (see n. 3). In short, it goes without saying that even if the text is less alluring from a stylistic or moralising perspective, this is no stable ground for denying its authenticity[187]. Therefore, before dealing with the problems of style and morality in *Quaestiones naturales*, let us first take a closer look at how Plutarch himself conceives of what constitues natural scientific discourse proper.

2. The rhetoric of scientific discourse according to Plutarch

Several passages in the *corpus Plutarcheum* indicate that Plutarch mostly identifies the rhetoric of ancient scientific discourses, and especially that of ancient meteorology, in terms of the rhetorical category of 'the sublime' (τὸ ὕψος). In *De prof. in virt.* 78E, for instance, Plutarch notes that practically all young, haughty philosophers pursue those forms of discourse that strive for repute (δόξα), and that some of them 'like birds, are led by their flightiness and ambition to alight on the resplendent heights of the natural phenomena' (οἱ μὲν ὥσπερ ὄρνιθες ἐπὶ τὴν λαμπρότητα τῶν φυσικῶν καὶ τὸ ὕψος ὑπὸ κουφότητος καὶ φιλοτιμίας καταίροντες). Clearly, the concept of ὕψος in this passage refers both to the style and content of the meteorological

---

[186] On the general stylistics of *Quaestiones naturales*, see K. Ziegler, 1951, col. 857, F.H. Sandbach, 1965, pp. 134–136, G.W.M. Harrison, 2000b and L. Senzasono, 2006, pp. 8–18. Regarding Plutarch's avoidance of hiatus, see J. Schellens, 1864, R. Flacelière, J. Irigoin, J. Sirinelli and A. Philippon, 1987, pp. ccxii–ccxiv, S.-T. Teodorsson, 1989, p. 128 (Plutarch did not avoid hiatus systematically, though: cf. *De vit. pud.* 534F and *Bellone an pace* 350DE). On Plutarch's use and avoidance of certain rhythmical clauses, see F.H. Sandbach, 1939, pp. 194–203 and M. Baldassari, 2000. On Plutarch's style and language more generally, see, e.g., K. Ziegler, 1951, cols. 931–938, S. Yaginuma, 1992 and L. Torraca, 1998.

[187] Let alone for doing away with it altogether. The same point was made for the modest literary merits of Plutarch's *Apophthegmata* by F. Fuhrmann, 1988, p. 4: "Mais peut-on ainsi rejeter tout ce qui, dans le corpus des oeuvres de Plutarque, n'est pas de haute tenue? A quel niveau situera-t-on la limite?"

type of discourse, which is, indeed, concerned with things of an elevated nature (viz. the sublime realm of τὰ μετέωρα or μετάρσια). A similar idea is present in Plutarch's account of Pericles' mode of speech in *Per.* 8, 1-2, where we read about Pericles' habit of adopting the style of Anaxagoras' natural scientific discourse in his own rhetorical speeches:

τῇ μέντοι περὶ τὸν βίον κατασκευῇ καὶ τῷ μεγέθει τοῦ φρονήματος ἁρμόζοντα λόγον, ὥσπερ ὄργανον, ἐξαρτυόμενος παρενέτεινε πολλαχοῦ τὸν Ἀναξαγόραν, οἷον βαφὴν τῇ ῥητορικῇ τὴν φυσιολογίαν ὑποχεόμενος. τὸ γὰρ 'ὑψηλόνουν τοῦτο καὶ πάντη τελεσιουργόν,' ὡς ὁ θεῖος Πλάτων φησί, 'πρὸς τῷ εὐφυὴς εἶναι κτησάμενος' ἐκ φυσιολογίας, καὶ τὸ πρόσφορον ἑλκύσας ἐπὶ τὴν τῶν λόγων τέχνην, πολὺ πάντων διήνεγκε.

Moreover, by way of providing himself with a style of discourse which was adapted, like a musical instrument, to his mode of life and the grandeur of his sentiments, he often made an auxiliary string of Anaxagoras, subtly mingling, as it were, with his rhetoric the dye of natural science. It was from natural science, as the divine Plato says (*Phdr.* 270a; cf. also *Them., Or.* 26, 329c), that he 'acquired his loftiness of thought and perfectness of execution, in addition to his natural gifts', and by applying what he learned to the art of speaking, he far excelled all other speakers.

There is a clear moralising, ethopoietic implication to this passage, according to which the rhetorical-scientific mode and grandeur of Pericles' speech is considered to be in conformity with his way of living[188] (contrast the haughtiness of young philosophers in *De prof. in virt.* 78E above). As such, Plutarch implies that the high ethical standard of Pericles' life is reflected in his stately and 'sublime' manner of speaking. Anaxagoras' influence on Pericles' character is also discussed a little earlier in *Per.* 4-5. At the end of this passage, Plutarch again alludes to the λόγος ὑψηλός of Anaxagoras' discourses, as reflected in Pericles' own character and speech (*Per.* 5, 1):

Τοῦτον ὑπερφυῶς τὸν ἄνδρα θαυμάσας ὁ Περικλῆς καὶ τῆς λεγομένης μετεωρολογίας καὶ μεταρσιολεσχίας ὑποπιμπλάμενος, οὐ μόνον ὡς ἔοικε τὸ φρόνημα σοβαρὸν καὶ τὸν λόγον ὑψηλὸν εἶχε καὶ καθαρὸν ὀχλικῆς καὶ πανούργου βωμολοχίας, ἀλλὰ καὶ προσώπου σύστασις ἄθρυπτος εἰς γέλωτα καὶ πρᾳότης πορείας καὶ καταστολὴ περιβολῆς πρὸς οὐδὲν ἐκταραττομένη πάθος ἐν τῷ λέγειν καὶ πλάσμα φωνῆς ἀθόρυβον καὶ ὅσα τοιαῦτα πάντας θαυμαστῶς ἐξέπληττε.

---

[188] On the influence of Anaxagoras' discourse on Pericles' rhetorical style, see P.A. Stadter, 1991, pp. 120-121.

This man (sc. Anaxagoras) Pericles extravagantly admired, and being gradually filled full of the so-called meteorology, he not only had, as it seems, a spirit that was solemn and a discourse that was lofty and free from plebeian and reckless effrontery, but also a composure of countenance that never relaxed into laughter, a gentleness of carriage and cast of attire that suffered no emotion to disturb it while he was speaking, a modulation of voice that was far from boisterous, and many similar characteristics which struck all his hearers with wondering amazement.

Arguably, Pericles' character and, connected with it, his manner of speaking elicited the same kind of wonder and amazement in its listeners as the meteorological phenomena would do of themselves. In any case, the main idea is again that Pericles' spiritual elevation, as engendered by Anaxagoras' meteorology, found its natural expression in his exalted speech[189].

What can we say about *Quaestiones naturales* in this regard, then? We can hardly speak of a λόγος ὑψηλός here, considering the collection's general lack of stylistic embellishment. Neither is it representative of Plutarch's own ethical persona and moralising ambitions as demonstrated elsewhere. The question is particularly intriguing, since several of the problems raised in this collection have specific meteorological interests (*Q.N.* 4, 24, 40). Senzasono has recently taken up a remarkable position, arguing that the general absence of stylistic embellishment in *Quaestiones naturales* is, in fact, an essential marker of the collection's scientific character. He argues that "l'adesione ai fenomeni fisici condiziona lo stile essenziale e sobriamente descrittivo del trattato naturalistico […]."[190] This point of view, however, tends to be biased by our modern conceptual standards of a sober and unembellished scientific discourse, which prescribes a type of phrasing that is clear and strives for a referential

---

[189] For a faint parallel to these two *Pericles* passages, see *Them.* 2, where Plutarch corrects Stesimbrotus' incorrect report that Themistocles was a student of Anaxagoras and a disciple of Melissus, the physicist (FGrHist 107, 1). Plutarch points out that Pericles, rather than Themistocles, was a pupil of Anaxagoras and that Melissus opposed him at the siege of Samos (in 440 BC). Themistocles, on the other hand, was a disciple of Mnesiphilus the Phrearrhian, who was a sophist, that is, 'neither a rhetorician nor one of the so-called natural philosophers' (οὔτε ῥήτορος ὄντος οὔτε τῶν φυσικῶν κληθέντων φιλοσόφων). Unlike Pericles, Themistocles only relied on his natural gifts (τῇ φύσει πιστεύων) – not also on a *philosophical* education –, which may explain why his youth essays were of a deplorable quality (*Them.* 2, 5: ἐν δὲ ταῖς πρώταις τῆς νεότητος ὁρμαῖς ἀνώμαλος ἦν καὶ ἀστάθμητος, ἅτε τῇ φύσει καθ' αὑτὴν χρώμενος κτλ.). For further commentary on this passage, see F.J. Frost, 1980, pp. 65–71.

[190] L. Senzasono, 2006, pp. 10–11.

description of the physical reality that is studied[191]. From a modern point of view one might, indeed, argue that the general discourse and register of *Quaestiones naturales* complies with the ethos of scientific objectivity, by demanding a purely descriptive, unembellished discourse and, in addition, also a depersonalisation of the authorial voice. It goes without saying, though, that such a view is anachronistic, at the least. In any case, as the passages above have shown, meteorological texts were rather marked by a highly rhetorical type of discourse.

The assumption, therefore, that any scientific discourse requires a type of diction that is equally objective and depersonalised as the object of study itself is assumed to be (i.e. nature), does not add up for a large number of ancient scientific texts, including those of Plutarch. In Antiquity, the focus on natural phenomena did not necessarily entail 1) a simple, unadorned style of discourse, let alone 2) an impersonal approach of the object studied.

1) As to the first point, problems rise regarding ancient scientific texts using poetry as a discursive medium, such as the didactic poems of Empedocles and Lucretius[192]. As we will see later on, Plutarch himself quotes again and again from the poets in his natural problems, not only from the didactic but also from the mimetic ones. Is this, then, some kind of an unscientific stain on his text? Plutarch's general attitude towards poetry is heavily influenced by his Platonism, but this does not, of course, imply that a poetical form automatically shrouds false content [see 4.1.2.3., n. 97]. Notably, regarding Empedocles' use of poetical epithets, Plutarch makes it clear in *Quaest. conv.* 683E that they do not complicate a suitably scientific approach to natural phenomena (DK31B80):

τὰ δὲ μῆλα καθ' ἥντινα διάνοιαν ὁ σοφὸς 'ὑπέρφλοια' προσειρήκοι, διαπορεῖν, καὶ μάλιστα τοῦ ἀνδρὸς οὐ καλλιγραφίας ἕνεκα τοῖς εὐπροσωποτάτοις τῶν ἐπιθέτων, ὥσπερ ἀνθηροῖς χρώμασι, τὰ πράγματα γανοῦν εἰωθότος, ἀλλ' ἕκαστον οὐσίας τινὸς ἢ δυνάμεως δήλωμα ποιοῦντος, κτλ.

But what puzzled me, I confessed, was what the philosopher meant by calling apples "succulent" (*hyperphloia*); especially since he was not in the habit of tricking out facts for the sake of elegant writing by

---

[191] The bias is obvious, for instance, in L. Senzasono, 2006, p. 9: "Questi nessi [c.q. structural markers in the text, such as γάρ, ὅθεν etc.] collegano rilievi di fenomeni naturali in una struttura essenziale, priva di qualsiasi abbellimento, come s'addice a un'esposizione che intenda essere scientifica, nel mondo Greco come in altre epoche fino ad oggi." For similar criticism of Senzasono's position, see L. Van der Stockt, 2011, p. 449.

[192] For the cases of Empedocles and Lucretius, see L. Taub, 2008 (see also *id.*, 2009 more generally).

using grandiose epithets, as if he were laying on gaudy colours, but in every case aimed at simple description of an essential fact or property. Etc.

As this passage shows, the scientificity of ancient scientific discourses cannot be properly evaluated on the basis of their form or register[193]. It rather illustrates that science was practiced in a wide range of literary genres, some, indeed, more literary than others.

2) The same conclusion can be drawn for the issue of depersonalisation in ancient scientific discourses. A comparison of *Quaestiones naturales* with *Quaestiones convivales* [see 2.2.] shows that the same kind of natural problems can be treated against the backdrop of 'real-life' table discussions, where each symposiast tries to defend his own personal theory in an eloquent fashion *vis-à-vis* that of his fellow symposiasts. There is not much authorial depersonalisation in *Quaestiones convivales* – to the contrary, Plutarch puts his closest friends in the scene along with his own literary *alter ego*, thus constructing scientific authority in a very personal way (see nn. 105–106). So even if Plutarch does not use the first person singular in *Quaestiones naturales*, this is not necessarily an indication of the collection's 'scientificity'. Indeed, as we will see later on, other personal forms, such as the second person singular and the first person plural, are still present, which is to be interpreted more likely in relation to the work's educational context and purpose[194] [see 3.1.4.]. Another point that is very important in view of the alleged objectivity of Plutarch's natural problems is the fact that ancient natural science did not necessarily complicate the author's attempt to assume a certain ethical persona and to communicate a moralising message through his text. Morality is, in fact, a relatively common aspect of ancient Greco-Roman scientific writing, as we saw in the case of Seneca's *Naturales quaestiones*[195] [see 1.1.6.]. Similarly, in Plutarch's writings on animal psychology, the scientific discourse serves as a means of promoting the author's moral agenda (see n. 14). It remains to be seen, then, why no such agenda is found in *Quaestiones naturales*.

---

[193] Cf. also, e.g., *Quaest. conv.* 658B for the idea that the poet Archilochus writes 'in accordance with nature' (φυσικῶς). For the 'very simple and antiquated' style of Solon's verses on natural science, cf. *Solon* 3, 6–8 (ἐν δὲ τοῖς φυσικοῖς ἁπλοῦς ἐστι λίαν καὶ ἀρχαῖος): see D. Leão, 2015. For the relation between science and poetry in Plutarch's *De sollertia animalium*, see also J. Bouffartigue, 2012, p. xii, n. 7.

[194] For the aspects of objectivity and subjectivity in ancient scientific and technical writing more generally, as conveyed respectively by an impersonal or personal discourse, see, e.g., G.E.R. Lloyd, 1987, pp. 56–70, H. von Staden, 1994, P. van der Eijk, 1997, pp. 115–119, H.M. Hine, 2009, V. Nutton, 2009, J. König, 2011.

[195] For the relation between science and ethics in Antiquity more generally, see G.E.R. Lloyd, 1985 and R.W. Sharples, 2000, pp. 14–22 and 2005, p. 2.

Before I will provide an alternative explanation for the problems of style and morality in *Quaestiones naturales*, let us first take a more detailed look at how, and to which extent, these problems precisely manifest themselves in our collection.

3. The problem of style

Although there are some notable exceptions, there is not much rhetorical liveliness in *Quaestiones naturales* and literary imagery is scarce. The collection's elliptical style renders the text obscure at times, and the mechanical structure of the problem chapters makes the whole collection relatively bloodless from a literary-aesthetic standpoint. Scholars speak of a 'matter-of-fact'[196] or 'referential'[197] style to denote the collection's descriptive approach. Generally speaking, the sentence structure remains simple throughout and contains a profusion of directive conjunctions and particles that mark the organisation and logic of the explanations in their most basic form. Furthermore, the explanations teem with a monotonous, but conveniently uniform, scientific vocabulary (which I will examine later on [see 4.3.4.]). In short, a clear and swift development of the explanations prevails: the aetiology is always *ad rem*, but conciseness often leads to obscurity[198] (as is the case, for instance, in the comparative ἀναγκαιότερα, said of spring rains falling before the summer in *Q.N.* 4, 913A: see the commentary *ad loc.*). As a rule, the information amassed is processed in an epitomic fashion, by which each and every recorded element plays a well-determined role in the development of the *causa* (as is the case, for instance, with Plutarch's silent adaptation of τομίας into μόνορχις in Aristotle's account about the infertility of wild boars in *Q.N.* 21, 917D: see the commentary *ad loc.*). Extensive details are generally avoided in this no-nonsense discourse, and what might seem to be a *fait divers* at first, will often end up playing a relevant role in the argumentation after a closer reading (as is the case, for instance, with the digression on magnetism in *Q.N.* 19, 916D: see the commentary *ad loc.*).

Even though the discourse of *Quaestiones naturales* is only seldom embellished, there are a few instances where we find exceptional glints of Plutarch's literary flair. The application of tropes and stylistic devices, such as metonymy and metaphor, is rather exceptional[199], but where they

---

[196] S.-T. Teodorsson, 1999a, p. 666. Cf. also J. Sirinelli, 2000, p. 365: "le caractère purement factuel de ses recueils".

[197] L. Senzasono, 2006, p. 10, n. 11 (with a reference to Jakobson's linguistic theory). The term is synonymous with 'denotative' or 'cognitive'.

[198] Cf. F.H. Sandbach, 1965, p. 135 and L Senzasono, 2006, pp. 11–12. G.W.M. Harrison, 2000b, p. 241 sees the positive side of it: "One of the great joys of Plutarch is his incisiveness and concision […]."

[199] See also L. Senzasono, 2006, pp. 12–15.

are used they make the argument more vivid. In *Q.N.* 2, 912A, rainwater is rather sacredly called τὸ ἐκ Διὸς ὕδωρ; in *Q.N.* 11, 914F, the tossing and upset of the soul (ἡ ψυχὴ σάλον ἔχουσα καὶ θορυβουμένη) stands in contrast to the calm of the sea; in *Q.N.* 34, the heat of the sun is compared to an enemy in battle chasing off the wind (*veluti hoste superatum*); and in *Q.N.* 39, the depth is called the mother of blackness (*mater nigritudinis*). Notably, in his study of Plutarch's general use of literary imagery, Fuhrmann marked only one instance in *Quaestiones naturales*, viz. in *Q.N.* 29, 919B: "Les météores éclatent comme des bulles"[200] (σέλα ῥηγνύμενα πομφόλυγος δίκην).

It is, indeed, true that an exceptional chapter is found in *Q.N.* 29, 919AB as a whole. Plutarch therein discusses the wondrous nature of hot springs *vis-à-vis* cold ones. His argument exhibits a remarkably vibrant case of rhetorical embellishment. Not so much the natural phenomenon of hot springs is at stake but rather the short-sighted marvelling for them by the common people[201]. Plutarch seems to be disillusioned by the fact that the common people's wonder for marvelous natural phenomena – as opposed, presumably, to that of the φυσικός – is not followed by an actual investigation into their 'nature', that is the natural causes of these phenomena (οὐ μέντοι θαυμάζουσιν οἱ πολλοὶ τὴν τούτων φύσιν). In order to emphasise his disillusion, Plutarch confronts the reader with a rhetorically substantiated invective against the common people for their superficial astonishment for rare natural phenomena (τὸ μὴ πολλάκις γινόμενον). Plutarch emphatically declares his own marvel for cosmic spectacles and the wonders of nature by scanning an evocative verse from Euripides (TGF 941: 'You see this infinite heaven up high / surrounding earth in a damp embrace') and by adding a two-part rhetorical question ('What a multitude of spectacles does it bring at night, how great is the beauty it exhibits by day?'). As we will see later on, the quotations from the poets are usually incorporated in Plutarch's natural problems primarily in order to fulfil a specific argumentative and illustrative function in the aetiologies [see 4.1.2.3.]. That is, they are not included for purely literary purposes. While they certainly add a literary accent to the text, this seems to be at the service of the main argument itself. Then again, the opposite seems to be the case for the Euripides quotation here in *Q.N.* 29, 919B, since it mainly contributes to Plutarch's rhetorical expression of wonder for nature rather than to a scientific theory about hot springs (these verses do, of course, have illustrative value in underlining Plutarch's sense of wonder, but they have no substantial argumentative significance). This is important for the rather exceptional character of this chapter in the

---

[200] F. Fuhrmann, 1964, p. 77.
[201] Unfortunately the ending of this chapter is lacunary (see the commentary *ad loc.* for further detail). For a separate discussion, see M. Meeusen, 2015b.

collection, which is why I will return to its significance for the intellectual agenda of *Quaestiones naturales* in further detail later on [see 3.2.2. and 4.1.1.1.].

What matters here is the fact that the general texture of *Q.N.* 29 differs considerably from that of the rest of the problem chapters in the collection. The exclamatory utterances and the overall style are clearly more rhetorical. As Oikonomopoulou suggests (in personal correspondence), "we may surmise a context of intellectual performance for this problem (perhaps sympotic?) – the question is whether this could be surmised for the collection as a whole". This is not unlikely if we bear in mind the link with the natural problems discussed in *Quaestiones convivales*[202] [see 2.2.]. As we will see later on, Plutarch's natural problems were also useful as philosophical exercises in the context of his own school in Chaeronea [see 3.1.]. But whether this means that the problems of *Quaestiones naturales* are the textual condensation or imitation of such 'performances' or were even intended to be 'performed' at a certain point remains uncertain. This is not, however, to reject the more generally 'performative' character of Plutarch's natural problems, at least in the sense that there are certain principles at work in them that strive for argumentative originality and for a creative refashioning of received knowledge [see 4.2.2.].

4. The problem of morality

The general absence and suppression of moralising dynamics in *Quaestiones naturales* is very prominent compared to many of Plutarch's other writings[203]. However, even if Plutarch does not provide any explicit moral advice here, there are still a few faint allusions to the field of ethics and moral conduct. These allusions are only made in light of a particular zetetic requirement, though, that is, for the further development or illustration of the physical aetiology at hand. A good example can be found in *Q.N.* 19, where the octopus' ability to change its colour is attributed, in the first *causa*, to the animal's cowardice. It is compared with the change of colour in cowardly persons (as is illustrated with a line from Homer, *Il.* 13, 279: 'the coward's complexion alters'). In the second *causa*, the

---

[202] Scholars have more often stressed the 'performative' character of the genre of *quaestiones*. In regards to Heraclitus' *Quaestiones Homericae*, cf., e.g., D.A. Russell and D. Konstan, 2005, p. xxix: "It may have been a showpiece, or, given its length, an earnest defense of Homer's piety. Anything more precise is guesswork". Cf. also D.A. Russell, 2003. See also K. Oikonomopoulou, 2013a, p. 147 (with n. 71 for further references).

[203] Notably, in *De tuenda* 133E, Plutarch draws a clear conceptual distinction between problems of natural science and stories that contain ethical considerations (ἀλλὰ πολλὰ μέν ἐστι τῶν φυσικῶν προβλημάτων ἐλαφρὰ καὶ πιθανά, πολλαὶ δὲ διηγήσεις ἠθικὰς σκέψεις ἔχουσαι) [see 3.1.3.].

context is again ethical: Plutarch quotes Pindar (fr. 43 Snell) and Theognis (215–216 West), who compare the octopus' ability to change its colour with the changeable character of human beings. Plutarch pokes fun at both poets by wondering whether they believe that the octopus treats its colour like a garment that can be easily changed whenever the animal wishes (ἢ καθάπερ ἐσθῆτι τῇ χρόᾳ νομίζουσι χρῆσθαι, ῥᾳδίως οὕτως ᾗ βούλεται μετενδυόμενον;)[204]. Thus, Plutarch suggests that both poets, with their exclusively moralising approach, do not actually have a correct insight into the physical mechanism behind the phenomenon at issue, and for this reason they make themselves into easy targets for derision. The underlying idea is that the octopus' colour change is not the effect of a deliberate choice but of deeper physical causes. This is why Plutarch at the beginning of the third *causa* points out that it is the octopus itself that initiates the effect by feeling fright, but that the determining factors of the cause lie elsewhere, viz. in the pores in its skin taking up the emanations that settle in them (ἆρ᾽ οὖν τὴν μὲν ἀρχὴν αὐτὸς ἐνδίδωσι τοῦ πάθους δείσας, τὰ δὲ κύρια τῆς αἰτίας ἐν ἄλλοις ἐστί;). Plutarch's interest in the verses of Pindar and Theognis in the second *causa*, thus, lies in the fact that they add something to the first *causa*. They do not only concern the change but also the adaptation of the octopus' colour to its surroundings. Eventually, the determining factor in the causation lies in the third and last *causa*, which is the most elaborate. It is introduced with the imperative σκόπει δή, and – most importantly – focuses on the physical mechanism behind the adaptation of the octopus' colour[205].

From a cluster of parallel passages in Plutarch's other works, we learn that the Chaeronean is prepared to exploit the topic of the octopus' metachrosis as a moral *exemplum* for people's changeable character, especially in the context of the opportunism of flatterers and politicians who adapt their character to always new situations (for a schematic presentation of this cluster [see 2.1.2.])[206]. The most relevant parallel in this regard is in *De am. mult.* 96F, where Plutarch advises against having

---

[204] The irony in this rhetorical question is obvious – a style of register that is very exceptional to the collection's discourse. For irony in Plutarch, see the appendix in J. Opsomer, 2000, pp. 328–329.

[205] However, Plutarch still considers the emotion of fear as an important factor for a proper explanation, since it initiates the change (τὴν μὲν ἀρχὴν αὐτὸς ἐνδίδωσι τοῦ πάθους δείσας). L. Senzasono, 2006, p. 201, n. 106 is correct that in this third *causa* "non si ha nemmeno un cenno al comportamento umano". Regarding the quotes from Pindar and Theognis, he notes (p. 198, n. 101): "Anche in questo tratto il mutamento di colore del polipo assume la connotazione d'un paradigma morale nonostante il contesto naturalistico di un opuscolo che tratta di fenomeni fisici: Plutarco è condizionato dai due testi poetici citati, che lo inducono a questa considerazione perché entrambi contengono un'esortazione a una certa condotta."

[206] For a separate study of this cluster, see M. Meeusen, 2012a, pp. 247–250.

many friends (πολυφιλία) by criticising the habit of certain individuals, who have no firmly grounded character of their own, to assimilate themselves to others. Plutarch compares this with the octopus' ability to change and adapt its colour to its surroundings, noting that these 'changes have no depth (i.e. they have nothing to do with the animal's character) but are generated entirely on the octopus' surface (i.e. the octopus' skin)' (αἱ μεταβολαὶ βάθος οὐκ ἔχουσιν, ἀλλὰ περὶ αὐτὴν γίγνονται τὴν ἐπιφάνειαν). I take this to imply that Plutarch saw the opportunistic adaptability of flatterers, as opposed to that of the octopus, not as superficial but as in keeping with their very personality (or rather the absence thereof). Plutarch further explains that the octopus' metachrosis is caused by the fact that emanations from nearby objects are taken up by its skin due to its alternate denseness and looseness of texture, which is, in fact, a synopsis of Plutarch's third *causa* in *Q.N.* 19, 916CF.

Even though Plutarch incorporates the natural phenomenon of the octopus' metachrosis in a moralising context in *De am. mult.* 96F, by providing it with a metaphorical connotation so that it is universally adaptable in the field of ethics (c.q. as a rejection of πολυφιλία), he makes it very clear that the phenomenon in itself is in fact free from any ethical connotation or 'depth' (βάθος). This, in turn, explains why the aetiology in *Q.N.* 19 remains for the most part on the 'surface' (ἐπιφάνεια) of physical aetiology. The lines from both Pindar and Theognis (as well as from Homer) do oblige Plutarch to include a certain degree of ethical 'depth' in quoting their verses, but he complies to this zetetic requirement only in view of a refutation of their accounts. After all, as Plutarch suggests in the third *causa*, the determining factors of the explanation lie elsewhere (916C: τὰ κύρια τῆς αἰτίας ἐν ἄλλοις ἐστί;). Plutarch thus shows that it is his primary objective in *Q.N.* 19 – and, hence, in *Quaestiones naturales* more generally – to scrutinise, what he calls, τὰ κύρια: these are clearly the most fundamental underlying physical principles. Therefore, any moral preoccupation is circumstantial or, at least, only of secondary importance to this work.

Plutarch avoids plunging into the 'depths' of morality again on several other occasions in *Quaestiones naturales*. In *De cap. ex inim.* 92B, for instance, he compares the improvement of roses and violets by planting garlic and onions beside them to the use of enemies and the joining of opposites. The same phenomenon is explained in *Q.N.* 41, but therein Plutarch remains on the 'surface' of physical aetiology (referring to the theories of emanation, attraction and motion). The best example by far is provided by *Q.N.* 36, where Plutarch manifestly passes over an opportunity for a moral digression. The natural problem at issue concerns the popular belief that bees are quicker to sting people who have just committed adultery (*stuprum*). The ethical depravation of this "immoral sexual act" (to use Sandbach's periphrastic translation)[207] turns out to be entirely subordinated to the physical aspect of the phenomenon itself (bees are cleanly insects, unable to endure the bad smell of adulterers). This becomes clear especially when confronted with the parallel account in *Coni. praec.* 144D, where Plutarch offers some 'moral support' to the groom. There is no need to treat this parallel in full detail here[208]. In short, Plutarch's advice to the groom would be to avoid adultery before (*sic!*) approaching his own wife, since the wife might react in the same way as a bee would react. Remarkably enough, there is a certain impulse towards a moralising diversion in *Q.N.* 36 (viz. in the words *adulteria* and *perfidia*), but this becomes far more explicit in *Coni. praec.* 144D. Passages like these show that Plutarch does not refrain from incorporating the same material from *Quaestiones naturales* in overtly moralising contexts elsewhere, but that in the collection itself there is a significant ethical disinterest.

Since elsewhere Plutarch would certainly approve of an ethically-laden view of the physical world, it is worthwhile to examine the relation between his natural and moral philosophy more closely[209]. Several key passages that deserve attention here concern Plutarch's general cosmology and theology (these are often related to the context of his theodicy, that is, his theological vindication of God's justice in respect of existing evil)[210]. In *De lat. viv.* 1129B, for instance, Plutarch objects to Epicurus' maxim λάθε βιώσας ('live unknown') by arguing that persons should not at all remain unknown to the outer world if they carry out their practice in the field of physics, ethics or politics in an honourable fashion (= fr. 524 Usener).

ἐὰν δέ τις ἐν μὲν φυσικοῖς θεὸν ὑμνῇ καὶ δίκην καὶ πρόνοιαν, ἐν δ' ἠθικοῖς νόμον καὶ κοινωνίαν καὶ πολιτείαν, ἐν δὲ πολιτείᾳ τὸ καλὸν ἀλλὰ μὴ τὴν χρείαν, διὰ τί λάθῃ βιώσας; ἵνα μηδένα παιδεύσῃ, μηδενὶ ζηλωτὸς ἀρετῆς μηδὲ παράδειγμα καλὸν γένηται;

But take one who in physics extols God and justice and providence, in ethics law and society and participation in public affairs, and in political life the upright and not the utilitarian act, what need has he to

---

[207] F.H. Sandbach, 1965, p. 219.

[208] For a separate study, see M. Meeusen, 2013c. Notably, Aelian also reports that bees chase off men wearing perfume and that they also recognise and pursue adulterers, which is considered a sign of their σωφροσύνη (*NA* 5, 11; for further parallels see the commentary *ad loc.*).

[209] In this regard, G. Soury, 1949, p. 323 speaks of "une sorte d'indifférenciation du physique et du moral, fréquente chez Plutarque".

[210] For further reading on Plutarch's theodicy, see, e.g., R. Del Re, 1950, J. Dillon, 2002 and F. Frazier, 2012, pp. 219–221. Cf. also D. Babut, 1969, pp. 287ff.

live unknown? In order to educate no one and become for no one an inspirer of virtuous emulation or a noble example?

What is remarkable here is that regarding the field of physics, Plutarch refers to God and providence as well as to justice (δίκη) – which would rather belong to the field of ethics. As opposed to the Epicureans, whose cosmology is based on the principles of chance and atomism [see 4.3.3.2.], Plutarch firmly believes that God can intervene in the material world, which is ordered in conformity with divine providence[211]. Roskam highlights the central importance of the passage at hand, and more specifically the mentioning of δίκη, for the argument in De latenter vivendo as follows:

> "That Plutarch also underlines the importance of justice in the domain of physics may come as a surprise, since one would a priori expect the mention of the element of δίκη rather in the field of ethics. Its presence in the domain of physics could perhaps be explained by Plutarch's thoroughly Platonic theology, in which justice and the good in general are inherently and inextricably bound up with God's essence. [...] Contrary to Epicurean physical thinking, which is ultimately based on atomism, Plutarch's general programme of physics, with its emphasis on divine providence and justice, rests on, and is fundamentally justified by, a theological perspective."[212]

Notably, in *Arist.* 6, 2–3, Plutarch again underlines the importance of δίκη in physical matters. This passage clarifies the actual range of the phrase ἐν φυσικοῖς as mentioned in *De lat. viv.* 1129B above. Plutarch writes that the inanimate part of nature shares its incorruptibility and power with the divine, but, in opposition to animate nature, it does not partake in virtue (ἀρετή), and hence does not participate in justice (δίκη καὶ θέμις), precisely because it is irrational.

> Καίτοι τὸ θεῖον [...] τρισὶ δοκεῖ διαφέρειν, ἀφθαρσίᾳ καὶ δυνάμει καὶ ἀρετῇ, ὧν καὶ σεμνότατον ἡ ἀρετὴ καὶ θειότατόν ἐστιν· ἀφθάρτῳ μὲν γὰρ εἶναι καὶ τῷ κενῷ καὶ τοῖς στοιχείοις συμβέβηκε· δύναμιν δὲ καὶ σεισμοὶ καὶ κεραυνοὶ καὶ πνευμάτων ὁρμαὶ καὶ ῥευμάτων ἐπιφοραὶ μεγάλην ἔχουσι· δίκης δὲ καὶ θέμιδος οὐδέν, ὅτι μὴ τῷ φρονεῖν καὶ λογίζεσθαι θεῖόν ἐστι, μεταλαγχάνει.

And yet divinity [...] is believed to have three elements of superiority, – incorruption, power, and virtue; and the most reverend, the divinest

---

[211] Cf., e.g., *De def. or.* 426DE. See G.E. Karamanolis, 2006, p. 106.
[212] G. Roskam, 2007, pp. 122–123. See also F. Frazier, 2012, p. 221. On the role of providence and chance in Plutarch's philosophical thinking, see the contributions in F. Frazier and D.F. Leão, 2010.

of these, is virtue. For vacuum and the ultimate elements partake of incorruption; and great power is exhibited by earthquakes and thunderbolts, and rushing tornadoes, and invading floods; but in fundamental justice nothing participates except through the exercise of intelligent reasoning powers.

Inanimate nature shares its incorruptibility and power with the divine, but it does not also partake in virtue because it is of a purely irrational and material order. Animate nature, by contrast, is constituted by more than just material components: it does have rational powers and, by implication, partakes in virtue. In the passage following the one just quoted, Plutarch notes that 'virtue is the only form of divine excellence within our grasp, but that it is often placed last', which is not right (τὴν δ' ἀρετήν, ὃ μόνον ἐστὶ τῶν θείων ἀγαθῶν ἐφ' ἡμῖν, ἐν ὑστέρῳ τίθενται). He also remarks that through injustice, life is made 'bestial' (ἡ δ' ἀδικία θηριώδη). This is presumably metaphorical, because in his writings on animal psychology, Plutarch emphatically marks the rational and, by implication, virtuous qualities of animals, which belong to the same animate realm in nature as human beings.

Context is important here: the general ethical purport of Plutarch's writings on animal psychology is very obvious[213]. The argument is of a polemical nature, aiming at undermining the Stoic tenet that animals have no rational abilities at all. In short, it seems that Plutarch is trying to show that animals actually live more in conformity with nature – another Stoic tenet – than human beings do, since the latter are often lead by vice rather than virtue. Arguably, Plutarch's works on animal psychology were intended to provide a 'noble example' for his readers to follow (i.e. a παράδειγμα καλόν, to use the wording of De lat. viv. 1129B above). To this end, Plutarch employs a moralising strategy by which he compares the ethics of human beings to the behavior of animals, often preferring the habits and attitudes of animals over those of humans[214]. In a highly rhetorical and moralising vein, he emphasises the unnatural behaviour of human beings (their ἀνθρωπίνη κακία: cf. De am. prol. 493B), as opposed to the natural incorruptibility of animals. As noted, this view strongly relativises the Stoic concern of living κατὰ φύσιν, showing that animals live more in accord with nature than humans do.

It is commonly accepted among scholars that Plutarch's sympathy for animals actually arises from his deep sense of humanity and φιλανθρωπία[215].

---

[213] For further discussion and reading, see S.T. Newmyer, 2014. The study of zoological phenomena more often served an ethical goal in Antiquity. Cf., e.g., H. Cherniss and W.C. Helmbold, 1957, p. 322, n. a.

[214] See, e.g., De am. prol. 493A–495A and Gryllus 989Cff.

[215] See, e.g., A. Barigazzi, 1992, p. 300, F. Becchi, 2002, p. 170, E. Lelli, 2010, p. 849,

In *De soll. an.* 966B, for instance, Aristotimus, one of Plutarch's students, remarks that the intelligence of animals is measured by philosophers according to several human ethical categories, such as purposefulness, memory, emotions, care for their offspring, courage, sociability, continence and magnanimity[216]. However, it seems only reasonable that the reduction from animal to human nature had a deeper motivation for Plutarch, although this is easily overlooked. If it is true that Plutarch 'loved' animals almost as much as he 'loved' humans, then he 'loved' what was rational and virtuous in them and, by implication, could be considered divine[217].

This last aspect is important in light of the role the animal kingdom plays in Plutarch's view on religion. In relation to the cosmological connection between the divine and the animate realms in nature (and more precisely in the context of divination), Aristotimus in *De soll. an.* 975B rather hyperbolically claims that he can produce thousands of signs and portents manifested by the gods through creatures of land and air (ἀλλὰ δὴ μυρίων μυριάκις εἰπεῖν παρόντων, ἃ προδείκνυσιν ἡμῖν καὶ προσημαίνει τὰ πεζὰ καὶ πτηνὰ παρὰ τῶν θεῶν). Plutarch's association between a rational, divine principle and the animal kingdom is also found, for instance, in *De Is. et Os.* 382AB, where we read that animals that are held in honour in Egyptian cults are actually the mirrors of the divine[218]. They are a natural instrument or medium for the God who orders all things: we should not honour these animals in themselves, but the divine *through* them (οὐ ταῦτα τιμῶντας, ἀλλὰ διὰ τούτων τὸ θεῖον ὡς ἐναργεστέρων ἐσόπτρων καὶ φύσει γεγονότων, ὥστ' ὄργανον ἢ τέχνην δεῖ τοῦ πάντα κοσμοῦντος θεοῦ νομίζειν κτλ.). This mirror metaphor ties in closely with Plutarch's Platonic world view, where nature

---

S.T. Newmyer, 2009, p. 501, J. Mossman and F. Titchener, 2011, p. 273 ("It is no surprise to us that a humane, compassionate, tolerant, and wise human like Plutarch wrote several essays specifically about animals"), J. Bouffartigue, 2012, p. xxxvi.

[216] *De soll. an.* 966B: Καθόλου δ', ἐπεὶ δι' ὧν οἱ φιλόσοφοι δεικνύουσι τὸ μετέχειν λόγου τὰ ζῷα, προθέσεις εἰσὶ καὶ παρασκευαὶ καὶ μνῆμαι καὶ πάθη καὶ τέκνων ἐπιμέλειαι καὶ χάριτες εὖ παθόντων καὶ μνησικακίαι πρὸς τὸ λυπῆσαν, ἔτι δ' εὑρέσεις τῶν ἀναγκαίων, ἐμφάσεις ἀρετῆς, οἷον ἀνδρείας κοινωνίας ἐγκρατείας μεγαλοφροσύνης, κτλ. On animal anthropomorphism in *De sollertia animalium*, see J. Mossman and F. Titchener, 2011, pp. 280–282. Cf. also, e.g., *De am. prol.* 493B: οἱ φιλόσοφοι τῶν προβλημάτων ἔνια διὰ τὰς πρὸς ἀλλήλους διαφορὰς ἐπὶ τὴν τῶν ἀλόγων φύσιν ζῴων ὥσπερ ἀλλοδαπὴν πόλιν ἐκκαλοῦνται, καὶ τοῖς ἐκείνων πάθεσι καὶ ἤθεσιν ὡς ἀνεντεύκτοις καὶ ἀδεκάστοις ἐφιᾶσι τὴν κρίσιν. The best example of animal anthropomorphism is probably provided in *Bruta animalia ratione uti* (*Gryllus*), where we find a lively discussion between Odysseus and Gryllus, the speaking boar (who has the last and most authoritative word in the discussion).

[217] See M. Meeusen, 2013c. See also L. Van der Stockt, 2005, p. 19 (who sees animals as "part of a world in which god, man and animals take care of each other"; *pace* S.T. Newmyer, 2009, p. 501). For further discussion see also S.T. Newmyer, 2006, pp. 17 ff.

[218] See R. Hirsch-Luipold, 2002, pp. 211–222 and J. Boulogne, 2005b.

is considered an inferior reflection of the intelligible realm[219]. Again in *Sept. sap. conv.* 163EF, Anacharsis shares his belief that God, insofar that he governs the inanimate world, uses the animate world to carry out his goals[220]:

δεινὸν γάρ, εἶπεν, εἰ πῦρ μὲν ὄργανόν ἐστι θεοῦ καὶ πνεῦμα καὶ ὕδωρ καὶ νέφη καὶ ὄμβροι, δι' ὧν πολλὰ μὲν σῴζει τε καὶ τρέφει, πολλὰ δ' ἀπόλλυσι καὶ ἀναιρεῖ, ζῴοις δὲ χρῆται πρὸς οὐδὲν ἁπλῶς οὐδέπω τῶν ὑπ' αὐτοῦ γιγνομένων. ἀλλὰ μᾶλλον εἰκὸς ἐξηρτημένα τῆς τοῦ θεοῦ δυνάμεως ὑπουργεῖν, καὶ συμπαθεῖν ταῖς τοῦ θεοῦ κινήσεσιν ἢ Σκύθαις τόξα λύραι δ' Ἕλλησι καὶ αὐλοὶ συμπαθοῦσιν.

For it is a dreadful mistake to assume that, on the one hand, fire is God's instrument, and breath and water also, and clouds and rain, by means of which He preserves and fosters many a thing, and ruins and destroys many another, but that, on the other hand, He never as yet makes any use whatever of living creatures to accomplish any one of His purposes. Nay, it is far more likely that the living, being dependent on God's power, serve Him and are responsive to His movements even more than bows are responsive to the Scythians or lyres and flutes to the Greeks.

If we return to *Quaestiones naturales* now, we already saw that Plutarch remains rather critical towards the rational abilities of animal beings (see n. 17). The link between natural phenomena (either animate or inanimate) and divine principles is generally absent. As outlined at the beginning of this chapter, Ziegler classified *Quaestiones naturales* alongside other specialised "naturwissenschaftliche Schriften". If ethics seems to figure into these writings, this is far less explicit or outspoken than is the case in the "tierpsychologische Schriften"[221].

In conclusion, ethical matters are never thematised in *Quaestiones naturales*[222]. They are only present insofar as they contribute to the central

---

[219] Cf. R. Hirsch-Luipold, 2002, p. 285: "Im sinnlichen wahrnehmbaren Kosmos hat die ewige, reine Gottheit sich selbst den Menschen als Leitschnur an die Hand gegeben. [...] Als Spiegel der göttlichen Wahrheit und Ordnung gewinnt der Kosmos Anteil am Göttlichen selbst. Auch untereinander haben die Phänomene der Welt diesen Verweischarakter, weil sie alle der einen Quelle entstammen."

[220] This does not imply, however, that there is no place at all for free will. Cf. also, e.g., *De Pyth. or.* 404B–405A.

[221] Cf. L. Senzasono, 2006, p. 33: "Ma mentre in opere come *De soll. an.* l'osservazione naturalistica è subordinate a una certa tesi che lo scrittore vuol dimostrare, nelle *Quaest. nat.* essa è fine a se stessa o, piú esattamente, è intesa a indagare le cause dei fenomeni in se stesse."

[222] I am loosely alluding here to the distinction between Plutarch's treatment of ethical topics as 'rhema' and as 'thema' made by L. Van Hoof, 2010, p. 39 (with n. 71). Ethics

focus on the natural explanations of the phenomena under discussion. Nevertheless, Plutarch is well aware of the fact that some of the problems treated in *Quaestiones naturales* can be reused in ethical discourses. So, to reformulate the initial problem: why does some kind of a sub-moralistic essentialism prevail in *Quaestiones naturales*? This question also relates to the sub-literary aspect of the work. Why does the aspect of physical aetiology fold back entirely on itself in order to be treated on its own terms, as the title of the collection clearly marks?

5. A 'generic' solution

I can briefly provide a solution to the problems of style and morality in *Quaestiones naturales*, which, as I will argue here, is probably dictated by the genre to which this text belongs. This means that a closer link should be drawn with the discourse of the Ps.-Aristotelian *Problems*, which served as a model for Plutarch's natural problems.

In regards to style, first of all, Ps.-Aristotle's model displays a very similar 'referential' discourse as found in *Quaestiones naturales*, and the same is true for Plutarch's other collections of *quaestiones* – with the exception of *Quaestiones convivales*. Fuhrmann, for instance, opposed "la forme rudimentaire des *Questions Naturelles, Romaines, Grecques*, et la froideur stéréotypée des recueils de cette espèce"[223] to the more literary and lively discourse of *Quaestiones convivales*. In the latter work, however, Plutarch is hybridising the genre of problems with that of the symposium, which explains the literary and dramatic elaboration of the content there [see 2.2.1.]. The genre of problems does display, at least, a minimal degree of formal elaboration (as will be further substantiated later on [see 2.3.2.]). Therefore, Senzasono is right that Fuhrmann "è giusto in generale, ma è inesatto parlare di "forma rudimentale": in realtà lo stile delle *Quaest. nat.* è elaborata nel senso dell'essenzialità denotativa."[224] Notably, Plutarch's

---

treated as 'rhema' is typical of Plutarch's works on practical ethics, where the author is concerned with a specific moral *practice*, whereas ethics is treated as a 'thema' in the technical treatises on philosophy. In these works, the consideration of ethical principles is related to the instruction and study of philosophical theory itself (i.e. the λόγοι rather than the ἄσκησις). In *Quaestiones naturales*, then, there is a certain impulse towards ethics as 'rhema', for instance, in the case of *Q.N.* 36, (viz. in the words *adulteria* and *perfidia*), but it is not further developed when compared to the parallel account in *Coni. praec.* 144D. Notably, the concept of *iniuria*, mentioned in *Q.N.* 33 and 37, has no ethical implications but refers to the physical damage or 'injury' done to waters by becoming stagnant (*Q.N.* 33), or to dogs by stones flung at them (*Q.N.* 37).

[223] F. Fuhrmann, 1972, p. xix, n. 2.

[224] L. Senzasono, 2006, p. 10. Senzasono adds (pp. 10–11 and p. 46, n. 70) that there is, nevertheless, a difference in style between *Quaestiones naturales*, on the one hand, and *Quaestiones Romanae* and *Graecae*, on the other (on the assumption that the style of

incorporation of literary (c.q. poetical and mythograpical) material in his natural problems lies significantly higher than in Ps.-Aristotle's *Problems* [see 4.1.2.2.–3.]. A plausible explanation is provided by the fact that Plutarch's natural problems, as opposed to those of Ps.-Aristotle, are part of the author's broader πολυμάθεια project. Even when dealing with matters pertaining to natural science, Plutarch aims to display and combine his all-round knowledge of several branches of Greek literature and learning, thus showing that he was a true πεπαιδευμένος.

Second, in regards to the lack of morality in *Quaestiones naturales*, a link can again be drawn with Ps.-Aristotle's model, where overt moralising dynamics are also absent. A few sections in Ps.-Aristotle's *Problems*, however, are generally related to the field of moral philosophy, viz. Books 27–30, which concern the topics of fear, courage, moderation, justice and intellectual virtue. The questions raised in these Books are never treated in a

---

the two latter collections is really the same). He bases his conviction on the idea that the discourse of *Quaestiones Romanae* and *Graecae* is more extensive, fluid and narrative than that of *Quaestiones naturales*, which he explains on the basis of the historical interest of the first two, whereas the essential and descriptive discourse of *Quaestiones naturales* would be conditioned by its natural scientific interests (but see above [1.2.2.]): "l'adesione ai fenomeni fisici condiziona lo stile essenziale e sobriamente descrittivo del trattato naturalistico, mentre lo stile dei due opuscoli d'argomento storico è talvolta piú disposto a un dettato esteso, fluido e sobriamente narrativo." As we will see later on, however, Plutarch reserves some space for natural science in the latter two collections as well, and also – the other way around – a cultural-antiquarian type of discourse in the former [see 2.4.2.]. In light of Plutarch's interest in natural scientific matters in *Quaestiones Romanae* and *Graecae*, C. Darbo-Peschanski, 1998, p. 27 has made the following conclusion: "Plutarque penserait donc une "cosmologie de l' histoire" comme prolongement et achèvement, sur le mode du redoublement analogique, de la cosmologie physique. [...] La conséquence en est qu' on peut s' interroger sur les causes (αἴτια) de ce que font et de ce que produisent historiquement les hommes comme on s' interroge sur les causes des phénomènes physiques." One should not, therefore, underestimate the 'historical' character of *Quaestiones naturales*, at least in light of the ancient concept of ἱστορία, 'inquiry' (indeed, Plutarch draws a great deal of his material from natural history, and also uses concepts related to the notion of ἱστορία passim: see *Q.N.* 1, 911E, 7, 914A, 9, 914B, 26, 918D, 34: *vulgo fertur*). It follows that at least from an ancient perspective, *Quaestiones naturales* is not more or less 'historical' than *Quaestiones Romanae* and *Graecae*, since the concept of ἱστορία includes both cultural-antiquarian and natural scientific types of inquiry (cf. LSJ, s.v.). See, e.g., Plu., *Cons. ad Apoll.* 119D (ἱστορία Ἑλληνική, Ῥωμαϊκή) and Pl., *Phd.* 96a (περὶ φύσεως ἱστορία). On natural scientific ἱστορία in Plutarch, see M. Battegazore, 1992, pp. 19–35, P. Donini, 1984, pp. 374 and J. Bouffartigue, 2012, pp. x–xii. See also L. Van der Stockt, 1987, p. 289 (for *Quaestiones Romanae*) and P. Payen, 2013 and 2014 (on Plutarch's antiquarianism, esp. in *Quaestiones Romanae* and *Graecae*). On Plutarch's general use of ἱστορία (esp. concerning historical narrative), see T. Duff, 1999a, pp. 17–21. For the relation between Herodotus' *Historiae* and natural history, see also R. Thomas, 2000, pp. 135–167 (esp. pp. 164–166 with nn. 93 and 99 for further literature).

strictly moralising way, though. In fact, they are often solved in a physical, if not in a more generally technical or specialised, fashion[225]. Most notably, the symptoms of fear and courage are generally explained in Book 27 in relation to a person's physical properties, esp. bodily coldness and heat[226]. The same explanatory scheme is found in *Quaestiones naturales*. For instance, in *Q.N.* 20, Plutarch explains the courageous character of the boar *vis-à-vis* the cowardly character of deer in terms of the hot (fiery) *vis-à-vis* the cold (watery) constitution of their bodies respectively. Similarly, in *Q.N.* 11, the emotion of fear is implicitly connected with coldness, since when people imagine some danger at sea, they tremble and shiver.

Now that we have analysed the close relationship between Plutarch's *Quaestiones naturales* and Ps.-Aristotle's *Problems*, it still remains to be seen, precisely why Plutarch opted for the genre of natural problems and not for another genre of writing. This, however, will be treated in the following chapters.

6. Conclusion and new questions

I conclude that it would prove of a doubtful insight in Plutarch's natural scientific project to claim that his study of natural phenomena concerns the examination of their physical causes alone. The fact, however, that it are precisely the αἰτίαι φυσικαί that receive central focus in *Quaestiones naturales* (as is marked by the title) does eventually define the collection's scientific outlook (as modelled after Ps.-Aristotle's *Problems*). However, Plutarch does incorporate the same or similar physical material in his other works, where he becomes the flamboyant moralist and litterateur we know so well. This raises questions about the position of *Quaestiones naturales* in the *corpus Plutarcheum* more generally. What kind of text is this, then, and which purpose did it serve? How does it relate to Plutarch's personal notes (ὑπομνήματα) and to his discussion of natural problems elsewhere (esp. in *Quaestiones convivales*)? These and related questions will be subject to debate in the following chapter.

---

[225] Book 29, most notably, deals with legal justice for the most part, which H. Flashar, 1962, p. 317 aptly describes as a "Fachwissenschaft" ("Wollte man den Titel "Problemata Physica" mit der Überlegung rechtfertigen, daß Ethik nicht an sich unbedingt den Bereich der Physis überschreitet, so ist zu bedenken, daß in den Probl. von Ethik im engeren Sinne kaum die Rede ist. Fast überall herrscht nämlich eine praktische Tendenz vor, die dazu führt, daß sich die behandelten Phänomene in den Zusammenhang einer Fachwissenschaft einordnen lassen."). For further reading and discussion, see R. Mayhew, 2011b. For the place of Books 27–30 in the *Problems* more generally (to be identified perhaps with the Aristotelian Προβλήματα ἐγκύκλια), see M. Meeusen, forthcoming e.

[226] See L.M. Castelli, 2011. There is a precedent for the mind-body relationship already in Plato, who writes that the diseases of the soul are due to the condition (ἕξιν) of the body (*Tim.* 86b).

2
The position of *Quaestiones naturales* in the *corpus Plutarcheum*

This chapter aims to fine-tune the position of *Quaestiones naturales* in Plutarch's oeuvre by studying which role the genre of natural problems more generally plays throughout his writings. I will argue that in *Quaestiones naturales* Plutarch creates an independent problematic framework for recording his aetiological speculations about particular natural phenomena in an autonomous way (i.e. free from other concerns such as stylistic embellishment and moralising dynamics [see 1.2.]). In an attempt to reject the traditional view that Plutarch's collections of *quaestiones* more generally were useful only as sets of notes (ὑπομνήματα) Plutarch drafted for personal use, I will try to demonstrate that they rather provide a medium for thematically sorting out, amassing and discussing all kinds of issues that struck Plutarch as being particularly problematic.

In examining how Plutarch incorporates the genre of natural problems, including specific *Quaestiones naturales* material, in his other writings, I will focus especially on the discursive effects procured by this technique. How does a parallel passage or a scientific digression function in the narrative or argumentative line of a particular text? Which methods are used to incorporate it in that specific discourse? And what are the similarities and differences with *Quaestiones naturales*? Once these and related questions have been clarified, I will zoom in on the compository relationship between *Quaestiones naturales* and *Quaestiones convivales* more in specific[1]. The results of these inquiries will form the basis for further research about the intellectual purpose of *Quaestiones naturales*. To this end, I will consider the likelihood of the work's publication, which I will further develop in chapter three.

2.1. *Scientific traits in the* corpus Plutarcheum

As an intellectual, a teacher, and a true paragon of ancient learning, Plutarch collected and reused any form of knowledge that attracted his personal attention. Any bit of information that interested him was jotted down in the form of personal notes (ὑπομνήματα) [see 2.3.2.]. These would surely serve him well one day, as is, indeed, suggested by the numerous parallel passages and clusters throughout his writings. In his reproduction

---

[1] A problem that, according to F.H. Sandbach, 1965, p. 138, however, "hardly admits of an answer".

of these materials, Plutarch often tried to iron out what, in his opinion, were the most problematic difficulties. He, therefore, sought explanations for many kinds of topics (e.g., cultural, antiquarian, literary, linguistic, philosophical, scientific etc.), but he also wanted his readers to acquaint themselves with these explanations and, thus, share in the richness of manifold learning (πολυμάθεια).

Even outside of his collections of *quaestiones* Plutarch often confronts his readers with discussions of numerous kinds of problems[2]. However, Plutarch's treatment of these topics is not always very closely related to the central narrative or argumentative line of the treatises at hand. On the contrary, they often divert the reader from what is really at issue. This does not imply, however, that Plutarch was just 'massing together useless material of research' in these passages, as he states himself (cf. *Nic.* 1, 5: οὐ τὴν ἄχρηστον ἀθροίζων ἱστορίαν). Yet, it remains to be seen what their use really was, then. In the *Vitae*, he refers to these kinds of digressions with the notion of παρεκβάσεις[3]; these are, in fact, abundant throughout his entire oeuvre, and are not restricted to the *Vitae* only [see 2.1.3.]. The length of these digressions varies from a single sentence to an entire paragraph, and they often display an 'aetiological climate'[4] that strongly reminds the reader of the genre of problems and its typical organisation of knowledge [see 1.1.4.].

The section at hand will mainly be concerned with Plutarch's natural scientific digressions, but I will also deal with his use of natural scientific *exempla* (i.e. metaphorical reinterpretations of natural phenomena). We have already briefly dealt with such *exempla* in the previous chapter (c.q. with their moralising intentions [see 1.2.4.]), but it will become clear here that this use of imagery was an intergral aspect of Plutarch's literary style.

1. Intellectual and literary interest of natural phenomena

As the examples below will show, Plutarch's scientific digressions concern very similar natural scientific topics as treated in *Quaestiones naturales*. By incorporating such problems in the narrative or argumentative line of several of his writings Plutarch clearly intended to promote his own research to the outer world and, thus, to demonstrate his own argumentative talent. This is not necessarily incompatible with the idea that by sharing this knowledge with his readers he intended to offer some kind

---

[2] Cf. R. Hirzel, 1912, p. 40: "Überall wird den Problemen nachgespürt, die sich in Wissenschaft und Leben darbieten, nicht bloß alten Problemen, sondern auch neuen, die der Augenblick, auch wohl nur der gesellige Scherz erfindet."

[3] See *Alex.* 35, 16 (regarding the scientific digression on naphtha; see further) and *Dion* 21, 9 (regarding the historical digression on Theste).

[4] This concept was coined by J.-M. Pailler, 1998, p. 80, who used it in light of the parallel material between the *Quaestiones Romanae* and the *Vitae*.

of learned diversion (as fellow πεπαιδευμένοι, they would surely appreciate such digressions). Yet, as we will see, apart from having an obvious intellectual interest, these scientific digressions often also serve a specific literary purpose in the organisation and development of Plutarch's text.

In order to illustrate this, I refer to the particularly intriguing digressions in the introductions to *De Pythiae oraculis* and *De defectu oraculorum*, which by their natural scientific interests deserve specific consideration here. The introduction to *De Pythiae oraculis* (395A–396C) is concerned with the patina of the bronze statues of Lysander and his admirals located near the entrance of the holy precinct in Delphi (cf. also *Lys.* 18, 1). In connection with this topic, the interlocutors discuss the problem of why oil covers bronze with rust[5]. In the introduction to *De defectu oraculorum* (410B–411D), on the other hand, the interlocutors deal with the ever-burning lamp at the shrine of Ammon in Egypt, which, so the local priests report, consumes less and less oil each year: does this imply that the years grow shorter and shorter[6]? I will not deal with these problems in detail here. What matters is that owing to the fact that these discussions are located at the very beginning of the treatises and are not emphatically connected with the main topic at issue (viz. that the oracles at Delphi are no longer given in verse and that many oracles in Greece have passed into disuse respectively)[7], I believe that this technique provides some kind of an introductory framework, some kind of a *Natureingang* perhaps, for the narrative (I will come back to this)[8].

The discursive value of these introductions lies in the fact that they at least for a while postpone the central issue that the author intends to treat in these writings by first providing a discussion of a completely different matter. As such, we can rightly speak of a literary and intellectual 'appetiser' in view of the intellectual σχολή that is required for the author to produce and for the reader to consume this kind of literature. This σχολή necessitates that the author and reader can take time to divert their attention a bit before getting to business. Starting off immediately with the central argument would not suit the decorum of this kind of literature, nor would it improve the literary verisimilitude that the author is trying to attain[9].

---

[5] Cf. Ps.-Arist./Alex. Aphr., *Suppl. probl.* 3, 17. For further detail, see J. Jouanna, 1975 and W.A. Franke and M. Mircea, 2005 (with an attempt towards a modern scientific explanation).

[6] For further discussion, see J. Hani, 1976, pp. 267–268 and E.G. Simonetti, forthcoming.

[7] See E. Valgiglio, 1992, p. 19 and A. Rescigno, 1995, p. 8, n. 1. Cf. also C. Kahle, 1912, pp. 93–95 and 103–104.

[8] Flacelière aptly speaks of a "lever de rideau" (in R. Flacelière, J. Irigoin, J. Sirinelli and A. Philippon, 1987, pp. ccxx and ccxxii).

[9] Cf. E. Valgiglio, 1992, p. 22: "abbiamo qui la scenografia, lo sfondo dell'azione

We are dealing here with dramatic dialogues between real-life persons (rather than with strictly systematic treatises), in the manner of Plutarch's *Quaestiones convivales* and generally inspired by Plato's dialogues.

Notably, the discussion about the bronze statues in *De Pyth. or.* 395A–396C begins and ends with a reference to the prearranged sight-seeing programme of the guides at Delphi. This does not seem to be of great interest to the interlocutors, who ask the guides to cut short their lengthy stories and readings of every single inscription (395A: Ἐπέραινον οἱ περιηγηταὶ τὰ συντεταγμένα μηδὲν ἡμῶν φροντίσαντες δεηθέντων ἐπιτεμεῖν τὰς ῥήσεις καὶ τὰ πολλὰ τῶν ἐπιγραμμάτων). Since no heed is given to their inquiries, the interlocutors start to discuss matters of greater personal interest (viz. the patina of the bronze statues of Lysander and his admirals), while the guides are left twiddling their thumbs. When afterwards there is a short moment of silence, the guides pursue their routine speeches (396C: Ἐκ τούτου γενομένης σιωπῆς πάλιν οἱ περιηγηταὶ προεχειρίζοντο τὰς ῥήσεις). Once they mention a certain oracle given in verse, the interlocutors interrupt again and start discussing the common quality of the verse in which oracles are delivered in their days – which is the main topic of the dialogue[10].

By the repeated interruption of the Delphic guides, Plutarch seems to suggest from the start that the treatise at hand will not just be a systematic, prearranged tour through the precinct of Delphi, but rather a more improvisatory, digressive promenade that leads the reader off-track – that is, off the trodden paths of the subject treated. As such, the purpose of the treatise is to look for new, original ways of approaching the very essence of the prophetic art, gradually unveiling the philosophical-religious power of the Delphic precinct. Indeed, the introductory scene of *De Pythiae oraculis* is literally set at the entrance of the holy precinct, where the very symbolic walk towards the Apollo-temple on the hill begins (cf. *De Pyth. or.* 394E, 402BC). The fact that a starting-point for the discussion is found in the genre of natural problems may be significant for the 'pre-philosophical' interest of this type of debate[11] [see 3.2.].

---

drammatica". A link can be drawn, for instance, with the introduction in *De gen. Socr.* 575B–588B, where Plutarch sets the scene of the discussion of Socrates' δαιμόνιον against the historical background of the Theban conspiracy against the Spartan tyrants.

[10] The guides cut a foolish figure again in *De Pyth. or.* 397DE, 400DE, 400F–401A, 401E. Cf. also *De E* 386B and the reactions to the accounts of the guide (ὁ περιηγητής) Praxiteles, in *Quaest. conv.* 675EF and esp. 723E–724D.

[11] The aetiology – which ends in aporetic silence – seems to suggest that the air at the precinct has unusual properties: it is at the same time dense and compact and tenuous and keen (*De Pyth. or.* 396AC). This may highlight the peculiar character of the precinct's natural environment (cf. *Q.N.* 23, 917EF) and may perhaps contain an implicit allusion to the prophetic exhalations released at the Delphic shrine (a topic treated at the very end of *De def. or.* 437C–438D).

The same can probably be presumed, then, for the problem of the lamp at the shrine of Ammon, which Plutarch treats in *De def. or.* 410B–411D in a very similar religious framework (viz. at the Delphic precinct, a short time before the Pythian games). Notably, this problem is put forward to the group by the travelling Cleombrotus of Sparta, 'a man fond of spectacles and learning' (ἀνὴρ φιλοθεάμων καὶ φιλομαθής), who, so Plutarch notes, 'was getting together a history to serve as a basis for a philosophy that had as its end and aim theology, as he himself named it' (συνῆγεν ἱστορίαν οἷον ὕλην φιλοσοφίας θεολογίαν ὥσπερ αὐτὸς ἐκάλει τέλος ἐχούσης). The fact that Plutarch mentions Cleombrotus' literary project seems very significant, since in a very similar way, natural history, and more precisely the discussion of a specific natural problem, serves as a preamble to the philosophical-religious discussion also in his own dialogue. This seems very significant in light of Plutarch's "effort to reconcile science and religion" in *De defectu oraculorum*[12].

Another suggestive means to provide some literary-intellectual diversion to the reader is found in Plutarch's frequent reinterpretation of natural phenomena in an expressive, rhetorical way as metaphorical *exempla*. In *Q.N.* 32, for example, Plutarch deals with the natural phenomenon of palm wood that rises against weight imposed upon it. This problem is explained in a purely physical way in *Q.N.* 32, whereas in a parallel passage in *Quaest. conv.* 724E, it is used as an *exemplum* for the athlete's well trained body and mind (notably, the palm tree's natural resilience remained a popular topic well beyond Antiquity in the form of a moral 'emblem': see the commentary *ad loc.*). In *Maxime cum principibus* 776F–777A, to give another example, Plutarch compares the teachings of philosophy with the natural powers of the sea-holly (*eryngium*):

> τὸ ἠρύγγιον τὸ βοτάνιον λέγουσι μιᾶς αἰγὸς εἰς τὸ στόμα λαβούσης, αὐτήν τε πρώτην ἐκείνην καὶ τὸ λοιπὸν αἰπόλιον ἵστασθαι, μέχρι ἂν ὁ αἰπόλος ἐξέλῃ προσελθών· τοιαύτην ἔχουσιν αἱ ἀπόρροιαι τῆς δυνάμεως ὀξύτητα, πυρὸς δίκην ἐπινεμομένην τὰ γειτνιῶντα καὶ κατασκιδναμένην. καὶ μὴν ὁ τοῦ φιλοσόφου λόγος, ἐὰν μὲν ἰδιώτην ἕνα λάβῃ, χαίροντα ἀπραγμοσύνῃ καὶ περιγράφοντα αὐτὸν ὡς κέντρῳ καὶ διαστήματι γεωμετρικῷ ταῖς περὶ τὸ σῶμα χρείαις, οὐ διαδίδωσιν εἰς ἑτέρους, ἀλλ' ἐν ἑνὶ ποιήσας ἐκείνῳ γαλήνην καὶ ἡσυχίαν ἀπεμαράνθη καὶ συνεξέλιπεν. ἂν δ' ἄρχοντος ἀνδρὸς καὶ πολιτικοῦ καὶ πρακτικοῦ καθάψηται καὶ τοῦτον ἀναπλήσῃ καλοκαγαθίας, πολλοὺς δι' ἑνὸς ὠφέλησεν, ὡς Ἀναξαγόρας Περικλεῖ συγγενόμενος καὶ Πλάτων Δίωνι καὶ Πυθαγόρας τοῖς πρωτεύουσιν Ἰταλιωτῶν. Κάτων δ' αὐτὸς ἔπλευσεν ἀπὸ στρατιᾶς ἐπ' Ἀθηνόδωρον κτλ.

---

[12] F.C. Babbitt, 1936b p. 349. For an excellent study of the passage in light of Plutarch's main argument, see E.G. Simonetti, forthcoming.

Of the plant *eryngium* they say that if one goat take it in its mouth, first that goat itself and then the entire herd stands still until the herdsman comes and takes the plant out, such pungency, like a fire which spreads over everything near it and scatters itself abroad, is possessed by the emanations of its potency. Certainly the teachings of the philosopher, if they take hold of one person in private station who enjoys abstention from affairs and circumscribes himself by his bodily comforts, as by a circle drawn with geometrical compasses, do not spread out to others, but merely create calmness and quiet in that one man, then dry up and disappear. But if these teachings take possession of a ruler, a statesman, and a man of action and fill him with love of honour, through one he benefits many, as Anaxagoras did by associating with Pericles, Plato with Dion, and Pythagoras with the chief men of the Italiote Greeks.

The popular belief about the goat and the sea-holly (which recurs in *De sera num.* 558E and *Quaest. conv.* 700D and probably originates from Arist., *HA* 610b29) clearly foregrounds the natural scientific subtext in this passage, and the same is true for the image of the spreading fire and the drawing of geometrical circles (as well as Plutarch's use of such technical terms as ἀπορροιαί and δύναμις). This type of natural scientific imagery is meant to serve a literary purpose in rendering Plutarch's argument more palatable to the reader, pointing out, in this case, that not only the philosopher but also the political ruler and, through him, the people can in fact benefit from philosophical teachings.

As already noted in the previous chapter, natural phenomena are often subject to a moralising type of exemplification in Plutarch's writings [see 1.2.4.]. A good example can be seen in *Phoc.* 3, 1–3, where Plutarch compares Cato the Younger's old-fashioned virtue in times of moral decay with fruits that grow out of season: these fruits are admired but not used. In the previous paragraph (*Phoc.* 2, 6–9), the non-rectilinear motion of the sun, which is said to provide an ideal temperature for all things on earth, is reinterpreted in light of a political precept, in order to suggest that a moderate government is best. This aspect of cosmic balance and harmony is considered to be in line with how the Platonic God rules the universe – that is, by means of reason and persuasion rather than by necessity (cf. *Tim.* 48a)[13]:

ὥσπερ οὖν τὸν ἥλιον οἱ μαθηματικοὶ λέγουσι μήτε τὴν αὐτὴν τῷ οὐρανῷ φερόμενον φοράν, μήτ' ἄντικρυς ἐναντίαν καὶ ἀντιβατικήν, ἀλλὰ λοξῷ καὶ παρεγκεκλιμένῳ πορείας σχήματι χρώμενον, ὑγρὰν καὶ εὐκαμπῆ καὶ παρελιττομένην ἕλικα ποιεῖν, ᾗ σῴζεται πάντα καὶ λαμβάνει τὴν ἀρίστην κρᾶσιν, οὕτως ἄρα τῆς πολιτείας ὁ μὲν ὄρθιος ἄγαν καὶ πρὸς ἅπαντα τοῖς δημοτικοῖς ἀντιβαίνων

---

[13] For further discussion of this topic, see L. Van der Stockt, 2012.

τόνος ἀπηνὴς καὶ σκληρός, ὥσπερ αὖ πάλιν ἐπισφαλὲς καὶ κάταντες τὸ συνεφελκόμενον οἷς ἁμαρτάνουσιν οἱ πολλοὶ καὶ συνεπιρρέπον· ἡ δ᾽ ἀνθυπείκουσα πειθομένοις καὶ διδοῦσα τὸ πρὸς χάριν, εἶτ᾽ ἀπαιτοῦσα τὸ συμφέρον ἐπιστασία καὶ κυβέρνησις ἀνθρώπων, πολλὰ πράως καὶ χρησίμως ὑπουργούντων, εἰ μὴ πάντα δεσποτικῶς καὶ βιαίως ἄγοιντο, σωτήριος, ἐργώδης δὲ καὶ χαλεπὴ καὶ τὸ σεμνὸν ἔχουσα τῷ ἐπιεικεῖ δύσμεικτον· ἐὰν δὲ μειχθῇ, τοῦτ᾽ ἔστιν ἡ πάντων μὲν ῥυθμῶν, πασῶν δ᾽ ἁρμονιῶν ἐμμελεστάτη καὶ μουσικωτάτη κρᾶσις, ᾗ καὶ τὸν κόσμον ὁ θεὸς λέγεται διοικεῖν, οὐ βιαζόμενος, ἀλλὰ πειθοῖ καὶ λόγῳ παράγων τὴν ἀνάγκην.

Now, the sun, as mathematicians tell us, has neither the same motion as the heavens, nor one that is directly opposite and contrary, but takes a slanting course with a slight inclination, and describes a winding spiral of soft and gentle curves, thus preserving all things and giving them the best temperature. And so in the administration of a city, the course which is too straight, and opposed in all things to the popular desires, is harsh and cruel, just as, on the other hand, it is highly dangerous to tolerate or yield perforce to the mistakes of the populace. But that wise guidance and government of men which yields to them in return for their obedience and grants them what will please them, and then demands from them in payment what will advantage the state, – and men will give docile and profitable service in many ways, provided they are not treated despotically and harshly all the time, – conduces to safety, although it is laborious and difficult and must have that mixture of austerity and reasonableness which is so hard to attain. But if the mixture be attained, that is the most concordant and musical blending of all rhythms and all harmonies; and this is the way, we are told in which God regulates the universe, not using compulsion, but making persuasion and reason introduce that which must be.

A myriad of such examples could be adduced to illustrate Plutarch's frequent reinterpretation of natural phenomena as rhetorical *exempla*[14]. What matters for us here is that this technique of comparing and unifying human affairs with natural phenomena is intelligent and often renders Plutarch's personal comments and criticism more enjoyable to read. As such, by using these images, Plutarch offers both a literary and an instructive, if not more philosophical and contemplative pleasure to his readers (as in the last case).

In *Quaestiones naturales*, by contrast, Plutarch's primary concern is to provide natural explanations for problematic phenomena. Therefore, he generally avoids referring to their metaphorical implications. Never-

---

[14] For more on Plutarch's natural metaphors, see A.I. Dronkers, 1892, pp. 102–142, F. Fuhrmann, 1964 (passim) and J. García López, 1991.

theless, several of these phenomena are exploited as rhetorical *exempla* elsewhere, viz. as images of fear (cf. *Q.N.* 11 ~ *De tranq. an.* 475F–476A), the lack of steadfast character (cf. *Q.N.* 19 ~ *De ad. et am.* 51D–53D, *De am. mult.* 96F–97A, *Alc.* 23, 4–5), democratic elections (cf. *Q.N.* 26, 918D ~ *Praec. ger. reip.* 801A), athletic strength (cf. *Q.N.* 32 ~ *Quaest. conv.* 724E), marital infidelity (cf. *Q.N.* 36 ~ *Coni. praec.* 144D), hostility (cf. *Q.N.* 37 ~ *De gar.* 514D), the joining of opposites (cf. *Q.N.* 41 ~ *De cap. ex inim.* 92B). Parallel passages of this sort nicely illustrate how Plutarch constantly refashions his scientific ideas in different literary contexts. The sheer amount of these parallels demonstrates that such physical themes play an important unifying role throughout the Chaeronean's oeuvre. As we will see in what follows, they are also very relevant for Plutarch's general writing method.

2. Cluster analysis in *Quaestiones naturales*

The most intriguing case in terms of parallel passages between *Quaestiones naturales* and Plutarch's other works is *Q.N.* 19, which concerns the octopus' change of colour. I have already partly discussed this problem in the previous chapter [see 1.2.4.], but will return to it here in light of Plutarch's writing technique.

What is important is that several key elements relating to Plutarch's discussion of the octopus' metachrosis as found in *Q.N.* 19 are repeated in a number of parallel passages, where the aspect of physical aetiology is not as central. The natural phenomenon is compared with the adaptable character of flatterers in *De ad. et am.* 51D–53D and *De am. mult.* 96F–97A, and with the opportunistic politics of Alcibiades in *Alc.* 23, 4–5. There is also a parallel concerning the animal's psychology in *De soll. an.* 978EF. Considering the topic's frequent recurrence, we can speak of a genuine 'cluster' of parallel passages here, which Van der Stockt has defined (in light of his method of 'cluster analysis') as "a repeated and structured collection of heterogeneous materials"[15]. Depending on the number of textual parallels, Van der Stockt makes a distinction between 'parallel passages' (two parallels) and 'clusters' (three or more parallels). Such parallels and clusters are often identified by a set of recurrent quotations, anecdotes, similes, concepts etc. In the case of the octopus cluster, these are found – as schematised below – in 1) the quotations from Pindar and Theognis, 2) a more ethical *vis-à-vis* more physical orientation, and 3) the reference to specific physical concepts (viz. emanations and breath)[16].

---

[15] L. Van der Stockt, 1999a, p. 580.
[16] For a separate analysis of this octopus cluster, see M. Meeusen, 2012a, pp. 247–250. There are several other such clusters with parallels in our collection, but it would bring us much too far to discuss each and every one of them in detail here. I will briefly discuss the

Octopus cluster	Quotations		Orientation		Concepts	
	Pind.	Theo.	Eth.	Phys.	Eman.	Breath
Alc. 23, 4-5			X	x		
De ad. et am. 51Dff.			X	x		
De am. mult. 96Ff.		X	X	x	X	
Q.N. 19, 916BF	X	X	X̶	X	X	X
De soll. an. 978EF	X	X	x	X		X

Key: X = clearly present; x = clearly present, but less strongly articulated than in *Q.N.* 19; X̶ = clearly present, but ruled out by Plutarch.

One of the most important advantages of Van der Stockt's method of cluster-analysis is that it, besides from being highly efficient and orderly, offers detailed insight into Plutarch's argumentative tactics and his writing and rewriting process, with a particular interest for his use of personal notes (ὑπομνήματα)[17] (I will deal with the specific nature of Plutarch's notes below [see 2.3.2.]). In the octopus cluster, then, it is not unlikely that Plutarch reuses and remodels the same material, drawing from (one or more of) his personal notes on zoological topics, and adapting this hypomnematic material to various contexts. Van der Stockt is well aware of the possibility, however, that one writing can be inspired by another, or that the parallelism in subject matter derives from mental, rather than textual, processes[18]. Indeed, also in the case of *Q.N.* 19 one cannot simply dismiss that the hypomnematic material at some point became an *idée fixe* in Plutarch's mind, such that the textual intermediation of a ὑπόμνημα in each and every case must eventually remain hypothetical.

Another important aspect regarding the parallel passages between *Quaestiones naturales* and Plutarch's other writings is that there are considerably few such parallels in the *Vitae*[19]. This is not at all the case

---

contents of these clusters in the commentary *ad loc.* (with specific attention also for the parallels in other authors). Cf. *Q.N.* 1 ~ *Quaest. conv.* 627AD, 695E; *Q.N.* 2-4 ~ *Quaest. conv.* 661BC, 663F, 664D-665C, 666A, 684E, 685BD; *Q.N.* 12 and 39 ~ *De prim. frig.* 950B; *Q.N.* 24 ~ *Quaest. conv.* 657F-659D, *De facie* 940A (see further); *Q.N.* 26 ~ *De soll. an.* 974BD, *Gryllus* 991E. The second cluster was analysed separately by L. Van der Stockt, 2011 (with a schematic representation on p. 451).

[17] For concrete applications of this method, see, e.g., L. Van der Stockt, 1999a, 1999b, 2004, 2011, B. Van Meirvenne, 1999, 2001, S.A. Xenophontos, 2012 and M. Meeusen, 2012a.

[18] Therefore, each case should be considered individually. See L. Van der Stockt, 1999a, p. 597 and 2004, p. 335, n. 10. See also S.A. Xenophontos, 2012, p. 87 and M. Meeusen, 2012a.

[19] Cf. *Q.N.* 19 ~ *Alc.* 23, 4-5; *Q.N.* 11 and 19 ~ *Arat.* 29, 6; *Q.N.* 11 ~ *Demetr.* 38, 4 and *Per.* 33, 5. These parallels are also rather weak in comparison to those in the *Moralia*.

with the *Moralia*, where we find an abundance of parallel material. The same material can be found especially in the more specialised natural scientific works (*De primo frigido*, *De facie*, *De sollertia animalium* and esp. *Quaestiones convivales*), but parallel passages are also present in Plutarch's non-scientific works. These parallels are not always entirely identical in form or content but often involve specific textual adaptations and rearrangements to suit the new context. A detailed analysis of each of these parallel accounts cannot be achieved here[20]. Instead, I will enumerate the most important procedures discernible in their incorporation.

First of all, a number of rather loose allusions to and weak reformulations of the same *Quaestiones naturales* material can be found throughout the *Moralia*, where the argumentation often relies on generally accepted scientific concepts and theories (e.g., the idea that salty seawater is naturally hot). The parallelism, however, is far more prominent in other cases, as can be seen, for instance, in the cluster of parallels passages concerning the production of dew by the moon (discussed in *Q.N.* 24, *Quaest. conv.* 659B and *De facie* 940A). In this cluster Plutarch repeats the same quotation from Alcman, where Dew is called the daughter of Zeus and Moon (43 Diehl: Διὸς θυγάτηρ Ἔρσα τρέφει καὶ Σελάνας δίας). Yet, there are also subtle differences in the arguments at hand. The most significant difference is that in *Q.N.* 24, Plutarch refers to the mechanism of attraction (ὁλκή) in explaining how dew comes to be, while in *Quaest. conv.* 659B and *De facie* 940A, he refers to the process of change (μεταβολή). This can be explained in light of the different aetiological contexts. In the latter two passages, Plutarch argues that the moon has a liquefying effect, and that the air (Ζεύς in Alcman's line) is liquefied by the moon into dew. The context of lunar liquefaction is absent, however, in *Q.N.* 24, where Plutarch argues (regarding the problem of why hunters are least successful in following animal tracks during full moons) that the moon draws the dew, which is a weak and impotent kind of rain, up from the earth like the sun does, but being unable to lift it to a height and to raise it, drops it again. Considering the clear Stoic overtones in this cluster (as attested in the allegorical reading of Alcman's verse and the allusion to exhalations as fuel for the moon and sun: see the commentary *ad loc.*), it is only likely that a certain compositional interference must have occured when Plutarch wrote down these passages (did he perhaps draw from his notes on a Stoic commentary on Alcman?). The slight differences in argumentative detail can be ascribed, then, to the different argumentative contexts in which the material was incorporated.

---

[20] For a list of the most obvious parallels between *Quaestiones naturales* and the *Moralia*, see the index (s.v. "Plutarch"), in F.H. Sandbach, 1965, pp. 239–240 (where *Quaestiones naturales* covers pp. 133–229). Sandbach's index records no parallels with the *Vitae* (see n. 19).

Second, when it comes to the number and detail of the explanations Plutarch provides in his natural problems, it seems that the aetiologies are far more systematic in *Quaestiones naturales* than in the parallel accounts. Indeed, one has the impression that Plutarch in this work aims to amass all of his knowledge on the natural phenomena at issue, while elsewhere he is more concerned with adapting only fractions of this material, often, indeed, in different contexts, so that an elaborate aetiology is not necessary. This is seen, for instance, in the octopus cluster, where the most exhaustive account is given in *Q.N.* 19 (see the scheme above). Only in one exceptional case does a specific argument take on greater rigor elsewhere than in *Quaestiones naturales*. This is the case in *Q.N.* 29, 919AB, where Plutarch argues that cold is a δύναμις in itself rather than a στέρησις of heat. This theory is elaborated in far greater detail in *De primo frigido* (946A–948A), a treatise in which the principle of cold is the main subject of inquiry. In *Q.N.* 29, however, the same theory is formulated in a very condensed fashion to serve as a starting-point for Plutarch's discussion of why we marvel at hot springs but not at cold ones.

What these clusters and parallels show is that the widely accepted unity and consisteny in the *corpus Plutarcheum* is considerably strengthened by the use of specific natural scientific topics, often identical or similar to those of *Quaestiones naturales*[21]. No wonder that Flacelière regarding Plutarch's digression on drinking water in *Aem. Paul.* 14 (discussed below) notes: "On croirait vraiment lire un paragraphe des *Causes physiques*."[22] In the following section, I will deal with the recurrent incorporation of scientific digressions in the *Vitae* more specifically and with the discursive role these digressions play in the biographical narratives at hand. Getting a clearer view of Plutarch's technique of incorporating such scientific digressions into the *Vitae* will be valuable to further study the position of *Quaestiones naturales* in the *corpus Plutarcheum*.

3. Scientific digressions in the *Vitae*

Plutarch's digressive writing method spans a wide range of topics in the *Vitae*. Most digressions in these writings deal with topics related to Greek

---

[21] For more on the unity between the *Vitae* and the *Moralia*, based on the scientific digressions in the former, see J. Boulogne, 2008, p. 748 (regarding the parallel on βουλιμία between *Quaest. conv.* 693E–695E and *Brut.* 25, 4–6; see further). Regarding the unity in the *corpus Plutarcheum* more generally, see J. Barthelmess, 1986, pp. 62–64 and the contributions in A.G. Nikolaidis, 2008.

[22] In R. Flacelière, J. Irigoin, J. Sirinelli and A. Philippon, 1987, p. lxxxii (see also p. xi, n. 2). Cf. also S. Saïd, M. Trédé and A. Le Boulluec, 1997, p. 444: "On retrouve partout la même érudition (la longue digression sur les eaux potables dans la *Vie de Paul-Émile* serait tout à fait à sa place dans les *Questions naturelles*) […]."

and Roman history and culture (viz. names, places and customs)[23]. This is only a logical consequence of Plutarch's basic intention with these writings, which is to portray the lives of illustrious Greek and Roman political figures. To this end, many cultural and historical realia require a detailed explanation for the reader to acquire an optimal understanding of the story-line and its broader context. Plutarch also incorporates more reflective digressions, in order to add a specific philosophical-theological layer – the ὕλη φιλοσοφίας – to his biographical discourses, pointing out, for instance, that God is capable only of doing good (*Per.* 39, 2–3), or that demons try to lead virtuous people astray (*Dion* 2, 4–7). As noted, there is often also a physical specification of the narratives by the incorporation of numerous scientific digressions. In the end, the physical world – or at least Plutarch's Platonic view of it – is the ultimate background against which these biographies are set.

These scientific digressions testify to Plutarch's intellectual concerns and desire to look for explanations. From a narratological perspective, however, the digressions do not always seem to have much relevance for the main story-lines, to which they often only bear indirect relevance. Barrow may well be right, therefore, that such digressions come in handy "as a means of suspending the interest of the reader" – after all, "Plutarch should not be read in a hurry"[24]. These digressions add a specific physical dimension to the text. Van der Stockt has convincingly argued, in this regard, that "some of Plutarch's scientific 'digressions' are no mere display of scholarship, but are quite functional: they explain the world in which the heroes are operating"[25]. Indeed, Plutarch's *Vitae* are often set against a specific geographical decor that plays a direct role in the development of the narrative[26]. Many historical events are, in fact, directly related to specific natural phenomena and their causes. At some points, nature

---

[23] J.-M. Pailler, 1998, p. 82, for instance, distinguishes three aetiological categories in the *Romulus*: aetio-etymology, aetio-toponymy, aetio-ethnology. On the digressions in the *Coriolanus*, see G. Roskam and S. Verdegem, forthcoming.

[24] R.H. Barrow, 1967, p. 65 (see also pp. 63–64). See Plutarch's own remark in *Timol.* 15, 11: ταῦτα μὲν οὖν οὐκ ἀλλότρια τῆς τῶν βίων ἀναγραφῆς οὐδ' ἄχρηστα δόξειν οἰόμεθα μὴ σπεύδουσι μηδ' ἀσχολουμένοις ἀκροαταῖς.

[25] L. Van der Stockt, 2013, p. 445. He also points at "Plutarch's endeavour to explore more or less virtuous human conduct in the world such as it is according to the *Platonist* Plutarch" (p. 438).

[26] Cf. J. Sirinelli, 2000, p. 363: "Dans les *Vies* il parle souvent des particularités géographiques des pays concernés. Il suffit de consulter la *Vie* d'Alexandre ou celle d'Antoine pour se rendre compte qu'il s'est beaucoup informé sur les régions traversées et avec beaucoup de discernement. On ne peut affirmer qu'il a une connaissance très poussée de toute la géographie de son temps, mais il semble clair que, chaque fois qu'il traite d'un sujet qui appelle des connaissances dans ce domaine, il fait le nécessaire pour se renseigner et sait où puiser ses informations."

even conditions human action[27] (solar or lunar eclipses, for instance, can engender fear in generals, thus causing military defeat)[28]. Moreover, the heroes of Plutarch's stories are, in a certain sense, presented as human products of nature, with a specific φύσις of their own. The link between a person's character and his bodily disposition becomes concrete, for instance, in Plutarch's reference to Lysander's melancholy in *Lys.* 2, 3. Plutarch there quotes Aristotle, who writes that 'great natures', like those of Socrates and Plato and Heracles, have a tendency to melancholy (τὰς μεγάλας φύσεις ἀποφαίνων μελαγχωλικάς), and that Lysander, not immediately, but when well on in years, was a prey to this affliction – this is a clear allusion to the famous chapter on melancholy in Ps.-Aristotle's *Problems* (953a10–955a40). Another likely allusion to the *Problems* is found in *Arat.* 29, 6 [quoted 3.1.1.], where Plutarch discusses Aratus' cowardice and its bodily manifestations (viz. heart palpitations, change in colour and looseness of the bowels), noting that such topics are popular points of discussion in the philosophers' schools (ἐν ταῖς σχολαῖς). By the recurrent link between ethics and physics in the *Vitae*, Plutarch's heroes can be considered the microcosmic pawns on the macrocosmic chessboard that is the world[29].

As the examples below will show, Plutarch knows very well that his natural scientific digressions might seem to contain rather redundant and heterogeneous materials in the context of the biographies of political figures. However, it turns out that these digressions, besides from serving as intellectual diversions, often also fulfil a specific literary function in the text, such as characterising the hero's personality or illustrating an important historical event (often in the context of divine intervention).

---

[27] See P. Desideri, 1992, pp. 77–81 and A. Ferreira, 2015.

[28] Cf. *Per.* 35, 2, *Nic.* 23 (with *De sup.* 169AB), *Aem. Paul.* 17, 7–13 (with P. Desideri, 1992, p. 83: "la conoscenza della causa scientifica del fenomeno naturale non esclude la possibilità di riconoscere in esso un segno divino"). On eclipses in Plutarch, see F.E. Brenk, 1977, pp. 41–45, A. Pérez Jiménez, 1992, L. Torraca, 1992, pp. 240–243, L. Lesage Gárriga, 2015.

[29] In view of Plutarch's doctrine of 'great natures' (μεγάλαι φύσεις), it could perhaps even be argued that the souls of Plutarch's heroes link up with the higher realm of the cosmos. For the relation between the world soul and the human soul in Plutarch, see P. Thévenaz, 1938, J. Opsomer, 1994b, and F. Ferrari and L. Baldi, 2002, pp. 52–54. On the Platonic concept of μεγάλαι φύσεις in the *Vitae*, see B. Bucher-Isler, 1972, pp. 79–81, T. Duff, 1999a, pp. 47–49, 1999b (p. 323 on *Lys.* 2, 3), F. Frazier, 2014, pp. 498–501. Cf. Pl., *Rep.* 491b–492a, 495b. For Platonic psychology in Plutarch's *Vitae*, see T. Duff, 1999a, pp. 72–98, esp. p. 91. At another level, the conceptual link between microcosmos and macrocosmos also figures, for instance, in *De facie*, where several cosmological principles are explained by means of concepts related to the human body (e.g., 928AC). See H. Görgemanns, 1970, pp. 107–111 and A. Pérez Jiménez, 1992, pp. 273–274. See also more generally L. Roig Lanzillotta, 2015.

Indeed, they are often cleverly woven into the overall narrative in such a way that they do not tip the work's unity out of balance. If a digression tends to deviate too far from the central story-line, Plutarch breaks it off in time (see the formulations of closure below)[30]. Apparently, he is well aware of the fact that a complete treatment of natural scientific topics is impossible in the *Vitae*, and that this should be reserved for a more specialised genre of writing.

Desideri was the first to devote a separate study to the natural scientific digressions in the *Vitae*[31], and there is also a more recent one by Boulogne[32]. Their overviews show that these digressions concern matters of physics, astronomy, geography, geometry, zoology, medicine, psychology and music. The aetiological structure and approach in these digressions, where Plutarch often provides several plausible explanations for a specific natural phenomenon, reminds the reader of the 'problematic' set-up of *Quaestiones naturales* [see 1.1.4.]. There is not enough space to analyse each and every one of the scientific digressions in the *Vitae* here – even Desideri notes that running through all the scientific passages in the *Vitae* may be a "cosa che probabilmente non avrebbe molto senso"[33] –, but the following examples may suffice to make things more concrete.

One of the most well-known scientific digressions is probably the one on the nature and origin of naphtha in *Alex.* 35[34]. Sansone has interpreted this passage in light of Alexander's character and physiognomy, arguing that "the volatile and flammable nature of naphtha is remarkably like the nature of Alexander as portrayed by Plutarch"[35]. As such, the digression is actually key to Plutarch's ethical portrait of Alexander. I will not provide an analysis of the entire passage here. It is worth mentioning that several parallels can be traced in *Quaestiones convivales* (viz. the marvellous phenomena of naphtha in 681C, the use of 'waterbeds' in 649EF, and Harpalus' failure to plant ivy in Babylonian soil in 648CD and 649E)[36]. The fact that some of these issues are treated in greater detail in

---

[30] Eight of the approximately 40 scientific digressions collected by J. Boulogne, 2008 contain a formulation of closure, meaning that Plutarch terminates these passages explicitly (see, p. 746, with n. 36).

[31] P. Desideri, 1992.

[32] J. Boulogne, 2008 (who on pp. 746–747 distinguishes four functions of these scientific digressions: "plaire", "instruire", "spécifier", "jugement personnel"). See also A. Ferreira, 2015.

[33] P. Desideri, 1992, pp. 73–74.

[34] Cf., e.g., J.R. Hamilton, 1969, p. 94: "The former passage is a good example of Plutarch's interest in science."

[35] D. Sansone, 1980, p. 63. Cf. also T. Whitmarsh, 2002, p. 190: "the heat of the East is inflaming Alexander, whose nature is already higly flammable". See also R. Caballero Sanchez, 1992, pp. 92–95, J. Mossman, 2006, pp. 290–291 and J. Boulogne, 2008, p. 737.

[36] According to J. Boulogne, 2008, p. 737, in this digression, Plutarch reuses pieces of

Quaestiones convivales is important, because at the end of the *Alexander* passage, Plutarch notes that if such digressions are kept within bounds, the impatient readers will perhaps complain about them less (*Alex.* 35, 16: τῶν μὲν οὖν τοιούτων παρεκβάσεων, ἂν μέτρον ἔχωσιν, ἧττον ἴσως οἱ δύσκολοι κατηγορήσουσιν). As the parallels in *Quaestiones convivales* show, the potential impatience on behalf of the reader does not so much involve the scientific contents of such digressions, but rather the fact that they tend to disrupt the fluency and coherence of the main story-line. Then again, Plutarch warned his reader in the introduction in *Alex.* 1, 2 that he is 'not writing history but biography' (οὔτε γὰρ ἱστορίας γράφομεν, ἀλλὰ βίους). Plutarch's digression on the fiery soil of Babylon, thus, turns out to be an important motive for the biographical narrative in that it, at least implicitly, illustrates Alexander's fiery character and physiognomy.

Another example can be found in *Lys.* 12, where Plutarch, after having mentioned that some people thought that Lysander's swift ending of the Peloponnesian war was the result of divine intervention (θεῖόν τινες ἡγήσαντο τοῦτο τὸ ἔργον), elaborates on the meteorite that fell in Aegospotami in 468–467 BC. Plutarch reports that this phenomenon was considered a divine portent in those days. He does not intend to reject this idea, but he gives a more physical motivation for it in the form of five explanations, including popular opinions, the theories of Anaxagoras and Daimachus, as well as his own criticisms and comments. The aetiology is relatively elaborate and occupies an entire paragraph. Importantly, Plutarch again abruptly concludes the aetiology with the remark that a more minute discussion of this subject belongs to 'another kind of writing' (*Lys.* 12, 7: ταῦτα μὲν οὖν ἑτέρῳ γένει γραφῆς διακριβωτέον). I will come back to this later.

The ending is even more abrupt in *Aem. Paul.* 14, where Plutarch illustrates Aemilius Paulus' superb leadership with a digression on drinking water. The story goes that Aemilius Paulus' troops were greatly overcome by thirst as there was no drinking water available. Aemilius Paulus saw green trees growing on the slopes of Mt. Olympus and inferred that drinking water must be present there. So he started digging at the foot of the mountain, yielding gallons of water for his soldiers to drink. As if intending to match Aemilius Paulus' practical ingenuity in these matters, albeit at a more theoretical level, Plutarch posits two theories in opposition to each other in order to explain where this drinking water exactly came from. The first theory is that water is generated when moist vapour and air under the earth are liquefied through compression and cooling. When the

---

"un dossier constitué autour de l'autorité de Théophraste et qu'il expose plus longuement dans les *Propos de Table*, où le nom du philosophe botaniste est cité, et il suggère d'induire que la Babylonie possède un sous-sol générateur de feu (πυριγόνον)". See Theophr., *HP* 4, 4, 1 (and *CP* 2, 3, 3; 2, 7, 3).

soil is manipulated by digging, in response, the water flows more freely. The same counts, by analogy, for women's breasts: it is only when a baby starts sucking them that they produce milk by converting the nourishment within them (this implies that the breasts are *not* like vessels filled with milk). Plutarch objects, however, that those who support this doctrine give the sceptical philosophers occasion to argue (οἱ δὲ ταῦτα λέγοντες ἐπιχειρεῖν δεδώκασι τοῖς ἀπορητικοῖς) – again by analogy – that living creatures do not have blood until the moment they are wounded, the blood then being generated through a transformation of some vapour or flesh, which causes its liquefaction. The alternative (and preferred) theory is that there are subterranean reservoirs and streams of water at hand, which under the weight and impulse of the pressure upon them (exerted by the mass of Mt. Olympus) discharge themselves into the vacuum afforded by the vents and wells. Plutarch closes the discussion rather inelegantly with the words ταῦτα μὲν περὶ τούτων ('that takes care of that!': *Aem. Paul.* 14, 11), as if to excuse himself for his somewhat schoolmasterly diligence[37].

A last example is in *Flam.* 10, 6, where Plutarch reports that during the Isthmian games, a flock of ravens fell from the sky due to very loud cheers from the crowd. Indeed, the Greeks must have been extremely cheerful the moment that Titus Flamininus declared them to be free. In this passage, Plutarch clearly builds towards a narrative climax, yet at the same time it seems that he is trying to keep his authorial cool (this may, indeed, have specific political dimensions in view of the altered political situation in his own days). Plutarch slows down the narrative pace by the repetition of Flamininus' proclamation – at first, the Greeks could not believe their ears. This is reinforced by the incorporation of three explanations for the natural anomaly of ravens falling from the sky:

> τὸ δὲ πολλάκις λεγόμενον εἰς ὑπερβολὴν τῆς φωνῆς καὶ μέγεθος ὤφθη τότε. κόρακες γὰρ ὑπερπετόμενοι κατὰ τύχην ἔπεσον εἰς τὸ στάδιον. αἰτία δ' ἡ τοῦ ἀέρος ῥῆξις· ὅταν γὰρ ἡ φωνὴ πολλὴ καὶ μεγάλη φέρηται, διασπώμενος ὑπ' αὐτῆς οὐκ ἀντερείδει τοῖς πετομένοις, ἀλλ' ὀλίσθημα ποιεῖ καθάπερ κενεμβατοῦσιν, εἰ μὴ νὴ Δία πληγῇ τινι μᾶλλον ὡς ὑπὸ βέλους διελαυνόμενα πίπτει καὶ ἀποθνήσκει. δύναται δὲ καὶ περιδίνησις εἶναι τοῦ ἀέρος, ἑλιγμὸν οἷον ἐν πελάγει καὶ παλιρρύμην τοῦ σάλου διὰ μέγεθος λαμβάνοντος.

And that which is often said of the volume and power of the human voice was then apparent to the eye. For ravens which chanced to be flying overhead fell down into the stadium. The cause of this was (1) the rupture of the air; for when the voice is borne aloft loud and strong, the air is rent asunder by it and will not support flying creatures, but lets them fall, as if they were over a vacuum, unless, indeed, (2) they

---

[37] For further commentary on this digression, see C. Liedmeier, 1935, pp. 162–167.

are transfixed by a sort of blow, as of a weapon, and fall down dead. It is possible, too, (3) that in such cases there is a whirling motion of the air, which becomes like a waterspout at sea with a refluent flow of the surges caused by their very volume.

The same phenomenon is mentioned in *Caes*. 63, 2 among several other bad omens witnessed the night before the Ides of March. It also recurs in *Pomp*. 25, 6–7 in greater detail, in the context of the *rogatio Gabinia* in the Roman senate and the people's impatient cry at the forum. In the latter passage Plutarch lists the same explanations and uses the same terminology as in the *Flamininus* passage, but the first solution is explicitly rejected and the third is less clearly distinguished from the second:

ἐπὶ τούτῳ λέγεται δυσχεράναντα τὸν δῆμον τηλικοῦτον ἀνακραγεῖν ὥστε ὑπερπετόμενον κόρακα τῆς ἀγορᾶς τυφωθῆναι καὶ καταπεσεῖν εἰς τὸν ὄχλον. ὅθεν οὐ δοκεῖ ῥήξει τοῦ ἀέρος καὶ διασπασμῷ κενὸν πολὺ λαμβάνοντος ἐνολισθαίνειν τὰ πίπτοντα τῶν ὀρνέων, ἀλλὰ τυπτόμενα τῇ πληγῇ τῆς φωνῆς, ὅταν ἐν τῷ ἀέρι σάλον καὶ κῦμα ποιήσῃ πολλὴ καὶ ἰσχυρὰ φερομένη.

At this, we are told, the people were incensed and gave forth such a shout that a raven flying over the forum was stunned by it and fell down into the throng. From this it appears (1) that such falling of birds is not due to a rupture and division of the air wherein a great vacuum is produced, but (2) that they are struck by the blow of the voice, which raises a surge and billow in the air when it is borne aloft loud and strong.

What these passages show is that Plutarch speaks of the same natural phenomenon on several occasions and in different contexts, where it always plays a specific discursive role. In the case of falling ravens it underlines the key-importance and extra-ordinary character of specific historical events. Even if the aetiology is not simply copy-pasted in these parallel accounts, specific conceptual and verbal reminiscences can still be detected, so that we may presume a certain intermediation in composition.

Notably, Plutarch had several such fixed theoretical and terminological schemes in the back of his mind that he could easily apply to different natural phenomena. An allusion, for instance, to the ὑπερβολὴ τῆς φωνῆς καὶ μέγεθος at the beginning of the *Flamininus* passage (and more precisely the physical impact of sounds on bodies) is found in *Quaest. conv.* 721EF, where Plutarch – in explaining a different problem, viz. why sounds carry better at night than during the daytime – defines sound as an impact on a sound-conducting body (ἡ δὲ φωνὴ πληγὴ σώματος)[38]. Similarly, the concept

---

[38] Cf. also *De fortuna* 98BC, *De genio Socr.* 588E, Pl., *Tim.* 67b and Arist., *DA* 420b29.

of impact (πληγή), in combination with that of surge (σάλον), is once again introduced, for instance, in the first *causa* in *Q.N.* 12, 914F, where Plutarch – in examining how sprinkling oil on the surface of the sea clears and calms the waters – gives Aristotle's explanation, according to which the wind, by its slipping off the smoothness so caused by the oil, makes no impact and raises no surge (Πότερον, ὡς Ἀριστοτέλης φησί, τὸ πνεῦμα τῆς λειότητος ἀπολισθαῖνον οὐ ποιεῖ πληγὴν οὐδὲ σάλον;). From parallels like these, we learn that Plutarch's natural problems, as known from *Quaestiones naturales*, do not hold an isolated position in the *corpus Plutarcheum* but actually stand in close dialogue with Plutarch's other works, where physical aetiology is concerned. Remarkably enough, though, Plutarch never refers to *Quaestiones naturales* in his other works in a direct way, which may suggest that the collection does not hold a very central position[39]. Even still, as the following section will show, several passages may qualify as indirect references.

4. Indirect references to *Quaestiones naturales*

Plutarch never directly refers to *Quaestiones naturales* throughout his writings, as he famously does to *Quaestiones Romanae* in *Cam.* 19, 8 (ταῦτα μὲν οὖν ἐν τῷ Περὶ αἰτίων Ῥωμαϊκῶν ἐπιμελέστερον εἴρηται) and in *Rom.* 15, 7 (Περὶ ὧν ἐπὶ πλέον ἐν τοῖς Αἰτίοις) [see 2.4.1.]. In the first passage, Plutarch deals with the *dies Alliensis*, a topic treated in *Quaest. Rom.* 25, 269F, while in the second passage, he explains three Roman customs that originated from the abduction of the Sabine women, viz. the exclamation of *Talassio*, the groom carrying the bride over the doorstep, and parting her hair with the head of a spear. These topics are treated in *Quaest. Rom.* 31, 271F–272B, 29, 271D and 87, 285BC, respectively. Clearly, a precise reference to one or more specific problem chapters in *Quaestiones Romanae* was not necessary and probably not possible either, considering the lack of any systematic organisation in the collection.

Even if there are no such direct references to *Quaestiones naturales*, several passages may still qualify as indirect references. It is not unreasonable to assume, for instance, that the phrase ταῦτα μὲν οὖν ἑτέρῳ γένει γραφῆς διακριβωτέον in *Lys.* 12, 7 (see above) indirectly refers to *Quaestiones naturales*, or at least to this 'kind of writing' – that is, to the *genre* of natural problem literature. As we saw, with this phrase Plutarch admits that the present discussion (of the meteorite that fell in Aegospotami) may seem somewhat out of context in the biographical narrative at hand. It is unclear, however, whether he is implying that this matter *should* (-τέον) either literally or figuratively be treated elsewhere, making it unclear

---

[39] See F.H. Sandbach, 1965, p. 133: "*Quaestiones Naturales*, however, are never cited by him".

whether the reference is to an existing work or not. However, seeing as this reference is not to one specific text but to an entire γένος γραφῆς, this does not really seem to matter anyway. My point is that this γένος γραφῆς would, indeed, comprise the basic characteristics of the genre of natural problems as known from *Quaestiones naturales* (that is, a formal disposition and enumeration of several plausible explanations for a specific natural problem, where the traditional doxography is critically evaluated). If one considers, moreover, that the problem of the meteorite should be examined *minutely* (διακριβωτέον)[40] elsewhere, it is not unlikely that the reference is in fact to *Quaestiones naturales*, where Plutarch's aim is to collect his knowledge on several natural problems in order to provide an aetiology that is as exhaustive as possible in each case [see 2.1.2.], but this remains uncertain. The specific problem of the meteorite in *Lys.* 12 cannot be retraced in our collection, but what is probably more important is that *Quaestiones naturales* would certainly have provided the right place for treating this problem. Due to its generality, though, the reference to 'another kind of writing' does not guarantee with absolute certainty that Plutarch is referring to *Quaestiones naturales*. It at least points out that the γένος γραφῆς of *Quaestiones naturales* is worth referring to as a distinct genre of natural scientific writing.

Other indirect references to *Quaestiones naturales* may be found elsewhere, for instance, in *De Is. et Os.* 352F, in the context of, what seems to be, a *Quaestio barbarica* about Egyptian priests. The problem can be reconstructed as follows: 'Why do Egyptian priests remove their hair, and why do they wear linen garments?'[41]. Plutarch explains that flax, as opposed to wool, is pure, and he also mentions that it is least apt to breed lice. At the end of this passage, Plutarch refers to his treatment of the subject of lice in 'another work' (περὶ ὧν ἕτερος λόγος). There is a clear parallel in *Quaest. conv.* 642BC, where Plutarch deals with the problem of why sheep bitten by wolves tend to have sweeter flesh but wool that breeds lice. Clearly, this topic could just as easily have been dealt with in *Quaestiones naturales*. The same is true for *Brut.* 25, 6, where Plutarch incorporates a digression on βουλιμία, a distemper (caused by fatigue and cold) from which Brutus suffered when he was near the city of Epidamnus. At the end of this digression Plutarch notes that the issue 'is discussed at greater length elsewhere' (ὑπὲρ ὧν ἐν ἑτέροις μᾶλλον ἠπόρηται). The reference is to *Quaest. conv.* 693E–695E, but given Plutarch's reference to Aristotle there (cf. Ps.-Arist., *Probl.* 887b38–888a23) and the subsequent criticism of the Stagirite's account by the symposiasts, one can imagine that the

---

[40] Cf. J. Boulogne, 2008, p. 746: "il [sc. Plutarque] pense qu'il s'agit d'un sujet important, qui mérite un traitement complet".

[41] Cf. J. Boulogne, 2005b, pp. 197–198. See T.S. Schmidt, 2008.

reference could just as well have been to our collection – or at least, again, to this 'kind of writing'.

Importantly, regarding the latter parallel and in light of the 'cross-fertilisation' between the *Vitae* and *Quaestiones convivales* more generally, Pelling is right that "we can rarely be sure that this '*Table Talk*-material' in the *Lives* is really informed by researches done 'for' the *Table Talk*, rather than drawn from material Plutarch had known for years"[42]. The issue is, indeed, intriguing, but as the problem about the Egyptian priests demonstrates it is not only relevant in light of the 'cross-fertilisation' between *Quaestiones convivales* and the *Vitae* but also the *Moralia*. One may wonder whether the whole of *Quaestiones naturales* (and Plutarch's other collection of *quaestiones* just as well) would perhaps count as "researches done 'for' the *Quaestiones convivales*", or if it is part of the "material Plutarch had known for years" (which I take to refer to his personal notes, his ὑπομνήματα). These two options are not necessarily incommensurable, if we may assume that the collection of *Quaestiones naturales* is itself a set of ὑπομνήματα drafted for the composition of *Quaestiones convivales*. The belief that *Quaestiones naturales* (and Plutarch's other collections of *quaestiones* just as well) were, indeed, composed as rough drafts was commonly accepted by traditional scholarship, but in what follows I will try to demonstrate that this assumption is untenable by showing that there is still an alternative explanation for the mutual correspondences with *Quaestiones convivales*.

In order to shed more light on the close relationship between the natural problems discussed in *Quaestiones naturales* and in *Quaestiones convivales*, then, the following section will provide a detailed comparison of the two works. On the basis of this comparison I will argue that the composition of these works must have been closely related. This, in turn, will provide further information for our study of the actual position of *Quaestiones naturales* in the *corpus Plutarcheum*.

2.2. *A comparative study of* Quaestiones naturales *and* Quaestiones convivales

Scholars have often argued, and rightly so, that the composition of the natural problems collected in *Quaestiones naturales* and in *Quaestiones convivales* must have been closely interrelated[43]. As I will try to demon-

---

[42] C. Pelling, 2011, p. 222. Regarding 'cross-fertilisation' as a central feature of Plutarch's method of composition in the Roman *Vitae*, see also C. Pelling, 1979, pp. 82–83.

[43] Cf. F. Klotz and K. Oikonomopoulou, 2011, p. 20. This is true perhaps also from a chronological perspective, see F.H. Sandbach, 1965, p. 138 [see the prologue]. Pace I. Gallo, 1998, p. 3527: "il confronto [sc. of *Quaestiones naturales*] con le 'quaestiones convivales', dove pure sono trattati problemi di vario genere, è solo apparente, perché diversa è la forma e l'elaborazione letteraria, quasi del tutto assente in questo […]."

strate here, both of these works are, indeed, tightly interwoven in several regards, even if there are also important divergences between the two. Regarding the style, organisation and content of the natural problems in *Quaestiones naturales* compared to those in *Quaestiones convivales*, Ziegler observes that:

> "Die Problemen [sc. in *Quaestiones naturales*] sind ganz in der Art derer, die in den Symposiaka zwischen P. und seinen Tischgenossen diskutiert werden, hier aber nicht literarisch-dialogisch ausgestaltet, sondern in der einfachen Kollektaneenform zusammengestellt"[44].

The three categories that Ziegler implicitly distinguishes in this short comparison are related to aspects of *elocutio* ("literarisch-dialogisch ausgestaltet"), *dispositio* ("in der einfachen Kollektaneenform zusammengestellt"), and *inventio* ("[die] Art [der Problemen]"). These are the three categories that I will also use in providing a more detailed comparison in the sections below[45].

As noted, some scholars have argued that we are dealing in *Quaestiones naturales* (and in Plutarch's other sets of *quaestiones*) with collections of personal notes, which Plutarch produced as the inferior textual substratum for composing his other writings (c.q. *Quaestiones convivales*). I will try to demonstrate that such a hypothesis not only tends to downplay the zetetic autonomy of Plutarch's collections of *quaestiones*, but also neglects the fact that these collections do not necessarily have the same didactic purpose as *Quaestiones convivales*. My first objection will be discussed further on in this chapter [see 2.3.3.], the second, in the following [see 3.1.4.]. Let us first consider where the ὑπομνήματα hypothesis precisely originates, so that the subsequent elaboration of my alternative theory, vindicating the independent status of *Quaestiones naturales*, gains in credibility.

1. The level of *elocutio*

From its early beginnings on, the symposium aimed at promoting social, political and cultural unity and interaction between (male) members of elite communities. Its main goal was to engender and strengthen the coherence of these communities, by means of both serious and more frivolous

---

[44] K. Ziegler, 1951, col. 857.
[45] I borrow these concepts (*elocutio*, *dispositio*, *inventio*) from the classical hermeneutical scheme set out by W. Babilas, 1961 [see the prologue, n. 82]. These categories are often closely interrelated to each other, so that they will not be analysed in strict separation from each other. Furthermore, some topics that will be dealt with here have already been examined earlier or will be later in further detail.

activities, ranging from discussing politics to deliberating over the wine, enjoying artistic performances, jointly singing skolia, solving riddles, etc. As a late representative of the literary genre of the symposium (the συμποτικὸν γένος, cf. *Quaest. conv.* 614A), Plutarch's *Quaestiones convivales* serves as a lively source for much of our knowledge about how such symposia were organised in elite milieus in the early Greco-Roman Empire and what was their binding function, in both social and intellectual terms.

With the *Symposia* of Xenophon and Plato the age-old sympotic institution poached on the preserves of the literary-philosophical tradition. Plutarch is proud to signal that he modelled his own *Quaestiones convivales* after these and related philosophical texts (cf. *Quaest. conv.* 612DE, 686D: see n. 80). This, of course, plays a determining role for the eventual outlook of Plutarch's own sympotic discussions and for their philosophical purpose. At the same time, the influence of Alexandrian scholarship is undeniable in Plutarch's *Quaestiones convivales*. There is proof that from the Hellenistic period on, the genre of the symposium was specifically associated with aetiological research. Some fragments that remain from Callimachus' Αἴτια are presumably set against a sympotic background[46]. The same scholarly approach lies at the basis of many, if not most, of the sympotic discussions recorded in *Quaestiones convivales*. The work as a whole can, thus, be seen as the product of a literary experiment, in which the Chaeronean tries to crossbreed the genre of problems with that of the dramatised, philosophical symposium. Therefore, Plutarch's ambition with this work was not only of a scholarly but also, and more primarily, of a philosophical kind.

The reader finds in *Quaestiones convivales* a fully-fledged work of literature, where Plutarch describes the lively discussions held at the table in his company. It is generally accepted that Plutarch in this work intended to elevate the somewhat profane genre of problems to a higher literary level by fusing it with that of the symposium (or, vice versa, to implant the problem format on the symposium genre). The lively *mise-en-scène* of the discussions aims to intensify the sense of dramatic and literary realism in the work. From a literary perspective, it is clear that Plutarch evokes a highly rhetorical discourse that echoes (and probably idealises [see 2.3.1.]) the real-life table discussions he held with his fellow symposiasts. The characters that Plutarch puts on stage – thus including his own literary *alter ego* – are mostly well-read and eloquent πεπαιδευμένοι, eager to deliver

---

[46] See frs. 43, 12–17 and 178 with A. Harder, 2012, p. 35 (in vol. 1) and pp. 301–302 and 955 (in vol. 2). See also A. Cameron, 1995, pp. 71–103. The pinnacle of this scholarly-sympotic tradition is reached in Athenaeus' *Deipnosophistae*.

on the spot deliberations on puzzling topics and capable of reproducing countless quotations by heart (from the poets, historians, philosophers etc.).

As we saw in the previous chapter, the general style of *Quaestiones naturales*, by contrast, remains at a rather sub-literary level [1.2.3.]. But even if the questions and answers are not dramatised so as to represent lively discussions, they still share the same learned and scholarly appeal of *Quaestiones convivales*[47]. Fuhrmann is exaggerating, then, when he writes:

> "Il faut signaler ici [sc. regarding *Quaestiones convivales*], en outre, l'extraordinaire foisonnement des citations, des récits, fables, apophtegmes, proverbes et images, qui fournissaient à eux seuls à Plutarque un moyen facile de dépasser à coup sûr la forme rudimentaire des *Questions Naturelles, Romaines, Grecques*, et la froideur stéréotypée des recueils de cette espèce."[48]

In fact, Plutarch does incorporate several citations, myths, stories, proverbs, and images in *Quaestiones naturales*, even if they appear in a more condensed form and are less numerous[49]. What is also important is the fact that these elements eventually serve the same discursive purpose as those recorded in *Quaestiones convivales*. They primarily contribute to a proper development of the problems and arguments themselves, so that their use in literary embellishment is only of secondary importance [see 1.2.3.].

To come back to Fuhrmann's account, and more precisely to what he adds directly after the passage just quoted, he is absolutely right that the

---

[47] Moreover, as scholars have effectively shown, at least a certain degree of elaboration went into the composition of *Quaestiones naturales* (see further). See G.W.M. Harrison, 2000b, esp. pp. 247–249 and L. Senzasono, 2006, p. 10: "è inesatto parlare di "forma rudimentale"".

[48] F. Fuhrmann, 1972, p. xix, n. 2.

[49] I have already dealt with the presence of literary images in *Quaestions naturales* [see 1.2.3.], and I will deal with the incorporation of myths and citations from both poets and prose authors later [see 4.1.2.2.–3. and 4.2.1.1. respectively]. There are also two proverbs (viz. in *Q.N.* 16, 915E and 21, 917B). In *Q.N.* 10, 914D, Plutarch refers to a story about the people of Halieis, who received an oracle ordering them to dip Dionysus in the sea. Regarding the style of Plutarch's "books of *problēmata* on antiquarian and scientific subjects, and the more technical philosophical treatises", see D.A. Russell, 1973, p. 34, who correctly observes that "[i]n all these, there is less scope for brilliant play of *exempla* or quotations: [but] the richness and the metaphorical style remain pervasive". Russell concludes that "Plutarch […] has *l'âme de la naïvité*; but in style, he has a sophistication and cunning which make interpretation a continuously exacting task". On Plutarch's method of citing in *Quaestiones convivales*, see J. König, 2010, esp. pp. 339–345 – we will later see that a similar method is applied in *Quaestiones naturales* [see 4.2.1.1.].

level of dramatic liveliness is not the same for each and every sympotic discussion in *Quaestiones convivales* (but that this is no reason to doubt the unity of the work altogether)[50]. It is not unimaginable, in this regard, that if *Quaestiones convivales* were stripped from its dramatic context, it would have the same 'matter-of-fact' style as *Quaestiones naturales*. Yet, it is not, therefore, a given fact that *Quaestiones naturales* was still awaiting a final veneer of literary polish, viz. by pouring it into the literary mould of the symposium[51]. In any case, the scientific parts are obviously not incorporated in an artless fashion in *Quaestiones convivales*, as if they are simply *patched* on the sympotic framework[52]. Therefore, it would be incorrect to speak of a genuine caesura between the more dramatic and the more aetiological types of discourse in *Quaestiones convivales*.

A nice way to illustrate this is by comparing the final explanation provided in *Q.N.* 3, 912EF with its parallel in *Quaest. conv.* 685DE: both passages concern the aphrodisiac properties of salt. In *Q.N.* 3, Plutarch examines why herdsmen put salt down for their cattle. He provides three explanations, arguing successively 1) that salt produces a bulk of food and fattens the cattle, 2) that it makes the cattle healthy and reduces their bulk, and 3) that it has generative and aphrodisiac properties. The formulation of the last *causa* is as clear as it is concise:

σκόπει δέ, μὴ καὶ γονιμώτερα καὶ προθυμότερα πρὸς τὰς συνουσίας· καὶ γὰρ αἱ κύνες κύουσι ταχέως τάριχος ἐπεσθίουσαι, καὶ τὰ ἁληγὰ τῶν πλοίων πλείους τρέφει μῦς διὰ τὸ πολλάκις συμπλέκεσθαι.

Consider, however, whether animals do not become more fertile and readier towards coition. Certainly, bitches conceive quickly when they eat salted meat after mating, and ships transporting salt harbour a larger number of mice, because they frequently copulate.

In *Quaest. conv.* 685DF, the same argument recurs at greater length, where it is attributed to Philinus, but the context is different. The problem at hand is why Homer calls salt divine (cf. *Il.* 9, 214: πάσσε δ' ἁλὸς θείοιο). The argument again closes off the discussion. Several new elements are added by Philinus, but the basic idea remains the same. Most notably, the account about bitches and mice recurs, albeit in a less abridged form.

---

[50] F. Fuhrmann, 1972, p. xix.

[51] For this theory, see K. Ziegler, 1951, col. 857, F.H. Sandbach, 1965, p. 135 (but more hesitative on p. 138), F. Fuhrmann, 1972, p. xiii, S.-T. Teodorsson, 2009, pp. 14–15.

[52] Cf. the so-called Στρωματεῖς (*Patchwork*), a doxographical miscellany attributed to Plutarch by Eus., *PE* 1, 7, 16 (= fr. 179 Sandbach; Lamprias catalogue nr. 62). Cf. also, e.g., Gell., *NA Praef.* 7. For the athetesis of this "puerile compilation", see F.H. Sandbach, 1969, pp. 324–327.

Philinus specifies regarding the popular belief that female mice become pregnant simply by licking salt that it is more likely that the saltiness serves as a kind of aphrodisiac (this is, indeed, closer to what Plutarch writes in *Q.N.* 3, 912EF).

Σιωπήσαντος δ' ἐμοῦ, Φιλῖνος ὑπολαβών 'τὸ δὲ γόνιμον οὐ δοκεῖ σοι' ἔφη 'θεῖον εἶναι, εἴπερ ἀρχὴ θεὸς πάντων;' ὁμολογήσαντος δ' ἐμοῦ 'καὶ μὴν' ἔφη 'τὸν ἅλ' οὐκ ὀλίγον πρὸς γένεσιν συνεργεῖν οἴονται, καθάπερ αὐτὸς ἐμνήσθης τῶν Αἰγυπτίων. οἱ γοῦν τὰς κύνας φιλοτροφοῦντες, ὅταν ἀργότεραι πρὸς συνουσίαν ὦσιν, ἄλλοις τε βρώμασιν ἁλμυροῖς καὶ ταριχευτοῖς κρέασι κινοῦσι καὶ παροξύνουσιν τὸ σπερματικὸν αὐτῶν ἡσυχάζον. τὰ δ' ἁληγὰ πλοῖα πλῆθος ἐκφύει μυῶν ἄπλετον, ὡς μὲν ἔνιοι λέγουσι, τῶν θηλειῶν καὶ δίχα συνουσίας κυουσῶν, ὅταν τὸν ἅλα λείχωσιν· εἰκὸς δὲ μᾶλλον ἐμποιεῖν τὴν ἁλμυρίδα τοῖς μορίοις ὀδαξησμοὺς καὶ συνεξορμᾶν τὰ ζῷα πρὸς τοὺς συνδυασμούς. διὰ τοῦτο δ' ἴσως καὶ κάλλος γυναικὸς τὸ μήτ' ἀργὸν μήτ' ἀπίθανον, ἀλλὰ μεμιγμένον χάριτι καὶ κινητικὸν ἁλμυρὸν καὶ δριμὺ καλοῦσιν. οἶμαι δὲ καὶ τὴν Ἀφροδίτην ἁλιγενῆ τοὺς ποιητὰς προσαγορεύειν καὶ μῦθον ἐπ' αὐτῇ πεπλασμένον ἐξενεγκεῖν, ὡς ἀπὸ θαλάσσης ἐχούσῃ τὴν γένεσιν, εἰς τὸ τῶν ἁλῶν γόνιμον αἰνιττομένους. καὶ γὰρ αὐτὸν τὸν Ποσειδῶνα καὶ ὅλως τοὺς πελαγίους θεοὺς πολυτέκνους καὶ πολυγόνους ἀποφαίνουσιν· αὐτῶν δὲ τῶν ζῴων οὐδὲν ἂν χερσαῖον ἢ πτηνὸν εἰπεῖν ἔχοις οὕτω γόνιμον, ὡς πάντα τὰ θαλάττια· πρὸς ὃ καὶ πεποίηκεν ὁ Ἐμπεδοκλῆς φῦλον ἄμουσον ἄγουσα πολυσπερέων καμασήνων.'

When I (sc. Plutarch) stopped speaking, Philinus took up the thread: "Don't you think that generation is divine, since the beginning of anything is always a god?" I said yes, and he went on: "Well, people hold that salt contributes not a little to generation, even as you yourself have said in talking about the Egyptians. Dog-fanciers, at any rate, whenever their dogs are sluggish towards copulation stimulate and intensify the seminal power dormant in the animals by feeding them salty meat and other briny food. Ships carrying salt breed an infinite number of mice, because, according to some authorities, the females conceive without coition by licking the salt. But it is more likely that the saltiness imparts a sting to the sexual members and serves to stimulate copulation. For this reason, perhaps, womanly beauty is called 'salty' and 'piquant' when it is not passive nor unyielding, but has charm and provocativeness. I imagine that the poets called Aphrodite "born out of the brine" and have spread the myth of her origin in the sea by way of alluding to the generative property of salt. For they also represent Poseidon himself and the sea gods in general as fertile and prolific. Even among the animals you cannot find one species of land or air that is so proliferous as are all the creatures of the sea. This is the point of Empedocles's line: Leading the mute tribe of fruitful fish (DK31B74)."

If we compare the two accounts, we see that a great deal of dramatic and rhetorical detail goes into the scientific argument in *Quaest. conv.* 685EF. Indeed, Plutarch does not incorporate any poetical quotations in *Q.N.* 3, 912EF. However, the poetical material from *Quaest. conv.* 685EF has the same basic purpose as it would have in *Quaestiones naturales*. It is not merely incorporated to embellish the discourse, but to serve as an illustration of the main argument itself (viz. that salt has a generative, and therefore divine, property). The divine character of salt is central to the debate in *Quaest. conv.* 684E–685F as a whole, but this really comes to a climax in Philinus' final explanation, where the divine principle of generation is at issue ('τὸ δὲ γόνιμον οὐ δοκεῖ σοι' ἔφη 'θεῖον εἶναι, εἴπερ ἀρχὴ θεὸς πάντων;'). Presumably, Philinus' argument is last in the aetiology, so as to provide some kind of theological closure to the physical aetiology. This aspect of theological closure is strengthened by the incorporation of mythological material (about Aphrodite, Poseidon and other sea gods) at the very end of the argument (μῦθον ἐπ' αὐτῇ πεπλασμένον ἐξενεγκεῖν). In addition to the allegorical value of these myths – understood as riddled allusions to the generative property of salt (εἰς τὸ τῶν ἁλῶν γόνιμον αἰνιττομένους) –, the passage seems to suggest that there is a higher dimension of philosophical truth that lies beyond the purely physical realm.

By contrast, in *Q.N.* 3, 912EF, Plutarch makes no allusion to the divine character of salt, but focuses exclusively on the αἰτίαι φυσικαί[53]. The introductory σκόπει δέ does, however, draw specific attention to this explanation. Perhaps, Philinus' more elaborate account in *Quaest. conv.* 685EF may explain why this is the case, as it draws a link with divine generation. The absence of mythological references here in *Q.N.* 3 does not imply, moreover, that Plutarch refrains from incorporating such material altogether in *Quaestiones naturales*, let alone that these myths do not provide a similar feature of closure to the physical aetiologies. In fact, these accounts are often placed at the very end of the aetiology in both *Quaestiones naturales* and *Quaestiones convivales* (as we will see later on [see 4.1.2.]). This is not only relevant for the issue of *elocutio* but also of *dispositio*.

2. The level of *dispositio*

The most basic ordering principle in *Quaestiones convivales* is the well-known organisation of the content into nine Books, each containing ten problem chapters each, with the deliberate exception of the last Book, which contains fifteen (see n. 118)[54]. With this considered limitation of

---

[53] The same observation was made by L. Van der Stockt, 2011, pp. 453–454.
[54] For the explicit formulation of this decimal system, see *Quaest. conv.* 612E, 629E, 660D, 697E, 736C. See G.W.M. Harrison, 2000a, p. 197, n. 21.

the content matter, Plutarch makes it absolutely clear that the work was not intended as an indefinite and boundless literary ἄπειρον. No such restrictions, though, are made for Plutarch's other collections of *quaestiones*[55], which highlights the open-ended character of the research projects at issue therein[56].

Regarding the macrostructural arrangement of Plutarch's collections of *quaestiones* in general and of *Quaestiones convivales* more specifically, I have argued earlier that the often chaotic and unpredictable surface of these texts can be explained from the perspective of *variatio* (as a basic feature of ancient miscellaneous literature more generally), while the at times close interconnection of the problem chapters and the recurrent themes and theories therein are the result of the principle of *concatenatio* [see 1.1.5.][57]. Indeed, the grouping together of different problems during one sympotic event is a relatively common feature in *Quaestiones convivales*, and the same structuring principle clearly recurs in Plutarch's other collections of *quaestiones*, especially in *Quaestiones naturales* (see the introduction to the commentary for a schematic overview). As to the internal arrangement of the problem chapters themselves, moreover, we see that the symposiasts in *Quaestiones convivales* put forth a variety of arguments and explanations to solve the problems, while the debate as a whole is guided by the principle of increasing plausibility (τὸ πιθανόν). Each symposiast personifies a specific position in the debate, leading to a combination of contending arguments. This organisation of the explanations is reminiscent of the development of the aetiologies in *Quaestiones naturales* and in Plutarch's other collections of *quaestiones* just as well.

In order to illustrate this, let us again turn to *Quaest. conv.* 684E–685F: two interconnected problems are treated there during one and the

---

[55] According to J. Opsomer, 1994a, p. 12, the collection of ten *Quaestiones Platonicae* may have been modelled on the same decimal system as found in *Quaestiones convivales*. Cf. also S.-T. Teodorsson, 1989, p. 38.

[56] This has led K. Oikonomopoulou, 2013a, p. 152 to conclude that *quaestiones* literature is, in fact, an integral part of the history and legacy of ancient encyclopaedic writing: "The *quaestiones* [...] are not collections of Plutarch's notes, but self-consciously fashion themselves as texts-in-progress for reasons in fact intrinsic to the kind of encyclopaedic function they envisage for themselves." She argues that there is "an underlying desire for encyclopaedic completeness, whose fulfilment can only be guaranteed through the continuation of research, perhaps *ad infinitum*" (p. 150). See also esp. pp. 152–153 for Oikonomopoulou's nuancing of the concept of 'encyclopaedism' in the context of *quaestiones* literature.

[57] On the aspect of structural order and disorder in *Quaestiones convivales*, see the introduction in F. Klotz and K. Oikonomopoulou, 2011, pp. 24–27. J. König, 2008, p. 97 describes "the symposium as an institution for sanctioned flirtation with disorder".

same sympotic event, viz. what is meant with the proverbial salt and bean friends and, connected with it, why Homer calls salt divine. Several of the explanations provided throughout the discussion recur in *Quaestiones naturales*, albeit in different forms and in different places. Florus is the συμποσίαρχος: he organises the dinner and also leads the accompanying discussion. He proposes the first problem and his interference and guidance recurs throughout the discussion as a structuring feature[58]. The only answer to the first problem is brought up by the scholar Apollophanes, who provides the obvious explanation (ἐκ προχείρου διέλυσεν) that the proverb – of salt and bean friends – refers to friends who are on very close terms, because they are prepared to have meals together. The symposiasts then raise the second problem (διηπορούμεν), which is closely connected to the preceding one via the topic of salt, by asking, more precisely, why it is considered divine. After an intermediate account which attests to the divine character of salt in the literature (viz. in Homer and Plato), and on the remarkable abstinence of Egyptian priests from salt, Florus urges his companions to leave the Egyptians out of the question and find a properly Greek explanation. The Egyptians do, indeed, complicate the problem (ἐπέτεινε δὲ τὴν ἀπορίαν), since if salt is divine, why then do Egyptian priests abstain from it on religious grounds? This kind of complication may not be appropriate in light of sympotic decorum, which demands that topics of discussion not become too complex [see 3.1.4.]. Plutarch (by means of his own character in the discussion) objects, however, that the Egyptians are not in conflict with the Greeks on this point (οὐδὲ τοὺς Αἰγυπτίους μάχεσθαι τοῖς Ἕλλησιν)[59]. He explains that Egyptian priests abstain from salt either for reasons of purity (because it has aphrodisiac properties owing to its heat – a point Philinus will elaborate upon later), or because it is delicious as a seasoning (making needful food enjoyable – this point is introduced with εἰκὸς δὲ καί). Florus then asks whether this is the reason why salt is considered divine. Plutarch affirms this and explains that salt is a basic need like water, daylight, the seasons and the earth (which is even generally considered to be a goddess) and that it is very useful for adapting food to our body and appetite (this point is paralleled in the first *causa* in *Q.N.* 3, 912DE). He also draws specific attention (σκόπει μή) to the fact that salt preserves bodies from decay (much like the soul preserves life in our body), as does the fire of lightning (these theories can be found also in *Q.N.* 1, 911D, 10, 914DE, 40 and *Quaest. conv.* 665C)[60].

---

[58] Cf. *Quaest. conv.* 684E (Ἐζήτει Φλῶρος, ἑστιωμένων ἡμῶν παρ' αὐτῷ), 684F–685A (Φλῶρος μὲν οὖν ἐᾶν ἐκέλευε τοὺς Αἰγυπτίους, Ἑλληνιστὶ δ' αὐτοὺς εἰπεῖν τι πρὸς τὸ ὑποκείμενον), 685AB ('Ἄρ' οὖν' ὁ Φλῶρος ἔφη 'διὰ τοῦτο θεῖον εἰρῆσθαι τὸν ἅλα φῶμεν;').

[59] For a separate study of Egyptian accounts in *Quaestiones convivales*, see M. Meeusen, forthcoming d.

[60] The imperative σκόπει μή may suggest that we are dealing here with Plutarch's

Philinus picks up the thread (ὑπολαβών) and adds a final point in agreement with Plutarch's previous position, by arguing that it seems likely (δοκεῖ) that generation is divine and that salt has a generative property (see above).

What this example nicely illustrates, then, is how two problems spontaneously cluster together in one and the same sympotic context. The principle of *concatenatio* is expressed by the prefix in διηπορούμεν, which is a stronger form of ἀπορέω and implies a notion of continuity, as it connects the second to the first problem[61]. The principle of probability, which functions as the main ordering principle in the development of the discussion, is expressed by the use of several concepts (such as δοκεῖ, εἰκός, ἴσως, οἶμαι etc.). Plutarch also recycles the same and similar material that he incorporates in *Quaestiones naturales*, albeit in a re-ordered fashion and in a new context. By the fact that these theories transgress the inter-textual boundaries of both collections, they testify to the adaptable and versatile nature of such scientific knowledge, as being applicable to very different problem contexts. This, in turn, is also relevant for the *inventio* of the scientific material in Plutarch's natural problems.

3. The level of *inventio*

The practice of solving natural problems allows for the efficient reuse, reordering and reinventing of numerous more or less fixed aetiological schemes in a multitude of always new problem contexts. When compared to the natural problems discussed in *Quaestiones convivales*, those collected in *Quaestiones naturales* would also make suitable topics for discussion during symposia. Even though the connection with a sympotic framework remains implicit at all times, most of the questions that Plutarch raises therein can be generally related to the thematic category of sympotic appetite (as is especially the case with the problems on wine)[62] and hence to more general sympotic themes. Oikonomopoulou makes the following conclusion in this regard:

---

personal theories [see 1.1.4., n. 110], but, then again, in *Quaest. conv.* 665C it is the rhetor Dorotheus who refers to the theory about lightning leaving corpses undecayed. As to Plutarch's source, see S.-T. Teodorsson, 1990a, p. 231: "Presumably the connection was first made in a Peripatetic work." Perhaps a Stoic tradition is not unlikely either, considering Plutarch's reference to the Stoic belief that the sow is dead flesh at birth but that the soul is implanted in it later, like salt, in order to preserve it (SVF 1, p. 116, fr. 516; 2, p. 206, frs. 722–723 and p. 333, 1154).

[61] Cf. LSJ, s.v. διαπορέω ii, 1 ("go through all the ἀπορίαι") and 2 ("commonly only a stronger form of ἀπορέω, raise an ἀπορία, start a difficulty").

[62] Cf. *Q.N.* 10, 27, 30–31 ~ *Quaest. conv.* 1, 6–7; 3, 3, 5, 7–9; 5, 3–4; 6, 7; 7, 3, 9–10. See also Book three of Ps.-Aristotle's *Problems*: ὅσα περὶ οἰνοποσίαν καὶ μέθην.

"Inspired by the physical reality of consumption at the symposium, they [i.e. the problems collected in *Quaestiones naturales*] prompted the exploration of topics such as the origin, nutritional benefits, and cultural value of sympotic staples such as wine, bread, water, fish, meat and vegetables (which could then ramify into the investigation of broader natural phenomena). They were also the result of curiosity about the material dimension of objects used at the symposium, or seen in religious locations such as Delphi: vessels, musical instruments, statues or sculptures."[63]

The relation with sympotic reality is, of course, much more palpable in the problems collected in *Quaestiones convivales*, where the discussions often directly arise from the circumstantial setting of the symposium (e.g., recent festivals, served meals or beverages, the place of the guests at the table, proper table talk itself etc.). As such, the wide variety of themes and subjects in this work is directly related to the miscellaneous organisation of the symposium itself[64] [see 1.1.5.].

Natural problems prove to be a popular topic of conversation in Plutarch's intellectual milieu. In fact, some of Plutarch's fellow symposiasts were well acquainted with the Ps.-Aristotelian *Problems*. In *Quaest. conv.* 734CD, most notably, Plutarch writes that a copy of the work was brought to Thermopylae, where Florus discussed it with his friends [quoted 3.2.1.]. The total amount of chapters in *Quaestiones convivales* that deal with natural scientific topics after the manner of *Quaestiones naturales*

---

[63] K. Oikonomopoulou, 2013a, pp. 146–147.

[64] Notably, in *Quaest. conv.* 629D, Plutarch makes a basic distinction between two types of problems: viz. συμποτικά and συμποσιακά. The category of συμποτικά covers problems *concerning* the symposium, whereas συμποσιακά are problems generally treated *at* the symposium. The first category is a subcategory of the latter, because it consists of meta-symposiac debates about the proper course and pragmatics of a symposium, which were also discussed *at* the symposium, such as whether philosophy is a fitting topic for conversation at a drinking party (*Quaest. conv.* 612E), or whether the host should arrange the placing of his guests or leave it to the guests themselves (*Quaest. conv.* 615C). Plutarch notes that both categories can be discussed *at* the symposium and can, therefore, be considered συμποσιακά (*Quaest. conv.* 629D). This probably explains the wording in the title of the collection (Συμποσιακῶν βιβλία Θ). H. Bolkestein, 1946, p. 7 has shown, however, that Plutarch is not always very conscientious in following the distinction between συμποτικά and συμποσιακά (cf., e.g., *Quaest. conv.* 645C, 660D vs. 686E, 717A, 736C). He adds that the distinction may be of Stoic origin, because these philosophers were very fond of grammatical issues and specifically of making subtle terminological distinctions. The Stoic Persaeus of Citium may have been the first to draw this distinction in his Συμποτικοὶ διάλογοι/Συμποτικὰ ὑπομνήματα (SVF I, pp. 100–101, frs. 451–453). See also F. Fuhrmann, 1972, p. xv, with n. 3 and J. König, 2007, p. 61.

covers approximately one third of the entire work[65]: the reader comes across problems that are related to ancient medicine[66], human physiology[67] (including sensations and affections)[68], zoology[69], botany[70], meteorology[71] etc.[72] By the fact that these natural problems are not concerned with highly complex issues in the field of natural philosophy, but, rather, deal with very concrete, 'everyday' phenomena[73], they bear a marked similarity in manner and style to the Ps.-Aristotelian *Problems*. The solutions that are provided have no direct practical use, but only serve the satisfaction of intellectual curiosity, which in itself, as we will see later on, has specific philosophical relevance for Plutarch[74] [see 4.1.1.].

2.3. *Hypomnematic text genetics of* Quaestiones naturales *and* Quaestiones convivales

From our previous comparison of the natural problems treated in *Quaestiones naturales* and *Quaestiones convivales*, we can safely conclude that there must be some text genetic tie between both collections, which deserves further study here. A study of the genesis of *Quaestiones convivales* in relation to that of *Quaestiones naturales* should clarify their relative compositional lineage. An initial problem that should be settled in this regard is the *vexata quaestio* of the historicity in the sympotic discussions recorded there, a controversial issue that still causes debate today[75].

---

[65] Cf. R. Flacelière, J. Irigoin, J. Sirinelli and A. Philippon, 1987, p. lxxxii and F.H. Sandbach, 1965, p. 138.

[66] Cf., e.g., *Q.N.* 26 ~ *Quaest. conv.* 6, 8; 8, 9.

[67] Cf., e.g., *Q.N.* 6, 36 ~ *Quaest. conv.* 2, 2; 3, 6; 4, 1, 10; 6, 8; 7, 1; 8, 10; 9, 11.

[68] Cf., e.g., *Q.N.* 8, 9, 11, 20, 22, 29 ~ *Quaest. conv.* 1, 8; 3, 3–4, 8; 5, 7; 6, 1–3.

[69] Cf., e.g., *Q.N.* 3, 17–22, 26, 28, 35–38 ~ *Quaest. conv.* 2, 3, 7–9; 4, 4; 8, 8.

[70] Cf., e.g., *Q.N.* 1, 2, 4–6, 14–16, 30–32, 41 ~ *Quaest. conv.* 2, 6; 3, 1–2; 4, 2, 10; 5, 3, 8, 9; 6, 10; 7, 2; 8, 4.

[71] Cf., e.g., *Q.N.* 2, 4, 7, 13, 18, 23–25, 34, 40 ~ *Quaest. conv.* 3, 10; 4, 2.

[72] For a similar categorisation, see R. Lopes, 2009, p. 419.

[73] See K. Oikonomopoulou, 2013a, p. 146: "the *QN*'s investigations do not emanate from a scientist's ivory tower, but are anchored in the economic and cultural parameters of practical life: agriculture, animal husbandry, hunting, fishing, sea-faring, swimming, feasting and drinking."

[74] See M. Meeusen, 2014. Similarly, for the aspect of intellectual curiosity in the *Supplementa problematorum* ascribed to Aristotle and Alexander of Aphrodisias, cf. S. Kapetanaki and R.W. Sharples, 2006, p. 1.

[75] For recent debate about the historicity of *Quaestiones convivales*, see esp. F. Titchener, 2009 (also 2011), G. Roskam, 2010, pp. 46–47 (with nn. 8 and 9 for further reading) and the introduction in F. Klotz and K. Oikonomopoulou, 2011, pp. 3–7.

1. Historicity and fiction in *Quaestiones convivales*

In the preface to the first Book of *Quaestiones convivales*, Plutarch addresses the dedicatee, Sossius Senecio, by declaring that the first three volumes that he sends to him present a set of conversations held at table in Rome and Greece[76]. In the preface to the second Book, he notes that he simply jotted down the conversations as each came to his mind[77]. On the basis of these accounts, scholars have accepted that we are dealing with genuine recollections in this work, and that the sympotic conversations that Plutarch records there are historical[78]. The discussions described in *Quaestiones convivales* would, thus, represent a development of personal notes that Plutarch took after the conversations in which he himself either actively or passively participated. The argument is further substantiated by the fact that many of the dinners recorded in *Quaestiones convivales* may very well have taken place at certain locations and during specific festive events, as they are often described in minute detail. The symposiasts that are put on stage are mostly close relatives, friends, students and acquaintances of Plutarch, rather than entirely fictitious characters[79]. The historicity of these settings and characters may, indeed, imply that the treatise is no *complete* literary fiction. However, this does not mean that the literary character of *Quaestiones convivales* should, therefore, be underestimated. Again in the preface to the first Book (*Quaest. conv.* 612DE, cf. also 686D), Plutarch places his work in the wider tradition of philosophical symposium literature, thus joining the line of several coryphaei in the genre (Plato, Xenophon, Aristotle, Speusippus, Epicurus, Prytanis, Hieronymus, and Dio from the Academy). The fact that Plutarch emphatically presents his work as a sample of this literary-philosophical tradition has cast considerable doubt on the historical character of its contents[80]. The view, however, that Plutarch is merely instrumentalising

---

[76] *Quaest. conv.* 612E: ᾠήθης τε δεῖν ἡμᾶς τῶν σποράδην πολλάκις ἔν τε Ῥώμῃ μεθ' ὑμῶν καὶ παρ' ἡμῖν ἐν τῇ Ἑλλάδι παρούσης ἅμα τραπέζης καὶ κύλικος φιλολογηθέντων συναγαγεῖν τὰ ἐπιτήδεια, πρὸς τοῦτο γενόμενος τρία μὲν ἤδη σοι πέπομφα τῶν βιβλίων.

[77] *Quaest. conv.* 629D: σποράδην δ' ἀναγέγραπται καὶ οὐ διακεκριμένως ἀλλ' ὡς ἕκαστον εἰς μνήμην ἦλθεν. This passage has already been discussed in light of the miscellaneous structure of *Quaestiones convivales* [see 1.1.5.].

[78] See, e.g., E. Graf, 1888, p. 59, H. Bolkestein, 1946, pp. 20–26, Z. Abramowiczówna, 1962, pp. 84–88. Cf. also the rather nonchalant ending of Book nine, *Quaest. conv.* 748D: Ταῦτα σχεδόν, ὦ Σόσσιε Σενεκίων, τελευταῖα τῶν ἐν τοῖς Μουσείοις τότε παρ' Ἀμμωνίῳ τῷ ἀγαθῷ φιλολογηθέντων (see F. Klotz, 2014, p. 210).

[79] *Quaestiones convivales* provides a great deal of prosopographical information about the symposiasts put on scene. See K. Ziegler, 1951, cols. 641–653 and 665–696 and B. Puech, 1992.

[80] For the Socratic symposia of Plato and Xenophon as literary models for Plutarch's *Quaestiones convivales*, see G. Roskam, 2010. For Plutarch and the genre of the symposium

the sympotic genre to disseminate his own variegated investigations to a wider public – that is, by using it as a purely literary fiction – does not hold against serious criticism either[81]. It remains to be seen, therefore, to what extent the lively descriptions of the sympotic discussions are faithful renderings or rather literary replicas of historic, real-life events.

Scholars have tried to reconcile both viewpoints[82], by arguing that in *Quaestiones convivales*, Plutarch intertwines sympotic authenticity with literary allusions. One can imagine that most of the talks are rooted in discussions that actually took place on a given occasion. Perhaps at certain points, they even contain the core positions defended by each of Plutarch's fellow symposiasts, but this is uncertain, and does not necessarily rule out the author's own interventions. Indeed, the sympotic discussions may very well contain a certain degree of additional aetiological elaboration and reorganisation of the arguments based on Plutarch's own research and reading. The following indications may support this theory.

First of all, Plutarch does not partake as a sympotic character in the discussion in 28 chapters. This has led many scholars to the suspicion that a considerable part of the work is fictional[83]. While this certainly casts doubt on the historicity of these chapters, however, from a narratological perspective perhaps Plutarch is making a certain Platonic gesture, viz. by stressing the authorial role of reportage in the recording of these sympotic discussions[84]. Indeed, Plato himself is also absent in his *Symposium*, and his absence in the *Phaedo* is illustrious.

Second, the attempt to maintain an aspect of historical verisimilitude does not seem to be equally successful in each and every sympotic discussion. Some of the chapters are less circumstantial when it comes to historical detail and sympotic liveliness (see n. 50). In these cases, the description of the sympotic setting does not receive a great deal of dramatic substantiation. In some cases, the portraits of the symposiasts, who are normally characterised by their personal interests, occupations and idiosyncrasies, remain rather vague (Plutarch there simply uses such generic situational markers as οἱ μέν, οἱ δέ, ἔνιοι, ἐδόκει, ἐλέχθη, ποτέ etc.)[85]. In some chapters, Plutarch does not mention the name of any of the

---

more generally, see, e.g., J. Martin, 1931, M. Vetta, 2000, S.-T. Teodorsson, 2009, F. Klotz, 2014. For the place of the symposium in Plato's philosophy, See M. Tecuşan, 1990.

[81] Cf., e.g., K. Hubert, 1911, p. 187 ("die Symposiaca [sind] ein durch und durch literarischen Werk"), J. Martin, 1931, pp. 173–179 (p. 173: "immer unter Beibehaltung der Fiktion").

[82] See, e.g., K. Ziegler, 1951, cols. 886–887, F. Fuhrmann, 1972, pp. vii–xix, S.-T. Teodorsson, 1989, pp. 12–15, J. Opsomer, 1994a, p. 8.

[83] See, e.g., A. Gudeman, 1927, col. 2526.

[84] Thanks are due to K. Oikonomopoulou for this suggestion.

[85] See, e.g., E.L. Minar, F.H. Sandach and W.C. Helmbold, 1961, p. 2.

symposiasts, but simply lines up a number of arguments anonymously without attributing them to specific persons (e.g., *Quaest. conv.* 619BF, 625AC). By putting the main focus on the development of the arguments, these chapters give the impression of being short expositions, rather than the condensation of real-life discussions[86] (even if the author attempts to maintain the illusion of reality by creating an artificial setting and evoking "un air de vérité"[87]). However, it may well be that these artificial table talks are simply rendered in summary or paraphrase due to selective or faulty recollection of the author[88].

Third, the use of lengthy monologues in the argumentations in *Quaestiones convivales* seem to betray the author's intervention, since it is unlikely that we are dealing in these cases with the symposiasts' *verba ipsissima* exactly as they were uttered (e.g., *Quaest. conv.* 629E–634F; the same is true for the use of indirect speech: e.g., *Quaest. conv.* 620A). If it is true, moreover, that these passages do contain at least a certain nucleus of authenticity[89], Plutarch may very well have made further elaborations and revisions to them. Since Plutarch may in some cases simply be using the literary characters of his fellow symposiasts as *porte-paroles* to voice his own opinions – a literary strategy with which he was not at all unfamiliar[90] –, it is not unimaginable that he intended to bring some kind of an intellectual tribute to his sympotic colleagues by labelling the explanations to the questions with their proper names and by staging them in, what, thus, turns out to be, a sympotic *liber amicorum*. In a way he, thus, immortalised his friends in this learned *Festschrift*.

Since the question of historicity and fiction in *Quaestiones convivales*, cannot be settled with any certainty or precision, it is safe to conclude with Titchener that "the *QC* do not need to be authentic to be real and true"[91]:

---

[86] A comparison with the fictitious *Septem sapientium convivium* is never far away in these cases. Cf. A. Gudeman, 1927, col. 2526 and F. Fuhrmann, 1972, p. ix. On the place of Plutarch's *Septem sapientium convivium* in the tradition of symposium literature, see J. Mossman, 1997b.

[87] F. Fuhrmann, 1972, p. xvii. Cf. also pp. ix–x and xviii: "nous pouvons au moins accepter comme historiques ceux auxquels se rapporte la relation la plus circonstanciée".

[88] Some symposia perhaps do have that effect on a person's memory. Thanks are again due to K. Oikonomopoulou for this suggestion.

[89] Cf. G. Roskam, 2010, p. 47.

[90] Plutarch's brother Lamprias, for instance, is often considered the literary delegate and mouthpiece of the Chaeronean's own opinions. For the role of Lamprias as narrator in *De defectu oraculorum*, see, e.g., F.C. Babbitt, 1936b, p. 349: "some have thought that Plutarch has used the person of Lamprias to represent himself, possibly because of the official position held by Plutarch at Delphi." Cf. also F. Ferrari, 1995, pp. 30–31. For Lamprias' role in *De facie*, see P. Donini, 2011, p. 36.

[91] F. Titchener, 2009, pp. 398–399 and 2011, p. 39.

"What the *QC* present us with is something a little in between: what at least conveys the texture of what MIGHT have happened, COULD have happened, and periodically HAD in fact happened. For Plutarch's purposes, this is really all the same thing [...]."[92]

As such, *Quaestiones convivales* vividly portrays the intellectual practice of sympotic debate as held in Plutarch's milieu, albeit in a dramatised and idealised fashion, with the goal of making these discussions accessible to posterity, much like Plato, Xenophon and other authors had done before (cf. *Quaest. conv.* 612DE, 686D). Scholars generally agree that a complex embroidery and reorganisation of hypomnematic material lies beneath the surface of the text in *Quaestiones convivales*[93]. This hypomnematic material was composed by the author on the basis of his own recollections, reading and research. As stated, this hypomnematic material has often been associated, and in some imprudent cases even identified, with Plutarch's collections of *quaestiones*, but this seems unlikely for several reasons, as will be set out in the following section[94].

2. Problems and personal notes

Contemporary Plutarch scholarship has devoted a great deal of its attention to the Chaeronean's use of personal notes (ὑπομνήματα) both in composing the *Moralia* and the *Vitae*[95]. In this section, I will examine the nature and function of these notes more closely with the goal of distinguishing them from Plutarch's collections of problems. A *locus classicus* is the ὑπομνήματα statement in the introduction to *De tranquillitate animi* (464E–465A)[96]. Close analysis of this passage will yield important information

---

[92] *Ibid.* A similar conclusion was made, e.g., for the discussion recorded in *De sollertia animalium* by J. Bouffartigue, 2012, p. xix: "On retiendra l'idée de "forme idéalisée", en ne perdant pas de vue que Plutarque n'écrit pas un reportage."

[93] See, e.g., J. Sirinelli, 2000, pp. 380–385 and p. 386: "On est tenté de donner comme sous-titre à cet ouvrage: un homme se penche sur son fichier!" The phrase ὡς ἕκαστον εἰς μνήμην ἦλθεν in *Quaest. conv.* 629D (discussed earlier [1.1.5.]) should perhaps be taken as an implicit allusion to Plutarch's reliance on his ὑπομνήματα.

[94] See, e.g., K. Hubert, 1911, pp. 174–176, 180, H. Bolkestein, 1946, p. 27 and F. Fuhrmann, 1972, p. xiii (quoted below).

[95] For the *Moralia*, see esp. L. Van der Stockt, 1999a, 2004, B. Van Meirvenne, 1999, 2001, M. Beck, 2010; for the *Vitae*, C. Pelling, 1979, pp. 94–95. See also, e.g., K. Mittelhaus, 1911, p. 23 and H. Martin, 1969, pp. 69–70. On the use of notes as a standard practice of literary composition in Antiquity, see J.P. Small, 1997, pp. 169–176, T. Dorandi, 1991, pp. 12–14 and 2000, pp. 28–50.

[96] For a detailed analysis of this passage, see esp. L. Van der Stockt, 1999a, pp. 577–580 (see also 1996, pp. 265–266 and 2004, p. 333).

about Plutarch's general writing technique and his use of personal notes more specifically.

The introduction to *De tranquillitate animi* (464E–465A) highlights the efficiency with which Plutarch accessed and used his personal notebooks, by highlighting the short period of time in which he completed the treatise on the basis of his ὑπομνήματα. Plutarch apologises to his correspondent Paccius for the haste with which the work was put together – 'I did not have the time I desired' (μήτε δὲ χρόνον ἔχων, ὡς προῃρούμην) – and explains that he only recently (ὀψέ, being the very first word of the treatise) received Paccius' petition urging him to write 'something on tranquility of mind, and also something on those subjects in the *Timaeus* which require more careful elucidation' (464E: παρεκάλεις περὶ εὐθυμίας σοί τι γραφῆναι καὶ περὶ τῶν ἐν Τιμαίῳ δεομένων ἐπιμελεστέρας ἐξηγήσεως). In *De tranquillitate animi*, Plutarch grants only the first part of Paccius' request (I will come back later to the second part about the subjects in the *Timaeus* [see 2.4.1.]). Plutarch adds that their beloved friend Eros, with whom he could send the treatise to Paccius, was in a hurry (ἐπιταχύνοντα) to get back to Rome. In the same breath, he admits that the hasty composition had a strong effect on the composition of the treatise, and that Paccius, therefore, should not expect to find a fully embellished literary work, but rather an edited sequence of rough material. Most importantly, Plutarch states that he 'extracted the topic of tranquility from the notes that I took for myself' (464F: ἀνελεξάμην περὶ εὐθυμίας ἐκ τῶν ὑπομνημάτων ὧν ἐμαυτῷ πεποιημένος ἐτύγχανον). As Van der Stockt notes, the translation of the phrase περὶ εὐθυμίας is not unproblematic: "he [sc. Plutarch] does *not* say that the actual theme of these hypomnemata was 'tranquility'! [...] On the other hand, neither does Plutarch *deny* that he consulted 'hypomnemata on tranquility'."[97] The phrase περὶ εὐθυμίας can be understood as a reference to the similar wording (περὶ εὐθυμίας τι) in Paccius' request, and can thus be interpreted as a periphrasis of the direct object ('I extracted <something> on tranquility from my notes'). It remains uncertain, therefore, whether Plutarch's notes were ordered thematically or not, but if his use of these notes was a standard practice for the composition of most of his writings – which is commonly accepted today –, this would certainly have facilitated the job[98].

In regards to the actual composition of *De tranquillitate animi*, one can vividly imagine Plutarch sitting at his writing desk, browsing through the personal notes that he amassed, perhaps over a fairly long period of time, from his own reading and research, and selecting the material that he found fit for transfer to his peer in Rome. When it comes to the precise purpose of

---

[97] L. Van der Stockt, 1999a, pp. 578–579.
[98] As for the bulk and systematisation of Plutarch's notebooks, cf. R.H. Barrow, 1967, p. 153.

drawing up notes, Plutarch indicates that he does this for personal reasons, viz. 'for myself' (*De tranq. an.* 464F: ἐμαυτῷ). Plutarch's notes were, therefore, primarily composed "by himself and for himself"[99]. With ἐμαυτῷ, Plutarch, thus, indicates that the notes were intended to serve his own memory. In this sense, they preserve material and thoughts that were dear to him, that is, in which he was personally interested. It was not Plutarch's intention to keep this knowledge to himself, though. On the contrary, they served a more practical goal in securing knowledge and personal reflections that could later be revisited. These notes are made accessible to Paccius in such a way as to, first and foremost, cater to his practical ethical needs. Paccius' main concern is, so Plutarch assumes, not one of literary 'calligraphy': he expects practical information and instruction on the topic (464F: ἡγούμενος καὶ σὲ τὸν λόγον τοῦτον οὐκ ἀκροάσεως ἕνεκα θηρωμένης καλλιγραφίαν ἀλλὰ χρείας βοηθητικῆς ἐπιζητεῖν καὶ συνηδόμενος).

Van der Stockt is probably correct in suggesting that Plutarch is not taking refuge in a literary τόπος here[100]. In fact, as a meticulous analysis of *De tranquillitate animi* would suggest, Plutarch's re-editing of his rudimentary notes does not so much aim at an upgrade of the literary stylistics, but rather at a reorganisation of that material by presenting it as a more or less continuous line of thought (with some inevitable defects

---

[99] L. Van der Stockt, 1999a, p. 579. Therefore, when speaking of Plutarch's ὑπομνήματα in this study, I do this in reference to his 'personal notes', as understood by Van der Stockt and the Leuven school of Plutarchists. I am well aware, however, that Plutarch – and with him many other ancient authors – used the term ὑπομνήματα (and other concepts derived from it, e.g. ὑπομνηματισμοί) not only for private documents, but also for published works: e.g., the journals of Sulla (*Sull.* 5, 5) or Caesar (*Comp. Dem. et Cic.* 3, 1, *Ant.* 15, 5). Some treatises on ethical matters are also designated as being ὑπομνήματα by Plutarch (*De Al. Magn. fort.* 328A). In *Adv. Col.* 1115B, Plutarch refers more specifically to Aristotle's physical and ethical ὑπομνήματα. The term is also used for other public records (e.g., *Sol.* 11, 2 and *De fort. Rom.* 326A). It is even used for certain institutions, like festivals (e.g., *Cam.* 33, 7). See L. Van der Stockt, 1999a, p. 576, with n. 18. See already A. von Premerstein, 1900, cols. 726–757 and F. Bömer, 1953, esp. pp. 215–226. Seeing that these are mostly published works, scholars speak of a genuine hypomnematic genre in ancient literature (A. von Premerstein, 1900, cols. 757–759, esp. col. 757 for scientific and technical *commentarii*, F. Bömer, 1953 and D. Amboglio, 1990, p. 503, esp. p. 506 for Plutarch's notion of ὑπομνήματα). The concept of ὑπομνήματα was also used, e.g., for technical commentaries on philosophical texts (see, e.g., F. Ferrari and L. Baldi, 2002, pp. 12–16).

[100] L. Van der Stockt, 1999a, p. 577, n. 9 and 1996b, p. 265, n. 3 (*pace* R. Flacelière, J. Irigoin, J. Sirinelli and A. Philippon, 1987, p. xxxv, n. 3: "coquetterie littéraire"). For similar formulations of Plutarch's intentional lack of literary embellishment, cf. *Reg. et imp. apophth.* 172BE and *Mul. virt.* 243A (with L. Van der Stockt, 1996b, pp. 266–272). See also M. Beck, 2010, p. 349: "Plutarch often was working under time constraints. His multiple duties as Delphic priest and town official in Chaeronea placed some limitations on the amount of time he could devote to literary pursuits."

in form and structure, though). As a consequence, the reader should not expect to find a highly embellished discourse, but rather one that remains relatively simple, concise, and contains information that is pertinent for direct instruction.

We do not know what Plutarch's personal notes looked like in terms of their level of composition and elaboration. Van der Stockt is inclined to conceive of a Plutarchan ὑπόμνημα "as a more or less elaborate train of thought, involving material previously gathered and certainly written in full syntactical sentences: we are beyond the stage of heuristics"[101]. Regarding its level of composition, he believes that a ὑπόμνημα "does not yet display literary finish" but "probably took the form of a rough draft"[102]. This brings us very close to the compositional level of the genre of problems, as described in the previous chapter [see 1.2.3.], but it remains to be seen whether the genre of problems can actually be considered hypomnematic, and, if so, to what extent.

In order to answer this question, let us return for a moment to the earlier mention of the complex embroidery and reorganisation of hypomnematic material in *Quaestiones convivales*, particularly in light of what Fuhrmann says:

> "Les *Propos de Table* sont, en grande partie, des développements [...] de notes prises par Plutarque sur ses lectures, notes tout à fait

---

[101] L. Van der Stockt, 1999a, p. 595. Plutarch's assembling and compiling (συνάγειν, συντάττειν) of material as a preparatory phase for the composition of his texts is explicitly marked, e.g., in *Cons. ad Apoll.* 121E, *Coni. praec.* 138C, *De coh. ira* 457D, and *Nic.* 1, 5.

[102] *Ibid.* Similarly, K. Ziegler, 1951, col. 787 argues that the term ὑπομνήματα in *De tranq. an.* 464F "ja nicht nur Auszüge aus Quellenschriften, sondern mindestens in gleichem Maße auch Niederschriften eigener Gedankengange bezeichnet". Cf. L. Van der Stockt, 1999a, p. 576: "we are not entitled to view hypomnemata as sources". Cf. also, e.g., R.H. Barrow, 1967, pp. 66–76, 109–110 and esp. p. 153: "Plutarch's notebooks contained not only quotations which seemed to him of appeal or of use, no doubt classified, but also summaries and abstracts, some at length, some little more than main headings, and no doubt the innumerable miscellaneous jottings which so assiduous a collector could not resist." This is not, of course, to reject the basic doxographical interests of Plutarch's notes. In *De coh. ira* 457DE, Fundanus (who is considered Plutarch's spokesman: see H. Martin, 1969, p. 69, with n. 30) states that he 'collects and peruses sayings and deeds of both philosophers and kings and tyrants' (συνάγειν ἀεὶ πειρῶμαι καὶ ἀναγινώσκειν οὐ ταῦτα δὴ νοῦν μόνα τὰ τῶν φιλοσόφων, οὕς φασι χολὴν οὐκ ἔχειν οἱ νοῦν οὐκ ἔχοντες, ἀλλὰ μᾶλλον τὰ τῶν βασιλέων καὶ τυράννων). Cf. also J. Opsomer, 1994a, p. 8 (with n. 15). The sayings and deeds of tyrants are collected in Plutarch's collections of *Apophthegmata*; those of philosophers are no longer extant. For further discussion on the use and status of Plutarch's collections of *Apophthegmata*, see P.A. Stadter, 2008 and M. Beck, 2010 (who distinguish the *Regum et imperatorum apophthegmata*, as opposed to the *Apophthegmata Laconica*, from Plutarch's 'primary' ὑπομνήματα).

semblables à ses *Questions Grecques, Romaines, Physiques* et aux *Problèmes* attribués à Aristote. Comme dans ces recueils, plusieurs réponses étaient données à chaque question, avec quelquefois une explication propre à l'auteur lui-même."[103]

Other scholars would not care as much to make any distinction between Plutarch's notes and problems as Fuhrmann does here (he writes "tout à fait semblables")[104]. It remains unclear, however, how much Fuhrmann himself would actually distance Plutarch's problems from his notes, since he agrees with Bolkestein (in what immediately follows after the quoted passage) that *Quaest. Rom.* 64, 279DE and 75, 281F are authentic ὑπομνήματα for the parallel accounts in *Quaest. conv.* 702Dff. (on the ancient Roman custom of not allowing a table to be removed empty after eating, nor to let a lamp be extinguished)[105]. Elsewhere, Fuhrmann describes these "notes" as "ces ébauches" or "recueils inférieurs"[106], presumably in light of the absence of καλλιγραφία mentioned in *De tranq. an.* 464F.

One may wonder, however, why Fuhrmann leaves the *Quaestiones Platonicae* unmentioned. Is it because these are ζητήματα rather than αἰτίαι and display a higher degree of elaboration, as can be inferred from their average length [see 1.1.6., n. 153]? Even then, other scholars include this collection among Plutarch's preparatory notes just as well, by arguing that it represents the raw material waiting to be incorporated in other writings, including *Quaestiones convivales*[107]. According

---

[103] F. Fuhrmann, 1972, p. xiii. Cf. also K. Hubert, 1911, pp. 174–176 and H. Bolkestein, 1946, p. 27. According to S.-T. Teodorsson, 2009, pp. 14–15, *Quaestiones convivales* is "based in part on his own [sc. Plutarch's] remembrances and notes and in addition on collections of *Problemata* and *Zetemata* and a great number of other sources".

[104] According to H.J. Rose, 1924, p. 51, *Quaestiones Romanae* is, in fact, "a series of selections from reading-notes" (cf. also pp. 48–49). Regarding *Quaestiones Romanae, Graecae* and *barbaricae*, R.H. Barrow, 1967, p. 66 argues in a similar way that these are drawn from Plutarch's reading-notes (cf. also p. 69). See also W.R. Halliday, 1928, p. 13: "The matter of both [sc. *Quaestiones Romanae* and *Graecae*] is derived from literary sources, and they consist essentially of a collection of notes, which Plutarch has put together, perhaps over a fairly long period, from his miscellaneous reading." Cf. also A. Carrano, 2007, p. 9.

[105] F. Fuhrmann, 1972, p. xiii and H. Bolkestein, 1946, pp. 27–35.

[106] F. Fuhrmann, 1964, p. 19 (cf. also 1972, p. xix, n. 2).

[107] Cf. H. Cherniss, 1976, p. 2. He believes that we are dealing in *Quaestiones Platonicae* with what we today would call 'collected notes', which he distinguishes from their literary equivalent, the symposium (alluding to *Quaestiones convivales*). He does not deny, however, that these collected notes and symposia could both be made available to interested readers [see 2.4.1.]. Cherniss notes, moreover (p. 3) that "Plutarch himself in his *Symposiacs* uses the term ζητήματα of the questions or problems there propounded and discussed, of which several without their literary embellishment could appropriately

to Opsomer, however, there can be absolutely no doubt that these are *not* rudimentary or simple notes:

> "Les *Questions Platoniciennes* ont une structure bien organisée et élaborée et il serait incorrect de penser qu'elles ne contiennent que le matériel brut. En outre, elles sont plus que de simples notes personnelles (ὑπομνήματα). La structuration et la construction méticuleuses des différentes *Questions* nous indiquent qu'elles ont été préparées pour être publiées."[108]

The same seems to be the case for Plutarch's *Quaestiones naturales*. In this work, the aspect of elaboration is confirmed by the author's ample collection of content matter and also by such formal characteristics as, for instance, the use of full syntactical sentences with hypotactic structures, the presence of rhetorical elements, including emphatic addresses to the reader, the global structuring of the aetiologies along the principle of increasing plausibility, and the thematic clustering of problem chapters around specific topics.

I will come back to the issue of the publication of Plutarch's collections of *quaestiones* later on [see 2.4.1.]. What is important here is the idea that Plutarch's simple notes only contain the rough and unfinished material (these are the ὑπομνήματα mentioned in *De tranq. an.* 464F), whereas Plutarch's collections of *quaestiones* display a higher level of elaboration. As a result, the level of composition of Plutarch's *quaestiones* is not situated at the primary level of the simple notes, but at a higher, secondary level of elaboration, where remnants of the original hypomnematic material are incorporated at certain points and further elaborated upon to fit the arguments at hand[109]. The same conclusion was made by Senzasono:

---

have been included in the *Platonic Questions*, just as all the latter could have been used as material for the *Symposiacs*." Such Platonic questions are raised in *Quaest. conv.* 697F, 700C, 718B and 739E. He also argues (p. 4, n. b) that *De def. or.* 421E–431A is perhaps an elaboration of a Platonic ζήτημα devoted to a passage in *Tim.* 55cd (on the number of worlds). Cf. also K. Ziegler, 1951, col. 834.

[108] J. Opsomer, 1996a, p. 83 (see also 1994a, p. 12 and 2010, p. 95).

[109] Opsomer (*ibid.*) does not explicitly deny the idea that we may be dealing with more complex notes in Plutarch's *Quaestiones Platonicae*, as opposed to more simple ones, but he does not explicitly assert this either (in 2010, p. 115 he notes, however, that the ὑπόμνημα Plutarch possibly used in composing the fifth *Quaestio Platonica* may have had a specific 'problematic' organisation itself). This distinction may be relevant here, as there may be some ground for assuming a substantial difference in gradation between rudimentary, simple notes and more elaborate, complex ones without further terminological discrimination being made by Plutarch himself (cf. also n. 99). The distinction between simple and complex notes may have some concrete basis in other, more or less, contemporary miscellanistic authors. For instance, Gellius' conception

"Cosí, nel complesso, le *Quaest. nat.*, nate forse come raccolta di appunti, presentano un ordine e una struttura stilistica che hanno le caratteristiche che abbiamo indicato, frutta d'indubia elaborazione formale." "Plutarcho ha rifinito lo stile, forse partendo da appunti [...]. Lo stile induce a supporre che Plutarco abbia formalizzato degli appunti presi per interesse scientifico [...]."[110]

Seeing that the level of elaboration of *Quaestiones naturales* oscillates between simple notes and a fully elaborated treatise[111], it is not so remarkable that some hypomnematic features are still noticeable in the collection. Even though there is a certain degree of elaboration of the content and structure in Plutarch's *quaestiones*, a specific element of, what can be labelled, 'hypomnematic negligence' has not been eliminated[112]. This negligence involves a compositional sloppiness at times that takes effect on several levels of the discourse. In the case of *Quaestiones naturales*, Plutarch's desire to be both precise and concise in his rendering of the arguments often ends up in obscurity (e.g., *Q.N.* 4, 913A: the comparative ἀναγκαιότερα is confusing). At times, he is also rather careless in using sources (e.g., *Q.N.* 1, 911E: Aristotle does not vindicate but, rather, reject the popular belief that seawater contains burnt earth) and is sometimes inaccurate and inexact in his claims (e.g., *Q.N.* 5, 913AB: Plutarch says that there are eight generic flavourings but sums up nine). Likewise, his cross-references between successive problems are not always successful (e.g., *Q.N.* 24, 917F contains a ghost-reference to what was previously said but cannot be clearly retraced). Sometimes, Plutarch does not answer a question in its entirety (e.g., *Q.N.* 21, 917B: he does not explain why all of the wild sows farrow at the same time, nor why

---

of 'notes' in his preface to the *Noctes Atticae* is confused or at least confusing. In *NA Praef.* 2–4 Gellius designates *Noctes Atticae* as being *commentarii, commentationes* or a *commentarius* in itself, but he also refers to the *annotationes* (*ad subsidium memoriae*) that provided the primary material for that work. See T. Dorandi, 2000, pp. 39–42 and L. Holford-Strevens, 2003, p. 33. There are four instances, moreover, where Gellius draws from Plutarch's *Quaestiones convivales*. In each of these passages, Gellius simply ignores the sympotic setting, so that the style of the *commentatio* comes close to that of a relatively unembellished problem. These passages are discussed by F. Klotz and K. Oikonomopoulou, 2011, pp. 235–236. See also L. Holford-Strevens, 2003, pp. 283–285. The passages at issue (with their parallel in *Quaestiones convivales*) are *NA* 3, 5 (~ *Quaest. conv.* 705E); 3, 6 (~ *Quaest. conv.* 724EF); 4, 11, 13 (~ *Quaest. conv.* 730B); 17, 11 (~ *Quaest. conv.* 697F–700B).

[110] L. Senzasono, 2006, pp. 18 and 45.

[111] L. Van der Stockt, 2011, p. 452, n. 30 reached a similar conclusion: "The relative lack of embellishment [...] of *QN* may reflect the intermediary stage between *hypomnema* and formal edition."

[112] For the aspect of compositional negligence in Plutarch's ὑπομνήματα, cf. also L. Van der Stockt, 1987, p. 287.

domesticated ones farrow at various moments), or the question itself is formulated badly (e.g., *Q.N.* 19, 916B: Plutarch criticises Theophrastus for explaining only the octopus' change of colour and not also its adaptation, but in the *quaestio* he himself mentions only the change). Arguably, these aspects of hypomnematic negligence can be attributed to the speed with which Plutarch composed his *quaestiones*. Signs of hasty composition have also been detected, for instance, in *Quaestiones convivales*[113], and the haste with which Plutarch made extractions from his personal notes is known from the ὑπομνήματα statement itself, discussed above (*De tranq. an.* 464E–465A).

From reading *Quaestiones naturales* one gets the impression that Plutarch not only tries to maintain the clarity but also the *momentum* of his streams of thought. The hypomnematic character of this work can be connected with the style of discourse of the Ps.-Aristotelian *Problems* (which underlines the relevance of the genre again [see 1.2.5.]). Indeed, scholars have argued that the *Problems* also have a specific hypomnematic disposition, to be explained in light of its educational origins[114]. In fact, already in the 6th century AD the hypomnematic character of the Aristotelian *Problems* was recognised on the basis of the collection's form and presentation. In the preface to his *Commentary on Aristotle's Categories*, the commentator Elias (David) draws up a classification of Aristotle's writings by distinguishing the so-called hypomnematic from the syntagmatic works[115]. As opposed to the syntagmatic works, the hypomnematic ones record only the gist of the matter (114, 2: ὑπομνηματικὰ μὲν λέγονται ἐν οἷς μόνα τὰ κεφάλαια ἀπεγράφησαν). They are further narrowed down into uniform and miscellaneous writings (114, 8: τῶν δὲ ὑπομνηματικῶν τὰ μὲν μονοειδῆ τὰ δὲ ποικίλα). Among the miscellaneous hypomnematic writings, Elias lists a collection of ἑβδομήκοντα βιβλία Περὶ συμμίκτων

---

[113] See F. Fuhrmann, 1972, pp. ix–x and J. Sirinelli, 2000, p. 369. Notably, in *Quaest. conv.* 612E, Plutarch promises to Sossius Senecio to send the rest of the work quickly (ταχέως). On signs of hasty composition also in the *Vitae* (attributed to the same Sossius Senecio), see C.T. Michaëlis, 1875, pp. 8–9 and C. Pelling, 1979, pp. 95–96.

[114] According to C. Jacob, 2004, pp. 43–44, for instance, Ps.-Aristotle's *Problems* are not as such "*hypomnémata* désordonnés et hétérogènes" but still "ébauches rédactionnelles": "On y trouverait non seulement un recueil de phénomènes, mais aussi les premières versions d'explications qui seront reprises, complétées ou remplacées dans les textes plus systématiques." Cf. also P. Louis, 1991, pp. xx, xxv and xxix: "Le style est révélateur de ce genre d'écrits. Les phrases sont souvent mal construites. Certaines sont incomplètes. Elles sont tantôt très courtes, tantôt exagérément longues, avec parfois plusieurs incises qui les rendent difficiles à comprendre. Il arrive même qu'elles se contredisent. N'est-ce pas là la marque de phrases rédigées ou copiées à la hôte? Mais ce qui fait justement l'intérêt de la plupart de ces problèmes, c'est la spontanéité du premier jet."

[115] For further discussion of Elias' account (and similar accounts in the works of other Aristotle commentators), see T. Dorandi, 2000, p. 85.

ζητημάτων (114, 13–14), which probably included the Ps.-Aristotelian *Problems* [see 1.1.3., n. 87]. Similar to the educational context in which Ps.-Aristotle's *Problems* were composed (c.q. Aristotle's Lyceum), the composition of *Quaestiones naturales* can be interpreted in light of the author's activities in his philosophical school (I will deal with Plutarch's school activity in the next chapter [see 3.1.4.]). Full elaboration of the style of the natural problems was not necessary, at least if the basic line of thought was sufficiently elucidated for a good understanding by the implied readership. It turns out that mainly for educational purposes, then, Plutarch, much like Ps.-Aristotle, was mainly concerned with the gist of the matter rather than the form. The fact that *Quaestiones naturales* still has certain hypomnematic traits should not necessarily come at the cost of the work's autonomous position in the *corpus Plutarcheum*, but should rather be explained in light of the eventual purpose of the collection as a school text[116]. As such, Plutarch's collections of *quaestiones* more generally created an independent space for the author to collect his personal research and findings in a thematic fashion, a point that will be further substantiated for *Quaestiones naturales* in the following section.

3. Zetetic autonomy in *Quaestiones naturales*

As argued previously, Plutarch did not compose *Quaestiones naturales* (or his other collections of *quaestiones*) as a hypomnematic *Fundgrube* of materials to be exploited for the redaction of his other writings. It is not just a collection of residual problems that Plutarch simply had no room for in *Quaestiones convivales* (or elsewhere)[117]. It may very well be the case, however, that, as a consequence of his decision to restrict each Book of *Quaestiones convivales* to ten chapters only, Plutarch ran out of space at certain points in the process of composing this work (see n. 54)[118].

---

[116] Cf. L. Senzasono, 2006, p. 47: "Può darsi che egli abbia scritto l'opera coll'intento di servirsene in seguito, ma concependola come fatto letterario autonomo e compiuto, oppure, com'è piú probabile, che abbia concepito il proposito di servirsene dopo la stesura ben elaborata dell'opera."

[117] Therefore, K. Ziegler, 1951, col. 857 (cf. also col. 887) rightly remains uncertain when he wonders: "Ist es [sc. *Quaestiones naturales*] vielleicht ein Rest von Materialien, die in die Symposiaka nicht mehr Aufnahme fanden?". In my opinion, the answer is negative. Notably, Ziegler vindicates the autonomy of *Quaestiones Romanae* (*vis-à-vis* the *Vitae*) with less doubt (col. 860).

[118] Plutarch's own established rule of a fixed number of ten chapters for each Book is in itself a severe restriction of the work's scope, but at the same time it is also a realistic decision on the side of the author *not* to chase encyclopaedic comprehensiveness. There are some ways to create extra space, though. Two problems can, for instance, merge under one heading (e.g., *Quaest. conv.* 664A, 684E, 700BC, 706E, 717A, 723A, 725F, 727A, 740F). Moreover, in Book nine Plutarch makes an explicit exception to his own rule of a

Remarkably enough, some of the problems that are only mentioned in passing in *Quaestiones convivales*, receive detailed elaboration in *Quaestiones naturales*. These I will discuss below in order to illustrate that the separate and meticulous treatment of these problems speaks to the autonomous character of *Quaestiones naturales*, rather than to their status of zetetic leftovers.

In *Quaest. conv.* 724EF, first of all, Plutarch compares the palm tree with a well-trained athlete, who possesses an unbendable vigour both in body and mind. A piece of palm wood is said to curve upward as though resisting a weight imposed upon it. It is only in *Q.N.* 32 that Plutarch provides an extensive aetiology for this natural phenomenon (amounting to three explanations in total). Similarly, in *Quaest. conv.* 700F, no one of the symposiasts ventures upon an explanation of the sweet tears of wild boars as opposed to the salty tears of deer, a detailed inquiry of which is offered in *Q.N.* 20[119] (which contains two explanations). The natural phenomenon of wild figs preventing domesticated figs from dropping their fruit, and thereby promoting their ripening, is mentioned in the same passage. This phenomenon is explained in the course of *Q.N.* 41 (containing one extensive explanation), but remains without an explanation in *Quaest. conv.* 700F.

Both phenomena of tears and figs (among others) serve as paradoxographical examples that Euthydemus and Patrocleas cite from their experience in farming and hunting (700E: οὐκ ὀλίγα τοιαῦτα τῶν ἀπὸ γεωργίας καὶ κυνηγίας προφέροντας) in order to support Plutarch's excuse for not accounting for the central problem of the so-called horncast seeds[120].

---

fixed number of ten chapters for each Book under the pretext of bringing an appropriate tribute to the nine Muses on their own festival (this is the setting of the fifteen chapters that follow). See *Quaest. conv.* 736C: ὁ δ' ἀριθμὸς ἂν ὑπερβάλλῃ τὴν συνήθη δεκάδα τῶν ζητημάτων, οὐ θαυμαστέον· ἔδει γὰρ πάντα ταῖς Μούσαις ἀποδοῦναι τὰ τῶν Μουσῶν καὶ μηδὲν ἀφελεῖν ὥσπερ ἀφ' ἱερῶν, πλείονα καὶ καλλίονα τούτων ὀφείλοντας αὐταῖς. According to S.-T. Teodorsson, 1996, p. 300, "[p]erhaps we may suppose that Plutarch, when setting out to write book IX, happened to see that he had a number of interesting questions left which he could not refrain from including." However, in this last Book no natural problems are discussed after the manner of *Quaestiones naturales*. Chapters 10 through 12 concerned natural problems in a more general sense, though, as we can learn from the titles of the lost chapters (talks 10 and 12 deal with more astronomical issues, and talk 11 with an ontological rather than a physical problem).

[119] The same observation was made by F.H. Sandbach, 1965, pp. 138 and 193, n. b. He adds that *Q.N.* 20 may, therefore, be composed subsequently to *Quaest. conv.* 700F, or that they are at least contemporaneous, but this is uncertain. Cf. S.-T. Teodorsson, 1996, pp. 40–41 (see also further).

[120] According to S.-T. Teodorsson, 1996, p. 41 the phenomena mentioned in *Quaest. conv.* 700F are only enumerated as examples here, because there existed well-known explanations for them (as *Q.N.* 20 and 41 show). Cf. also L. Senzasono, 2006, p. 202, n. 108.

In what follows, Florus asserts that these (and similar) problems are not childish nonsense, and that they should not be given up as *insoluble* (701A: Ἐπεὶ δὲ τοῦτο μὲν ὁ Φλῶρος ᾤετο παιδιὰν εἶναι καὶ φλύαρον, ἐκείνων δ' οὐκ ἄν τινα τῆς αἰτίας ὡς ἀλήπτου προέσθαι τὴν ζήτησιν)[121]. This is, indeed, proven by the aetiologies in *Q.N.* 20 and 41, where extensive explanations are found. The fact, then, that the two problems at issue are only mentioned in passing and remain "notoriously unsolved"[122] in *Quaestiones convivales*, whereas they do receive a detailed aetiology in *Quaestiones naturales*, is very significant, since it suggests that *Quaestiones naturales* does not necessarily 'need' the sympotic framework of *Quaestiones convivales* to accomplish its own zetetic goal. Or in other words: Plutarch does not necessarily require a boost of wine to solve these problems (cf. *Quaest. conv.* 700E: τὰς ζητήσεις πολὺ προθυμοτέρας καὶ θρασυτέρας τὰς ἀποφάνσεις τοῦ οἴνου ποιοῦντος). If the two chapters in *Quaestiones naturales* where these problems are solved (i.e. *Q.N.* 20 and 41) are to be considered editorial remains – on the uncertain assumption, *nota bene*, that their composition is chronologically prior to or contemporary with that of the discussions in *Quaestiones convivales* –, why are they worked up in such aetiological detail? Why did Plutarch not simply delete them altogether? If it is because they were composed subsequently to *Quaestiones convivales*, we cannot speak of them as leftovers at all, but rather as a continuation of the same zetetic project (but then again, matters of chronology remain unclear [see the prologue]).

Rather than assuming that the research conducted in *Quaestiones naturales* is actually done 'for' *Quaestiones convivales*, I firmly believe that it is at least equally plausible that Plutarch relies on, incorporates and elaborates the same or similar hypomnematic material into the problematic framework of *Quaestiones naturales* as well as into the more dramatised and literary context of *Quaestiones convivales*[123]. This explains why there are numerous parallel passages between the two works, including their

---

[121] What Florus presumably *does* consider to be παιδιά (if ἐκείνων in his reply, indeed, refers back to ταῦτ' in Euthydemus' account – as it does in the translation of E.L. Minar, F.H. Sandach and W.C. Helmbold, 1961, p. 25) is what Euthydemus adds: these are the popular beliefs that celery grows better if it is trampled and crushed as it grows, and that the same is true for cumin if it is sown with curses and maledictions.

[122] F.H. Sandbach, 1965, p. 193, n. b (Sandbach does not, however, mention the second parallel on fig trees). Solving these and similar problems is not considered a sinecure, though (cf. 700D: τὴν αἰτίαν ἀνεύρετον, 700DE: πρᾶγμα πίστιν ἔχον ὅτι γίγνεται, τὴν δ' αἰτίαν ἔχον ἄπορον ἢ παγχάλεπον).

[123] Cf. F.H. Sandbach, 1965, p. 138: "*A priori* it might be guessed that the former [sc. *Quaestiones naturales*] provided raw material that was worked up into a literary form in the latter [sc. *Quaestiones convivales*]. [...] The facts in general do not [however] seem to exclude the possibility that material found for the *Symposiacs* was used in composing the *Quaestiones Naturales* and vice versa."

slight divergences in argument at times. It would also account for the fact that certain topics, as the cases above have shown, are touched upon only superficially in *Quaestiones convivales*, while they receive a separate and circumstantial aetiology in *Quaestiones naturales*. It is only reasonable, then, that *Quaestiones naturales* is not a zetetic appendix to *Quaestiones convivales*. In fact, one could turn the tables and argue, the other way round, as Harrison did, that:

> "[*Quaestiones convivales*] allowed Plutarch to examine customs and phenomena that might not have found comfortable places within other of his Αἰτίαι and the exegesis of which did not warrant a separate essay."[124]

Whatever may be the case, *Quaestiones naturales* clearly has a high degree of zetetic autonomy, offering to its author plenty of space for the treatment of natural problems mostly on their own terms (i.e. in view of their physical causality without much further consideration of matters regarding style, morality etc.). The same aspect of zetetic autonomy can also be presumed, then, for Plutarch's other collections of *quaestiones* (viz. *Quaestiones Romanae*, *Graecae* and *Platonicae*), which, in this logic, portray several specialised fields of interest that reflect the author's versatile research occupations[125]. Let it be absolutely clear, however, that the notion of compository autonomy does not, of course, imply that Plutarch's collections of *quaestiones* were composed in complete intellectual isolation from each other. As we will see in the following section, sometimes there are clear thematic overlaps between these works, meaning that they are part and parcel of a larger, overarching research project, a project inspired by Plutarch's quest for all-round πολυμάθεια [see 2.4.2.].

Another point is that the differences in form between Plutarch's *Quaestiones convivales* and his other collections of *quaestiones* involves a difference in implied reading and readership. I will argue in the next chapter that there are, indeed, good reasons to believe that there is a difference in authorial intention in these works [see 3.1.4.]. Before this is

---

[124] G.W.M. Harrison, 2000a, p. 197. In a similar vein, F. Klotz and K. Oikonomopoulou, 2011, p. 20 have pointed out that "the *Table Talk*'s inquiries are, in the overwhelming majority of cases, similar or approximate, but never identical to those of the other collections, suggesting that Plutarch self-consciously avoided close replication of material across the corpus. And, whereas its answers often employ scientific theories and arguments that are encountered in the other collections as well, they never do so with the same ends in view."

[125] As G.W.M. Harrison, 2000a, p. 198 has argued, therefore, "[o]ne hardly feels constrained any more to state that these essays were meant to stand on their own [...]."

possible, however, we must first focus on the issue of publication itself of Plutarch's *quaestiones*.

2.4. Opening up Plutarch's zetetic archive

The aim of this section is to investigate the possibility of a publication of Plutarch's collections of *quaestiones*, and of *Quaestiones naturales* more specifically. What indications do we have that these works were published or were at least intended to be made public one way or another? A detailed answer to this question will, in turn, provide further food for thought in the next chapter about the educational context in which these collections were useful.

1. The issue of publication: problems as functional literature

When it comes to the issue of publication of Plutarch's collections of *quaestiones*, only little is known with any certainty, leaving much room for conjecture. We will see here that there is reason to assume that Plutarch probably disclosed his collections of *quaestiones* to his students and close peers who presumably shared his interest for their different strands of inquiry. Plutarch's philosophical school was, in all likelihood, situated in his own house in Chaeronea [see 3.1.2.]. Perhaps his students even took up residence there. Considering that Plutarch complains about the lamentable availability of books in small towns like his own[126], one can very well imagine that he, in his role as a helpful teacher, not only allowed them access to his own library, but also to his thematically ordered collections of *quaestiones* that he had composed on the basis of his own readings and discussions. But of course, it is only likely that he also granted requests made by friends living abroad (as the ὑπομνήματα statement nicely illustrates; see above). Plutarch's intellectual repute, although centered in Chaeronea, must have radiated throughout the entire Mediterranean region. As we will see in what follows, there are several indications to make these points more concrete.

Regarding Plutarch's *Quaestiones Platonicae*, first of all, scholars have argued that they "might be made available to interested readers"[127]. Notably, in *Quaest. Plat.* 1003A, Plutarch reports that a specific issue 'has already been frequently discussed by us' (τὸ πολλάκις ὑφ' ἡμῶν λεγόμενον)[128]. This phrase, and especially the use of the first person plural, would not have much sense unless a certain audience, presumably situated in the context

---

[126] Cf. *De E* 384E (quoted n. 133) and *Quaest. conv.* 675B. See D.A. Russell, 1973, pp. 42–43 (with J. Sirinelli, 2000, p. 365). On the smallness of Chaeronea, see *Dem.* 2, 2 (with R. Flacelière, J. Irigoin, J. Sirinelli and A. Philippon, 1987, pp. lix–lx).

[127] H. Cherniss, 1976, p. 2.

[128] See J. Opsomer, 1994a, p. 12 (see also 1996a, p. 83).

of Plutarch's classroom, is implicitly being addressed by it[129] (as we will see later on, similar personal forms can be found throughout *Quaestiones naturales* [see 3.1.4.]). Another indication of the publishable character of *Quaestiones Platonicae* is found in the ὑπομνήματα statement in *De tranq. an.* 464E [see 2.3.2.]. As seen previously, this passage speaks of Paccius' petition to Plutarch to write him something not only περὶ εὐθυμίας but also περὶ τῶν ἐν Τιμαίῳ δεομένων ἐπιμελεστέρας ἐξηγήσεως. The part about the 'passages in *Timaeus* that require futher elucidation' can, and has, been linked to the exegetical chapters on the *Timaeus* in *Quaestiones Platonicae*[130]. Cherniss argued, in this regard, that the ὑπομνήματα that Plutarch mentions in *De tranq. an.* 464F probably contained "such things as our ζητήματα [sc. *Quaestiones Platonicae*] or the material for them"[131]. The second seems more plausible, though[132]. What is particularly unlikely, however, is that Plutarch eventually sent this *Timaeus* exegesis to Paccius in its rough hypomnematic form (after all, *De tranquillitate animi* are no rough ὑπομνήματα either). Therefore, Plutarch probably elaborated this exegetical material, presumably in the form of one or more *quaestiones*, before handing it over to his friend Eros. If this is true, the passage at hand implies a transfer of one or more chapters from *Quaestiones Platonicae* to Rome. It would be absurd to claim, of course, that the geographical distance is a prerogative for the disclosure of such *quaestiones*. If Plutarch was prepared to disclose this knowledge to acquaintances living abroad, why would he not do the same for those living closer to home, viz. to his close students and friends[133]?

---

[129] J. Glucker, 1978, p. 264, n. 27 notes, moreover, that "[f]rom *De An. Procr.* 1012B, with its obvious reference to *Plat. Quaest.* it appears that it is his students in this particular lecture who demand an exposition of this problem. It thus seems that *Plat. Quaest.* owe their origin largely to things said (εἰρημένα) in the classroom" (also cited by J. Opsomer, 1994a, p. 13).

[130] *Quaest. Plat.* 2, 1000E–1001C, 4, 1002E–1003B, 5, 1003B–1004C, 7, 1004D–1006B, 8, 1006B–1007E. Scholars have considered it unlikely that Paccius is referring here to *De animae procreatione in Timaeo*, because this treatise is composed (much like *De tranquillitate animi* itself) in the form of an open letter to the author's sons Autobulus and Plutarch (1012A). However, *De animae procreatione in Timaeo* may in itself be based on one or more *Quaestiones Platonicae*, or rather on similar underlying notes. See H. Cherniss, 1976, p. 133 and F. Ferrari and L. Baldi, 2002, p. 9.

[131] H. Cherniss, 1976, p. 4, n. b.

[132] See J. Opsomer, 2010, pp. 94–95. Notably, Plutarch only mentions that he extracted περὶ εὐθυμίας from his personal notes, so the same is not necessarily true for the exegetical material of the *Timaeus* also, although this is not unlikely, of course. Paccius' twofold petition can perhaps be taken to imply, then, that he asks Plutarch to send him something on tranquility and – since he presumably has to browse through his personal archive anyway – to attach some exegetical material regarding the *Timaeus* as well.

[133] J. Glucker, 1978, p. 264 even argues that "it is not unlikely that his [sc. Plutarch's]

Another indication of the publishable character of Plutarch's collections of *quaestiones* is found in Plutarch's aforementioned self-references to *Quaestiones Romanae* in *Cam.* 19, 8 (ταῦτα μὲν οὖν ἐν τῷ Περὶ αἰτίων Ῥωμαϊκῶν ἐπιμελέστερον εἴρηται) and in *Rom.* 15, 7 (Περὶ ὧν ἐπὶ πλέον ἐν τοῖς Αἰτίοις) [see 2.1.4.]. These references would be meaningless if the collection was meant for personal use only, that is, as a private set of inquiries into Roman antiquities that was not made accessible to the intended reader of the *Vitae*. Therefore, scholars have accepted that *Quaestiones Romanae* was published by Plutarch himself and made accessible to a broader circle of readers who were interested in such antiquarian matters[134]. In this regard, it is important to note that these self-references explicitly indicate the aetiology to be more detailed in *Quaestiones Romanae* than in the *Vitae* (ἐπιμελέστερον ~ ἐπὶ πλέον). Plutarch, thus, openly promotes the antiquarian research from *Quaestiones Romanae* to the reader in these passages, by declaring that he has collected this kind of knowledge in a separate and more specialised way there. Arguably, due to the fact that the *Romulus* reference (considering the vagueness of the phrase ἐν τοῖς Αἰτίοις) is less precise than the *Camillus* reference (which specifically refers to *Quaestiones Romanae*), it is not unlikely that by the former, Plutarch is referring to his aetiological works in general, rather than to one specific collection of *quaestiones*. There is debate among scholars as to whether the *Romulus* reference refers specifically to *Quaestiones Romanae* or, more generally, to the entire triptych collection of *Quaestiones Romanae*,

---

Platonicae Quaestiones were written among other things, in response to such [c.q. Paccius'] inquiries", but it is just as likely that Plutarch already had the material on hand before such requests came to his address. Scholars have argued, moreover, that Plutarch himself also had access to other people's archives and libraries. See J. Sirinelli, 2000, p. 283: "Plutarque a sans doute beaucoup travaillé sur des notes ou des documents qu' on lui avait préparés." According to R.H. Barrow, 1967, p. 152, Plutarch perhaps even contacted his Roman friends for information via letters, just as Paccius does in the introduction to *De tranquillitate animi*. Such a request is, indeed, found in the introduction to *De E*, albeit in an address to Plutarch's Athenian comrad Sarapion (384E): ὅρα δ' ὅσον ἐλευθεριότητι καὶ κάλλει τὰ χρηματικὰ δῶρα λείπεται τῶν ἀπὸ λόγου καὶ σοφίας, <ἃ> καὶ διδόναι καλόν ἐστι καὶ διδόντας ἀνταιτεῖν ὅμοια παρὰ τῶν λαμβανόντων. ἐγὼ γοῦν πρὸς σὲ καὶ διὰ σοῦ τοῖς αὐτόθι φίλοις τῶν Πυθικῶν λόγων ἐνίους ὥσπερ ἀπαρχὰς ἀποστέλλων ὁμολογῶ προσδοκᾶν ἑτέρους καὶ πλείονας καὶ βελτίονας παρ' ὑμῶν, ἅτε δὴ καὶ πόλει χρωμένων μεγάλῃ καὶ σχολῆς μᾶλλον ἐν βιβλίοις πολλοῖς καὶ παντοδαπαῖς διατριβαῖς εὐπορούντων. The circulation of knowledge in this way (i.e. in the form of problems, notes, excerpts etc.), was in fact very customary among intellectuals in Plutarch's time, which certainly testifies to a high degree of intellectual freedom and promiscuity. See, e.g., E. Lao, 2008, p. 36 (with a discussion of Pliny, *Ep.* 3, 5, 17).

[134] See, e.g., H.J. Rose, 1924, pp. 47–48, K. Ziegler, 1951, cols. 857 and 860, R.H. Barrow, 1967, p. 66, J. Boulogne, 1992, p. 4687, 1994, p. 126, 1998, p. 31, J.-M. Pailler, 1998, p. 77 (*pace* W.R. Halliday, 1924, p. 13 with W. Nachstädt, W. Sieveking and J.B. Titchener, 1935, p. 274).

Graecae and the lost *barbaricae*[135]. If it is true, however, that Plutarch is referring to his aetiological writings in a more generic way, it is not unlikely that the reference also covers his other collections of Αἰτίαι, thus including *Quaestiones naturales*[136] [see 1.1.4., n. 91].

Interestingly, on the basis of these two self-references in the *Vitae* Boulogne designated Plutarch's *Quaestiones Romanae* to be an autonomous 'work of reference'[137], that is, so I take it, a depository of specialised antiquarian inquiries that is worth referring to whenever suitable. In line with Boulogne, I believe that *Quaestiones naturales* can be considered a 'work of reference' also. Indeed, the aetiologies in *Quaestiones naturales* are often more detailed than in their parallel accounts [see 2.1.2.]. Moreover, in *Lys.* 12, 7, Plutarch concludes the digression on the meteorite that fell in Aegospotami by referring the interested reader to the γένος γραφῆς of *Quaestiones naturales*, if it is not simply this work that Plutarch had in mind [see 2.1.4.]. Of course, Boulogne's designation of collections of *quaestiones* as 'reference works' should not be understood in a modern sense (i.e. as systematic encyclopaedias, handbooks, manuals or the like). In the end, there is no clear structure to these collections nor a table of contents, by the use of which material can be easily traced in the collection. By contrast, it seems that these collections provide a more general 'frame of reference', that is, a general explanatory framework in accordance with which problems could be properly solved.

As we will see in the next chapter, this 'referential framework' proves specifically useful for didactic purposes [see 3.2.1.]. Plutarch's collections of *quaestiones* could be consulted and used by the reader to retrieve clearly shaped but roughly finished information whenever this was needed. It is perhaps not inappropriate, therefore, to consider the genre of problems as an integral part of ancient 'functional literature' (*Gebrauchsliteratur*, *letteratura di consumo*)[138]. The aspect of utilisation and consumption

---

[135] See R.H. Barrow, 1967, pp. 66–67, esp. n. 1 (on p. 184) and J. Boulogne, 1998, p. 31 and 2002, p. 91. Cf. also T.S. Schmidt, 1999, p. 10, n. 38 and R. Preston, 2001, p. 95 (with n. 44).

[136] If the reference is, indeed, a generic one, this may explain the neutral form ἐν τοῖς Αἰτίοις – perhaps to be emended in the feminine ἐν ταῖς Αἰτίαις [see 1.1.6., n. 146]?

[137] J. Boulogne, 1992, p. 4686 and 1998, p. 31 ("un ouvrage de référence"); cf. also 2002, p. 92 ("un livre de référence sur Rome"). K. Oikonomopoulou, 2013a, pp. 144–147 has recently suggested that each of Plutarch's collections of *quaestiones* function as encyclopaedic-style reference works. Note, moreover, that Aristotle also explicitly refers to his *Problems* on several occasions throughout his writings: see *Mete.* 363a24 (cf. also 381b13), *Somn. vig.* 456a29, *Iuv.* 470a18, *PA* 676a18, *GA* 747b5, 772b11, 775b37. See P. Louis, 1991, pp. xxv–xxvii.

[138] See the contributions in O. Pecere and A. Stramaglia, 1996. J. Opsomer, 2010, p. 95 suggests (somewhat hesitantly) to speak of the implied reader in Plutarch's *quaestiones* as "un 'utilisateur', en présupposant un contexte didactique". Cf. also p. 115: "Le lecteur –

of ancient problem literature can, indeed, be illustrated from several papyri that were edited in the question-and-answer format, some of which contain extracts from Plutarch's *quaestiones*[139]. The existence of these papyrological sources, the origin of which is generally linked to the context of schooling, at least testifies to the importance of question-and-answer literature in ancient education. Arguably, the kind of theories, concepts and argumentative turns Plutarch uses in his collections of *quaestiones* could be reused and remoulded in new discussions concerning similar problems in any given situation (for instance during symposia, as described in *Quaestiones convivales*). As such, one could even compare these collections of *quaestiones* to, say, instruction manuals, atlases, books of recipes, or other kinds of 'open source' literature, that is, literature that is 'open' for free use and re-use by the reader. By its interrogative structure, the content of this kind of literature is, indeed, very dynamic, and the response triggered by it will differ from reader to reader in a considerably idiosyncratic fashion.

In conclusion, it is not unlikely that *Quaestiones naturales* did not remain locked up in Plutarch's office, but that it was made accessible to interested readers, or that the Chaeronean at least envisaged to prepare it for publication at some point, presumably in an educational context. In light of the idea that Plutarch's collections of *quaestiones* were probably made accessible to a group of interested readers that stood in close contact with Plutarch himself, Sirinelli has appropriately called the Chaeronean "un homme-ressource" (rather than that he was "un esprit véritablement encyclopédique")[140]. Regarding Plutarch's transfer of knowledge in the problem format, Sirinelli's conclusion is worth quoting in full:

"Plutarque a pu céder aux sollicitations de ses amis qui ont fait valoir l'intérêt que présenteraient ces collections [de problèmes] pour un

---

ou utilisateur – idéal devait être familier avec les règles du genre, qui n'incluent pas seulement l'usage idiomatique de certaines particules et phrases, mais aussi l'usage de certaines stratégies textuelles et argumentatives."

[139] Some papyrus fragments remain from *Quaestiones convivales* (PSI inv. 2055, PL III 543 A; see G. Messeri Savorelli and R. Pintaudi, 1997, pp. 174–177) and also (possibly) from *Quaestiones Graecae* (P. Oxy. 2688 and 2689; see W. Morel, 1969, p. 219). P. Oxy. 2744 (2nd century AD) contains Ps.-Arist./Alex. Aphr., *Suppl. probl.* 2, 156 (= *Probl. ined.* 2, 153) and attributes it to Aristotle's ἀπορήματα. See S. Kapetanaki and R.W. Sharples, 2006, p. 231, n. 466. In addition, several fragments from a papyrus codex of Books two to five from Ps.-Plutarch's/Aëtius' *Placita* have also survived (P. Antinoopolis 85 and 213). See also more generally R. Cribiore, 2001, pp. 208–209 and 212. In addition, there is a Greek papyrus from the 1st century AD (P. Berol. inv. 9764), which mentions the 'well-known study by problem' (lines 17–18: ἐπὶ τὸν πολυθρύλητον τὸν προβληματικὸν [...] λόγον). See M.-H. Marganne, 1998, pp. 13–34.

[140] J. Sirinelli, 2000, pp. 365–366.

public intéressé par les questions traitées; c'était du reste la mode des collections de citations, de pensées ou de faits marquants. [...] Ce n'était pas une tâche mineure et indigne d'un écrivain de qualité mais même, à une époque où les bibliothèques demeuraient rares, un instrument intellectuel très prisé. [...] Cette pratique prouve seulement une conception différente du métier et des devoirs d'écrivain et, dans le cas de Plutarque, nous éclaire un peu mieux sur ce qu'on attendait de lui, sur les sources du savoir à cette époque et ce qu'on entendait alors par la fonction de communication et d'information."[141]

Now that we have considered the publishable character of Plutarch's collections of *quaestiones* and their general usability as school texts, I will shortly reflect on their actual method of storage, with specific attention to aspects of thematic classification and overlap. Further indications of Plutarch's school context in *Quaestiones naturales* will be discussed in the next chapter [see 3.1.].

2. Classification and overlap

Jacob identified the genre of problems with sets of index cards ("fiches"), which are further categorised into thematic folders ("dossiers de travail")[142]. Even though, nothing is known with any certainty about the actual form and organisation of Plutarch's archive, scholars have often also conceived of it in a very concrete, physical way, imagining it as some sort of a systematised card-index box[143]. In a very similar way, Plutarch's collections of *quaestiones* serve as a discursive medium for the storage of several kinds of inquiries. They provide an accumulative textual format for the author's progressive research, where new problems and answers could always be added or older ones revised. The thematic categorisation of Plutarch's collections of *quaestiones*, by distinguishing several subsections (viz. antiquarian, scientific, literary, philosophical etc.), certainly improved their efficient usability, even if they are not catalogued in a fully

---

[141] J. Sirinelli, 2000, p. 365.

[142] C. Jacob, 2004, pp. 43–44 (with K. Oikonomopoulou, 2013a, p. 134). Jacob also points out that these problems are no simple notes because of their coherent language and style. Furthermore, in regards to the composition of *Quaestiones convivales*, J. Sirinelli, 2000, p. 386, n. 1 notes: "Il serait passionnant de pouvoir déceler dans les *Propos de table* tels qu'ils se présentent aujourd'hui les traces d'un classement des archives de Plutarque. On sent parfois dans telle ou telle séquence de plusieurs dîners, enjambant même la division en livres [...], un air de parenté qui suppose une classification originelle par thèmes et non par convives."

[143] See already A. Gudeman, 1927, col. 2526, who speaks of "Zettelkasten". See also T. Dorandi, 2000 more generally. On ancient conceptions of memory in itself as an 'archive', see J.P. Small, 1997, pp. 81–137, 224–239.

systematic fashion [see 1.1.4.]. Yet, at the same time, the technique of thematic categorisation seems to have had specific disadvantages, since on certain occasions, there may have been difficulties in classification.

Take, for instance, *Quaest. Plat.* 7, 1004D-1006B, which deals with the mechanism of ἀντιπερίστασις and its operation in several natural phenomena as discussed in *Tim.* 79e-80c[144]. Plutarch chose to classify this problem with *Quaestiones Platonicae*, considering the close link with Plato's text. But because of its focus on physics and natural phenomena, it would perhaps not have been out of place in *Quaestiones naturales* either[145]. There is, in fact, a close parallel on magnetism between *Quaest. Plat.* 1005BD and *Q.N.* 19, 916D (ἐν κύκλῳ περιιών ~ περιέλευσις), and the theory of ἀντιπερίστασις also recurs in *Q.N.* 13, 915B (moreover, Plato is quoted in *Q.N.* 1, 911D and 5, 913CD [see 4.2.1.1.]). I believe that this type of thematic overlap contributes to a sense of mutual coherence between Plutarch's collections of *quaestiones* more generally, which is also seen at work elsewhere.

Notably, there is also room for physical aetiology in *Quaestiones Romanae*[146], and the same is obviously true for the Αἰτίαι τῶν Ἀράτου Διοσημιῶν (frs. 13-20 Sandbach). Another interesting example can be found in *Quaest. Graec.* 7, 292CD, which deals with the so-called floating clouds. The link with Greek culture is not very clear in this chapter. In fact, it is rather problematic. Perhaps this chapter ended up in the wrong folder in Plutarch's archive by mistake. According to Halliday, "Plutarch would more tidily have placed [this 'alien'] among his *Aetia Physica*"[147]. This is certainly supported by the fact that Plutarch quotes from the fourth Book of Theophrastus' *Meteorology* (192 FHSG). The wrong (re?)location of this problem can perhaps speak to the practical, but at times, indeed, hasty and messy, use and consultation of Plutarch's collections of *quaestiones* (perhaps by his students?). However, at a more conceptual level, it is not unlikely that these examples rather demonstrate how physical aetiology is an important unifying factor that effectively contributes to the general coherence of Plutarch's corpus of *quaestiones*, and hence of his oeuvre more generally[148] [see 2.1.2.].

---

[144] See J. Opsomer, 1999 [see 4.3.1.2., n. 176].

[145] Cf. K. Oikonomopoulou, 2013a, p. 144.

[146] Cf. *Quaest. Rom.* 1, 263E, 2, 264B, 19, 268CD, 24, 269CD, 38, 273E, 77, 282CD, 78, 282EF, 101, 288B, 102, 288C, 106, 289C (ἢ φυσικώτερον ἔχει λόγον τὸ πρᾶγμα καὶ φιλοσοφώτερον), 111, 290AB. See J. Boulogne, 1992, pp. 4704-4706, 1994, p. 130 and 2002, p. 98.

[147] W.R. Halliday, 1928, p. 14. Other physical material is found in *Quaest. Graec.* 10, 293A (on the small plant called 'sheep-escaper') and to a lesser degree 9, 292E (on the month called 'Bysios', wrongfully associated with the word φύσιος, 'growth').

[148] Cf., e.g., C. Darbo-Peschanski, 1998, p. 28 for the 'cosmological' connection

Of course, physical aetiology is not the only unifying factor. Also the other way around, a number of chapters treated in *Quaestiones naturales* express a sensitivity for cultural-antiquarian inquiry, which reminds the reader of the problems treated in *Quaestiones Graecae*. This is the case, for instance, in *Q.N.* 10, 914D, where Plutarch refers in parenthesis to a story told about the people of Halieis, who received an oracle instructing them to dip Dionysus in the sea. Similarly, in *Q.N.* 14, 915C, Plutarch wonders why the people of Doris pray for a bad harvest of hay. And in *Q.N.* 23, 917F, he explains why people do not hunt in the vicinity of Mt. Etna in Sicily. Moreover, the method of incorporation of quotations from the poets, such as Homer or Aratus, is reminiscent of the exegetical-aetiological approach known from Plutarch's fragmentary Ὁμηρικαὶ μελέται and Αἰτίαι τῶν Ἀράτου Διοσημιῶν (Homer is quoted in *Q.N.* 5, 913D, 19, 916B, 20, 917A, 21, 917D, 34 and Aratus in *Q.N.* 2, 912D [see 4.1.2.3.])[149]. Arguably, these overlaps do not only testify to the, at times, very close affiliation between the different research projects in Plutarch's collections of *quaestiones*, but also reveal the openness and all-round applicability of many kinds of knowledge to different contexts – a dynamic that lies at the heart of Plutarch's πολυμάθεια project[150].

3. Conclusion and new questions

In conclusion, Plutarch's interest in natural problems is an important unifying factor throughout his entire oeuvre, both in the *Vitae* and the *Moralia*. The collection of *Quaestiones naturales* holds an important position in this regard, because Plutarch there aims to treat such problems mostly on their own terms. As such, the collection provides an independent textual medium that allows for Plutarch to autonomously store and retrieve the results of his research on specific natural scientific topics. He could refer the interested reader to this work whenever necessary, making it a work that was certainly publishable. From the perspective of composition, *Quaestiones naturales* oscillates between simple notes and

---

between *Quaestiones Romanae/Graecae* and the *Vitae*: "Les Αἴτια ne seraient donc pas un simple recueil de curiosités sur lesquelles un esprit érudit s'exercerait à des tentatives d'explication pour le plaisir de spéculer. Ils semblent s'inscrire dans la logique des oeuvres jugées les plus importantes de Plutarque et, comme les *Vies* ou les traités physiques, mettre au centre de leur propos la rationalité du devenir et du *cosmos* ainsi que les limites de la connaissance qu'on peut avoir de celle-ci."

[149] Notably, Plutarch's collection of Ὁμηρικαὶ μελέται (from which only frs. 122–127 Sandbach remain) also had specific physical interests: fr. 127 Sandbach concerns the atmospheric influence on the consistency of the shoots of plants.

[150] Indeed, the aspect of diversity (πολυειδία), as a typical feature of Hellenistic poetics, is brought to a climax in the miscellaneous *Quaestiones convivales*. See I. Gallo and C. Moreschini, 2000, pp. 14–15.

a fully elaborated treatise. The relationship with *Quaestiones convivales* is important, because both works are closely related from a text genetic perspective, probably being based on the same and similar hypomnematic material. It remains to be seen, however, to which degree Plutarch's intentions with these two works coincide in terms of their educational goals. New questions emerge, especially concerning the concrete reading of *Quaestiones naturales* (vis-à-vis *Quaestiones convivales*). Who are the intended readers of Plutarch's natural problems? What is their function and how do they relate to Plutarch's overarching educational project? These questions will demand a closer examination of the socio-cultural and intellectual-philosophical context from which Plutarch's natural problems emerge. These and related issues will be treated in the following chapter.

3
Quaestiones naturales and zetetic παιδεία

Throughout the *corpus Plutarcheum*, we find important information about the intellectual value of Plutarch's natural problems and their popularity in the author's social milieu. The aim of this chapter will be to focus on the socio-cultural and intellectual-philosophical backdrop of *Quaestiones naturales*. I will start by discussing the collection's social *Sitz im Leben* by reconstructing its implied readership and educational context. Afterwards, I will zoom in on the work's educational interests, arguing that the search for physical causes is inspired not only by generally scholarly but also genuine philosophical motives. Plutarch's natural problems, thus, promote some kind of 'intellectual gymnastics' and serve as a preamble to higher, meta-physical speculations.

3.1. Sitz im Leben: *readership and educational context*

Even though there is no proof that Plutarch self-handedly published *Quaestiones naturales*, we have seen in the previous chapter that the work is, in any case, publishable and that Plutarch presumably also prepared it for publication [see 2.4.1.]. There is debate, however, as to what kind of readership the collection is meant to address[1]. According to Sandbach, the work was intended "for circulation among interested friends, but not for a general public"[2]. Similarly, Fuhrmann believes that "[c]es recueils de notes [sc. *Questions Grecques, Romaines, Physiques*] étaient destinés à l'usage privé, peut-être à des exercices d'école, mais non promis à l'édition publique"[3]. Senzasono, by contrast, has argued that Plutarch prepared the publication of *Quaestiones naturales* not for a select group of friends but for a large readership[4]. For most ancient works, however,

---

[1] C. Hubert, 1960, p. v left this question open: "Has quaestiones, [...] utrum Plutarchus ipse in volgus ediderit an tantum in suum usum conscripserit, ut materiam praeberent qua in futuris libris condendis (velut Quaestionibus Convivalibus) uteretur, litigari potest [...]."

[2] F.H. Sandbach, 1965, p. 135.

[3] F. Fuhrmann, 1972, p. xiii. Cf. already H. Bolkestein, 1946, p. 27: "ad privatum usum conscripta esse videntur [sc. "[i]psa haec quaestionum corpuscula Problematis illis non dissimilia [...] quae sub Aristotelis nomine circumferuntur"] aut ut scholas habenti fortasse ad manum essent, certe non ut ab omnibus legerentur, tamquamsi elegantium litterarum essent monumenta".

[4] L. Senzasono, 2006, pp. 45–46: "Se Plutarco ha elaborato lo stile, come appare

the standard avenue of publication was to first circulate a text among peers, who, in their turn, made copies and spread them among their own peers, thus causing an exponential growth of the text's readership[5]. One may wonder, therefore, who the initial, intended readers of *Quaestiones naturales* were and what we can learn about the collection as "exercises d'école".

Unfortunately, there is no formal guide, no preface to *Quaestiones naturales* which could shed a light on the collection's aetiological organisation and programme. In any case, *Q.N.* 1 does not have any programmatic value (though it may contain a subtle Platonic σφραγίς, as we will see later on [see 4.2.1.1.]). One seminal passage that attests to the intellectual agenda of *Quaestiones naturales* as a whole can be found in *Q.N.* 29, 919AB. Here, Plutarch's vituperation of people's superficial marvel for rare natural phenomena (as discussed earlier on [see 1.2.3.]) can be seen as an appeal to a serious and mature study of nature. Plutarch's tirade *against* the common people, of course, does not imply that he is writing *for* the common people – they would, in any case, draw little benefit from it –, but rather, for people who have already begun their studies in (natural) philosophy or are, in any case, informed about its basic principles, concepts and procedures. The implied reader, then, is invited to carry out his physical inquiries, in order to distinguish himself from the common plebs who are unfamiliar with the finer elements of physical aetiology. Bearing in mind that Plutarch's intended reader was probably an informed reader, in what follows, I will try to clarify in what precise contexts such natural problems were useful, then.

1. Natural problems and philosophical σχολή

In regards to the exemplary cowardice of Aratus (the famous Greek general and statesman) in the presence of seeming peril, Plutarch notes (*Arat.* 29, 6) that the physiological symptoms of heart palpitation, change in the colour of one's skin, and looseness of the bowels, are popular questions

---

chiaro, l'ipotesi della destinazione a un circolo di amici cade o almeno appare improbabile: agli amici egli poteva dare degli appunti inconditi ma chiari o tutt'al piú una serie di problemi non elaborati stilisticamente ad uso pratico. L'elaborazione induce senz'altro a propendere per l'ipotesi della pubblicazione destinata al grande pubblico. [...] [S]i può ritenere altamente probabile che Plutarco, una volta portato a termine l'opuscolo, lo affidasse, subito o qualche tempo dopo, ai copisti per una pubblicazione destinata al grande pubblico."

[5] See R. Starr, 1987, who discusses the circulation of (literary) texts in the Roman world on the basis of widening concentric circles depending on the varying degrees of friendship (he distinguishes circles of friends from those of strangers, i.e. friends of friends etc.). Thanks are due to K. Oikonomopoulou for this reference.

in the philosophers' schools (τοὺς φιλοσόφους ἐν ταῖς σχολαῖς ζητοῦντας)⁶. He explains that these topics are generally accounted for in a twofold manner, viz. by reference to Aratus' cowardice (i.e. in an ethical sense) and some defective disposition and coldness in the body (i.e. in a physical sense):

> τοὺς φιλοσόφους ἐν ταῖς σχολαῖς ζητοῦντας, εἰ τὸ πάλλεσθαι τὴν καρδίαν καὶ τὸ χρῶμα τρέπεσθαι καὶ τὴν κοιλίαν ἐξυγραίνεσθαι παρὰ τὰ φαινόμενα δεινὰ δειλίας ἐστὶν ἢ δυσκρασίας τινὸς περὶ τὸ σῶμα καὶ ψυχρότητος, ὀνομάζειν ἀεὶ τὸν Ἄρατον, ὡς ἀγαθὸν μὲν ὄντα στρατηγόν, ἀεὶ δὲ ταῦτα πάσχοντα παρὰ τοὺς ἀγῶνας.

> In the schools of philosophy, when the query arises whether palpitation of the heart and change of colour and looseness of the bowels, in the presence of seeming peril, are the mark of cowardice, or of some faulty temperament and chilliness in the body, Aratus is always mentioned by name as one who was a good general, but always had these symptoms when a contest was impending.

Parallels can be found for this in Book 27 of Ps.-Aristotle's *Problems* (entitled ὅσα περὶ φόβον καὶ ἀνδρείαν), which deals with the physiological manifestation of fear and courage in the human body more generally⁷. Notably, some of the symptoms that are mentioned in the *Aratus* passage are also recorded or alluded to in *Quaestiones naturales*. In *Q.N.* 11, 914EF, for instance, the natural phenomenon of people's bowels turning to water (ἐξυγραίνονται) when they are seasick is again related to fear (φόβος) and implicitly also to cold, as the persons suffering from this condition tremble and shiver (τρέμουσι καὶ φρίττουσι). In *Q.N.* 19, 916BF, moreover, Plutarch deals with the change of colour in the octopus' skin colour, arguing that this physical process is triggered by fear (τὴν μὲν ἀρχὴν αὐτὸς ἐνδίδωσι τοῦ πάθους δείσας). What we can learn from the *Aratus* passage, then, is that the kind of natural problems treated in *Quaestiones naturales* must have been particularly useful for discussions in Plutarch's philosophical school context also (ἐν ταῖς σχολαῖς).

As seen before, the practice of solving natural problems goes back on Aristotle's teaching in the Lyceum [see 1.1.3.]. It is interesting to see, then, how the genre became well-entrenched also in Plutarch's own

---

⁶ A similar passage is found in *Per.* 35, 2, where Plutarch refers to the story that is often told in the philosophers' schools (ταῦτα μὲν οὖν ἐν ταῖς σχολαῖς λέγεται τῶν φιλοσόφων) about Pericles' attempt to overcome the fear of his skipper for an eclipse of the sun by keeping his cloak before the man's eyes, thus drawing an analogy with the sun's obscuration by the moon.

⁷ See L.M. Castelli, 2011. As to fear turning people's bowels to water, cf. Ps.-Arist., *Probl.* 948b35–949a8 (with Gell., *NA* 19, 4).

philosophical school[8]. Arguably, the Ps.-Aristotelian *Problems* themselves were actually read, studied, and discussed in Plutarch's philosophical school, thus, serving as some kind of a text-book model by which Plutarch and his peers were inspired to perform similar research[9]. There is no direct evidence for this, but it may be relevant that L. Mestrius Florus is found discussing a copy of the *Problems* with *his* friends – including Plutarch – in *Quaest. conv.* 734CD [quoted 3.2.1.]. Similarly, we also have evidence from Gellius that the *Problems* were read and discussed in another Platonic school context, viz. in the school of his teacher L. Calvenus Taurus, who was a Platonist and may have even been one of Plutarch's students[10]. How does Plutarch's own research into natural problems fit in with his educational programme, then, and what can we learn, in this regard, about his industriousness as a philosophy tutor?

2. Plutarch's academy

Plato's Academy had ceased to exist as a philosophical institution for more than two centuries by the time of Plutarch's death (it closed with the sack of Athens by Sulla in 86 BC). In his home town of Chaeronea, however, and presumably in his own house, Plutarch directed "eine Art Filiale der

---

[8] Cf. M. Schuster, 1917, p. 37: "Mit gutem Grunde dürfen wir die aetia physica als eine Sammlung von Problemen ansprechen, die in Plutarchs Schule zur Erörterung gekommen sind; es sind die alten bekannten Probleme, an denen die philosophischen Schulen seit Aristoteles gearbeitet haben und die den Schülern reichen Stoff zu Disputierübungen boten." See also K. Ziegler, 1951, col. 664 (quoted below). On the use of Ps.-Aristotle's *Problems* as a school handbook in the Lyceum, see, e.g., H. Flashar, 1962, pp. 341–346. On the use of problems more generally in pedagogical contexts, see already A. Gudeman, 1927, col. 2529: "Weit lebenskräftiger erwies sich diese Methode [sc. of λύσεις] sowohl in der Praxis der Schule wie in der Literatur, insofern sie dazu diente, allerlei Kenntnisse zu vermitteln [...]." See also, e.g., C. Jacob, 2004.

[9] Cf., e.g., F. Fuhrmann, 1972, p. xxi: "Plutarque s'inspire surtout [...] du Corpus aristotélicien, particulièrement des *Problèmes*, qu'il croyait authentiques et qui représentaient un manuel de prédilection pour les gens cultivés".

[10] In *NA* 19, 6, Gellius reports that when he was a student in Athens he read and discussed a passage from the *Problems* (fr. 243 Rose) together with his teacher L. Calvenus Taurus. K. Oikonomopoulou, 2013a, p. 137 connects this with the school practice of συνανάγνωσις ('reading together'). Again in *NA* 20, 4, Gellius informs us that the same Taurus assigned the daily reading of a specific passage from Aristotle's προβλήματα ἐγκύκλια to a student to divert him from his company with actors (see M. Meeusen, forthcoming e). On Taurus' role in the *Noctes Atticae* more generally, see M.L. Lakmann, 1995 (esp. pp. 216–220 on his "Schulpraxis"); see also pp. 227–228 for Taurus' relationship with Plutarch (Lakmann remains sceptical about Taurus being a student of Plutarch, though). On Taurus' sympotic teaching methods, see also G. Roskam, 2009, pp. 377–379 (with n. 30 for further reading).

athenischen Akademie"[11], where he, after the manner of his own teacher in Athens, Ammonius[12], taught philosophy with a particular interest and specialisation in the writings of Plato and the Academic tradition[13]. Apart from the existence of this school under the directorship of Plutarch, not much is known about its actual organisation and daily routines. Most scholars agree that it was not a very official institution[14]. We do not, in any case, have any account of Plutarch's school practices. There aren't any course schedules nor are there any lists of students who attended Plutarch's classes. Thus, we simply do not know how many students Plutarch tutored, where they came from, or what social or intellectual background they may have had. Nevertheless, we are not entirely bereft of information, because throughout his oeuvre, Plutarch hints at the general school-protocols in his academy and informs us about his teaching methods, the subjects of his courses, and, most notably, how students should behave during lectures.

In his 1917 study of *De sollertia animalium*, Schuster goes into the details of Plutarch's tutorship often with a great amount of imagination (at times, perhaps, even too much)[15]. From Plutarch's texts, we can deduce only more general institutional customs and educational practices. We learn that Plutarch's method of instruction entails a number of different approaches. Thus, Schuster distinguishes between Plutarch's acroamatic and erotematic teaching[16]. Among Plutarch's writings, we come across *ex cathedra* course-lectures, such as *Adversus Colotem* and *De communibus*

---

[11] K. Ziegler, 1951, col. 663. Cf. F. Fuhrmann, 1972, p. xviii: "une manière de petite université". Plutarch refers to his school and instruction with the words σχολή and διατριβή (cf., e.g., *De aud.* 42A, *De E* 385A, *Quaest. conv.* 613C, 613F, 702A, 705B, 713C, *De facie* 929B, 942C). See M. Schuster, 1917, p. 2 and J. Opsomer, 1998, p. 25 (with n. 61). J. Boulogne, 2003, p. 36, nn. 94, 95 even translates διατριβή in *Non posse* 1086D as "la sale de conférences".

[12] For a philosophical profile of Ammonius, see J. Opsomer, 2009. See also P. Donini, 1986b.

[13] In *De soll. an.* 964D, Plutarch's father Autobulus says that his son, under the direction of Plato, instructs inquisitive persons who have no love of wrangling (Πλάτωνος ὑφηγουμένου, δείκνυσιν [...] τοῖς μὴ φιλομαχεῖν ἕπεσθαι δὲ καὶ μανθάνειν βουλομένοις). On Plutarch's school pragmatics, see, e.g., J. Boulogne, 2003, pp. 34–37. For further reading on the Academy and Plutarch's school, see J. Opsomer, 1998, pp. 21–26.

[14] For the informal character of Plutarch's academy, cf., e.g., K. Ziegler, 1951, cols. 662–663, R.H. Barrow, 1967, pp. 18–19, D.A. Russell, 1973, p. 13, J. Glucker, 1978, pp. 257–280, J. Boulogne, 2003, p. 34 and J. Bouffartigue, 2012, p. xix. Pace M. Schuster, 1917, pp. 19–21.

[15] Take, for instance, his argument that the lessons took place at fixed hours and that the ancient Greeks were early risers (M. Schuster, 1917, pp. 20–21). Did Plutarch use a school bell, by the way (cf. Cic., *De or.* 2, 5, 21)?

[16] M. Schuster, 1917, p. 27.

notitiis adversus Stoicos[17], where the students play a rather passive role and just have to listen to what is said. However, there are also philosophical discussions that require much more active input from the students in formulating their own well-reasoned opinions on certain subjects, as can be seen in *Non posse*[18].

Interestingly, Plutarch at times reports from his own years as a student under Ammonius. In *Quaest. conv.* 645D–646A, for instance, we read that after a dinner party in Athens flower garlands were offered to the symposiasts. Ammonius ridicules the custom of wearing such garlands and considers it a girlish habit, so the young men who were present at the event took them off out of shame. Plutarch, still a pupil himself at the time (and a diligent one), knew, however, that Ammonius actually threw this problem in the middle of the group not so much to offend the young men as to provide an exercise in inquiry (γυμνασίας ἕνεκα καὶ ζητήσεως). And so the sympotic discussion about the natural properties of flower garlands and ivy begins.

A similar passage to illustrate how the students' own zetetic skills are put to the test in a mostly competitive situation, when prompted to answer to a problem put forward by the teacher, is seen in *De sollertia animalium*. The second part of this work is presented as a school discussion about the intelligence of animals. In the preceding discussion between Autobulus and Soclarus, we read that a symposium took place the other day, during which someone read an encomium of hunting. The symposiasts declared that all animals are, to some extent, rational and partake in διάνοια and λογισμός. In light of this understanding, Aristotimus and Phaedimus, two of Plutarch's students[19], received their formal assignment for their discussion (ἀγών), viz. to determine whether land or sea animals are more clever. In the discussion that follows, each of them defends the opposite case. The use of such contradictory discussions, in which topics are argued

---

[17] In *Adversus Colotem* Plutarch gives the course himself (cf. 1108Bff.), but in *De communibus notitiis adversus Stoicos* (cf. 1060Bff.) this task is allotted to Diadumenus (presumably Plutarch's double: cf. D. Babut, 1969, p. 38).

[18] In a recent study of Plutarch's conceptualisation of teacher-student communication, G. Roskam, 2004 has analysed the level of independence that students were allowed. He thus distinguishes between a propaedeutic stadium (under the guidance of the παιδαγωγός, the διδάσκαλος, and the γραμματικός) and the actual philosophical παιδεία (under the guidance of a καθηγητής). The first stadium is generally passive-receptive for the student and monologic for the teacher, and the second stadium more active and dialogic. More generally useful in the context of ancient education are H.-I. Marrou, 1948 (esp. pp. 252–255 and 289–299 regarding to the use of problems), T. Morgan, 1998, R. Cribiore, 2001 and the contributions in Y.L. Too, 2001.

[19] See M. Schuster, 1917, pp. 57ff. and H. Cherniss and W.C. Helmbold, 1957, p. 312, n. b.

from both sides, was a common practice in philosophical schools (i.e. the ἐπιχείρησις εἰς ἑκάτερον or *disputatio in utramque partem* [see 4.3.3.1.])[20].

In most of Plutarch's dialogical writings, however, more than two persons participate in the discussion, so that a multiplicity of answers is given, each of which illuminates a specific aspect of the complex problem at hand. It is the task of the teacher or the presiding moderator of the debate to steer the discussion in the right direction, then[21]. He can do this by adducing a range of questions and problems (as marked with the verbs ζητεῖν, προβάλλειν and διαπορεῖν). In the prologue to *De E* (385AB), for instance, we read that the problem of the enigmatic E at Delphi had been brought up many times in school discussions (ἐν τῇ σχολῇ), but that thus far, Plutarch had avoided it. Now that he is confronted with the same problem in the presence of his sons and some strangers who are about to leave Delphi, he feels embarrassed to divert the discussion any longer. Plutarch then creates a schoolish setting by finding a place for himself and his audience to sit near the temple and by looking for answers himself and asking questions (καθίσας παρὰ τὸν νεὼν τὰ μὲν αὐτὸς ἠρξάμην ζητεῖν τὰ δ' ἐκείνους ἐρωτᾶν)[22].

Of course, the student was also allowed to ask questions himself, often after the lectures[23]. However, Plutarch writes in *De aud.* 42E–44A that the student should follow certain rules in doing so. We read, for instance, that those individuals who lead the speaker to digress to other topics, interject questions or raise new difficulties are not pleasant or agreeable company at a lecture (*De aud.* 42F: οἱ γὰρ εἰς ἄλλας ὑποθέσεις ἐξάγοντες καὶ παρεμβάλλοντες ἐρωτήματα καὶ προσδιαποροῦντες, οὐχ ἡδεῖς οὐδ' εὐσυνάλλακτοι πρὸς ἀκρόασιν ὄντες). Additionally, one should wait for the right time to ask a question and only ask questions that are useful and relevant. The questions should also be adapted to the speaker's competence and

---

[20] The same format also lies at the basis, e.g., of Plutarch's *Aqua an ignis sit utilior*, which may originate from a similar school exercise (for the issue of authenticity [see 1.1.1., n. 8]).

[21] For Plutarch's ἡγεμονία, e.g., in *Adversus Colotem* and *Non posse*, cf. H. Adam, 1974, p. 6. The debates often take place under the guidance of the teacher, but another leading figure can also fulfil this task, e.g. the συμποσίαρχος in case of a symposium (see *Quaest. conv.* 620A–622B). In *De soll. an.* 965DE, Optatus is appointed as 'judge' (βραβευτής) in the discussion, because he is considered to be an expert on Aristotle's writings (Δεῦρο δὴ καθίζου πρὸς ἡμᾶς, ὅπως, εἰ δεήσει μάρτυρος, μὴ τοῖς Ἀριστοτέλους πράγματα βιβλίοις παρέχωμεν, ἀλλὰ σοὶ δι' ἐμπειρίαν ἑπόμενοι τοῖς λεγομένοις ἀληθῶς τὴν ψῆφον ἐπιφέρωμεν).

[22] Indeed, sometimes the teacher gives the solution himself. Therefore, he should develop personal insights (cf. *De virt. mor.* 440E, *De an. procr.* 1012B).

[23] Gellius also reports that his Platonic teacher L. Calvenus Taurus often held question time style debates after his lectures (*NA* 1, 26: *dabat enim saepe post cotidianas lectiones quaerendi, quod quis uellet, potestatem*). Cf. C. Jacob, 2004, p. 30. On the practice of ἐρώτησις with the neo-Platonists, see H.G. Snyder, 2000, pp. 111–118.

should not be too high in number. Therefore, one should not ask natural scientific or mathematical questions to those who are more concerned with the ethical side of philosophy or problems related to logic to natural scientists (*De aud.* 43C: μὴ παραβιάζεσθαι τὸν μὲν ἠθικώτερον φιλοσοφοῦντα φυσικὰς ἐπάγοντα καὶ μαθηματικὰς ἀπορίας, τὸν δὲ τοῖς φυσικοῖς σεμνυνόμενον εἰς συνημμένων ἐπικρίσεις ἕλκοντα καὶ ψευδομένων λύσεις). Asking the right question, thus, turns out to be some kind of a regulated art in itself with which the student should acquaint himself.

Clearly, the method of questioning and answering obliges the participants to actively partake in the discussion. As to the right way to obtain an answer to a question, Plutarch often insists on providing a *personal* response, which he considers a useful means for sharpening a person's zetetic skills[24]. A relevant passage for this insistence on providing personal responses to problems can be found in the finale of *De audiendo* (48BC). Herein, Plutarch encourages lazy people to do some thinking of their own instead of continually asking questions about the same things to the lecturer:

τοὺς δ' ἀργοὺς ἐκείνους παρακαλῶμεν, ὅταν τὰ κεφάλαια τῇ νοήσει περιλάβωσιν, αὐτοὺς δι' αὑτῶν τὰ λοιπὰ συντιθέναι, καὶ τῇ μνήμῃ χειραγωγεῖν τὴν εὕρεσιν, καὶ τὸν ἀλλότριον λόγον οἷον ἀρχὴν καὶ σπέρμα λαβόντας ἐκτρέφειν καὶ αὔξειν. οὐ γὰρ ὡς ἀγγεῖον ὁ νοῦς ἀποπληρώσεως ἀλλ' ὑπεκκαύματος μόνον ὥσπερ ὕλη δεῖται, ὁρμὴν ἐμποιοῦντος εὑρετικὴν καὶ ὄρεξιν ἐπὶ τὴν ἀλήθειαν.

But as for those lazy persons whom we have mentioned, let us urge them that, when their intelligence has comprehended the main points, they put the rest together by their own efforts, and use their memory as a guide in thinking for themselves, and, taking the discourse of another as a germ and seed, develop and expand it. For the mind does not require filling like a bottle, but rather, like wood, it only requires kindling to create in it an impulse to think independently and an ardent desire for the truth.

A zetetic agenda is implied in the 'stimulus and desire for the truth' mentioned here (ὁρμὴν εὑρετικὴν καὶ ὄρεξιν ἐπὶ τὴν ἀλήθειαν). Instead of implanting knowledge from without, an appeal is made to activate the knowledge already present in a person's memory in his search for answers (τῇ μνήμῃ χειραγωγεῖν τὴν εὕρεσιν)[25]. This emphasis on activating independent

---

[24] On the stimulation of the zetetic attitude of Plutarch's students, see G. Roskam, 2004, p. 103.

[25] A parallel can be drawn with Socrates' maieutical method in philosophy, cf. *Quaest. Plat.* 1000E (with J. Opsomer, 1998, pp. 193-212): οὐδὲν ἐδίδασκε Σωκράτης, ἀλλ' ἐνδιδοὺς ἀρχὰς ἀποριῶν ὥσπερ ὠδίνων τοῖς νέοις ἐπήγειρε καὶ ἀνεκίνει καὶ συνεξῆγε τὰς ἐμφύτους νοήσεις·

thought ties in closely with Plutarch's more general conceptualisation of personal ingenuity in intellectual debate [see 4.2.2.2.].

3. Digestive discussions and problematic promenades

Zetetic education is not restricted to the discussions in Plutarch's school context but extends to other social settings that have a less schoolish character (σχολή being an elastic concept). Such a setting is found in the discussions at sympotic events[26], where for Plutarch a healthy balance between seriousness and play is even required by protocol[27]. Another occasion is the spontaneous discussion that arises during the so-called 'peripatos'[28], that is, a stroll often after dinner or after a lecture (as it was

---

καὶ τοῦτο μαιωτικὴν τέχνην ὠνόμαζεν, οὐκ ἐντιθεῖσαν ἔξωθεν, ὥσπερ ἕτεροι προσεποιοῦντο, νοῦν τοῖς ἐντυγχάνουσιν, ἀλλ' ἔχοντας οἰκεῖον ἐν ἑαυτοῖς ἀτελῆ δὲ καὶ συγκεχυμένον καὶ δεόμενον τοῦ τρέφοντος καὶ βεβαιοῦντος ἐπιδεικνύουσαν. On Plutarch's view of Socrates as an exemplary teacher, see G. Roskam, 2004, pp. 104–105, 108 and 2011c, pp. 421–425. See also C. Pelling, 2005.

[26] On the educational agenda of the symposia described in *Quaestiones convivales* and on convivial teaching more generally, see, e.g., M. Schuster, 1917, pp. 51–52, J. König, 2007, G. Roskam, 2009, E. Kechagia, 2011a, L. Van der Stockt, 2000b and 2011.

[27] Intellectual entertainment and instruction were important aspects of the symposium (see, e.g., G. Roskam, 2009). Even if the notion of σπουδογέλοιον is not indicated by Plutarch with this precise term, its basic idea can be found throughout *Quaestiones convivales*. Cf., e.g., *Quaest. conv.* 621DE (ἔστι γὰρ καὶ γέλωτι χρῆσθαι πρὸς πολλὰ τῶν ὠφελίμων καὶ σπουδὴν ἡδεῖαν παρασχεῖν), 629F (οὐ γάρ τι μικρόν […] τῆς ὁμιλητικῆς μόριον ἡ περὶ τὰς ἐρωτήσεις καὶ τὰς παιδιὰς τοῦ ἐμμελοῦς ἐπιστήμη καὶ τήρησις), 686D (τὰ δὲ φιλοσοφηθέντα μετὰ παιδιᾶς σπουδάζοντες εἰς γραφὴν ἀπετίθεντο). Cf. also *De soll. an.* 960B (σὺν οἴνῳ καὶ παρὰ πότον οὐ μετὰ σπουδῆς). On the aspect of play in *Quaestiones convivales*, see F. Frazier, 1998. For the notion of σπουδογέλοιον in ancient Greek literature more generally, see D. Arnould, 1990, pp. 113–122.

[28] For the connection between the peripatos and the discussion of problems, see, e.g., R. Hirzel, 1895, vol. 1, p. 364 (with notes for further references), M. Schuster, 1917, pp. 48–51, A. Gudeman, 1927, col. 2522, K. Ziegler, 1951, col. 664 (quoted below). In several of Plutarch's writings, the peripatos serves as some kind of a hodological framework for the dialogue's development. In *Non posse*, for instance, Plutarch records the discussions that spontaneously arose during the customary peripatos after school hours (1086D: ἐπεὶ δὲ καὶ τῆς σχολῆς διαλυθείσης ἐγένοντο λόγοι πλείονες ἐν τῷ περιπάτῳ πρὸς τὴν αἵρεσιν, ἔδοξέ μοι καὶ τούτους ἀναλαβεῖν). In *De sera* the entire discussion is framed within a peripatos (548AB: ὥσπερ ἐτυγχάνομεν περιπατοῦντες). In the introduction to *Septem sapientium convivium*, Plutarch describes the walk on the way to Periander's banquet, offering the participants the opportunity for free, leisurely conversation and discussion of several problems, as it was often held (146E: ἐβαδίζομεν οὖν ἐκτραπόμενοι διὰ τῶν χωρίων, καθ' ἡσυχίαν). From *De facie* 937CD, we learn that the interlocutors had, until thus far, been walking, but they now sit down upon the steps, where they remain seated until the end of the discussion (εἰ δοκεῖ, καταπαύσαντες τὸν περίπατον καὶ καθίσαντες ἐπὶ τῶν βάθρων ἑδραῖον αὐτῷ παράσχωμεν ἀκροατήριον). A similar hodological framework is also present in *De Pythiae oraculis*, where the discussion is set against the background of the Delphic sanctuary [see 2.1.1.].

traditionally made in the Lyceum)[29]. In spite of their less formal character, the symposium and the peripatos can be considered a supplement to Plutarch's school education[30].

Interestingly, several passages from Plutarch's *De tuenda sanitate praecepta* show that there lies not just an intellectual but also a specific physiological agenda behind these sympotic and peripatetic discussions. *De tuenda* comes in the form of a dialogue between the physician Moschion and his friend Zeuxippus (but surely many of the author's own ideas and convictions on this topic must be present as well)[31]. Although there are clear reminiscenses to medical literature in this work, Plutarch is not highly concerned with the technical side of medicine as such[32]. Rather, the work is about the practical side of intellectual living. At some point, we read that the discussions during the symposium and peripatos have a direct impact not only on the mind, but also on the body of its participants, and that they, thus, contribute to a person's mental and physical well-being at the same time.

Regarding the 'exercises suitable for scholars' (*De tuenda* § 16, 130A: περὶ γυμνασίων φιλολόγοις ἁρμοζόντων), Zeuxippus, by reference to ancient

---

Cf. also *De def. or.* 412D: Ἤδη δέ πως ἀπὸ τοῦ νεὼ προϊόντες ἐπὶ ταῖς θύραις τῆς Κνιδίων λέσχης ἐγεγόνειμεν.

[29] For the peripatos as an emblematic occupation of Peripatetic philosophers, cf. *De Al. Magn. fort.* 328A, where ἐν Λυκείῳ περίπατον συνέχειν is opposed to ἐν Ἀκαδημείᾳ θέσεις λέγειν. Cf. also *Quaest. conv.* 734CD [quoted 3.2.1.] and *Alex.* 7, 4 (Ἀριστοτέλους ὑποσκίους περιπάτους). According to Cic., *Ep. ad Attic.* 7, 1, 1 the slow *ambulatio* is typical for philosophers. For more on walking in Roman culture, see T.M. O'Sullivan, 2011. Cf. also E. Lao, 2008, pp. 179–180 and J. Scheid, 2012, p. 147.

[30] See K. Ziegler, 1951, col. 664: "Die diskussion wurde öfters, gemäß peripatetischer Tradition, im Umhergehen geführt, und die Symposien, von denen natürlich die schwierigeren, volle Konzentration erfordernden Themen ausgeschlossen wurden, bildeten eine Ergänzung des eigentlichen Unterrichts und so gewissermaßen einen Teil der Schule. [...] [A]uch Fragen aus der Physik – in dem weiten antiken Sinne, wonach sie auch Biologie und Medizin umfaßt – [wurden] nicht selten behandelt [...]." For the distinction between the symposium and a more strict school context, cf., e.g., *Quaest. conv.* 712A (with J. Sirinelli, 2000, p. 385): ὥστε γραμματοδιδασκαλεῖον ἡμῖν γενέσθαι τὸ συμπόσιον. Then again, it was not at all inappropriate to bring along students to the symposium in order to introduce them as novices in the intellectual milieu (cf., e.g., *Quaest. conv.* 660D). The other way round, a certain degree of informality in the private milieu of Plutarch's academy, cannot be excluded either, as we saw earlier on (see n. 14).

[31] In fact, the ἑταῖρος from *De tuenda* 122F has been identified with Plutarch himself. See F.C. Babbitt, 1928, pp. 215 and 220, n. a.

[32] As F.C. Babbitt, 1928, p. 214 points out, Plutarch's "advice is meant for men whose work is done with their heads rather than their hands". For an interpretation of *De tuenda sanitate praecepta* in light of Plutarch's practical ethics (or 'diet-ethics'), see L. Van Hoof, 2010, pp. 211–254 (with further useful literature). For the influence of Plutarch's Platonism on this treatise, see L. Senzasono, 1992, pp. 19–25.

πνεῦμα-theory, recommends the daily use of the voice by conversing, reading aloud or declaiming (*De tuenda* 130BD). He asserts that speaking during the peripatos (after rubbing oneself with oil, thus equalising the bodily breath) is an appropriate exercise enabling one to 'simultaneously instruct, question, learn, and use one's memory' (*De tuenda* 130F: ἅμα διδάσκῃ τι καὶ ζητῇ καὶ μανθάνῃ καὶ ἀναμιμνῄσκηται γυμναζόμενος). He also points out that rhetorical and sophistical debates, by contrast, are unhealthy, due to the improper use of the voice (*De tuenda* 130F–131A: μόνον ἐκεῖνο φυλακτέον, ὅπως μήτε πλησμονὴν μήτε λαγνείαν μήτε κόπον ἑαυτοῖς συνειδότες ἐντεινώμεθα τῇ φωνῇ τραχύτερον, ὃ πάσχουσι πολλοὶ τῶν ῥητόρων καὶ τῶν σοφιστῶν).

A little later, Zeuxippus indicates a number of remarkable physiological advantages and disadvantages provided by the peripatos after dinner[33]. He opposes two theories on this topic (including Aristotle's) and concludes that pleasant discussion in itself may suffice for the proper digestion of food (*De tuenda* § 21, 133F–134A):

Ἐπεὶ δ' Ἀριστοτέλης οἴεται τῶν δεδειπνηκότων τὸν μὲν περίπατον ἀναρριπίζειν τὸ θερμόν, τὸν δ' ὕπνον, ἂν εὐθὺς καθεύδωσι, καταπνίγειν, ἕτεροι δὲ τὴν μὲν ἡσυχίαν οἴονται τὰς πέψεις βελτίονας ποιεῖν, τὴν δὲ κίνησιν ταράττειν τὰς ἀναδόσεις, καὶ τοῦτο τοὺς μὲν περιπατεῖν εὐθὺς ἀπὸ δείπνου τοὺς δ' ἀτρεμεῖν πέπεικεν, ἀμφοτέρων ἂν οἰκείως ἐφάπτεσθαι δόξειεν ὁ τὸ μὲν σῶμα συνθάλπων καὶ συνέχων μετὰ τὸ δεῖπνον, τὴν δὲ διάνοιαν μὴ καταφερόμενος μηδ' ἀργῶν εὐθὺς ἀλλ' ὥσπερ εἴρηται διαφορῶν ἐλαφρῶς τὸ πνεῦμα καὶ λεπτύνων τῷ λαλεῖν τι καὶ ἀκούειν τῶν προσηνῶν καὶ μὴ δακνόντων μηδὲ βαρυνόντων.

Aristotle (fr. 233 Rose) holds that walking about on the part of those who have just dined revives the bodily warmth, while sleep, if they go to sleep at once, smothers it; but others hold that quiet improves the digestive faculties, while movement disturbs the processes of assimilation; and this has persuaded some to walk about immediately after dinner, and others to remain quiet. In view of the two opinions a man might appear properly to attain both results who after dinner keeps his body warm and quiet, and does not let his mind sink at once into sleep and idleness, but, as has been previously suggested, lightly diverts and enlivens his spirits by talking himself and listening to another on one of the numerous topics which are agreeable and not acrimonious or depressing.

The reference to what 'has been previously suggested' (ὥσπερ εἴρηται) in this passage is to the preceding paragraph (*De tuenda* § 20, 133BF),

---

[33] For the connection between the peripatos and the digestion of food, cf. also *De comm. not.* 1071D: οὐ σπουδάζομεν εὐκαίρως περιπατεῖν ἕνεκα τοῦ πέττειν τὴν τροφὴν ἀλλὰ πέττειν τὴν τροφὴν ἕνεκα τοῦ περιπατεῖν εὐκαίρως.

which deals with the topics that are fit and agreeable for discussion. There, Zeuxippus first attacks the athletic trainers and teachers of gymnastics, who state that scholarly conversation at dinner spoils the food and makes the head heavy. He admits that problems in the field of dialectics are not very pleasant because 'they bring on a headache and are extremely fatiguing' (*De tuenda* 133C: διαλεκτικὴ δὲ "τρωγάλιον" ἐπὶ δείπνῳ "γλυκὺ" μὲν οὐδαμῶς κεφαλαλγὲς δὲ καὶ κοπῶδες ἰσχυρῶς ἐστιν)[34], but one should not, therefore, do away with the discussion of problems after dinner altogether. The rest is worth quoting in full (*De tuenda* 133DF):

αὐτοὶ δὲ πειθόμενοι τοῖς ἰατροῖς παραινοῦσιν ἀεὶ τοῦ δείπνου καὶ τοῦ ὕπνου λαμβάνειν μεθόριον καὶ μὴ συμφορήσαντας εἰς τὸ σῶμα τὰ σιτία καὶ τὸ πνεῦμα καταθλίψαντας εὐθὺς ὠμῇ καὶ ζεούσῃ τῇ τροφῇ βαρύνειν τὴν πέψιν ἀλλ' ἀναπνοὴν καὶ χάλασμα παρέχειν, ὥσπερ οἱ τὰ σώματα κινεῖν μετὰ δεῖπνον ἀξιοῦντες οὐ δρόμοις οὐδὲ παγκρατίοις τοῦτο ποιοῦσιν ἀλλὰ βληχροῖς περιπάτοις καὶ χορείαις ἐμμελέσιν, οὕτως ἡμεῖς οἰησόμεθα δεῖν τὰς ψυχὰς διαφέρειν μετὰ τὸ δεῖπνον μήτε πράγμασι μήτε φροντίσι μήτε σοφιστικοῖς ἀγῶσι πρὸς ἅμιλλαν ἐπιδεικτικὴν ἢ κινητικὴν περαινομένοις. ἀλλὰ πολλὰ μέν ἐστι τῶν φυσικῶν προβλημάτων ἐλαφρὰ καὶ πιθανά, πολλαὶ δὲ διηγήσεις ἠθικὰς σκέψεις ἔχουσαι καὶ τοῦτο δὴ τὸ "μενοεικές," ὡς Ὅμηρος ἔφη, καὶ μὴ ἀντίτυπον. τὰς δ' ἐν ἱστορικαῖς καὶ ποιητικαῖς ζητήσεσι διατριβὰς οὐκ ἀηδῶς ἔνιοι δευτέρας τραπέζας ἀνδράσι φιλολόγοις καὶ φιλομούσοις προσεῖπον. εἰσὶ δὲ καὶ διηγήσεις ἄλυποι καὶ μυθολογίαι, καὶ τὸ περὶ αὐλοῦ τι καὶ λύρας ἀκοῦσαι καὶ εἰπεῖν ἐλαφρότερον ἢ λύρας αὐτῆς φθεγγομένης ἀκούειν καὶ αὐλοῦ. μέτρον δὲ τοῦ καιροῦ τὸ τῆς τροφῆς καθισταμένης ἀτρέμα καὶ συμπνεούσης τὴν πέψιν ἐγκρατῆ γενέσθαι καὶ ὑπερδέξιον.

But as for ourselves, we shall follow the advice of the physicians who recommend always to let some time intervene between dinner and sleep, and not, after jumbling our victuals into our body and oppressing our spirit, to hinder our digestion at once with the food that is still unassimilated and fermenting, but rather to provide for it some respite and relaxation; just as those who think it is the right thing to keep their bodies moving after dinner do not do this by means of foot-races and strenuous boxing and wrestling, but by gentle walking and decorous dancing, so we shall hold that we ought not to distract our minds after dinner either with business or cares or pseudo-learned disputations, which have as their goal an ostentatious or stirring rivalry. *But many of the problems of natural science are light and enticing*, and there are many stories which contain ethical considerations and the "soul's

---

[34] Cf. also *De aud.* 43A: μᾶλλον δ' ἄν τις ἀκροατοῦ καταγελάσειεν εἰς μικρὰ καὶ γλίσχρα προβλήματα τὸν διαλεγόμενον κινοῦντος, οἷα τερθρευόμενοί τινες τῶν νέων καὶ παρεπιδεικνύμενοι διαλεκτικὴν ἢ μαθηματικὴν ἕξιν εἰώθασι προβάλλειν κτλ.

satisfaction," as Homer has phrased this, and nothing repellent. The spending of time over questions of history and poetry some persons, not unpleasingly, have called a second repast for men of scholarship and culture. There are also inoffensive stories and fables, and it is less onerous to exchange opinions about a flute and a lyre than to listen to the sound of the lyre and the flute itself. The length of time for this is such as the digestion needs to assert itself and gain the upper hand over the food as it is gradually absorbed and begins to agree with us.

The idea that 'many of the problems of natural science are light and enticing', that is, 'easy and persuasive' (πολλὰ μέν ἐστι τῶν φυσικῶν προβλημάτων ἐλαφρὰ καὶ πιθανά), is very significant for the type of problems collected in *Quaestiones naturales* and for their intellectual appeal. Also, the fact that this type of problems is placed next to other topics that are considered 'satisfactory to the soul' (μενοεικές, cf. also *Phoc.* 2, 3), 'not repellent' (μὴ ἀντίτυπον) and 'inoffensive' (ἄλυποι), is relevant for the purpose of Plutarch's own natural problems. Indeed, even if the words ἐλαφρὰ καὶ πιθανά qualify πολλὰ – rather than πάντα – τῶν φυσικῶν προβλημάτων, it is not implausible that this qualification applies to the genre of natural problems more generally, as known from the Ps.-Aristotelian *Problems*. Gellius, for one, confirms that these problems are 'most delightful and filled with choice knowledge of all kinds' (*NA* 19, 4: *Aristotelis libri sunt, qui* Problemata Physica *inscribuntur, lepidissimi et elegantiarum omnigenus referti*). Plutarch's qualification, then, seems to imply that natural problems are not very complex but relatively easy (ἐλαφρά), because anyone can solve them: learned men and lesser intellectuals alike (cf. *Quaest. conv.* 613E, quoted below). Their persuasiveness (πιθανά), on the other hand, probably has a deeper philosophical implication, because in light of Plutarch's Platonic epistemology, solving natural problems is a conjectural science, where the plausibility of the arguments is of the utmost importance [see 4.3.2.2.].

The opposition with complex, dialectical problems cannot be any more obvious. These should be avoided for physiological reasons, because, as we saw, 'they induce headaches and are very fatiguing' (*De tuenda* 133C: κεφαλαλγὲς δὲ καὶ κοπῶδες ἰσχυρῶς ἐστιν). The same is true for pseudo-learned, 'sophistic disputations, which only strive for rivalry' (*De tuenda* 133E: σοφιστικοῖς ἀγῶσι πρὸς ἅμιλλαν ἐπιδεικτικὴν ἢ κινητικὴν περαινομένοις). One must, therefore, be cautious about 'passionate and convulsive vociferations, because spasmodic expulsion and straining of the breath produces ruptures and sprains' (*De tuenda* 130D: αἱ γὰρ ἀνώμαλοι προβολαὶ καὶ διατάσεις τοῦ πνεύματος ῥήγματα καὶ σπάσματα ποιοῦσιν). Straining the voice too hard is, in fact, an experience 'of rhetoricians and sophists' (*De tuenda* 131A: τῶν ῥητόρων καὶ τῶν σοφιστῶν). It is precisely for this reason that excessive rivalry and sophistic arguments are excluded from the decorum of sympotic discussions.

This is not to say, of course, that there is no place for rhetorical manoeuvres at all at Plutarch's table – quite to the contrary –, but it does clearly imply that sympotic discussions should be more philosophical in nature, albeit not overtly philosophical either (as we will see in a moment). It is presumably in this sense, then, that the topics that Zeuxippus recommends for discussion, including natural problems, are not only 'beautiful and useful but also contain an element of pleasurable allurement and sweetness' (*De tuenda* 133C: τῶν ἐν τῷ καλῷ καὶ ὠφελίμῳ τὸ ἐπαγωγὸν ὑφ' ἡδονῆς καὶ γλυκὺ μόριον ἐχόντων). These discussions have immediate psychosomatic effects and function as a kind of intellectual *digestif* that takes as long as the assimilation of the food requires.

In *Quaestiones convivales*, many topics such as these are treated by Plutarch and his fellow symposiasts, and the general sympotic protocols from this work are very similar, if not identical, to those described in *De tuenda* (and *vice versa*). For instance, in the preface to the fifth Book (*Quaest. conv.* 672D–673A), Plutarch considers sympotic topics to be 'extraordinary' (ζητεῖν τι τῶν περιττῶν) and part of the so-called 'second repast' (δεύτερα τραπεζά), which consists of 'delights set aside for the soul' (τῇ ψυχῇ ταμιεῖον εὐπαθειῶν)[35]. Regarding the easy and persuasive character of natural problems – these are ἐλαφρὰ καὶ πιθανά, as we just saw (*De tuenda* 133E) – a great deal of relevant information can be found in the very first chapter of the first Book. There, Plutarch and his fellow symposiasts discuss the programmatic problem of 'whether philosophy is a fitting topic for conversation at a drinking party' (*Quaest. conv.* 612E: Εἰ δεῖ φιλοσοφεῖν παρὰ πότον). In approving this, Plutarch insists that the topics of inquiry at the symposium should be 'easy to handle' (*Quaest. conv.* 614E: ἐλαφραὶ ζητήσεις). Indeed, 'the matter of inquiry should remain rather facile, the topics familiar and the subjects suitable and not complex, since, in this way, the less intellectual guests may neither be stifled nor turned away' (*Quaest. conv.* 614D: εἶναι δὲ δεῖ καὶ αὐτὰς τὰς ζητήσεις ὑγροτέρας καὶ γνώριμα τὰ προβλήματα καὶ τὰς πεύσεις ἐπιεικεῖς καὶ μὴ γλίσχρας, ἵνα μὴ πνίγωσι τοὺς ἀνοητοτέρους μηδ' ἀποτρέπωσιν). Complex and abstruse topics lead

---

[35] Cf. also Gell., *NA* 7, 13, where 'sympotic questions' (*quaestiunculae sympoticae*) are considered 'ingenious' (*argutiae*), 'neither weighty nor serious' (*non gravia nec reverenda*), but 'pleasant and neat' (*lepida et minuta*); they are no 'pointless or idle sophisms' (*captiones* [...] *futtiles atque inanes*), no 'trifling amusements' (*nugarum aliquem ludum*), but the 'sweetmeats of the desserts' (*mensarum secundarum* [...] τραγημάτια). Cf. also Macrob., *Sat.* 7, 3, 23: *quod genus* [sc. *quaestiones convivales*] *veteres ita ludicrum non putarunt ut et Aristoteles de ipsis aliqua conscripserit et Plutarchus et vester Apuleius, nec contemnendum sit quod tot philosophantium curam meruit.*

to irritation (*Quaest. conv.* 614E: ἐν πράγμασι γλίσχροις καὶ δυσθεωρήτοις τούς τε παρατυγχάνοντας ἀνιῶσιν). Engaging in complex argumentation over one's wine is a 'sophistical' thing to do, and should therefore be avoided at the symposium (*Quaest. conv.* 615B: λόγοις δὲ γλίσχροις παρὰ πότον κεχρῆσθαι σοφιστικὸν μέν, οὐ καλὸν δ' οὐδὲ συμποτικόν). Plutarch's main concern is that everyone be able to participate in the discussion, learned men and men without erudition alike (*Quaest. conv.* 613E: ἂν ὀλίγοι τινὲς ἰδιῶται παρῶσιν, ὥσπερ ἄφωνα γράμματα φωνηέντων ἐν μέσῳ πολλῶν τῶν πεπαιδευμένων ἐμπεριλαμβανόμενοι φθογγῆς τινος οὐ παντελῶς ἀνάρθρου καὶ συνέσεως κοινωνήσουσιν)[36].

This clearly reflects on the level of complexity of sympotic discussions, among which natural problems take a prominent place (they make up approximately one third of *Quaestiones convivales*). Still in the first chapter of the first Book, Plutarch says that the dilemma, of course, is that philosophy is mostly concerned with 'subtle and disputatious problems' (*Quaest. conv.* 614F: λεπτὰ καὶ διαλεκτικὰ προβλήματα), that is, problems that are annoying for guests who are not philosophers. Thus, 'the height of sagacity is to talk philosophy without seeming to do so' (*Quaest. conv.* 614A: συνέσεως ἄκρας φιλοσοφοῦντα μὴ δοκεῖν φιλοσοφεῖν – I will come back to this later). Furthermore, it is not so much the 'compulsion of the arguments', but rather their 'persuasiveness', that matters for Plutarch (*Quaest. conv.* 614C: διὰ τοῦ πιθανοῦ μᾶλλον ἢ βιαστικοῦ τῶν ἀποδείξεων). The emphasis on the persuasiveness of the arguments recurs throughout *Quaestiones convivales* and can be linked not only with the social dimension of sympotic debates but also with Plutarch's underlying method of argumentation itself (cf. also, e.g., 697D: ἐνῇ τι καὶ πιθανόν) [see 4.3.2.2.].

The social dynamics of Plutarch's convivial milieu are very important for gaining a better insight into his sympotic protocols. Considering the proliferation of secondary literature on this topic, I will limit myself to outlining this topic only in general terms here[37]. We can see, then, that for Plutarch, the intellectual satisfaction provided by the discussion after dinner is inextricably bound with the social function of the symposium itself. In the preface to the seventh Book, Plutarch writes that subjects of

---

[36] Cf., e.g., P. Donini, 1992, p. 110.

[37] For a socio-political analysis of Plutarch's symposia, see, e.g., the contributions in J.R. Ferreira, D. Leão, M. Tröster and P. Barata Dias, 2009 (*sub* "Section 2: The *Symposion* as a Space for Social and Political Gatherings"), J.G. Montes Cala, M. Sanchez Ortiz de Landaluce and R. Gallé Cejudo, 1999, F. Klotz and K. Oikonomopoulou, 2011 and also M. Vamvouri Ruffy, 2012. Cf. F. Fuhrmann, 1972, p. xix, for the view that *Quaestiones convivales* evokes "tout un tableau de la vie sociale sous l'Empire romain". On the realities of public banqueting in *Quaestiones convivales*, see, e.g., J.C. Relihan, 1992, pp. 232, P. Schmitt-Pantel, 1992, pp. 471–482 and F. Pordomingo Pardo, 1999.

discourse, like friends, should be admitted to dinner only if they are of proven quality. If not, they should be refused entrance to the select group (*Quaest. conv.* 697DE: ἄξιόν ἐστι μηδὲν ἧττον λόγους ἢ φίλους δεδοκιμασμένους παραλαμβάνειν ἐπὶ τὰ δεῖπνα). Guests are invited to the symposium for the sake of drinking together rather than for drinking per se[38]. This means that drinking is only a means of gathering people who are mostly of a distinguished social standing, in order to foster learned discussions and social networking. Plutarch's attitude towards drinking and inebriation is, indeed, heavily influenced by the intellectual practices of his social milieu. He strongly disapproves of a symposium that ends up in 'a lascivious carousal and drunken jabber' (*Quaest. conv.* 716F), but sympotic conversation in itself is important, since it 'steadies' those who drink (*Quaest. conv.* 660C, 643AB)[39]. Personal participation in the sympotic conversations is, therefore, essential: every symposiast should contribute to the discussion and share his own – preferably original – opinion to the debate.

Even though Plutarch attributes the ideals of 'community' (κοινωνία), 'cordiality' (φιλοφροσύνη), 'humanity' (φιλανθρωπία), 'goodwill' (εὔνοια) and 'mildness' (πραότης) to sympotic companionship[40], and even stresses its 'democratic' nature[41], its aristocratic character should not, of course, be underestimated. In any case, Plutarch's discourse tends to exhibit a bias of elitist exclusivity. This is not, however, a goal in itself for Plutarch. As seen above, less intellectual people were also allowed at his table (*Quaest. conv.* 614D: ἵνα μὴ πνίγωσι τοὺς ἀνοητοτέρους μηδ' ἀποτρέπωσιν, cf. also 613E: ἰδιῶται). Moreover, when Plutarch says that he shuns 'banquets

---

[38] On heavy drinking and alcoholism in Antiquity, see J.D. Rolleston, 1927 and J.H. D'Arms, 1995.

[39] For a discussion of Plutarch's approval of a moderate consumption of wine and for its capacity to encourage friendly interaction, see H.G. Ingenkamp, 1999, A.G. Nikolaidis, 1999, P.A. Stadter, 1999 and S.-T. Teodorsson, 1999b. On Plato and wine, see P. Boyancé, 1951.

[40] See L. Van der Stockt, 2000b, p. 94 and F. Frazier and J. Sirinelli, 1996, pp. 180 ff. Cf. Also A.M. Scarcella, 1998, pp. 14–20.

[41] Cf. *Quaest. conv.* 616F (δημοκρατικόν ἐστι τὸ δεῖπνον), 657B, *Sept. sap. conv.* 154DF. In *Quaest. conv.* 621B the democratic character is placed in a negative light (νῦν μὲν ἐκκλησίαν δημοκρατικὴν νῦν δὲ σχολὴν σοφιστοῦ γινομένην). On the political agenda of Plutarch's *Quaestiones convivales*, see also P. Schmitt-Pantel, 1992, pp. 471–482 and M. Vamvouri Ruffy, 2012. On the social ἰσότης of the symposiasts in *Quaestiones convivales*, see J.H. D'Arms, 1990, p. 313. At the same time, the guests preserve their αὐτονόμια, because they have their personal interests and idiosyncrasies (cf. *Quaest. conv.* 643A). The good συμποσίαρχος should, therefore, know his guests and their physical and mental dispositions, viz. how much they can drink, their character, and – most of all – their particular field of expertise (cf. *Quaest. conv.* 613D, 613E, 621A, 643C, 644D).

for citizens and foreigners', he does so not because he is a snob or a xenophobe of any kind, but because he prefers to be among 'intellectuals' (φιλόλογοι)⁴².

For Plutarch, social elitism is intrinsically bound with intellectual pre-eminence rather than parvenu arrogance⁴³. It is more specifically distinguished from regular, more popular, symposia by the introduction of philosophy – as recurrently opposed to egotistic, self-absorbed sophistry – as the primary concern of the age-old institution of the symposium⁴⁴. Plutarch finds his examples for this in authors like Plato and Xenophon [see 2.3.1., n. 80]. At times, his adherence to the philosophical symposium receives a polemical, anti-Epicurean dimension, as is the case, for instance, in *Non posse* 1095CD, where he blames Epicurus for allowing no place, not even over wine, for 'scholarly enquiries' (φιλόλογα ζητήματα) or 'problems' (προβλήματα) concerning music and poetry. In fact, Plutarch sees the Epicureans as finding no pleasure in the contemplative part of the soul whatsoever but only in the pleasure of the belly⁴⁵.

There is a tendency in recent scholarship to connect the socio-intellectual dynamic of Plutarch's symposia with the Second Sophistic's culture of παιδεία – that is, a culture were overt demonstration of personal

---

⁴² As is the case, e.g., in *Quaest. conv.* 723A: Ἰσθμίων ἀγομένων ἐν τῇ δευτέρᾳ τῶν Σώσπιδος ἀγωνοθεσιῶν τὰς μὲν ἄλλας ἑστιάσεις διεφύγομεν, ἑστιῶντος αὐτοῦ πολλοὺς μὲν ἅμα ξένους πάντας δὲ πολλάκις τοὺς πολίτας· ἅπαξ δὲ τοὺς μάλιστα φίλους καὶ φιλολόγους οἴκοι δεχομένου καὶ αὐτοὶ παρῆμεν.

⁴³ In *Quaest. conv.* 615CD, for instance, Plutarch draws a bead on the sympotic protocol of placing guests at fixed places at the table in relation to their social status. We read that when his brother Timon once hosted a symposium and decided that the guests could choose their place themselves, an ostentatious stranger turned up at the door and immediately left again, because there was no place left that was worthy of him (οὐκ ἔφη τὸν ἄξιον ἑαυτοῦ τόπον ὁρᾶν λειπόμενον). As a result, the symposiasts were indignant about the stranger's arrogance and bid him goodbye.

⁴⁴ See especially the programmatic opening problem in the first Book of *Quaestiones convivales* (612E: Εἰ δεῖ φιλοσοφεῖν παρὰ πότον;). Cf. also, e.g., *Quaest. conv.* 716D: οἱ φιλοσοφίαν, ὦ Σόσσιε Σενεκίων, ἐκ τῶν συμποσίων ἐκβάλλοντες οὐ ταὐτὸ ποιοῦσι τοῖς τὸ φῶς ἀναιροῦσιν, ἀλλὰ χεῖρον. In addition, Plutarch's distinction between philosophical problems and (popular) trivial riddles (cf., e.g., *Quaest. conv.* 673A) also has specific socio-intellectual implications (see S. Beta, 2009). On the omnipresence of philosophy in *Quaestiones convivales*, see R. Lopes, 2009.

⁴⁵ For Plutarch, on the contrary, the discovery of solutions to entangled problems provides a genuine intellectual pleasure. See *Non posse* 1096BC: λύσεις ἀποριῶν ἐν τῷ πρέποντι καὶ γλαφυρῷ τὸ οἰκεῖον ἅμα καὶ πιθανὸν ἔχουσαι τὸ τοῦ Ξενοφῶντος ἐκεῖνό μοι δοκοῦσι (*Cyn.* 5, 33) καὶ τὸν ἐρῶντα ποιεῖν ἐπιλανθάνεσθαι· τοσοῦτον ἡδονῇ κρατοῦσιν. The entertainment value and joy (ἡδονή) that is provided by solving problems is also appreciated by Plutarch throughout *Quaestiones convivales*. Cf., e.g., 646B (δέδοκται μηδεμίαν ἡδονὴν ἀσύμβολον δέχεσθαι), 673A (ἐπὶ τὰς αὑτῆς ἡδονὰς τρέπεται, λόγοις εὐωχουμένη καὶ μαθήμασι καὶ ἱστορίαις καὶ τῷ ζητεῖν τι τῶν περιττῶν).

knowledge and learning serves as a means to confirm and sharpen elite identity, especially by distinction from the common, un(der)educated plebs[46]. However, this elitist ambition for intellectual distinction from the common plebs is a well-known τόπος throughout ancient literature and is not restricted to Second Sophistic literature only (as we will see, its presence in *Q.N.* 29, 919AB serves an underlying philosophical programme [see 3.2.2.]). Moreover, Plutarch's emphasis on the centrality of philosophy as a collective search for the truth clearly distinguishes his symposia from those recorded in a work like Athenaeus' *Deipnosophistae*, where the link between the sympotic genre and the Second Sophistic adoration of παιδεία is far more discomforting. In Plutarch's case, moreover, the consolidation of elite identity is reinforced by the friend-making (φιλοποιόν) character of the dinner-table, which requires a friendly sociability on behalf of the symposiasts at all times, rather than an continuous self-promotion of one's own education[47]. The unification of the intellectual upper-class is realised at these occasions by the emphasis on social networking between people who enjoyed the same education and can, therefore, call themselves 'well-educated' (*Quaest. conv.* 634F: πεπαιδευμένοι καλῶς). One may wonder, then, how high this degree of παιδεία actually is in the case of Plutarch's natural problems and what is their level of technicality and complexity.

4. *Quaestiones naturales* as school text: technicality and complexity

As seen in the previous section, Plutarch's golden rule at the symposium is to 'discuss philosophy without seeming to do so' (*Quaest. conv.* 614A: συνέσεως ἄκρας φιλοσοφοῦντα μὴ δοκεῖν φιλοσοφεῖν). Indeed, in *Quaestiones*

---

[46] For the connection between *Quaestiones convivales* and the Second Sophistic, see, e.g., J. König, 2008, p. 88, F. Klotz and K. Oikonomopolou, 2011, p. 3 and M. Vamvouri Ruffy, 2012, pp. 218–220. Cf. also S. Goldhill, 2009, pp. 109–110. Regarding παιδεία as a key concept in scholarship on the Second Sophistic, see M. Gleason, 1995, T. Schmitz, 1997 and T. Whitmarsh, 2005. Regarding the symposium as providing a framework for consolidation of elites already in Archaic and Classical Greece, see O. Murray, 1990b and M. Griffith, 2001, pp. 56–59. G.W.M. Bowersock, 1985, p. 665 doubts about the connection between Plutarch and the Second Sophistic. For a more systematic analysis of Plutarch's conceptualisation of sophists and sophistry, see T. Schmitz, 2014. I side with his conclusion that "Plutarch's world is not the world of the Second Sophistic" (pp. 32, 40), neither in a chronological nor in a geographical sense. Yet, even if Plutarch, much like Plato, depicts sophistic argumentation in a negative fashion, as being diametrically opposed to proper philosophical conduct, he was well acquainted with the agonistic debates of the sophists, and at times even employs overt rhetorical strategies in his own arguments – albeit never, so it should be added, for purely sophistical but for higher, philosophical reasons.

[47] Cf., e.g., *Quaest. conv.* 612D, 621C (διαγωγὴ γάρ ἐστιν ἐν οἴνῳ τὸ συμπόσιον εἰς φιλίαν ὑπὸ χάριτος τελευτῶσα) 660AB, *Sept. sap. conv.* 156D, *Ca. Ma.* 25, 4. On Plutarch's authorial self-presentation in *Quaestiones convivales* as a complex mean between self-promotion and self-effacement, see J. König, 2011. See also F. Klotz, 2007.

convivales Plutarch at several points shows to his reader how the discussion of seemingly futile topics can eventually evolve into more, albeit much concealed, philosophical inquiries. This higher, philosophical layer is generally absent from *Quaestiones naturales*, where the main focus is on the natural causes of the phenomena under discussion (think of the parallel, discussed earlier on, about the generative property of salt and its divine nature in *Quaest. conv.* 685DE vis-à-vis *Q.N.* 3, 912EF [see 2.2.1.]). As such, the natural problems collected in *Quaestiones naturales* are more 'technical' in kind than those treated in *Quaestiones convivales*, that is, they are more concerned with the type of knowledge and terminology that belongs to the τεχνῖται, such as doctors, farmers etc., rather than to the φιλόσοφοι (for this distinction, cf. *De prim. frig.* 948BC, which I will discuss later on [see 4.3.2.1.]). Scholars have differentiated the intended reading of the two works accordingly.

It is generally accepted that in *Quaestiones convivales* Plutarch aims to portray how intellectual discussions should be held in real-life situations. Therefore, the educational value of *Quaestiones convivales* is to offer some kind of a manual for proper socio-philosophical conduct[48]. *Quaestiones naturales*, on the other hand, by its primary focus on the physical aetiologies rather than on the form and dialogical set-up of the arguments, represents the actual 'theory' for such discussions. This, in turn, reflects the different educational purposes of the two texts, representing the argumentative dynamics of intellectual discussion of natural problems 'in practice' and 'in theory', respectively. In this sense, *Quaestiones convivales* adds a social dimension to the discussion of problems: it not only shows how they should be solved, but how they should be solved in a real-life situation among friends and strangers (or potentially new friends). Van der Stockt is right in writing:

> "The *QN*, then, are an introduction to the art of philosophical conversation about causes in the *QC*: to its required 'zetetic' attitude, and to the 'polite' conversation at a symposium. As a result, the *QC* offer us a more clear sight on the criteria applied for the scientific validation of opinions, explanations, and assertions [...]."[49]

---

[48] Cf. L. Van der Stockt, 2000b, pp. 93–98. See also S. Goldhill, 2009, pp. 109–110, J. König, 2007, p. 62, 2010, p. 332 and F. Klotz, 2014, p. 209. On *savoir-vivre* and conviviality in *Quaestiones convivales*, cf. also, e.g., J. Sirinelli, 2000, pp. 376–379. For the revival of the conversational ethics and etiquettes of *Quaestiones convivales* during the Renaissance, see M. Jeanneret, 1991, pp. 65–68.

[49] L. Van der Stockt, 2011, p. 453. In the same way, C. Jacob, 2004, p. 46 considers Plutarch's *Quaestiones convivales* as "l'un des modes d'emploi" of Ps.-Aristotle's *Problems*.

In a following step, one may wonder how the aspect of technicality relates to that of complexity in Plutarch's natural problems. As seen before, the sympotic decorum of *Quaestiones convivales* prescribes that discussions at the table may not become too complex or sophisticated. One of the key-words here is that of γλίσχρον, which literally means 'sticky' but is often used by Plutarch in a metaphorical sense to qualify an idea that is 'complex', 'all too subtle' or 'far-fetched'[50]. It remains to be seen, then, whether a higher degree of complexity and sophistication is allowed perhaps in the more 'technical' *Quaestiones naturales*. I believe that this is rather unlikely, though, and that for Plutarch, natural problems are essentially ἐλαφρὰ καὶ πιθανά (as suggested in *De tuenda* 133E [see 3.1.3.]).

A close analysis of the parallel passage between *Q.N.* 4, 912F–913A and Plutarch's λόγος in the second chapter of Book four (*Quaest. conv.* 664D–665A) will allow us to further substantiate these points. Both passages deal with the generative property of rainwaters that are accompanied by thunder and lightning (these are called ἀστραπαῖα, 'lightning waters', by farmers)[51]. In the passage from *Quaestiones convivales* (664A–665A), Plutarch and his fellow symposiasts deal with the popular belief that truffles are produced by thunder. Agemachus, who hosts the dinner in Elis, defends this popular belief (ἰσχυρίζετο τῇ ἱστορίᾳ) against those symposiasts who remain sceptical and reject it. The latter argue that thunder does not

---

[50] For the metaphorical implications of Plutarch's concept of γλίσχρος and its medical overtones, see M. Vamvouri Ruffy, 2012, pp. 67–75. The notion of γλίσχρον has several meanings: viz. 'sticky', 'tough', 'importunate', 'penurious', 'niggardly', 'mean', 'shabby', 'poor', 'carefully', 'detailed' or 'difficult to see' (see LSJ, s.v.). It is closely connected to the notion of σοφιστικόν in Plutarch's writings (cf. *Quaest. conv.* 615B; quoted above). The adjective γλίσχρος is used synonymously with δυσθεώρητος, meaning 'abstruse' or 'hard to understand' (cf. *Quaest. conv.* 614E, quoted above; cf. also *Quaest. conv.* 731A). In *Quaest. conv.* 614D (quoted above), Plutarch says that the matter of sympotic inquiry should remain 'rather easy' (τὰς ζητήσεις ὑγροτέρας), the topics 'familiar' (γνώριμα τὰ προβλήματα) and the subjects 'suitable and not complex' (τὰς πεύσεις ἐπιεικεῖς καὶ μὴ γλίσχρας), so that the 'less intellectual' guests (τοὺς ἀνοητοτέρους) may neither be stifled nor turned away. The meaning of 'abstruse' or 'hard to understand' is clearly present in the concept of γλίσχρας here. The pejorative connotation of 'petty' or 'frivolous' may also be implied, but not in the sense of 'easy' (cf. *De aud.* 43A, where problems in the field of dialectics and mathematics are considered μικρὰ καὶ γλίσχρα). It can also be used in connection with philosophical issues (*De aud.* 47B: φιλοσοφίας ἐχούσης τι καὶ γλίσχρον ἀμέλει καὶ ἀσύνηθες), and it is even related to play and unconvincing inventiveness (*De aud. poet.* 31E: γλίσχρος ἐστίν, οὐ παίζων ἀλλ' εὑρησιλογῶν ἀπιθάνως), as well as to ridiculous explanations (cf. *Quaest. conv.* 670B: αἰτίαις γλίσχραις, ἐνίων δὲ καὶ πάνυ γελοίαις).

[51] This same parallel is discussed by F. Klotz and K. Oikonompoulou, 2011, pp. 20–21, albeit not in light of what is γλίσχρον for Plutarch, nor of the subtle philosophical agenda in *Quaestiones convivales*, but in view of Plutarch's flexible reuse of specific bits of knowledge in different contexts. Cf. also L. Van der Stockt, 2011, pp. 450–454.

actually generate truffles but simply makes them visible in the cracks in the earth. Agemachus, by contrast, lists several other natural *mirabilia* that are related to thunder and lightning, which, so he says, are 'difficult if not completely impossible to solve' but are not, therefore, unworthy of belief (χαλεπὰς καταμαθεῖν ἢ παντελῶς ἀδυνάτους τὰς αἰτίας ἔχοντα κτλ.). Among these *mirabilia* he mentions the farmers' account of the generative properties of lightning water (τὰ δ' ἀστραπαῖα τῶν ὑδάτων εὐαλδῆ καλοῦσιν οἱ γεωργοὶ καὶ νομίζουσιν). Here, Plutarch enters the discussion and argues that '*at least at the time*' he cannot provide 'a *more plausible* explanation' (οὐδὲν γὰρ ἔν γε τῷ παρόντι φαίνεσθαι πιθανώτερον) for the fertility of lightning water than that the heat of the rain, so produced by the fire of lightning, enriches the soil. This explanation runs parallel to the second *causa* in *Q.N.* 4, 913A, where Plutarch, regarding the same problem of lightning water, argues that the heat of lightning concocts the moisture of rainwater and makes it agreeable and useful to the growth of things.

Notably, at the end of his explanation in *Quaest. conv.* 665A (after adding a number of other sub-arguments), Plutarch says that his fellow symposiasts may find his λόγος somewhat γλίσχρος[52]. He assures them, however, – and this is the very essence of his λόγος – that most of the effects of thunder and lightning are like that, that is, they have a generative property (γόνιμον), and this testifies to their ostensibly *divine* character (εἰ δέ γε γλίσχρος [...] ὁ λόγος ὑμῖν δοκεῖ, τοιαῦτά τοι τὰ πλεῖστα τῶν βρονταῖς καὶ κεραυνοῖς συνεπομένων· διὸ καὶ μάλιστα τοῖς πάθεσι τούτοις δόξα θειότητος πρόσεστι). This is followed by an account of the divine character of thunder and lightning by the rhetorician Dorotheus, which closes off the problem at hand (and opens up the discussion of a new and closely related popular belief, viz. that sleepers are never struck by thunderbolts). In the parallel account in *Q.N.* 4, however, Plutarch adds yet another natural explanation (in the third and final *causa*) that can also be considered γλίσχρος. He argues that it is not so much the thunder and lightning, which particularly occur in spring due to uneven temperature in the air, that make the rain more fertile. Rather, the rainwater itself is useful to plants, because it comes 'before the summer heat' (πρὸ τοῦ θέρους), and, thus, protects the plants against this heat. Now, if the second *causa* can already be considered γλίσχρος to a certain degree – as Plutarch suggests in *Quaest. conv.* 665A –, this is perhaps even more the case in this third *causa* in *Q.N.* 4, since a subtle

---

[52] The adjective γλίσχρος qualifies Plutarch's entire λόγος here. This means that it does not only refer to Plutarch's sub-arguments about the special charateristics and flavour imparted by such rains on vegetation (e.g., the belief that dew makes grass sweeter and that rainbows fill trees with fragrances); nor is it restricted to the theory that a truffle is a being that is constituted by itself (Plutarch argues that it is not a normal φυτόν, because it bears no fruit and has no roots and is made entirely of earth that is slightly altered by the rain).

aetiological distinction is made there. Plutarch suggests that the influence of lightning and thunder on the increase in fertility is an incidental adjunct rather than the main cause of this phenomenon. In other words, thunder and lightning are rather circumstantial phenomena that have nothing to do with the generative properties of rainwater as such (compare the initial, sceptical argument of the symposiasts who reject the popular belief about the generation of truffles by lightning). As we will see later on, there is often, indeed, an increase in specificity and detail in the development of Plutarch's aetiologies [see 4.3.3.3.]. One should not, therefore, jump to conclusions, though, and claim that, in light of the aforementioned technical character of *Quaestiones naturales*, our collection admits, or is even reserved for γλίσχρον argumentations.

Those who would, nevertheless, stick to such a belief might come to assert that the particle γε in the phrase ἔν γε τῷ παρόντι ('*at least* at this moment'; see above) implies that Plutarch would be able on other occasions – c.q. in the more technical discussion in *Q.N.* 4 (where the specific natural phenomenon of the lightning water is discussed separately and on its own physical terms) – to provide another explanation that is not only πιθανώτερον but also even more γλίσχρον (viz. the third and final *causa* with its subtle aetiological distinction). Such a claim would run the risk of neglecting the different argumentative contexts of the two discussions, though. The main problem at hand in *Quaest. conv.* 664B–665A is not why lightning water is fertile for plants but why truffles are thought to be produced by thunder. The phrase οὐδὲν γὰρ ἔν γε τῷ παρόντι φαίνεσθαι πιθανώτερον should be interpreted in this specific context: Plutarch cannot find an explanation that is more probable *at this very moment*, that is, in the current discussion – as opposed to other discussions, such as *Q.N.* 4, where the main problem is not that of the generation of truffles, though. It is not unlikely, therefore, that Plutarch with the phrase ἔν γε τῷ παρόντι is subtly alluding to *Q.N.* 4 [see 2.1.4.]. He must have been aware of the fact, then, that the contexts were very different. Plutarch would probably have acknowledged that other explanations for the problem about the lightning water are possible, as the third *causa* in *Q.N.* 4 proves. Arguably, then, his explanation in the discussion at issue in *Quaest. conv.* 664B–665A is very *ad hoc*, since it is, indeed, the *most probable* one, *at least in the present context*.

If Plutarch had made the same subtle aetiological distinction in his λόγος in *Quaest. conv.* 664D–665A as he does in the third *causa* in *Q.N.* 4, he would actually have sided with the camp of the sceptical non-believers (who reject the popular belief that thunder is really generative of itself and argue that it merely brings the truffles to light). In that case, there would have been no opportunity for further discussion but only an awkward silence (notably, in *Quaest. conv.* 641CE and also in 642AB, similar subtle aetiological distinctions are made, which close off the discussions at hand [see 4.3.3.3.]). Therefore, we should probably cut Plutarch some argumentative slack in his physical aetiologies, especially since, from his Platonic epistemological perspective, nothing in nature can be captured in terms of certainty [see 4.3.3.1.]. It is presumably for this reason, then, that he considers natural problems to be essentially πιθανά (*De tuenda* 133E; quoted above). If Plutarch had also added the third *causa* from *Q.N.* 4 in his λόγος in *Quaest. conv.* 664B–665A – disregarding whether or not it is even πιθανώτερον – this would have entirely disrupted the rationale of his own argument. It is precisely for this reason that Plutarch says that *in this specific context, nothing seems more plausible* than the explanation that he provides[53].

Another important – if not the most important – point that should be added is that at the very end of his λόγος in *Quaest. conv.* 665A, Plutarch shifts from speaking of the physical properties of lightning towards its divine character (διὸ καὶ μάλιστα τοῖς πάθεσι τούτοις δόξα θειότητος πρόσεστι). This point is further elaborated upon in the subsequent argument by the rhetorician Dorotheus, who notes that the divine theory is accepted 'not only by ordinary people but also by some philosophers' (οὐ γὰρ μόνον οἱ πολλοὶ καὶ ἰδιῶται τοῦτο πεπόνθασιν, ἀλλὰ καὶ τῶν φιλοσόφων τινές). Arguably, the incorporation of such religious-theological topics in *Quaestiones convivales* is indicative of the philosophical goal of the sympotic discussions at hand. It nicely elucidates Plutarch's golden rule, as stated above, that philosophy should be discussed in guarded terms during symposia, viz. 'without seeming to do so' (*Quaest. conv.* 614A). No such divine interpretation is found in the parallel passage in *Q.N.* 4, though, which, in turn, illustrates the more 'technical' approach of *Quaestiones naturales* as a

---

[53] As F. Klotz and K. Oikonomopoulou, 2011, p. 21 have rightly concluded, therefore: "Seen together then, the two chapters vividly showcase the remarkable flexibility with which Plutarch can engage with a core of factual knowledge, adapting it to different contexts and objectives: what the fictional Plutarch in the *Table Talk* cites as a self-sufficient explanation (rain following from thunder is highly nourishing for plants, because of the admixture of heat) in the *Natural Questions* features as a partial answer to an open question (what is it that makes the rain following from thunder nourishing?), which admits different scientific explanations – all of which are exhaustively explored. The difference is not just of focus, but also concerns context and register. What the *Table Talk* especially underscores is the way such knowledge can naturally spring up in the relaxed context of learned conversation, blending in with folk wisdom, oscillating between seriousness and play (chapter 4.2's main topic of discussion is after all truffles!), and serving the needs of a speaker's selected argumentative strategy. It thus brings before us the fact that scientific inquiry is never a value-free exercise, but one influenced by situations and objectives often extraneous to a search for scientific 'truth'. Moreover, it allows us to place scientific knowledge in a wider cultural framework, by drawing attention to the kinds of cultural preoccupation – in this instance, the marvellous – which it can address." Cf. also L. Van der Stockt, 2011, pp. 450–454.

whole. As seen previously, Plutarch in this work primarily focuses on the physical causes of the phenomena studied therein without much further ado[54] [see 1.1.6.].

This does not necessarily speak to the complexity of the natural problems as such, though, but rather to their different intellectual dispositions. So the key question still remains about the actual level of παιδεία that is required on behalf of the reader. How much knowledge is actually needed beforehand for a proper understanding of these problems? In order to answer this question, an analogy can be drawn with *Quaestiones Platonicae*.

For an 'esoteric' school text (see n. 73), the aetiologies in *Quaestiones naturales* never seem to become too complex. If the text is in any way obscure, this is a matter of style rather than content [see 2.3.2.]. In fact, a minimal acquaintance with the theoretical and terminological apparatus seems to suffice for an adequate understanding of the text [see 4.3.4.]. It is only likely, then, that the implied reader did not necessarily have to consider himself a specialist in ancient physics to be able to follow the general lines of argumentation. Dörrie reached a similar conclusion regarding *Quaestiones Platonicae*, by arguing, *mutatis mutandis*, that no profound philosophical foreknowledge was required for a good understanding of the problems collected therein. Thus, it should not be considered specialist literature, but rather the work of an author aiming to popularise specific debates in the Platonic tradition[55]. Opsomer, however, refuted this theory by arguing that a serious and profound degree of philosophical foreknowledge actually was required for a proper understanding of the text[56]. This obviously tells us something about its implied readership, but what holds true for *Quaestiones Platonicae* does not necessarily hold true for *Quaestiones naturales* just as well.

If we bear in mind the opposition between the notions of αἰτίαι and ζητήματα [see 1.1.6.], it is clear that the kind of research conducted in *Quaestiones naturales* is less specialised than in *Quaestiones Platonicae*, in that it is not exegetical (at least in a strict sense). Our collection is

---

[54] This was also observed by L. Van der Stockt, 2011, p. 454: "*QN* offers no leg up here. But that doesn't mean Plutarch's physical world is without god! It only means that we have to look for Plutarch's *philosophia prima* elsewhere in his oeuvre."

[55] H. Dörrie, 1959, p. 4: "Auch in anderer Hinsicht fand der Leser, was er erwarten durfte: er durfte erwarten, daß die Fragen von vielfältiger Thematik – eben vermischter Art waren; er durfte erwarten, sich nicht in dornige Spezialuntersuchungen, die Vorkenntnisse erfordern, verstrickt zu sehen, sondern ein auch dem Laien übersehbarer Beweisgang mußte zu plausiblen Ergebnißen führen." According to some scholars, Plutarch's *De facie*, for instance, was also destined for a readership of non-specialists (see H. Cherniss and W.C. Helmbold, 1957, p. 19; cf. also P. Donini, 1984, p. 369). See, e.g., *De facie* 938C.

[56] J. Opsomer, 1994a, pp. 12–13 (see also 1996a, pp. 82–83).

not concerned with the clarification of specific enigmatic passages in a text or with the tradition of that text with which the reader had to be acquainted first. Rather, it deals with explanations of more general, everyday problems, that are, in a certain sense, universal and common to everyone's experience. As Oikonomopoulou rightly states:

> "[T]he *QN*'s investigations do not emanate from a scientist's ivory tower, but are anchored in the economic and cultural parameters of practical life: agriculture, animal husbandry, hunting, fishing, seafaring, swimming, feasting and drinking."[57]

In some way, 'nature', understood as a conglomerate of countless natural phenomena and the current beliefs about them, is the 'text' being examined here – though, of course, Plutarch himself generally relies on the scientific literature at hand to make sense of it. Bearing in mind, however, that Plutarch's collections of *quaestiones* are all set in the educational context of his philosophical school[58], it is not unimaginable that the actual readerships of *Quaestiones naturales* and *Quaestiones Platonicae* were identical. Thus, we should think of Plutarch's readers here as philosophers, not only experts, but also students who are advancing in their philosophical παιδεία, and also well-educated πεπαιδευμένοι with an informed notion of the philosophical tradition in general. The bottom-line, then, is that *Quaestiones naturales* is more generally comprehensible than *Quaestiones Platonicae*, but that both texts originate from the very same school context. In the end, natural problems remain essentially ἐλαφρὰ καὶ πιθανά for Plutarch (although a natural explanation may sometimes become somewhat – albeit never excessively – complex or sophisticated). This is not necessarily irreconcilable, then, with the view that *Quaestiones naturales* presents the 'theory' of scientific discussions, whereas *Quaestiones convivales* presents such discussions 'in action'.

In order to receive a more concrete idea of Plutarch's readership and of the kind of audience that was interested and actually participated in such inquiries, we can think of the types and characters put on stage in *Quaestiones convivales*. These are Plutarch's intimi and belong to his more or less direct intellectual milieu: viz. his students, family-members, friends, but also friends of friends, both παιδευόμενοι and πεπαιδευμένοι, young and old, philosophers and all-round intellectuals, specialists and laymen alike, all of whom had their personal interest and occupations

---

[57] K. Oikonomopoulou, 2013a, p. 146.

[58] Also historical-antiquarian problems, as those treated in *Quaestiones Romanae* and *Graecae*, belong to the so-called 'second repast' (cf. *De tuenda* 133E: ἐν ἱστορικαῖς ζητήσεσι), and this is also the case with the more philosophical-exegetical ones (cf., e.g., *Quaest. conv.* 718B: Πῶς Πλάτων ἔλεγε τὸν θεὸν ἀεὶ γεωμετρεῖν;).

in day to day life (either politics, rhetoric, poetry, literature, medicine, agriculture, philosophy etc.), but shared a general interest in a broad array of intellectual matters, thus by no means despising the lower strata of natural philosophy[59].

5. The dialogue between author and reader: vivacity and historicity

Even if *Quaestiones naturales* represents only the 'theory' of natural scientific discussions, there is still some degree of argumentative vivacity to it, as can be seen in its dialogical organisation (albeit to a far lesser degree than is the case in *Quaestiones convivales*, of course). This is not so exceptional to the genre of problems more generally. Regarding Ps.-Aristotle's *Problems*, some scholars have argued that the question-and-answer format is, in fact, all about argumentative liveliness, and that the problems are actually based on genuine philosophical debates (in the Lyceum), or at least resemble the conditions of such debates. At the same time, however, it is generally acknowledged that the problem format had become a standard method of composition in dealing with scientific topics already by the time of Aristotle, which underlines the 'literary' character

---

[59] According to L. Van der Stockt, 2011, p. 453, "*QN* offers philosophical fuel for the philosophical discussion and is destined to people in need of that fuel, namely all those who are not philosophers [*cave*], who uphold only popular opinion, but want to take part in philosophical discussion. These are the young students in Chaeronea and older friends of Plutarch: people who controlled their agenda and had had some training in rhetoric and grammar (poetry)." Cf. also K. Ziegler, 1951, cols. 662–665, esp. col. 663: "Der Schülerkreis besteht teils aus älteren Leuten, die zu dem Leiter mehr in einem Freundschafsverhältnis stehen (οἱ συνήθεις), teils aus den jungen Studenten (νέοι, νεανίσκοι, μειράκια), von denen uns eine Anzahl namentlich vorgestellt wird". For the diversity of the interlocutors in *Quaestiones convivales*, cf., e.g., F. Klotz, 2007, p. 653 and G. Roskam, 2009, p. 376. See also J. Sirinelli, 2000, p. 380: "En somme, ainsi que Plutarque le précisera, il s'agit de banquets de personnes cultivées et non pas de spécialistes, qu'ils soient hommes d'affaires, politiciens ou sophistes, ou même philosophes, ou plus exactement, les conversations (car ce sont les conversations qui sont ainsi qualifiées) ne doivent pas être des conversations de spécialistes. Chacun, [...] et mêmes les philosophes, doit abandonner les questions techniques pour ne discuter que de questions intéressant des hommes simplement cultivés." If required by some social protocol, specialists can, of course, discuss non-specialist matters, or at least converse in a non-specialist way. On the intended readership of *Quaestiones convivales*, see also F. Klotz and K. Oikonomopoulou, 2011, pp. 27–29. Moreover, regarding the readers of Plutarch's *Vitae*, P.A. Stadter, 1988, pp. 292–293 argues that they were "male, upper class, and leisured. [...] Plutarch's audience were also intellectuals, well-read and familiar with the science of their day". According to D.A. Russell, 1973, p. 43, Plutarch addresses himself in his entire oeuvre to "the highly-trained, the imaginative, the leisured", but H.G. Ingenkamp, 1976, p. 547 has put this into perspective: "Sicher war P. ein solcher Leser besonders willkommen, aber seine Gelehrsamkeit hindert den durchschnittlich Gebildeten nicht am Zugang."

of the problem genre⁶⁰. It is not unimaginable, therefore, that some parts of Plutarch's *Quaestiones naturales* may originate from actual dialogues in his school, or that Plutarch at least composed this collection with a specific school context in mind. It would not be unrealistic to assume that Plutarch, as the head of his own school, participated in the school discussions and actually took notes on what was said⁶¹. These notes he later incorporated and remodelled in his *quaestiones*, adding his own views and comments, based on his own reading and research. If this is correct, *Quaestiones naturales* represents the proceedings of Plutarch's school seminars on natural scientific topics, providing a *status quaestionis* of the discussions held, after a thorough review by Plutarch himself. However, except from a few indications, presented below, there is no solid proof for the historicity of *Quaestiones naturales* (as is also the case for *Quaestiones convivales*, as we saw earlier on [2.3.1.]).

Scholars have identified the enigmatic figure of Λαῖτος, whose answers to two specific meteorological problems are recorded in *Q.N.* 2, 911F and 6, 913E, with Ofellius Laetus, a contemporary Platonic philosopher with whom Plutarch perhaps had a real-life discussion on specific meteorological phenomena once [see 4.2.1.1., n. 115]. This has been inferred from the imperfect ἔλεγε which introduces Laetus' two accounts, and which, according to scholars, may suggest a personal encounter with Plutarch⁶². Notably, Plutarch again uses ἔλεγε to introduce the opinion of another contemporary Platonic philosopher, the Lacedaemonian Tyndares, in *Quaest. conv.* 728E. This man also makes his appearance in *Quaest. conv.* 717E and 718C, where his accounts are introduced differently, though, with ἔφη and εἶπεν respectively. The evidence is, indeed, uncertain, since Plutarch uses ἔλεγε not only of contemporaries but also of past authorities,

---

⁶⁰ Cf. H. Flashar, 1962, p. 341, C. Jacob, 2004 and K. Oikonomopoulou, 2013b. For some aspects of counter-liveliness in Ps.-Aristotle's *Problems*, see, however, H. Flashar, 1962, p. 345: "Nicht immer sind die Problemata der unmittelbare Reflex der lebendige Diskussion, denn in zahlreichen Fällen können wir nachweisen, daß das Frage- und Antwortschema erst nachträglich einem in Aussageform bereits fixierten Gedanken aufgesetzt ist. Dieser Umsetzungsprozeß ist überhaupt ein wesentlicher Bestandteil der Quellenverarbeitung. Immerhin braucht sich die Tatsache eines in einer Quelle vorgegebenen Gedankens und die Möglichkeit einer Schuldiskussion nicht unbedingt auszuschließen [...]. Bei der Beurteilung der Entwicklung der Problemata-Form, die von der lebendigen Diskussion zum Handbuchschema führt, muß man auch berücksichtigen, daß die für die Probl. charakterische Frage- und Antwortform schon bei Ar. in starkem maße 'literarisch' geworden ist".

⁶¹ Cf., e.g., *Quaest. Plat.* 1003A: τὸ πολλάκις ὑφ' ἡμῶν λεγόμενον. Cf. also *Non posse* 1086D: ἐπεὶ δὲ καὶ τῆς σχολῆς διαλυθείσης ἐγένοντο λόγοι πλείονες ἐν τῷ περιπάτῳ πρὸς τὴν αἵρεσιν, ἔδοξέ μοι καὶ τούτους ἀναλαβεῖν.

⁶² Cf. J. Opsomer, 2008, p. 586. For further reading, see also G.W.M. Bowersock, 1982 and M. Meeusen, 2013a.

such as Plato (*Cor.* 15, 4; *Non posse* 1091D), Euripides (*Alc.* 1, 5), and Cleanthes (*Alc.* 6, 2). It remains to be seen, then, whether any aspect of historicity can be deduced from the two Laetus quotes, a problem that hardly admits of an answer. This is not, of course, to reject that some parts of *Quaestiones naturales* may contain the argumentative nucleus of the responses that Plutarch's fellow discussants gave to the problems at hand and which he then recorded in a primarily anonymous fashion (perhaps, indeed, with the exception of Laetus). In *Q.N.* 3, 912EF, to give just one example, Plutarch anonymously records a solution that is ascribed to Philinus in *Quaest. conv.* 685D, albeit in a different context [see 2.2.1.].

The issue of historicity is very difficult, if not impossible, to settle, since we should always bear in mind Plutarch's potential editorial interferences in the text. What is perhaps more important, therefore, is that Plutarch's general school context has clearly left its marks on the collection's discourse, disregarding whether this influence occured directly or not. In the end, the possibility that *Quaestiones naturales* is simply exploiting the question-and-answer format, precisely because of the latter's roots in school practice, cannot be excluded[63]. The bottom line would, then, be that *Quaestiones naturales* does not necessarily have to originate from historical school discussions to have a basic educational intention. In either case, the question-and-answer format operates as a useful educational tool for the systematisation and communication of scientific knowledge between the author and his readership. It is also an efficient tool in Plutarch's attempt to convince the reader that certain physical explanation are plausible and deserve specific consideration. As such, in each natural problem, we find the condensation of some kind of a virtual dialogue, where the author poses a question and, at the same time, suggests several answers to the reader in an interrogative way. A dramatised version of such dialogues is found in *Quaestiones convivales*, where Plutarch shows how these problems provided popular topics of conversation during the social event of the symposium. This is not, however, to claim that the

---

[63] According to K. Oikonomopoulou (in personal correspondence), readers of *Quaestiones naturales* of a more advanced age would have perhaps enjoyed learning from a text written in a format familiar to them from the years of their school instruction. This may, indeed, be the case, but whether this implies that discussing natural problems is normally a practice suitable for young students only is not a given fact. In any case, both younger and older symposiasts are found discussing natural problems in *Quaestiones convivales*. In fact, the fictional Plutarch himself makes his appearance both as a young student and as a more mature philosopher [see the prologue, n. 25]. Therefore, the philosophical value of Plutarch's natural problems is not necessarily age-related. In fact, they provide a useful intellectual exercise for more mature philosophers just as well, as we will see further on [3.2.2.].

genre of problems *mimics* that of the dialogue: at the very least there are similar structural principles at work in both genres.

On closer inspection, the problem format has several structural similarities with that of the (philosophical) dialogue[64]. It allows for critical discussion, refutation and correction of traditional view-points and for a tentative proposition of personal theories in a generally zetetic fashion. Moreover, the solution to one question may give rise to another question that is closely related. This may account for the concatenative process that has been noted for the development of many problem chapters in our collection at a macrostructural level[65] [see 1.1.5.]. But then again, this concatenated structure is not necessarily caused by the dynamics of real-life discussions; it can also be explained on the basis of the author's editorial interventions, or of the coherence of the content in the source text on which the chapters are sometimes based [see 4.2.1.2.].

At a microstructural level [see 1.1.4.], the profusion of question marks and the absence of aetiological closure, testify to the author's awareness that the explanations he provides do not necessarily fully exhaust the entire aetiological potential of a specific problem. As such, the almost incessant formulation of explanations in an interrogative and anti-dogmatic fashion in the aetiologies is not so much a rhetorical technique inspired by feigned uncertainty (analogous to rhetorical questions), but a sign of scientific caution, both on the side of the teacher and the student[66]. In fact, it is one of Plutarch's firm epistemic convictions that it is impossible to formulate natural explanations that are certain and definite (ἐποχή). As a consequence, the φυσικός has to manage with probable arguments and leave space for the potential addition of new solutions, corrections and criticisms [see 4.3.2.]. When it comes to the interrogative and open-ended structure of the explanations, one could even argue that the role of the reader, in a certain sense, merges with that of the author. The reader is invited, then, to evaluate the positions the author proposes and, if possible, to formulate his own explanations. This certainly makes the reading process more active and engaging, and for that matter, more attractive from a didactic perspective.

Closely connected to this point is the idea that, from an epistemological perspective, Plutarch, as a teacher, actually places himself on par with his students, since they are both looking for what are essentially plausible

---

[64] See, e.g., C. Jacob, 2004, J. Opsomer, 2005, pp. 198–199, 2010, p. 115 ("une forme rudimentaire de dialogue") and K. Oikonomopoulou, 2013b. For a study of Plutarch's general use of the dialogue format, see L. Van der Stockt, 2000b (with further references).

[65] The same argument was made for Ps.-Aristotle's *Problems* by H. Flashar, 1962, p. 301.

[66] A similar conclusion was made for Ps.-Aristotle's *Problems* by H. Flashar, 1962, p. 341.

natural explanations. Therefore, obvious authorial self-promotion in providing such arguments is unnecessary. This probably explains why Plutarch's personal statements and remarks are *not* formulated in an authorial fashion in *Quaestiones naturales*[67]. In fact, Plutarch effaces everything personal by avoiding the use of the first person singular ('I') in favour of the less emphatic first person plural ('we'). In some cases[68], though, the first person plural may function as a substitute for the first person singular (scholars call this the "authorial we")[69]. Yet, this is rather unlikely, if not impossible, in other cases[70]. The use of the first person plural can be considered a suggestive means, then, of involving the student in a joint search for physical causes (scholars call this the "sociative we")[71]. It

---

[67] Cf. G.W.M. Bowersock, 1985, p. 666: "But spread throughout his [sc. Plutarch's] works is a genuine and irresistible humanity, unfettered by egotism or pretence. In this respect too he is highly unusual." The rather impersonal authorship is also typical of Ps.-Aristotle's *Problems* (see C. Jacob, 2004, p. 41, V. Nutton, 2009, pp. 58–59, K. Oikonomopoulou, 2013b). Similarly, in regards to *Quaestiones Platonicae*, J. Opsomer, 2010, p. 93 notes that "l'intervention de l'auteur est minimale". This general impersonal authorship can perhaps be connected, in the case of Plutarch, with the ἀφασία of the Academic Sceptics [see 4.3.2.1.], that is, the avoidance of formulating theories of one's own, a practice that goes back on Socrates' aporetic attitude in philosophical dialogues (cf. *Quaest. Plat.* 1000B: τῶν ὥσπερ Σωκράτης ὁμολογούντων μηδὲν ἴδιον λέγειν). A relevant passage for this ἀφασία is found in *De facie* 922E–923A, where the Stoic Pharnaces attacks Lamprias for employing τὸ περίακτον ἐκ τῆς Ἀκαδημείας, viz. μὴ παρέχειν ἔλεγχον ὧν αὐτοὶ λέγουσιν. Lucius speaks in defense of Lamprias and declares that ἡμεῖς (i.e. the Academic philosophers) μὲν οὖν οὐδὲν αὐτοὶ παρ' αὑτῶν λέγομεν. The fact that on several occasions, they *do* employ the first person, might complicate the consistency of this sceptical feature in *De facie* (cf. H. Görgemanns, 1970, p. 86), but P. Donini, 2011, pp. 38–39 is probably right that "semplicemente, in quelle occasioni i suoi personaggi si identificano con la teoria che difendono non perché improvvisamente se ne siano fatti conquistare e abbiano abbondonato ogni riserva a proposito della sua veridicità definitiva così tradendo le premesse dell'epistemologia 'academica' e del *Timeo*, ma soltanto in quanto la considerano la migliore delle spiegazioni disponibili". Arguably, then, the same counts, *mutatis mutandis*, for Plutarch's personal statements (by means of his own literary *alter ego*) in *Quaestiones convivales* (see further). Note, moreover, that the aspect of ἀφασία or avoidance to beget personal doctrines does not withhold Plutarch from proposing his own original contributions to the problems, again, at least within the limits of his Platonic-Academic epistemology [see 4.2.2.2.].

[68] Cf. *Q.N.* 1, 911E (παραινοῦμεν) and 15, 915E (εἰρήκαμεν).

[69] See H. von Staden 1994, p. 108 and H. Hine, 2009, p. 21.

[70] Cf. *Q.N.* 5, 913B (ὁρῶμεν), 8, 914B (ὁρῶμεν), 29, 919A (θαυμάζομεν), 30, 919B (λέγομεν).

[71] F. Slotty, 1928. Regarding the most common uses of the first person plural in ancient scientific writing, see H. von Staden, 1994, pp. 108–109 and H. Hine, 2009, pp. 21–22. For the didactic relationship expressed by the first person plural in *Quaestiones convivales*, see J. König, 2011, p. 185. On Plutarch's use of the first person plural in *Quaestiones Romanae*,

evokes a certain sense of zetetic community (κοινωνία)⁷² and, in most cases, hints at some kind of didactic pact between the teacher and the student by invoking, and at times even insisting on, shared believes, customs and experience. For instance, in *Q.N.* 8, 914B, Plutarch wonders why *we* observe that the sea grows warmer when it is agitated, whereas all other liquids – paradoxically – become colder when disturbed (Διὰ τί, τῶν ἄλλων ὑγρῶν ἐν τῷ κινεῖσθαι καὶ στρέφεσθαι ψυχομένων, τὴν θάλατταν ὁρῶμεν ἐν τῷ κυματοῦσθαι θερμοτέραν γιγνομένην;). Rather than discussing whether or not this is really true, Plutarch, by use of the first person plural in ὁρῶμεν, insists on assuming that this is the case, if only for the sake of argument [see 4.1.1.2.]. It can be inferred from this aspect of authorial self-effacement that Plutarch deliberately aims to temper his authorial voice, so that the text never becomes too obtrusive or self-promoting.

This is far less the case in *Quaestiones convivales*, presumably for reasons of 'exotericism' – *vis-à-vis* the 'esotericism' of *Quaestiones naturales*⁷³. Arguably, in *Quaestiones convivales*, Plutarch aims to pro-

---

cf. also R. Preston, 2001, p. 114 ("[Plutarch's] use of the first person plural [...] suggests the possibility of a unified, undifferentiated humanity") and P. Payen, 2014, pp. 245. For the *Vitae*, see also C. Pelling, 2004, pp. 411–412: "It is indeed often unclear exactly how that category of 'us' is envisaged: 'we Greeks', 'we cultured beings', 'we people of humane sensibility', 'we who are interested in the past'? Does it include real readers in subsequent generations as well as those 'in our day', i.e. Plutarch's own? But in any case, it is evidently a category that includes narratee as well as narrator. [...] The blurring is important in insinuating that *of course* narrator and narratee are people who think along similar lines." Cf. also *id.*, 2011, p. 208: "that characteristic Plutarchan 'we' and 'us' [...] contrives to suggest a large happy family of readers". See also T. Duff, 2014, pp. 340–342.

⁷² In *Praec. ger. reip.* 816DE, Plutarch reports his father's advice that it is more polite and unambitious to use the first person plural instead of the first person singular as a way of reporting on an association with colleagues (οὐ γὰρ μόνον ἐπιεικὲς τὸ τοιοῦτον καὶ φιλάνθρωπόν ἐστιν, ἀλλὰ καὶ τὸ λυποῦν τὸν φθόνον ἀφαιρεῖ τῆς δόξης).

⁷³ If *Quaestiones naturales* can rightly be considered an 'esoteric' text, which Plutarch presumably composed for the inner circle in his own school, *Quaestiones convivales* can be considered a more 'exoteric' work, promoting Plutarch's own research activities and that of his intellectual milieu to the outer world (the work was dedicated to Sossius Senecio). The esoteric nature of *Quaestiones naturales* can be generally linked, then, with its theoretical character, and the exoteric nature of *Quaestiones convivales* with the social pragmatics that it promotes by describing the discussion of problems 'in action'. This distinction may not remain unproblematic, though. In any case, scholars have also treated the issue of exotericism *vis-à-vis* esotericism with regard to Ps.-Aristotle's *Problems*, but at first sight, their claims seem irreconciliable. This incongruity mainly rests in a different perspective of interpretation. Some scholars especially focus on the form and organisation of the *Problems* in considering them esoteric, whereas others consider them exoteric on the basis of their content. According to C. Jacob, 2004, p. 44, n. 32, for instance, Ps.-Aristotle's collection of *Problems* is an esoteric work to be situated in the Lyceum: "[L]es *Problèmes* seraient un exemple parmi d'autres des différentes formes d'écriture à l'oeuvre

mote the practices and achievements of his research to a wider audience by placing his own literary *alter ego* in the sympotic spotlights at several points in the discussions (often allowing him the final and, thus, most authoritative word on a given topic [see 1.1.4., n. 104]). To this end, Plutarch often presents his literary *alter ego* as some kind of *primus inter pares* in the discussions with his peers[74]. This is not to say, however, that *Quaestiones convivales* is an ego-document, since the views of Plutarch's sympotic peers often have even more weight in the discussion. What is perhaps more important, however, is that a certain aspect of argumentative caution and undecidedness is never abandoned (cf. n. 67).

Another discursive means of involving the reader in a joint search for physical causes is found in the use of the second person singular in imperatives like σκόπει δέ/δή (μή)[75]. Such phrases can be interpreted as a "metadirective"[76] address to the reader, meant to emphatically encourage

---

dans la partie ésotérique du corpus aristotélicien, qui nous est parvenue: 'textes d'école' comprenant des notes de cours, des textes de conférences, des brouillons, mais aussi des instruments de travail partagés par les membres de la communauté, pouvant faire l'objet de révisions et d'enrichissements." Cf. also W.S. Hett, 1936, p. viii ("The form of the book suggests that it originated as a lecture notebook containing problems for discussion.") and H. Flashar, 1962, p. 341. On the contrary, P. Louis, 1991, pp. xix–xx has argued that the *Problems* are exoteric, because they are sometimes labelled ἐγκύκλια (see M. Meeusen, forthcoming e), implying that they are 'in the hands of everyone', making it a work of philosophical vulgarisation (see also A. Garzya and R. Masullo, 2004, p. 12). Louis does not as such reject the school context for the *Problems*, though, since he describes them as "des notes prises au cours de lectures ou à la suite d'observations personnelles" (p. xxix). If we now transpose this problem on Plutarch, we see that the natural problems treated in *Quaestiones naturales* and *Quaestiones convivales* are very similar from the perspective of content to those treated in Ps.-Aristotle's *Problems*, so that they can be considered exoteric [see 2.2.3.]. From a formal perspective, though, *Quaestiones naturales* is closer to Ps.-Aristotle's *Problems* than is *Quaestiones convivales*, so that the former can be considered more esoteric, at least according to these standards [see 2.2.2.]. For Plutarch's acquaintance with the distinction between Aristotle's exoteric and esoteric writings, cf. *Adv. Col.* 1115B and *Alex.* 7, 2.

[74] On the intricacy of Plutarch's authorial self-promotion and self-effacement in *Quaestiones convivales*, see J. König, 2011 and F. Klotz, 2007. As J. König, 2011, p. 188 observes, Plutarch "was on the whole [...] uninterested in confrontational Galenic debunking of intellectual rivals and predecessors". Cf. also J. König, 2008, p. 97.

[75] Cf. *Q.N.* 3, 912E, 12, 915A, 19, 916C. The use of the second person singular is also found in the verb ἐμβαίης in Heraclitus' river-statement in *Q.N.* 2, 912A and in ὁρᾷς in Euripides' quote in *Q.N.* 29, 919B.

[76] See R. Risselada, 1993, pp. 258–278 and H. Hine, 2009, p. 19: "[The] point is that it is part of the unspoken contract between writer and reader that the reader normally pays attention, believes statements the writer makes, considers arguments presented by the writer, and so on; so 'listen to this', 'believe me' and the like are foregrounding and making explicit what is normally implicit."

him to take a specific point that has been made into consideration and to be attentive and thoughtful towards what has been said ('Consider this!', 'Attention please!'). As such, these imperatives certainly make the zetetic process more engaging for the reader, as seen before [see 1.1.4., n. 110].

Bearing in mind that we are dealing in *Quaestiones naturales* with a 'functional' type of literature, as we saw [2.4.1.], there is much reason to assume that Plutarch composed the work as some sort of a technical manual, setting an example that could be followed by the reader when discussing natural problems himself[77]. As such, the collection provides a general aetiological framework or frame of reference to be followed and reactivated when confronted with similar natural problems, a point that deserves further consideration in the following section.

3.2. Quaestiones naturales *as a preamble to metaphysics*

For Plutarch, physics, as the causal study of natural phenomena, has no practical use in itself (contrary to, say, politics)[78], but has a more intellectual goal, in as far as the search for natural explanations provides an efficient means of making sense of the immediate world around us. In contrast to Plutarch's other natural philosophical writings, the research conducted in *Quaestiones naturales* is, as we saw, of a mainly 'technical' kind, meaning that not much place is left in it for a 'higher' type of causality. Van der Stockt is probably right, therefore, that the collection is "a leg up for philosophy at its best"[79]. In what follows, I intend to explore precisely how we should read *Quaestiones naturales* as a pre-philosophical text, then, and what is the intellectual and propaedeutic function of physical aetiology more generally according to Plutarch.

1. Natural problems as a means of exercising the mind

As suggested before, the discussion of natural problems is intrinsically bound up with Plutarch's intellectual πολυμάθεια project [see 2.4.2.]. An important passage illustrating this is found in *Quaest. conv.* 734CD, where L. Mestrius Florus is reading a copy of Aristotle's *Problems*, which he shared (μετεδίδου) with his friends for pleasant conversation (οὐκ ἄχαριν

---

[77] For Plutarch's *Moralia* more generally as an exponent of the 2nd century Greek handbook movement, cf. W.H. Stahl, 1962, p. 133.

[78] Cf., e.g., Plutarch's account in *Per.* 16, 7 of Anaxagoras' theoretical way of living *vis-à-vis* Pericles' practical one (οὐ ταὐτὸν δ' ἐστὶν οἶμαι θεωρητικοῦ φιλοσόφου καὶ πολιτικοῦ βίος, ἀλλ' ὁ μὲν ἀνόργανον καὶ ἀπροσδεῆ τῆς ἐκτὸς ὕλης ἐπὶ τοῖς καλοῖς κινεῖ τὴν διάνοιαν, τῷ δ' εἰς ἀνθρωπείας χρείας ἀναμειγνύντι τὴν ἀρετήν ἐστιν οὗ γένοιτ' ἂν οὐ τῶν ἀναγκαίων μόνον, ἀλλὰ καὶ τῶν καλῶν ὁ πλοῦτος, ὥσπερ ἦν καὶ Περικλεῖ, βοηθοῦντι πολλοῖς τῶν πενήτων).

[79] L. Van der Stockt, 2011, p. 452.

διατριβήν) during the daytime strolls[80]. We read that Florus was full of questions, as is natural for a philosopher[81], thus confirming Aristotle's saying that great learning (πολυμάθεια) provides many starting-points:

> Προβλήμασιν Ἀριστοτέλους φυσικοῖς ἐντυγχάνων Φλῶρος εἰς Θερμοπύλας κομισθεῖσιν αὐτός τε πολλῶν ἀποριῶν, ὅπερ εἰώθασι πάσχειν ἐπιεικῶς αἱ φιλόσοφοι φύσεις, ὑπεπίμπλατο καὶ τοῖς ἑταίροις μετεδίδου, μαρτυρῶν αὐτῷ τῷ Ἀριστοτέλει λέγοντι τὴν πολυμάθειαν πολλὰς ἀρχὰς ποιεῖν. τὰ μὲν οὖν ἄλλα μεθ᾽ ἡμέραν οὐκ ἄχαριν ἡμῖν ἐν τοῖς περιπάτοις διατριβὴν παρέσχεν.

Florus, who was engaged in reading a copy of Aristotle's *Problems* that had been brought to Thermopylae, was himself full of questions, as is natural for a philosophical spirit, and shared them with his friends too, proving Aristotle's own statement that "great learning gives many starting-points." Most of the questions raised provided us with a pleasant pastime during our daytime walks.

Aristotle's quote, according to which 'great learning provides many starting-points' (fr. 62 Rose: πολυμάθεια πολλὰς ἀρχὰς ποιεῖ), originates from his lost Περὶ παιδείας. In light of Plutarch's description of Florus as a philosopher full of questions who shares the *Problems* with his friends for discussion, the quote does, indeed, seem to have specific educational meaning. I take it to imply that great learning (πολυμάθεια) functions as a useful starting-point for philosophical research (notably, the chapter at hand inquires into the nature of dreams and ends with a reference to divination). If great learning provides many starting-points, it is obviously not a τέλος in itself. The ultimate τέλος for Plutarch, as a philosopher, is to look for philosophical knowledge – however unattainable this may

---

[80] The pleasant character of natural problems is also acknowledged, e.g., in *De tuenda* 133B, where they are included among many pleasant scholarly diversions at table (καλὰς καὶ ἡδείας ἀπόψεις καὶ ἀποστροφάς). For the intellectual pleasure provided by natural problems in *Quaestiones convivales*, cf. also F. Fuhrmann, 1972, p. xxiv: "Pour les questions d'ordre scientifique, il est particulièrement grave de ne pas discuter les problèmes dans leur fond. Or c'est ce qui se passe ici. Au lieu de chercher les causes véritables [quid?] des phénomènes, Plutarque se contente en général de la vraisemblance, en citant plusieurs théories qui s'y rapportent, ou en rappelant ce que divers auteurs en ont dit. Les différentes opinions se succèdent ainsi sans aucune analyse et le plus souvent sans solution, comme si ceux qui sont chargés de les défendre s'amusaient avec elles." For a more positive evaluation, see F. Frazier and J. Sirinelli, 1996, p. 206: "Sans jamais tourner à la leçon de philosophie – ce qui serait un manquement inacceptable à l'atmosphère détendue du banquet –, celui-ci témoigne donc d'un esprit curieux et toujours en éveil chez ses participants et d'un véritable plaisir d'exercer ensemble son esprit." See also F. Frazier, 1998.

[81] Elsewhere, Florus is called a 'lover of antiquities' (*Quaest. conv.* 702D: φιλάρχαιος).

be. It remains to be seen, then, how natural problems can serve as useful starting-points to achieve this more philosophical goal. Before we can do this, we should first look into their more basic intellectual appeal, as distinct from their higher, philosophical purpose.

In regards to the basic intellectual appeal of Plutarch's natural problems, we learn from several passages throughout his oeuvre that they provide a virtual training court for mental exercise (γυμνασία)[82]. This is generally interpreted in light of their function as scholarly-rhetorical exercises, that is, a training in natural scientific ζήτησις, to be situated in the wider context of the Chaeronean's philosophical school. As scholars have argued in light of the natural problems discussed in *Quaestiones convivales*, the practice of solving such problems by means of looking for plausible natural explanations is a scholarly endeavour, suitable for all-round πεπαιδευμένοι, to whom it offers an incentive for rhetorical-argumentative creativity. When read in this way, the problem-format offers an agonistic (but amusing and friendly) framework for rhetorical demonstration and the ingenious display of multifarious παιδεία. Indeed, the symposiasts generally try to show off their knowledge of traditional authorities in combination with their proficiency to remodel this knowledge in an original fashion to ever new problem contexts. However, further nuance is necessary, insofar that the eventual utility and purpose of such exercises remains unexplained. Were they merely scholarly, but for the rest entirely noncommittal, games played by learned people, or is there more to them? I will deal with this in the next section [3.2.2.].

When it comes to the implied reading of Plutarch's natural problems and their role as scientific literature, scholars agree that they intend to activate the reader's attentiveness for the kind of theories and terminologies used in solving such problems. As such, they promote, what has been called, an 'active reading'[83], meaning that they provide some kind of theoretical model that presents a general aetiological method and design for the reader to follow when dealing with natural problems himself.

---

[82] See, e.g., *De tuenda* 130A (περὶ γυμνασίων φιλολόγοις ἁρμοζόντων), *Quaest. conv.* 628D (ἐγγυμνάσασθαι [...] ὁ λόγος παρέξει), 646A (γυμνασίας ἕνεκα καὶ ζητήσεως). See S.-T. Teodorsson, 1989, p. 290: "The discussions at the drinking-parties were pursued as a sport and training in εὑρησιλογία [...]." The idea that natural inquiry is an opportunity for intellectual exercise is a widespread τόπος in ancient literature. For the conception of natural science as an intellectual exercise (mostly with an ethical finality), see P. Hadot, 2002, pp. 207–211.

[83] This hypothesis has especially been discussed in view of the problems treated in *Quaestiones convivales*. See J. König, 2007 and E. Kechagia, 2011a. See also C. Jacob, 2004 and K. Oikonomopoulou, 2013a more generally. I have argued elsewhere that the same idea also applies to Ps.-Alexander's *Medical puzzles and natural problems*, where the situation seems far less hypothetical (see M. Meeusen forthcoming f).

These scientific concepts and theories – in short, the general aetiological framework – could be reused and remoulded in new discussions concerning similar problems, for instance, during symposia[84]. This does not necessarily imply, however, that the intended reader should simply learn these problems by heart in order to reproduce or 'perform' them during discussions. Extemporisation and on the spot ingenuity was, after all, a much valued intellectual virtue by Plutarch and his peers [see 4.2.2.2.]. If the main educational purpose of this branch of research is not so much achieved by the exact reproduction and 'performance' of the problems that have been recorded[85], then, it is presumably found in the implicit and indirect acquisition of a more general aetiological sensitivity, some kind of an aetiological *Fingerspitzengefühl* for explaining natural phenomena. By means of solving specific case studies in his problems, Plutarch, thus, intends to outline a general aetiological *modus operandi* for the implied reader to absorb and to reactivate whenever suitable. A similar notion of the reader's acquisition of a general "explanatory schema for natural philosophy" has recently been discussed by Kechagia for the natural problems treated in *Quaestiones convivales*[86]:

---

[84] Interestingly, Plutarch does, indeed, inform us that there is a certain sympotic protocol that guests should prepare themselves intellectually to participate in the discussion at table. See, e.g., *Sept. sap. conv.* 147EF: ἡ γὰρ οὐκ οἴει, καθάπερ ἑστιάσοντός ἐστι τις παρασκευή, καὶ δειπνήσοντος εἶναι; κτλ. In *Quaest. conv.* 629C, the sympotic conversation is considered part of τὰ εἰς τὰ δεῖπνα καὶ τὰ συμπόσια παρασκευαζόμενα. Similarly, Gellius reports that his teacher L. Calvenus Taurus often invited some of his students (viz. the *iunctiores*) to dinner at his house and that each diner was obliged to prepare a light and entertaining topic for discussion, 'suitable for a mind enlivened with wine' (*NA* 7, 13: *Quaerebantur autem non grauia nec reuerenda, sed* ἐνθυμημάτια *quaedam lepida et minuta et florentem uino animum lacessentia*). This contribution to the discussion is considered the 'tax' or συμβολή, that is, some kind of an intellectual entrance-fee to the dinner (cf. LSJ, s.v. iv). The same imagery recurs passim throughout *Quaestiones convivales* (cf. 664D, 668D, 682A, 694B, 719EF).

[85] This cannot be refuted with full certainty either, though. The bite-size format of such problems would, in any case, facilitate memorisation. On the memorability and re-usability of miscellaneous knowledge (as collected in Plutarch's *Quaestions convivales*), see S. Goldhill, 2009, p. 109. See also F. Klotz and K. Oikonomopoulou, 2011, p. 16 and K. Oikonomopoulou, 2013a, pp. 144–145. For further reading on ancient mnemotechnics and its importance in ancient education, see J.P. Small, 1997, R. Cribiore, 2001, T. Morgan, 1998. For the link between compilation literature and teaching practices more generally, see M. Horster and C. Reitz, 2010, p. 11.

[86] E. Kechagia, 2011a, pp. 97–99. Cf. also J. Sirinelli, 2000, pp. 382–383: "Ce [sc. *Quaestiones convivales*] ne sont pas les plus anodines. Elles représentent assez exactement le niveau moyen des problèmes posés mais ce n'est pas à cette aune que se mesure pour Plutarque leur intérêt: c'est à la nature et à la diversité des réponses. Il convient que l'on puisse, pour répondre, invoquer l'autorité d'auteurs connus, la discuter, confronter les réponses et, éventuellement, en retenir une pour des raisons plausibles. Ce n'est donc pas l'importance du sujet traité qui fait l'importance de la question, mais c'est plutôt

"In other words, what the reader effectively gets to learn through reading the zeteseis one after another are not specific (and firm) explanations of a certain phenomenon; he rather learns how to use the basic tools available, with the help of which he can embark on an inquiry into nature. Just as philosophy can be employed in every question needing an answer, so this particular explanatory schema can be tried out in search of an answer to almost any question relevant to natural philosophy."[87]

If the main objective of the active reading process is, indeed, found in the reader's mastery of this explanatory scheme and in his ability to remodel and reactivate this scheme in other problem contexts, the active reading in itself sharpens the zetetic potential of the reader, who is challenged not only to remain critical and attentive in reading the aetiologies and mining them for useful insights but also to be inventive in constructing his own solutions to the problems that are raised. As such, Plutarch's natural problems not only appeal to a good understanding by the reader, but also demand a critical response from him in return. This appeal is found both at the micro- and at the macrostructural level of the discourse in *Quaestiones naturales*.

At the microstructural level, an appeal is done to the reader by means of the interrogative and open-ended formulation of the explanations, as well as by the metadirective addresses in the text that closely engage the reader

---

l'ingéniosité et la force de la démonstration. Il s'agit d'une sorte de rituel: évoquer les avis des poètes et des savants, les discuter, discuter des raisonnements présentés ou sous-entendus. Le ressort de ce jeu de société est la question posée (*problèma*) et la recherche de la solution (*zetema*), autrement dit c'est une question de méthode et non pas un jeu de devinettes qui fait l'intérêt de ces discussions. Sous cet angle il n'y a pas de petits sujets, mais une activité de l'esprit et du jugement qui satisfait les *philologoi* présents, lesquels en retirent non seulement des informations mais surtout des travaux pratiques et des confrontations de procédure." See also p. 387: "C'est bien là que réside l'intérêt de l'ouvrage aujourd'hui encore: il nous livre beaucoup de renseignements sur la manière dont un contemporain cultivé de Trajan se représentait le monde, mais surtout il nous fait connaître les mécanismes intellectuels de cet homme, ce qui lui paraissait constituer les règles d'un raisonnement scientifique et le "critère" de la vérité."

[87] E. Kechagia, 2011a, p. 99. The same conclusion has also been made for Ps.-Aristotle's *Problems*. See, e.g., C. Jacob, 2004, pp. 40–48 (*sub* "Les Problèmes aristotéliciens: savoir ou pratique?"). Cf. also, e.g., A. Blair, 1999, p. 174: "The resolution of *problemata* involves the manipulation of the common pool of Aristotelian and Hippocratic notions about nature and human physiology: humors and qualities, phenomena of antiperistasis (or opposition), concoction, sympathy, and the like." Cf. also p. 175: "*Problemata* are best understood as exercises in manipulating concepts of physics and medicine, using methods of argumentation acquired earlier. The goal is perhaps less to reach a single, "true" answer than to display mastery and ingenuity in the use of fundamental principles."

in the zetetic process (see n. 76). By formulating the physical aetiologies in this way, Plutarch, in a certain sense, breaches the fiction of his own text and taps into the reader's reality. He extends a hand to the reader and invites him to actively participate in the virtual discussion, in order to evaluate the proposed solutions and to formulate personal answers to the problems that have been raised[88]. In the end, Plutarch leaves it up to the reader to decide which explanation he considers most convincing, and, if possible, to formulate his own solution to the problem so as to challenge the author's own problem solving ability. Of course, the author tries to convince the reader by formulating plausible explanations to the problems himself, but at the same time, on the side of the intended reader, there is also a certain willingness to be convinced (cf. *Quaest. conv.* 665E: εὐπείθεια).

At the macrostructural level, on the other hand, we have seen before that the mostly chaotic and disorganised line-up of the problem chapters also stimulates active reading [see 1.1.5.]. In regards to the generally unstructured variation of themes and topics in *Quaestiones convivales*, König argues that "Plutarch embeds the requirement for personal response in the very form of his writing, forcing us to take up the provocative challenges of interpretation precisely through his arrangement of material."[89] This means that the implied reader should remain attentive to the repetition of specific recurring themes and theories in the text, in order to deduce a certain sense of structural unity for himself in spite of the work's thematic diversity and miscellaneous arrangement. As such, the collection's chaotic organisation challenges the implied reader to "read things disjointedly and out of context or not" and at the same time "to experience the way in which disparate material can begin to resolve itself into unity if only we read carefully enough"[90]. These claims also apply to Plutarch's *Quaestiones*

---

[88] In this sense, the function of the reader resembles that of an *umbra* accompanying someone to a symposium, without being directly invited by the host, but who is nevertheless welcome to join in (cf. *Quaest. conv.* 706E). Regarding the concrete reading of *Quaestions convivales*, G.W.M. Harrison, 2000a, p. 196, n. 19 also notes that "as self-aware as self-effacing, Plutarch certainly had to know that we are his grateful *umbrae*".

[89] J. König, 2007, p. 50. See also pp. 45–46: "The *Sympotic questions* prompts us to read actively – in other words to respond creatively and philosophically for ourselves to the many different questions under discussion, and to stay alert to the recurring themes and patterns of the texts. Plutarch also shows us his fellow dinner-guests learning that style of active response for themselves, using the topics they discuss as springboards for personal response, as stepping-stones in their philosophical lives. The work demonstrates, in other words, how processes of universally relevant philosophical enquiry can start from frivolous snatches of conversation."

[90] J. König, 2007, p. 61. Scholars have, indeed, argued that a similar overarching scheme lies behind the disordered problems in *Quaestiones Romanae* (see, e.g., J. Boulogne, 1992, pp. 4698–4707, who analysed it in terms of politics, Greekness and

naturales. In this work too, subtle continuities of thematic and theoretic leitmotifs stimulate the reader's response to the text (this is the principle of *concatenatio*). As such, the author encourages and challenges the reader to form a coherent picture of the text and, by implication, also of the physical world behind it, in terms of the natural processes and powers that operate in it.

Clearly, the reader's direct involvement in the zetetic process underlines the intellectual appeal of this kind of literature. However, as noted before, a deeper, philosophical motive still lies at the heart of Plutarch's interest in natural problems, a feature that has been neglected by scholars thus far and that can put the hypothesis about the active reading in a broader perspective[91].

2. Natural problems as a means of easing the mind

The aim of this section is to demonstrate that the eventual goal of Plutarch's natural problems was not just of a generally scholarly-intellectual but of a more elevated and philosophical kind. These two features do not necessarily exclude one another, but even so the second seems to outstrip the first. This will be very seminal in determining how Plutarch's natural problems relate to his overarching natural philosophical programme, an issue that will be further examined in the next chapter.

An important passage to start with is *Q.N.* 29, perhaps the most significant chapter to interpret the eventual goal of *Quaestiones naturales* (it has already partly been discussed earlier on [see 1.2.3. and 3.1.]). Initially, the reader of *Q.N.* 29 might expect Plutarch to simply treat yet another natural problem here, but the tone of the discourse rapidly changes. Plutarch wonders why we marvel at hot springs but not at cold ones, while it is clear that heat is the cause of the former and cold of the latter (Τίς ἡ αἰτία, δι' ἣν τὰ ψυχρὰ τῶν ὑδάτων οὐ θαυμάζομεν ἀλλὰ τὰ θερμά; καίτοι δῆλον ὅτι θερμότης αἰτία τούτων ὡς ψυχρότης ἐκείνων). From the subsequent argument, we learn that it is not so much the natural phenomenon of hot or cold springs

---

anthropology), *Quaestiones Graecae* (see, e.g., P. Payen, 1998b, pp. 49–55, who analysed it in terms of geographical categories) and *Quaestiones naturales* (K. Oikonomopoulou, 2013a, p. 152 [see 1.1.5.]).

[91] J. König, 2007, p. 47 has actually considered Plutarch's "insistence on personal response [in *Quaestiones convivales*] as a central part of philosophy". He argues that "the frivolous joys of ingenious speculation are shown to embody the most important principles of philosophical education" (p. 56). He also says (p. 61) that "Plutarch thus repeatedly emphasises the requirement that the philosopher should be able to use any conversation as a starting point for philosophy, by applying his or her own distinctive skills of reading." Unfortunately, König does not specify how this leap towards philosophy – which I take to be *philosophia prima* and not a generally scholarly-intellectual mode of conversation – can be made then.

as such, but rather the aspect of marvelling itself that is at issue here[92]. Plutarch explains that it appears that nature attributes marvellousness to rarity and stimulates the research of how a phenomenon comes to be only if it occurs infrequently (ἀλλ' ἔοικε τῷ σπανίῳ τὸ θαυμάσιον ἡ φύσις νέμουσα πῶς γίνεται ζητεῖν τὸ μὴ πολλάκις γινόμενον). In what follows, he starts to target the common people, whom he accuses of not feeling any wonder for the nature of celestial phenomena that can be seen during night and day (οὐ μέντοι θαυμάζουσιν οἱ πολλοὶ τὴν τούτων φύσιν). Their attention only goes to rare phenomena such as rainbows, the variety of clouds by day, meteors bursting like bubbles, and comets ... – and then the text breaks off (ἴριδες δὲ καὶ ποικίλματα νεφῶν ἡμέρας καὶ σέλα ῥηγνύμενα πομφόλυγος δίκην καὶ κομῆται ****). I take this to imply that the wonder of the common people remains superficial, not only because they are not interested in 'less marvellous' phenomena, but also – and more importantly so – because they do not look into the nature (φύσις) of these phenomena. Unlike a genuine φυσικός, who would also be astonished about these phenomena, albeit not in the same superficial way, the common people have no intention towards natural philosophical insight, so that they do not really wonder about the actual nature of wonder-inducing phenomena. In other words, the kind of wonder they experience does not lead on to natural philosophical contemplation. As I will try to show in what follows, Plutarch considers this lack of a rational, physical approach a fertile ground for superstition (δεισιδαιμονία).

As just indicated, the end of *Q.N.* 29 is lacunary and open to conjecture. Notably, a similar polysyndetic enumeration of wonder-inducing celestial spectacles can be found in the conclusion of *De Pyth. or.* 409CD (in the context of the prophetic art). Plutarch there reprimands children's selective amazement at marvellous phenomena in the heavens: 'It is a fact that children take more delight and satisfaction in seeing rainbows, haloes, and comets than in seeing moon and sun etc.' (καὶ γὰρ οἱ παῖδες ἴριδας μᾶλλον καὶ ἅλως καὶ κομήτας ἢ σελήνην καὶ ἥλιον ὁρῶντες γεγήθασι καὶ ἀγαπῶσι κτλ.). Another parallel for this childish marvelling is found in *Amatorius* 766A (in the context of love): 'It is like the eagerness of children to catch the rainbow in their hands, attracted by its mere appearance' (ὥσπερ οἱ παῖδες προθυμούμενοι τὴν ἶριν ἑλεῖν τοῖν χεροῖν, ἑλκόμενοι πρὸς τὸ φαινόμενον). From these parallels, one can induce that Plutarch originally concluded his invective in *Q.N.* 29 with the same topic, namely that the childish astonishment of the ignorant plebs for such celestial phenomena – and hence also for other astonishing phenomena, such as the hot springs – is motivated on irrational grounds[93]. As we learn from *Per.* 6, 1, these

---

[92] If wonder is the beginning of philosophy, why not wonder about wonder itself [see 4.1.1.1., n. 23]?

[93] P.R. Hardie, 1992, pp. 4747–4748 also highlights people's "foolish wonder at meteorological marvels" in *Q.N.* 29 (n. 21), and he correctly interprets the parallels in

irrational grounds can be identified with superstition (δεισιδαιμονία), which Plutarch describes as a 'feeling which is produced by amazement at what happens in regions above us':

Οὐ μόνον δὲ ταῦτα τῆς Ἀναξαγόρου συνουσίας ἀπέλαυσε Περικλῆς, ἀλλὰ καὶ δεισιδαιμονίας δοκεῖ γενέσθαι καθυπέρτερος, ὅσην τὸ πρὸς τὰ μετέωρα θάμβος ἐνεργάζεται τοῖς αὐτῶν τε τούτων τὰς αἰτίας ἀγνοοῦσι καὶ περὶ τὰ θεῖα δαιμονῶσι καὶ ταραττομένοις δι' ἀπειρίαν αὐτῶν, ἣν ὁ φυσικὸς λόγος ἀπαλλάττων ἀντὶ τῆς φοβερᾶς καὶ φλεγμαινούσης δεισιδαιμονίας τὴν ἀσφαλῆ μετ' ἐλπίδων ἀγαθῶν εὐσέβειαν ἐργάζεται.

These were not the only advantages Pericles had of his association with Anaxagoras. It appears that he was also lifted by him above superstition, that feeling which is produced by amazement at what happens in regions above us. It affects those who are ignorant of the causes of such things, and are crazed about divine intervention, and confounded through their inexperience in this domain; whereas the doctrines of natural philosophy remove such ignorance and inexperience, and substitute for timorous and inflamed superstition that unshaken reverence which is attended by a good hope.

If my conjecture is correct, Plutarch's message in *Q.N.* 29 seems to be that, unlike a genuine φυσικός, the plebs have no intention of developing natural philosophical insight, meaning that they do not really wonder about the actual nature (φύσις) of these phenomena. They only focus on the miraculous and supernatural character of some rare phenomena without any serious attempt at understanding even these in a proper physical way, being 'attracted by their mere appearance' (cf. *Amatorius* 766A above). By contrast, the more everyday phenomena, such as the positions of the sun, the movement of the stars and the phases

---

De Pyth. or. 409CD and *Amatorius* 766A in their broader Platonic context, where the childish people do not aim to reach the ultimate intelligible truth. In the present context of *Q.N.* 29, however, it seems that the phrase οὐ μέντοι θαυμάζουσιν οἱ πολλοὶ τὴν τούτων φύσιν primarily implies that the people do not feel any wonder for the *nature* – i.e. for the natural causes – of these phenomena. Then again, in light of Plutarch's dualistic causality, the higher, intelligible causes are always closely related to the natural ones [see 4.1.2.], and in this sense, the meteorological phenomena dealt with here presumably also have a divine motivation for Plutarch – which he perhaps mentioned in the lost part of *Q.N.* 29 (cf. also *Lys.* 12: οἱ δὲ καὶ τὴν τοῦ λίθου πτῶσιν ἐπὶ τῷ πάθει τούτῳ σημεῖόν φασι γενέσθαι). But even so, the central focus is clearly on natural causes in *Quaestiones naturales*, so that a divine motivation may only be implied. Cf. also, e.g., *Quaest. conv.* 720E with S.-T. Teodorsson, 1996, p. 183: "The causes that should be investigated are those of the physical processes which are the objects of the scientist, in contrast to the ultimate cause, the will of Providence, which lies beyond his competence." Cf. Pl., *Tim.* 68e.

of the moon, are much more 'regular' phenomena, and, thus, deserve attention also. Notably, Plutarch elsewhere emphasises that contemplation of astronomical phenomena has actually empowered us to acquire philosophical knowledge, an idea that ties in closely with Platonic epoptics [see 4.1.2.2., n. 65].

A similar idea recurs on several occasions in *Quaestiones convivales*, especially at those points where one or more symposiasts are reprimanded because they do not look for physical explanations of natural *mirabilia* but simply remain perplexed about their wondrous character (they often relate these phenomena to the cosmic antipathy or sympathy of the Stoics, without further aetiological specification or detail [see 4.1.1.3., n. 45]). This does not mean, of course, that Plutarch sides with the other type of symposiasts who do not believe in such natural marvels or popular beliefs altogether but simply reject them. His position seems to be more moderate, giving wonderful popular beliefs the benefit of the doubt (see, for instance, the discussion of the belief that thunder generates truffles in *Quaest. conv.* 664A–665A[94] [see 3.1.4. and 4.1.1.1.]). This benefit of the doubt, in turn, ties in closely, as we will see in the following chapter, with Plutarch's scientific method, and more precisely with his Platonic-Academic caution (εὐλάβεια) [see 4.3.2.1.].

As we saw at the beginning of this chapter, Plutarch's vituperation of people's short-sighted marvel at natural phenomena may be considered an actual appeal towards a more serious and mature study of nature. To this end, Plutarch aims to lure the reader into an intellectual contemplation of natural causes, as it allows him to distinguish himself from the common, superstitious plebs, that are unfamiliar with the subtleties of such a study. In this sense, physical aetiology serves as a means for achieving intellectual distinction and can be considered a first step towards genuine philosophy. It is a weapon against superstition and a useful instrument in attaining an intelligent devotedness to the gods, that is, εὐσέβεια – being a means between the religious extremes of atheism and superstition[95] [see 4.1.3.].

An interesting passage for the underlying philosophical-religious motivation of Plutarch's natural scientific project is found in the conclusion of *De tranquillitate animi* (477CD). Plutarch there says that 'tranquillity of mind' (εὐθυμία) can be achieved by contemplating the world's divinely inspired spectacles. The Chaeronean here takes position as a Platonic

---

[94] Plutarch here actually aims to save a traditional belief in a physical way. See the discussion by A. Setaioli, 2009.

[95] Cf. D. Babut, 1969, p. 517: "Son idéal est de trouver le juste milieu entre la crédulité naïve, qui fait prendre le moindre fait insolite pour un signe, et a vite fait de sombrer dans la superstition, et, de l'autre côté, l'étroitesse rationaliste, qui récuse tout ce dont elle ne peut rendre compte." See also M. Meeusen, 2014 and 2015b.

contemplator mundi, portraying human beings as spectators of a divinely governed universe:

Ἄγαμαι δὲ τοῦ Διογένους, ὃς τὸν ἐν Λακεδαίμονι ξένον ὁρῶν παρασκευαζόμενον εἰς ἑορτήν τινα καὶ φιλοτιμούμενον 'ἀνὴρ δ᾽" εἶπεν 'ἀγαθὸς οὐ πᾶσαν ἡμέραν ἑορτὴν ἡγεῖται;' καὶ πάνυ γε λαμπράν, εἰ σωφρονοῦμεν. ἱερὸν μὲν γὰρ ἁγιώτατον ὁ κόσμος ἐστὶ καὶ θεοπρεπέστατον· εἰς δὲ τοῦτον ὁ ἄνθρωπος εἰσάγεται διὰ τῆς γενέσεως οὐ χειροκμήτων οὐδ᾽ ἀκινήτων ἀγαλμάτων θεατής, ἀλλ᾽ οἷα νοῦς θεῖος αἰσθητὰ μιμήματα νοητῶν, φησὶν ὁ Πλάτων, ἔμφυτον ἀρχὴν ζωῆς ἔχοντα καὶ κινήσεως ἔφηνεν, ἥλιον καὶ σελήνην καὶ ἄστρα καὶ ποταμοὺς νέον ὕδωρ ἐξιέντας ἀεὶ καὶ γῆν φυτοῖς τε καὶ ζῴοις τροφὰς ἀναπέμπουσαν. ὧν τὸν βίον μύησιν ὄντα καὶ τελετὴν τελειοτάτην εὐθυμίας δεῖ μεστὸν εἶναι καὶ γήθους κτλ.

And I am delighted with Diogenes, who, when he saw his host in Sparta preparing with much ado for a certain festival, said, "Does not a good man consider every day a festival?" And a very splendid one, to be sure, if we are sound of mind. For the universe is a most holy temple and most worthy of a god; into it man is introduced through birth as a spectator, not of hand-made or immovable images, but of those sensible representations of intelligible things that the divine mind, says Plato (*Tim.* 92c, *Epinom.* 984a), has revealed, representations which have innate within themselves the beginnings of life and motion, sun and moon and stars, rivers which ever discharge fresh water, and earth which sends forth nourishment for plants and animals. Since life is a most perfect initiation into these things and a ritual celebration of them, it should be full of tranquillity and joy etc.

As Hirsch-Luipold argues, this passage is very relevant for Plutarch's view on religion: "With this statement the cult is transferred into daily life, as a quotidian form of divine service"[96]. At a natural philosophical level, the contemplation of the intelligible things (νοητά) exposes the meta-physical and divine aspect behind material reality, that is, what lies behind the face of nature, which in itself is only a sensible representation thereof (αἰσθητὰ μιμήματα). In this sense, the study of nature paves the way for more abstract, philosophical contemplations, and as such, Plutarch's physics dovetails with his higher Platonic philosophy[97]. The

---

[96] R. Hirsch-Luipold, 2014, p. 170. He adds: "as humans we are supposed to celebrate every day of our lives as a festival of the gods. Plutarch, in connecting the traditional religious world of symbols with philosophical interpretation, renders religion the basis of striving for understanding and personal happiness."

[97] The distinction between αἰσθητά and νοητά can be taken as a reference to the philosophical ἀνωτάτω πορεία Plutarch discusses in *De prim. frig.* 948BC [see 4.3.2.1., n. 195]. Cf. also, e.g., *Quaest. conv.* 718DE (in the context of geometry).

natural phenomena that Plutarch is dealing with in his natural problems belong to the lowest regions of the universe (c.q. 'the rivers which ever discharge fresh water, and earth which sends forth nourishment for plants and animals'), but they are equally subject to 'the innate beginnings of life and motion' (ἔμφυτον ἀρχὴν ζωῆς καὶ κινήσεως) as is the case for the higher, celestial phenomena (c.q. 'sun and moon and stars'). This means that there is no strict conceptual distinction for Plutarch between natural phenomena here on earth or at a distance in the heavens. In the end, they are all sensible representations of intelligible things.

An interesting parallel for this can be found in *De cur*. 517CE, where Plutarch advises the 'busybody' (πολυπράγμων) to turn his soul to 'better and more pleasant subjects' than those suitable for baser forms of curiosity (ἐπὶ τὰ βελτίω καὶ τὰ ἡδίω τρέψαντι τὴν ψυχήν)[98]. One can, for instance, shift and divert one's inquisitiveness towards natural phenomena, either great or small.

τὰ ἐν οὐρανῷ πολυπραγμόνει, τὰ ἐν γῇ τὰ ἐν ἀέρι τὰ ἐν θαλάττῃ. μικρῶν πέφυκας ἢ μεγάλων φιλοθεάμων; εἰ μεγάλων, ἥλιον πολυπραγμόνει ποῦ κάτεισι καὶ πόθεν ἄνεισι· ζήτει τὰς ἐν σελήνῃ καθάπερ ἀνθρώπῳ μεταβολάς, ποῦ τοσοῦτον κατανήλωσε φῶς πόθεν αὖθις ἐκτήσατο, πῶς

ἐξ ἀδήλου πρῶτον ἔρχεται νέα
πρόσωπα καλλύνουσα καὶ πληρουμένη,
χὤταν περ αὑτῆς εὐγενεστάτη φανῇ,
πάλιν διαρρεῖ κἀπὶ μηδὲν ἔρχεται.

καὶ ταῦτ' ἀπόρρητ' ἐστὶ φύσεως, ἀλλ' οὐκ ἄχθεται τοῖς ἐλέγχουσιν. ἀλλὰ τῶν μεγάλων ἀπέγνωκας; πολυπραγμόνει τὰ μικρότερα, πῶς τῶν φυτῶν τὰ μὲν ἀεὶ τέθηλε καὶ χλοάζει καὶ ἀγάλλεται παντὶ καιρῷ τὸν ἑαυτῶν ἐπιδεικνύμενα πλοῦτον, τὰ δὲ νῦν μέν ἐστιν ὅμοια τούτοις νῦν δ' ὥσπερ ἀνοικονόμητος ἄνθρωπος ἀθρόως ἐκχέαντα τὴν περιουσίαν γυμνὰ καὶ πτωχὰ καταλείπεται, διὰ τί δὲ τὰ μὲν προμήκεις τὰ δὲ γωνιώδεις τὰ δὲ στρογγύλους καὶ περιφερεῖς ἐκδίδωσι καρπούς. ἴσως δὲ ταῦτ' οὐ πολυπραγμονήσεις, ὅτι τούτοις οὐθὲν κακὸν ἔνεστιν.

Direct your curiosity to heavenly things and things on earth, in the air, in the sea. Are you by nature fond of small or of great spectacles? If of great ones, apply your curiosity to the sun: where does it set and whence does it rise? Inquire into the changes in the moon, as you

---

[98] For a study on ancient *curiositas*, see A. Labhardt, 1960 (esp. pp. 210–216 on natural scientific curiosity). For the passage at issue, cf. A.M. Battegazzore, 1992, pp. 48–49.

would into those of a human being: what becomes of all the light she has spent and from what source did she regain it, how does it happen that

> When out of darkness first she comes anew,
> She shows her face increasing fair and full;
> And when she reaches once her brightest sheen,
> Again she wastes away and comes to naught? (Soph., fr. 787)

And these are secrets of nature, yet nature is not vexed with those who try to find them out. Or suppose you have renounced great things. Then turn your curiosity to smaller ones: how are some plants always blooming and green and rejoicing in the display of their wealth at every season, while others are sometimes like these, but at other times, like a human spendthrift, they squander all at once their abundance and are left bare and beggared? Why, again, do some plants produce elongated fruits, others angular, and still others round and globular?

In what follows, Plutarch even attaches a moral advantage to this type of scientific curiosity by noting that physical inquiry diverts the busybody's curiosity from malicious subjects towards relatively innocent ones. But he is well aware that the busybody might not be interested in these natural scientific matters precisely because there is nothing morally depraved (κακόν) in them[99].

As suggested in the previous passage (*De tranq. an.* 477CD), Plutarch's distinction between grandiose, heavenly phenomena (c.q. the rising of the sun and the stages of the moon) and the rather small, profane ones (c.q. those on the earth, in the air and in the sea) is not strict. Moreover, as appears from the examples that Plutarch gives of the latter phenomena (viz. the loss of leaves in plants and the shapes of different fruits), these clearly belong to the category of 'everyday' natural phenomena studied in *Quaestiones naturales*. The fact that these problems evoke the same sense of intellectual curiosity as the grandiose phenomena do is very significant for their shared intellectual appeal.

The idea that nature is not vexed with those who try to find out its 'secrets' (ἀπόρρητα) is very important. Nature is represented here as an oracle, whose mysterious 'utterances' should be investigated and interpreted in order to properly understand their underlying meaning. Plutarch makes it very clear that natural philosophical inquisitiveness is not an act of sacrilegious profanation. After all, in line with his dualistic view on causality, the search for physical causes in explaining marvellous

---

[99] For the ethical influence of natural philosophy on the character and intellect of its practitioners, cf. also *Per.* 5, 1 and 8, 1–2 [quoted 1.2.2.].

natural phenomena, either great or small, is not an alternative to a more religious-philosophical interpretation [see 4.1.2.1.]. Rather, the concept of ἀπόρρητα implies a feature of philosophical-religious caution towards the divine here (εὐλάβεια), the workings of which we are unable to fathom with our human intellect [see 4.3.2.1.].

The bottom line is that Plutarch's world is constituted in such a way that there is no rigid border between the natural and intelligible realms in the cosmos. Natural phenomena are firmly rooted in the category of the divine, so that their natural causes go firmly hand in hand with the final ones [see 4.1.2.]. As such, Plutarch's search for physical causes is not intended as an alternative but rather as a confirmation of his religious outlook on the world. Therefore, the contemplation of the world's divinely inspired spectacles provides εὐθυμία. And it is in this sense that natural science not only challenges but also eases Plutarch's mind.

3. Conclusion and new questions

I conclude that *Quaestiones naturales*, as a school text, conveys only the theory of scientific discussions, whereas in *Quaestiones convivales* Plutarch adds a supplementary social and philosophical dimension to the problems, showing how such problems could be solved in real-life situations. Notwithstanding the 'technical' nature of *Quaestiones naturales*, Plutarch does not consider the genre of natural problems highly complex. On the contrary, they are 'easy and persuasive', which implies that we are not dealing with specialist literature (*De tuenda* 133E: ἐλαφρὰ καὶ πιθανά). I have argued that the genre of natural problems provides a scholarly-rhetorical exercise in argumentative creativity ('mental gymnastics'), presenting an aetiological framework for the reader to absorb by an 'active reading', but that its eventual goal is of a more philosophical and even religious kind for Plutarch. Physical aetiology offers a useful instrument for doing away with superstition (δεισιδαιμονία) and for attaining a well-reasoned devotion towards the divine (εὐσέβεια)[100]. Needless to say that both options are fully commensurable with one another, in that the practice of physical aetiology can, thus, be considered an intellectual *exercise* in eradicating *superstitious* beliefs. As such, the genre of natural problems provides some kind of a preamble to more meta-physical speculations, and, thus, can be considered a light version of philosophy.

A detailed analysis of the aetiological design and scientific context of *Quaestiones naturales* in the following chapter will yield valuable insights into Plutarch's scientific *modus operandi*. This will help us to conjure the physical world system that is silently propagated in it. Such a study

---

[100] Cf. P. Desideri, 1992, p. 81: "è a quest'ultima [sc. alla superstizione], e non alla religiosità, che la scienza si contrappone". See also M. Meeusen, 2015b.

embraces questions related to Plutarch's Platonic view on causality, his reliance on traditional sources and his urge for argumentative creativity, and the general explanatory scheme employed by him. These issues will be treated in the following chapter.

4
Plutarch's Platonic world view: the aetiological design of *Quaestiones naturales* and its scientific context

4.1. Science and its foes? The ancient scientific value of **Quaestiones naturales**

In order to interpret *Quaestiones naturales* in light of Plutarch's overarching natural philosophical programme and Platonist convictions, this final chapter will focus on the peculiarities of the method the Chaeronean applies both in raising and solving natural problems. To this end, I will first examine the wonder-inducing nature of the phenomena that receive Plutarch's attention and how he approaches them [4.1.1.]. A more overarching question will then concern Plutarch's well-known dualistic view on causality. In this regard, I will also examine the presence of a more mythological and poetic type of discourse in Plutarch's scientific writings [4.1.2.]. This will provide us with the necessary background information for further inquiry into the issues of scientific authority and its discursive construction in Plutarch's natural problems [4.2.], and also into more specific aspects related to Plutarch's scientific methodology in the sections to follow [4.3.].

4.1.1. Saving popular beliefs: the wonders and paradoxes of nature

I will start by examining the often paradoxographical and truly 'enigmatic' nature of the natural phenomena that Plutarch investigates in his natural problems. As regards Plutarch's custom of solving natural problems, it is clear that their main purpose is to satisfy intellectual curiosity[1]. At several points in *Quaestiones naturales*, the reader finds himself confronted with rather bizarre natural problems, such as the sweet taste of the tears of wild boars as opposed to the salty and ordinary ones of deer (*Q.N.* 20), or the palm tree's ability to rise against a weight imposed upon it (*Q.N.* 32). The kind of scientific research conducted in such problems might seem to be rather playful if not absurd, and as such might remind one

---

[1] The same conclusion was reached by S. Kapetanaki and R.W. Sharples, 2006, p. 1 for the *Supplementary problems* ascribed to Aristotle and Alexander of Aphrodisias, in which even the medical problem chapters have no immediate practical use.

of the sympotic σπουδογέλοιον described in *Quaestiones convivales*[2] [see 3.1.3., n. 27]. Teodorsson has, in fact, distinguished these "examples of fanciful beliefs in *Quaestiones naturales* which Plutarch accepts without hesitation" from other problems that are of a more "respectable, scientific kind"[3]. This distinction, however, seems unnecessary, since Plutarch does not make it himself. Seeing that Plutarch makes a considerable attempt to explain these and similar fanciful phenomena in an intelligent, physical way, the question is legitimate as to why he takes such curiosa seriously and how this ties in with his Platonic philosophy. In other words, what is Plutarch's actual intention for accounting for such problems in a serious way, and what does he see as constituting proper scientific conduct, then[4]?

The philosophical-educational context of Plutarch's school has clearly left its mark on the scientific outlook of his natural problems, to the effect that they are for a great part detached from nature itself. For Plutarch, scientific demonstration (ἀπόδειξις) remains, to a considerable extent, a theoretical practice, that is, a rhetorical-argumentative exercise fuelled by scholarly πολυμάθεια (viz. his acquaintance with the scientific tradition and literature). In order to balance this view, we saw that these problems are not merely trivial *Spielereien*, which Plutarch did not attach any philosophical relevance to. Indeed, Plutarch did not believe or at least ascribe credibility to such phenomena *only* for reasons of rhetorical exercise[5] [see 3.2.2.]. Of

---

[2] See, e.g., the evaluation by J. Sirinelli, 2000, p. 382: "Cet ouvrage, dont les sujets sont pour nous souvent déconcertants et parfois fastidieux, a été paradoxalement composé avec allégresse. Plutarque trouve du plaisir à nous rapporter ces propos que le lecteur d'aujourd'hui se laisse aller à juger comme des "curiosités" parfois puériles, dignes plutôt d'Élien ou d'Athénée."

[3] S.-T. Teodorsson, 1999a, p. 666.

[4] Pliny the Elder's predilection for natural *mirabilia* serves as a nice parallel. See, e.g., V. Naas, 2011, pp. 66–67, who first negatively remarks that "Pliny's attitude, between rationality and *mirabilia*, is also typical of the evolution of science at his time. The rise of the paradox goes with a decline of the sciences." A little further on, though, she draws a more reasonable conclusion: "We must not seek in the *mirabilia* a proof of the decline of knowledge. They simply reflect a different kind of knowledge [...]."

[5] Solving natural problems is more than simply a matter of sophistic bravura for Plutarch. But even so, there are certainly links with contemporary *suasoriae*, where it was a common rhetorical practice and strategy to defend the most absurd cases first. Cf. *De aud.* 44F, where Plutarch mentions a panegyric on vomiting, on fever and on the kitchen-pot, and says that they actually have a certain degree of plausibility (see R. Flacelière, J. Irigoin, J. Sirinelli and A. Philippon, 1987, pp. ccvi–ccvii). Sophists already declaimed encomia on τὰ φαυλότατα τῶν ὄντων in Isocrates' days (*Panath*.135; cf. *Paneg*. 188). Isocrates mentions a praise of the fly and one of salt (*Hel.* 12). The encomium on salt is also mentioned in Pl., *Symp.* 177b, and Lucian's comic *Praise of the Fly* shows that this type of declamatory pyrotechnics was still very alive after Plutarch's death. Cf. also, e.g., the pieces by Favorinus (frs. 1–2 Barigazzi and L. Holford-Strevens, 1997,

course, it is only likely that some element of intellectual entertainment and rhetorical persuasion was intrinsically bound up with Plutarch's science of natural problems, but this should not necessarily interfere with the claim that it can be taken seriously as an intrinsic part of his natural philosophical programme (or of ancient science more generally)[6]. Similarly, in regards to the discussion in *De facie* of the problem of the face appearing in the moon, Gianakaris argues that:

> "Man always seems to accept strange phenomena more easily when seen as part of a system intelligible to human experience. [...] Plutarch's age, like our own, apparently was willing to contemplate science-fiction possibilities in the universe but not to pursue all theories indiscriminately."[7]

The border between 'science' and 'fiction' was probably less fixed for Plutarch than it is for us today. This is presumably due to the fact that Plutarch had a different, more religious, outlook on the world. He obviously lived in the same physical reality as we do, but he looked at it in a different way and, therefore, saw different things. Because of his religious devotion, Plutarch was inclined to accept more fanciful phenomena and popular beliefs. He primarily did this in order to explain the world he lived in on a theological basis and in view of his Platonic-Academic philosophy. The following sections are intended to further subtantiate this point.

1. *Natural problems and the fabric of strangeness*

The strangeness and paradox of Plutarch's natural problems are central to the wider genre of natural history that embraces paradoxology and *mirabilia* literature. Scholars have shown that a great deal of ancient natural scientific writing actually relies on pre-scientific data, including bizarre and unreliable popular beliefs[8]. Plutarch's discussion of natural

---

p. 200). For further reading on this type of 'adoxographical' literature, see A.S. Pease, 1926. For its relation wih the 2nd century culture of the Second Sophistic, see G. Anderson, 1993, pp. 171–199 (*sub* "*Adoxa paradoxa*: the *pepaideumenos* at play").

[6] For a similar discussion, see M. Sassi, 1993, pp. 465–468 (*sub* "Credevano gli antichi ai loro *mirabilia*?" – with the allusion to the classic work of P. Veyne, 1983, pp. 126–137).

[7] C.J. Gianakaris, 1970, p. 104.

[8] For the role of *mirabilia* and the marvellous in ancient scientific literature, see, e.g., G.E.R. Lloyd, 1983, M. Sassi, 1993 (esp. pp. 454 ff. regarding Peripatetic natural history), J.F. Healy, 1999, pp. 63–70, T. Murphy, 2004, pp. 87 ff., V. Naas, 2011, M. Beagon, 2011 (esp. p. 85 regarding *Quaest. conv.* 680D; see further). For more background and further literature on ancient paradoxography, see G. Schepens and K. Delcroix, 1996 and C. Jacob, 1983. For the role of θῶμα already in Herodotus' work, see G.E.R. Lloyd, 1979, pp. 29–32 and R. Thomas, 2000, pp. 135–167. Cf. *De Her. mal.* 855EF.

problems ties in well with the more common intellectual endeavour from Hellenistic and Roman times of saving such popular beliefs through scientific and philosophical interpretations[9]. In this sense, the study of nature is designed, not so much to deny common-sense beliefs but to vindicate them as being an intrinsic aspect of natural scientific research. At the beginning of his work on *Science, Folklore and Ideology*, Lloyd writes that many Greek scientists "often remain deeply influenced by such [popular] beliefs"[10]. Although they try to understand the physical world according to a safe, rational basis, they still incorporate these beliefs into their scientific treatises on a large scale. This does not mean, of course, that their inquiries remain unsubstantiated from an intellectual perspective. On the contrary, in their attempts to rationalise such popular beliefs, ancient scientists mostly try to assimilate them with a critical analysis, just as can be seen in Plutarch's natural problems. There, Plutarch records many folk traditions concerning fauna, flora and meteorology, albeit never without any sense of aetiological thoroughness and detail. After all, he is prepared to deal with these popular beliefs in an intelligent way – at least within the conceptual scope and limits of ancient physics – and indistinctly from the other, perhaps to modern readers, more serious problems (see above).

Plutarch's interest in natural *mirabilia* is based on a belief that what is contrary to expectation (παράδοξον) is not necessarily beyond reason (παράλογον). This idea is formulated in *Sept. sap. conv.* 163D by Pittacus in the context of the wondrous myth of Enalus:

εἴ τις εἰδείη διαφορὰν ἀδυνάτου καὶ ἀσυνήθους καὶ παραλόγου καὶ παραδόξου, μάλιστ' ἄν, ὦ Χίλων, καὶ μήτε πιστεύων ὡς ἔτυχε μήτ' ἀπιστῶν τὸ 'μηδὲν ἄγαν' ὡς σὺ προσέταξας διαφυλάττοι.

---

[9] For the notion of 'saving' popular beliefs through physical explanation, see, for instance, the conclusion in Ammonius' account of the ever-burning lamp at Ammon's shrine in *De def. or.* 411D: εἰ δεῖ τοῖς Ἀμμωνίοις ἀνασῴζειν καίπερ ἄτοπον καὶ ἀλλόκοτον οὖσαν τὴν ὑπόθεσιν. (For the pun on Ammonius' name in this phrase, see J. Opsomer, 2009, p. 143, n. 97.) In a similar way, in *De def. or.* 420C, the same Ammonius quotes Theophrastus (fr. 263 FHSG) regarding the contested existence of demigods: ὀρθῶς ἔφη μοι δοκεῖ Θεόφραστος ἀποφήνασθαι τί γὰρ κωλύει φωνὴν δέξασθαι σεμνὴν καὶ φιλοσοφωτάτην; καὶ γὰρ ἀθετουμένη πολλὰ τῶν ἐνδεχομένων ἀποδειχθῆναι δὲ μὴ δυναμένων ἀναιρεῖ, καὶ τιθεμένη πολλὰ συνεφέλκεται τῶν ἀδυνάτων καὶ ἀνυπάρκτων.

[10] G.E.R. Lloyd, 1983, p. 1. This is the case, e.g., with Aristotle's biological writings [see 4.1.1.3.]. See also, for instance, Lloyd's account on the informants of Theophrastus, pp. 119-135, esp. 133: "He [sc. Theophrastus] does not merely record many folk traditions concerning the use of plants, but is prepared to take some of them seriously. But his admirable openness occasionally tips over into uncritical or naïve acceptance, and his reserving judgement becomes the expression of a bafflement that he was unlikely to be able to resolve."

If a man realises a difference between the impossible and the unfamiliar, and between what is beyond reason and what is contrary to expectation, such a man, Chilon, who would neither believe nor disbelieve at haphazard, would be most observant of the precept, 'Avoid extremes,' as you have enjoined.

The epistemological value of this μηδὲν ἄγαν is seminal also for Plutarch's sceptical, Academic attitude towards natural *mirabilia*, mediating between belief and disbelief, as we will see [4.3.2.1.]. The paradox in itself, as something 'contrary to expectation', can be considered a subcategory of the *mirabilia*, because it just as well arouses a certain sense of wonder for what is said and responds to a sentiment of initial disbelief (I will deal with the actual phrasing of natural paradoxes in *Quaestiones naturales* later on [see 4.3.3.1.]).

The often surprising character of the natural phenomena under discussion in Plutarch's natural problems can be explained by the fact that a great amount of traditional, and often, indeed, bizarre and unreliable lore (ἱστορία, δόξα, φήμη) was incorporated from an early stage into the scientific tradition. In the case of *Quaestiones naturales*, Plutarch's sources can be identified, either directly or indirectly[11], with un(der)educated informants like travellers or merchants (e.g., *Q.N.* 1, 911E on plants growing in the Indian Ocean)[12], women (e.g., *Q.N.* 6, 913F on dew making overweight women thinner, and *Q.N.* 26, 918D on pregnant women eating stones and dirt), sponge-divers (e.g., *Q.N.* 12, 915A on oil producing illumination and transparency in the depths of the sea), farmers (e.g., *Q.N.* 41 on better flowering in plants), shepherds (e.g., *Q.N.* 3, 912DF on the aphrodisiac effects of salt on cattle and other animals), hunters (e.g., *Q.N.* 20, 916F on the taste of the tears of boars and deer, and *Q.N.* 22, 917D on the taste of the bear's fore-paws), fishermen (e.g., *Q.N.* 17, 915F on the fabrication of fishing lines from the hair of horses), bee-keepers (e.g., *Q.N.* 35-36 on the behaviour of bees towards pungent smells) etc.

Clearly, Plutarch did not reject these popular beliefs[13]. In fact, wonders and miracles more generally are an inherent aspect of his world view.

---

[11] For instance, the paradoxographical account about she-wolves in *Q.N.* 38 is borrowed from the *De animalibus* of Antipater (presumably the Stoic philosopher of Tarsus: see n. 113), who probably relies on traditional hunter lore himself.

[12] F.H. Sandbach, 1965, p. 151, n. g is probably correct in pointing out that "[t]ravellers tales, both of mangroves and of seaweed, seem to lie behind these reports". Cf., e.g., Theophr., *HP* 4, 7, 3 (which mentions an expedition of those returning from India sent out by Alexander).

[13] Perhaps, his sympathy for popular beliefs can even be linked with his devotion towards traditional belief more generally, that is, the πάτριος πίστις, which, as Flacelière notes, is one of the two "constantes de sa pensée" – the other one being philosophy. See

In *Q.N.* 29, 919B, as we saw, Plutarch notes that 'it appears that nature attributes marvellousness to rarity and stimulates the research of how a phenomenon comes to be only if it occurs infrequently' (ἀλλ' ἔοικε τῷ σπανίῳ τὸ θαυμάσιον ἡ φύσις νέμουσα πῶς γίνεται ζητεῖν τὸ μὴ πολλάκις γινόμενον). I have taken Plutarch's subsequent complaint against the common people for not feeling any wonder for the nature of more 'everyday' celestial phenomena that can be seen during night and day (οὐ μέντοι θαυμάζουσιν οἱ πολλοὶ τὴν τούτων φύσιν) to imply that their amazement remains superficial, because they do *not* look into the actual causes of natural phenomena [see 3.2.2.]. In other words, the kind of wonder they experience does not lead on to natural philosophical contemplation. As Oikonomopoulou rightly notes, this passage "may well be a reference to the genre of paradoxography, which flourished in the period of the empire"[14]. It is indeed remarkable that the paradoxographical genre is not, or not at least greatly, concerned with formulating explanations for the natural *mirabilia* that it collects. By contrast, this genre of writing is mostly preoccupied with simply listing wonder-inducing phenomena and with preserving their wondrous nature by intentionally abandoning any attempts of formulating a reasonable explanation for them[15]. In many cases, the paradoxographer even omits the explanations that have already been provided for the *mirabilia* in the scientific literature, from which they are mostly drawn. Therefore, if the link with

---

R. Flacelière, J. Irigoin, J. Sirinelli and A. Philippon, 1987, pp. cli–clii: "Les deux constantes de sa pensée, depuis sa jeuneusse jusqu'à sa mort, à savoir un double attachement, une double fidélité à la πάτριος πίστις, d' une part, et de l' autre à la philosophie, [...] ont toujours été les mêmes. Jamais sa conception de la vie et du monde ne s' écarta sensiblement de ces deux pôles." For Plutarch's notion of πίστις, see D. Babut, 1969, p. 517 ("la foi ne supprime pas la raison, elle la dépasse et l' intègre") and 1994, pp. 580–581. F. Frazier, 2008, p. 61 does not believe, however, that there is an "émergence d' un concept de 'foi' dans la pensée de Plutarque, qui servirait de terme alternatif au *logos*: la *patrios pistis* pose davantage le problème de la place de la tradition, si importante dans la philosophie impériale". As M. Bonazzi, 2014, p. 129 has argued, moreover, "Plutarch does not anticipate the fideism of modern times". See also J. Opsomer, 1996b, p. 187, 1998, pp. 178–179 (see also pp. 131–132 on Plutarch's σεμνότης) and G. Van Kooten, 2012. Cf., e.g., *De Pyth. or.* 402E: δεῖ γὰρ μὴ μάχεσθαι πρὸς τὸν θεὸν μηδ' ἀναιρεῖν μετὰ τῆς μαντικῆς ἅμα τὴν πρόνοιαν καὶ τὸ θεῖον, ἀλλὰ τῶν ὑπεναντιοῦσθαι δοκούντων λύσεις ἐπιζητεῖν τὴν δ' εὐσεβῆ καὶ πάτριον μὴ προΐεσθαι πίστιν. Cf. also *Amatorius* 756B: ἀρκεῖ γὰρ ἡ πάτριος καὶ παλαιὰ πίστις, ἧς οὐκ ἔστιν εἰπεῖν οὐδ' ἀνευρεῖν τεκμήριον ἐναργέστερον [...] ἀλλ' ἕδρα τις αὕτη καὶ βάσις ὑφεστῶσα κοινὴ πρὸς εὐσέβειαν, ἐὰν ἐφ' ἑνὸς ταράττηται καὶ σαλεύηται τὸ βέβαιον αὐτῆς καὶ νενομισμένον, ἐπισφαλὴς γίνεται πᾶσι καὶ ὕποπτος.

[14] K. Oikonomopoulou, 2013a, p. 146, n. 69. In personal correspondence, she rightly adds that "[i]n a way, Plutarch re-claims the notion of θαῦμα from the paradoxographers".

[15] Cf. G. Schepens and K. Delcroix, 1996, pp. 390–394. C. Jacob, 1983 appropriately speaks of "la fabrication du merveilleux" as a main preoccupation of the genre (see esp. p. 133, *sub* "Le privilège du fait brut. La disparition des causes").

the genre of paradoxography is, indeed, legitimate in *Q.N.* 29, Plutarch's acclamation is not so much to the paradoxographical kind of natural phenomena recorded by this tradition, but rather to the fact that the common people – much like the paradoxographers themselves – make no attempt to explain and understand such phenomena in a proper physical way.

As an intellectual and a philosopher, Plutarch has no objections against the use of natural *mirabilia* and paradoxes as starting-points for learned and philosophical discussion, at least if the explanations applied to them remain plausible. For instance, regarding the popular belief that lightning produces truffles (*Quaest. conv.* 664A–666D [see 3.1.4.]), Agemachus argues against the non-believers (who object that lightning is not generative of itself but merely brings the truffles to light) that 'the miraculous should not be regarded as unworthy of belief, even if it is very difficult if not impossible to find the causes'[16]. Plutarch (as a character in the discussion) follows Agemachus in not rejecting the popular belief and makes an attempt to explain it in a physical way (by arguing that lightning water is fertile owing to its heat), thus implying that it is incorrect to dismiss popular, pre-scientific beliefs without at least trying to find suitable theories for explaining them.

Démarais has conducted a more general study of the *mirabilia* discussed in *Quaestiones convivales*[17]. She has demonstrated that many of these discussions concern an apparent contradiction or an unusual report, and that the symposiasts are repeatedly confronted with something 'illogical' (ἄλογον), 'strange' (ἄτοπον), 'surprising' (θαυμαστόν, θαυμάσιον), 'improbable' (οὐκ εἰκόν), 'paradoxical' (παράδοξον), 'unbelievable' (ἄπιστον), if not 'impossible' (ἀδύνατον)[18]. Thus, she shows that the genre of problems has a clear preference for extraordinary topics and in fact supposes a "fabrique de l'étrange"[19]. By highlighting the enigmatic aspect

---

[16] *Quaest. conv.* 664C: ὁ δ᾽ Ἀγέμαχος ἰσχυρίζετο τῇ ἱστορίᾳ καὶ τὸ θαυμαστὸν ἠξίου μὴ ἄπιστον ἡγεῖσθαι. καὶ γὰρ ἄλλα πολλὰ θαυμάσια βροντῆς ἔργα καὶ κεραυνοῦ καὶ τῶν περὶ ταῦτα διοσημιῶν εἶναι, χαλεπὰς καταμαθεῖν ἢ παντελῶς ἀδυνάτους τὰς αἰτίας ἔχοντα. Similarly, further on in *Quaest. conv.* 665C, Dorotheus regarding the natural phenomenon of lightning never striking people who are asleep (and related phenomena) remarks that such *mirabilia* can be believed or rejected (καὶ ταῦτα μὲν ἔξεστι πιστεύειν καὶ μή). He adds an even more astonishing phenomenon (πάντων δὲ θαυμασιώτατον), noting that 'we all *know*, so to say, that the bodies of those struck by lightning leave undecayed' (ὃ πάντες ὡς ἔπος εἰπεῖν ἴσμεν, ὅτι τῶν ὑπὸ κεραυνοῦ διαφθαρέντων ἄσηπτα τὰ σώματα διαμένει). If the phrase ὡς ἔπος εἰπεῖν, indeed, applies to ἴσμεν, rather than to πάντες here (*pace* P.A. Clement and H.B. Hoffleit, 1969, p. 325), it may have a certain relativising (perhaps even humoristic) effect, implying *as if* this is a commonly known fact.

[17] L. Démarais, 2005.

[18] See L. Démarais, 2005, p. 161 (with references).

[19] L. Démarais, 2005, p. 161: "La présentation du problème, comme celle des *mirabilia*, suppose une 'fabrique de l'étrange'."

of the *mirabilia* and their ability to be reformulated as problems, Plutarch stresses the aetiological challenge that they give rise to[20]. To push these *mirabilia* into the realm of the ἄπιστον (without further ado) would, in fact, be an intellectual refusal of turning them into problems for debate[21].

Plutarch did not dismiss such *mirabilia*, nor should his custom of turning them into natural problems necessarily be taken as an attempt to extricate their wondrous character altogether. In fact, Plutarch approaches the *mirabilia* in the exact sense as they are recorded in the tradition, so that an important aspect of the initial wonder remains when reformulating them as problems. Moreover, the aetiological preoccupation of natural problems does not necessarily undermine the wondrous character of these phenomena. As can be inferred from Plutarch's inquisitive and anti-dogmatic approach in his physical aetiologies and from his custom of postponing final judgement (ἐποχή), wonder is, in fact, key to his natural scientific method [see 4.3.2.]. A relevant passage in this regard can be found in *Quaest. conv.* 680CD, which discusses the problem of those who are said to cast an evil eye. Some symposiasts consider this matter to be completely silly and laugh at it, but L. Mestrius Florus does not see it this way[22]:

ὁ δ' ἑστιῶν ἡμᾶς Μέστριος Φλῶρος ἔφη τὰ μὲν γιγνόμενα τῇ φήμῃ θαυμαστῶς βοηθεῖν, τῷ δ' αἰτίας ἀπορεῖν ἀπιστεῖσθαι τὴν ἱστορίαν, οὐ δικαίως, ὅπου μυρίων ἐμφανῆ τὴν οὐσίαν ἐχόντων ὁ τῆς αἰτίας λόγος ἡμᾶς διαπέφευγεν· 'ὅλως δ'' εἶπεν 'ὁ ζητῶν ἐν ἑκάστῳ τὸ εὔλογον ἐκ πάντων ἀναιρεῖ τὸ θαυμάσιον· ὅπου γὰρ ὁ τῆς αἰτίας ἐπιλείπει λόγος, ἐκεῖθεν ἄρχεται τὸ ἀπορεῖν, τουτέστι τὸ φιλοσοφεῖν· ὥστε τρόπον τινὰ φιλοσοφίαν ἀναιροῦσιν οἱ τοῖς θαυμασίοις ἀπιστοῦντες. δεῖ δ''' ἔφη 'τὸ μὲν διὰ τί γίγνεται τῷ λόγῳ μετιέναι, τὸ δ' ὅτι γίγνεται παρὰ τῆς ἱστορίας λαμβάνειν.'

Mestrius Florus, our host, declared that actual facts lend astonishing support to the common belief. Yet the reports of such facts are commonly rejected because of the want of an explanation; but this is not right, in view of the thousands of other cases of indisputable

---

[20] L. Démarais, 2005, p. 163: "Défi pour l'étiologie, les *mirabilia* représentent une forme de quintessence du questionnement 'problématique', c'est-à-dire formulé (ou formulable) en *problema*."

[21] L. Démarais, 2005, p. 166: "[R]eléguer les *mirabilia* dans le domaine de l'ἄπιστον, c'est renoncer à en faire des *problemata*."

[22] Later on in *Quaest. conv.* 701A, Florus shares a similar opinion regarding other wondrous beliefs: ἐπεὶ δὲ τοῦτο μὲν ὁ Φλῶρος ᾤετο παιδιὰν εἶναι καὶ φλύαρον, ἐκείνων δ' οὐκ ἄν τινα τῆς αἰτίας ὡς ἀλήπτου προέσθαι τὴν ζήτησιν [see 2.3.3., n. 121]. Similarly, in *Quaest. conv.* 698E, he calls for a vindication of Plato's contested theory that drink passes through the lungs (οὕτως ὑφησόμεθα τοῦ Πλάτωνος ἐρήμην ὀφλισκάνοντος;).

fact in which the logical explanation escapes us. "In general," he went on, "the man who demands to see the logic of each and every thing destroys the wonder in all things. Whenever the logical explanation for anything eludes us, we begin to be puzzled, and therefore to be philosophers. Consequently, in a way, those who reject marvels destroy philosophy. The right method," he maintained, "is to search out the reason for facts by means of logic, but to take the facts themselves as they are recorded."

This is meant not so much as an incentive to accept the phenomenon without further ado, but to look for plausible explanations while maintaining an inquisitive attitude in the debate. Two points are important here for Plutarch's method of raising and solving problems (as voiced in this passage through the figure of Florus).

As to Plutarch's method of raising natural problems, first of all, the last sentence of Florus' account, where he promotes the aetiological study of popular beliefs exactly as they are put on record, is very important: 'the right method is to search out the reason for facts by means of logic, but to take the facts themselves as they are recorded' (δεῖ τὸ μὲν διὰ τί γίγνεται τῷ λόγῳ μετιέναι, τὸ δ' ὅτι γίγνεται παρὰ τῆς ἱστορίας λαμβάνειν). I will further examine this point in the next section.

Second, as to Plutarch's method of solving natural problems, the passage nicely illustrates his aporetic attitude in (natural) philosophy. Its formulation may, however, seem somewhat paradoxical in itself. The point is that by explaining the logic behind *mirabilia*, wonder disappears, and similarly, but the other way around, by rejecting these *mirabilia*, there is no wonder to begin with, so that in both cases there can, in fact, be no philosophy. After all, wonder is the beginning of philosophy, but a logical explanation is the end of wonder and, thus, of philosophy[23]. As Opsomer indicates in this regard, "[ἀ]πορία is [...] vital to philosophy" for Plutarch[24]. This means that his philosophy is an essentially zetetic search for the truth which attempts to, but cannot eventually

---

[23] For the ancient τόπος that wonder triggers philosophy, cf. also, e.g., *De E* 385C with Pl., *Tht.* 155d and Arist., *Met.* 982b11–15. For the opposite idea that philosophy is the end of wonder, cf., e.g., *De aud.* 44BC, with the interpretation of G. Roskam, 2005, p. 352, who holds (in his italics) that: "*Whereas Plato considers wonder to be the beginning of philosophy, Plutarch considers philosophy to be the end of wonder.*" However, considering Plutarch's aporetic attitude in philosophy, other scholars (see n. 24) are more inclined to underline the harmony between Plato and Plutarch in this regard – at least in the context of *natural* philosophy. The ambiguity related to the concept of wonder either as an impetus or an obstruction to attaining knowledge (in Antiquity and later) is outlined by M. Beagon, 2011, pp. 80–88.

[24] J. Opsomer, 1998, p. 80. See also M. Meeusen, 2014, p. 325.

culminate in genuine 'science' (ἐπιστήμη). When it comes to natural philosophy, we will see further on that from Plutarch's epistemological perspective, it is impossible to acquire steadfast knowledge in the study of natural phenomena, so that an aspect of uncertainty and wonder will always remain in physical research [see 4.3.2.1.]. So even if this is Florus speaking, Plutarch's own approach is much in line with this account.

2. *Democritus and the cucumber*

As is to be expected in light of Plutarch's Platonism, the stipulation of empirical verification – which is a premise in Aristotelian science – is of secondary importance for the Chaeronean's scientific project. This is not to say, though, that the role of αἴσθησις is absent or of no significance at all in Plutarch's natural problems [see 4.3.2.1.], but even so he is not only interested in those natural phenomena that have been positively proven to occur in nature. A relevant passage is found in *Quaest. conv.* 628BD, which illustrates Plutarch's choice of a more theoretical model of natural scientific research. Here, in regards to the veracity of an antiquarian problem (viz. 'Why is the chorus of the phylè Aiantis at Athens never judged last?'), Philopappus tells the following story about Democritus (DK68A17a):

εἰπόντος δὲ τοῦ ἑταίρου Μίλωνος 'ἂν οὖν ψεῦδος ᾖ τὸ λεγόμενον;' 'οὐδέν' ἔφη 'δεινόν' ὁ Φιλόπαππος 'εἰ ταὐτὸ πεισόμεθα Δημοκρίτῳ τῷ σοφῷ διὰ φιλολογίαν. καὶ γὰρ ἐκεῖνος ὡς ἔοικε τρώγων σίκυον, ὡς ἐφάνη μελιτώδης ὁ χυμός, ἠρώτησε τὴν διακονοῦσαν, ὁπόθεν πρίαιτο· τῆς δὲ κῆπόν τινα φραζούσης, ἐκέλευσεν ἐξαναστὰς ἡγεῖσθαι καὶ δεικνύναι τὸν τόπον. θαυμάζοντος δὲ τοῦ γυναίου καὶ πυνθανομένου τί βούλεται, 'τὴν αἰτίαν' ἔφη 'δεῖ με τῆς γλυκύτητος εὑρεῖν, εὑρήσω δὲ τοῦ χωρίου γενόμενος θεατής.' 'κατάκεισο δή' τὸ γύναιον εἶπε μειδιῶν, 'ἐγὼ γὰρ ἀγνοήσασα τὸ σίκυον εἰς ἀγγεῖον ἐθέμην μεμελιτωμένον'· ὁ δ' ὥσπερ ἀχθεσθεὶς 'ἀπέκναισας' εἶπεν 'καὶ οὐδὲν ἧττον ἐπιθήσομαι τῷ λόγῳ καὶ ζητήσω τὴν αἰτίαν, ὡς ἂν οἰκείου καὶ συγγενοῦς οὔσης τῷ σικύῳ τῆς γλυκύτητος.' οὐκοῦν μηδ' ἡμεῖς τὴν Νεάνθους ἐν ἐνίοις εὐχέρειαν ἀποδράσεως ποιησώμεθα πρόφασιν· ἐγγυμνάσασθαι γάρ, εἰ μηδὲν ἄλλο χρήσιμον, ὁ λόγος παρέξει.'

His companion Milo said, "What if actually the information is false?" "No matter!" said Philopappus. "It is not bad if the same thing does happen to us that happened to the wise Democritus because of love for learning. It seems that the juice of a cucumber he was eating appeared to have a honeylike taste, and he questioned his serving-woman about where she had bought it. When she indicated a certain garden, he got up and told her to take him and show him the place. The woman was astonished and asked what he had in mind. 'I must find,' he replied, 'the explanation for the sweetness, and I shall find it if I see the place.' 'Sit down,' said the woman with a smile, 'the fact is I accidentally put the cucumber in a honey-jar.' 'That was very annoying of you,' said Democritus with pretended anger, 'and I shall apply myself not the less to the problem and seek the explanation as if sweetness were proper and natural to this cucumber.' Let us not, then, make Neanthes's indifference in some items a pretext for running away, for this discussion will be a good exercise, if nothing else useful."

It is said of Democritus that he wrote about 'the most wondrous and paradoxical things of nature'[25]. The passage at hand subscribes to this idea. According to Abramowiczówna[26], the irony of this anecdote aims to subvert the zetetic attitude of philosophers in general towards fictitious problems. She finds it remarkable, however, that the symposiasts are not affected by the irony of the story, but consider it as encouragement to simply go on explaining the problem. It remains to be seen, however, whether the anecdote is really ironic to begin with. Of course, a certain humorous effect cannot be denied in it, but the actual moral of the story is, I take it, that, for Plutarch (and for Democritus in the story), explaining natural scientific problems should not so much be concerned with the ὅτι of the phenomenon at hand but with the διὰ τί (to allude to Florus' conclusion in *Quaest. conv.* 680D, quoted above)[27]. In other words, not the reality of natural phenomena as such is the main topic of inquiry but their causes. A phenomenon that *can* lay a hypothetical claim to being real or empirically verifiable, disregarding whether it also actually *does*, requires an appropriate explanation[28]. Indeed, in his natural problems, Plurach does

---

[25] DK68A99a: τὰ θαυμαστὰ καὶ τὰ παραλογώτατα τῆς φύσεως. Cf. also M. Sassi, 1993, p. 449. I have discussed Democritus' aetiological activities earlier on [see 1.1.6.]. For a study of Democritus' model of causal research, see P.-M. Morel, 1996.

[26] Z. Abramowiczówna, 1960, pp. 113–114 (passage also discussed in 1962, p. 83). Cf. also S.-T. Teodorsson, 1989, pp. 158–159.

[27] For the conceptual distinction between the ὅτι and the διὰ τί, cf. also, e.g., *Quaest. conv.* 700D: πρᾶγμα πίστιν ἔχον ὅτι γίγνεται, τὴν δ' αἰτίαν ἔχον ἄπορον ἢ παγχάλεπον.

[28] The concept of 'indifference' (εὐχέρεια) in the passage at hand refers to the irresponsibility of historians to falsify historical facts (cf. LSJ, s.v.), but it also applies to the field of natural history, as is the case in the Democritus anecdote [see 1.2.5., n. 224]. For the issue of historical accuracy in the *Vitae*, Plutarch's account in *Alex.* 1 is well-known (cf. also *Galba* 2, 5 and *Fab.* 16, 6). The conclusion reached for this passage by C. Pelling, 1980, p. 135 is relevant in light of Plutarch's natural science also: "It is simply that the boundary between truth and falsehood was less important than that between acceptable and unacceptable fabrication, between things which were "true enough" and things which were not. Acceptable rewriting will not mislead the reader seriously; indeed readers will grasp more of the important reality if they accept what Plutarch writes than if they do not.

not so much emphasise the 'actuality' of the natural phenomena but their 'potency'. 'Potential phenomena' can and should be studied – these are phenomena that possibly exist in principle (since they are recorded in the tradition), but are not, therefore, empirically proven to occur in nature and may, thus, remain without a parallel in actual physical reality[29]. The notion of 'what can and might be' is formulated explicitly in *De facie* 938C (regarding the possibility of habitation on the moon):

καὶ γὰρ ὡς ἀληθῶς τῶν σφόδρα πεπεισμένων τὰ τοιαῦτα διαφέρουσιν οὐδὲν οἱ σφόδρα δυσκολαίνοντες αὐτοῖς καὶ διαπιστοῦντες ἀλλὰ μὴ πράως τὸ δυνατὸν καὶ τὸ ἐνδεχόμενον ἐθέλοντες ἐπισκοπεῖν.

It is, moreover, a fact that there really is no difference between those who in such matters are firm believers and those who are violently annoyed by them and firmly disbelieve and refuse to examine calmly what can be and what might be[30].

The fact that Plutarch firmly opposes those who either blindly reject or blindly accept such possibilities is very significant for his own attitude towards natural *mirabilia*. Even though for Plutarch there is no strong commitment to what actually occurs in physical reality in order to explain natural phenomena, a certain degree of scientific caution (εὐλάβεια) towards the actual existence of these *mirabilia* obviously remains. It seems that Plutarch often gives such phenomena the benefit of the doubt. This is, I believe, seminal for the theoretical and sceptical character of Plutarch's

---

Truth matters; but it can sometimes be bent a little." Cf. also, e.g., *Adv. Col.* 1115C for Colotes' εὐχέρεια in his attacks on Plato.

[29] The Aristotelian view of 'actualisation', where potential phenomena become actual one way or another, is not at issue here. Rather, I am alluding to the wording of S. Sambursky, 1963, pp. 234–235, who argues, by contrast, that the Greeks in general were only able to study natural phenomena in their actuality (i.e. as they occur directly in nature), and that they were, therefore, unable to execute laboratory-like experiments (viz. regarding artificially recreated phenomena in an unnatural environment, i.e. phenomena that are isolated and extracted from their direct natural context). See p. 235: "The latter [sc. potential phenomena] become actual only in the laboratory. In such a sense we may call an experiment unnatural. This, no doubt, is how it seemed to the Greeks, who would have thought it paradoxical to study natural phenomena by unnatural methods." However, Plutarch's theoretical and non-experimental model in explaining natural phenomena (and that of many other ancient scientific authors) can also be called 'unnatural' according to Sambursky's standards (see also n. 244).

[30] It seems that Plutarch does not intend to make a technical distinction here between 'the potential' (τὸ δυνατόν) and 'the contingent' (τὸ ἐνδεχόμενον) but uses both concepts as synonyms, basically denoting 'what is possible'. See H. Cherniss and W.C. Helmbold, 1957, p. 163, n. e and P. Donini, 2011, p. 329, n. 306.

science of natural problems. Plutarch does not necessarily intend to study nature as it is (empirically), but as it could be (theoretically), that is, if the postulated premises are accepted *a priori* as they are formulated in the *quaestiones*. In his natural problems, Plutarch, much like Democritus in the story (though perhaps to a lesser degree), does not ascribe much value to empirical verification (or falsification) of the natural phenomena, but, indifferent to their claim to being real or not, intends to provide a more or less reasonable explanation of the problem, if only for the sake of intellectual exercise [see 3.2.1.]. However, this 'if only' (εἰ μηδὲν ἄλλο χρήσιμον) does not necessarily imply that each and every problem that Plutarch deals with would simply be false, or, *a fortiori*, that Plutarch does not really believe them, or does not at least ascribe some credibility to them. In any case, it puts the often absurd character of the phenomena discussed into perspective, which, at the very least, appear to be worthy of a plausible explanation.

As we saw earlier, the Greek title of *Quaestiones naturales* – viz. Αἰτίαι φυσικαί – hints at a distinction between *quaerenda* and *quaestiones*, that is, between explanations and problems respectively[31] [1.1.6.]. From this title it is clear that the primary focus of the collection is on the physical causes of the problems, that is, the διὰ τί, rather than the natural phenomena themselves that are questioned, that is, the ὅτι[32]. The formulation of each *quaestio* is motivated on the basis of a conviction that the natural phenomenon functions as a manifestation, a σημεῖον, of material forces at work in nature – these are the *quaerenda* that are decipherable and fathomable by means of physical aetiology[33]. However difficult or even impossible it may sometimes seem, it is *a priori* assumed that the natural problems are explicable and that there are plausible physical αἰτίαι to support them[34]. The

---

[31] This distinction is echoed, e.g., in *Quaest. conv.* 641C: Ἐγὼ δὲ τοῦτο μὲν ἔφην ἀπόδρασιν εἶναι τῆς ἐρωτήσεως μᾶλλον ἢ τῆς αἰτίας ἀπόδοσιν.

[32] The διὰ τί in the *quaestiones* inquires specifically into the material, not the higher, type of causes of the natural phenomena at hand. Moreover, the recurrent use of the preposition διά/δι' in the explanations links up closely with this interrogative διὰ τί in the *quaestiones*. In *Q.N.* 29, 919B, the phrase πῶς γίνεται implies basically the same as the introductory τίς ἡ αἰτία δι' ἥν (= διὰ τί): *pace* L. Senzasono, 2006, pp. 230–231, n. 163. Cf. P. Donini, 1992, p. 107: "Ben lontano dal preoccuparsi di ricordare sempre anche le cause "divine", quando si domanda il perché di un fatto del mondo sensibile o di un fenomeno fisico Plutarco si limita per lo più nelle *Questioni* [*conviviali*] a parlare delle sole cause che altrove chiama subordinate, meno importanti, naturali o necessarie […]."

[33] For the conceptualisation of natural σημεῖα in *Quaestiones naturales*, cf. *Q.N.* 2, 912D, 18, 916A, 30, 919C (25, 918A).

[34] Cf., e.g., *Quaest. conv.* 641C (τούτων γὰρ ἐμφανῆ τὴν πεῖραν ἐχόντων, χαλεπὸν εἶναι τὴν αἰτίαν, εἰ μὴ καὶ παντελῶς ἀδύνατον, καταμαθεῖν), 664C (χαλεπὰς καταμαθεῖν ἢ παντελῶς ἀδυνάτους τὰς αἰτίας ἔχοντα), 690F (ἔστι γὰρ μάλιστα δυσθεώρητος), 700D (quoted n. 27), 701A (Ἐπεὶ δὲ τοῦτο μὲν ὁ Φλῶρος ᾤετο παιδιὰν εἶναι καὶ φλύαρον, ἐκείνων δ' οὐκ ἄν τινα τῆς αἰτίας ὡς ἀλήπτου

quaestio itself is not questioned: it actually has a rationale, a *raison d'être*, of itself, and the problematic natural phenomenon is, by implication, a virtual given only in the guise of a problem.

It is incorrect, therefore, to identify the *quaestiones* with 'hypotheses', for the simple fact that the latter need to be verified by further research: hypotheses are accepted until the opposite has been proven to be true (by falsification), whereas the *quaestiones* in *Quaestiones naturales* are accepted without further ado, so that a certain aspect of objectivity and credibility is ascribed to them. The concept of hypotheses more appropriately applies to the arguments in the aetiologies, which are tentative by their anti-dogmatic and interrogative formulation. There is no urge to fully prove or demonstrate things in Plutarch's physical aetiologies (this is, in fact, considered impossible from his Platonic-Academic point of view [see 4.3.2.1.]). In other words, the *quaestio* is not only an 'unknown factor' in the development of the problem, but also an 'invariable' or a 'given' in itself, the validity of which is not generally put to question. The *quaestio* functions as a stepping-stone on which the entire aetiology is founded. In a certain sense, the explanations to these problems are the real 'variables' in the problems. They are not formulated dogmatically and are, therefore, always a potential subject for criticism and revision.

3. *Plutarch's popular beliefs: anti-Aristotelian and anti-Stoic dynamics*

As seen in the previous section, empirical verification in dealing with natural problems was not of great concern to Plutarch's scientific methodology, and, as we will see later on, he had good epistemological reasons for it [see 4.3.2.3.]. What is important, and what has also been flagged before, is the anti-Aristotelian position Plutarch is implicitly taking. It is worth digging a bit deeper regarding Plutarch's subordination of the veracity of the natural phenomena to their physical causes in light of Aristotle's concept of science, which famously departs from what is positively given in nature. Afterwards, I will also confront Plutarch's approach with that of the Stoics.

Indeed, judging from his recurrent statement that natural science should progress from a consideration of what actually appears to the senses (see below), Aristotle has often been hailed by historians of science as the first empiricist. However, on closer inspection, much of the material gathered in his biological writings derives from popular hearsay rather than from personal observations. Even though Aristotle probably sought empirical verification for at least some of his assertions in these writings

---

προέσθαι τὴν ζήτησιν). In *Quaest. conv.* 700D Plutarch refers to a work from Theophrastus ἐν οἷς πολλὰ συναγήοχεν καὶ ἱστόρηκεν τῶν τὴν αἰτίαν ἀνεύρετον ἡμῖν ἐχόντων (probably Περὶ τῶν λεγομένων ζῴων φθονεῖν: fr. 175 Wimmer = 362A FHSG). Cf. S.-T. Teodorsson, 1996, p. 35.

(he mentions dissections, for instance), scholars have convincingly shown that, to a great extent, his approach relies on reported observation rather than on autopsy[35]. In a famous passage concerning his scientific method in biology, Aristotle states that 'first the phenomena should be grasped [...], then their causes discussed' (PA 640a14–15: πρῶτον τὰ φαινόμενα ληπτέον [...] εἶτα τὰς αἰτίας λεκτέον)[36]. This means that for Aristotle, natural science should start from a consideration of what actually appears to the senses, and that examining the reasons of a reported phenomenon without knowing whether it actually exists or not leads to bad science (in principle). Aristotle's injunction to first grasp the observable phenomena (τὰ φαινόμενα) and to discuss their causes afterwards (τὰς αἰτίας) is a genuine appeal towards empirical verification. However, as just noted, Aristotle's actual scientific practice is often at odds with this injunction. Despite his official empiricist concern, Aristotle often violates his own methodological rule by basing his inquiry on popular hearsay and on doubtful assumptions rather than on empirical observations.

Even though it is uncertain as to whether Plutarch was acquainted with the empirical injunction from PA 640a14–15 just quoted, he is clearly not following Aristotle's advice. In fact, a clear anti-Aristotelian attitude speaks from Plutarch's explicit subordination of the reality of the natural phenomena (i.e. the ὅτι = τὰ φαινόμενα) to their causes (i.e. the διὰ τί = τὰς αἰτίας). Notably, Aristotle's account in PA 640a14–15 is very close from a formal perspective (λέγειν – λαμβάνειν) to Quaest. conv. 680D, quoted above, where Florus argues, on the contrary, that 'the right method is to search out the reason for facts by means of logic, but to take the facts themselves as they are recorded' (δεῖ τὸ μὲν διὰ τί γίγνεται τῷ λόγῳ μετιέναι, τὸ δ' ὅτι γίγνεται παρὰ τῆς ἱστορίας λαμβάνειν). Florus' wording is

---

[35] E.g., I. Düring, 1961, pp. 218–221 was right that: "we should be careful not to over-emphasize Aristotle's own rôle as a pioneer of the empirical sciences. [...] [H]e was, perhaps to a far greater extent than we sometimes are inclined to believe, a desk-work scholar. [...] Yet he is always quick to castigate those who do not start from the φαινόμενα. 'Facts', 'data', did not mean the same for him as for us; practically every statement in which he himself believed was classed as a 'fact', or as 'true'." Cf. also H. von Staden, 1989, p. 118. For a general outline of the aspect of empirical research in Aristotle's science, see G.E.R. Lloyd, 1979, pp. 200–225.

[36] A similar empiricist concern is found, e.g., in Herophilus, fr. 50a (and b) von Staden: 'Let the phenomena be said first, even if they are not first' (λεγέσθω δὲ τὰ φαινόμενα πρῶτα, καὶ εἰ μὴ ἔστιν πρῶτα). In his well-known article on Aristotle's conception of what constitutes the φαινόμενα, G.E.L. Owen, 1961 argued that Aristotle used this concept not in the sense of observable phenomena here, but of τὰ ἔνδοξα or τὰ λεγόμενα (cf. also I. Düring, 1961). This was rejected by H. von Staden, 1989, pp. 117–119, who argued, more convincingly, that the concept of φαινόμενα really denotes the observable phenomena in Aristotle's biological writings, but that this does not eventually preclude the incorporation of τὰ ἔνδοξα or τὰ λεγόμενα in these writings (see also the review by R.J. Hankinson, 1990, pp. 213–215).

very close also to another Aristotelian account about the proper method of scientific inquiry: 'when we know the fact, we seek the reason why' (*APo.* 89b29–30: ὅταν δὲ εἰδῶμεν τὸ ὅτι, τὸ διότι ζητοῦμεν). Even though it remains unclear as to whether Plutarch is intentionally alluding to these Aristotelian accounts (Aristotle's name and authority remain unmentioned), it is clear that Aristotle's appeal towards an empirical science did not greatly affect him in his natural problems. However, when push comes to shove, it turns out that the actual scientific practice of both philosophers was not always that different after all.

Plutarch's interest in explaining marvellous natural phenomena is more generally germane to the genre of natural problems. A large amount of paradoxes and *mirabilia* found their way into the Ps.-Aristotelian *Problems*[37] (we know that Florus owned a copy of this work: cf. *Quaest. conv.* 734CD [quoted 3.2.1.]). As such, the scientific outlook of Plutarch's natural problems is not that different from those attributed to Ps.-Aristotle, which served as his model. Notably, the explanations (διὰ τί) in the Ps.-Aristotelian *Problems* also receive priority to the veracity of the phenomena themselves (ὅτι). As Mayhew observes, the *Problems* "never ask whether something exists or whether some proposition is true."[38]

---

[37] The *Problems* have a clear predilection for explaining popular beliefs (ἔνδοξα). See H. Flashar, 1962, p. 299 and pp. 342–343, M. Sassi, 1993, p. 455, A. Blair, 1999, p. 173, R. Mayhew, 2011a, pp. xxi–xxii. See, for instance, the compilation by O. Thomas in his BMCR review (21 august 2012) of R. Mayhew's 2011 Loeb edition of Ps.-Aristotle's *Problems*: "Why do warthogs find each other attractive? Why do certain noises send a chill down the spine? Why do we get more enjoyment from tunes that we already know? Why do you yawn if I yawn? And why, while we yawn, do we lose our hearing? Why do children get more nits (and runny noses, and nosebleeds) than adults? Why does holding one's breath cure hiccups? Why can't one tickle oneself? Why is sex the highest pleasure? Why do drunks see double, or see the room spinning? Why does cutting an onion make you cry? Why does fear loosen the bowels? Why is it more shocking to kill a woman than a man? Why are most professional performers odious? Why do we count in base ten? (Is it because of the Pythagorean tetraktys? Or because we have ten fingers?) Why do some people feel sleepy the moment they open a book?" Ps.-Aristotle's *De mirabilibus auscultationibus* is an exponent of the genre of ancient paradoxography, but this work is generally considered spurious. This is not the case for Ps.-Aristotle's *Problems*, which contains at least a certain, albeit undefinable, nucleus of authentic Aristotelian problems [see 1.1.3.].

[38] R. Mayhew, 2011a, p. xxii. Cf. also, e.g., A. Blair, 1999, p. 173 (and pp. 187–188): "The question, in διὰ τί, asks not about the existence or nature of a fact, but about the cause of a fact that is presumed so well known that it is not even stated before it is explained. However bizarre the 'fact' may seem to us, the *problema* never includes discussion of its veracity but only of its cause." Cf. also J. Mansfeld, 2010, p. 44: "Aristotle is quite clear that one need not always put all the questions. As to the ὅτι and the διότι, he says [...] that when you already know the 'that' you immediately ask for the 'why'." Cf. also C. Jacob, 2004, p. 45: "Ces questions donnent une réalité objective aux phénomènes

This may explain, then, why Plutarch did not bother to verify the popular beliefs in his own natural problems but simply took them for granted, perhaps even considering this procedure a characteristic feature of the Aristotelian genre of natural problems. This would, in turn, put Plutarch's alleged anti-Aristotelianism into perspective.

A good example is found in *Q.N.* 38, where Plutarch examines why she-wolves give birth to their young at a fixed time of the year within twelve days. As to the belief that they do this within the timespan of twelve days, Plutarch in the second *causa* mentions the story about Leto, 'to which certain people refer' (*Quidam ad fabulam Latonae referunt*). When Leto became pregnant from Zeus, she could not find a safe haven from Hera anywhere. Thus, Zeus transformed her into a wolf for a period of twelve days, during which she travelled to Delos. In this way, she procured that all wolves should be able to litter in that same period from then on. Among the *quidam* mentioned at the beginning of the *causa*, Aristotle certainly comes first, since the same mythological account is also recorded in *HA* 580a14–19. The Stagirite remains sceptical, however, about the popular belief that she-wolves (and dogs) mate and litter within twelve days. He writes: 'whether this really is the time for their pregnancy or not, has not yet been definitely established by observation; that is merely what is asserted' (*HA* 580a20–22: εἰ δ' ἐστὶν ὁ χρόνος οὗτος τῆς κυήσεως ἢ μὴ ἔστιν, οὐδέν πω συνῶπται μέχρι γε τοῦ νῦν, ἀλλ' ἢ ὅτι λέγεται μόνον). The fact *that* this phenomenon 'is merely asserted' (ὅτι λέγεται μόνον) is important, since it does not withhold Plutarch from addressing the problem anyhow. He may be relying directly on Aristotle or on an intermediary source, perhaps a lost natural problem, where this very assertion was supported and explained (see the commentary *ad loc.*).

As such, the fundamental difference between Aristotle's and Plutarch's scientific projects does not so much lie in the authority they ascribe to popular hear-say – in fact, they seem to be relatively convergent at this point –, but rather in the epistemological relevance they attribute to empirical observation. Aristotle openly vindicates an empirical approach in natural scientific research, while Plutarch does not see observational data as procuring reliable knowledge. In other words, while the Stagirite neglects to empirically double-check what is put on record in popular natural ἱστορία, the Chaeronean is not primarily concerned with verifying (or falsifying) these popular beliefs in the first place, since he is convinced, as a faithful Platonist, that data pertaining to sense perception are

qu' elles problématisent: loin d' être des faits aléatoires, ils sont présentés comme des phénomènes généralisables et récurrents dont il est légitime de chercher les causes. Ces dernières peuvent relever de principes physiques ou de principes psychologiques, de schémas scientifiques ou de la sagesse populaire, non dénués parfois d' une certaine forme d' humour, du moins aux yeux du lecteur moderne [...].''

essentially deceitful in kind [see 4.3.2.1.]. This does not imply that Plutarch ascribes no value whatsoever to empirical knowledge, but it is quite clear that his main concern lies elsewhere. Considering their authoritative statute, Plutarch assumes that popular beliefs must contain at least a certain aspect of reliability. Perhaps more important even, he believes that these often marvellous beliefs hint at the workings of a higher type of causality.

A relevant passage to illustrate this is found in *Cor.* 38 which speaks of the popular belief that statues of deities can cry, sweat, bleed or even speak, or at least give the appearance of doing so (the context is the story of the goddess Fortuna expressing words of thanks to Coriolanus' mother and wife via her newly erected statue). Plutarch points out that this is not, in fact, impossible for both natural and super-natural reasons:

ἰδίοντα μὲν γὰρ ἀγάλματα φανῆναι καὶ δακρυρροοῦντα καί τινας μεθιέντα νοτίδας αἱματώδεις οὐκ ἀδύνατόν ἐστι· καὶ γὰρ ξύλα καὶ λίθοι πολλάκις μὲν εὐρῶτα συνάγουσι γόνιμον ὑγρότητος, πολλὰς δὲ καὶ χρόας ἀνιᾶσιν ἐξ αὑτῶν, καὶ δέχονται βαφὰς ἐκ τοῦ περιέχοντος, οἷς ἔνια σημαίνειν τὸ δαιμόνιον οὐδὲν ἂν δόξειε κωλύειν. δυνατὸν δὲ καὶ μυγμῷ καὶ στεναγμῷ ψόφον ὅμοιον ἐκβαλεῖν ἄγαλμα κατὰ ῥῆξιν ἢ διάστασιν μορίων βιαιοτέραν ἐν βάθει γενομένην. ἔναρθρον δὲ φωνὴν καὶ διάλεκτον οὕτω σαφῆ καὶ περιττὴν καὶ ἀρτίστομον ἐν ἀψύχῳ γενέσθαι παντάπασιν ἀμήχανον, εἰ μηδὲ τὴν ψυχὴν καὶ τὸν θεὸν ἄνευ σώματος ὀργανικοῦ καὶ διηρμοσμένου μέρεσι λογικοῖς γέγονεν ἠχεῖν καὶ διαλέγεσθαι.

For that statues have appeared to sweat, and shed tears, and exude something like drops of blood, is not impossible; since wood and stone often contract a mould which is productive of moisture, and cover themselves with many colours, and receive tints from the atmosphere; and there is nothing in the way of believing that the Deity uses these phenomena sometimes as signs and portents. It is possible also that statues may emit a noise like a moan or a groan, by reason of a fracture or a rupture, which is more violent if it takes place in the interior. But that articulate speech, and language so clear and abundant and precise, should proceed from a lifeless thing, is altogether impossible; since not even the soul of man, or the Deity, without a body duly organised and fitted with vocal parts, has ever spoken and conversed.

Plutarch, in what follows, notes that 'history forces our ascent with numerous and credible witnesses' (ἡμᾶς ἡ ἱστορία πολλοῖς ἀποβιάζεται καὶ πιθανοῖς μάρτυσιν). He does not, however, accept these beliefs blindly but tries to save them by providing a natural explanation (see the parallel in *Cam.* 6, 5–6 below). Even so, Plutarch is clear that it is 'altogether impossible' (παντάπασιν ἀμήχανον) for these statues to really produce articulate speech. He concludes that 'an experience *different* from that of sensation arises in the imaginative part of the soul, and persuades men to think it sensation' (ἀνόμοιον αἰσθήσει πάθος ἐγγινόμενον τῷ φανταστικῷ τῆς ψυχῆς συναναπείθει τὸ δόξαν). This is very relevant in light of the deceitful nature of data pertaining to sense perception more generally, underlining how this deceit works exactly in the context of natural *mirabilia*.

In what follows, Plutarch explains that too eager acceptance of such miracles is a sign of superstition. He formulates this in rather shrouded and euphemistic terms, calling superstitious people 'those who cherish strong feelings of good-will and affection for the deity, and are therefore unable to reject or deny anything of this kind' (τοῖς ὑπ᾽ εὐνοίας καὶ φιλίας πρὸς τὸν θεὸν ἄγαν ἐμπαθῶς ἔχουσι καὶ μηδὲν ἀθετεῖν μηδ᾽ ἀναίνεσθαι τῶν τοιούτων δυναμένοις). And he says of them that they 'have a strong argument (sc. as provided by their belief in talking statues) for their faith in the wonderful and transcending character of the divine power' (μέγα πρὸς πίστιν ἐστὶ τὸ θαυμάσιον καὶ μὴ καθ᾽ ἡμᾶς τῆς τοῦ θεοῦ δυνάμεως). Plutarch shatters this illusion by pointing out that 'the deity has no resemblance whatever to man, either in nature, activity, skill, or strength' (οὐδενὶ γὰρ οὐδαμῶς ἀνθρωπίνῳ προσέοικεν οὔτε φύσιν οὔτε κίνησιν οὔτε τέχνην οὔτ᾽ ἰσχύν). This is, of course, no cause to resort to radical atheism. On the contrary, so Plutarch concludes, 'most of the deity's powers, as Heraclitus says, "escape our knowledge through incredulity" (DK22B86)' (τῶν μὲν θείων τὰ πολλά, καθ᾽ Ἡράκλειτον, ἀπιστίῃ διαφυγγάνει μὴ γινώσκεσθαι). In short, this means that God is able to communicate with us, albeit not by employing human speech but by other means strange to us (e.g., via dreams: cf. *Per.* 13, 7–8). What the passage shows, then, is not that natural *mirabilia* (as recorded in popular ἱστορία) should be rejected but should be approached with the necessary circumspection and caution.

As we know from a work like *De superstitione*, Plutarch more often invites his reader to avoid religious immoderation by following a 'middle course' between atheism (ἀθεότης) and superstition (δεισιδαιμονία) – these he considers the extreme antipodes in the spectrum of impiety (ἀσέβεια)[39]. Plutarch at times connects the concept of a 'middle course' with the Delphic imperative to 'avoid extremes' (μηδὲν ἄγαν). This is the case, for instance, in *Cam.* 6, 5–6, where a parallel account is found about the marvellous phenomena connected to divine statues (the broader context is that of the statue of Juno assenting to its transfer from Veii to Rome after the sack by Camillus). Again with a reference to traditional ἱστορία, Plutarch reports that 'not a few historians wrote that statues of gods could talk, drip with sweat, utter audible groans, turn away their faces and close

---

[39] Cf. *De sup.* 169F. For the opposition between atheism and superstition, cf., e.g., *De sup.* 164E and esp. 165C: ἡ μὲν ἀθεότης λόγος ἐστὶ διεψευσμένος, ἡ δὲ δεισιδαιμονία πάθος ἐκ λόγου ψευδοῦς ἐγγεγενημένον (for their mutual relation, cf. *De sup.* 171AB, F). This opposition recurs passim throughout Plutarch's writings: cf., e.g., *De ad. et am.* 66CD, *De Is. et Os.* 355D, 378A, 379E etc.

their eyes' (*Cam.* 6, 4: ἱστορήκασιν οὐκ ὀλίγοι τῶν πρότερον κτλ.). Plutarch warns, however, that certainty is unattainable in such cases for the human intellect:

πολλὰ δὲ καὶ τῶν καθ' ἡμᾶς ἀκηκοότες ἀνθρώπων λέγειν ἔχομεν ἄξια θαύματος, ὧν οὐκ ἄν τις εἰκῇ καταφρονήσειεν. ἀλλὰ τοῖς τοιούτοις καὶ τὸ πιστεύειν σφόδρα καὶ τὸ λίαν ἀπιστεῖν ἐπισφαλές ἐστι διὰ τὴν ἀνθρωπίνην ἀσθένειαν, ὅρον οὐκ ἔχουσαν οὐδὲ κρατοῦσαν αὑτῆς, ἀλλ' ἐκφερομένην ὅπου μὲν εἰς δεισιδαιμονίαν καὶ τῦφον, ὅπου δ' εἰς ὀλιγωρίαν τῶν θείων καὶ περιφρόνησιν· ἡ δ' εὐλάβεια καὶ τὸ μηδὲν ἄγαν ἄριστον.

And we ourselves might make mention of many astonishing things which we have heard from men of our own time, – things not lightly to be despised. But in such matters eager credulity and excessive incredulity are alike dangerous, because of the weakness of our human nature, which sets no limits and has no mastery over itself, but is carried away now into vain superstition, and now into contemptuous neglect of the divine. Caution is best, and to go to no extremes.

For Plutarch, an unreasoned rejection of the possibility of divine intervention through miraculous natural phenomena allows for radical disbelief and atheism, much in the same way as unreasoned acceptance thereof does for credulity and superstition. In order to avoid the religious extremes of atheism and superstition, Plutarch adheres to some kind of rational devotion to the gods (εὐσέβεια)[40], which, as we learn from *De sup.* 171F, lies in between both extremes (ἐν μέσῳ κειμένην τὴν εὐσέβειαν)[41]. I will later return to the epistemological value of the μηδὲν ἄγαν and εὐλάβεια at the end of this passage in light of Plutarch's penchant for Academic philosophy in more detail [see 4.3.2.1.]. Important for the argument at hand is that Plutarch's attempt to save popular *mirabilia* relates to issues of theology and world view. For Plutarch each natural phenomenon can be considered a sign not only of the working of physical forces in nature, but also of a higher entity that orders the world in a providential way [see 4.1.2.]. This allows Plutarch to draw a close affiliation between the natural and the divine realms in nature, which, in turn, closely alines his world system with Plato's and distinguishes it from Aristotle's.

---

[40] See R. Flacelière, J. Irigoin, J. Sirinelli and A. Philippon, 1987, p. clxii: "Mais sa foi, certes, n' était pas celle du charbonnier; elle était raisonnée, réfléchie, 'éclairée' comme il sied à un philosophe fermement convaincu du primat de la raison."

[41] Cf. also, e.g., *Comp. Nic. et Crass.* 5, 3: χαλεπὴ μὲν ἐν τούτοις ἡ ἀσφάλεια καὶ δύσκριτος· ἐπιεικέστερον δ' αὐτῆς τοῦ παρανόμου καὶ αὐθάδους τὸ μετὰ δόξης παλαιᾶς καὶ συνήθους δι' εὐλάβειαν ἁμαρτανόμενον.

Aristotle's God, as perceived by Plutarch, is more divorced from the world that we live in. He has only set the world in motion (in his capacity of ὃ οὐ κινούμενον κινεῖ) and for the rest concentrates his mind exclusively upon himself (in his capacity of νοήσεως νόησις)[42]. This implies that he does not intervene in nature in a providential way and does not care about the animate world that we, human beings, belong to. By consequence, the incorporation of popular and wonderful beliefs in the Stagirite's natural scientific writings is not motivated on religious grounds, but is rather symptomatic of the author's urge for scholarly-encyclopaedic comprehensiveness. If Aristotle is interested in collecting 'raw data' from traditional natural history, it is not because they hint at the existence of a higher level of reality, but because they require further scientific investigation. In the case of Plutarch, by contrast, a deeper religious-philosophical motive does play an important role in his interest in natural *mirabilia*.

Notably, the Stoics also had this custom of saving natural *mirabilia* in their attempt to support a providential ordering of the world[43]. Again, however, their motives are fundamentally different from Plutarch's. Stoic cosmology is based on the assumption of a universally predetermined fate (εἱμαρμένη), which is seen as a predetermined sequence of causes, where every cause is considered the predetermined effect of a previous cause, which in turn is the effect of yet another previous cause, and so on. It follows that the world's natural ordering, from its very outset until its very end, is predetermined from a causal perspective (which also enables us to make predictions about the future by means of divination). Importantly, for the Stoics the world's providential ordering is essentially linked to a physical interpretation of all causality, whereas in the case of Plutarch there is room also for a higher type of causality, which transcends physical reality (but still stands in close contact with it). According to the Stoics, a divine fire steers the cosmos, and the world's providence is based on this material-logical principle. As such, the providential ordering of nature is founded on an essentially material basis, since the divine principle is immanent in the natural world. Plutarch's God, by contrast, has a more transcendental nature, although his influence can still be felt in the natural world (this means that God is not simply detached from the lower natural realm as is the case rather in Aristotle's theology)[44].

---

[42] Cf. *De def. or.* 426DE: ὁ δ' ἀληθινὸς (sc. Ζεὺς) ἔχει καλὰς καὶ πρεπούσας ἐν πλείοσι κόσμοις μεταβολάς, οὐκ ἐπὶ κενὸν ἄπειρον ἔξω βλέπων οὐδ' ἑαυτὸν ἄλλο δ' οὐδὲν (ὡς ᾠήθησαν ἔνιοι) νοῶν, ἀλλ' ἔργα τε θεῶν καὶ ἀνθρώπων πολλὰ κινήσεις τε καὶ φορὰς ἄστρων ἐν περιόδοις καταθεώμενος κτλ.

[43] See, e.g., Sen., *De prov.* I, 2–4.

[44] Plutarch's Platonic God arranges nature in a providential fashion, but his essence is located in the intelligible realm. For the difference between Plutarch's concept of divine

As a consequence, the religious motivation for Plutarch's attempt to save natural *mirabilia* is very different from that of the Stoics. Whereas Plutarch does not reject the relevance of physical aetiology in interpreting what are essentially divinely inspired phenomena, the Stoics attribute everything in nature to a predetermined chain of material causes. In so doing, they invoke the concepts of natural sympathy and antipathy, by which they assume a permanent interaction of every phenomenon in the cosmos, in a positive or negative way respectively[45]. Plutarch does not seem to be a great enthusiast of these theories. In fact, it seems that for him the sympathy/antipathy argument is a non-explanation. At least, it is presented as not being very plausible or convincing from an aetiological perspective. Presumably by its lack of explanatory detail and elaboration the sympathy/antipathy argument remains rather close to the realm of fable and superstition, for Plutarch, and cannot, therefore, be considered an adequate physical explanation. This may explain why it is not mentioned in *Quaestiones naturales*[46], and why on several occasions in *Quaestiones convivales*, Plutarch even objects to it openly[47]. In *Quaest. conv.* 641B, for

---

providence *vis-à-vis* that of the Stoics, see G.E. Karamanolis, 2006, p. 108: "Against the Stoic view that providence is immanent in Nature in particular, Plutarch argues that nature has been arranged in a certain way, and it is in this arrangement that the essence of divine providence lies (*De facie* 927C–D)." See also p. 106: "Plutarch regards God as being constantly involved with the world, exercising providence over everything in it." For further reading, see also J. Opsomer, 2014. For the place of divine πρόνοια in Plato's cosmos, cf., e.g., *Tim.* 30c. On Plutarch's view of the workings of providence and chance in the development of history, see S. Swain, 1989 (his appendix of terms on pp. 298–302 is useful in light of Plutarch's conceptualisation of providence). Cf. also F. Titchener, 2014.

[45] For the strong Stoic connotation of the cosmological theories of antipathy and sympathy, see, e.g., T. Weidlich, 1894, pp. 4–11, K. Reinhardt, 1926, pp. 178–186 and G. Soury, 1949, pp. 322–323. Bolus of Mendes (Ps.-Democritus), a contemporary of Callimachus, wrote an influential work Περὶ ἀντιπαθειῶν καὶ συμπαθειῶν (see M. Wellmann, 1897). See also S.-T. Teodorsson, 1989, pp. 255–256 and 1999a, pp. 667–668. In addition, Book seven of Ps.-Aristotle's *Problems* is entitled ὅσα ἐκ συμπαθείας, but this type of sympathy concerns medical contagions (in a broad sense). See R. Mayhew, 2011a, pp. 228–229. The theme of *Q.N.* 6 perhaps goes back on this topic (see the commentary *ad loc.*).

[46] However, there may be an implicit allusion to this Stoic theory in the sympathetic account about she-wolves that Plutarch borrows from Antipater in the first *causa* of *Q.N.* 38 (presumably the Stoic philosopher of Tarsus: see n. 113). There are also a number of instances where the theories of attraction and motion (ὁλκή and φορά) are mentioned, often in combination with that of emanation, but this is not necessarily an allusion to the Stoic theory of natural sympathy/antipathy in a strict sense, bur rather to a more general mechanical theory of natural movements. Cf. *Q.N.* 7, 914A, 19, 916D, 24, 918A, 26, 918C, 41. Notably, Plato rejected the theory of ὁλκή for the working of several natural phenomena (including magnetism) in *Tim.* 80c (cf. *Quaest. Plat.* 1005BD and *Q.N.* 19, 916D).

[47] For a list of problems pertaining to sympathy and antipathy in *Quaestiones*

instance, the antipathy adepts are described as 'babblers' (θρυλοῦντες), and in *Quaest. conv.* 664CD such antipathies are considered mere 'chatter' – significantly, this is meant as an emphatic invitation to search for a theory that will *explain* such phenomena (ἀδολεσχῶ παρακαλῶν ὑμᾶς ἐπὶ τὴν ζήτησιν τῆς αἰτίας)[48]. Arguably, then, Plutarch in his natural problems aims to take an anti-Stoic position by further investigating and providing detailed physical aetiologies of the natural *mirabilia* at hand, that is, by suggesting more plausible natural explanations for them (than simply assuming the workings of natural sympathy and antipathy), without moreover forsaking a higher type of causality[49].

We can safely conclude that Plutarch's positive attitude towards natural *mirabilia* ties in closely with his philosophical-religious convictions about the natural world and its providential ordering. When Plutarch carefully attaches credence to natural *mirabilia* – giving them the benefit of the doubt –, he does this with necessary epistemic caution (εὐλάβεια) and with an underlying philosophical-religious motive in the back of his mind (εὐσέβεια), according to which divine providence is active and permanent in the natural world. I will further specify this in the next section in light of Plutarch's dualistic view on causality, which lies at the very basis of his Platonic world view.

---

conviales and in other works by Plutarch, see M. Wellmann, 1928, pp. 25–26 and T. Weidlich, 1894, pp. 53–58.

[48] I take this to imply that according to the Stoic approach of *mirabilia* the natural causality of these phenomena should be accepted without further aetiological specification (presumably they were seen as the factual consequences of universal faith, understood as a predetermined sequence of natural causes). Regarding the popular belief that thunder produces truffles, for instance (as discussed in *Quaest. conv.* 664A–665A [see 3.1.4.]), A. Setaioli, 2009, pp. 442–443 draws a link with Stoic divination theory and argues that "the area in which the Stoics tried hardest to reconcile popular traditions with their own philosophy was of course divination. This form of prediction of the future was theoretically founded on the doctrine of συμπάθεια, the mutual connection and reciprocal influence of all natural phenomena, stemming from the basic ideas of πρόνοια ("providence") and εἱμαρμένη ("fate", conceived as an uninterrupted chain of causes), but the need to save the pre-philosophical folkloric traditions connected with divination forced the Stoics to assume a link between the facts traditionally considered as signs and the ensuing phenomena considered to be announced by them – *which restricted them to an empirical observation admitting of no experimental test or rational ascertainment of causal sequences*" (my Italics). On the role of divination in Plutarch's own philosophical thinking, see J. Opsomer, 1996b.

[49] For Stoic aversion to Aristotelian-style aetiology (more precisely that of Posidonius), cf. Strabo, *Geogr.* 2, 3, 8 [quoted 1.1.6., n. 167]. For Plutarch's general anti-Stoic attitude, see D. Babut, 1969, pp. 22–69. See also J. Opsomer, 2014, pp. 92–93.

4.1.2. Plutarch's dualistic causality: rationalising the divine and the use of myth and poetry

Plutarch's outlook on the world prescribes that natural phenomena are not fathomable in terms of physical causes alone, since they also allow for an alternative interpretation based on their deeper and underlying significance. Thus, Plutarch firmly believes that the order of nature has a divine basis[50]. In light of this understanding, the section at hand aims to discuss Plutarch's dualistic view on causality and how this was influenced by Plato's philosophy[51].

1. Plato's scientific revolution

For Plutarch, as a Platonist, the world of contingent natural phenomena is an image of a higher, intelligible model. He believes that the natural world of becoming mirrors the divine reality and provides a faint reflection of it in a material form[52]. By consequence, Plutarch's world system is constituted in such a way that there is no rigid border between physical and meta-physical causes[53]. From numerous passages throughout the *corpus Plutarcheum*, we learn, rather, that the reverse is true.

The *locus classicus* is *Per.* 6, where Plutarch describes Pericles' association with Anaxagoras and the man's natural science (φυσικὸς λόγος). As we saw earlier on, this acquaintance provided Pericles with the advantage of rising above superstition (δεισιδαιμονία) and paved the way for genuine devotion to the gods (εὐσέβεια)[54] [quoted 3.2.2.]. In what follows, Plutarch does not specify how this occurred precisely but tells a legendary story instead about a one-horned ram whose head was brought to Pericles from his farm. Plutarch's attempt to reconcile the diametrically opposed interpretations of this marvellous natural phenomenon by the

---

[50] This natural order is the work of God. Cf., e.g., *De def. or.* 430E: οὐ γὰρ ὁ θεὸς διέστησεν οὐδὲ διώκισε τὴν οὐσίαν, ἀλλ' ὑπ' αὐτῆς διεστῶσαν αὐτὴν καὶ φερομένην χωρὶς ἐν ἀκοσμίαις τοσαύταις παραλαβὼν ἔταξε καὶ συνήρμοσε δι' ἀναλογίας καὶ μεσότητος· κτλ.

[51] For a seminal study of Plutarch's view on causality, see P. Donini, 1992. See also, e.g., F. Ferrari, 1995, p. 79 and J. Opsomer, 1998, pp. 181–184.

[52] Cf. *De Is. et Os.* 372F: εἰκὼν γάρ ἐστιν οὐσίας ἐν ὕλῃ γένεσις καὶ μίμημα τοῦ ὄντος τὸ γινόμενον. This conception ties in closely with Plutarch's Platonic cosmology. For Plutarch's view of the physical world as an image of the divine, see R. Hirsch-Luipold, 2002, esp. pp. 174–222 (and pp. 284–285 for a general synopsis).

[53] G.E.R. Lloyd, 1979, pp. 32, 51, 57 would speak of "double determination", where it is believed that natural phenomena are brought about both by the gods and by natural causes.

[54] Cf. also, e.g., *De sup.* 169EF: Ὅθεν ἔμοιγε καὶ θαυμάζειν ἔπεισι τοὺς τὴν ἀθεότητα φάσκοντας ἀσέβειαν εἶναι, μὴ φάσκοντας δὲ τὴν δεισιδαιμονίαν. καίτοι γ' Ἀναξαγόρας δίκην ἔφυγεν ἀσεβείας ἐπὶ τῷ λίθον εἰπεῖν τὸν ἥλιον, Κιμμερίους δ' οὐδεὶς εἶπεν ἀσεβεῖς ὅτι τὸν ἥλιον οὐδ' εἶναι τὸ παράπαν νομίζουσι. See F. Brenk, 1977, p. 39 and A. Pérez Jiménez, 1996.

seer Lampon, on one side, and the natural philosopher Anaxagoras, on the other, is revealing for his own understanding of causality in general (*Per.* 6, 2–4):

λέγεται δέ ποτε κριοῦ μονόκερω κεφαλὴν ἐξ ἀγροῦ τῷ Περικλεῖ κομισθῆναι, καὶ Λάμπωνα μὲν τὸν μάντιν, ὡς εἶδε τὸ κέρας ἰσχυρὸν καὶ στερεὸν ἐκ μέσου τοῦ μετώπου πεφυκός, εἰπεῖν ὅτι δυεῖν οὐσῶν ἐν τῇ πόλει δυναστειῶν, τῆς Θουκυδίδου καὶ Περικλέους, εἰς ἕνα περιστήσεται τὸ κράτος παρ' ᾧ γένοιτο τὸ σημεῖον· τὸν δ' Ἀναξαγόραν τοῦ κρανίου διακοπέντος ἐπιδεῖξαι τὸν ἐγκέφαλον οὐ πεπληρωκότα τὴν βάσιν, ἀλλ' ὀξὺν ὥσπερ ᾠὸν ἐκ τοῦ παντὸς ἀγγείου συνωλισθηκότα κατὰ τὸν τόπον ἐκεῖνον ὅθεν ἡ ῥίζα τοῦ κέρατος εἶχε τὴν ἀρχήν. καὶ τότε μὲν θαυμασθῆναι τὸν Ἀναξαγόραν ὑπὸ τῶν παρόντων, ὀλίγῳ δ' ὕστερον τὸν Λάμπωνα, τοῦ μὲν Θουκυδίδου καταλυθέντος, τῶν δὲ τοῦ δήμου πραγμάτων ὁμαλῶς ἁπάντων ὑπὸ τῷ Περικλεῖ γενομένων.
ἐκώλυε δ' οὐδέν, οἶμαι, καὶ τὸν φυσικὸν ἐπιτυγχάνειν καὶ τὸν μάντιν, τοῦ μὲν τὴν αἰτίαν, τοῦ δὲ τὸ τέλος καλῶς ἐκλαμβάνοντος· ὑπέκειτο γὰρ τῷ μέν, ἐκ τίνων γέγονε καὶ πῶς πέφυκε θεωρῆσαι, τῷ δέ, πρὸς τί γέγονε καὶ τί σημαίνει προειπεῖν. οἱ δὲ τῆς αἰτίας τὴν εὕρεσιν ἀναίρεσιν εἶναι τοῦ σημείου λέγοντες οὐκ ἐπινοοῦσιν ἅμα τοῖς θείοις καὶ τὰ τεχνητὰ τῶν συμβόλων ἀθετοῦντες, ψόφους τε δίσκων καὶ φῶτα πυρσῶν καὶ γνωμόνων ἀποσκιασμούς· ὧν ἕκαστον αἰτίᾳ τινὶ καὶ κατασκευῇ σημεῖον εἶναί τινος πεποίηται. ταῦτα μὲν οὖν ἴσως ἑτέρας ἐστὶ πραγματείας.

A story is told that once upon a time the head of a one-horned ram was brought to Pericles from his country-place, and that Lampon the seer, when he saw how the horn grew strong and solid from the middle of the forehead, declared that, whereas there were two powerful parties in the city, that of Thucydides and that of Pericles, the mastery would finally devolve upon one man, – the man to whom this sign had been given. Anaxagoras, however, had the skull cut in two, and showed that the brain had not filled out its position, but had drawn together to a point, like an egg, at that particular spot in the entire cavity where the root of the horn began. At that time, the story says, it was Anaxagoras who won the plaudits of the bystanders; but a little while after it was Lampon, for Thucydides was overthrown, and Pericles was entrusted with the entire control of all the interests of the people.
Now there was nothing, in my opinion, to prevent both of them, the naturalist and the seer, from being in the right of the matter; the one correctly divined the cause, the other the object or purpose. It was the proper province of the one to observe why anything happens, and how it comes to be what it is; of the other to declare for what purpose anything happens, and what it means. And those who declare that the discovery of the cause, in any phenomenon, does away with the meaning, do not perceive that they are doing away not only with divine portents, but

also with artificial tokens, such as the ringing of gongs, the language of fire-signals, and the shadows of the pointers on sundials. Each of these has been made, through some causal adaptation, to have some meaning. However, perhaps this is a matter for a different treatise.

Stadter rightly argues that: "By introducing examples of signals made by men, P. implies by analogy that the gods make natural phenomena function as signs also"[55]. As seen in the previous section, a natural phenomenon is some kind of a σημεῖον for Plutarch, albeit not only of the manifestation of physical causes, but also of a deeper meaning[56] (see n. 33). The distinction in the explanation between these two aspects, viz. between the physical αἰτίαι (ἐκ τίνων γέγονε καὶ πῶς πέφυκε) and the higher τέλος (πρὸς τί γέγονε καὶ τί σημαίνει), is not absolute for Plutarch but of a complementary kind, because they form a tight explanatory unity. An adequate scientific explanation of natural phenomena should, therefore, take into account both of these aspects.

Plutarch's dualistic causality ties in closely with his Platonic philosophy, as can be deduced, for instance, from *Nic.* 23, 2–4, where Plutarch reports on Nicias' superstitious reaction when witnessing a lunar eclipse. In this passage, Plutarch outlines some kind of a micro-development in the history of ancient Greek science, where Anaxagorean physics became subjected (ὑπέταξε) to divine and meta-physical principles in Plato's cosmology[57]:

ὁ γὰρ πρῶτος σαφέστατόν τε πάντων καὶ θαρραλεώτατον περὶ σελήνης καταυγασμῶν καὶ σκιᾶς λόγον εἰς γραφὴν καταθέμενος Ἀναξαγόρας οὔτ' αὐτὸς ἦν παλαιὸς οὔτε ὁ λόγος ἔνδοξος, ἀλλ' ἀπόρρητος ἔτι καὶ δι' ὀλίγων καὶ μετ' εὐλαβείας τινὸς ἢ πίστεως βαδίζων. οὐ γὰρ ἠνείχοντο τοὺς φυσικοὺς καὶ μετεωρολέσχας τότε καλουμένους, ὡς εἰς αἰτίας ἀλόγους καὶ δυνάμεις ἀπρονοήτους καὶ κατηναγκασμένα πάθη διατρίβοντας τὸ θεῖον, ἀλλὰ καὶ Πρωταγόρας ἔφυγε,

---

[55] P.A. Stadter, 1989, p. 87. He interprets the concluding phrase (ταῦτα μὲν οὖν ἴσως ἑτέρας ἐστὶ πραγματείας) as a reference to *De defectu oraculorum* (see below). For the influence of Anaxagoras' philosophy on Pericles' statesmanship, see *id.*, 1991, pp. 120–122. For further commentary on this passage, see also J.P. Hershbell, 1982, pp. 141–142.

[56] For the idea that natural phenomena contain a deeper, divine significance, cf. also, e.g., *Sept. sap. conv.* 149CE (on the monster brought to Periander) and *Cor.* 38, 2 (regarding talking statues; quoted above).

[57] Plutarch also considers Plato's philosophy to be a turning point in other, more exact scientific disciplines. It initiated a theoretical distinction (διεκρίθη) between abstract geometry and technical mechanics, as practiced by Archimedes, Eudoxus, Archytas and Menaechmus (cf. *Marc.* 14; 17 and *Quaest. conv.* 718EF, with G.E.R. Lloyd, 1973, pp. 93–95, A. Georgiadou, 1992, P. Culham, 1992, J. Sirinelli, 2000, pp. 356–357). In addition, for Plutarch's idea of historical progress in the field of astronomy, cf. *Arist.* 19, 7 (with J. Boulogne, 2008, p. 741).

καὶ Ἀναξαγόραν εἰρχθέντα μόλις περιεποιήσατο Περικλῆς, καὶ Σωκράτης, οὐδὲν αὐτῷ τῶν γε τοιούτων προσῆκον, ὅμως ἀπώλετο διὰ φιλοσοφίαν. ὀψὲ δ᾽ ἡ Πλάτωνος ἐκλάμψασα δόξα διὰ τὸν βίον τοῦ ἀνδρός, καὶ ὅτι ταῖς θείαις καὶ κυριωτέραις ἀρχαῖς ὑπέταξε τὰς φυσικὰς ἀνάγκας, ἀφεῖλε τὴν τῶν λόγων τούτων διαβολήν, καὶ τοῖς μαθήμασιν εἰς ἅπαντας ὁδὸν ἐνέδωκεν.

The first man to put in writing the clearest and boldest of all doctrines about the changing phases of the moon was Anaxagoras. But he was no ancient authority, nor was his doctrine in high repute. It was still under seal of secrecy, and made its way slowly among a few only, who received it with a certain caution rather than with implicit confidence. Men could not abide the natural philosophers and "visionaries," as they were then called, for the reason that they reduced the divine agency down to irrational causes, blind forces, and necessary incidents. Even Protagoras had to go into exile, Anaxagoras was with difficulty rescued from imprisonment by Pericles, and Socrates, though he had nothing whatever to do with such matters, nevertheless lost his life because of philosophy. It was not until later times that the radiant repute of Plato, because of the life the man led, and because he subjected the compulsions of the physical world to divine and more sovereign principles, took away the obloquy of such doctrines as these, and gave their science free course among all men.

The same notion of a scientific revolution *avant la lettre* that arose from Plato's criticism of Anaxagorean natural science is found in De def. or. 435F–436A, where Plutarch describes the final and the efficient causes (τὸ οὗ ἕνεκα καὶ ὑφ᾽ οὗ) as *better* causes (βελτίονας) than the physical causes and the law of necessity (φυσικαὶ αἰτίαι καὶ τὸ κατ᾽ ἀνάγκην), although he still speaks of them as being closely interrelated:

ἀπολογήσομαι δὲ μάρτυρα καὶ σύνδικον ὁμοῦ Πλάτωνα παραστησάμενος. ἐκεῖνος γὰρ ἀνὴρ Ἀναξαγόραν μὲν ἐμέμψατο τὸν παλαιόν, ὅτι ταῖς φυσικαῖς ἄγαν ἐνδεδεμένος αἰτίαις καὶ τὸ κατ᾽ ἀνάγκην τοῖς τῶν σωμάτων ἀποτελούμενον πάθεσι μετιὼν ἀεὶ καὶ διώκων, τὸ οὗ ἕνεκα καὶ ὑφ᾽ οὗ, βελτίονας αἰτίας οὔσας καὶ ἀρχάς, ἀφῆκεν αὐτὸς δὲ πρῶτος ἢ μάλιστα τῶν φιλοσόφων ἀμφοτέρας ἐπεξῆλθε, τῷ μὲν θεῷ τὴν ἀρχὴν ἀποδιδοὺς τῶν κατὰ λόγον ἐχόντων, οὐκ ἀποστερῶν δὲ τὴν ὕλην τῶν ἀναγκαίων πρὸς τὸ γιγνόμενον αἰτίων, ἀλλὰ συνορῶν, ὅτι τῇδέ πῃ καὶ τὸ πᾶν αἰσθητὸν διακεκοσμημένον οὐ καθαρὸν οὐδ᾽ ἀμιγές ἐστιν, ἀλλὰ τῆς ὕλης συμπλεκομένης τῷ λόγῳ λαμβάνει τὴν γένεσιν.

I shall defend myself by citing Plato as my witness and advocate in one. That philosopher found fault with Anaxagoras, the one of early times, because he was too much wrapped up in the physical causes and was always following up and pursuing the law of necessity as it was worked out in the behaviour of bodies, and left out of account the purpose and the agent, which are better causes and origins. Plato himself was the first of the philosophers, or the one most prominently engaged in prosecuting investigations of both sorts, to assign to God, on the one hand, the origin of all things that are in keeping with reason, and on the other hand, not to divest matter of the causes necessary for whatever comes into being, but to realise that the perceptible universe, even when arranged in some such orderly way as this, is not pure and unalloyed, but that it takes its origin from matter when matter comes into conjunction with reason.

The reference is probably to Pl., *Phd.* 97b–99d, where Socrates vents his disappointment about Anaxagoras' almost exclusive focus on physical causes. At first, Anaxagoras' mention of the all-embracing νοῦς seemed promising to Socrates, because he assumed that it would arrange everything 'in such a way as it is best for it to be' (ταύτῃ ὅπῃ ἂν βέλτιστα ἔχῃ). Eventually, though, Anaxagoras' theory did not meet up to Socrates' initial expectations (I have discussed this passage in further detail in the prologue). For a clearer account of causality in Plato we must turn to the *Timaeus*, where a basic distinction is drawn between necessary and divine causes (68e). Plato here describes natural causes as 'contributory causes' (46ce, 76d: συναιτίαι) and natural necessity as the 'wandering cause' (48a: πλανωμένη αἰτία): these are auxiliary or secondary to the primary cause[58], which Plato varyingly calls 'the best cause' (29a, cf. *Quaest. conv.* 720B), 'the maker and father of the universe' (28c), 'demiurge' (29a), 'mind' (47e), 'God' (30a, 53b) etc.

Accordingly, a little bit further in *De def. or.* 436DE, Plutarch returns to the issue of causality. Even though his preference clearly goes to the primary cause (τὸ δι' οὗ καὶ ὑφ' οὗ), he still approves of a synthesis thereof with natural causes (τὸ ἐξ ὧν καὶ δι' ὧν)[59]:

---

[58] Cf. also *Tim.* 29e, 68e. For the distinction between primary and secondary causes in Plato's *Timaeus*, see, e.g., W. Scheffel, 1976, pp. 118–139. For the influence of this dichotomy on Plutarch's philosopy, see, e.g., P. Donini, 1984, p. 374, 1992a and J. Opsomer, 1998, p. 183.

[59] In an attempt to bring Plutarch's terminology in this passage more in line with the traditional (Aristotelian) scheme of causes, P. Donini, 1992a, p. 101 proposed to emend δι' οὗ (= instrumental cause) in δι' ὅ (= final cause). This is clever, because it is only reasonable that the φυσικοί, rather than the θεολόγοι, are more likely to neglect the final cause (τὸ δι' ὅ), but, then again, all the manuscripts read δι' οὗ (cf. also *Quaest. conv.* 698B: τὸ οὗ ἕνεκα […] καὶ πρὸς ἥν). I believe Donini is right in asserting that Plutarch allows "una certa libertà" in his formulation of the different causes. Therefore, it is not unlikely that he is *generally* alluding to the traditional scheme of causes without having the intention of becoming too technical or precise. It remains to be seen, then, whether τὸ δι' οὗ is really simply equal to τὸ δι' ὧν for Plutarch in this passage (τὸ δι' + gen. of the relative pronoun being the

Καθόλου γάρ, ὥς φημι, δύο πάσης γενέσεως αἰτίας ἐχούσης οἱ μὲν σφόδρα παλαιοὶ θεολόγοι καὶ ποιηταὶ τῇ κρείττονι μόνῃ τὸν νοῦν προσεῖχον τοῦτο δὴ τὸ κοινὸν ἐπιφθεγγόμενοι πᾶσι πράγμασι 'Ζεὺς ἀρχὴ Ζεὺς μέσσα, Διὸς δ' ἐκ πάντα πέλονται·' ταῖς δ' ἀναγκαίαις καὶ φυσικαῖς οὐκέτι προσῄεσαν αἰτίαις. οἱ δὲ νεώτεροι τούτων καὶ φυσικοὶ προσαγορευόμενοι τοὐναντίον ἐκείνοις τῆς καλῆς καὶ θείας ἀποπλανηθέντες ἀρχῆς ἐν σώμασι καὶ πάθεσι σωμάτων πληγαῖς τε καὶ μεταβολαῖς καὶ κράσεσι τίθενται τὸ σύμπαν. ὅθεν ἀμφοτέροις ὁ λόγος ἐνδεὴς τοῦ προσήκοντός ἐστι, τοῖς μὲν τὸ δι' οὗ καὶ ὑφ' οὗ τοῖς δὲ τὸ ἐξ ὧν καὶ δι' ὧν ἀγνοοῦσιν ἢ παραλείπουσιν. ὁ δὲ πρῶτος ἐκφανῶς ἁψάμενος ἀμφοῖν καὶ τῷ κατὰ λόγον ποιοῦντι καὶ κινοῦντι προσλαβὼν τὸ ἀναγκαίως ὑποκείμενον καὶ πάσχον ἀπολύεται καὶ ὑπὲρ ἡμῶν πᾶσαν ὑποψίαν καὶ διαβολήν.

To sum up, then: while every form of creation has, as I say, two causes, the very earliest theological writers and poets (Orph. fr. 168) chose to heed only the superior one, uttering over all things that come to pass this common generality: "Zeus the beginning, Zeus in the midst, and from Zeus comes all being"; but as yet they made no approach towards the compelling and natural causes. On the other hand the younger generation which followed them, and are called physicists or natural philosophers, reverse the procedure of the older school in their aberration from the beautiful and divine origin, and ascribe everything to bodies and their behaviour, to clashes, transmutations, and combinations. Hence the reasoning of both parties is deficient in what is essential to it, since the one ignores or omits the intermediate and the agent, the other the source and the means. He who was the first to comprehend clearly both these points and to take, as a necessary adjunct to the agent that creates and actuates, the underlying matter, which is acted upon, clears us also of all suspicion of wilful misstatement.

The first person to have done this, that is, to endorse both aetiological approaches towards natural phenomena, is obviously Plato. As such, Plato's reconciliation of both types of causality is depicted as a real turning point in the history of ancient Greek philosophy.

---

instrumental cause). In my opinion, the most important aspect of Plutarch's formulation of the different types of causes is the shift in number in the relative pronouns οὗ – ὧν: viz. τὸ δι' οὗ καὶ ὑφ' οὗ (= the primary cause) vs. τὸ ἐξ ὧν καὶ δι' ὧν (= the physical causes). This can be taken to imply, then, that whereas nature is subject to a plurality of lower causes (cf. De def. or. 435F–436A; quoted above), the higher cause is essentially a singularity, viz. God (or synonymous concepts, which basically imply one and the same divine entity or principle). Whatever may be the case, Plutarch's main point is clear and comprehensible with or without Donini's emendation.

Plutarch is clearly self-fashioning himself in these passages as a faithful son of the Platonic revolution. For him, as for Plato, natural phenomena are grounded in divine principles. Therefore, one's devotedness to the gods does not necessarily result in bad science. At the same time, one's attempt to save the phenomena by means of natural explanations is not an act of impiety, since natural causality does not provide an alternative to religion[60]. This raises questions about how to precisely trace these two approaches – viz. the physical and the theological – in Plutarch's own natural scientific works[61]. To this end, in what follows, I will focus on how Plutarch embeds the voices of the παλαιοὶ θεολόγοι καὶ ποιηταί (as mentioned in *De def. or.* 436DE) in his scientific works. Afterwards, I will turn to the category of οἱ νεώτεροι καὶ φυσικοί, examining Plutarch's discursive method of incorporating the authority of authors of scientific prose into his natural problems [see 4.2.1.1.]. In the next two sections, I will start by analysing the categories of the θεολόγοι and the ποιηταί successively.

2. *Science, religion and mythology*

As seen in the previous section, the demarcation line between what belongs to the realms of physics and religion is not rigid for Plutarch, but osmotic. There is, in fact, a clear preeminence of theology over physics in Plutarch's natural philosophy[62], since his God is the cornerstone and culmination point of his world system[63]. As we will see in the section at hand, this

---

[60] Cf., e.g., *De cur.* 517D: καὶ ταῦτ' ἀπόρρητ' ἐστὶ φύσεως, ἀλλ' οὐκ ἄχθεται τοῖς ἐλέγχουσιν [quoted 3.2.2.]. Cf. also, e.g., *De facie* 923A, where Lucius' sarcasm in his reference to Cleanthes' past accusation of Aristarchus' attempt to save the phenomena as an act of impiety is very clear (ὡς κινοῦντα τοῦ κόσμου τὴν ἑστίαν, ὅτι τὰ φαινόμενα σῴζειν ἀνὴρ ἐπειρᾶτο). See, e.g., J. Opsomer, 1998, p. 181: "For Plutarch there is no conflict between 'rationalism' and faith." *Pace* E. Teixeira, 1992, p. 214.

[61] I make this distinction – viz. myth/poetry vs. prose – mainly for practical reasons. Seeing that Plutarch makes this distinction himself (in *De def. or.* 436DE just quoted), it seems useful to do the same, if only for the sake of a transparent development of the analysis in what follows. This is not to claim, however, that these distinctions are entirely water-tight, let alone that for Plutarch the poets and mythographers have no authority in the field of natural science. Empedocles, for instance, belongs to both categories of ποιηταί and φυσικοί strictly speaking (see n. 92). Nevertheless, a fundamental difference between these categories is found in the fact that the poetical and mythographical accounts in Plutarch's natural problems not only have argumentative value in the aetiologies, but also add an extra layer to the technical-physical discourse (e.g., by providing a literary illustration of a specific argument [see 1.2.3.]).

[62] Cf. P. Donini, 1986a, p. 210: "ma la fisica non è la scienza suprema per Plutarco: c'è ovviamente un'altra e superiore dimensione della filosofia teoretica".

[63] See P. Donini, 1986a, p. 211 (cf. also *id.*, 1994a, p. 48, n. 32, 2011, p. 96) and J. Opsomer, 1998, p. 214.

theological preeminence recurs throughout Plutarch's natural scientific writings, even if it is not always strongly emphasised.

The Platonic endnote in *Aqua an ignis* 958E is particularly worth quoting in order to illustrate this, if only because it is often neglected (presumably for reasons of the work's controversial authorship [see I.I.I., n. 8]). The passage deals with the function of water and fire in sensory perception and argues for the pre-eminence of sight over the other senses[64]:

> Καὶ μήν, οὗ πλεῖστον ἡ κρατίστη τῶν αἰσθήσεων μετείληφεν, οὐκ ἂν εἴη λυσιτελέστατον; οὐχ ὁρᾷς οὖν, ὡς τῇ μὲν ὑγρᾷ φύσει οὐδεμία τῶν αἰσθήσεων καθ' αὑτὴν προσχρῆται χωρὶς πνεύματος ἢ πυρὸς ἐγκεκραμένου, τοῦ δὲ πυρὸς ἅπασα μὲν αἴσθησις, οἷον τὸ ζωτικὸν ἐνεργαζομένου, μετείληφεν, ἐξαιρέτως δ' ἡ ὄψις, ἥτις ὀξυτάτη τῶν διὰ σώματός ἐστιν αἰσθήσεων, πυρὸς ἔξαμμα οὖσα; καὶ ὅτι θεῶν πίστιν παρέσχηκεν· ἔτι τε, ἣ Πλάτων φησί, δυνάμεθα κατασχηματίζειν πρὸς τὰς τῶν ἐν οὐρανῷ κινήσεις τὴν ψυχὴν διὰ τῆς ὄψεως.

> And, to be sure, will not that (sc. substance, i.e. either water or fire) be the most advantageous of which each of the senses has the greatest proportion? Do you not perceive, then, that there is no one of the senses which uses moisture by itself without an admixture of air or fire; and that every sense partakes of fire inasmuch as it supplies the vital energy; and especially that sight, the keenest of the physical senses, is an ignited mass of fire and is that which has made us believe in the gods? And further, through sight, as Plato says (*Tim.* 47ab), we are able to conform our souls to the movements of the celestial bodies.

In this passage, Plutarch portrays human beings as spectators of a divinely governed universe[65]. According to Plato, quoted here (*Tim.* 47ab), the divine gift of sight has led us to inquire into astronomical phenomena, which in turn has empowered us to acquire philosophical knowledge, an idea that lies at the very basis of Platonic epoptics. From a parallel account in *De sera num.* 550DE, we learn that the contemplation of the heavenly motions and cosmic ordering is a useful means of doing away

---

[64] On the pre-eminence of sight over the other senses, cf. also, e.g., *Quaest. conv.* 654D, Pl., *Phdr.* 250d, Arist., *Met.* 980a26–27, Cic., *De or.* 2, 86–87, SVF 2, pp. 232–233, frs. 863 and 866.

[65] Cf., e.g., *Quaest. conv.* 718D (in the context of geometry enabling us to *see* the intelligible realm): γεωμετρίαν ὡς ἀποσπῶσαν ἡμᾶς προσισχομένους τῇ αἰσθήσει καὶ ἀποστρέφουσαν ἐπὶ τὴν νοητὴν καὶ ἀίδιον φύσιν, ἧς θέα τέλος ἐστὶ φιλοσοφίας οἷον ἐποπτεία τελετῆς;. Cf. E.L. Minar, F.H. Sandach and W.C. Helmbold, 1961, p. 120: "the philosopher passes, with the help of geometry, from study of physical objects to the vision of the ideas."

with errant passions and allows us to assimilate our soul to God's beauty and goodness. What we should try to attain, then, from contemplating the ordered universe is 'likeness to God' (ὁμοίωσις θεῷ, cf. *De Is. et Os.* 351CD). This is the eventual τέλος for Plutarch as a middle-Platonic philosopher[66]. And it is probably to this idea that Plutarch is alluding in the above passage when he writes that sight enabled us to believe in the gods (θεῶν πίστιν παρέσχηκεν).

The idea of a divine ordering of the cosmos, as perceived in the cosmic balance of opposite forces at work in nature, is most clearly formulated in *De prim. frig.* 946EF, where Plutarch explains why God is called a cosmic harmoniser and musician:

ἡ μὲν γὰρ κατὰ στέρησιν καὶ ἕξιν ἀντίθεσις πολεμικὴ καὶ ἀσύμβατός ἐστιν, οὐσίαν θατέρου τὴν θατέρου φθορὰν ἔχοντος· τῇ δὲ κατὰ τὰς ἐναντίας δυνάμεις καιροῦ τυχούσῃ πολλὰ μὲν αἱ τέχναι χρῶνται, πλεῖστα δ᾽ ἡ φύσις ἔν τε ταῖς ἄλλαις γενέσεσι καὶ ταῖς περὶ τὸν ἀέρα τροπαῖς, καὶ ὅσα διακοσμῶν καὶ βραβεύων ὁ θεὸς ἁρμονικὸς καλεῖται καὶ μουσικός, οὐ βαρύτητας συναρμόττων καὶ ὀξύτητας οὐδὲ λευκὰ καὶ μέλανα συμφώνως ὁμιλοῦντα παρέχων ἀλλήλοις, ἀλλὰ τὴν τῆς θερμότητος καὶ ψυχρότητος ἐν κόσμῳ κοινωνίαν καὶ διαφοράν, ὅπως συνοίσονταί τε μετρίως καὶ διοίσονται πάλιν, ἐπιτροπεύων καὶ τὸ ἄγαν ἑκατέρας ἀφαιρῶν εἰς τὸ δέον ἀμφοτέρας καθίστησι.

For the opposition of a negation to a positive quality is an irreconcilable hostility, since the existence of the one is the annihilation of the other. The other opposition, however, of positive forces, if it occurs in due measure, is often operative in the arts, and very often indeed in various phenomena of nature, especially in connexion with the weather and the seasons and those matters from which the God derives his title of harmoniser and musician, because he organises and regulates them. He does not receive these names merely for bringing sounds of high and low pitch, or black and white colours, into harmonious fellowship, but because he has authority over the association and disunion of heat

---

[66] See J. Dillon, 2014, p. 62. This τέλος is connected with the aspect of εὐθυμία in *De tranq. an.* 477CD, where the sense of sight is again very central, viz. in the word θεατής (passage discussed earlier on [see 3.2.2.]). P. Donini, 1988, p. 132 argues that everything in Plutarch's world "is at the god's service and designed for his worship" (cf. also M. Battegazzore, 1992, p. 25). However, as J. Opsomer has objected (in personal correspondence), it seems useful to make an Aristotelian distinction here between two forms of teleology, viz. between that which is strived after, and the instance to whose benefit this goal is strived after (*DA* 415b2–3: τὸ δ᾽ οὗ ἕνεκα διττόν, τὸ μὲν οὗ, τὸ δὲ ᾧ). Plutarch emphasises that providence is executed in the interest of human beings and the world. Cf. J. Opsomer, 2014, p. 91: "the gods exercise providence for our benefit (and not just *to* our benefit)".

and cold in the universe, to see that they observe due measure in their combination and separation, and because, by eliminating the excess of either, he brings both into proper order.

The idea that there is a divine agent and ruler who procures cosmic order is made somewhat in passing here. Yet, its mere presence clearly indicates that *De primo frigido*, in dealing with a specific natural problem (viz. 'Which, if any, is the active principle or substance of cold?': cf. 945F), not only relies on physical causality but is ultimately based on theological principles[67].

The idea that natural phenomena have a divine motivation to which their physical causality is ultimately subordinated is not very emphatically argued for in Plutarch's natural problems collected in *Quaestiones naturales* and *Quaestiones convivales*. According to Van der Stockt, there are in fact no divine causes in *Quaestiones naturales*: "This is strange because physics should end in *philosophia prima*, supreme philosophy"[68]. In several discussions in *Quaestiones convivales*, by contrast, we saw earlier on that there are several allusions to the divine aspect of natural phenomena (such as the generative properties of salt or lightning in *Quaest. conv.* 665A, 684E–685F). One may presume that the sympotic decorum would probably not allow more overt philosophical ruminations, which explains why these divine aspects are only seldom elaborated upon (cf. *Quaest. conv.* 614A [quoted 3.1.4.]).

An important passage where Plutarch alludes to the higher causal motivation of natural phenomena is found in *Quaest. conv.* 699B (concerning Plato's contested view that drink passes through the lungs). Here, Plutarch says that 'the ingenious organisation of nature's activities is beyond the range of words, and it is impossible to explain adequately the exact working of the agencies it employs – that is breath and heat' (ἡ γὰρ φύσις οὐκ ἐφικτὸν ἔχει τῷ λόγῳ τὸ περὶ τὰς ἐνεργείας εὐμήχανον, οὐδ' ἔστι τῶν ὀργάνων αὐτῆς τὴν ἀκρίβειαν οἷς χρῆται (λέγω δὲ τὸ πνεῦμα καὶ τὸ θερμόν) ἀξίως διελθεῖν). Scholars have taken this to imply that the exact way in which nature actually works cannot be captured in physical terms or by the use of any human discourse, considering the divine powers at work in it[69]. There seems to be a concrete Platonic background to this idea in *Tim.* 28c, where Plato writes that it is not as such impossible (ἀδύνατον) to personally discover

---

[67] Similar references to a divinely governed cosmology can be found, e.g., in *De def. or.* 430E–431A and *Aqua an ignis* 957B. For further study, see P. Donini, 1986a, pp. 207–208.

[68] L. Van der Stockt, 2011, pp. 453–454. For the general absence of higher causes also in the natural problems treated in *Quaestiones convivales*, cf. P. Donini, 2011, p. 96, n. 203.

[69] As P. Donini, 1986a, pp. 208–209 argues, "sono infatti implicite le operazioni della demiurgia che non sono però completamente esplicabili dal discorso umano".

(εὑρεῖν) God – even if this is a hard task –, but to *declare* (λέγειν) him unto all men[70]. Plutarch knows very well that divine, intelligible truth is, indeed, extremely hard, if not altogether impossible, to reach, considering the epistemological weakness and limits of our human understanding[71]. To capture his essence in words is simply impossible.

As the natural problems discussed in *Quaestiones convivales*, thus, show, a deeper philosophical-theological style of discourse is not absent in them, although it is only seldom foregrounded. It remains to be seen, then, whether similar implicit allusions to the category of the divine are also present in *Quaestiones naturales* despite the work's main attention for natural explanations [see 1.1.6.]. Indeed, in line with Ps.-Aristotle's *Problems*, Plutarch's *Quaestiones naturales* is primarily concerned with material-mechanical causes. But even so, I believe that there still are some hints in the work towards a higher type of causality[72]. I will try to corroborate this view by examining the passages, 1) where Plutarch hints at a providential ordering of nature, and 2) where he incorporates mythological material in his physical discourse.

1) First of all, there are a couple of instances in *Quaestiones naturales* where nature's providential ordering is central to the physical aetiologies. This is the case most notably for two passages, where Plutarch relies on meteorological lore. In *Q.N.* 2, 912CD, Plutarch argues that when frogs croak louder, this is a sign of impending rain (σημεῖον ὑετοῦ μέλλοντος). In *Q.N.* 18, 916AB, he explains that when the calamary leaps out of the sea or when the octopus hurries back to the shore and grasps small rocks, this forecasts a great storm (σημεῖόν ἐστι μεγάλου χειμῶνος [...] σημεῖόν πνεύματος ὅσον οὔπω παρόντος). Even though Plutarch does not explicitly say so, these meteorological signs imply that there is a providential ordering in nature which makes weather predictions possible. Indeed, elsewhere, Plutarch openly refers to the divine inspiration and the mantic powers of animals[73] (as well as of other things in nature), by which the gods transmit a message

---

[70] The same Platonic idea is formulated at several places in Plutarch's oeuvre. Cf., e.g., *De Is. et Os.* 381B (φωνῆς γὰρ ὁ θεῖος λόγος ἀπροσδεής ἐστι). Compare the concept of ἀπόρρητα (natural science as some kind of a 'mystery cult') in *De cur.* 517D [quoted 3.2.2.]. Cf. also, e.g., *Quaest. conv.* 728F: ὁ μὲν ἀληθὴς ἴσως λόγος καὶ νῦν ἀπόθετος καὶ ἀπόρρητος εἴη, τοῦ δὲ πιθανοῦ καὶ εἰκότος οὐ φθόνος ἀποπειρᾶσθαι. On "mystical silence" in Plutarch, see P. Van Nuffelen, 2007.

[71] Cf. also *De aud. poet.* 17DF. At times, Plutarch says that the human intellect cannot attain the divine truth (cf. *De Is. et Os.* 351C) but only a share of it (cf. *Ad princ. iner.* 781A).

[72] Notably, in Ps.-Aristotle's *Problems* there is an "occasional interest in teleological explanation", but this is very exceptional and not of the order of Plutarch's divine causes (see R. Mayhew, 2011a, p. xxiii, who notes only two cases). See also B.J. Stoyles, 2015.

[73] Cf., e.g., *De soll. an.* 975AC, 976C, *De Is. et Os.* 382AB, *Sept. sap. conv.* 163EF [see 1.2.4.].

to the world⁷⁴. Moreover, the account about frogs is illustrated at the end of *Q.N.* 2 with a quotation from Aratus, the poet of meteorological signs. As to Plutarch's interest in Aratus' *Phaenomena*, the conclusion of Negri is very significant:

> "[I]l Plutarco fedele sacerdote dell'Apollo delfico, cioè del dio che per eccellenza comunica con gli uomini attraverso 'segni', dovette sentirsi particolarmente in sintonia con il poeta dei 'segni' benevolmente offerti da Zeus agli uomini, le διοσημεῖαι appunto."⁷⁵

Notably, Plutarch composed an actual collection of Αἰτίαι τῶν Ἀράτου Διοσημιῶν of which some fragments remain (frs. 13–20 Sandbach; Lamprias catalogue nr. 119). Plutarch's aetiological-exegetical approach of Aratus' verses in this work – and also at the end of *Q.N.* 2 – seems to indicate that these verses require further physical explanation in order to reveal their full causal extent, Aratus adducing only the higher cause (c.q. divine providence).

2) Even if Plutarch's primary concern in *Quaestiones naturales* is with the physical causes of natural phenomena, the Olympic pantheon is not dispelled from the physical scene. There are seven references to mythological figures and gods in our collection (see the scheme below). The mention of their names may appear to be somewhat casual in some cases (*Q.N.* 2, 912A, 10, 914D, 21, 917B), while in other cases, we find a short allusion to a more extended myth (*Q.N.* 24, 918A, 36) or a short paraphrase of it (*Q.N.* 23, 917F, 38). As Hardie has argued in regards to the mythographical material Plutarch incorporates in *Quaestiones convivales*, "[t]he interpretation of myth is often handled as an exercise in solving

---

⁷⁴ Similarly, the two mythological accounts recorded at the very end of the aetiology in *Q.N.* 36 open up onto the broader context of divine punishment (c.q. by means of bee-stings). The first one is recorded by Theocritus and concerns the infidelity of Anchises and Aphrodite, the second by Pindar concerning Rhoecus' unfaithfulness (or insult?) to a nymph. Cf. n. 85 below. See also M. Meeusen, 2013c.

⁷⁵ M. Negri, 2004, p. 288. Plutarch quotes Aratus only here in *Q.N.* 2, 912D (Arat., *Phaen.* 946–947) and in *De soll. an.* 967F (Arat., *Phaen.* 956). In composing his Αἰτίαι τῶν Ἀράτου Διοσημιῶν, Plutarch may very well have consulted one of the many commentaries that existed on Aratus' work (cf. W.C. Helmbold, 1957, pp. 370–371, n. c and F.H. Sandbach, 1969, p. 89; for the overlap between his aetiological and exegetical approach, see [2.4.2., n. 149]). It is worth mentioning, in this regard, that the Stoics Boethus of Sidon and Posidonius also composed collections of αἰτίαι φυσικαί on the basis of Aratus' prognostics (cf. Gem. 17, 48 and Cic., *De div.* 1, 8, 13; 2, 21, 47). Whether the Ἄρατος, a work listed in the Lamprias catalogue (nr. 40 vs. nr. 24: Ἄρατος καὶ Ἀρταξέρξης), is a biography of the Greek didactic poet or of the Greek statesman is unclear, but the second seems more likely (*pace* K. Ziegler, 1951, col. 698).

problems."[76] Myths often directly contribute to the arguments at hand and, as such, demonstrate the symposiast's argumentative ingenuity in activating his scholarly πολυμάθεια in an original way. However, as I will try to show, there may be more to these mythological accounts, in that they may hint at a higher level of causality.

Clearly, such mythographical accounts in Plutarch's natural problems provide an important alternative for the purely physical discourse. If myths, indeed, hint at higher philosophical truth[77], it is not unlikely that those recorded in *Quaestiones naturales* contribute to the physical aetiologies in a meta-physical way. In fact, rather than reducing the myths to a purely physical explanation (after the manner of the allegorical interpretations of the Stoics), they more often confirm and corroborate the main argument at hand or even serve as an explanation in their own right. This means that Plutarch does not incorporate mythological tales in his natural problems so as to reduce the gods and their actions to physical principles. It seems that the exact opposite dynamic is at work here (though he does not, therefore, shun such allegorical interpretations altogether: see below).

At several points, Plutarch seems to incorporate mythological material so as to provide some kind of a 'mystifying' extension to the physical aetiology, in much the same way as this is the case, for instance, with the closing myth in *De facie*[78]. In this work, the astrophysical inquiry into the substance and nature of the moon, does not as such insist on a denial of the moon's divine and spirited nature. In fact, the opposite is true, since an

---

[76] P.R. Hardie, 1992, p. 4751. He notes, moreover, that "[s]ome of the physical allegorizations in the 'Qu. conv.' have the air of *ad hoc* improvisations, examples of the εὑρησιλογία which such discussions are designed to promote" (p. 4772).

[77] For Plutarch, myth is not, of course, an explicit record of the truth as such, but contains a deeper meaning. P.R. Hardie, 1992, p. 4754 notes that in many passages, Plutarch, indeed, "describes myth as a faint reflection of a transcendental truth. [...] Myths act as ladders to the truth, which may then be kicked away." See also pp. 4746–4749 for the relationship between myth and truth in Plutarch more generally. In *De Iside et Osiride*, for instance, Plutarch interprets the Isis myth in light of Plato's philosophy in the *Timaeus* (see R.M. Jones, 1980, p. 25). Cf. also F. Ferrari, 1995, p. 174: "si può dire che il mito fornisce il quadro filosofico-metafisico entro il quale contestualizzare le argomentazioni 'scientifiche'". For Plutarch's ambivalent attitude towards myths in general, see L. Van der Stockt, 1992a, pp. 88–97.

[78] The same conclusion was made for the scientific-mythological account about lunar eclipses in *De genio Socr.* 591C by R. Flacelière, 1951, esp. pp. 213–214: "Il me semble que Plutarque s'est complu à entretenir dans ses mythes une atmosphère de vague, de pénombre, de mystère, par différents moyens qui vaudraient la peine d'être étudiés [...]." For the relation between "science et mystique" in the light of Plutarch's eschatology, see also Y. Vernière, 1977, pp. 164–178. In fr. 156 Sandbach, Plutarch calls myths a 'mystic theology' (μυστηριώδης θεολογία).

eschatological, Platonic myth is appended to the treatise about the purpose of the moon in the universe, explaining its importance for the life-cycle of human souls (*De facie* 940F–945D). I see no reason why Plutarch's motivation for incorporating such Platonic type of eschatological myth should essentially differ from the more traditional type – in any case, Plutarch does not make a clear conceptual distinction between the two. Clearly, the annexation of such mythological material does not remain without consequences for the eventual outlook and scope of Plutarch's natural scientific discourse[79].

Notably, in Plutarch's natural problems, the mythographical accounts are never disparaged or even hinted at as being dispensible *faits divers*. What may perhaps arrest the reader's attention the most is the fact that these accounts tend to be located at the very end of the aetiology, which, as we saw earlier on [1.1.4., n. 104], is the most significant *locus* in Plutarch's aetiologies (*Q.N.* 23, 24, 36, 38 – at the end of *Q.N.* 23, Plutarch may have actually re-edited the text in order to obtain a mythological finale to the problem: see the commentary *ad loc.*). This can be taken to imply, then, that Plutarch incorporated these myths in order to hint at some kind of a *causa finalis*, in the sense of a higher, divine cause. In any case, numerous discussions in *Quaestiones convivales* also contain such a mythological ending[80], and the same is true for *De facie* as we just saw. Yet, as Donini argues in regards to the closing myth of *De facie*, "the explanations of the myth are not literally true: they are only an example and a suggestion of how matters could otherwise stand"; therefore, they must "simply be understood as a hint of another truth, different from physical truth"[81]. If Senzasono is right, then, that the myth about Kore and Pluto at the end of *Q.N.* 23 "non aveva

---

[79] H. Görgemanns, 1968, pp. 10–11 (cf. also *id.*, 1970, p. 85) rightly connects this in the case of *De facie* with Plutarch's human aim towards *Sinngebung*: "Für Plutarch ist das übergreifende Thema offenbar [...] die kosmische Theologie. Das "Wozu?" der so fremd und fern erschienenen Himmelshelft bewegt ihm, und er versucht in verschiedener Weise, wissenschaftlich und mythisch, darauf zu antworten. Er folgt damit eine menschliche Bedürfnis nach Sinngebung, das von der strengen Wissenschaft nicht befriedigt wird. [...] Mechanistisches Funktionieren wird ausgeschlossen; statt dessen wird eine sinvolle Ordnung durch einen göttlichen "Werkmeister" angenommen." Cf. also F. Ferrari, 1995, pp. 173–175. L. Van der Stockt, 2011, pp. 454–455 adds that "[i]f that kind of science was able, to quote Görgemanns, 'to make people feel at home in the cosmos', it remains to be seen if modern science is equally successful in that respect." Plutarch's science is, indeed, incommensurable with modern science in this regard. For a study of the philosophical myth of Thespesius in *De sera num.* 563B–568A, see F. Frazier, 2010.

[80] See *Quaest. conv.* 657E, 671BC, 679DE, 685EF, 714C, 716BC, 718AB, 720C, 739D, 743BC, 747A. On the telling of tales as an essential passtime at the symposium, see M.V. Ruffy, 2011, pp. 140–142.

[81] P. Donini, 1988, pp. 138–139. On the relation between myth and reason (μῦθος – λόγος), see also R. Hirsch-Luipold, 2014, p. 174.

per Plutarco lo stesso peso delle osservazioni naturalistiche"[82], it should be added that Plutarch does not explicitly prefer one mode of explanation to the other. This can be explained, then, by the fact that for him there is no rigid border between the natural and higher causes.

Hardie has also paid attention to the structural use of myth in Plutarch's writings: he notes that "Plutarch uses myth to highlight the structure of an essay or dialogue, especially at the beginning and the end"[83]. It remains to be seen, however, whether such mythological material is, as Hardie believes, "used essentially as ornament"[84]. If my hypothesis is correct, it is not unlikely that Plutarch incorporates mythological material into his physical aetiologies in order to hint, in a very concealed way, at the higher level of causality behind natural phenomena. After all, for Plutarch, the world we live in is the work of a divine cosmic ruler who organises nature in a providential way, so that natural causes do not exclude a higher, divine principle. In this sense, Zeus, is responsible for fixing the period during which she-wolves litter (*Q.N.* 38). The cosmic ruler also cares about a proper punishment for improper behaviour as can be implied by the myths recorded in *Q.N.* 36 (he punishes adulterers – and even the lower gods – for their *stuprum* by means of bee-stings)[85]. There are also divine precincts in nature (*Q.N.* 23, 917F: ἄσυλον, 'sanctuary'), where nature behaves in a somewhat unnatural (read: super-natural) fashion, since once upon a time, some mythological event took place there (*Q.N.* 23, 917F: Pluto's abduction of Kore near Mt. Etna). Additionally, Plutarch mentions several gods in a more allegorical fashion in relation to the physical attributes over which they preside, that is, over which they are κύριος[86]: viz. Zeus and heaven (*Q.N.* 2, 912A and 24, 918A), Dionysus and wine (*Q.N.* 10, 914D) and Cypris and love (*Q.N.* 21, 917B).

As previously noted, Plutarch exercises considerable restraint in endorsing allegorical interpretations of myths, probably because this reduces the gods to natural phenomena and material categories, an interpretative strategy that was common to the Stoics due to their monistic view of the world [see 4.1.1.3.]. This interpretive strategy is not, however, entirely absent from Plutarch's writings, and is also seen at work in *Quaestiones naturales*[87]. In *Q.N.* 24, 918A, he quotes Alcman's line where

---

[82] L. Senzasono, 2006, p. 215, n. 135.

[83] P.R. Hardie, 1992, p. 4783.

[84] *Ibid.*

[85] For fabulous stories about (late) divine punishment, cf. *De sera num.* 556F–557F.

[86] Cf. P.R. Hardie, 1992, p. 4768.

[87] For the Stoic practice of allegorising myths and its influence on Plutarch, see D. Babut, 1969, pp. 367–440 and J. Opsomer, 2014, p. 92. For physical allegory in Plutarch's mythological accounts, see W. Bernard, 1990, pp. 218–222 (and pp. 183–275 more generally) and P.R. Hardie, 1992, pp. 4766–4772. Notably, Plutarch reports (in fr. 157 Sandbach) that

Dew (Ersa) is allegorically called the daughter of Zeus (air) and Moon (Selene). In general, however, Plutarch's physical aetiologies are not concerned with rationalising the myth in such an allegorical way, where the mythological packaging and imaginative content is done away with once the myth's metaphorical value has been brought to light by a physical interpretation[88]. In many cases, it seems that Plutarch actually tries to save these traditional myths in the form in which they are recorded[89], much in the same way as he does with many other popular beliefs about the natural world, as we saw [4.1.1.].

The following table shows where the mythological material is located in *Quaestiones naturales* and in which cases we can speak of a *causa finalis* (*CF*), in the meaning attributed to it above. In the remaining cases where there is no *causa finalis*, we are dealing with more casual references to deities in connection with the physical attribute over which they reside (i.e. over which they are κύριος).

No.	*Q.N.*	**Mythological god/figure**	*CF*
1	2, 912A	Zeus	
2	10, 914D	Dionysus	
3	21, 917B	Cypris	
4	23, 917F	Kore, Pluto	X
5	24, 918A	Zeus, Ersa, Selene	X
6	36	Anchises, Aphrodite, Rhoecus	X
7	38	Leto, Hera, Zeus	X

'ancient physics among both Greeks and barbarians took the form of a scientific account hidden in mythology' (Ὅτι μὲν οὖν ἡ παλαιὰ φυσιλογία καὶ παρ' Ἕλλησι καὶ βαρβάροις λόγος ἦν φυσικὸς ἐγκεκαλυμμένος μύθοις κτλ.).

[88] Elsewehere, Plutarch sometimes does vindicate such a process of *Entmythologizierung*, though, by removing the fictitious parts of myths. Cf., e.g., *Thes.* 1, 3. See L. Van der Stockt, 1992a, pp. 140-141 and C. Pelling, 2002, pp. 171-195.

[89] For a general study of how ancient philosophers (including Plutarch) 'saved' myths, see L. Brisson, 2004. The notion of 'saving' myths can be seen, for instance, in the digression on naphtha in *Alex.* 35, 10-12, where Plutarch refers to the tragic story of Medea and says that 'some people reasonably wish to bring fables into conformity with truth' (εἰκότως οὖν ἔνιοι τὸν μῦθον ἀνασῴζοντες πρὸς τὴν ἀλήθειαν). J.R. Hamilton, 1969, p. 94 connects this with the Stoic allegorical interpretation of myths and argues that "Plutarch doubtless came across the reference to Medea in one of the many Stoic works he read". According to J. Boulogne, 2008, p. 737, however, in this digression, Plutarch reuses pieces of "un dossier constitué autour de l'autorité de Théophraste" [see 2.1.3., n. 36]. Both theories do not necessarily exclude one another, if we may assume that both traditions were already combined, or that Plutarch did this himself.

I conclude that Plutarch's incorporation of mythological elements in his scientific discourse forms an inherent, if not essential, aspect of it. If my hypothesis is correct, these elements not only illustrate Plutarch's scholarly approach in explaining natural problems but are also fundamentally in line with his dualistic view on causality, in that they may indicate that there are divine forces at work in nature. As such, they hint at an intrinsic consistency in Plutarch's general scientific project and its underlying world view. Arguably, these mythographical elements articulate what remains 'beyond words' in a purely physical discourse (by which I allude to *Quaest. conv.* 699B, quoted above). Indeed, they somehow enliven the mainly physical approach of the natural problems and thus also their sub-literary style [see 1.2.3.]. Notably, the mythographical accounts in *Q.N.* 24, 918A and 36 are in verse, which sharpens the contrast with the purely physical discourse even more. It will be useful to further consider the relationship between physics and poetry in Plutarch's natural scientific discourses in the following section, by taking a closer look at Plutarch's incorporation of poetic material in *Quaestiones naturales*.

3. Science and poetry

Apart from relying on the θεολόγοι and their myths, Plutarch also recurrently calls on the authority of the ποιηταί in explaining natural problems. The relatively high number of poetical quotations in *Quaestiones naturales* – 22 in total – makes it clear that the poets play a significant role in Plutarch's scientific discourse. As the scheme below will show, several poets are cited by name (apart from two anonymous metrical proverbs). Empedocles is the most frequently quoted authority in the entire collection (seven times) and is closely followed by Homer (five times)[90]. Pindar and Euripides are each quoted twice, and Alcman, Aratus, Theocritus and Theognis once.

A first important observation is that the link between physics and poetry is clear from the fact that some ancient φυσικοί composed their works in verse (see n. 61)[91]. This is the case with Empedocles (whom Plutarch calls a φυσικός in *De cur.* 515C)[92] and with Aratus, who composed his work

---

[90] In *De facie*, Homer and Empedocles are also the most frequently cited poets (see P. Raingeard, 1935, p. xxviii).

[91] For a more general account of ancient natural science treated in verse, see L. Taub, 2008 and 2009.

[92] In *Quaest. conv.* 683E, Plutarch writes that Empedocles (described as a σοφός) did not pursue literary embellishment (καλλιγραφίας ἕνεκα) in using literary epithets but aimed at a referential description of essential facts or properties [quoted 1.2.2.]. For Plutarch's evaluation of Empedocles' verses, see also *De aud. poet.* 16C. Notably, Hippol., *Haer.* 5, 20, 5 mentions Plutarch as a composer of ten Books of Ἐμπεδοκλέα (= fr. 24 Sandbach), which is nr. 43 in the Lamprias catalogue (Plutarch's authorship remains uncertain for this

on astronomical and meteorological matters in verse⁹³. But Alcman also wrote cosmogonic poetry, and the Poet (Homer) is even considered 'a sensitive observer' of natural phenomena in *Quaest. conv.* 627E (ὑπερφυῶς τοῦ ποιητοῦ τὸ γινόμενον συνεωρακότος)⁹⁴. In *Quaest. conv.* 699A, Plutarch also says of Euripides that he 'has keener eyes' than Erasistratus, the famous physician (βλέπων τι ὀξύτερον). It is clear, therefore, that the poets have a great deal of authority in the field of natural science, or that Plutarch at least ascribes it to them. In this respect, it is important to note that in Plutarch's scientific writings, we cannot make a clear distinction between the quotations from the didactic and the mimetic poets, since they are all treated in the same way by him. It seems unlikely, then, that this distinction really mattered to Plutarch himself, who indiscriminately interprets the poets in a physicalist fashion, thus often ingeniously stripping their verses from their original contexts (I will come back to this point later).

Another seminal issue, which has already been highlighted earlier on [see 1.2.3.], is that the poetical accounts clearly enliven the referential style of the purely physical discourse, but that this is not necessarily their primary objective. Plutarch mostly incorporates this poetical material in order to contribute to the main arguments themselves or to illustrate them in the sub-arguments, meaning that they do not fulfil a simply aesthetic function in the text. Therefore, Barrow's "rough and ready rule" according to which citations in Plutarch's *Moralia* are only recorded for the purpose of superfluous illustration – in the sense that they do not contribute to the main arguments themselves – does not seem to be apposite⁹⁵.

When it comes to Plutarch's heuristical method, it remains uncertain as to whether he extracted the poetical material directly from the works of the poets or relied on intermediary sources (e.g., commentaries or florilegia). Some of the quotations may suggest first-hand extraction, as is the case with the one from Aratus in *Q.N.* 2, 912D, but Plutarch may just as well rely on an intermediary source here (e.g., a commentary: see n. 75). It goes without saying, moreover, that Plutarch must have been

---

lost work, though: see F.H. Sandbach, 1969, p. 103). The discussion of the poetical value of Empedocles' verses is also found in Aristotle (cf. *Poet.* 1447b18 and fr. 70 Rose).

⁹³ On Plutarch's acquaintance with Aratus' oeuvre, see n. 75.

⁹⁴ Cf. also, e.g., *Quaest. conv.* 698EF, 684F. Plutarch composed a set of Ὁμηρικαὶ μελέται in four Books (Lamprias catalogue nr. 42), of which six fragments remain (frs. 122–127 Sandbach). We learn from fr. 127 Sandbach that physical aetiology was an inherent aspect of Plutarch's Homer exegesis [see 2.4.2., n. 149]. A genuine Homeric question is found in *Q.N.* 34, which concerns *Il.* 19, 415–416 (on the swiftness of the west wind). On physical allegory and scientific explanations in Heraclitus' *Quaestiones Homericae*, see D.A. Russell and D. Konstan, 2005, pp. xxi–xxii [see 1.1.3., n. 50].

⁹⁵ R.H. Barrow, 1967, p. 156. The same conclusion was reached for *Quaestiones Romanae* by L. Van der Stockt, 1987, p. 291. On Plutarch's general method of citing poets, see C. Bréchet, 2007.

well acquainted with Homer's works, but even so, this does not guarantee that he went directly to the source text (Homer is quoted via Aristotle in *Q.N.* 21, 917D). In addition, there may be reason to assume that the quotation from Alcman in *Q.N.* 24, 918A originates from a Stoic tradition, considering the allegorical context in which it is situated. It would not be an easy undertaking, if not an impossible one, to make an attempt towards generalisation here. But what is perhaps more important is the fact that Plutarch's interpretive strategy implies that the accounts of the poets – disregarding whether the poets themselves were aware of it or not – are in conformity with the natural problem at hand, or are at least bent to fit in the new context. As a rule, Plutarch simply extracts the poetical material from its original context and adapts it to the new one (some of these contexts were originally physical, others not). In this way, it often receives the necessary syntactic adaptations, and more importantly, a new semantic appropriation [see 4.2.1.1.].

Plutarch does not always simply accept what he reads, though. He knows very well, after all, that the poets tell many lies, sometimes even intentionally, for the purpose of giving 'pleasure and gratification to the ear' (*De aud. poet.* 16A: πρὸς ἡδονὴν ἀκοῆς καὶ χάριν). His literary criticism, which becomes explicit (only) in *Q.N.* 19, 916BC, involves the idea that while the poets (c.q. Pindar and Theognis) certainly do have authority in the field of natural science, they do not always have good insight into the mechanism that lies behind the natural phenomena they mention (which makes them an easy target of derision)[96]. As the accounts of Pindar and Theognis in *Q.N.* 19, 916BC show, the octopus has the ability to adapt and assimilate its colour to its surroundings, much like certain people do by imitating their neighbours. Plutarch, however, ridicules the poets and asks whether 'they believe that the octopus treats its colour like a garment that it can easily change whenever it wishes' (ἢ καθάπερ ἐσθῆτι τῇ χρόᾳ νομίζουσι χρῆσθαι, ῥαδίως οὕτως ἣ βούλεται μετενδυόμενον;). He concludes, to the contrary, that the octopus' change in colour is not the effect of a deliberate choice but of underlying physical causes [see 1.2.4.].

The scheme below sets out all the poetical quotations from *Quaestiones naturales* with special attention to their place and ranking in the aetiologies. No steadfast rules can be detected in this ranking, though, which seems to suggest that Plutarch is simply quoting the poets whenever he finds the proper occasion for it. Homer, for instance, is quoted once in the *quaestio*, twice in the first *causa* and again twice in the final *causa* of several problems. As to the poetical quotes that are recorded in the final *causa*, it is not unlikely that they, at least in some cases – viz. where they

---

[96] There is some disdain towards the poets (c.q. Alcaeus) also, e.g., in *Quaest. conv.* 698A: οὐδὲν ἔφη θαυμαστόν, εἰ ποιητικὸς ἀνὴρ Ἀλκαῖος ἠγνόησεν ὃ καὶ Πλάτων ὁ φιλόσοφος. καίτοι τὸν μὲν Ἀλκαῖον ἀμωσγέπως εὐπορήσειν βοηθείας κτλ.

actually conclude the *causa* –, allude to a higher level of causality. This is the case most clearly in *Q.N.* 2, 912D, with the quote from Aratus, and in *Q.N.* 36, with the mythological accounts of Theocritus and Pindar [see 4.1.2.2.]. This remains hypothetical, of course, but it is not, in any case, unlikely, since, for Plutarch, "poetry is the bridesmade of philosophy", as Gianakaris puts it, insofar that it is often considered to shroud an essential part of the intelligible truth[97].

No.	*Q.N.*	Poetical source	Rank
1	2, 912C	Emp., DK31B81	*c.* 4©
2	2, 912D	Arat., *Phaen.* 946–947	*c.* 5©†
3	5, 913D	Hom., *Od.* 5, 322–323	*c.* 4°†
4	16, 915E	Proverb	*q.*
5	19, 916B	Hom., *Il.* 13, 279	*c.* 1©
6	19, 916BC	Pind., 43 Snell	*c.* 2
7		Theog., 215–216 West	
8	19, 916CD	Emp., DK31B89	*c.* 3
9	20, 917A	Hom., *Od.* 19, 446	*c.* 1
10	20, 917A	Emp., DK31A78	*c.* 2
11	21, 917B	Proverb	*q.* ©
12	21, 917B	Eur., TGF 895	*c.* 1°
13	21, 917C	Emp., DK31B64	*c.* 3°
14	21, 917D	Hom., *Il.* 9, 539	*c.* 4°†
15	23, 917E	Emp., DK31B101	*c.* 1°
16	24, 918A	Alcm., 43 Diehl	*c.* 1°†?
17	29, 919B	Eur., TGF 941	*c.* 1

---

[97] C.J. Gianakaris, 1970, p. 128. Plutarch's attitude towards poetry is ambivalent. As a scholar and an all-round πεπαιδευμένος, he openly displays his knowledge of and attachment to the poets, arguing for the didactic function of their verses. But as a Platonist, he takes a more polemical position (see L. Van der Stockt, 1992a, p. 85, see also pp. 88–97 for the relation between myth and poetry). This explains why Plutarch writes in *De aud. poet.* 17DE that 'the art of poetry is not *greatly* concerned with the truth' (ποιητικῇ μὲν οὐ πάνυ μέλον ἐστὶ τῆς ἀληθείας). In *De aud. poet.* 16C, he says that Socrates – the philosopher *par excellence* – was 'the champion of truth all his life rather than a plausible or naturally clever workman in falsehood', that is, a poet (γεγονὼς ἀληθείας ἀγωνιστὴς τὸν ἅπαντα βίον, οὐ πιθανὸς ἦν οὐδ' εὐφυὴς ψευδῶν δημιουργός). So, when he was induced by certain dreams to take up poetry, he only put Aesop's fables into verse instead of writing poetry of his own, assuming that 'there can be no poetic composition without falsehood' (ποίησιν οὐκ οὖσαν ᾗ ψεῦδος μὴ πρόσεστι). Notably, Plato banned the poets and, with them, all mimetic artists from his ideal state (*Rep.* 603ab), but morally uplifting art was still very welcome (*Rep.* 401cd, 607be). See A.G. Nikolaidis, 2013, pp. 170 and 176–178.

No.	Q.N.	Poetical source	Rank
18	31, 919CD	Emp., DK31B81	c. 2°
19	34	Hom., *Il.* 19, 415	q. ©
20	36	Theocr. 1, 105–107	c. 1©†
21		Pind., 252 Snell	
22	39	Emp., DK31B94	c. 1©

Key for rank: *c.* = *causa*; *q.* = *quaestio*; ° = part of the argument; © = concludes *c./q.*; † = final *c.*

4.2. *Constructing scientific authority: between continuity, ingenuity and innovation*

As seen in the previous section, Plutarch's approach to natural problems is for the most part of a scholarly and literate kind. The Chaeronean mostly extracts scientific knowledge from books: folklore, mythology and poetry proclaim aspects of the 'truth', and it is precisely therein that their authority lies[98]. In order to gain further insight into the intellectual traditions to which *Quaestiones naturales* adheres, the section at hand will offer an analysis of Plutarch's scientific prose sources. Specific attention will also be directed toward Plutarch's method of composing and solving natural problems on the basis of his source material [4.2.1.]. Afterwards, I will try to balance this analysis by studying the underlying dynamics of argumentative creativity and ingenuity in Plutarch's physical aetiologies, showing that *Quaestiones naturales* is no mere accumulation of purely doxographical material [4.2.2.]. By taking these issues into consideration, the section at hand can be seen as a companion to the previous one on the popular, mythological and poetical traditions on which Plutarch relies, since each of these traditions were dealt a great deal of authority in the field of natural science by him (see n. 61).

---

[98] Cf. the observation of A. Gudeman, 1927, col. 2523 that we are dealing in collections of problems more generally "mit am Schreibtisch entstandenen Werken". However, the oral context of the discussion of problems should not be neglected. Considering the educational background of *Quaestiones naturales*, we should remain cautious in assuming Plutarch's *direct* acquaintance with the sources he apparently relies on, because in many cases, we may just as well be dealing with the (argumentative core of the) explanations that his fellow interlocutors gave to the problems at hand, and which Plutarch then incorporated into his own text [see 3.1.4.].

4.2.1. Character and use of the scientific tradition

In what follows, I will provide an analysis of Plutarch's scientific prose sources[99]. To this end, I will mainly focus on the explicit, that is, the nominatim quotations from prose authors. Arguably, an analysis of these quotations is representative of Plutarch's more general knowledge and use of the scientific tradition, as well as of his appreciation of the authorities on which he relies. Therefore, it will provide a great deal of useful information for a subsequent study of Plutarch's method of composition in his natural problems on the basis of his source material. This is not, however, the place for an all-embracing analysis of Plutarch's incorporation of traditional material in *Quaestiones naturales*, that is, an analysis that would also include Plutarch's countless allusions and implicit references to the scientific literature more generally (occasion for this will be provided in the commentary). Nevertheless, I will start by stressing more general aspects of Plutarch's reliance on the ancient scientific tradition, shedding a few ideas on this topic also.

1. Quotations from scientific prose authors

First of all, as just noted, many of Plutarch's references and allusions to the scientific literature remain vague and implicit. As seen before, Plutarch probably alludes to the Ps.-Aristotelian *Problems* a number of times without mentioning Aristotle's name [see 1.1.3., n. 83]. For instance, several of the accounts on animal diseases in *Q.N.* 26, 918BC can be retraced to Aristotle's *Historia animalium*, but this is not the case for each

---

[99] As K. Ziegler, 1951, cols. 857–858 noted, Plutarch relies on the following authorities ("Gewährsmänner") in *Quaestiones naturales*: Heraclitus, Empedocles, Plato, Aristotle, Theophrastus, the physicians Apollonius and Mnesitheus, Dionysius ὁ ὑδραγωγός, Antipater (author of *De animalibus*), a number of poets and the obscure Laetus. Cf. also R. Flacelière, J. Irigoin, J. Sirinelli and A. Philippon, 1987, p. lxxxii. For a concise analysis of Plutarch's "fonti letterarie", see also L. Senzasono, 2006, pp. 24–25. Notably, Senzasono argues (p. 25) that in those cases where a parallel can be found in a Latin author, this is probably due to the use of a common source that is now lost, but he does not exclude Plutarch's direct knowledge of those Latin authors either: "In questo opuscolo le concordanze sono probabilmente frutto in massima parte di dipendenze da una fonte comune, spesso perduta; in qualche caso non è da escludere una conoscenza diretta, come nel caso delle *Nat. quaest.* di Seneca e, soprattutto, date le frequenti concordanze, di Plinio il Vecchio." It is true that Plutarch had a notion of Latin later on in his life (as he notes himself in *Dem.* 2, 2–4; cf. A. De Rosalia, 1991 and J. Scheid, 2012, p. 7), but it seems rather unlikely that his limited knowledge enabled him to read technical scientific works in Latin (by which I leave unmentioned the clear inconsistencies with these authors at points), which is why I will leave Seneca and Pliny the Elder aside here. The latter is not explicitly quoted throughout Plutarch's entire œuvre, the former only once (in *De coh. ira* 461F–462A; cf. also *Galba* 20, 1).

and every one of them, and Aristotle is not explicitly quoted by name. It is not always clear in such cases whether Plutarch directly relies on a specific source text where a parallel is found, which he then perhaps connected with material from other sources, or on an intermediary tradition, such as lost natural problems, where they were already combined. The same uncertainty counts, for instance, for the material that can be related to Theophrastus' lost *De aquis* (see *Q.N.* 1, 911D, 5, 913CD, 7, 914A, 13, 915B). It is not unlikely that some accounts in *Quaestiones naturales* – and especially a number of the explicit quotations from Theophrastus – originate from this work, but the intermediation of lost natural problems cannot be excluded, especially if one considers that much Theophrastean material is still present in the Ps.-Aristotelian *Problems* that came down to us.

Additionally, the potential influence of a largely anonymous tradition, either popular or more specialised, should not be underestimated. In regards to the incorporation of ancient medical knowledge, most notably, we see that Plutarch not only cites 'accomplished physicians' (*Q.N.* 26, 918D: οἱ χαρίεντες ἰατροί), such as Mnesitheus of Athens or Apollonius the Herophilean, but also mentions several more popular beliefs in an anonymous fashion. He reports, for instance, that dew makes fat people thinner by imbibing it or by soaking it up on their cloths (*Q.N.* 6, 913F; cf. Cael. Aur., *Tard. pass.* 5, 139), or that pregnant women eat stones and dirt (*Q.N.* 26, 918D; cf. Hipp., *De superfetat.* 18).

As to the parallels with the Hippocratic writings, these remain rather vague, and at some points there are also obvious divergences[100]. This does not imply, however, that these parallels are merely coincidental, since the Hippocratic tradition, including its general theoretical and terminological framework, was incorporated into the Ps.-Aristotelian *Problems* from an

---

[100] For the parallels between *Quaestiones naturales* and the *corpus Hippocraticum* (esp. *Aer.* 7–8 ~ *Q.N.* 2, 912A; 3, 912E; 5, 913C; 9, 914C; 33), see R.M. Aguilar, 1994, pp. 42–43 and V. Andò, 2004, pp. 177–178 (who, however, assume Plutarch's direct reliance on the Hippocratic writings). The same problem applies to Plutarch's *De tuenda sanitate praecepta*. F.C. Babbitt, 1928, p. 214, believes that the "body of Hippocratic medical writings, along with others, was in circulation, and had undoubtedly been read by Plutarch". Also according to L. Senzasono, 1992, p. 11, the Hippocratic writings were "più o meno conosciuti da Plutarco". He notes (n. 9) that "la presenza latente degli scritti ippocratici è probabilmente diffusa, più o meno, in tutto l'opuscolo [c.q. *De tuenda*]". I remain hesitant towards Plutarch's firm acquaintance with the Hippocratic tradition, though. Cf. also K. Ziegler, 1951, col. 925: "Bei dem ziemlich breiten Raum, den medizinische Dinge bei P. einnehmen, möchte man glauben, daß er allerlei medizinische Literatur gelesen hat. Aber Zitate daraus sind nich zahlreich […]. Für sonstige naturwissenschaftliche Frage sind Aristoteles und Theophrastos P.s Hauptquellen gewesen." For Plutarch's quotes from the Hippocratic writings, see W. Helmbold and E.N. O'Neil, 1959, p. 19. See also the general study on "Plutarque et la médecine" by J. Boulogne, 1996.

early stage onwards[101]. It seems rather unlikely, therefore, that Plutarch is relying on the Hippocratic writings directly. In any case, Hippocrates is not cited explicitly in *Quaestiones naturales*. The idea, for instance, that fever turns moisture into bile in *Q.N.* 1, 911E can be generally related to the humoral theory of the Hippocratics. However, seeing that this idea illustrates a quote from Aristotle that cannot be retraced in the extant Aristotelian writings, it perhaps more likely originates from a lost problem, so that the context is more generally Peripatetic (see the commentary *ad loc.*).

The following scheme sets out the explicit quotations from scientific prose authors in *Quaestiones naturales* with special attention to their place and ranking in the aetiologies. In what follows, I will zoom in on individual (groups of) intellectuals, viz. the Presocratics, Plato, Aristotle etc., and more specifically on the ranking of their authority in Plutarch's explanations.

No.	*Q.N.*	Authority	Rank
1		Pl., *Tim.* 90a; *Rep.* 491d, 546a	
2	1, 911D	Anaxag., DK59A116	*c.* 1°
3		Democritus[102]	
4	1, 911E	Arist., *Mete.* 358a?	*c.* 4°†
5	2, 911F	Laetus	*c.* 1°
6	2, 912A	Arist., fr. 215 Rose	*c.* 2°
7	2, 912A	Heracl., DK22B12	*c.* 2
8	3, 912DE	Apollonius, fr. 33 von Staden	*c.* 1
9	5, 913C	Pl., *Tim.* 59e	*c.* 3°
10	5, 913D	Pl., *Tim.* 65de	*c.* 4†
11	6, 913E	Laetus	*c.* 1°
12	7, 914A	Theophr., 161 W. = 214C FHSG	*c.* 2†
13	9, 914B	Dionysius	*q.* ©
14	12, 914F	Arist., *Pr.* 935b?	*c.* 1°
15	13, 915B	Theophr., 163 W. = 173 FHSG	*c.* 1°
16	19, 916B	Theophr., 188 W. = 365C FHSG	*c.* 1°
17	21, 917D	Arist., *HA* 578a?	*c.* 4°†

---

[101] The Ps.-Aristotelian *Problems* contain a great deal of Hippocratic theories and concepts [see 1.1.3., n. 75]. For the sources of the *Problems* in general, see H. Flashar, 1962, pp. 333–341.

[102] This fragment is not recorded among the Democritus fragments collected by DK (nor is it mentioned in the list of Plutarch's quotations by W.C. Helmbold and E.N. O'Neil, 1959, p. 22). Cf. also J.P. Hershbell, 1982b, p. 81, n. 2.

No.	Q.N.	Authority	Rank
18	26, 918DE	Mnesith., fr. 16 Bertier	c. 2†
19	38	Antip., SVF 3, p. 251, fr. 48	c. 1°
20	40	Arist., fr. 218 Rose	c. 1©†

Key for rank: c. = *causa*; q. = *quaestio*; ° = part of the argument; © = concludes c./q.; † = final c.

For Plutarch, the Presocratics were essentially masters in natural scientific matters (in their capacity of φυσικοί), which explains why he recurrently relies on their authority in his natural problems. In *Quaestiones naturales*, Heraclitus, Anaxagoras and Democritus are cited only once each, but this number rises sensitively with Empedocles, who is quoted no less than seven times, making him the most frequently quoted authority in the collection. In regards to Plutarch's reliance on the Presocratics (throughout his entire oeuvre), Sambursky claimed that it "is not only typical of the eclectic spirit of Plutarch's time, but demonstrates the closeness of its intellectual viewpoint to the scientific approach of the great pre-Socratic philosophers"[103]. The issue of eclecticism has, of course, led to much debate among historians of philosophy and has long been abandoned by Plutarchists today[104]. Also the second part of Sambursky's claim is problematic: it is somewhat dubious for the contemporary value of Plutarch's scientific writings, as it implies an intellectual setback. Plutarch does, indeed, show great interest in Presocratic natural philosophy and in the Presocratics themselves (e.g., underlining Anaxagoras' importance as Pericles' philosophy tutor). Yet, the role that he attributes to them in the history of science remains inferior to the intellectual agenda of Plato and the scientific revolution lead by him (outlined above [see 4.1.2.1.]).

Interestingly, Aristotle also frequently engages with Presocratic physics (e.g., in the doxographical sections of *De generatione et corruptione* or the *Parva naturalia*), albeit often in such a way that his own scientific theories come out best[105]. Such self-promoting dynamics, though, are generally absent from Plutarch's treatment of the Presocratics in his natural problems. Instead of engaging in a polemic with the Presocratics, Plutarch generally appeals to their authority in order to underpin and illustrate his own explanations. As such, they are not rival φυσικοί but welcome allies

---

[103] S. Sambursky, 1963, pp. 211–212. For Plutarch's acquaintance with the Presocratics in general, see A. Fairbanks, 1897, A.M. Battegazzore, 1992, p. 54, n. 50, C. Santaniello, 2004 and S.-T. Teodorsson, 2011. See also the separate case studies by J.P. Hershbell.

[104] See J.M. Dillon, 1988.

[105] For further detail on Aristotle's review of the Presocratics, see C. Collobert, 2002.

in Plutarch's scientific project. In *Q.N.* 19, for instance, Plutarch refers to Empedocles' theory of emanations in order to back up his (presumably) personal explanation of the octopus' metachrosis [see 4.2.2.2.].

However, seeing that Plato – in defiance of a correct chronology – is mentioned *before* Anaxagoras and Democritus in the very first *causa* of *Q.N.* 1, 911D, it is clear where Plutarch's heart eventually lies. This point should not go unnoticed here. Arguably, this chronological inversion has implicit programmatic value for *Quaestiones naturales*, as it reveals Plutarch's personal preference for Plato – who, thus, carries of the privilege of being quoted as the very first authority at the very beginning of the collection, *before* the Presocratics. This does not imply, of course, that Plutarch does not esteem the Presocratics at all, since Plato is mentioned *along with* two Presocratic eminencies (viz. Anaxagoras and Democritus). Therefore, the chronological inversion does not strictly separate Plato from the φυσικοί[106]. In fact, the phrase οἱ περὶ Πλάτωνα καὶ Ἀναξαγόραν καὶ Δημόκριτον may imply one happy family of natural philosophers, thus *including* Plato. However, if it is true that Plutarch does, indeed, show himself to prefer Plato (or more periphrastically, οἱ περὶ Πλάτωνα) at the beginning of *Quaestiones naturales*, it is not unlikely that we are dealing with Plutarch's philosophical signature here, that is, some kind of a subtle Platonic σφραγίς ('seal', 'signet'), for the entire collection. Plutarch's philosophical loyalty to Plato does not require illustration, nor does a subtle intellectual signature seem to be out of place at the beginning of a collection of natural problems, particularly considering the strong Peripatetic tradition in which it stands [see 1.1.3.]. We know from elsewhere that Plutarch does not as such aim to reject the approach of the φυσικοί and their search for natural causes but aims to complete it with a teleological perspective, after the model of Plato [see 4.1.2.1.]. Perhaps the chronological inversion is, thus, meant as a subtle reminder or a silent

---

[106] A similar conclusion was reached by D. Babut, 1994, pp. 574–475 (with n. 138) for *De prim. frig.* 948C, where Plato is also mentioned along with Democritus: "Platon n'est pas entièrement dissocié ici des autres physiciens"; n. 138: "[l]e passage n'oppose pas, en effet, Platon *en tant que métaphysicien* à tous les autres *en tant que physiciens*". Democritus is named again in association with Plato (and with other philosophers) in *De tranq. an.* 472D, *Adv. Col.* 1108BC and 1124C (cf. J.P. Hershbell, 1982b, p. 96 with n. 51). In regards to the reference to Anaxagoras and Democritus here in *Q.N.* 1, 911D, J.P. Hershbell, 1982a, pp. 146–147 notes: "In itself, the report is of little value, but the mention of Anaxagoras (also Democritus) together with Plato, suggests Plutarch's esteem for Anaxagoras as a student of the natural world." See also p. 153: "Because of Plutarch's own personal interest in the workings of nature, it is not surprising that he has regard for Anaxagoras' views […]. In Plutarch's eyes, Anaxagoras was also a precursor of his own fight against superstitious explanations of the world's happenings. But however sympathetic Plutarch may have been to Anaxagoras' beliefs, the latter's views did not really explain the purposive activity of nature, a doctrine dear to both Plutarch and to his master Plato."

invitation to read the work in the spirit of Plato's natural philosophical project and in light of the scientific revolution that it caused according to Plutarch (what the reader gets to read are, indeed, only the Αἰτίαι φυσικαί [see 1.1.6.]). In that case, the rough rule that the first *causa* is usually reserved for *communis opinio* may be less operative here (at the very least, this passage may reveal the elasticity in such a theory). Plato is quoted two more times, viz. in *Q.N.* 5, 913CD, where he closes off the aetiology. There can be little doubt, moreover, that Plutarch was able to quote Plato by heart, but he may just as well have consulted the Platonic text directly, or indirectly via an intermediary source (e.g., a commentary). In any case – and this is perhaps not so remarkable –, the three Platonic quotations can be retraced to the *Timaeus* (among several other parallel passages in Plato's oeuvre).

While the intellectual signature may be Platonic, the general atmosphere of the collection is clearly Peripatetic [see 1.1.2.]. There are five quotes from Aristotle, which make him a very important authority in the collection (on a par with Homer, and only surpassed by Empedocles)[107]. This can be further illustrated by the fact that if Aristotle's account is not explicitly considered to be 'plausible' (cf. *Q.N.* 2, 912A, 12, 914F), it closes off the aetiology (cf. *Q.N.* 1, 911E, 21, 917D, 40). In *Q.N.* 40, Aristotle is identified as a φιλόσοφος and is considered one of 'the best scientists' (οἱ κρείττους τῶν φυσικῶν – it is uncertain, however, whether this praise is original or part of the reformulation of the chapter by Psellus). Moreover, as would appear from the ranking of the quotes, Plutarch seems to appreciate Aristotle more than his successor, Theophrastus, whom he quotes three times. Theophrastus kicks off the aetiology in *Q.N.* 13, 915B and 19, 916B, and is ranked in the second *causa* in *Q.N.* 7, 914A[108]. Plutarch considers his solution 'plausible but insufficient' in *Q.N.* 19, 916B (πιθανῶς [...] οὐχ ἱκανῶς). In many cases, though, a more attractive alternative follows upon the accounts of Aristotle and Theophrastus, but they are never openly rejected, since their explanations are not altogether implausible according to Plutarch[109].

A small group of authorities remains, among which only the Greek doctors Apollonius the Herophilean (*Q.N.* 3, 912DE)[110] and Mnesitheus

---

[107] For a separate analysis of Aristotelian quotes in Plutarch's natural problems, see M. Meeusen, 2011 and 2016.

[108] For a general study of Plutarch as a reader of Theophrastus, see J. Boulogne, 2005c. Boulogne rightly notes that the first *causa* in the Chaeronean's aetiological system is not "synonyme de rang préférentiel, mais de premier pas vers la vérité" (pp. 292–293, with further references).

[109] *Pace* F.H. Sandbach, 1965, p. 134.

[110] Surnamed 'Mys', presumably because he wrote a treatise Περὶ μυῶν (ca. 60 BC). Herophilus of Chalcedon, the leader of the school to which Apollonius belonged, worked

of Athens (Q.N. 26, 918DE)[111] can be identified with certainty[112]. These doctors are not quoted anywhere else by Plutarch, which makes a direct consultation of their works implausible. While Plutarch quotes Apollonius in the first *causa* on a par with οἱ πολλοί, he references Mnesitheus in the last one and as a representative of οἱ χαρίεντες ἰατροί (their accounts are located in the sub-argument). Furthermore, in the first *causa* of Q.N. 38 (concerning the period of twelve days during which she-wolves litter), Plutarch quotes a passage from Antipater's *De animalibus*, probably the Stoic philosopher of Tarsus[113]. We do not know who the Dionysius from Q.N. 9, 914B is,

---

as an anatomist in Alexandria around 300 BC (cf. also *De cur.* 518D). For further reading on Herophilus and his followers, see H. von Staden, 1989, with pp. 540–554 on Apollonius (esp. p. 544 regarding the quote in Q.N. 3, 912DE).

[111] Mnesitheus of Athens (ca. 350 BC) wrote several works on dietetics (Περὶ τῶν ἐδεσμάτων, Περὶ παιδίου τροφῆς). For further reading, see J. Bertier, 1972.

[112] For Plutarch's acquaintance with Hellenistic medicine, see S. Grimaudo, 2004, R.M. Aguilar, 2005 and I. Rodríguez Alfageme, 2005. The literature on Plutarch and medicine mostly concerns specific case studies. On popular medicine in Plutarch, see I. Rodríguez Alfageme, 1999a. For a more general overview, see J. Boulogne, 1996 (esp. p. 2766, with n. 31 for Plutarch's treatment of medical topics in *Quaestiones naturales*).

[113] = SVF 3, p. 251, fr. 48 (von Arnim refers to Cic., *De div.* 2, 33 and notes: "possunt haec Stoico Antipatro vindicari de συμπαθείᾳ disserenti, sed certi nihil affirmare licet"). For Plutarch's quotations from Antipater of Tarsus, see W.C. Helmbold and E.N. O'Neil, 1959, p. 5. In *De soll. an.* 962F (= SVF 3, p. 251, fr. 47), Plutarch probably relies on Antipater's Περὶ ζῴων again (concerning asses and sheep and their lack of cleanliness). The scholiast on Apoll. Rh., *Arg.* 2, 88–89a refers to a similar phenomenon as in Q.N. 38 (concerning the mating habits of bulls) in the same work (Ἀντίπατρος ἐν τῷ περὶ ζῴων). C. Wendel, 1942, pp. 216–217 dismisses the hypothesis that the Antipater from this passage can be identified with a doctor from the time of Augustus. This was considered uncertain already by M. Wellmann, 1894, but no longer by G. Lachenaud, 2010, p. 213, n. 27. According to A. Dyroff, 1897, p. 403, however, the work should be attributed to Antipater of Tyre and was mainly directed against Antipater of Tarsus. This was shown to be unlikely by G. Tappe, 1912, pp. 19, 1 and 52, 1 (cf. also M. Schuster, 1917, p. 77 and D. Babut, 1969, pp. 211–214). C. Wendel, 1942, pp. 216–217, by contrast, remains sceptical, and argues that we are dealing in Q.N. 38 with an unknown chronicler of animal συμπάθεια and παράδοξα (cf. also L. Senzasono, 2006, pp. 247–248, n. 208). The Stoic predilection for paradoxes and *mirabilia* is well-known [see 4.1.1.3.]. Cf. also J. Schmidt, 1949 and F.H. Sandbach, 1965, pp. 222–223, n. b: "Yet marvels did not come amiss to Stoics, as being evidence of the workings of Providence or of the unity of all things in the universe." Cf., e.g., Cleanthes' story about ants in *De soll. an.* 967E (= SVF 1, p. 116, fr. 515). Another relevant passage is found in *Quaest. conv.* 626EF, where Theon the critic asks why Chrysippus never gave an explanation for any of the strange and extraordinary things he frequently mentions (SVF 3, p. 146, fr. 546: τί δήποτε Χρύσιππος ἐν πολλοῖς τῶν παραλόγων καὶ ἀτόπων ἐπιμνησθείς κτλ.). The Stoic Themistocles answers that Chrysippus does so by way of example, because people are easily and irrationally trapped by what appears likely, and contrariwise disbelieve what appears unlikely (626F: ῥᾳδίως ἡμῶν καὶ ἀλόγως ὑπὸ τοῦ εἰκότος ἁλισκομένων καὶ πάλιν ἀπιστούντων τῷ παρὰ τὸ εἰκός).

quoted in *oratio obliqua* in the *quaestio*, where he is surnamed ὁ ὑδραγωγός ('the designer of aqueducts', 'the hydraulic engineer')[114]. Lastly, a certain Λαῖτος is quoted twice, viz. in *Q.N.* 2, 911F and 6, 913E, both times in the first *causa* and regarding specific meteorological phenomena related to precipitation (viz. rain and dew respectively). Plutarch's references to the man's explanations are kept fairly short, and in *Q.N.* 2, 911F, he explicitly rejects an explanation proposed by him (ἢ τοῦτο μὲν οὐκ ἀληθές;). Scholars have identified this Λαῖτος with Ofellius Laetus, a contemporary Platonic philosopher and author of a meteorological hymn (a μετάρσιος ὕμνος), whom we know from two inscriptions: one from Athens (*I.G.* II², 3816), the other from Ephesus (*I.Eph.* VII, 2, 3901). It is not unlikely that Plutarch personally knew this man as a fellow (or perhaps rival?) Platonist, but this remains uncertain[115] [see 3.1.5., n. 62].

As to Plutarch's general method of citing, it is his habit to keep the quotations fairly brief, but even so, he does not seem to show tremendous concern for precision, as he often renders the original text in his own words[116]. The adaptation of the traditional material is often

---

[114] Throughout the *corpus Plutarcheum*, several persons are named Dionysius, which is a commonly used Greek name (cf. *De def. or.* 421E). The Dionysius from *Q.N.* 9, 914B remains obscure, though. Notably, in *Amatorius* 761B, Dionysius of Corinth is mentioned as a poet and author of a collection of Αἴτια. In *Quaest. conv.* 744F, a certain farmer (ἡμεῖς οἱ γεωργοί) called Dionysius of Melitè participates in one of the discussions during the festival of the Muses in the Academy (S.-T. Teodorsson, 1996, p. 356 notes that "[t]his Attic farmer is the only representative of his profession appearing in the *Talks*"). This Dionysius was perhaps a close friend or acquaintance of Plutarch's, but even so it remains uncertain whether he can be identified with the Dionysius from *Q.N.* 9, 914B. The same can be said of Dionysius of Delphi, named in *De soll. an.* 965C, who was the father of Aeacides and Aristotimus. According to L. Senzasono, 2006, p. 179, n. 64, the Dionysius from *Q.N.* 9, 914B must have been a sufficiently known person (like Laetus; see n. 115), even if Plutarch cites him indirectly, but this seems unlikely precisely for that reason. In addition, the epitheton ὑδραγωγός may perhaps refer to Dionysius' local (?) nickname (rather than to his profession): 'the waterdrinker', 'the sufferer of dropsy' (cf. LSJ, s.v.). This remains speculative but is not, in any case, implausible in the context of the question at issue in *Q.N.* 9 (Διὰ τί τοῦ χειμῶνος ἧττον πικρὰ γίνεται γευομένοις ἡ θάλαττα;). For a list of ancient scientists named Dionysius, see P.T. Keyser and G.L. Irby-Massie, 2008a, pp. 258–265.

[115] For further reading on Ofellius Laetus, see G.W.M. Bowersock, 1982, J. Opsomer, 2008 and M. Meeusen, 2013a.

[116] See already A. Fairbanks, 1897, pp. 78–79 and also D.A. Russell, 1980, p. 14: "Plutarch's quotations and allusions are often loose and inexact". Whether this happens on purpose or not is not always clear. It remains to be seen, moreover, whether precision in quoting authorities κατὰ λέξιν was really a prerequisite for Plutarch. In *De prim. frig.* 947F, for instance, Plutarch notes that his paraphrase of Anaximenes (DK13B1) comes close to the original wording, which, apparently, was good enough for him to make his point (οὕτω πως ὀνομάσας καὶ τῷ ῥήματι). The lack of precision in quoting can perhaps be explained by the vagaries of Plutarch's memory or by the inexactness of his personal notes. See,

more conspicuous, though, especially when Plutarch changes the specific wording of the citation to fit the context (as is the case, for instance, with the adaptation of τομίας into μόνορχις in the quote from Aristotle in *Q.N.* 21, 917D; see the commentary *ad loc.*). Moreover, Plutarch often extracts material from a very specific context, so that it needs to be adapted in several ways to the new context, that is, not only by syntactic or verbal but also by semantic modifications (as is the case, for instance, with Plutarch's claim in *Q.N.* 5, 913B that salt is 'not generated', whereas Theophrastus, who is his probable source, implies that it 'does not generate'; see the commentary *ad loc.*). Plutarch's custom of lifting specific accounts from their original context is widely observed throughout his entire oeuvre and testifies to his desire and ability to use the available sources in an original fashion, so that a simple reproduction is deliberately avoided. This is especially the case with the quotations from the poets, since they are often reinterpreted in a somewhat exegetic fashion, thus receiving a new physical appropriation, as we saw [4.1.2.3.]. Plutarch also applies this interpretive strategy, for instance, to Heraclitus' famous river statement in *Q.N.* 2, 912A: 'you could not step into the same rivers twice, because other waters flow upon you'. Plutarch reinterprets this in a very literal, physical fashion in order to support the theory that river water has a fresh and new-born property. As such, the deeper and original pan-cosmic meaning of Heraclitus' saying is no longer relevant. Such a literal interpretation, which is, indeed, much in keeping with the general referential stylistics of the collection [see 1.2.3.], illustrates how Plutarch freely and often very playfully processes received knowledge and traditional authority in his natural problems, twisting it where he personally finds possible and fit[117].

---

e.g., W.C. Helmbold and E.N. O'Neil, 1959, p. ix ("[h]is memory was prodigious, and his confidence in it no less so") and R. Flacelière, J. Irigoin, J. Sirinelli and A. Philippon, 1987, p. lix. In some cases, though, the adaptation seems deliberate, which can be explained, then, in light of Plutarch's urge to process traditional authority in an original way, even if this means that the original text should be bent a little. In *Quaest. conv.* 718C, for instance, the 'misquotation' from Plato is certainly deliberate: Plutarch is very well aware that Plato nowhere literaly says that 'the demiurge is always practicing geometry', but the saying is not, therefore, essentially unplatonic (γέγραπται μὲν ἐν οὐδενὶ σαφῶς τῶν ἐκείνου βυβλίων, ἔχει δὲ πίστιν ἱκανὴν καὶ τοῦ Πλατωνικοῦ χαρακτῆρός ἐστιν). Even if Plato did not actually say this, it is not, therefore, an 'unorthodox' saying, and that is what really matters for Plutarch. Or formulated more negatively, in the words of S.-T. Teodorsson, 1996, pp. 162–163: "Plut. clearly indicates that he is aware of the fact that the subject to be discussed goes beyond Plato's doctrine, and he implicitly warns that the opinions are not entirely orthodox."

[117] For Plutarch's similar method of citing in *Quaestiones convivales*, see J. König, 2010, esp. pp. 339–345 (see also *id.*, 2011 for an account of how the symposiasts stage and re-enact the voices of past authorities by quoting them). On Plutarch's general method of citing the poets, see C. Bréchet, 2007. See also E. Bowie, 2008.

2. Problematisation of scientific knowledge

One of the central aims of the genre of natural problems, besides from raising questions about problematic passages in the scientific literature, is to communicate received knowledge in, what can be called, a 'problematised' fashion, meaning that the author often adapts and remoulds certain passages from a specific source text into the problem format, in order to reframe the source text and/or to reopen it for discussion, a procedure seen at work throughout Ps.-Aristotle's *Problems*[118]. In doing so, the author often adds his own criticisms and further remarks, suggestions, specifications, so as to advance or review traditional theories. The same procedure is clearly present also in Plutarch's natural problems, as we will see here. Despite the sophisticated use of source material, the result comes across quite organically, because the borrowed material ties in neatly with the development of the problem chapters[119]. As such, the transformation of traditional material into the problem format – both in the *quaestiones* and in the *causae* – testifies to the author's attempt at a problematisation of a wide array of ancient Greek scientific learning. Nature and, by implication, the field of natural scientific knowledge, is cut up into small problem units and is transmitted in a piecemeal fashion. Each of these problem units usually focuses on one specific natural phenomenon. Therefore, as Blair argues in light of Ps.-Aristotle's *Problems*, "*Problemata* are one of the ways of attaching particulars to the universals of *scientia* developed in systematic treatises, through commonsensical but often sophisticated reasoning."[120]

---

[118] In some cases, not the subject for the *quaestio* is found in a specific source text, but one or more of the explanations themselves. For more detail on this technique in Ps.-Aristotle's *Problems*, see H. Flashar, 1962, p. 334: "Aber es handelt sich keineswegs um mechanische Exzerpte, vielmehr ist die Art der Quellenbenutzung im einzelnen gans verschieden: teils geht nur die Frage, nicht aber die Antwort, teils nur die Antwort, nicht aber die Fragestellung aus Ar. zurück." For a general account on the problematisation of source material into the problem format, see C. Jacob, 2004, pp. 48–53.

[119] The same technique can be found in Plutarch's other collections of *quaestiones*, esp. *Quaestiones Romanae* and *Graecae*. For *Quaestiones Graecae*, see K. Giesen, 1901, p. 449 and W.R. Halliday, 1928, pp. 14–15; for *Quaestiones Romanae*, see L. Van der Stockt, 1987, pp. 287–288 and J. Boulogne, 1992, p. 4686, n. 27: "Il est peu vraisemblable que Plutarque ait commencé par dresser une liste de problèmes à résoudre, pour ensuite procéder aux recherches nécessaire. L'ordre des opérations a dû être inverse."

[120] A. Blair, 1999, p. 175. Cf. also R. Mayhew, 2011a, pp. xxi–xxii: "In the broadest terms, the purpose of the chapters of the *Problems* is to raise questions – about passages in the works of Aristotle or Theophrastus or other Peripatetic philosophers and scientists, about passages in the works of medical writers (and especially the Hippocratic treatises), and in general about *endoxa* (the reputable opinions in the air at the time, on any number of subjects)."

A manageable case for illustrating this procedure in Plutarch's *Quaestiones naturales* is the cluster of problems on hunting in *Q.N.* 23–25, which discusses the influence of the meteorological conditions of spring, winter, dew and full moons on tracks and trails left behind by animals. There is a dense cluster of parallel passages with Xen., *Cyn.* 5, 1–5, as set out in the following table:

Xen., *Cyn.* 5, 1–5	Plut., *Q.N.* 23–25, 917E–918B
5, 1–2 **χειμῶνος** μὲν οὖν πρῴ οὐκ ὄζει αὐτῶν, ὅταν πάχνη ᾖ ἢ παγετός· [...] καὶ αἱ κύνες μαλκίουσαι τὰς ῥῖνας οὐ δύνανται **αἰσθάνεσθαι** ὅταν ᾖ τοιαῦτα, πρὶν ἂν ὁ ἥλιος **διαλύσῃ** αὐτὰ ἢ προϊοῦσα ἡ ἡμέρα· τότε δὲ καὶ αἱ κύνες ὀσφραίνονται καὶ αὐτὰ ἐπαναφερόμενα ὄζει.	*Q.N.* 25, 918AB: Διὰ τί τὸ δρόσιμον γενόμενον διὰ τοῦ ψύχους δυστίβευτον; [...] ἢ δεῖ μὴ μόνον ἔχειν ἴχνη τὸν στιβευόμενον τόπον ἀλλὰ κινεῖν τὴν ὄσφρησιν, κινεῖ δὲ **λυόμενα** καὶ χαλώμενα μαλακῶς ὑπὸ θερμότητος, ἡ δ' ἄγαν περίψυξις πηγνύουσα τὰς ὀσμὰς οὐκ ἐᾷ ῥεῖν οὐδὲ κινεῖν τὴν **αἴσθησιν**· ὅθεν καὶ τὰ μύρα καὶ τὸν οἶνον ἧττον ὄζειν ψύχους καὶ **χειμῶνος** λέγουσιν· ὁ γὰρ ἀὴρ πηγνύμενος ἴστησι τὰς ὀσμὰς ἐν αὐτῷ καὶ οὐκ ἐᾷ ἀναδίδοσθαι.
5, 3–4 ἀφανίζει δὲ καὶ ἡ πολλὴ **δρόσος** καταφέρουσα αὐτά [...] οἱ δὲ ὑετοὶ κατακλύζουσι καὶ αἱ ψακάδες, καὶ ἡ σελήνη ἀμαυροῖ τῷ θερμῷ, μάλιστα δὲ ὅταν ᾖ **πανσέληνος** κτλ. 5, 5 τὸ δὲ **ἔαρ** κεκραμένον τῇ ὥρᾳ καλῶς παρέχει τὰ ἴχνη λαμπρά, πλὴν εἴ τι ἡ γῆ ἐξανθοῦσα βλάπτει τὰς κύνας, εἰς τὸ αὐτὸ **συμμιγνύουσα τῶν ἀνθῶν τὰς ὀσμάς**.	*Q.N.* 24, 917F–918A: Διὰ τί περὶ τὰς **πανσελήνους** ἥκιστα ταῖς ἰχνοσκοπίαις ἐπιτυγχάνουσιν; Ἢ διὰ τὴν εἰρημένην αἰτίαν; **δροσοβόλοι** γάρ αἱ πανσέληνοι κτλ. *Q.N.* 23, 917E: Διὰ τί δυστίβευτος ἡ τοῦ **ἔαρος** ὥρα; Πότερον αἱ κύνες [...] τὰς ἀπορροίας ἀναλαμβάνουσιν, ἃς ἐναπολείπει τὰ θηρία τῇ ὕλῃ, ταύτας δὲ τοῦ ἔαρος ἐξαμαυροῦσι καὶ **συγχέουσιν αἱ πλεῖσται τῶν φυτῶν καὶ τῶν ὑλημάτων ὀσμαί**, καὶ ὑπὲρ τὴν ἄνθησιν ὑπερχεόμεναι καὶ κεραννύμεναι περισπῶσι καὶ διαπλανῶσι τὰς κύνας τῆς τῶν θηρίων ὀσμῆς ἐπιλαβέσθαι; κτλ.

Two further remarks should be made here. First of all, the parallel passages in Xenophon are very relevant for the concatenative clustering of the distinct problem chapters around the same topic in Plutarch. Indeed, this problem cluster nicely illustrates how the obvious connective threads between two or more chapters can sometimes be explained on the basis of the coherent structure of the source text which they problematise [see 1.1.5.]. Second, even though there may be reason to assume that Plutarch was acquainted with Xenophon's *Cynegeticus*[121], it is not clear that this text served as a direct subtext for the three problem chapters at hand. After all, the intermediation of another text (e.g., one or more lost natural problems) and/or an oral report cannot be excluded (e.g., during a sympotic dicussion: cf. the encomium on hunting mentioned in *De soll. an.* 959B). Hardie correctly states that "[b]y Plutarch's time much scholarly and philosophical discussion had been reduced to the status of the προβλήματα"[122]. But this is not, of course, to underestimate Plutarch's own authorial intermediation in composing such problems. In any case, each of the chapters in *Q.N.* 23–25 contains more than one demonstrable source, since Plutarch also relies on *communis opinio*, incorporates mythological material, and quotes Empedocles and Alcman by name (these are located in the passages that are bracketed in the table above). It is only likely, then, that Plutarch draws on a mixture of sources for each problem chapter and, in doing so, adds his own findings and comments by developing or adding new material or by nuancing older material (cf., e.g., the allegorical interpretation of Alcman's line in *Q.N.* 24, 918A, where the moon's ability to draw up moisture from the earth is emphasised, *vis-à-vis* the parallels in *Quaest. conv.* 658B and *De facie* 929A, where the moon's liquefying ability is at issue; see the commentary *ad loc.*).

Sandbach is probably correct, then, in concluding that it is unlikely that Plutarch drew *all* of his information from one and the same book in *Quaestiones naturales*[123]. Even if Plutarch incorporates a great deal of received knowledge in his natural problems, we should not lose track of his personal intuition for scientific speculation and zetetic ingenuity.

---

[121] See J. Mossman and F. Titchener, 2011, p. 277. Cf. *Non posse* 1096C with the quote from *Cyn.* 5, 33.

[122] P.R. Hardie, 1992, p. 4751. He adds that "the form is characterized by the massive deployment of argument and learning on questions which are selected for their inherent trickiness rather than by any externally determined standard of importance; by a corresponding obsession with detail at the expense of wider connections; and, not so obviously, by the readiness to let go of the question once a number of plausible explanations have been found, which may be ranked in order of likelihood but need not be".

[123] F.H. Sandbach, 1965, p. 135: "Notes of this sort [sc. *Quaestiones naturales, Romanae* and *Graecae*] can arise from summarizing and abstracting from a single book, but they may be drawn from diverse sources, and include suggestions and criticisms made by the note-taker himself. Any contention that Plutarch took all his questions from a single source [...], does not admit of profitable discussion. If the questions referred to Aristotle, Theophrastus, and the unknown Laetus implied first-hand consultation of these authors, the contention would be untenable; but there is no better reason for asserting than for denying that Plutarch went directly to them."

Quaestiones naturales is not a mere doxography of traditional authorities, since, as Teodorsson remarks, Plutarch in this work "did not follow his sources slavishly but developed the material further in his own way"[124]. This leads the way to a further examination of the actual innovative dynamics in Plutarch's natural problems in the following section.

4.2.2. Scientific innovation and performance

1. A note on the sociology of knowledge and παιδεία

Plutarch's eagerness to examine and reprocess received knowledge can be regarded as one of the unifying factors throughout his entire oeuvre[125]. Regarding Plutarch's scholarly and literate approach in scientific matters, Sirinelli observes that: "Le nombre des citations qu'il propose et leur diversité sont destinés souvent, plus qu'à fortifier son argumentation, à faire admirer la force de sa culture et l'étendue de sa documentation."[126] From a sociological perspective, one can, indeed, argue that profound knowledge of the tradition testifies to a person's degree of πολυμάθεια and παιδεία in a specific field of expertise (therefore, presumably, Plutarch's quotations are mostly nominatim)[127]. However, restricting things to matters of intellectual ostentation – to be measured only in terms of Bourdieu's

---

[124] S.-T. Teodorsson, 1999a, p. 666. *Pace* E. Lelli, 2010, p. 849: "Ancora di taglio dossografico sono gli *Aitia physica*, elenco di dottrine su disparati argomenti di scienza naturale."

[125] Cf. G. Roskam, 2011c, p. 420. On the unity in Plutarch's work more generally, see J. Barthelmess, 1986, pp. 62–64 and the contributions in A.G. Nikolaidis, 2008.

[126] J. Sirinelli, 2000, p. 363. He adds: "Il est dans toute l'extension du terme un *philologos*. Il suffit de le comparer à cet égard à Sénèque, Épictète et Dion Chrysostome pour discerner clairement où il veut situer son originalité et sa supériorité." Cf. also F. Klotz, 2014, p. 208. For fierce criticism of an exclusively sociological reading of (serious) philosophical texts, see J. Opsomer, 2014, pp. 90–91: "It is true that knowledge and power are often closely intertwined. Yet it would be a vast exaggeration to claim – as some theorists in their awe for Foucault do – that it would invariably be illicit to study philosophical ideas without looking at their relation to cultural power structures. Just as the history of philosophy is not eliminatively reducible to rhetoric, social and cultural contexts alone cannot account for why persons hold certain views. Intellectual traditions, philosophical argument, hermeneutics, and theoretical constraints linking various ideas across philosophical sub-domains are much more important in accounting for the views of serious philosophers."

[127] In a sympotic context, the πεπαιδευμένος should be able to flag his sources by name, and if he fails, this is at the risk of becoming an object of ridicule (cf. *Quaest. conv.* 675DC). Then again, as C. Pelling, 2011, pp. 216–217 remarks (regarding *De se ipsum laud.* 544A), forgetfulness is "one of the engaging weaknesses one can readily admit to" (cf. also D.A. Russell, 1993, p. 431 and F. Klotz, 2007, p. 661). On the aspect of memory and its vagaries in *Quaestiones convivales* more generally, see K. Oikonomopoulou, 2011, pp. 108–123.

cultural capitalism – would clearly impoverish the actual historical value of Plutarch's scientific project, as it would fail to appropriately account for its higher philosophical goals (discussed earlier on [see 3.2.2.]). Nevertheless, recent studies have especially highlighted the importance of rhetorical display (ἐπίδειξις) in scientific works produced in the early Roman Empire, often drawing links with the Second Sophistic's culture of παιδεία, and this feature has often – and increasingly so – also been discussed for Plutarch's *Quaestiones convivales*, as we saw[128] [see 3.1.3., n. 46]. However, with respect to the sympotic discussions described in this work, one may wonder what there really is to gain in terms of socio-cultural prestige, when – not just from a social, but also from an underlying epistemological perspective – the symposiasts are, in fact, each other's intellectual peers and equals [see 4.3.2.]. The relative plausibility of the arguments is, indeed, an important criterion in the participants' personal investments in the discussion, but the only common currency that is really at stake, in the end, is the philosophical truth (πολυμάθεια being just a means to an end for Plutarch). So why should we doubt Plutarch's sincere and honest interest in the subject matter if at the heart of his research project lies a higher philosophical motive? In this light, a person's acquaintance with traditional knowledge is important to the more introvert development, rather than extravert ostentation, of personal research and reflection, in that it underpins any personal arguments or theories that sprout from it, and in which the author personally believes, or to which he at least ascribes a certain measure of plausibility (bearing in mind the principle of charity [see the prologue, n. 3]).

Let it be clear, moreover, that close familiarity with the tradition was a quintessential predisposition for an eager scholar and intellectual in Plutarch's time. However, the ability to handle this knowledge in a personal and original way, by assessing it critically or by adapting it where possible and where necessary to the new context, is even more important than the mere reproduction or 'performance' of received knowledge *tel quel* [see 1.2.3., n. 202]. In this sense, a person's acquaintance with the tradition plays an essential role in the search for plausible explanations[129]. If it is true, moreover, that one can only be original and be fully aware of it if one 'knows' the tradition, then it is equally true that great learning is not a τέλος in itself, but provides many 'starting-points' (cf. the notion of ἀρχαί in *Quaest. conv.* 734D [quoted 3.2.1.]). Accordingly, Plutarch is well aware that firm acquaintance with the tradition only provides a

---

[128] See, e.g., M. Gleason, 2009 for the performative dimension of Galen's anatomy demonstrations (ἐπιδείξεις). Notably, in *De ad. et am.* 71A, Plutarch vituperates the shamelessness of physicians to conduct operations in theatres in order to gain new clients.

[129] Cf. *De virt. mor.* 440E: Βέλτιον δὲ βραχέως ἐπιδραμεῖν καὶ τὰ τῶν ἑτέρων, οὐχ ἱστορίας ἕνεκα μᾶλλον ἢ τοῦ σαφέστερα γενέσθαι τὰ οἰκεῖα καὶ βεβαιότερα προεκτεθέντων ἐκείνων.

stable foothold for his research, because inventiveness requires knowledge to be 'performative', that is, open to rhetorical-argumentative reuse and adaptation. In *De prim. frig.* 952C–955C, for instance, the theory that earth is the principle of cold is commonly understood as Plutarch's "original contribution to theoretical physics"[130]. Nevertheless, it is clear from the preceding doxography that Plutarch only managed to formulate this theory by critically assessing several other traditional opinions (viz. of Aristotle, the Stoics, Empedocles and Strato). Therefore, Plutarch's inventiveness and originality does not arise *ex nihilo*, but from a thorough acquaintance with the tradition. The following section will examine this point in further detail for Plutarch's natural problems.

2. *The pragmatics of Plutarch's scientific ingenuity and creativity*

Throughout the *corpus Plutarcheum*, improvised ingenuity is designated by the term εὑρησιλογία. The verb derived from it, εὑρησιλογέω, means "to invent ingenious arguments, explanations or pretexts"[131], but it often has a pejorative connotation, as it often applies to rhetorical and slightly sophistical practices. Plutarch considers it a most valuable asset when used in the right proportion, but if an explanation becomes too ingenious or rhetorical, he criticises it as constituting bad philosophy[132].

We learn a great deal from *Quaestiones convivales* regarding the pragmatics of Plutarch's conception of ingenuity and originality in solving problems. As we saw earlier on, for Plutarch, sympotic events provide an occasion for social and intellectual interaction, where each symposiast tries to contribute and bring his own share to the debate [see 3.1.3.]. Notably, even if a solution has already been deemed adequate, this is no reason to stop looking for new arguments. In *Quaest. conv.* 656D,

---

[130] H. Cherniss and W.C. Helmbold, 1957, p. 227. The originality of Plutarch's theory was also acknowledged by O. Longo, 1992, p. 228, but G. D'Ippolito and G. Nuzzo, 2012, p. 64 remain uncertain.

[131] LSJ, s.v.

[132] Plutarch places the term in a good light in *De def. or.* 414A and in *Quaest. conv.* 642A and 656A, but it more often receives a negative connotation (it is even related to inebriety in *Quaest. conv.* 682BC; cf. also 700E). Plutarch notes that verbal ingenuity in poetic texts can corrupt the young (*De aud. poet.* 28A), and that it tends to constitute bad philosophy (*De aud. poet.* 31E, *De comm. not.* 1070F, 1072F). In the worst case, it can even be directly opposed to serious philosophy, being synonymous rather with vainglorious παιδιά (*De Stoic. rep.* 1033B). From *Quaest. Rom.* 283C and *Quaest. conv.* 625C, 656A, 682B, we also learn that even though the author considers a *causa* to be an unsound conjecture, he may still write it down in a collection of problems, albeit simply to disprove it or to provide an alternative explanation for it. G. Roskam, 2009, p. 373 has argued that Stoic philosophers, in particular, are blamed for their sophistical ingenuities. See also the section on εὑρησιλογία in K. Oikonomopoulou, 2011, pp. 120–123 and M. Meeusen, 2012a.

for instance, Plutarch's father reproduces Aristotle's explanation for the problem of why the so-called tipsy are more deranged than very drunk people. Aristotle argued that tipsy people judge badly because they follow illusory appearances. Plutarch's father is not very enthusiastic about the explanation, though, and invites his fellow symposiasts to say 'something of their own' on the subject (τι ἴδιον ἐπιχειρήσομεν εἰπεῖν). Before reproducing Aristotle's explanation, he notes that, 'even though Aristotle is normally very sharp in such investigations, he does not seem to have sufficiently examined the cause' of the problem at hand (οὐ γὰρ ἱκανῶς μοι δοκεῖ, καίπερ ὀξύτατος ὢν ἐν τοῖς τοιούτοις ζητήμασι, διηκριβωκέναι τὴν αἰτίαν)[133]. On his father's request, Plutarch then examines Aristotle's explanation and concludes that 'it is sufficient as far as causality is concerned' (ἀποχρῶν οὗτος ἦν πρὸς τὴν αἰτίαν ὁ λόγος). But, even so, he still feels prompted to add 'something of his own' (ἴδιόν τι), which he does by arguing that the power of wine is variable in proportion to its quantity. Apparently, Aristotle's explanation attains a certain level of sufficiency (ἀποχρῶν) from a causal perspective, but this does not preclude the formulation of yet other explanations, presumably because for Plutarch (both as author and as sympotic character) the aspect of plausibility is the only criterion that matters in these discussions, rather than the singling out of one ultimately correct explanation[134].

Another relevant passage, where Aristotle's authority is again central, is found in the previous talk from *Quaest. conv.* 655D–656B, which is set in the same sympotic context. The problem at hand is why γλεῦκος ('must'; i.e. sweet, new wine) is least intoxicating. Two young philosophers make their own attempt at an explanation: Hagias argues that the excessive sweetness of the wine prevents people from drinking a quantity that is sufficient for intoxication, and Aristaenetus of Nicaea argues that the sweetness blunts the intoxicating effects of the wine[135]. Their ingenuity is heartily approved by the group, because, so Plutarch writes, 'they did not fall upon the evident arguments but attained their personal explanations' (Σφόδρ' οὖν ἀπεδεξάμεθα τὴν εὑρησιλογίαν τῶν νεανίσκων, ὅτι τοῖς ἐμποδὼν οὐκ ἐπιπεσόντες ἰδίων ηὐπόρησαν ἐπιχειρημάτων). This does not withhold Plutarch from also adding some arguments which he describes as 'at hand and easy to comprehend' (τά γε πρόχειρα καὶ ῥᾴδια λαβεῖν): these are the heaviness

---

[133] Parallels are found for this problem in Ps.-Arist., *Probl.* 871a8–16 and 875a29–40.

[134] Cf. F. Frazier and J. Sirinelli, 1996, p. 200 (see also pp. 197–200 more generally): "l'originalité n'est pas imagination débridée, mais élaboration d'une théorie plausible".

[135] The phrase ἔν τισιν ἐνίοις γράμμασιν ἀνεγνωκὼς ἔφη μνημονεύειν, ὅτι γλεῦκος μιχθὲν οἴνῳ παύει μέθην in Aristaenetus' account is an implicit reference to Ps.-Arist., *Probl.* 872b32–873a4 (see S.-T. Teodorsson, 1989, p. 370). Perhaps, Aristotle is not named explicitly as Aristaenetus' source, in order to (falsely) underline the aspect of personal ingenuity in his argument (see below).

of the wine, which, as Aristotle – presumably in one of his lost natural problems – says, breaks through the stomach[136], and the large quantity of pneumatic and watery substances that are mixed with the wine (= fr. 220 Rose). Plutarch thus shows that the formulation of personal ingenuities should not necessarily go at the cost of mentioning also the more obvious, traditional (c.q. Aristotelian) explanations, seeing that these also bear direct relevance to the discussion at hand and to its search for plausible arguments. It is precisely herein, then, that Plutarch's subtle criticism of the young philosophers presumably lies.

A most relevant passage to illustrate the importance of received knowledge as a starting-point for personal creativity is found in *Quaest. conv.* 694D, where the symposiasts are looking for the cause of 'ox-hunger' (βουλιμία). This problem is treated in Ps.-Arist., *Probl.* 887b38–888a23, where the same processes are central as described in Plutarch's explanation. After an introduction of the problem, Plutarch in an interior monologue reflects on the importance of argumentative creativity in the context of intellectual inquiry. Notably, Plutarch very seldom uses this technique of interior monologue, as attested here, which only highlights the importance of the passage at issue[137].

Γενομένης δὲ σιωπῆς, ἐγὼ συννοῶν ὅτι τὰ τῶν πρεσβυτέρων ἐπιχειρήματα τοὺς μὲν ἀργοὺς καὶ ἀφυεῖς οἷον ἀναπαύει καὶ ἀναπίμπλησι, τοῖς δὲ φιλοτίμοις καὶ φιλολόγοις ἀρχὴν ἐνδίδωσιν οἰκείαν καὶ τόλμαν ἐπὶ τὸ ζητεῖν καὶ ἀνιχνεύειν τὴν ἀλήθειαν

There was a silence during which I reflected that it suits the dull and unschooled to accept and be full of the solutions provided by our predecessors, whereas to the ambitious and learned it provides a familiar beginning and an encouragement to search and track down the truth.

The idea that traditional knowledge provides a beginning (ἀρχή) for zetetic discussions may very well be an echo of Aristotle's quote in *Quaest. conv.* 734D, according to which 'great learning provides many starting-points' (ἀρχαί [quoted 3.2.1.])[138]. In what follows, Plutarch suits the action

---

[136] See S.-T. Teodorsson, 1989, p. 371. R. Mayhew, 2011a, p. 111, n. 23 draws a parallel with Ps.-Arist., *Probl.* 872b25–32 and 874b13–21.

[137] See S.-T. Teodorsson, 1990a, p. 289: "This example of inner soliloquy is unique in the *Talks* and rare in Ancient literature on the whole." (With the well-known exception, of course, of Marcus Aurelius' Τὰ εἰς ἑαυτόν.)

[138] See J. König, 2007, p. 57: "This passage [i.e. *Quaest. conv.* 734D] is typical of patterns which are repeated over and over again throughout the *Sympotic questions*: the use of past authority to provide a stimulus for present discussion; explicit recommendation

to the word. He first brings Aristotle's account to mind (ἐμνήσθην τῶν Ἀριστοτελικῶν), where the cause of the disease of βουλιμία is found in the processes of heating and colliquation. The discussion then proceeds, some persons attacking Aristotle's theory, others advocating it, and this is 'only reasonable', so Plutarch writes (Ὅπερ οὖν εἰκός, τοῦ λόγου λεχθέντος ἐπεραίνετο, τῶν μὲν ἐπιφυομένων τῷ δόγματι τῶν δ' ὑπερδικούντων). The emphasis on εἰκός here, implies that the continuation of the discussion by the formation of two camps is a logical consequence in the context of the debate, since those symposiasts advocating Aristotle will have to come up with new arguments against those attacking him. This clearly indicates that received (c.q. Aristotelian) knowledge only provides an incentive (ἀρχή) for discussion and should not simply be taken for granted, as was, indeed, highlighted in the interior monologue.

What we learn from these passages, then, is that traditional authority often functions as a starting-shot in the race for the truth[139]. It provides an ἀρχή for zetetic ingenuity, that is, a first step in the direction of an innovative explanation. However, originality and creative, improvised speculation are valued more than a person's mere acquaintance with past authorities. Therefore, a true intellectual should by no means be content with the tradition, but when he eventually ventures upon original speculation, it will often be in a progression and advancement of received knowledge[140].

The same ambivalence can be seen throughout *Quaestiones naturales*, especially in those passages where Plutarch relies on traditional authorities and at the same time displays a strong sense for aetiological originality. In these cases, Plutarch neatly balances past authorities with personal speculations. To this end, he sometimes marks his criticism of traditional authorities in a very explicit way. This is the case, for instance, in *Q.N.* 12, 914F regarding the problem of why oil that is sprinkled on sea-water causes

---

of independent thought [...]; and use of the language of contribution to describe individual attempts at explanation." Cf. also *De aud.* 48C: τὸν ἀλλότριον λόγον οἷον ἀρχὴν καὶ σπέρμα λαβόντας κτλ. [quoted 3.1.2.].

[139] For a more inclusive analysis of the construction and deconstruction of Aristotle's authority in Plutarch's natural problems, see M. Meeusen, 2016.

[140] As J. König, 2011, p. 190 observes, "[p]ast and present speak with each other particularly within the all-embracing framework of the symposium." Plutarch recurrently deploys the metaphor of "entering into conversation with the past" (*id.*, 2008, p. 90, with n. 18): cf., e.g., *Quaest. conv.* 651F, 653B, 718C. The imagery used in staging traditional authorities often becomes very plastic, so that we come across "vivid metaphors which depict the quoted text as an object in its own right, to be controlled and mastered by the symposiasts" (*id.*, 2011, pp. 200–201). In *Quaest. conv.* 734F–735A, most notably, Democritus' theory of the 'spectral films' (εἴδωλα) is compared with an old weapon Florus brushes up in his own argument, and in 735C the language from the world of boxing and wrestling is used.

clearness and calm (καταφάνεια καὶ γαλήνη). According to Aristotle, so Plutarch writes in the first *causa*, the wind, slipping off the smoothness (so caused by the oil), makes no impact and raises no surge. Plutarch criticises this view in the second *causa*, where he highlights the incompleteness of Aristotle's theory: the explanation is plausible, but only so regarding the external aspect of the phenomenon (ἢ τοῦτο μὲν πιθανῶς εἴρηται πρὸς τὰ ἐκτός). He then draws attention to the internal (c.q. submarine) aspect of the problem by referring to diver lore: divers take oil into their mouth and blow it out in the depths, so that they may have light and transparency when under water. As Plutarch notes, it is impossible to adduce the cause here to slippage of the wind, as Aristotle did (οὐκ ἔστιν ἐκεῖ πνεύματος ὄλισθον αἰτιάσασθαι). By reason of its denseness, so Plutarch further explains, the oil (in its movement out of the divers' mouth), pushes and forces the sea aside, which is earthy and irregular (and thus cannot mix with the oil). Afterwards, when the sea flows back to itself and draws together, intermediate passages are left, which provide transparency and clearness to the eyes. What this argument nicely illustrates is how Plutarch aims to criticise Aristotle's theory not necessarily by rejecting it (he considers it a plausible point of departure, after all), but by further elaborating upon it. Aristotle only explained the external aspect of calm (γαλήνη), as mentioned in the *quaestio*, whereas Plutarch's second *causa* deals with the internal clearness (καταφάνεια) caused by the oil. In the third *causa*, Plutarch will eventually try to combine these two aspects (viz. of καταφάνεια καὶ γαλήνη) in an attempt to formulate a complete solution to the problem (I will come back to this argument later [see 4.3.3.1.]).

Another good example where Plutarch makes a genuine attempt at a hybridisation of traditional material with his own innovative insights is found in *Q.N.* 19, where the octopus' change of colour is central[141]. In the first *causa*, Plutarch writes that Theophrastus ascribes this change to the octopus' cowardice: fear triggers a physiological process in the body under the influence of the animal's breath. Plutarch considers this theory to be 'plausible but insufficient' (πιθανῶς […] οὐχ ἱκανῶς), since Theophrastus only explains the change of the colour but not its adaptation to the animal's surroundings (this was not, however, mentioned in the *quaestio*). The incompleteness of Theophrastus' account is illustrated in the second *causa*, where Plutarch provides two poetical quotations. Pindar and Theognis mention the adaptation of the octopus' colour in their verses, but they do not provide an apposite explanation for the physical mechanism behind this phenomenon. This is not the effect of a deliberate choice, after all, but of underlying physical causes. Plutarch's own theory, which follows in the third *causa*, sets out on explaining what Theophrastus

---

[141] For a separate case study of *Q.N.* 19 in light of Plutarch's argumentative creativity in his natural problems, See M. Meeusen, 2012a.

left unsolved (that is the aspect of the colour's adaptation). In so doing, Plutarch refers to Empedocles' theory of emanations and argues that the emanations from nearby objects interlock in the pores of the octopus' skin. The pores contract when the animal feels fear, and, thus, change and adapt the animal's colour. Importantly, at the beginning of his (in all likelihood) personal explanation in the third *causa*, Plutarch alludes back to Theophrastus' initial theory (viz. fear triggers a pneumatic process in the body) and suggests that it contains 'the starting-point but not the most important aspect of the explanation, which lies elsewhere' (ἆρ' οὖν τὴν μὲν ἀρχὴν αὐτὸς ἐνδίδωσι τοῦ πάθους δείσας, τὰ δὲ κύρια τῆς αἰτίας ἐν ἄλλοις ἐστί;). This clearly indicates how a traditional (c.q. Theophrastean) theory again provides an ἀρχή for Plutarch's own original contribution to the problem. Plutarch, thus, shows that he aims to complete what Theophrastus left unsolved. To this end, he creatively calls on the authority of the poets and of Empedocles to introduce his own original contribution to the problem.

I conclude that by remodelling traditional theories in a problematised fashion in his natural problems (viz. by remoulding them in the problem format) and simultaneously by looking for original viewpoints, Plutarch, somehow advanced the science of his day one problem at a time. It is at least so that by underlining the innovativeness of some of his explanations he clearly shows that what is to be avoided in solving natural problems is intellectual lazyness (his insistence on providing personal responses to problems more generally is also stressed in *De aud*. 48BC, as we saw previously [quoted 3.1.2.]). Plutarch's concept of originality may not, however, be as adventurous as some modern scholars may have hoped [see prologue, n. 79]. In fact, his idea of zetetic originality is rather ambivalent: it departs from the tradition, to which it permanently looks back, and at the same time looks forward to new, innovative perspectives[142]. As such, the innovative dynamic in Plutarch's zetetic project is strongly intertwined with the incorporation of traditional material and authority. In almost half of the natural problems collected in *Quaestiones naturales*, and often more than once in the same problem chapter, Plutarch quotes a wide variety of authorities by name. However, the quotations not only testify to the author's scholarly acumen and πολυμάθεια, but they also come in handy for the sake of heuristics itself, in that they provide a useful 'starting-point' (ἀρχή) for further discussion.

We can now return to where we started from, by stressing Plutarch's scholarly and literate approach in his natural problems. As we have already seen before, the Chaeronean's study of natural phenomena remains situated on a theoretical level, so that it is for a great part detached from what is

---

[142] For the close relationship between tradition and innovation as a common feature of ancient Greek scientific literature more generally, see, e.g., G.E.R. Lloyd, 1987, pp. 50–108, T.S. Barton, 1994a, pp. 149–152 and J. König, 2011, p. 182.

positively given in nature itself. As we will see in the following section, this has great repercussions for Plutarch's scientific methodology, and also for a proper understanding of it in light of contemporary philosophical debates. The remainder of this chapter will, therefore, be devoted to an analysis of several aspects of Plutarch's scientific method in solving natural problems in relation to his other natural scientific writings. This will further reveal the world view that Plutarch is promoting in these writings.

4.3. *Plutarch's scientific methodology: a rough guide to explaining natural phenomena*

The main goal of this section is to demonstrate that there are significant correspondences between the aetiological design of *Quaestiones naturales* and the overall method Plutarch employs in his other natural scientific writings. In analysing Plutarch's scientific approach in his natural problems, I will pay special attention to the following topics: his main attention for the material side of natural phenomena [4.3.1.], the epistemological limits of this type of inquiry [4.3.2.], its logical and rhetorical dynamics [4.3.3.], and the scientific terminology that Plutarch employs in his aetiologies [4.3.4.]. It seems appropriate to treat these more 'technical' aspects of Plutarch's scientific methodology under a separate heading. Nevertheless, the two preceding sections in this chapter are still very relevant for the analysis at hand. After all, a good understanding of Plutarch's dualistic view on causality is seminal for a proper demarcation of the ontological and epistemological backdrop of Plutarch's scientific project. Likewise, the aspect of scientific authority is important for examining the intellectual backdrop of the physical theories and concepts Plutarch employs in his aetiologies. The question as to whether he really envisages a reconciliation between the Aristotelian/Peripatetic and the Platonic/Academic tradition in his natural problems by blending both traditions into his general explanatory scheme will be revisited at the end of this chapter as a means to conclude the first part of this study (the question itself was already raised at the beginning of the first chapter [see 1.1.2.]).

4.3.1. **Material principles and natural processes**

In his natural problems, Plutarch has no ambition to be very precise or 'exact', at least in the sense that his approach is not of a quantitative but of a qualitative kind. He is mainly concerned with theoretical speculations regarding natural substances, their properties and the processes to which they are subject[143]. These substances are often described without any

---

[143] For an account of the "carattere prevalentemente qualitativo delle relazioni strutturali" in *Quaestiones naturales*, cf. also L. Senzasono, 2006, pp. 41–44, esp. pp. 41–42:

consideration of what lies at the very heart of them, in terms, for instance, of geometrical solids, atoms or other principles. Moreover, some things in nature simply are 'by nature', which probably hints at an aspect of natural necessity and in most cases does not appear to require any further explanation[144].

In his natural problems, Plutarch does not carry out any mathematical measurements in order to abstract quantifiable data from physical reality. This *modus operandi* of taking mathematical measurements was common in other, more exact ancient scientific disciplines, such as geometry, astronomy, acoustics, harmonics, optics, catoptrics, statics, hydrostatics, mechanics, but not in ancient physics, meteorology biology and medicine. In *De E* 387F, Plutarch famously states that he was enthusiastically devoted to mathematics during his youth but began to hold to the adage 'avoid extremes' soon after entering the Academy (τηνικαῦτα προσεκείμην τοῖς μαθήμασιν ἐμπαθῶς, τάχα δὴ μέλλων εἰς πάντα τιμήσειν τό 'μηδὲν ἄγαν' ἐν Ἀκαδημείᾳ γενόμενος)[145]. Even so, the astrophysical section in *De facie*, with its references to astronomical theory and calculation (e.g., *De facie* 935DE), shows that Plutarch's interest in mathematical matters was not completely doused by his conversion to the Academy. Nevertheless, the physical-aetiological parts of his scientific writings clearly outweigh the exact mathematical ones[146]. Notably, in *Quaest. conv.* 720E, Plutarch's

---

"Si tratta infatti soprattutto di problemi che oggi chiameremmo biologici o fisio-patologici e nel mondo antico in tale ambito di ricerca scientifica non interessavano le relazioni quantitative, diversamente da quanto accadeva in campo astronomico e geografico o comunque dove fosse ritenuta possible una geometrizzazione o matematizzazione della realtà fisica." Senzasono is primarily concerned with the opposition between the qualitative and the, at times, indeed, more quantitative approach of the phenomena in *Quaestiones naturales*, but unfortunately the former category is not greatly substantiated in his analysis, and the accounts he considers to be exceptions to the qualitative approach may not be that exceptional, as we will see further on. Senzasono does not, moreover, explain Plutarch's qualitative approach in light of his more general scientific method.

[144] See *Q.N.* 2, 911F (πέφυκε), 6, 913F (φύσει), 10, 914D (πεφυκότα), 12, 915A (φύσει), 18, 916A (φύσει), 19, 916B (φύσει), 31, 919D (φύσιν ἔχων [...] πέφυκεν), 41 (φυσικῶς – the translation of F.H. Sandbach, 1965, p. 227 is rather pregnant: "by a law of nature"). The phrase παρὰ φύσιν occurs only once (*Q.N.* 31, 919D). Naturally, such phrases are very common in ancient Greek science (cf., e.g., *De def. or.* 424C: πῇ μὲν ἐν ταῖς κατὰ φύσιν χώραις ὑπάρχειν, πῇ δ' ἐν ταῖς παρὰ φύσιν). Therefore, a strict connection with Aristotelian terminology seems unlikely (*pace* L. Senzasono, 2006, pp. 236–237, n. 182).

[145] For discussion of this passage in light of Plutarch's philosophical career, see D. Babut, 1994, pp. 556–558 (cf. also J. Opsomer, 1998, p. 130). For Plutarch's attitude towards Platonic mathematics more generally, see M. Isnardi Parente, 1992. Cf. also L. Senzasono, 2006, p. 42 (with n. 63).

[146] The distinction between the mathematical sciences and knowledge related to the sensible world is, indeed, very Platonic. Cf., e.g., *Quaest. conv.* 718DE and 744D.

Platonic teacher, Ammonius, says that 'the proper task of the φυσικός is to study material and organic principles' (τὰς ὑλικὰς καὶ ὀργανικὰς ἀρχὰς); he also calls for an investigation of 'the causes which operate by the inevitable process of nature' (τὰ δι' ἀνάγκης φύσει περαινόμενα τῶν αἰτίων ἀνευρίσκειν)[147]. This allows us to analyse in more detail how these aetiological categories figure in Plutarch's *Quaestiones naturales*.

1. Material principles

As to the material principles, which concern the constitutive organisation of physical bodies and their attributes, Plutarch in his natural problems traces all of physical matter to the well-known Empedoclean scheme of the four primary elements (fr. DK31B17), viz. earth, water, air and fire, to which breath (πνεῦμα) is added[148]. These primary elements are pure and unmixed (cf. *De prim. frig.* 955A: γῆν [...] αὐτὴν καθ' αὐτὴν ἀποκεκριμένην τῶν ἄλλων). As such, they function as the elementary building blocks for all composite material bodies that appear in nature (such as blood, seawater, wood etc.). As always, πνεῦμα is a special case. Being the basic result of πνεῖν ('to blow', or 'breathe'), πνεῦμα is essentially air containing a specific motive force (i.e. air in motion)[149], but it is also related to fire (*Q.N.* 32: *ignea et spirabilis facultas*)[150]. Considering its motive force, πνεῦμα has a lot in common with wind (*Q.N.* 12, 914F–915A, 14, 915D, 18, 916A) and is closely affiliated with ἄνεμος (*Q.N.* 8, 914B, 34: *ventus*). However, it has a more 'elementary' value, as it retains its motive force *ad infinitum*[151]. Therefore, rainwater falling from the sky is imparted not just with air, but with 'air in motion', that is, πνεῦμα (*Q.N.* 2, 912AC, 4, 912F). Similarly, in wine, the pneumatic (and the watery) substances are considerably unstable and, thus, inclined to suffer change (*Q.N.* 10, 914D). Plutarch mainly uses

---

[147] *Quaest. conv.* 720E: δεῖ καὶ τὰ δι' ἀνάγκης φύσει περαινόμενα τῶν αἰτίων ἀνευρίσκειν καὶ τοῦτο τοῦ φυσικοῦ ἴδιόν ἐστιν, ἡ περὶ τὰς ὑλικὰς καὶ ὀργανικὰς ἀρχὰς πραγματεία.

[148] The lack of a fully adequate English equivalent for the Greek πνεῦμα is well-known (see F.H. Sandbach, 1965, p. 141 and p. 187, n. f, who at times prefers to transliterate the concept). For the scheme of the four traditional elements, cf., e.g., *De prim. frig.* 947E and *Aqua an ignis* 957A. For the five worlds theory and the scheme of the five primary bodies related to it, see *De E* 389F–390A and *De def. or.* 430A–431A (with αἰθήρ instead of πνεῦμα as the fifth primary body). See G.E. Karamanolis, 2006, pp. 104–105.

[149] Cf. F.H. Sandbach, 1965, pp. 140–141 and L. Senzasono, 2006, pp. 151–152, n. 18.

[150] Notably, Aristotle describes πνεῦμα as being 'analogous to the element of the stars', that is, αἰθήρ (*GA* 736b37–737a1: ἀνάλογον οὖσα τῷ τῶν ἄστρων στοιχείῳ). Therefore, even if αἰθήρ is also a fiery substance, it is not simply identical to πνεῦμα (cf., e.g., *De prim. frig.* 951D and Pl., *Epin.* 981c).

[151] By contrast, in his distinction between ἀήρ and πνεῦμα in *Top.* 127a3–13 (οὐκ ἔστιν ὅλως ἀὴρ τὸ πνεῦμα), Aristotle defines the latter as the 'motion of air' rather than 'air in motion' (which has a more elementary value).

the concept in a material sense in his natural problems[152]. Therefore, the 'breath of life' in the lungs, probably has the same properties as regular πνεῦμα (*Q.N.* 36: *spiritus vitalis*). In *Q.N.* 19, 916BE, the πνεῦμα acts as a connector between the octopus' soul and body: the passive impulse of fear in the animal's ψυχή triggers an active movement and change in the body by the intermediation of the πνεῦμα[153]. In this case, Plutarch refers to Peripatetic pneumatology (he quotes Theophr., fr. 188 Wimmer = 365C FHSG), but πνεῦμα theory had become very common in ancient Greek physics and medicine by Plutarch's time. The pneumatic school of medicine (1st century AD) deserves specific mention here; the doctrines of this school (if we may call it that) were congenial to the πνεῦμα theory of the Stoics. The same connection can perhaps also be drawn for the concept of 'transpiration', which implies an active passage of air through an object or body (διαπνέω: cf. *Q.N.* 22, 917D and 27, 918EF)[154].

In line with traditional physical theory, Plutarch further specifies the four primary elements (earth, water, air and fire) by the four primary qualities (cf. *De prim. frig.* 947E: ποιότητες πρῶται), viz. heat, cold, dryness and moistness. Some of these qualities are specifically linked to a physical substance: heat, for instance, is 'innate' to seawater (*Q.N.* 8, 914B: σύμφυτον)[155]. To these primary qualities, Plutarch adds a vast number

---

[152] Yet, in other contexts, πνεῦμα has a more 'super-natural' implication, as being "something midway between the material and the spiritual" (H.W. Parke and D.E.W. Wormell, 1956, p. 23). Notably, in *De def. or.* 437C–438D, we read that the δύναμις of the hallucinogenic πνεῦμα that is released in the vicinity of the Delphic oracle 'comes from the gods and demigods, but, for all that, it is not unfailing nor imperishable nor ageless, lasting into that infinite time by which all things between earth and moon become wearied out, according to our reasoning' (438CD: ἔστι δὲ θεία μὲν ὄντως καὶ δαιμόνιος, οὐ μὴν ἀνέκλειπτος οὐδ' ἄφθαρτος οὐδ' ἀγήρως οὐδὲ διαρκὴς εἰς τὸν ἄπειρον χρόνον ὑφ' οὗ πάντα κάμνει τὰ μεταξὺ γῆς καὶ σελήνης κατὰ τὸν ἡμέτερον λόγον).

[153] Cf. G. Verbeke, 1945, p. 266: "Cette explication [sc. the third *causa* in *Q.N.* 19] est intéressante à notre point de vue parce qu'elle révèle le pneuma comme principe de la vie et du mouvement." Cf. also L. Senzasono, 2006, p. 196, n. 96. Aristotle famously describes πνεῦμα as an ὄργανον of the soul, by which movement is imparted to the body (*DA* 433b18, *GA* 789b8–9; cf. also esp. *MA* 703a3–b1). Cf. F.H. Sandbach, 1965, p. 141: "a living being contains air in motion that performs some of the functions that our physiology gives to the nervous system and the hormones".

[154] On Plutarch's conceptualisation of πνεῦμα more generally, see G. Verbeke, 1945, pp. 260–287 (esp. p. 267 for the link with Stoic and medical theory). On pneumatic processes in Ps.-Aristotle's *Problems*, see M. Meeusen, forthcoming g.

[155] This idea recurs several times in the first section of problems on salt and water in *Quaestiones naturales* (cf. *Q.N.* 1, 7–10, 13). Cf. also, e.g., Arist., *Mete.* 358b6–9. The Aristotelian doctrine of σύμφυτον θερμόν as a life-sustaining principle is already present in the Hippocratic writings. Cf. also, e.g., *Quaest. conv.* 635C (with S.-T. Teodorsson, 1989, p. 208).

of secondary qualities[156]: physical substances can be heavy, light, large, small, transparent, obscure, smooth, rough, firm, loose etc. Moreover, from the primary elements, specific elementary qualities are derived, viz. earthy, watery, airy, fiery and pneumatic. If one or more of these elementary qualities is ascribed to a certain substance, this means that the substance not only contains the qualities of that specific element (either primary or secondary; cf., e.g., *Q.N.* 1, 911D: ἐμβριθές ἐστι καὶ γεῶδες), but also a share of that element itself. This means that physical qualities have a specific material value for Plutarch.

The elements and qualities are the material fundaments on which Plutarch's natural world, as depicted in his natural problems, is built. They provide a terminological arsenal of concepts that can be freely used in explaining all different kinds of natural phenomena. One issue that deserves special attention here (and that has just been highlighted) concerns the direct association between physical matter and its qualities in Plutarch's natural science. In this regard, Sandbach rightly points out that Plutarch "tends to think of a quality as specifically linked to a substance"[157]. The opening sentence of *De primo frigido* illustrates this nicely. Here, Plutarch wonders whether there is a principle of cold, just as fire is of what is hot: nothing has the quality of heat without containing fire, just as nothing can become cold without containing the principle of cold (*De prim. frig.* 945F: Ἔστι τις ἄρα τοῦ ψυχροῦ δύναμις, ὦ Φαβωρῖνε, πρώτη καὶ οὐσία, καθάπερ τοῦ θερμοῦ τὸ πῦρ, ἧς παρουσίᾳ τινὶ καὶ μετοχῇ γίνεται τῶν ἄλλων ἕκαστον ψυχρόν;). Plutarch will eventually accept this thesis by rejecting the opposite (Aristotelian) theory that cold is a 'privation' (στέρησις) of heat, and by arguing that earth is the principle of cold (as opposed to what the Stoics, Empedocles and Strato believed). This is not just a metonymical ingenuity, but a genuine physical scheme that pervades Plutarch's natural scientific writings.

There are numerous instances in *Quaestiones naturales* where the same link between physical matter and its qualities is clearly present. Plutarch does not, however, apply it to the primary elements and qualities only but also to the compound material bodies constituted by them. In *Q.N.* 17, 915F, for instance, the hairs of male horses are considered stronger than those of female horses, by analogy with the general strength in the horses' other body parts: the idea is that the male animal is 'essentially' strong so that it must also have strong hairs, whereas the female is 'essentially' weak so that its hairs must be weak. Similarly, in *Q.N.* 32, it is argued that the great strength of the palm tree as a whole is also present in its

---

[156] For the distinction between primary and secondary qualities, cf., e.g., Arist., *PA* 646a15.

[157] F.H. Sandbach, 1965, pp. 139–140. Cf. also L. Senzasono, 2006, p. 38: "la causa materiale o proprietà potenziale deviene termine di una relazione unitaria".

separate parts, that is, in its trunk but also even in its soft and tender branches and twigs. Seeing that there is no clear distinction between physical matter and its qualities in Plutarch's natural problems, the difference between the particular phrasing of material substances and the qualities is not always crystal-clear either. The ancient Greek language would enable one to distinguish, for instance, between τὸ ὑγρόν (i.e. 'a substance characterised by moistness', 'moisture': e.g., *Q.N.* 1, 911E) and ἡ ὑγρότης (i.e. 'the quality of moistness', 'moistness': e.g., *Q.N.* 14, 915D). In general, Plutarch complies with this distinction, but his phrasing is not always consistent, because the adjectival phrase (τὸ ὑγρόν) occasionally seems to indicate a quality and the common noun (ἡ ὑγρότης) a substance (e.g., *Q.N.* 4, 912F, 31, 919D).

Sandbach made the same observation regarding the notions of τὸ γλυκύ (i.e. "a substance characterized by sweetness") and ἡ γλυκύτης (i.e. "the quality of sweetness")[158]. In fact, Plutarch's general concept of the χυμοί ('flavours') is also illustrative of his tendency to conceive of matter and its qualities in close connection to each other. A flavour can be seen as some kind of liquid (matter) that possesses an intrinsic taste (quality). In *Quaestiones naturales*, the χυμοί are often described as being present in living beings (not so much in animal beings, though, but in plants: cf. *Q.N.* 5, 913A, C, D, 27, 918E)[159]. Despite the fact that rainwater is counted among the ἄχυμα in *Q.N.* 2, 912B, it is considered sweet, since it contains water with an inherent sweet taste (*Q.N.* 2, 912C, 9, 914C). Sandbach understands from *Q.N.* 2, 912C that "the 'sweetness' of rainwater is a separable constituent, though doubtless far the largest, which can be caused to leave it and enter into a plant"[160]. Sandbach presumably puts sweetness in inverted commas to hint at the material-in-combination-with-the-qualitative aspect of the sweet flavour present in rainwater (indeed, Plutarch speaks of τὸ γλυκὺ τῶν ὀμβρίων). However, rainwater also contains several other material elements and qualities, so it is not 'pure' (i.e. it is not simply identical to a sweet flavour, but just contains it). A little bit

---

[158] F.H. Sandbach, 1965, pp. 139–140.

[159] In *Q.N.* 5, 913C, Plutarch refers to Plato's description (*Tim.* 59e) of a χυμός as 'water that is strained through a plant' (ἢ χυμὸς μέν ἐστιν, ὡς Πλάτων εἶπεν, ὕδωρ ἠθημένον διὰ φυτοῦ). See LSJ, s.v. i: "juice of plants". Notably, in *Quaestiones naturales*, Plutarch does not use this concept in relation to Hippocratic humoral theory, that is, in relation to the presence of 'humours' in animal bodies (cf. W.H.S. Jones, 1923, pp. xlvi–li). This does not imply, however, that he has no notion of humoral theory whatsoever (in *Q.N.* 1, 911E, for instance, he reports that fever turns moisture into bile). F.H. Sandbach, 1965, p. 140 translates χυμοί not simply as 'flavours', but as 'flavourings', by which he presumably refers to liquids giving off a certain taste (rather than to actual spices or aromas; *pace* L. Senzasono, 2006, pp. 168–169, n. 40 and p. 172, n. 45).

[160] F.H. Sandbach, 1965, p. 140.

earlier in *Q.N.* 2, 912A, for instance, Plutarch argues that 'rainwater is light, airy, and mixed with breath' (κοῦφόν ἐστι τὸ ἐκ Διὸς ὕδωρ καὶ ἀερῶδες, καὶ πνεύματι μεμιγμένον, cf. also *Q.N.* 4, 912F).

Similar to sweet rainwater, salty seawater, is not 'pure' either. In *Q.N.* 1, 911CF, Plutarch draws attention to several material aspects of seawater in an attempt to explain why it does not provide nourishment to trees: he argues that seawater is thick, heavy and earthy (*causa* 1), that it has a drying effect (*causa* 2), contains a fatty, oily content (*causa* 3) and is undrinkable and bitter by an admixture of burnt earth (*causa* 4). From other accounts in *Quaestiones naturales* we learn that seawater has numerous other attributes. It contains not only bitter but also sweet drinkable water, because many rivers empty out into the sea, but the sweet water, which is light and lies at the surface, evaporates by the heat of the sun (*Q.N.* 9, 914C). Likewise, when a thunderbolt strikes seawater, the sweet water is expelled immediately, leaving only salt crystals behind (*Q.N.* 40). Seawater also contains innate heat (see n. 155), it is transparent, earthy and heavy (*Q.N.* 8, 914B), and a certain amount of air is also mixed with it (*Q.N.* 12, 915A).

Clearly, for Plutarch, there is no material purity in the natural phenomena we perceive around us, since physical bodies contain more than only one element or quality. The material substances that appear in nature are rather conceived of as being complex compounds of different primary elements and a multitude of primary and secondary qualities connected to them. This idea has direct repercussions for the aetiological design of Plutarch's physical aetiologies. It allows him to apply a variegated focus in the explanations that he provides and to freely differentiate between a myriad of physical attributes and properties, with the feature of argumentative plausibility being the only criterion to be taken into account. There is at some points, however, also a certain aspect of regularity in Plutarch's explanations of natural phenomena, insofar that specific elements and qualities are more or less consequently and repetitively attributed to specific material bodies. Seawater, for instance, contains a heavy, earthy component and is hot, as we saw (cf. *Q.N.* 1, 911D, 5, 913C, 7, 914AB, 8, 914B, 10, 914D). The air in the atmosphere is also hot and transparent because it is full of sunlight and fire (cf. *Q.N.* 12, 915A, 39, *De facie* 930F). Yet, despite there being fixed concepts of this sort, Plutarch still allows a degree of aetiological flexibility in applying them in his explanations (I will come back to this later [see 4.3.3.1.]). For instance, in *Quaest. conv.* 652B, he is found extemporising (αὐτοσχεδιάσαι) that the δύναμις of wine is actually cold, whereas, normally a hot δύναμις is attributed to it[161].

---

[161] Cf. *Q.N.* 10, 914D, 31, 919C, *Quaest. conv.* 701F. Cf. also, e.g., Ps.-Arist., *Probl.* 871a2. The argument in *Quaest. conv.* 652B is inspired by Pyrrhonic scepticism.

Another important point is that Plutarch often draws specific attention in his explanations to opposite elements and qualities in a material body or natural process. Not infrequently, he focuses on specific binary oppositions in the natural phenomena that he tries to explain. In *Q.N.* 20, for instance, he deals with the sweet taste of the tears of wild boars as opposed to the salty and ordinary ones of those of deer. In the first *causa*, Plutarch explains this on the basis of the hot and cold character of both animals respectively: the natural heat in wild boars melts the salty particles in their tears so that they become sweet, whereas no such process takes place in deer, due to their natural coldness. The second *causa* focuses on the composition of the blood of both animals, from which the tears (according to Empedocles) are discharged: the blood of wild boars is rough and black owing to their heat, whereas that of deer is thin and watery, and the taste of the tears, which are secreted from the blood, depends on these opposite qualities. The focus on such binary oppositions is a relatively common explanatory strategy in ancient science more generally[162]. In Plutarch's case, it can be related to his belief that there are opposite forces at work in natural bodies[163]. This polarity of natural forces does not, of course, fragment Plutarch's world view. By contrast, he accepts that the order of nature is actually based on such oppositions, which are themselves kept in check by the demiurgic ordering of the universe[164].

2. *Natural processes*

While the material principles are concerned with the constitutive organisation of physical bodies in terms of elements and qualities, natural processes, on the other hand, concern the changes and movements to which these bodies are subject. The process of change is expressed in *Quaestiones naturales* with such terms as μεταβάλλω (μεταβολή), ἐξίστημι or τρέπω. The terminology, of course, tends to be more specific in many cases. Apart from the basic alterations and phase transitions of material bodies (as expressed with a range of concepts, such as χυλόω, τήκω, ψύχω, περίψυξις, λεπτύνω, μαλάσσω, πήγνυμι, πῆξις, ἐξατμίζω, ἀναξηραίνω, ἀναπνοή, πυκνόω, πύκνωσις)[165], the processes of generation (γίγνομαι, γεννάω, γένεσις), corrup-

---

[162] See G.E.R. Lloyd, 1964 and 1966, pp. 15–171.

[163] Cf. *De prim. frig.* 947F: ἐν τοῖς φυσικοῖς σώμασιν ἀντιστοιχίαν ὑποληπτέον ὑγρῶν πρὸς ξηρὰ καὶ ψυχρῶν πρὸς θερμά, τὸ κατὰ λόγον ἅμα καὶ τὰ φαινόμενα διαφυλάττοντας.

[164] Cf. *De prim. frig.* 946EF [quoted 4.1.2.2.], 951D: οὔτε τὴν φύσιν ἔχει λόγον ἐφεξῆς τῷ φθείροντι τάξαι τὸ φθειρόμενον, ὥσπερ οὐ κοινωνίας οὖσαν οὐδ' ἁρμονίας ἀλλὰ πολέμου καὶ μάχης δημιουργόν. χρῆται μὲν γὰρ ἐναντίοις εἰς τὰ ὅλα πράγμασι· χρῆται δ' οὐκ ἀκράτοις οὐδ' ἀντιτύποις, ἀλλ' ἐναλλάξ τινα θέσιν καὶ τάξιν οὐκ ἀναιρετικὴν ἀλλὰ κοινωνικὴν δι' ἑτέρων καὶ συνεργὸν ἐν μέσῳ παρεμπλεκομένην ἔχουσι.

[165] Some material substances appear in nature in several physical states: the substance

tion (φθείρω, φθορά), concoction (πέττω, πέψις), mixture (κεράννυμι, κρᾶσις, ἀνάμιξις) and concentration of opposites (ἀντιπερίστασις) are also relevant here. The latter three concepts may require some further elucidation.

Generally speaking, concoction (πέψις) is a physical process by which a material substance is generated out of another substance by means of heat. In the case of plants or animals (Q.N. 33: *stirpibus et animantibus* [...] *concoquatur*)[166], it is often the nourishing subtance that is concocted into an assimilate substance by means of natural heat (but external heat can also trigger this process: Q.N. 4, 913A, 27, 918E). Concoction does not only refer to the digestion of food (Q.N. 2, 912B, 22, 917D), but also, for instance, to the conversion of grape juice into wine (Q.N. 27, 918E). It is identified with the process of putrefaction (in Q.N. 2, 912C: σῆψις), but this last point is, as Sandbach points out, "contrary to the standard Aristotelian opinion that distinguishes these two processes, the former [πέψις] being the formation of a substance, effected by natural internal heat, the latter [σῆψις] its destruction, due to alien external heat"[167]. Even so, there are obvious parallels between Plutarch's and Aristotle's notion of concoction (and, similarly, also that of non-concoction, ἀπεψία: Q.N. 2, 912B). Yet, the concept is not, therefore, specifically Aristotelian in kind, since concoction is a standard process in ancient Greek physical theory more generally (in fact, Aristotle himself borrowed this concept from medical literature)[168].

The same is true for the notion of κρᾶσις[169], which basically refers to a process of mixture or blending (Q.N. 23, 917E, 27, 918E). It is often used in a more specific sense, viz. as the material result of such a process[170]. When

---

of salt, for instance, is constitutive of salt liquids (e.g., the salt flavour in seawater and in tears), but it also occurs in a solid state (e.g., salt licking stones and salt crystals).

[166] Cf. L. Senzasono, 2006, p. 240, n. 191: "Il verbo *concoquere*, dato il suo uso tecnico, del quale il Longolio era certo cosciente [he was a physician], con ogni probabilità corrisponde a πέττειν del testo originale perduto."

[167] F.H. Sandbach, 1965, p. 157, n. a. See Arist., Mete. 379b10–381b23 for an account of concoction and non-concoction and their various species. On decay, see Mete. 379a11–b9.

[168] See F.H. Sandbach, 1965, p. 138. For concoction in the Hippocratic writings, see W.H.S. Jones, 1923, pp. li–lii (cf. esp. *De prisc. med.* 18–19). According to L. Senzasono, 2006, p. 153, n. 20, Plutarch may rely on both Aristotle and medical authors for this concept.

[169] See A.L. Peck, 1965, pp. lxxv–lxxvii. Cf. also J. Boulogne, 2006/7, pp. 3–4. The notion of κρᾶσις (δι' ὅλων) had become common currency in physical theory by Plutarch's time (cf. *Coni. praec.* 142F–143A: δεῖ δέ, ὥσπερ οἱ φυσικοὶ τῶν ὑγρῶν λέγουσι δι' ὅλων γενέσθαι τὴν κρᾶσιν, οὕτω τῶν γαμούντων καὶ σώματα καὶ χρήματα καὶ φίλους καὶ οἰκείους ἀναμειχθῆναι δι' ἀλλήλων: cf. Antipater of Tarsus, SVF 3, p. 255, fr. 63, 15).

[170] The concept of κρᾶσις is not simply synonymous with μίξις. Cf. M. Vamvouri Ruffy, 2012, p. 125: "la *mixis* est le mélange des éléments, la *krasis* la fusion qui s' ensuit". Cf. *Quaest. conv.* 626D, 648D, 620E and esp. *De prim. frig.* 946DE: καὶ γὰρ ἕξεως μὲν οὐκ ἔστι μῖξις πρὸς στέρησιν οὐδ' ἀναδέχεται δύναμις οὐδεμία τὴν ἀντικειμένην αὐτῇ στέρησιν ἐπιοῦσαν οὐδὲ

used, for instance, in reference to the soil (*Q.N.* 16, 915E), κρᾶσις refers to the composition of the ground in terms of the different material elements and qualities that are blended in it (not simply earth). In regards to the mixing of the principles of heat and cold (e.g., in the air), the concept of κρᾶσις can be translated as our modern 'temperature' or, more specifically, 'climate' (*Q.N.* 4, 913A, 31, 919D)[171]. The notion of κρᾶσις is also used in regards to the bodies of animals (*Q.N.* 26, 918DE), thus referring to the blending of their bodily constituents and qualities[172]. In this sense, κρᾶσις is best translated as 'constitution' or 'composition', which is synonymous with 'condition' (*Q.N.* 26, 918E: διάθεσις). This constitution is variable in living beings, and it is not the same in sickness as in health[173].

The process of ἀντιπερίστασις involves the concentration of one of two opposites (ἀντί-) by the other, which surrounds and encloses it (περί-). Only the *nomen actionis* (ἀντιπερι(σ)ταται) is referred to in *Quaestiones naturales*, viz. in *Q.N.* 13, 915B[174], where Plutarch relies on Theophrastus (fr. 163 Wimmer = 173 FHSG) in arguing that coldness in the winter season overwhelms the heat that is present in the seawater, so that the latter is concentrated and enclosed at the bottom of the sea (the phrase κατακλείεται

ποιεῖ κοινωνὸν ἀλλ' ἀντεξανίσταται· θερμὰ δ' ἔστιν ἄχρις οὗ κεραννύμενα ψυχροῖς ὑπομένει, καθάπερ μέλανα λευκοῖς καὶ βαρέσιν ὀξέα καὶ γλυκέσιν αὐστηρά, παρέχοντα τῇ κοινωνίᾳ ταύτῃ καὶ ἁρμονίᾳ χρωμάτων τε καὶ φθόγγων καὶ φαρμάκων καὶ ὄψων προσφιλεῖς πολλὰς καὶ φιλανθρώπους γενέσεις.

[171] Notably, Book 14 of Ps.-Aristotle's *Problems*, entitled ὅσα περὶ κράσεις, mostly deals with problems concerning climate. As F.H. Sandbach, 1965, p. 139 notes, "whereas we do not normally remember that by etymology 'temperature' means 'blending', and regard a temperature as a point on a scale, the word κρᾶσις implies a blending in certain proportions of absolute heat with absolute cold". The use of thermometers, as we know them today, did not exist in Antiquity, and temperatures were not measured in a quantitative fashion (i.e. in units of K, °F, °C etc.), if they were measured at all. It is not unlikely that such a quantitative approach would even have seemed outlandish to Plutarch, who himself conceives of κρᾶσις as a blending of proportionate amounts of heat and cold (cf., e.g., *Quaest. conv.* 649D and *De prim. frig.* 946DE; quoted n. 170). For a history of the thermometer and of thermometry more generally, see W.E.K. Middleton, 1966. Galen was the first to describe heat and cold in a numeric fashion. The earliest Greek writings about the expansion of air by heat were composed by Philo of Byzantium and Heron of Alexandria.

[172] Notably, Plutarch does not explicitly refer to the Hippocratic blending of humours (χυμοί) in the body in *Quaestiones naturales* (see n. 159).

[173] Cf., e.g., Galen, *Temp.* I, 509, 1–4 Kühn. P.J. van der Eijk, 2013, p. 188 (with n. 19 for further literature) describes κράσεις as "les combinaisons proportionnées des quatre qualités et les modalités de leur variation, qui déterminent les différences dans la constitution physique entre les corps humains et les corps des animaux et les variations entre les individus humains".

[174] See L. Senzasono, 2006, pp. 187–188, n. 78 (this term is not recorded under the heading of "scientific vocabulary" by F.H. Sandbach, 1965, pp. 138–141, probably because it is not "recurrent" in our collection). It seems that the physical process of ἀντιπερίστασις is not always mentioned by name (cf., e.g., *Q.N.* 27, 918EF and *Quaest. conv.* 694DE).

[…] ἡ θερμότης ὑπὸ τοῦ ψυχροῦ κρατήσαντος summarises what happens). The concept of ἀντιπερίστασις has a strong Peripatetic connotation (it is also attested in Ps.-Aristotle's *Problems*)[175], but has its roots, as Opsomer has shown, in the Platonic theory of περίωσις[176]. Additionally, the concept of περίψυξις is closely related to that of ἀντιπερίστασις, by the notion of enclosure or surrounding (περί-). By this process, a substance is solidified by the influence of cold around it (*Q.N.* 7, 913F, 10, 914D, 25, 918B; cf. *De prim. frig.* 949B).

Apart from the natural processes that involve material change, Plutarch also mentions an abundance of processes that involve an aspect of spatiality and are often mechanical in kind. These again embrace a multitude of concepts, relating to natural movement (κινεῖν, κίνησις, φεύγειν, φέρειν, φορά, διεκθέω, ῥεῖν, σάλον, ἀναπέμπω, ἀναδίδωμι etc.), attraction (ἕλκειν, ὁλκή), contraction (συγκοπή), dilatation (διαστέλλω), collision (πληγή), dissipation (διαφορέω), dispersion (διαχέω), loosening (λύω, χαλάω), expulsion (ἐκβάλλω, ἐκπέμπω, ἐξωθέω), insertion (ἐμβάλλω) etc. The theories of emanations and effluences (ἀπόρροιαι, ἀπορροαί) and, often linked with them, that of the pores (πόροι), deserve specific consideration here. These concepts have an Empedoclean touch but had become more generally used in physical theory by Plutarch's time[177] (also especially via the atomist movement)[178].

---

[175] See H. Flashar, 1962, pp. 328–329. Cf. also, e.g., Arist., *Mete.* 348b2 (with H.D.P. Lee, 1952, pp. 82–83, n. b).

[176] J. Opsomer, 1999. Aristotle in *De respir.* 472b6–473a14 comments on the Platonic theory of breathing as formulated in *Tim.* 79a–80c, where the process of περίωσις is central. In order to remain as close as possible to the wording of his source, Aristotle mostly uses the original Platonic terminology (περίωσις, περιωθεῖν), but in 472b16 the Aristotelian concept occurs (ἀντιπεριισταμένων). Plutarch comments on the same Platonic passage in *Quaest. Plat.* 7, 1004D–1006B, where we find a combination of Platonic (1004E: περιωθεῖν) and Aristotelian terminology (1004D: ἀντιπερίστασιν), which at times results in peculiar hybrid neologisms (1005D: ἀντιπεριώσεως, 1005F: ἀντιπεριωθουμένοις).

[177] Cf., e.g., *Quaest. conv.* 649D (= Emp., DK31B77): Ἐμπεδοκλῆς δὲ πρὸς τούτῳ καὶ πόρων τινὰ συμμετρίαν αἰτιᾶται κτλ. Cf. also Pl., *Men.* 76c (= Emp., DK31A92): Οὐκοῦν λέγετε ἀπορροάς τινας τῶν ὄντων κατὰ Ἐμπεδοκλέα; […] Καὶ πόρους εἰς οὓς καὶ δι᾽ ὧν αἱ ἀπορροαὶ πορεύονται;. For the concept of 'Zwischenräume' in Ps.-Aristotle's *Problems*, see H. Flashar, 1962, p. 330. For the use of the concept of πόροι in a medical context, cf., e.g., *Quaest. conv.* 687BC (with S.-T. Teodorsson, 1990a, pp. 241–243). In *Quaest. conv.* 689BC, Plutarch criticises 'those who advocate the theory of passages' (οἱ τοὺς πόρους ὑποτιθέμενοι). This is probably an allusion to the followers of Asclepiades of Bithynia, who founded the methodic school in medicine.

[178] Cf., e.g., Democr., DK68A165: ὁ Δημόκριτος δὲ καὶ αὐτὸς ἀπορροίας τε γίνεσθαι τίθεται καὶ τὰ ὅμοια φέρεσθαι πρὸς τὰ ὅμοια, ἀλλὰ καὶ εἰς τὸ κενὸν πάντα φέρεσθαι. Notably, Plutarch does not connect the concept of pores with a notion of κενόν (*vacuum*) in *Quaestiones naturales*. He emphatically rejects the existence of the void in *Quaest. Plat.* 1004DE (following Plato) and in *De def. or.* 424D (following Aristotle). Its existence is not necessarily implied, moreover, by a phrase like τὰς τῶν πόρων κενώσεις καὶ ἀναπληρώσεις (*Quaest. conv.* 689AB).

In *Q.N.* 19, 916D, Plutarch relies on Empedocles in arguing that there are particles emanating from all objects, either animate or inanimate (DK31B89: πάντων εἰσὶν ἀπορροαὶ ὅσσ᾽ ἐγένοντο). These emanations play an important role, for instance, in processes related to sense perception (*Q.N.* 23, 917E, *Quaest. conv.* 680F–681A) and in bodily processes more generally, such as the octopus' metachrosis (*Q.N.* 19, 916CF). Pores, on the other hand, are empty interstices in animate or inanimate bodies through which matter can pass. In the case of animals and plants they refer to the pores in the skin (*Q.N.* 19, 916EF) or in the body (*Q.N.* 3, 912DE, 5, 913D, 31, 919D). When they are mentioned in relation to the earth (*Q.N.* 2, 911F) or seawater (*Q.N.* 12, 915A), they are more naturally translated as 'passages', 'ducts' or 'channels'.

The emanations hover freely and can pass trough the pores of certain substances or lodge themselves in the ones with which they are commensurate in form (*Q.N.* 19, 916F: ταῖς ἀπορροίαις πόρους συμμέτρους ἔχουσιν)[179]. If there is no commensuration, the emanations are blocked off or slip off the surface of the substance. There is a more abstract, geometrical motivation here, which takes into account the form of both the emanations and the pores. It is important to add, however, that Plutarch phrases this in approximate rather than in exact mathematical terms. For instance, in regards to the octopus' metachrosis (*Q.N.* 19, 916BF), he argues that the skin of the creature contains many pores in which many *minute* particles (μέρη καὶ θραύσματα πολλὰ καὶ λεπτά) that are continually detached from the rocks by the seashore can settle. These fragments slip off the surface of animals that have *narrower* pores or pass quickly through those that have *more open* ones (λανθάνει περιολισθάνοντα τῶν πυκνοτέρους ἐχόντων πόρους ἢ διεκθέοντα τῶν μανοτέρους), but the octopus has a flesh which is obviously *honeycombed* (ἀνθρηνιώδης), so that it offers places for these particles to lodge. To give another example, the pungency (δριμύτης) of salt opens up the passages in the bodies of animals and thus better prepares the way for the food to be distributed (*Q.N.* 3, 912D). By contrast, such passages in plants are too *narrow* (διὰ λεπτότητα) for the *large-sized* (παχυμερές) earthy constituents of salt to pass through (*Q.N.* 5, 913D). In the same way, unmixed wine penetrates the roots of vines that are sprinkled with it, contracting and clogging the passages (τοὺς πόρους συναγαγὼν καὶ πυκνώσας), so that water cannot enter into the plant (*Q.N.* 31, 919D). Thus, we see that the basic geometrical motivation of the emanations and the pores never becomes very exact. This ties in closely with Plutarch's more generally qualitative rather than quantitative approach in his natural problems, as highlighted earlier on.

---

[179] Cf. also, e.g., *Quaest. conv.* 689B: πᾶσι γὰρ ὄντων πόρων, ἄλλας πρὸς ἄλλα συμμετρίας ἐχόντων κτλ.

Indeed, when the description of natural processes involves a variable amount (ποσότης) of material qualities, Plutarch mostly formulates this in relative terms (c.q. by means of comparatives). In *Q.N.* 9, 914BC, for instance, he connects the relative bitterness and sweetness of seawater with the variable effect of the sun's heat in the summer and in the winter, respectively. He explains that seawater is *more bitter* in the summer, insofar that the heat of the sun removes the sweet parts of the seawater. Likewise, it is *less bitter* (and thus sweeter) in the winter, when the sun has a gentler effect. To give another example, in *Q.N.* 8, 914B, Plutarch argues that seawater becomes *warmer* when agitated insofar that heat is innate to it, but that this is not true of all other liquids, which grow *colder*. As Senzasono notes, there is no attempt to mathematise this proportion in terms of numerical quantities[180]. Only in an exceptional case does Plutarch refer to a physical change in an absolute fashion, that is, in terms of numeric quantities. This is the case in *Q.N.* 7, 914A, where Plutarch records a hydrostatic account from Theophrastus (fr. 161 Wimmer = 214C FHSG). He writes that the weight ratio of a vessel filled with water from a certain source on Mt. Pangaeum in Thrace depends on the season: the vessel weighs *twice* as much in the winter as it does in the summer (ἐν δὲ Θράκῃ περὶ τὸ Πάγγαιον ἱστορεῖ Θεόφραστος εἶναι κρήνην, ἀφ' ἧς ταὐτὸ γέμον ἀγγεῖον ὕδατος ἱστάμενον χειμῶνος ἕλκειν διπλάσιον σταθμὸν ἢ θέρους). Seeing, however, that Plutarch simplifies the numbers from Theophrastus' account – the original weight ratio would be 96 to 46 (cf. Ath., *Deipn.* 2, 42b = fr. 159, 15–21 Wimmer = 214A, 13–17 FHSG) –, it is clear that he is primarily referring to the water's relative quality of heaviness *vis-à-vis* lightness in relation to the seasonal temperature rather than to its exact mass density. Therefore, this account should not be seen as an exception to Plutarch's qualitative approach[181].

In conclusion, for Plutarch, natural science as the causal study of natural phenomena is not an exact, but rather a conjectural science[182]. Explaining natural phenomena in the immediate world around us – that is, in the sublunary region, as opposed to those situated in the astronomical realm – is a matter of estimation rather than calculation, and of gentle persuasion rather than rigorous proof and demonstration. Therefore, the exact mathematisation of physical reality was not a common procedure in Plutarch's physical aetiologies, and the same is true for the Ps.-Aristotelian

---

[180] L. Senzasono, 2006, p. 43: "non c'è tentativo di matematizzare in termini numerici il rapporto".

[181] *Pace* L. Senzasono, 2006, p. 44.

[182] It is comparable, at least from a methodological perspective, to the στοχαστικαὶ τέχναι, such as ancient medicine or astrology. Cf. LSJ, s.v. στοχαστικός 2: "proceeding by guesswork" (e.g., Pl., *Philebus* 55e). Cf. T. Barton, 1994a, p. 16 (with n. 57) and 1994b, p. 7. For medicine as a 'stochastic art', see K. Ierodiakonou, 1995.

Problems, in which he found his model[183]. Plutarch does not carry out any exact measurement of physical data in his natural problems[184]. Even in those cases in *Quaestiones naturales* which tend toward mathematical quantification, the quality of the material substances and natural processes remains central. Notably, there are no references in Plutarch's natural problems to Plato's solids and their geometric formations[185]. Presumably, no further abstraction of the natural elements and their qualities was necessary in order to properly explain what is happening in the immediate world around us. In the end, a speculative consideration of the working of natural processes in the world suffices to attain a certain level of plausibility in the aetiologies. The idea that natural science provides a conjectural body of knowledge is very seminal for Plutarch's scientific method and will be further substantiated in the following section in light of his adherence to Platonic natural philosophy. In line with Plato, Plutarch did not consider the object itself that was being studied – that is, nature – to be ontologically stable. In this way, it could not provide indisputable and steadfast knowledge from an epistemological perspective either.

4.3.2. Towards the limits of natural science

As is well-known, the study of natural phenomena is for Plutarch intrinsically bound with his Platonic world view[186]. Accordingly, the ontological distinction Plutarch makes between the sensible and intelligible realms

---

[183] Book 15 of Ps.-Aristotle's *Problems* does concern mathematical topics, albeit in a physical-aetiological framework (ὅσα μαθηματικῆς μετέχει θεωρίας ἁπλῶς καὶ ὅσα περὶ τὰ οὐράνια). For the distinction between natural and mathematical problems, cf., e.g., *De aud.* 43C.

[184] For the natural philosopher's (Platonic) disdain for the use of geometrical instruments, cf., e.g., *Per.* 16, 7: ὁ μὲν ἀνόργανον καὶ ἀπροσδεῆ τῆς ἐκτὸς ὕλης ἐπὶ τοῖς καλοῖς κινεῖ τὴν διάνοιαν. However, in an anti-dogmatic context, Plutarch still vindicates mathematical calculation by means of geometrical instruments, because he prefers this approach to plain assumption. This is the case with Xenagoras' calculation οὐ παρέργως, ἀλλὰ μεθόδῳ καὶ δι' ὀργάνων of the height of Mt. Olympus as opposed to the dogmatic attitude of the geometers, who believe that no mountain or sea can be higher or deeper than ten stadia (*Aem. Paul.* 15, 9–11). See J. Boulogne, 2008, p. 748: "Ce rationalisme pondéré d'un scepticisme de méthode dénote une méfiance indiscutable à l'égard de la raison et de ses spéculations théoriques, susceptibles à la fois d'errements et de progrès. C'est pourquoi Plutarque incline à accorder, de façon pragmatique, plus volontiers sa confiance à l'expérience qu'à la spéculation pure […]". For Plato's influence on Plutarch in this regard, see R. Flacelière, J. Irigoin, J. Sirinelli and A. Philippon, 1987, pp. lxx–lxxiii and M. Isnardi Parente, 1992. In *Quaest. conv.* 718E, Plutarch quotes Philolaus (DK44A7a) and calls geometry ἀρχὴ καὶ μητρόπολις of the other scientific disciplines.

[185] On Plutarch's concept of geometric atomism, see J. Opsomer, 2015.

[186] The literature on this topic is vast and has accumulated exponentially in the last decades. Generally useful are J. Opsomer, 1998 and 2005. For a recent overview with further literature, see also G.E. Karamanolis, 2010 and P. Donini, 2011, pp. 27–40.

has an immediate epistemological implication, taking effect in an analogous distinction of corresponding levels of explanation [see 4.1.2.]. In Plato's cosmology, the sensible world we live in is seen as a world of becoming that is never truly existent (*Tim.* 27d–28a). Sensible objects are created, perpetually in motion, subject to the processes of generation and corruption, and they are only apprehensible by 'opinion in addition to sense' (*Tim.* 28a: δόξῃ μετ' αἰσθήσεως). These sensible objects are the second-rate likenesses or images of the real, intelligible world of the forms, which, themselves, are self-identical, ungenerated, indestructible, and perceived by mind alone (νόησις: *Tim.* 52a, 92c). In the *Timaeus*, Plato famously gives probabilistic reasoning an important place within the field of natural philosophy. Timaeus describes his own account as a 'likely account' or a 'likely story' (εἰκὼς λόγος, εἰκὼς μῦθος: e.g., *Tim.* 29d), which is often interpreted as being programmatic for Plato's own study of natural phenomena[187]. In order to foreground the same epistemological backdrop in Plutarch's natural problems, this section will draw a closer link between Plutarch's 'sceptical' approach in natural science and what influence it draws from Plato and the Platonic tradition. This will also bring into consideration Plutarch's evaluation and use of data pertaining to sense perception and the role of autopsy.

1. A 'sceptical' Plutarch: ἐμπειρία, ἐποχή and εὐλάβεια

In accordance with Plutarch's dualistic view on causality [see 4.1.2.], the world of natural phenomena is subordinated to a higher realm of intelligibles, and the same holds true for the kind of knowledge that the study of these different realms yields. Even though Plutarch occasionally highlights the importance of ἐμπειρία and αἴσθησις in studying natural phenomena and even prefers serious observation to theoretical speculation, this can by no means be generalised[188]. Plutarch's problem with data pertaining to sense perception is that they often procure unreliable knowledge. This idea shines through on several occasions in the natural

---

[187] According to F.M. Cornford, 1937, pp. 28–29 (see also pp. 28–32 more generally) this implies "that there can be no exact, or even self-consistent, science of Nature. [...] There is [...] no exact truth to which our account of physical things can ever hope to approximate". See also more recently D.J. Zeyl, 2000, pp. xxxii–xxxiii (p. xxxii: "Probably what Plato means is that *within the constraints in which the story must be told* something like this account is the most plausible one can hope for."), T.K. Johansen, 2004, M.F. Burnyeat, 2009 and L. Brisson, 2012.

[188] Cf., e.g., *Quaest. conv.* 641C (with D. Lehoux, 2003), 699D, 725C, *De facie* 933A, C, *De soll. an.* 975DE. For the idea that theory should be squared with fact according to Plutarch, cf. also, e.g., *De prof. in virt.* 75F: οἱ δὲ μὴ τιθέμενοι τὰ δόγματα πρὸς τοῖς πράγμασιν ἀλλὰ τὰ πράγματα πρὸς τὰς ἑαυτῶν ὑποθέσεις ὁμολογεῖν μὴ πεφυκότα καταβιαζόμενοι πολλῶν ἀποριῶν ἐμπεπλήκασι τὴν φιλοσοφίαν.

problems treated in *Quaestiones convivales*[189]. For instance, in regards to Chaeremonianus' report that he actually saw a tiny fish, the remora, slowing down a ship once (θεάσασθαι γὰρ πλέων ἐν τῷ Σικελικῷ καὶ θαυμάσαι τὴν δύναμιν), Plutarch lists a number of similar *mirabilia* and states that 'these phenomena are, indeed, obvious to the senses but it is not easy, if not entirely impossible to explain them' (*Quaest. conv.* 641C: τούτων γὰρ ἐμφανῆ τὴν πεῖραν ἐχόντων, χαλεπὸν εἶναι τὴν αἰτίαν, εἰ μὴ καὶ παντελῶς ἀδύνατον, καταμαθεῖν). In what follows, he explains that we should not mistake the effect for the cause in this case: it is not so much the remora that slows down the ship, but rather the seaweed that sticks to the hull of the ship and that attracts this tiny fish. This reflects on the value of Chaeremonianus' initial report (i.e. his autopsy claim), revealing its basic unreliability. What the passage shows, then, is that sensory data are very valuable for scientific research, but that they are at least equally unreliable.

Another relevant passage is found at the end of *Quaest. conv.* 697F–700B, where Plutarch personally defends Plato's contested view that drink passes through the lungs (*Tim.* 70c, 91a)[190]. Plutarch ascribes great probability to a set of arguments that he draws from sense perception (*Quaest. conv.* 699D: ἐκ τῆς αἰσθήσεως), among which he especially emphasises the bodily function of the lungs and bladder in processing liquids. He remains cautious, though, and adds that such knowledge is uncertain, especially because the subject itself is obscure (*Quaest. conv.* 700B):

εἰκότα γὰρ μακρῷ ταῦτα μᾶλλον ἐκείνων. τὸ δ' ἀληθὲς ἴσως ἄληπτον ἔν γε τούτοις, καὶ οὐκ ἔδει πρὸς φιλόσοφον δόξῃ τε καὶ δυνάμει πρῶτον οὕτως ἀπαυθαδίσασθαι περὶ πράγματος ἀδήλου καὶ τοσαύτην ἀντιλογίαν ἔχοντος.

This (sc. corroboration from sense perception) is far more probable than the other accounts. Certainty, however, is doubtless[191] unattainable in questions of this sort; and it was wrong to make such a rash attack, in a matter which is obscure and admits of so many contrary arguments, against a philosopher pre-eminent in reputation and in influence (i.e. Plato).

This passage clearly illustrates Plutarch's high opinion of Plato, Platonic doctrine and Platonic epistemology (c.q. his sceptical attitude towards natural phenomena and observational data). The idea that natural phenom-

---

[189] See, e.g., Z. Abramowiczówna, 1962, pp. 83–84.
[190] Cf. the parallel in *De Stoic. rep.* 1047CD.
[191] F. Frazier and J. Sirinelli, 1996, p. 22 translate ἴσως as "peut-être", but the "sans doute" of D. Babut, 1994, p. 573 is stronger.

ena of this kind are inexplicable (τὸ δ' ἀληθὲς ἴσως ἄληπτον) and obscure (ἄδηλα) is of great significance, since Plutarch makes a very similar conclusion at the end of his argumentation in *De prim. frig.* 955C. Here we find the famous ἐποχή statement, which is a *locus classicus* in the debate on Plutarch's epistemological framework[192].

In *De primo frigido*, Plutarch discusses the problem of whether cold is a privation of heat or has a principle of its own, and prefers the second option (cf. also *Q.N.* 29, 919AB). After criticising several theories on the principle and primary element of cold in a doxographical fashion, he expounds his own view according to which the element of earth is the most likely candidate. Plutarch concludes his study with an exhortation to Favorinus, to whom the treatise is dedicated[193]. In order to anticipate Favorinus' evaluation of the new theory, he declares that suspension of judgement (ἐποχή) is the right philosophical position in such matter (*De prim. frig.* 955C):

Ταῦτ', ὦ Φαβωρῖνε, τοῖς εἰρημένοις ὑφ' ἑτέρων παράβαλλε· κἂν μήτε λείπηται τῇ πιθανότητι μήθ' ὑπερέχῃ πολύ, χαίρειν ἔα τὰς δόξας, τὸ ἐπέχειν ἐν τοῖς ἀδήλοις τοῦ συγκατατίθεσθαι φιλοσοφώτερον ἡγούμενος.

Compare these statements, Favorinus, with the pronouncements of others; and if these notions of mine are neither less probable nor much more plausible, say farewell to opinion, in the belief that it is more philosophic to suspend judgement when the truth is obscure than to take sides.

A very fundamental, philosophical dynamic lies at the basis of the finale of *De primo frigido*, which originates from Plutarch's sincere epistemological conviction that inferior knowledge (δόξα) springs from an ontologically inferior object (φύσις)[194]. Plutarch aims to demonstrate that firm knowledge cannot be attained from natural phenomena, because they pertain to the

---

[192] See, e.g., J. Opsomer, 1998, pp. 213-221, and more recently P. Donini, 2011, pp. 31-35, esp. p. 32 (with further references) and G. D'Ippolito and G. Nuzzo, 2012, pp. 65-68.

[193] Favorinus, the famous philosopher of Arelate, dedicated a work to the Chaeronean, entitled 'Plutarchos or on the Academic disposition' (Πλούταρχος ἢ Περὶ τῆς Ἀκαδημαϊκῆς διαθέσεως). The Lamprias catalogue mentions a 'Letter to Favorinus on friendship' (nr. 132: Ἐπιστολὴ πρὸς Φαβωρῖνον περὶ φιλίας). For further reading, see L. Holford-Strevens, 1997 and J. Opsomer, 1997 and 1998, pp. 213-240.

[194] It is unlikely that in the finale of *De primo frigido* Plutarch is playing along with some τόπος of feigned modesty undeserving of attention. There is, of course, a certain feature of modesty at play here, but this is not just a rhetorical strategy that Plutarch is deploying *ad hoc* in an attempt to render his own theory more digestible for Favorinus. Cf., e.g., the ending of *De defectu oraculorum*, where the author suggests to postpone the discussion to another time (438D: ταῦθ' ὑπερκείσθω).

level of sense perception, and are, therefore, essentially obscure and uncertain. Thus, they cannot grant any evident comprehension.

For a better understanding of the concept of ἐπέχειν in this closing comment, we should take a closer look at *De prim. frig.* 948BC earlier on. In this passage, Plutarch (after leaving the question of whether cold has a principle of its own, and before starting to inquire which principle this may be) incorporates a seminal paragraph, where he demarcates the domain of natural philosophy from that of the crafts on the basis of the different procedures followed by their practitioners. The passage is worth quoting in full:

> οἱ μὲν οὖν, τῶν σκαληνῶν καὶ τριγωνοειδῶν σχηματισμῶν ἐν τοῖς σώμασι κειμένων, τὸ ῥιγοῦν καὶ τρέμειν καὶ φρίττειν καὶ ὅσα συγγενῆ τοῖς πάθεσι τούτοις ὑπὸ τραχύτητος ἐγγίνεσθαι λέγοντες, εἰ καὶ τοῖς κατὰ μέρος διαμαρτάνουσι, τὴν γοῦν ἀρχὴν ὅθεν δεῖ λαμβάνουσι· δεῖ γὰρ ὥσπερ ἀφ' ἑστίας τῆς τῶν ὅλων οὐσίας ἄρχεσθαι τὴν ζήτησιν. ᾧ καὶ μάλιστα δόξειεν ἂν ἰατροῦ καὶ γεωργοῦ καὶ αὐλητοῦ διαφέρειν ὁ φιλόσοφος. ἐκείνοις μὲν γὰρ ἐξαρκεῖ τὰ ἔσχατα τῶν αἰτίων θεωρῆσαι· τὸ γὰρ ἐγγυτάτω τοῦ πάθους αἴτιον ἂν συνοφθῇ, πυρετοῦ μὲν ἔντασις ἢ παρέμπτωσις, ἐρυσίβης δ' ἥλιοι πυριφλεγεῖς ἐπ' ὄμβρῳ, βαρύτητος δὲ κλίσις αὐλῶν καὶ συναγωγὴ πρὸς ἀλλήλους, ἱκανόν ἐστι τῷ τεχνίτῃ πρὸς τὸ οἰκεῖον ἔργον. τῷ δὲ φυσικῷ θεωρίας ἕνεκα μετιόντι τἀληθὲς ἡ τῶν ἐσχάτων γνῶσις οὐ τέλος ἐστὶν ἀλλ' ἀρχὴ τῆς ἐπὶ τὰ πρῶτα καὶ ἀνωτάτω πορείας. διὸ καὶ Πλάτων ὀρθῶς καὶ Δημόκριτος αἰτίαν θερμότητος καὶ βαρύτητος ζητοῦντες οὐ κατέπαυσαν ἐν γῇ καὶ πυρὶ τὸν λόγον ἀλλ' ἐπὶ τὰς νοητὰς ἀναφέροντες ἀρχὰς τὰ αἰσθητὰ μέχρι τῶν ἐλαχίστων ὥσπερ σπερμάτων προῆλθον.

Now those who affirm that there are certain uneven, triangular formations in our bodies and that shivering and trembling, shuddering and the like manifestations, proceed from this rough irregularity, even if they are wrong in the particulars, at least derive the first principle from the proper place; for the investigation should begin as it were from the very hearth, from the substance of all things. This is, it would seem, the great difference between a philosopher and a physician or a farmer or a flute-player; for the latter are content to examine the causes most remote from the first cause, since as soon as the most immediate cause of an effect is grasped – that fever is brought about by exertion or an overflow of blood, that rusting of grain is caused by days of blazing sun after a rain, that a low note is produced by the angle and construction of the pipes – that is enough to enable a technician to do his proper job. But when the natural philosopher sets out to find the truth as a matter of speculative knowledge, the discovery of immediate causes is not the end, but the beginning of his journey to the first and highest causes. This is the reason why Plato and Democritus, when they were inquiring into the causes of heat and heaviness, were right not to stop their investigation with earth and fire, but to go on carrying back sensible phenomena to rational origins until they reached, as it were, the minimum number of seeds.

The study of natural phenomena, as conducted either by the natural philosopher (φυσικός/φιλόσοφος) or the 'technician' (τεχνίτης, i.e. a craftsmen, such as a doctor, farmer or musician), departs from the material object. But while the 'technician' limits his research to the immediate, natural causes, the natural philosopher continues his intellectual pursuit further upwards (ἀνωτάτω πορεία), viz. from the 'technical' data pertaining to sense perception (τὰ αἰσθητά) towards the intelligible principles (αἱ νοηταὶ ἀρχαί)[195]. What this passage makes clear, then, is that the discovery of the natural causes provides only the beginning (ἀρχή) for an investigation into the higher, intelligible causes[196]. Therefore, 'technical' knowledge can be considered a first step towards natural philosophical contemplation (θεωρία)[197] [see 3.2.2.]. This is, of course, seminal in light of Plutarch's 'technical' inquiry into the natural causes in his natural problems. Even though each type of knowledge – viz. of τὰ αἰσθητά or of αἱ νοηταὶ ἀρχαί – is important for natural philosophical contemplation (θεωρία), the former category remains subordinated to the latter. Even the use of a more abstract and theoretical approach – viz. by invoking Democritus' atoms and Plato's triangles[198] –

---

[195] For this intellectual ἀνωτάτω πορεία, cf. also *De E* 393D (ἀνώτερω προάγειν) and *Adv. Col.* 1115E (ἀνωτέρω δ' οὐ προῆλθον). These passages are rightly interpreted in light of Plutarch's adherence to Plato by G. Roskam, 2011b, p. 60 (see also G.E. Karamanolis, 2006, p. 99). Cf. also *Quaest. conv.* 718EF (ἐπανάγει [...] μὴ φερομένης ἄνω) with Pl., *Rep.* 527b (πρὸς τὸ ἄνω σχεῖν), 532bc (ἐπάνοδος [...] ἐπαναγωγή) and *Phdr.* 249c (ἀνακύψασα).

[196] See P. Donini, 1986a, pp. 210–211, J. Opsomer, 1998, pp. 215–216. For the distinction between τέχναι and ἐπιστῆμαι in this passage and its importance for Plutarch's philosophical thinking, see L. Van der Stockt, 1992b, pp. 291–293, esp. p. 292: "It seems that all that is needed here is a further distinction between what we call science and ... philosophy!" In this passage, Plutarch indeed demarcates the φυσικός-in-alliance-with-the-φιλόσοφος from the τεχνίτης. Therefore, the distinction between φυσικός and φιλόσοφος is not as strict in this passage as G.E. Karamanolis, 2010 suggests: "This is what, for Plutarch, demarcates the philosopher from the mere natural scientist [...], a distinction much exploited by later Platonists [...]."

[197] For Plutarch's use of θεωρία in the sense of theoretical contemplation, with specific applications in the fields of physics, mathematics, geometry, theology and psychology, cf. E. Kechagia, 2011a, p. 87, n. 22. For Aristotle's ruminations on the relation between ἐμπειρία, τέχνη and ἐπιστήμη, see, e.g., *Met.* 981aff.

[198] The reference is, indeed, to Plato's geometric atomism (see n. 185), not to his intelligible forms (see G. Boys-Stones, 1997a, p. 227, n. 2, whose attack on Donini was countered by D. Babut, 2007, p. 79, n. 49). It may seem to some that Plato's theory of triangular formations is implicitly criticised in *De prim. frig.* 948B (Plutarch initially writes οἱ μέν, but Plato's name eventually appears in 948C). In any case, this strategy of

will not eventually suffice, in Plutarch's opinion, to get at the bottom of things, because he knows that there still is a higher, primary, and, by implication, divine seed: this is 'the godly harmoniser and musician', who, as we saw earlier on, orders all things in the universe (*De prim. frig.* 946F: ὁ θεὸς ἁρμονικὸς καλεῖται καὶ μουσικός [see 4.1.2.2.]).

Notably, in the doxography on the principle of cold that immediately follows, Plutarch takes a deliberate step downwards on the ladder of ontology/epistemology, by arguing that 'it is better to first attack things perceptible to the senses in which Empedocles, Strato and the Stoics locate the substances of the qualities' (*De prim. frig.* 948D: Οὐ μὴν ἀλλὰ καὶ τὰ αἰσθητὰ ταυτὶ προανακινῆσαι βέλτιόν ἐστιν, ἐν οἷς Ἐμπεδοκλῆς τε καὶ Στράτων καὶ οἱ Στωικοὶ τὰς οὐσίας τίθενται τῶν δυνάμεων)[199]. A little bit later, though, Plutarch reminds us that the senses are unreliable, because 'they often provide false information to us' (*De prim. frig.* 952A: ἡ μὲν αἴσθησις πολλάκις ἡμᾶς ἐξαπατᾷ)[200]. What Plutarch is implying in the finale of *De primo frigido* (955C), then, is that one should remain cautious when confronted with physical δόξαι – including the one he puts forward on his own personal account –, because these concern a sensible object that is essentially obscure (ἄδηλον). Natural phenomena are inherently obscure (ἄδηλα), because they are ever-changing. As we can learn from several passages, for instance, throughout *De E apud Delphos*, only God *is* in the full meaning of the word. Everything else in this world is subject to a continuous ontological flux of becoming and perishing (*De E* 393A: γιγνόμενα πάντα καὶ φθειρόμενα). To 'know' this God equals true and perfect science (i.e. ἐπιστήμη), which is, however, unattainable by the human intellect, or at least not in this life. True philosophy is primarily concerned with genuine knowledge (ἐπιστήμη) and not with mere opinion (δόξα), which is more relevant in the field of natural science. Therefore, the φυσικός has to manage with probable accounts and explanations, but what is plausible from one perspective, may seem less plausible, or even implausible from another. This probably explains why Plutarch presents his own personal theory about the principle of cold (identifying it with the element of earth) from the very outset as being persuasive and plausible, while Chrysippus

---

implicit criticism occurs on several other occasions in the *Moralia* (see D. Babut, 1994, p. 574, with n. 137, and 1969, p. 95, with nn. 5–6; cf. also 2007, p. 79). However, Plutarch's criticism of Plato is not fundamental. After all, what matters most for Plutarch's argument is that Plato (and Democritus with him), *as opposed to the technicians*, who simply settle for an examination of the immediately perceptible natural causes, sets a good example by pursuing a higher mode of explanation. This is an issue of scientific method rather than orthodoxy. Cf. J. Opsomer, 1998, pp. 217 and 221, with n. 26.

[199] For this deliberate 'step downwards', see also P. Donini, 1986a, p. 211 and J. Opsomer, 1998, p. 216, n. 11.

[200] Cf. also, e.g., *De E* 392E: ψεύδεται δ' ἡ αἴσθησις ἀγνοίᾳ τοῦ ὄντος εἶναι τὸ φαινόμενον.

rejected it as being utterly unadmissable and absurd (*De prim. frig.* 952D: ὡς ἀδόκιμόν τινα παντελῶς τοῦτον καὶ ἄτοπον ἀπορρίψας τὸν λόγον, ἐγώ μοι δοκῶ μηδὲ τὴν γῆν ἄμοιρον εἰκότων καὶ πιθανῶν ἀποφαίνειν). In other words, it does not really matter whether Favorinus considers Plutarch's δόξα 'neither less probable nor much more plausible than that of others' (κἂν μήτε λείπηται τῇ πιθανότητι μήθ' ὑπερέχῃ πολύ). After all, in both cases, he would be dealing with an inferior category of relative probability, and not with the absolute category of steadfast knowledge suitable for a true philosopher. Therefore, in matters pertaining to natural science, that is, in matters that are essentially unclear and conjectural, suspension of judgement (ἐποχή) is a more philosophical attitude than taking sides.

Plutarch's notion of suspension of judgement in natural scientific matters (ἐποχή) ties in closely with his intellectual caution in divine matters (εὐλάβεια, ἀσφάλεια)[201]. These concepts are closely affiliated with each other in that they both originate from an aporetic awareness that the philosopher cannot eventually attain certain or true knowledge in physical or divine matters respectively[202]. Plutarch often relates this cautious attitude to the Delphic imperatives to 1) 'avoid extremes' (μηδὲν ἄγαν)[203] and to 2) 'know yourself' (γνῶθι σαυτόν)[204], which, as Opsomer has shown, have both an Academic and a religious connotation for him[205]. 1) As we saw earlier on, the extremes that should be avoided

---

[201] J. Glucker, 1978, p. 268 argued (*pace* P.H. De Lacy, 1953, pp. 83–84) that the concept of εὐλάβεια occurs more often in Stoic and Peripatetic than in Platonic texts, but that Plutarch may have "lifted" the word and some of its meanings from the Stoic terminology in order to use it in a Platonic context.

[202] P. Donini, 1986a, pp. 205 (with n. 11), 209 and 212–213 (cf. also 2011b, pp. 34–35) has argued that there is a strict difference in meaning between εὐλάβεια (ἀσφάλεια) and ἐποχή. He distinguishes 1) ἐποχή, which concerns opinions based on evidence from the senses, from 2) caution in the domain of the sciences and 3) εὐλάβεια, which implies a cautious attitude towards divine matters. The distinction between 1) and 2) was later rejected by D. Babut, 1994, p. 573 and J. Opsomer, 1998, p. 218 (cf. also pp. 178–186). Donini remarks, moreover, that "[q]uesta cautela [εὐλάβεια] potrebbe infatti anche apparire come un'estensione o una conseguenza dell'ἐποχή" (p. 213). However, the other way around, ἐποχή can also be considered a direct effect, if not a sub-category, of εὐλάβεια (cf. also P.H. De Lacy, 1953, p. 83). Either way, Donini was the first to show that both concepts are closely connected in Plutarch's epistemology and that they have seminal importance for his philosophical method.

[203] See *Cam.* 6, 6 [quoted 4.1.1.3.], *Sept. sap. conv.* 163D [quoted 4.1.1.1.], 164B, *De E* 387F, *De Pyth. or.* 408E, *De def. or.* 431A, *De gar.* 511B.

[204] See *De ad. et am.* 49AB, 65E, *De cap. ex inim.* 89A, *Cons. ad Apoll.* 116D, *Sept. sap. conv.* 164B, *De E* 385D, 392A, 394C, *De Pyth. or.* 408E, *De gar.* 511B, *Adv. Col.* 1118C, Lamprias catalogue nr. 177.

[205] J. Opsomer, 1998, p. 185. For the proverbial value of these Delphic maxims, see also J.A. Fernández Delgado, 1991, pp. 207–208.

are those of eager credulity and incredulity in the context of things that are unclear (mostly divinely inspired *mirabilia* [see 4.1.1.3.]). 2) Self-knowledge, on the other hand, should lead a true philosopher to maintain an aporetic attitude in his inquiries, much like Socrates did in Plato's dialogues[206]. Philosophy is, after all, a continuous zetetic pursuit of the truth, for Plutarch, whose only philosophical dogma is the infallibility of Platonic doctrine. Accordingly, there are obvious sceptical overtones in Plutarch's writings, but, as recent scholarship has shown, this does not necessarily imply that Plutarch considered himself a full-blooded Academic Sceptic, rather than a true-hearted Platonist[207]. Plutarch certainly appreciated the Academic Sceptics' anti-dogmatic attitude in philosophy, but at the same time he still uncritically valued Plato's doctrines as well.

Plutarch's attitude towards the Academic Sceptics and their scientific method is, indeed, positive: he is well affiliated, for instance, with their notions of ἐποχή (suspension of judgement), ἀκαταληψία (impossibility of infallible apprehension) and ἀφασία (abstention from expressing a personal opinion)[208]. He even personally defends the Academic Sceptics and the method that they employ against the Epicureans and the Stoics, which is not a mere polemical strategy to assail his personal adversaries[209]. On the contrary, he actually considers the Academic Sceptics important authorities in the wider Platonic tradition, to which he himself adheres[210]. Furthermore, there is an obvious sceptical undertone in Plutarch's own writings, as the ἐποχή statement in *De prim. frig.* 955C clearly illustrates.

---

[206] Cf., e.g., *De ad. et am.* 72A, *Adv. Col.* 1117D with Pl., *Phdr.* 229e, *Charm.* 164e–165a.

[207] Some scholars have seen an Academic Sceptic in Plutarch, or at least noted a great influence of the New Academy on his thinking (especially so on the basis of the ἐποχή statement in *De prim. frig.* 955C). See, e.g., J. Schroeter, 1911, J.J. Hartman, 1916, pp. 253–254, K. Ziegler, 1951, col. 856, H. Cherniss and W.C. Helmbold, 1957, p. 227, H. Görgemanns, 1970, pp. 86–89, J. Glucker, 1978, pp. 287–288, J. Dillon, 1988, p. 107, n. 9, D. Babut, 1994. More recent scholarship emphasises that Plutarch's debt towards the Academic Sceptics should not be exaggerated in favour of a more basic Platonic stance. See, e.g., P.H. De Lacy, 1953, pp. 83–84, R.M. Jones, 1980, pp. 18–19, P.R. Hardie, 1992, p. 4754, A.M. Ioppolo, 2004, p. 310, F. Ferrari, 2005, p. 384, J. Opsomer, 1998, 2009, p. 169, P. Donini, 2002, pp. 250–251, 271–272, 2011, pp. 31–35, M. Bonazzi, 2014, J. Dillon 2014.

[208] For ἐποχή, see Sext. Emp., *HP* I, 31ff.; for ἀκαταληψία, *HP* I, 200–201; for ἀφασία, *HP* I, 192–193.

[209] Cf., e.g., *Adv. Col.* 1124B. See, e.g., D. Babut, 1969, pp. 276–284, G. Boys-Stones, 1997b, J. Opsomer, 1998, pp. 84–105, 2005, p. 175, E. Kechagia, 2011b, pp. 305–311.

[210] Plutarch vindicated the unity of the Platonic Academy in a lost treatise listed in the Lamprias catalogue (nr. 63: Περὶ τοῦ μίαν εἶναι τὴν ἀπὸ τοῦ Πλάτωνος Ἀκαδήμειαν). See P. Donini, 1994b, pp. 5064–5073, D. Babut, 1994, p. 550, n. 9, 2007, J. Opsomer, 1998, pp. 59–60, 171 ff., 2005, p. 169.

In the following section, I will examine this in further detail for the natural problems treated in *Quaestiones naturales* by focusing on how Plutarch's (generally) sceptical attitude features in the aetiologies at hand. To this end, I will first analyse the categories of τὸ πιθανόν and τὸ ἀληθές. Afterwards, I will also take into consideration the argumentative role he ascribes to knowledge pertaining to sense perception and also the issue of autopsy.

2. *Truth and probability in* Quaestiones naturales

As seen previously, Plutarch believes that the φυσικός, in studying natural phenomena, can only formulate plausible opinions and should, therefore, suspend his judgement on things that are essentially uncertain (ἐποχή). He should remain cautious and acknowledge that the ultimate truth is unfathomable and remains hidden beyond nature (εὐλάβεια). In the natural problems treated in *Quaestiones convivales*, Plutarch shows how this approach takes effect in the context of sympotic deliberations[211]. In these discussions, Plutarch and his fellow symposiasts are in search of a correct physical explanation without ever explicitly claiming to have reached it. The symposiasts formulate only plausible explanations, and they remain cautious not to show too much confidence in their own contributions to the discussion. This aspect of plausibility is not only an essential feature of the social decorum of sympotic table talk, as we saw, where argumentative egotism is considered a practice suitable for sophists rather than philosophers[212]. It also reveals Plutarch's underlying methodological concerns and is more fundamentally bound with the *genre* of natural problems itself, which is essentially concerned with

---

[211] E. Kechagia, 2011a, p. 101 is absolutely right that in his natural problems in *Quaestiones convivales* "Plutarch is no outright sceptic; he is, rather, a dedicated Platonist for whom the workings of the physical world can never be fully apprehended *unless one appealed to a higher level of explanations* [...]". In a similar vein, A.H. Armstrong, 1967, p. 61 notes that "Plutarch [...] has kind words for the scepticism of the New Academy, though, for himself, scepticism means simply cautiousness in committing oneself to a definite solution of a difficult matter." Cf. also, e.g., J. Boulogne, 1992, pp. 4695–4696 (regarding *Quaestiones Romanae*): "Un tel aveu d'impuissance n'est pas, ici, à confondre avec le doute des sceptiques. La perspective de Plutarque relève plutôt du probabilisme. [...] Par sa terminologie, il ne cesse de rappeler que nous sommes dans le royaume de la doxa et non pas de l'épistème."

[212] See, e.g., L. Van der Stockt, 2000b, p. 94 (with n. 9 for references). See also J. König, 2007, pp. 57–58: "The requirement of being an entertaining conversationalist, and to be generous with one's own interventions, seem to outweigh any requirement to aim for a single correct answer." Cf. also, e.g., J. Opsomer, 2009, pp. 131 and 133. For εἰκὼς λόγος and its Platonic background in *Quaestiones convivales*, see E. Kechagia, 2011a, 99–104.

argumentative persuasion (cf. *De tuenda* 133E: ἐλαφρὰ καὶ πιθανά [see 3.1.3.]). It is only likely, then, that the notion of plausibility in Plutarch's natural problems has a deeper, epistemological motivation for him, which is also foregrounded in the problems collected in *Quaestiones naturales*, as we will see.

Indeed, genre bears concrete meaning: it represents a certain attitude toward reality and the world, and does not come into existence by hazard, as a merely artificial game played by learned authors[213]. In the case of *Quaestiones naturales*, the fundamental motivation behind Plutarch's choice for the genre of natural problems relies on his aporetic awareness that it is hard, if not altogether impossible, to attain the highest measure of accuracy and certainty in matters pertaining to natural science. The genre of natural problems provides a convenient format in which a given natural phenomenon is explained in a variety of more or less plausible explanations [see 1.1.4.]. These explanations are phrased interrogatively, and none of the physical aetiologies receive any closure, that is, a response that removes all possible doubt[214]. In this way, place is always left for potential criticism and further investigation, and final judgement is ultimately suspended (ἐποχή)[215]. Also the use of the categories of τὸ πιθανόν and τὸ ἀληθές in the aetiologies tell us much about Plutarch's argumentative strategies. The Chaeronean has a tendency to resort to correct explanations (τὸ ἀληθές), but this never becomes a strict aetiological compulsion. In the end, the explanations are mainly concerned with an

---

[213] Cf., e.g., G.B. Conte, 1992, p. 120: "Genre functions as a mediator, permitting certain models of reality to be selected and to enter into the language of literature; it gives them the possibility of being 'represented'." Cf. also *id.*, 1994, p. 132: "genre must be thought of as a discursive form capable of constructing a coherent model of the world in its own image".

[214] Cf. already J. Schroeter, 1911, pp. 23–24: "Die praktische Anwendung dieser Lehre [sc. of ἐποχή] bieten die Abhandlungen: quaest. nat. und aqua an ign. util. In diesen Schriften tritt Plutarch keiner von den vorgetragenen Ansichten bei, sondern überlässt es dem Leser, die wahrscheinlichste zu wählen."

[215] H. Görgemanns, 1970, p. 87 also connected Plutarch's sceptical attitude with the "zetematische Form der Aet. phys. und der Quaest. conv.". Cf. also R.M. Jones, 1980, p. 19 and P. Donini, 1986a, p. 205. For a similar aporetic approach in *Quaestiones Romanae*, see J. Boulogne, 2002, p. 94. Notably, R. Preston, 2001, p. 112 opposed the tentative explanations of *Quaestiones Romanae* with the dogmatic ones of *Quaestiones Graecae*, arguing that for Plutarch, "Roman culture will elude definitive explanation". It is not unreasonable that Greek culture was, indeed, better known or, in any case, more familiar to Plutarch than the more exotic Roman one. Perhaps, it is not so remarkable, therefore, that interpretative pluralism is scarce in *Quaestiones Graecae*, where we are mostly dealing with an attempt at defining, rather than explaining, specific Greek cultural phenomena [see 1.1.4., n. 95]. Nevertheless, Preston's theory was rejected by T. Morgan, 2011, p. 72, n. 77.

anti-dogmatic formulation of plausible explanations (τὸ πιθανόν)[216]. Three further observations can be made in this regard.

1) First of all, several passages show that the feature of aetiological sufficiency plays an important role in Plutarch's search for plausible explanations. In *Q.N.* 12, for instance, Plutarch examines why oil sprinkled on seawater causes clearness and calm. The aetiology opens with Aristotle's explanation, according to which the wind, slipping off the smoothness (so caused by the oil), makes no impact and raises no surge. In the second *causa*, Plutarch considers Aristotle's explanation to be plausible with regard to the external aspect of the phenomenon (ἢ τοῦτο πιθανῶς εἴρηται πρὸς τὰ ἐκτός;), but he adds that it gives no internal explanation of how this phenomenon works in the sea (ἐν τῷ βυθῷ), where it is of course impossible to adduce slipping of the wind as a cause (οὐκ ἔστιν ἐκεῖ πνεύματος ὄλισθον αἰτιάσασθαι). As such, Aristotle's explanation can only be considered partially correct at best, but this does not, of course, make it *im*plausible in principle, so that it is still worth mentioning. The association between the aspects of aetiological plausibility and sufficiency is rendered more explicit in *Q.N.* 19, where Plutarch examines the octopus' colour-changing ability. In the first *causa*, Plutarch provides Theophrastus' explanation, according to which the change of colour is due to the octopus' natural cowardly nature (δειλόν ἐστι φύσει ζῷον) and the working of its πνεῦμα. This explanation is considered plausible but insufficient in the second *causa*, because it only accounts for the change of the octopus' colour and not also for its adaptation to its surroundings (ἢ τοῦτο πρὸς τὴν μεταβολὴν πιθανῶς λέλεκται πρὸς δὲ τὴν ἐξομοίωσιν οὐχ ἱκανῶς;). As such, Theophrastus provides only a partial explanation of the problem (as was also the case with Aristotle's explanation in *Q.N.* 12). Nevertheless, it is still considered to be a plausible account, since, so Plutarch adds in the third *causa*, it adduces an important 'starting-point' in the aetiology (ἀρχήν [...] ἐνδίδωσι). In both of these cases, Plutarch shows that what is plausible from one perspective may turn out to be insufficient from another[217].

---

[216] The conceptualisation of τὸ πιθανόν in *Quaestiones naturales* (see *Q.N.* 2, 912A, 12, 915A, 19, 916B, 26, 918E, 39) clearly outweighs that of τὸ ἀληθές (see *Q.N.* 2, 911F–912A and 21, 917D).

[217] Cf. L. Senzasono, 2006, p. 197, n. 98 (cf. also pp. 18–19): "l'opposizione del persuasivo (o probabile) e del sufficiente non esclude, anzi implica la possibilità del vero." "La spiegazione sufficiente è appunto quella adeguata alla realtà di fatto del fenomeno e quindi vera dal punto di vista gnoseologico. La spiegazione probabile sotto un certo aspetto può a sua volta essere insufficiente o inadeguata sotto un altro". However, for Plutarch, the ultimately 'true' (or better: 'correct'; see n. 228) explanation cannot be attained nor formulated in a physical discourse (considering the epistemic restriction imposed by ἐποχή).

2) Second, the category of τὸ πιθανόν involves an aspect of relativity, in the sense that one explanation can be more or less πιθανόν than another (cf. Q.N. 39: *probabilius*)[218]. One may wonder, however, what Plutarch exactly means in the last *causa* of Q.N. 21 (regarding the problem of the higher fertility of domesticated sows *vis-à-vis* wild ones) when he asks whether Aristotle's account (of the fertility of wild boars) is, literally, 'also true' (ἢ καὶ τὸ λεγόμενον ὑπ' Ἀριστοτέλους ἀληθές;). Indeed, the category of τὸ ἀληθές is used in an interrogative and anti-dogmatic context here and is, thus, ultimately pushed back into the region of plausibility and verisimilitude[219]. Thus, even if Aristotle's account has the appearance of providing a correct explanation to the problem, it does not surpass the realm of plausibility. This makes it possible for Plutarch to use the category of τὸ ἀληθές in a relative rather than an absolute sense here. Arguably, the use of the copulative καί (in ἢ καὶ τὸ λεγόμενον ὑπ' Ἀριστοτέλους ἀληθές;) implies that the three preceding explanations may *also* have the appearance of being correct (ἀληθές), and may lay an equally legitimate claim to it as the fourth (see n. 262). Alternatively, the καί may be used adverbially, expressing "emphatic assertion or assent" (see LSJ, s.v.); the translation, then is: 'Or, *indeed*, is Aristotle's account true?'. This does not, of course, alter the relative value of the truth claim at hand.

3) Third, at some points, Plutarch explicitly relates the criterion of τὸ πιθανόν to that of τὸ ἀληθές, without, therefore, identifying the two as being completely equivalent. For instance, in the first *causa* of Q.N. 2 (regarding the problem of why trees and seeds naturally receive more nourishment from rainwater than from irrigational water), Plutarch records Laetus' explanation that focuses on the rainwater's impact. He does not consider the explanation very convincing, though, and suggests, in an interrogative fashion (in the second *causa*), that it is incorrect (ἢ τοῦτο μὲν οὐκ ἀληθές, ἀλλ' ἔλαθε τὸν Λαῖτον κτλ.;). Plutarch then wonders whether Aristotle's explanation, which draws attention to the the rainwater's freshness, is correct (τὸ δὲ τοῦ Ἀριστοτέλους ἀληθές;). Yet, insofar that

---

[218] Cf. L. Senzasono, 2006, p. 249, n. 212: "*probabilius* è quasi certamente la traduzione di πιθανώτερον, data la discreta frequenza di questo termine qui e altrove nei *Moralia*". Senzasono refers to *Quaest. conv.* 629A, 664D, 741C, *De facie* 930C, 933CD, *De prim. frig.* 951F. Longolius translates πιθανῶς in Q.N. 12, 915A and 19, 916B as *probabiliter*. Furthermore, it is unlikely that the original Greek text read ἢ μᾶλλον, which Longolius elsewhere translates as *an potius* (Q.N. 3, 912E, 7, 913F, 28, 919A).

[219] A similar conclusion was reached for *Quaestiones Romanae* by J. Boulogne, 1992, p. 4695: "Le mode interrogatif de la formulation des causes examinées ne tient pas, toutefois, seulement au respect de la forme traditionnelle du genre littéraire des Q.R. Il correspond surtout à une réflexion personnelle, consciente de l'impossibilité de parvenir à des certitudes. Il est, d'ailleurs, remarquable que les explications qualifiées de vraies ne le sont, pour ainsi dire, jamais que sur le mode interrogatif."

running waters do, indeed, share their freshness with rainwater but are less nourishing, he argues that this explanation is 'also/indeed plausible rather than true' (ἢ καὶ τοῦτο πιθανὸν μᾶλλον ἢ ἀληθές ἐστι;). Notably, this pattern of speech (viz. πιθανὸν μᾶλλον ἢ ἀληθές) recurs in *De def. or.* 424C and in *Quaest. conv.* 627B, again regarding a specific Aristotelian account[220]. Scarcella may be right in pointing out that this phrase has a playful and ironic connotation[221]. In my opinion, this does not, however, make it irrelevant in light of Plutarch's underlying epistemology. What the phrase probably implies, then, is that what is at first suggested to be a correct explanation, after close examination is considered to have falsely aroused that appearance, and is, therefore, degraded as being plausible (rather than true). The basic idea here is that the relative plausibility of a physical explanation, rather than its absolute correctness, is the main criterion to be taken into consideration in evaluating this type of theories. Or as Teodorsson has put it: "These expressions seem at first sight to indicate that Plut. was a thorough researcher, eager to arrive at the truth, but in reality he was prone to be content with τὸ πιθανόν or τὸ εἰκός [...]."[222] This can be linked, then, with Plutarch's philosophical attitude in postponing final judgement in natural scientific matters. Moreover, the phrase also indicates that there is a clear epistemological distinction between what is plausible and what is correct for Plutarch, but that it is not always easy to draw the line in reality. Therefore, these categories are not strictly opposed to each other at a conceptual level[223]. After all, what is plausible is not necessarily *in*correct[224]. Then again, at a conceptual level, presumably only one plausible explanation can be the correct one (but this cannot be said of Aristotle's explanation here in *Q.N.* 2). It is from this overlapping of the categories of τὸ πιθανόν and τὸ ἀληθές that misunderstandings can eventually arise, and this is why the φυσικός should remain pensive in his judgement[225].

---

[220] It can also be seen in an adapted form in *De prim. frig.* 952B (cf. also *Quaest. conv.* 687DE). See also, e.g., Galen, *SMT* II, 471, 13–14 Kühn. A similar idea is expressed in Herod., *Hist.* 2, 22: Ἡ δὲ τρίτη τῶν ὁδῶν πολλὸν ἐπιεικεστάτη ἐοῦσα μάλιστα ἔψευσται.

[221] A.M. Scarcella, 1998, p. 340, n. 352: "un gradevole tono ironico". *Pace* L. Senzasono, 2006, pp. 19–20, with n. 29, who seems to link Scarcella's irony too strictly with Socratic irony.

[222] S.-T. Teodorsson, 1989, p. 150 (translated by L. Senzasono, 2006, p. 20 without reference). For the conceptualisation of the category of τὸ εἰκός in *Quaestiones naturales*, see *Q.N.* 2, 912C, 5, 913D, 7, 914B, 19, 916D, 20, 917A, 29, 919B.

[223] *Pace* L. Senzasono, 2006, pp. 19 ("opposizione", "distinti", "contrapposizione"), 20 ("contrapposti") and 149, n. 15 ("opposizione").

[224] Cf. L. Senzasono, 2006, p. 22: "Plutarco possa attribuire al probabile una possibilità di verità".

[225] This also works the other way around, insofar that the categories of the 'plausible'

In conclusion, Plutarch never claims to know what is ostensibly true or false in a natural scientific context. Nevertheless, in spite of suspending judgement, he still shows that he has a certain opinion about these aspects. If he suggests an explanation to *not* be true, he does not take it to be strictly implausible (*Q.N.* 2, 911F–912A). Conversely, what is plausible sometimes turns out to be only partially correct (*Q.N.* 12, 915A, 19, 916B). In this sense, the notion of plausibility can be considered a heuristic stimulus to further complete the aetiology[226]. It functions as a useful argumentative criterion (often, indeed, ranked in the first *causa*: *Q.N.* 2, 911F, 12, 915A, 19, 916B), by means of which the φυσικός is exhorted to look for an even more plausible or more comprehensive explanation. In the end, Plutarch is satisfied by formulating plausible arguments, since in some cases, what is plausible is also potentially correct (ranked in the final *causa*: *Q.N.* 26, 918E, 39).

We can safely assume from these observations that the natural problems collected in *Quaestiones naturales* comprise some kind of *status quaestionis* of an ongoing zetetic search for plausible arguments, thus embodying a collection of open-ended natural scientific ἐπιχειρήματα (i.e. endeavours, attempts, essays)[227]. In the end, the search for the 'truth' in the sensible world leads to more or less plausible opinions (δόξαι), which cannot attain the cognitive status of absolute science (ἐπιστήμη). The physical aetiologies Plutarch provides are only indicative of τὸ ἀληθές. Therefore, ἐποχή is the most appropriate method in natural scientific inquiry, even if it does not offer a clear way out of the problems that nature confronts us with. When all is said and done, many blind spots will inevitably remain on the map of nature[228]. As we saw previously, from an epistemological

---

and the 'incorrect' do not necessarily exclude one another, as can be inferred from Plutarch's criticism of Laetus in combination with that of Aristotle here in *Q.N.* 2, 911F–912A. Their explanations are suggested to be οὐκ ἀληθές but still πιθανόν. What is untrue may sometimes seem plausible (cf. *De aud. poet.* 16C: μεμιγμένον πιθανότητι ψεῦδος). There may even be some truth to what seems improbable (cf. *Quaest. conv.* 626F: ῥᾳδίως ἡμῶν καὶ ἀλόγως ὑπὸ τοῦ εἰκότος ἁλισκομένων καὶ πάλιν ἀπιστούντων τῷ παρὰ τὸ εἰκός).

[226] Cf. L. Senzasono, 2006, p. 22, n. 31: "la stessa probabilità è un criterio euristico".

[227] The concept of ἐπιχείρημα is mentioned on several occasions in the context of the natural problems discussed in *Quaestiones convivales* (cf. 651C, 656B, 662D, 694D, 701F). For the link between Plutarch's aetiological originality and the notion of "essai", cf. also F. Frazier and J. Sirinelli, 1996, p. 198. In Aristotelian logic, an ἐπιχείρημα is more technical in meaning (cf. LSJ, s.v.), referring to a dialectical syllogism (cf. *Top.* 162a17: συλλογισμὸς διαλεκτικός).

[228] I here add a critical note on Senzasono's interpretation of Plutarch's (generally) sceptical stance in *Quaestiones naturales* (L. Senzasono, 2006, esp. pp. 20–21). Senzasono argues that Plutarch's scepticism is not fundamental in his natural problems but 1) methodological and 2) partial or limited. 1) The methodological aspect implies that "si tratta d'un atteggiamento che guida la ricerca, non d'una convizione gnoseologica che si radica

perspective, natural phenomena are immanently unclear (ἄδηλα) since they belong to the domain of sense perception, and therefore cannot grant

---

in una concezione scettica sistematica, sia pure probabilistica" (p. 21, n. 31; see also p. 149, n. 15). Senzasono therefore prefers to speak of "agnosticismo metodologico, inspirato allo scetticismo accademico" (p. 150, n. 15). There are, indeed, clear sceptical overtones in Plutarch's natural scientific method, which does not, however, imply, as we saw above, that Plutarch would consider himself to be an Academic Sceptic *pur sang* rather than an enthusiast of Plato. Therefore, Plutarch is not following Carneades' 'probabilism' in a strict way, as Senzasono knows (p. 21, n. 31): "È probabile che un intellettuale come Plutarco, antidogmatico e aperto a istanze problematiche, sia stato stimolato dalla gnoseologia di Carneade e indotto da essa all'uso frequente del termine [sc. πιθανόν], tanto piú che aveva fatto parte dell'Accademia; ma [...] basta Platone a spiegare l'atteggiamento di Plutarco e l'uso del termine è originariamente platonico." However, regarding the ἐποχή statement in *De prim. frig.* 955C, Senzasono seems to overdo things a bit when he speaks of "un cauto scetticismo di convenienza, da adottarsi quando le idee non sono né inferiori né superiori alla probabilità (πιθανότης)" (p. 150, n. 15). Plutarch's scepticism in natural scientific matters is more than simply a methodological convenience (to use Senzasono's wording), since it does not just imply an *ad hoc* approach, applicable only on occasion (*pace* also the conclusion drawn for *Quaestiones convivales* by F. Fuhrmann, 1972, p. xxv: "Le scepticisme n'est jamais présenté que comme une diversion."). On the contrary, Plutarch's anti-dogmatic and aporetic approach towards natural phenomena is applied consistently all throughout his scientific writings in general, thus including his natural problems more in specific, and it is firmly rooted in his Platonic epistemological conviction that the φυσικός should manage with plausible arguments. 2) The partial or limited aspect of Plutarch's scepticism, on the other hand, implies, to use Senzasono's own words again, that Plutarch examines "certi aspetti e non altri della realtà che è oggetto di conoscenza" (p. 21), which means – if I understand Senzasono correctly – that Plutarch's approach "è limitato a ciò che non è chiaro, non investe la totalità del reale; la stessa probabilità è un criterio euristico e non esclude la conoscibilità del vero in generale" (p. 22, n. 31); "la sua origine può essere scettica solo limitatamente all'uso del termine 'probabile', non per la concezione della verità in generale" (p. 21). In regards to the intelligible, divine truth in itself, however, one could object that, contrary to what Senzasono writes, Plutarch believed that it *cannot* be reached in its full extent by the human intellect (which is why he advocates εὐλάβεια instead). Regarding the phrase πιθανὸν μᾶλλον ἢ ἀληθές (in *Q.N.* 2, 912A; see above), Senzasono (p. 19; cf. also p. 24) argues that it actually *is* possible for Plutarch to attain the truth ("si ammette la possibilità di conoscere il vero"; "non esclude l'approdo alla verità"), but, then again, it is a mistake to assume that τὸ ἀληθές can be attained because it is merely mentioned there. In his commentary to this passage, Senzasono, does, indeed, draw attention to the interrogative and non-assertoric formulation of the phrase, but he still seems to go too far when he adds that "in generale si ha sempre l'ammissione implicita della possibilità del vero non identificabile col probabile: dunque qui non c'è riduzione di esso [i.e. il vero] al probabile, e quindi scetticismo" (p. 149, n. 15). As we saw earlier on, however, the concept of τὸ ἀληθές in Plutarch's physical aetiologies does *not*, in fact, surpass the realm of verisimilitude. Moreover, it seems that there is a category mistake in Senzasono's account, obfuscating the conceptual border between the logical category of truth and the psychological-subjective category of certainty. This idea applies specifically reliable knowledge[229]. In what follows, I will take a closer look at the use and role of sensory knowledge in *Quaestiones naturales*, including the controversial issue of Plutarch's personal observations.

3. *Sense perception and the issue of autopsy in* Quaestiones naturales

In accordance with Plato, Plutarch firmly believes that natural phenomena have something that we apprehend by opinion and something that we apprehend by intellect[230]. He, therefore, draws a sharp distinction between sensible and intelligible knowledge. Knowledge of intelligibles is superior to that of the senses, since the latter is situated at the level of opinion and conjecture[231]. In light of Plutarch's (generally) sceptical method in natural science, it should be noted that the Academic Sceptics themselves did not intend to do away with the working of αἴσθησις as such but with the unstable δόξαι that it yields[232]. Even though Plutarch accordingly rejects a scientific method that is entirely based on αἴσθησις, this does not imply that sense perception is of no importance in his study of natural phenomena. On the contrary, Plutarch ascribes a great deal of argumentative value to the data pertaining to sense perception in his aetiologies. The overall function of this empirical knowledge is often restricted, though, to somewhat indecisive inductions, associations, analogies, examples etc., that help corroborate the

---

to Plutarch's use of the concept of τὸ ἀληθές in *Quaestiones naturales*. One should draw a functional distinction between the object language (i.e. in Plutarch's text) and the descriptive-analytical language (i.e. how we should interpret it): on the second level the opposition is not plausible-*true* but plausible-*certain*, i.e. 'with certainty true' (thanks are due to J. Opsomer for this remark). Of course, the intelligible truth does, indeed, exist for Plutarch, but certainty not, or at least not in a domain like physics. This is precisely why Plutarch's physical aetiologies are formulated in an anti-dogmatic, fashion. For further criticism of the alleged partialness of Plutarch's scepticism (as put forward by P. Donini, 1986a), see also D. Babut, 1994, pp. 570ff.

[229] Cf., e.g., *De prim. frig.* 955C and *Quaest. conv.* 700B. The conceptual distinction between ἄδηλα and ἀκατάληπτα, as made by Carneades (see J. Opsomer, 1998, pp. 164-165), is not of importance here.

[230] Cf. *Adv. Col.* 1114C: ἔχει τι δοξαστὸν ἡ φύσις ἔχει δὲ καὶ νοητόν κτλ. See M. Bonazzi, 2004, pp. 67-68.

[231] Cf. *Quaest. Plat.* 1001D: τοῖς δ' αἰσθητοῖς πίστιν, εἰκασίαν δὲ τοῖς περὶ τὰ εἴδωλα καὶ τὰς εἰκόνας. See J. Opsomer, 1998, p. 196.

[232] Cf. *Adv. Col.* 1122F: ὁ γὰρ τῆς ἐποχῆς λόγος οὐ παρατρέπει τὴν αἴσθησιν, οὐδὲ τοῖς ἀλόγοις πάθεσιν αὐτῆς καὶ κινήμασιν ἀλλοίωσιν ἐμποιεῖ διαταράττουσαν τὸ φανταστικόν, ἀλλὰ τὰς δόξας μόνον ἀναιρεῖ χρῆται δὲ τοῖς ἄλλοις ὡς πέφυκεν. See J. Opsomer, 1998, pp. 88 and 101: "The reason why we should distrust the senses is not in the first place that our sensory cognitive faculty is defective as such, but rather that the world is not the sort of place which it is possible to know with total clarity." For the idea that sight and hearing are not the result of chance but of reason, cf. *De fortuna* 98BC.

physical aetiologies in an irresolute fashion (see the examples below)[233]. As such, these empirical data are highly indicative and supportive of the plausibility of Plutarch's arguments, but in the end, they are only slightly compelling for producing argumentative certainty[234].

For instance, in *Q.N.* 1, 911D, Plutarch argues that salt water supports boats and swimmers, from which he induces that it has an earthy quality. Regarding this earthy component in salt water, he argues that it cannot provide trees with nourishment, since it is unable to enter the roots and rise in the plant (on account of its thickness and heaviness). Similarly, in *Q.N.* 2, 912B, Plutarch argues that rainwater falling from the sky makes bubbles, from which he induces that it contains light air and breath. Thus, rainwater is more nutritive for plants than irrigational water, since it can be guided and transmitted more quickly into the plant by its tenuity (an attribute of air and breath). These inductions are based on the association of the quality of the water with its corresponding localities (and the primary elements situated there): if it comes from the sky (ἀήρ), it is pure and light; if it is mingled with earth (γῆ), impure and heavy. A case of analogy is found, for instance, in *Q.N.* 1, 911E, where the natural phenomenon of burning heat that destroys drinkable water is compared with fever in the body, which is said to turn moisture into bile. A direct link is drawn here between processes that occur in nature and in living bodies, suggesting that these are based on the same or similar physical principles. An interesting case of exemplification is found, for instance, in *Q.N.* 12, 915A: in order to illustrate that oil causes transparency in the sea, Plutarch adduces a popular account, according to which divers take oil into their mouth and blow it out into the water, so that they – literally – get light and transparency in the depths of the sea (cf. *De prim. frig.* 950BC). Numerous other instances of inductions, associations, analogies and examples of this sort (c.q. relating to empirical knowledge) could be added. What is important, however, is that Plutarch is relying, in most of these cases, on relatively common physical schemes and received beliefs that were already current on the contemporary scientific scene. For instance, in regards to the popular account about divers and oil, the introductory verb φασι clearly indicates that Plutarch is relying on a traditional belief, probably originating from diver lore (perhaps via the intermediation of a lost problem; see the commentary *ad loc.*). Plutarch incorporated and adapted such data to the new problem contexts. Considering that Plutarch elsewhere emphatically underlines the inferior epistemological value of this kind of knowledge, it seems rather unlikely

---

[233] This approach is very common in ancient physical theory. E.g., for the use of analogy, see G.E.R. Lloyd, 1966, pp. 172–420. See also L. Taub, 2003, 98–102 (on Aristotle).

[234] Cf., e.g., E. Teixeira, 1992, p. 213 (regarding the natural problems treated in *Quaestiones convivales*): "Plutarque use de comparaisons pour mieux convaincre, bien que ce qu'il avance parfois ne corresponde pas toujours à la réalité."

that he is relying on personal experience in these cases, that is, on empirical data that he personally gathered or verified in the field.

This has not, however, discouraged some scholars from thinking otherwise with regard to a number of natural phenomena recorded in *Quaestiones naturales*, where Plutarch could perhaps be relying on his own personal observations. The arguments in support of this theory are not very convincing, though[235]. The meticulous and detailed description of some natural phenomena was seen as an important indication of Plutarch's personal observations by Senzasono. However, descriptive detail alone is not a necessary, let alone sufficient, condition for assuming personal observation[236] (think of Plato's minute description of Atlantis). Senzasono also regards the apparent lack of traditional sources in some cases as an additional indication of Plutarch's autopsy[237]. However, in these cases Plutarch may just as well be relying on an oral tradition, or some of his sources may have gone lost – or are we perhaps dealing with an aspect of aetiological ingenuity? Whatever may be the case, the examples that Senzasono provides are not convincing.

In regards to a number of problem chapters on hunting (viz. *Q.N.* 20, 22–25 and 28), Senzasono argues that Plutarch relies on personal experience. Plutarch would also have had personal knowledge of navigation and fishing (as would appear from *Q.N.* 11 and 13; see also *Q.N.* 17–19). However, in these cases, we are dealing with rather weak indications of Plutarch's alleged personal observation. Notably, the natural problem at issue in *Q.N.* 20 about the tears of wild boars tasting sweet (as opposed to the salty and ordinary ones of deer) is cited by Euthydemus and Patrocleas from *their* acquaintance with farming and hunting (in *Quaest. conv.* 700EF: ἀπὸ γεωργίας καὶ κυνηγίας προφέροντας)[238]. In addition, if Plutarch was cultivated in botany, as Senasono believes (considering the names of specific marsh plants in *Q.N.* 2, 912A)[239], this would have only been indirectly and not by his own experience in gardening. It seems far more reasonable to assume that in his capacity as an armchair scientist, Plutarch draws most of his information from the tradition, which he intends to

---

[235] In his study on 'Plutarch and autopsy', J. Buckler, 1992 concludes "that autopsy was neither of particular nor central interest to him" (p. 4829). Regarding Plutarch's *de visu* observation in *Quaestiones Romanae*, see L. Van der Stockt, 1987, p. 287 and also J. Boulogne, 2002, p. 100: "Si la plupart des sujets paraissent fournis par l'expérience même du voyageur qu'est Plutarque, au titre de choses vues, plusieurs portent sur le passé et présentent manifestement une provenance livresque."

[236] L. Senzasono, 2006, pp. 25–28, esp. p. 25: "In certi casi Plutarco sembra aver osservato direttamenti i fenomeni, tanto è puntuale e minuta la notizia che li tramanda."

[237] L. Senzasono, 2006, p. 27.

[238] *Pace* L. Senzasono, 2006, p. 26, n. 39 and p. 203, n. 110.

[239] L. Senzasono, 2006, p. 148, n. 13.

recycle and 'problematise' for intellectual discussion (as we saw earlier on [4.2.1.2.]). Therefore, Plutarch the hunter, sailor, fisherman, gardener, but also the cattle breeder (*Q.N.* 3, 21), winegrower (*Q.N.* 30–31) and bee-keeper (*Q.N.* 35–36) will remain the Plutarch of our imagination.

In a number of cases, the use of *oratio obliqua* clearly betrays Plutarch's reliance on received knowledge (I have already mentioned the use of the verb φασι in *Q.N.* 12, 915A in the context of diver lore). Regarding the question of why the sea becomes less bitter to the taste (γευομένοις) in winter, Plutarch in *Q.N.* 9, 914B indicates that he does not speak from his own experience but relies on what people say to be recorded by Dionysius, the designer of aqueducts (τοῦτο γάρ φασι καὶ Διονύσιον ἱστορεῖν τὸν ὑδραγωγόν: this is *oratio obliqua* to the second degree). In *Q.N.* 22, 917D, to give another example, Plutarch wonders why people say (φασί) that the bear's fore-paws have the sweetest flesh and is the most delicious to eat. This means that Plutarch, in raising the problem, has not necessarily tasted the flesh himself but can just as well be relying on the popular tradition (c.q. hunting lore)[240]. Indeed, in regards to the observations in *Q.N.* 26 on animal diseases, Senzasono this time rightly underlines Plutarch's dependence on the tradition, because many parallels can be found in the literature[241].

Among the 'technicians' (τεχνῖται) mentioned in *De prim. frig.* 948BC [quoted 4.3.2.1.], doctors and farmers especially (not so much musicians)[242] have specific authority in the context of the natural problems treated in *Quaestiones naturales*, since several of these problems are of specific medical or agricultural interest. Ancient doctors sought empirical verification for most of their assertions (e.g., when diagnosing their patients or when performing surgery), as did farmers (e.g., when confronted with an abnormality in the growth of their crops or in the behaviour of their cattle).

---

[240] *Pace* F.H. Sandbach, 1965, p. 198, n. a. As Sandbach notes, Xenophon reports that bears could be caught ἐν ξέναις χώραις (*Cyn.* 11, 1), which may perhaps imply that they no longer lived on Greek territory by Xenophon's time. The appearance of bears in Greece was again reported in the 2nd century AD, though (cf. Paus., *Graec. descr.* 1, 32, 1; 3, 20, 4; 7, 18, 12; 8, 23, 9; cf. also L. Senzasono, 2006, pp. 210–211, n. 126; see the commentary *ad loc.*). Bears were also used in the battles in the arena (cf., e.g., *De soll. an.* 977D, Pliny, *NH* 8, 130, Sen., *Dial.* 5, 30, 1, Martial., *Spect.* 22). Considering his vegetarian leanings, one may doubt that Plutarch ever tasted bear meat himself.

[241] L. Senzasono, 2006, p. 28. This means that these observations "possono suscitare, se non la certezza, almeno il forte sospetto che Plutarco non abbia verificato personalmente quanto riferisce. [...] È probabile che egli [...] si sia rifatto solo a fonti letterarie che tramandavano le abitudine di certi animali, senza un controllo sperimentale." Senzasono draws the same conclusion for the observations in *Q.N.* 5 on the filtration of seawater (see further).

[242] For musical problems, see *Quaest. conv.* 657BE, 672C, 704C–706E, 710C–711A, 741B and Book 19 of Ps.-Aristotle's *Problems* (ὅσα περὶ ἁρμονίαν).

The same is true for other specialists, like hunters (e.g., when distinguishing the trails of different kinds of animals) or bee-keepers (e.g., when trying to find out ways to sedate their bees). This technical know-how must have been a welcome source of inspiration for Plutarch to rely on. It seems rather unlikely, however, that Plutarch, considering his Platonic and Academic leanings, ventured to verify his sources empirically, or – *a fortiori* – that he even considered to do so[243].

In those instances where one may be inclined to speak of 'experiments' or of 'experimental observations' (see the examples below), it turns out that Plutarch frequently relies on the scientific literature – mostly from the Peripatetics – and accepts the eventual outcome without any attempt towards verification (or falsification). Plutarch often incorporates such 'experimental' data in the sub-arguments in order to validate or support the main arguments, but the procedure of controlled experimentation, where the repeatability of the results is essential, is clearly absent in his natural problems[244]. In *Q.N.* I, 911E, for instance, Plutarch advises (παραινοῦμεν) people not to throw seawater onto flames, because it contains a fatty, oily constituent that would only fan the fire. He adduces this point in order

---

[243] Cf., e.g., S.T. Newmyer, 2014, p. 225. In fact, this lack of empirical and experimental verification was not uncommon in ancient science. The same conclusion was made, for instance, for Seneca's *Naturales quaestiones* by T.H. Corcoran, 1971, p. xiv: "[Seneca's] age evaluated theories by the arguments of analogy. To test a theory by controlled experiment was not a standard reflex and it would be inappropriate to expect it." See also, for instance, the account of G.E.R. Lloyd, 1983, pp. 119–135 on the informants of Theophrastus (esp. p. 133; quoted n. 10). See also more generally *id.*, 1991, pp. 70–99 and L. Taub, 2003, pp. 102–103 (with n. 135 for further literature). On ancient empirical arguments, see G.E.R. Lloyd, 1979, 126–225. *Pace* L. Senzasono, 2006, p. 27: "Certo, in molti casi nei quali la presenza di citazioni esplicite o di concordanza con altri testi s'impone, non si può quasi mai escludere del tutto la diretta osservazione o almeno l'esperienza del fenomeno."

[244] According to L. Senzasono, 2006, pp. 30–31 the aspect of empirical observation in *Quaestiones naturales* is, indeed, insufficient due to a lack of systematic experimentation, but he still maintains that in most cases, Plutarch relies on personal observation. Senzasono would, therefore, connect this lack of experimental repetition with Plutarch's observation of sensible phenomena in their direct natural context, that is, without any attempt towards generalisation: "Cosí qui questa mancanza di ripetizione sperimentale è connessa al fatto che Plutarco tendenzialmente, lungi dall'isolare i fenomeni di cui indaga le cause, li vede piú o meno sempre legati all'ambiente entro il quale si manifestano, quindi nella loro immediatezza." The idea that ancient Greek science in general was not concerned with a systematic experimental repetition of its findings ("carenza statistica"), and that it did not intend to artificially isolate particular natural phenomena by excerpting them from their natural environment – which is a common approach in our modern laboratories – is borrowed by Senzasono from S. Sambursky, 1963, p. 233. In my opinion, however, there is an obvious aspect of artificial isolation of these phenomena out of their direct natural context in Plutarch's natural problems in that they are approached from a theoretical rather than empirical standpoint (see also n. 29).

to explain why seawater cannot nourish plants. Plutarch's argument is that the oil in the seawater is hostile to plants and destroys those that are smeared with it. The idea that seawater is highly combustible because it contains a fatty, oily constituent is paralleled on several occasions in Ps.-Aristotle's *Problems* (see the commentary *ad loc.*). Thus, it is unlikely that Plutarch tested or personally experienced what he dissuades others from doing here[245]. To give another example, in *Q.N.* 5, 913C, Plutarch asserts that when seawater is boiled, it loses its saltiness and tang. This would explain why the salty flavour, which is initially present, one way or another, in ripening fruits, loses its natural character by the action of heat caused by the ripening process. Clear parallels can again be drawn with Aristotle's writings for this belief (see the commentary *ad loc.*). A little bit further in *Q.N.* 5, 913CD, Plutarch argues that seawater loses it saltiness, which is earthy and has large particles, when it is filtered. We read that drinkable moisture can be reached by digging on the seashore, that sweet filtered water can be drawn from the sea by a jar made of wax, and that the passage of seawater through white clay renders it drinkable[246].

---

[245] See the doubt, however, in L. Senzasono, 2006, p. 143, n. 6: "Quanto all' "esortazione" a non gettare acqua marina sulle fiamme, sembra che si riferisca, anche per l'uso della prima persona plurale (παραινοῦμεν), all'esperienza diretta all'autora, o almeno a un uso di cui egli spesso ha sentito parlare."

[246] Modern attempts to repeat the experiment with the wax jar have failed – in spite of the fact that Arist., *HA* 590a22 writes: Ἤδη γὰρ εἰληφέναι τούτου συμβέβηκε πεῖραν (Aristotle mentions this experiment also on other occasions: see the commentary *ad loc.*). Some scholars have tried to correct Aristotle's text: instead of κηρίνοις they suggested to read κεραμίνοις, κεδρίνοις or κισσίνοις, but this does not solve the problem. H. Diels, 1905, pp. 310–316, esp. 310–311 already considered the experiment to be false, and argued that perhaps a certain amount of seawater permeated through the cork into the vessel and obtained a sweet taste from the honey that was still sticking to the wax. He adds, p. 314: "Ebenso unrichtig ist die Behauptung (913CD), das Meerwasser werde, durch eine Tonschicht geleitet, seines Salzes beraubt, da diese als Filter wirke." G.E.R. Lloyd, 1991, p. 90 upholds Aristotle's experiment by arguing that water vapour perhaps condensed in the jar once it was submerged. F.H. Sandbach, 1965, p. 165, n. d remains sceptical: "Salt can be extracted from seawater by filtering through certain resins, but not through clay or wax. The mistake over the origin of fresh water in a well by the sea-shore [...] is understandable, but not that over the story of the wax vessel [...]." L. Senzasono, 2006, pp. 27–28 considers this experiment "un caso limite", and on the basis of its "falsità" he argues that "l'intervento verificatore di Plutarco appare del tutto improbabile o è addirittura da escludarsi": "Sarebbe bastato un esperimento elementare mediante un recipiente di cera o d'argilla per verificare che l'acqua di mare non perde affatto la salinità in virtú di tali recipienti, perché evidentemente non viene filtrata da essi. Dunque Plutarco ha accolto da fonti letterarie, o forse da tradizione orali, questa notizia senza curarsi di verificarla." However, A. Stückelberger, 1996, p. 379 made an attempt to explain the phenomenon by invoking modern chemical theory. He speaks of a specific "Ionenaustauscheffekt": "Ton hat namlich die Fahigkeit, Cl-Ionen und Na-Ionen der Salzlosung gegen andere im Ton vorkommende Ionen auszutauschen und

There are again obvious parallels to these accounts in Aristotle (see the commentary *ad loc.*). To give a last example, in *Q.N.* 7, 914A, Plutarch argues that the winter cold compresses the water of rivers and makes it heavy and solid. In support of this argument he refers to the phenomenon of clepsydrae drawing up water more slowly in winter than in summer, and to Theophrastus' account that a certain amount of water from a source on Mt. Pangaeum in Thrace would weigh twice as much in winter as in summer. It seems only likely that Plutarch borrowed these data from the scientific tradition and adapted them to the new problem context without verifying or falsifying them (Theophrastus' lost *De aquis* served as a probable source)[247]. Since empirical knowledge cannot lead on to steadfast science, according to Plutarch, it does not really matter how many times these claims have been checked or re-checked [see 4.1.1.3.]. These 'experimental' data are only incorporated in view of underpinning the plausibility of the arguments at issue. As such, it is perhaps better not to speak of 'experiments' or of 'experimental observations' at all in Plutarch's natural problems. It turns out that these data are often treated on a par with other paradoxographical phenomena, which *a priori* receive Plutarch's benefit of the doubt, as we saw [4.1.1.2.]. Thus, they fall into the same category as traditional knowledge and *idées reçues*.

Even in those few instances where Plutarch employs terminology related to human sense perception, this does not necessarily imply that he directly relies on his own personal observation. When in *Q.N.* 5, 913AB, for instance, he wonders why we observe (ὁρῶμεν) that only one of the eight generic flavours, namely the salty, is not produced by any fruit, whereas the others actually are, it is clear that he aims to evoke a sense of shared, communal experience, as is expressed by the use of the first person plural [see 3.1.4.]. But Plutarch is not necessarily speaking from his personal experience here. In any case, it seems unlikely that he collected all kinds of fruits to do the test, rather than that he just accepted this *a priori*. In fact, the same problem was already raised and solved by Theophrastus (*CP* 6, 10, 1–2), so it is probably Plutarch's primary intention here to provide a few insightful theoretical speculations of his own about the old problem of flavours. This is also the case, in *Q.N.* 9, 914B, where Plutarch wonders

---

damit Salzwasser bis zu einem gewissen Grade zu entsalzen". L. Taub, 2003, p. 103 (with further references) is probably right, however, that "[w]hile Aristotle may have done the experiment as described, it seems possible that he is reporting hearsay evidence, which literally does not hold water".

[247] See L. Senzasono, 2006, pp. 43–44: "È assai probabile che Plutarco non abbia eseguito alcuna verifica sperimentale [...]." F.H. Sandbach, 1965, p. 169, n. b is sceptical about Theophrastus' report: "An experiment involving an exact quantitative observation is unusual in Antiquity, and this one cannot have been correctly performed, if the result is correctly reported."

why the sea becomes less bitter to the taste (γευομένοις) in winter – this he knows indirectly from Dionysius, as we saw –, and in *Q.N.* 22, 917D, where the sweet taste (φαγεῖν ἡδίστην) of the bear's fore-paws is under scrutiny – he probably knows this from popular hunter lore, as we saw.

To give another example, in *Q.N.* 8, 914B, Plutarch wonders why we observe (ὁρῶμεν) that the sea grows warmer when it is agitated, whereas all other liquids become colder when moved and disturbed. By the use of the first person plural he is again evoking a certain sense of shared experience. According to Sandbach, this "phenomenon [is] familiar to bathers, but not registered by thermometers." He explains that "[t]he cause of the illusion is to be sought mainly in the relative temperatures of sea and air, and the chilling effect of the wind that often accompanies a rough sea."[248] This may well be true, but one may still wonder why Plutarch adds that the same effect does *not* take place in other liquids (e.g., the fresh water in lakes or rivers, but perhaps also such liquids as oil, milk, wine, etc. – the *quaestio* speaks of ὑγρόν, rather than ὕδωρ). Plutarch does not necessarily have to rely on personal observation to raise this problem, let alone to solve it in a plausible way. One can very well imagine that it is precisely in the paradoxical element in the *quaestio* that the problematic aspect of this phenomenon is situated, rather than in Plutarch's alleged personal observation, or, for that matter, in the empirical attestation of this phenomenon in the actual physical reality to begin with. If Plutarch had *not* denied that this particular phenomenon takes place in other liquids just as well, this would have entirely eroded the paradox and thus the problematic aspect of the *quaestio* at hand.

Indeed, Plutarch's explanation of the problem in *Q.N.* 8 is clearly based on this paradoxical element: he *a priori* treats the property of heat as an intrinsic feature of seawater (heat is innate to it: σύμφυτον), as opposed to all other liquids (where it is considered an incidental and alien intrusion: ἐπεισόδιον καὶ ἀλλοτρίαν). In Plutarch's world, seawater contains innate heat and other liquids do not (see n. 155). This did not require empirical verification for him, an idea that clearly highlights the theory-ladenness of empirical knowledge[249]. As such, the explanation that he provides is,

---

[248] F.H. Sandbach, 1965, pp. 170–171, n. b. See also L. Senzasono, 2006, p. 29: "Plutarco quindi in questo caso assume un dato sensoriale soggettivamente reale considerandolo direttamente oggettivo, senza sottoporre a una mediazione critica la sensazione per verificare quale sia la sua origine e se essa abbia una corrispondenza nella realtà oggettiva o si tratti di un'impressione (sua o di altri). Così la vera causa del fenomeno gli sfugge [...]." Cf. also p. 178, n. 60: "Qui il dato sensoriale è direttamente assunto come dato sperimentale senza riflessione critica [...]."

[249] On the secondary importance of empirical verification in the formation of ancient scientific theories, but also in our modern evaluation of them, see D. Lehoux, 2003. Lehoux discusses *Quaest. conv.* 641C, where it is argued that there is tangible experience (πεῖραν)

indeed, plausible, albeit only in this specific context, that is, when all premises are accepted as stated. However, in *Q.N.* 18, 916B, Plutarch argues, on the contrary, that the calamary leaps out of the sea in an attempt to escape the cold and the disturbance in the depths (τὸ ψῦχος καὶ τὴν ἐν βάθει ταραχὴν τῆς θαλάττης). This implies that cold can, in fact, be accompanied by movement in the sea – in opposition to what Plutarch says in *Q.N.* 8, 914B (the sea grows warmer when it is agitated). Clearly, then, Plutarch's world is not constituted such that it leaves no room for variations of such physical theories (c.q. the innate heat of seawater), at least within the broad limits of τὸ πιθανόν. These theories are not *idées fixes* in a strict sense, because nothing in the sublunary region is ontologically stable. Thus, all possibilities should also remain open at an aetiological level[250]. Therefore, we should probably cut Plutarch some aetiological slack in physical matters, especially in the context of sense perception (I will come back to this point later [see 4.3.3.2.]).

Similarly, close to the end of the aetiology in *Q.N.* 8, Plutarch refers to the seawater's transparency in order to illustrate its innate heat (transparent water contains light, and by implication fire and heat). One may wonder, though, if this transparency does not apply to other kinds of water just as well, such as that of lakes, rivers or fountains. Plutarch does not mention this, precisely because it would disturb the logic of his argumentation.

---

for the antipathetic phenomenon of magnets losing their attractive force when rubbed with garlic. Who would even consider putting this ancient belief to the test today in his rejection of it? On the basis of this question, Lehoux develops his thesis according to which shifts in ontological categories can occur over time, and that the evaluative significance of empirical verification is *a priori* subjected to these categories. He argues (p. 334) that "the category of 'experience' is heavily intertwined with the category of 'theory'" (c.q. the ancient theory of cosmic sympathy and antipathy *vis-à-vis* the modern theory of magnetic attraction). He, therefore, notes that "obviousness has more to do with the [epistemological] classifications of facts than it does with the experiences of those facts. But the epistemologies on both sides [i.e. in ancient and modern evaluations of such phenomena] try to tie that obviousness of kind to obviousness of experience, by surreptitiously including classification under the rubric of experience" (p. 327). For further (critical) discussion of Lehoux' account, see the conclusion in M. Meeusen, 2014.

[250] Cf. J. Opsomer, 2009, p. 131: "The very fact that opposed principles can apparently be applied at will in physical contexts show something important about what we can know about the physical world according to Plutarch […]." "[T]he physical world is by nature such as to elude a firm epistemic grasp". The same conclusion was made for the (alleged) contradictions in *De facie* by P. Donini, 1988, p. 138: "The contradictions […] have the purpose of insisting on the nondefinitive, not fully certain, nonabsolute nature of scientific explanations. The Platonic philosopher suggests that when one engages in the science of nature or of the heavens, one must always remember that in a wider vision […] the explanations could be different, involving metaphysical forces or entitites which are not even exactly perceptible by science."

Nevertheless, Sandbach tries to redeem Plutarch by suggesting that "[t]he sea is no more transparent than fresh water, but some Greeks, perhaps as not familiar with any deep fresh waters, thought that it was"[251]. Sandbach's claim seems to imply that Plutarch never saw the water of a pond in his entire life – however, he is not *that* much of an armchair scientist either.

In *Q.N.* 39, Plutarch deals with the problem of why water is seen as white in its upper layer but black at the bottom. In the explanation he draws attention to the visual aspect of both colours (*spectatur*), but only in order to emphasise the paradoxical opposition between them. Again, we are dealing with a classic paradox. The main problem is that if water is *essentially* black (a physical theory that recurs, for instance, in *De Is. et Os.* 364B and in *De prim. frig.* 950A), why is it seen as white in its upper layer but black only at the bottom? Apparently, it did not really matter for Plutarch – for the sake of the argument, presumably – that if one takes a portion of the deep water, it will turn out to have the same colour as that in the upper layer[252].

A similar conclusion can be drawn for *Q.N.* 11, 914EF. Here, Plutarch addresses the problem of seasickness, which, according to what he says in the *quaestio*, occurs on the sea rather than on rivers, even if the weather is calm. Plutarch argues that of all sensations, the sense of smell induces sickness most (ἣ μάλιστα ναυτίαν κινεῖ τῶν αἰσθήσεων ἡ ὄσφρησις). On the basis of this reference to human sensation (αἴσθησις)[253] and a lack of exact sources, Senzasono suggests that we are dealing here with an instance of Plutarch's personal experience in navigation[254]. Surely, though, one does not necessarily have to suffer from a certain affliction (c.q. seasickness), in order to speculate about its origins – otherwise, doctors would find themselves in a most distressful situation. Of course, one cannot rule out the possibility that Plutarch personally experienced this ailment during

---

[251] F.H. Sandbach, 1965, p. 171, n. c. He refers to Ps.-Arist., *Probl.* 932b8 and 935b17.

[252] Cf. W. Capelle, 1910, p. 333, n. 2: "Daß das Wasser von Natur weder schwarz noch weiß, sondern farblos ist, scheint man im Altertum nicht erkannt zu haben, offenbar, weil man es nicht in durchsichtigem Glase geprüft hat." One could just as well use one's hands to test this, so it is only likely that the aspect of experimentation was not that important an issue to begin with for Plutarch.

[253] Notably, it is only in this passage in *Quaestiones naturales* that Plutarch relates the concept of αἴσθησις to human sense perception. In other passages, it refers to the sensation in animals: *Q.N.* 18, 916A (ταχὺ δὴ προαισθάνεται δι' εὐπάθειαν τοῦ χειμῶνος), 25, 918B (ἡ δ' ἄγαν περίψυξις πηγνύουσα τὰς ὀσμὰς οὐκ ἐᾷ ῥεῖν οὐδὲ κινεῖν τὴν αἴσθησιν), 26, 918D (οὐ λογισμῷ τοῦ συμφέροντος ἀγούσης τῆς αἰσθήσεως).

[254] L. Senzasono, 2006, p. 27: "Cosí il *Problema* XI [...] presuppone senza dubbio l'esperienza diretta della navigazione e la riflessione personale sui malesseri che essa può comportare [...]." Cf. also p. 185, n. 73: "qui Plutarco si riferisca a sue esperienze personali e rifletta su di esse in modo originale".

a sea voyage once, but it is just as likely that he heard or read about it. In either case, I would argue – perhaps *ad nauseam* at this point – that this element of sense perception is subordinated to the logical development of the aetiology itself. The *quaestio* is again formulated as a traditional paradox: even if the weather is calm, people are liable to suffer from seasickness, and more so when sailing on the open sea than on rivers. One may wonder whether this is really so. Modern scientists would ascribe the cause of seasickness to the rocking motion of the ship on the water over which it sails, regardless of whether it is that of the sea or of rivers. They hold that the illness originates from the conflicting signals that the eyes and the equilibrium sensors in the ears send to the brain. Plutarch, however, links this phenomenon to our perception of smell in conflict with that of sight. Once again, he conjures up a physical opposition in the aetiology, this time between the strange (ἀηθείᾳ) smell of the undrinkable seawater and the customary (συνήθης) smell of the drinkable fresh water of rivers. There is no point in putting this to the test for Plutarch, because human αἴσθησις is fundamentally deceptive anyway (he emphatically speaks of φαντασίαν κινδύνου). In fact, this is actually what the problem in *Q.N.* 11 is all about, effectively illustrating how our senses (c.q. smell) trigger our base emotions (c.q. fear), which can eventually fool our souls and sicken our bodies (c.q. by causing seasickness).

In conclusion, Plutarch's relative nonchalance towards observatory knowledge and empirical verification in his natural problems can be connected not only with the intellectual context of problem solving itself, where the theoretical aspect of argumentative ingenuity is considered an important criterion in the debates, but also, and probably more fundamentally so, with his Platonically inspired depreciation of the low epistemological value of sensible data. In accordance with Platonic-Academic philosophy, Plutarch accepts that natural phenomena have a low ontological statute and, thus, yield inferior knowledge. Therefore, the logical persuasiveness of the explanations is more important for Plutarch than the actual verification or even verifiability of sensory data. The only thing that matters for Plutarch is providing a set of more or less plausible explanations for each natural problem. In the end, the logical development of the explanations takes priority, and when it is necessary, the data pertaining to sense perception are adapted to the logical development of the arguments themselves, rather than the other way around. Van der Stockt is correct, therefore, that:

> "there is, in ancient science and Plutarch's practice, undeniably a certain degree of negligence in the gathering and classifying of physical data, of testing hypotheses against those data, let alone of experimentation;

the criterion for validating scientific opinions seems to have been more the comprehensiveness of "logical" explanations and the coherence of the theory"[255].

In the following section, I will examine in further detail how Plutarch organises his natural problems in a generally logical fashion, and which methodological dynamics are playing along in this process.

4.3.3. Logical-rhetorical dynamics

In his natural problems, Plutarch is not concerned with formulating cogent proofs but probable arguments, by means of which he tries to convince the reader of what is said[256]. The emphasis on argumentative persuasion has specific rhetorical overtones – rhetoric being the ability to see what is possibly persuasive in every given case (cf. Arist., *Rh.* 1355b26–27 and *Top.* 149b25) –, but it is also an intrinsic aspect of Plutarch's natural scientific method, as we saw previously. By providing several plausible explanations for each problem and by postponing final judgement (ἐποχή), Plutarch intentionally avoids taking a dogmatic position and leaves space for further examination. We know from *De tuenda* 133CE that problems in the field of dialectics are not very pleasant, according to Plutarch, because they 'bring on a headache and are extremely fatiguing', whereas natural problems are 'easy and persuasive' (ἐλαφρὰ καὶ πιθανά [quoted 3.1.3.]). Plutarch thus distinguishes physical aetiology from purely logical demonstration (ἀπόδειξις), that is, a type of reasoning which, in line with Aristotelian dialectics, departs from premises that are strictly true and certain[257].

---

[255] L. Van der Stockt, 2011, pp. 454–455.

[256] For the distinction between plausible *vis-à-vis* cogent demonstration, see, e.g., *Quaest. conv.* 614CD, where preference goes to the former category (διὰ τοῦ πιθανοῦ μᾶλλον ἢ βιαστικοῦ τῶν ἀποδείξεων ἄγουσι τὸν λόγον). In this passage, Plutarch finds an example in Plato, who in his *Symposium*, even when discussing serious philosophical matters such as the final cause and the primary good, does not labour his proof (οὐκ ἐντείνει τὴν ἀπόδειξιν), but employs simple premises, examples and myths to make his point (ὑγροτέροις λήμμασι καὶ παραδείγμασι καὶ μυθολογίαις). The same Platonic strand of argumentation is also present throughout *Quaestiones convivales*, and in Plutarch's other collections of *quaestiones*, including *Quaestiones naturales*.

[257] In Aristotelian logic, demonstrative reasoning (ἀπόδειξις) is deductive proof by syllogism, which is based on primary and true premises (as opposed to dialectical, contentious and false reasoning: cf. Arist., *Top.* 100a25–101a24). It can also apply to a rhetorical demonstration, though, viz. an ἐνθύμημα. See LSJ, s.vv. Cf. Arist., *Rh.* 1355a6–8: ἔστι δ' ἀπόδειξις ῥητορικὴ ἐνθύμημα, καὶ ἔστι τοῦτο ὡς εἰπεῖν ἁπλῶς κυριώτατον τῶν πίστεων, τὸ δ' ἐνθύμημα συλλογισμός τις κτλ. For the syllogism as a logical means of demonstrating the truth, cf. *De E* 387A: ἐπεὶ τοίνυν φιλοσοφία μὲν περὶ ἀλήθειάν ἐστιν ἀληθείας δὲ φῶς ἀπόδειξις ἀποδείξεως δ' ἀρχὴ τὸ συνημμένον, εἰκότως ἡ τούτο συνέχουσα καὶ ποιοῦσα δύναμις ὑπὸ σοφῶν ἀνδρῶν τῷ μάλιστα τὴν ἀλήθειαν ἠγαπηκότι θεῷ καθιερώθη.

This does not imply, of course, that the arguments in Plutarch's natural problems would be strictly illogical. As I will try to show in this section, the laws of logic and dialectics remain operative in Plutarch's natural problems, albeit in a broader sense, viz. by being applied to the overall plausibility, not the veracity, of the arguments at hand[258]. Indeed, Plutarch's custom to argue from probabilities impedes his physical aetiologies from developing into fully logical demonstrations (ἀποδείξεις). Even still, the basic principles and rules of logic and dialectics are observed throughout the entire collection, which is true at least within the epistemological confines of physical aetiology. More concretely, I will examine how Plutarch deploys several such dialectical-rhetorical strategies in order to bolster the persuasiveness of his arguments[259]. I will also try to explain why – that is, for which methodological motives – they are deployed in his natural problems.

1. Contradiction, non-contradiction and aetiological freedom

First of all, the principle of non-contradiction is operative throughout *Quaestiones naturales*, albeit not always in a very strict sense. At some points, Plutarch intentionally avoids logical contradictions among the different problems. This is the case, most clearly, in *Q.N.* 5, 913B, where he (regarding the problem of why the salty flavour is not produced by any fruit) in the first *causa* notes that salt is not nutritious for animals that feed on seeds and plants (a similar point is made in the first *causa* of *Q.N.* 1, 911C). At the end, though, he adds that it still functions as a relish for some animals and takes away their satiety (ἥδυσμα δ' ἐνίοις γίνεται τῷ τὸ πλήσμιον ἀφαιρεῖν τῶν τρεφόντων). This addition does not directly contribute to the argument at hand, but it is still useful as an allusion to *Q.N.* 3, 912DE earlier on, where the same point was made, albeit regarding another problem (viz. why herdsmen put down salt for their cattle). Plutarch there argues in the first *causa* that salt stimulates the appetite in animals (τήν τε γὰρ ὄρεξιν ἡ δριμύτης ἐκκαλεῖται) by opening up the pores and thus ameliorating the passage of the food for distribution into the body. By implicitly alluding in *Q.N.* 5, 913B to what was previously argued in *Q.N.* 3, 912DE Plutarch, thus, avoids contradiction.

Such back-references between different problems can also be marked more explicitly, for instance, with phrases like ὡς εἰρήκαμεν (*Q.N.* 16, 915E) or διὰ τὴν εἰρημένην αἰτίαν (*Q.N.* 24, 917F; also used within the same

---

[258] Notably, in *Q.N.* 15, 915D the argument is actually called a λόγος, in the sense of a logical account. Cf. also, e.g., *Quaest. conv.* 680CD [quoted 4.1.1.1.], *De facie* 932D.

[259] For the operation of several such logical principles also in *Quaestiones Romanae*, see J. Boulogne, 1992, pp. 4690–4694 ("raisonnement logique") and pp. 4696–4698 ("esprit dialectique").

aetiology in *Q.N.* 4, 913A). The avoidance of contradiction is not, however, attested consistently throughout the entire collection (at times, not even within one and the same aetiology: see further)[260]. Take, for instance, the aforementioned case concerning the relation between the motion and the temperature of seawater in *Q.N.* 8, 914B vis-à-vis *Q.N.* 18, 916B. In the former passage, Plutarch argues that motion causes warmth in the sea, but in the latter it is said to produce coldness. As noted previously regarding this type of contradiction, we should probably allow some aetiological freedom in Plutarch's natural problems, since nothing in the realm of natural phenomena and sensory data is certain from an epistemological perspective, so that all possibilities remain open also on an aetiological level (see n. 250).

The principle of non-contradiction is also operative within each problem chapter separately, where it applies to the validity of both the *quaestiones* and the *causae*. As to the *quaestiones*, first of all, the natural phenomena are often formulated on the basis of a specific paradox, which aims to evoke a sense of wonder for a given 'contradiction' in nature. These paradoxes are most commonly formulated (with the usual ellipses in phrasing in *Quaestiones naturales*) as follows: 'Why does phenomenon X have the effect Y, while phenomenon Xa does not have the effect Ya?' (e.g., *Q.N.* 7, 8, 13) or 'Why does phenomenon X have the effect Y, while not-Y is in fact expected?' (e.g., *Q.N.* 11, 26, 28). The paradox can be considered a subcategory of the *mirabilia*, because it just as well arouses a sense of wonder for what is said, and responds to a sentiment of initial disbelief [see 4.1.1.1.]. With the aetiology that follows, the author then tries to demonstrate that the contradictory aspect in this paradoxical phenomenon is only a matter of appearance and can be solved by providing pertinent explanations, thereby revealing the actual logic behind the paradox. This is usually achieved when a subtle conceptual distinction is made in the problematic natural phenomena itself. This distinction is often already hinted at in the specific phrasing of the *quaestio*, more precisely by the mentioning of or allusion to opposite substances, qualities, powers or processes[261]. Take, for instance, *Q.N.* 13, 914BC, where Plutarch wonders why fishermen's nets decompose more in the winter than in the summer, whereas the opposite is true for other objects. The abnormality of this paradoxical phenomenon is twofold: why does this process of decomposition occur more in the winter than in the summer and not the other way around, as could be expected

---

[260] *Pace* H. Flashar, 1962, p. 370, who ensures us that *Quaestiones naturales* is entirely free from contradictions (and doublets). On Plutarch's contradictions more generally, see A.G. Nikolaidis, 1991. On contradictions in Ps.-Aristotle's *Problems*, see C. Prantl, 1852, pp. 358–359 (with E.S. Forster, 1928, p. 165).

[261] For the idea that there are, indeed, opposite forces at work in nature, cf. *De prim. frig.* 947F (quoted, n. 163).

(since heat, which is an attribute of the summer rather than the winter season, has a putrefactive quality)? And why is the opposite true for other objects (presumably on land), but not for the nets (in the sea)? In other words: why does nature seem to make an exception for the nets? In his aetiology, Plutarch will draw specific attention to these opposite attributes in order to solve the paradox (viz. winter vs. summer, and nets vs. other objects). The first *causa* draws attention to the opposition between cold and heat. The heat of the seawater withdraws under the atmospheric cold in winter and is concentrated by it (ἀντιπερίστασις), so that the water in the depths of the sea becomes warmer and, by consequence, decomposes the nets more easily. It is implied that the same process (ἀντιπερίστασις) does not occur in the case of other objects on land in winter, on account of the presence of atmospheric cold. Plutarch makes a more subtle distinction in the second *causa*, where he argues that the nets do not actually decompose, but undergo a process similar to it (see further).

At the level of the *causae*, secondly, we see that within one and the same explanation no contradictions occur, probably because this would disturb the internal logic of the argument at hand. Nevertheless, such contradictions sometimes do occur among the separate *causae* in one and the same aetiology. It is perhaps useful for our further analysis to draw a functional distinction here between two different types of *causae*, based on the underlying logic they each have in the aetiologies. 1) At times, the aetiology consists of a number of *causae* that are complementary to each other, in the sense that they complete one another. In this case, the natural phenomenon is conceived of as the result of a complex natural process, and the different explanations each mention a cause that explains part of this process (cf., e.g., *Q.N.* 12, 19, 26, 34). For instance, regarding the problem in *Q.N.* 26 of why animals seek and pursue substances that have remedying properties when they are ill, and often restore themselves to health by using them, Plutarch gives two complementary explanations. According to the first *causa*, it is by the attractive qualities of odours that animals find the proper cure for their disease. However, this does not explain why animals are attracted by these odours only when they are ill. Plutarch explains this point in the second *causa*, where he argues that the animal's bodily constitution (κρᾶσις) follows the disease, and as such influences the appetite. The two explanations are complementary to each other and must be read together for a proper understanding of the problem. At times, the aspect of aetiological complementarity is formulated more periphrastically with a phrase like ἅμα συνημμένον (*Q.N.* 21, 917C: συμμένον?) or more emphatically with a phrase like ἢ δεῖ μὴ μόνον κτλ. (*Q.N.* 25, 918B). It may also be implied in the introductory ἢ καί at the beginning of the *causae* (*Q.N.* 2, 912A, 21, 917C, 28, 919A (?), 31, 919D)[262]. 2) In other

---

[262] In this sense, ἢ καί signals a connection between two or more solutions (implying

cases, the *causae* provide alternative interpretations for the same problem, which are not linked to each other and can be seen as completely distinct approaches to the problem. This is the most common type of explanation. Two sub-categories deserve specific consideration here. A) Sometimes the explanations exclude each other by mentioning causes that cannot occur at the same time. B) On other occasions, Plutarch tries to cast doubt on the natural phenomenon itself, as mentioned in the *quaestio* (I will deal with the second sub-category later [see 4.3.3.3.]).

As regards the first sub-category (2A), apparently, Plutarch's well-reasoned undecidedness in natural science allows contrary arguments to be taken into consideration. This is the case, for instance, in *Q.N.* 3, 912DE, concerning the problem of why herdsmen put down salt for their cattle. In the first *causa*, Plutarch argues that salt fattens the cattle, whereas according to the second it reduces their bulk. Plutarch here marks the contradictory value of his second argument with the introductory ἢ μᾶλλον, but on other occasions the contradiction is more explicit with a phrase like ἢ τοὐναντίον (e.g., *Q.N.* 27, 918E). Let it be absolutely clear, however, that these 'contradictions' are *not* logical errors per se. On the contrary, they aim to provide an alternative explanation that is, in all regards, opposed to the preceding one but is at least equally, if not more, plausible. It is, in fact, a well-planned rhetorical strategy and challenge for Plutarch to reverse a certain argument in order to argue in favour of its exact opposite[263]. This argumentative technique ties in closely with the

---

something like οὐ μόνον [...], ἀλλὰ καί [...]: see LSJ, s.v. καί B, 2, a), rather than that it introduces an alternative explanation that must be seen as fully distinct from the other *causae* ('or rather', 'or indeed'). For the complementary value of ἢ καί, see J.D. Denniston, 1966, p. 299: "B is true as well as A". This is not to suggest that, in the case of Plutarch's problems, two explanations can be true for one and the same phenomenon (cf. F.C. Babbitt, 1936a, p. 2: "presumably not more than one can be right"); rather, it implies that B may lay an equally legitimate claim to being 'true' as A does, the connection bearing on the underlying epistemological conviction that certainty is eventually unattainable. This very meaning is also attested in Plutarch's use of ἢ καί throughout the *Quaestiones Romanae* (267E (ἢ καὶ ἄλλως), 270B (ἢ καὶ τὸ τοῦ Θεμιστοκλέους ἔχει λόγον;), 273E, 275C (ἢ καὶ τοῦτο λύεται τῇ ἱστορίᾳ;), E, 277D (ἢ καὶ τοῦτο λυτέον τῇ ἱστορίᾳ;), 279E, F, 280E, 281F, 283A, 286B, 288B, 291A).

[263] Cf. J.R. Hamilton, 1969, p. xxx (in the context of rhetorical school exercises): "one should not rest content with refuting one's opponents, but should seek to prove the exact opposite". In regards to the reversing (ἀναστρέφειν) of arguments, compare the analogy in *De Pyth. or.* 396E, where the following story is told by the Epicurean Boëthus. Pauson the painter was commissioned to paint a rolling horse, but painted it galloping. The patron was displeased, but Pauson laughed and turned the painting upside down, so that the horse appeared to be rolling instead of galloping. According to Bion, this is also what happens with some arguments when they are inverted (τοῦτό φησιν ὁ Βίων ἐνίους τῶν λόγων πάσχειν, ὅταν ἀναστραφῶσι).

'contradictory discussion' (ἐπιχείρησις εἰς ἑκάτερον, *disputatio in utramque partem*) that was a common educational practice in the schools of rhetoric and philosophy at the time, where topics were argued from both sides, pro and con[264]. As is well-known, several of Plutarch's works are based on this antithetic scheme[265]. Babut has even argued that this is a typical procedure in Plutarch's scientific writings in general:

> "Et c'est cette méthode de la *disputatio in utramque partem* que l'on voit mise en œuvre dans les écrits de l'auteur qui traitent de problèmes de physique, qu'il s'agisse d'excercices purement scolaires, comme les *Aetia physica* et la dissertation *Aquane an ignis sit utilior*, ou de travaux plus élaborés et susceptibles de refléter davantage ses propres options philosophiques, comme le *De facie* ou le *De primo frigido* [...]."[266]

This method of contrary discussion can be connected with the concept of 'equipollence' (ἰσότης λόγου or ἰσοσθένεια) often invoked by the Academic Sceptics[267], that is, the argumentative strategy, where a person, in order to dispute his opponents' claims to the truth, substantiates his own theory with arguments that are equally convincing or plausible, a procedure that eventually results in ἐποχή (as is the case, e.g., in *De prim. frig.* 955C [quoted 4.3.2.1.]). A similar dynamic is seen at work in the contradictory arguments in Plutarch's natural problems, since they each contribute an equal share of plausibility to the aetiology, often without any clear value judgement being made.

Strictly speaking, the concept of contrary discussion involves an opposition of two essentially irreconcilable explanations[268]. However,

---

[264] The technique of contrary discussion can be retraced to the Δισσοὶ λόγοι (on which, see P. O'Grady and D. Silvermintz, 2008, pp. 147–151). It was also well-entrenched in the philosophy of Plato, Aristotle and the Academic Sceptics (cf., e.g., *De Stoic. rep.* 1035F–1036A, *Adv. Col.* 1124A). See, e.g., J. Boulogne, 1992, p. 4697 (with n. 100), D. Babut, 1994, pp. 566–567, J. Opsomer, 1998, pp. 186–190, E. Kechagia, 2011a, pp. 96–97.

[265] It can be seen both in Plutarch's scientific and non-scientific writings. For the contrary discussion in *Quaestiones convivales*, see, e.g., 667C–669E. See also esp. the contrary argumentations in *De sollertia animalium* and *Aqua an ignis sit utilior*, but also, e.g., in *Animine an corporis affectiones sint peiores*, *De fortuna*, *De fortuna Romanorum*, *De Alexandri Magni fortuna aut virtute*, *Bellone an pace clariores fuerint Athenienses* etc. The Lamprias catalogue mentions a work Περὶ τῆς εἰς ἑκάτερον ἐπιχειρήσεως in three Books (nr. 45). Cf. R. Flacelière, J. Irigoin, J. Sirinelli and A. Philippon, 1987, pp. ccvii–ccx.

[266] D. Babut, 1994, p. 567.

[267] Cf. Sext. Emp., *HP* I, 202–206. See J. Opsomer, 1994a, pp. 77, 79 and 1998, pp. 186, 214–216 (with n. 13 for further references).

[268] See, e.g., those problems where Plutarch alternately draws attention to the external

Plutarch often opposes more than two explanations in his aetiologies, and the opposition is not always that strict either. Even though he does often employ an aetiological parting into two opposite *causae*, in some cases yet other alternative *causae* are added (see below). Moreover, in those cases where only two explanations are given, these are not always strictly opposed to each other[269]. There is reason to assume, therefore, that yet another, more complex, logical principle plays an important role in Plutarch's physical aetiologies. Two further observations should be made in this regard.

First of all, this 'contradictory' approach, where an argument ('thesis') and a counterargument ('antithesis') are directly opposed, can be subsumed by some kind of combinatory argument ('synthesis' – the Hegelian scheme of *Aufhebung* is never strictly applied, though). Such an aetiological scheme is found, for instance, in *Q.N.* 12, where the external and the internal effects of the phenomenon of oil poured on seawater are opposed to each other in the first two explanations, while in the third solution, both of these effects are combined. The first two *causae* are complementary to each other, explaining the seawater's γαλήνη and καταφάνεια provided by the oil respectively, but the last one tries to comprise and combine the preceding *causae* by explaining both the effects of καταφάνεια καὶ γαλήνη at the same time (as mentioned in the *quaestio*), so that it is the most complete in the aetiology. The second point, which is more complex, has to do with the aspect of 'aetiological comprehensiveness' in Plutarch's natural problems[270]. I will deal with this issue in the following section.

2. Aetiological comprehensiveness and pluricausality

In Plutarch's natural problems, each *causa* is meant to adduce a more or less plausible element to the aetiology, without final judgement being made (ἐποχή). The combination of a number of plausible explanations – either as alternatives for or complementary to each other – can be considered an implicit attempt at aetiological comprehensiveness, in the sense that an

---

and internal explanations of a specific phenomenon (cf. *Q.N.* 7, 12, 21, 26 – the alternation also occurs the other way around: cf. *Q.N.* 16, 17: see G. Roskam, 2011c, pp. 424–425).

[269] A similar observation was made for *Quaestiones Romanae* by J. Boulogne, 1992, p. 4697.

[270] For this, I – partly – rely on a theory Boulogne has proposed in view of the logic behind Plutarch's aetiologies in *Quaestiones Romanae* (J. Boulogne, 1992; quoted below), and which has also been applied to Plutarch's other collections of *quaestiones* by other scholars. For its application to *Quaestiones Platonicae, Romanae, naturales, convivales*, and also *De E* and *De Iside et Osiride*, see G. Roskam, 2011c, pp. 424–425. For its application to *Quaestiones convivales* more in specific, see J. König, 2009, p. 88 and F. Klotz and K. Oikonomopoulou, 2011, p. 26.

effort is made to exhaust the principle of plausibility as much as possible. Boulogne has drawn a similar conclusion for Plutarch's *Quaestiones Romanae*:

> "Face à la complexité du monde, Plutarque applique la méthode capable, à son avis, de lui en procurer la saisie la plus totale possible. D'où une interrogation systématiquement multiforme, fondée sur la conviction que seule la pluricausalité ouvre, dans le champ d'étude parcouru, le chemin de la compréhension."[271]

The aspect of 'pluricausality' (to use Boulogne's wording) is not necesarily the same as 'causal overdetermination', which implies that of several explanations that are given each may be sufficient in itself to account for the problem at hand[272]. Rather, Plutarch's attempt towards aetiological comprehensiveness reveals an important aspect of his natural scientific method, viz. that of epistemic weakness, according to which the φυσικός should acknowledge that the ultimately correct explanation cannot be grasped by means of physical aetiology. The formulation of several more or less plausible explanations in the aetiologies is not, of course, an obstacle in the search for a valid explanation: it rather promotes it. At the same time, it testifies to the author's impulse to put into words the diversified range of causal attributes that may be operative in a specific natural phenomenon,

---

[271] J. Boulogne, 1992, p. 4690. Boulogne adds (p. 4697): "Dans ces conditions, la multiplication des interrogations [...] tend [...] vers une espèce de démonstration du bien-fondé de la promotion d'une solution regardée comme la plus valable. Plus l'addition des disjonctions s'accroît, plus celles-ci s'enrichissent de la confrontation des origines possibles ainsi rapprochées les unes des autres, plus nombreuses, du coup, deviennent les associations et les combinaisons qu'elles provoquent et plus grande leur chance d'emprisonner la réalité dans leurs rets." I am not so sure, however, whether the ultimately correct explanation will be reached by doing the sum of the plausible elements found in the distinct *causae*, each containing a complementary 'part of the truth'. In the end, the aetiology as a whole will provide only a plausible account. Cf. *id.*, 1994, p. 129: "le domaine étudié n'offre aucune certitude absolue. Néanmoins, plus que le doute ou la prudence liée négativement à une impuissance intellectuelle, ce qui sous-tend cette démarche demeure principalement la conviction que la réponse la plus plausible est celle qui parvient à réunir de façon cohérente le maximum des parcelles de vérité dispersées au sein de la multiplicité des explications connues et possibles".

[272] Cf. *De E* 387B, for the idea that, since nothing in the cosmos becomes without a cause (οὐδενὸς ἀναίτιος ἡ γένεσις), the connection of individual causes leads to profound knowledge of everything (ὁ τὰς αἰτίας εἰς ταὐτὸ συνδεῖν τε πρὸς ἄλληλα καὶ συμπλέκειν φυσικῶς ἐπιστάμενος οἶδε καὶ προλέγει 'τά τ' ἐόντα τά τ' ἐσσόμενα πρό τ' ἐόντα'). This should not be interpreted in light of 'causal overdetermination' but in light of the Stoic theory of fate (εἱμαρμένη), which is based on a predetermined sequence of causes (the context is that of the prophetic art) [see 4.1.1.3.].

and thus in nature as a whole. The search for logical comprehensiveness is, thus, motivated on a firm epistemic basis.

Notably, the aspect of 'pluricausality' was a common argumentative procedure also in the scientific method of the Epicureans, especially in matters pertaining to meteorology[273]. However, as König has argued in light of the aetiological pluralism of the natural problems treated in *Quaestiones convivales*, it has a very different motivation in Plutarch:

> "Epicurean theory holds that all explanations are equally valuable, the main aim of explanation being to remove superstition by showing that a number of plausible rational explanations exist; in some of his works Plutarch rejects that assumption, tending to hierarchise his alternative explanations according to plausibility [...], but the *Quaest. conv.* in places comes close to endorsing that Epicurean view, albeit for very different reasons, by the suggestion that all responses may be equally valid because of their equal capacity to inspire philosophical reflection."[274]

In what follows, I will zoom in on 1) the aspect of superstition, and 2) the value and validity of the aspect of plausibility in Plutarch's aetiologies.

1) König is absolutely right in pointing out that Plutarch's custom of listing several plausible explanations for one and the same natural phenomenon is fundamentally different from that of Epicurus. This can be explained in light of the completely different outlooks on the world these two thinkers had, and especially in light of their dissimilar views on theology. As is well-known, Epicurus did not reject the existence of the gods, but situated them in the *intermundia*, thus generally separating them from our world. These gods do not really care about us, but focus all of their attention on their own blissful state of being. In an attempt to attain ἀταραξία, we should strive to emulate their tranquil way of living, but we should not fear their wrath. By providing several equally valid natural explanations, then, for the at times very frightening natural phenomena around us, Epicurus aims to show that these can be explained by the working of natural causes alone (only one of these causes being valid in our world: cf. Lucr., *De rer. nat.* 5, 526–532). Although for Plutarch physical aetiology also serves as a useful tool to do away with irrational feelings of superstition, as discussed previously [see 3.2.2.], he would probably consider Epicurus' approach atheistic, at least in the sense

---

[273] See E. Asmis, 1984, pp. 321–330, F. Jürss, 1994, L. Taub, 2003, pp. 127–137, F. Bakker, 2010, pp. 8–68.

[274] J. König, 2007, p. 54, n. 38. Cf. also J. Boulogne, 1992, p. 4694 (with n. 77), 1994, p. 129 and P.R. Hardie, 1992, p. 4761: "One may compare the Epicurean attitude to meteorology; curiosity is satisfied when a number of possible materialist accounts have been given, and the danger of superstition has been removed".

that it does not account for the providential ordering of the world. This world is based, rather, on the principles of chance and atomism instead[275]. Even if it seems that the eventual purpose of Epicurus' science and its search for 'pluricausality', which is to procure ἀταραξία, is not so different from the εὐθυμία ('tranquillity of mind') Plutarch's science envisages, the fundamental difference is that Plutarch's concept of εὐθυμία is based on a firm belief that God's benevolent influence is very palpable in this world, whereas Epicurus' concept of ἀταραξία is established rather on the absence of the gods. As opposed to Epicurus, Plutarch draws a much closer link between the material and the divine realms. Accordingly, he is convinced that only by taking into account *both* the natural and the intelligible aspects of the natural spectacle the φυσικός will be able to attain a proper devotion towards the divine (εὐσέβεια)[276] [see 4.1.1.3.]. So if this type of research is to "inspire philosophical reflection" (to use König's words), it will only be by shifting towards a higher level of explanation.

2) In regards to the value and validity of the aspect of plausibility in Plutarch's physical aetiologies, it is true that one natural explanation is often presented as being more or less plausible than another, hence the possibility of hierarchising them (cf. the finale in *De prim. frig.* 955C [quoted 4.3.2.1.]). However, eventually each of the natural explanations cannot be *more* than plausible – it cannot attain the level of τὸ ἀληθές [see 4.3.2.2.] –, so that they are, at least from a strictly epistemological perspective, equally invalid (as opposed to König's equal

---

[275] This is also why the exclusion of mythology from natural scientific discourse is a necessary condition for Epicurus (see D.L. 10, 104: μόνον ὁ μῦθος ἀπέστω· ἀπέσται δὲ ἐάν τις καλῶς τοῖς φαινομένοις ἀκολουθῶν περὶ τῶν ἀφανῶν σημειῶται). This, of course, is very different from Plutarch's approach [see 4.1.2.2.].

[276] Arguably, the distinction Plutarch makes between superstitious and atheistic reactions to natural phenomena ties in closely with his (at times, indeed, somewhat misguided) representation of his sworn philosophical adversaries, viz. the Stoics and the Epicureans respectively. Superstition applies to the world view of the Stoics, in the sense that they accept the working of divine providence in nature on a rather unscientific basis (they rely on their blind faith in a universally predetermined fate, which is immanent in nature [see 4.1.1.3.]), whereas the Epicureans – always according to Plutarch – openly promote atheism in their natural scientific inquiries by rejecting divine providence. For Plutarch, however, the natural causality of our world relies on a divine order, and both of these features go firmly hand in hand in his opinion, so that the φυσικός should take both of these aspects into consideration. A similar observation was made by G.E. Karamanolis, 2006, p. 109: "Plutarch's criticism of the Epicurean and the Stoic positions on the divine providence is a clear contrast. [...] In his view, the Epicurean denial of divine providence arises from their atheism, while the Stoic view on providence rests on a materialist conception of God and their assumption of a universally pre-determined fate, since both Epicureans and Stoics are fundamentally mistaken about God's nature." See also J. Opsomer, 2014, pp. 90 and 93.

validity). It is this idea of equal invalidity that stimulates Plutarch's search for multiple natural explanations for one and the same natural phenomenon in an attempt to attain aetiological comprehensiveness. Several explanations are plausible in principle, so that the φυσικός should always leave space for the formulation of new explanations by postponing final judgement (ἐποχή). As a result, Plutarch's quest for aetiological comprehensiveness is a very sophisticated process that often involves a high level of aetiological subtlety. I will further examine this in the following section in light of the logical and rhetorical strategies Plutarch employs in his physical aetiologies more generally.

3. Aetiological subtlety and sophistication

Each and every element that Plutarch adduces in his physical aetiologies performs a significant and meaningful function in the development of the explanations. Therefore, a careful reading is often required in order to fully seize the precise logic that lies behind an explanation. This is not only true for the specific wording Plutarch uses and their precise meaning (as is the case, for instance, with the meaning of ἀναγκαιότερα in *Q.N.* 4, 913A; see the commentary *ad loc.*), but also, more generally, for the formulation of the explanations themselves. These are aften formulated in a deductive fashion, where the stated premises should be accepted as stated. In *Q.N.* 22, 917D, for instance, the explanation develops in a syllogistic fashion. In this problem, Plutarch wonders why people say that the bear's fore-paws (or 'hands') have the sweetest flesh and are most delicious to eat. He argues that those body parts that concoct the food the most provide the most delicious meat (this is the first premise). In the sub-argument, he adds that the best concoction is by what 'transpires', being that part of the body that is most in motion and exercised. Plutarch continues that the bear makes most movements with its fore-paws, which it uses as feet when walking or running and as hands when grasping (this is the second premise). Therefore, as the reader can deduce for himself, the bear's fore-paws 'transpire' most and, by implication, provide the best concoction. For this reason, they are the most delicious (this is the suggested conclusion). From a structural point of view the deductive-syllogistic development of Plutarch's argument remains implicit in the *causa* and the reader has to deduce the conclusion for himself. From a logical point of view, moreover, the argumentation is not strictly valid, because the first premise is formulated interrogatively – that is, as a hypothetical argument –, meaning that the deduction proceeds from an uncertain starting-point. Therefore, we are not dealing with a fully logical demonstration (ἀπόδειξις) but rather with a rhetorical one. But even so, it is clear that the explanation forms a tight and concise logical unity which is firmly captured in a deductive frame. This is what eventually determines its argumentative power and plausibility.

Notably, a certain increase in specificity and detail in the explanations can often be noticed as the aetiology proceeds. It would be wrong to assume, however, that the aetiological development in Plutarch's natural problems is necessarily bound up with an increase in complexity. The last *causa* of *Q.N.* 2, 912C, for instance, is actually introduced as being the easiest and most evident explanation of the entire aetiology (ἡ πάντων ἑτοιμώτατον ἐστι καὶ ῥᾷστον αἰτιάσασθαι κτλ.)[277]. Nevertheless, the criterion of increasing plausibility, which is the main structuring principle in the development of Plutarch's physical aetiologies [see 1.1.4.], often involves an according increase in aetiological subtlety. For instance, in *Q.N.* 17, 915F–916A, Plutarch deals with the problem of people using the (tail) hairs of stallions rather than mares in order to manufacture fishing lines. He gives two explanations for this: in the first one, he draws an analogy with the relative strength in other body parts of horses in arguing that the stallion has stronger hair than the mare. In the second explanation, a more subtle distinction is made between the male *vis-à-vis* the female body on the basis of their anatomical dissimilarity. Plutarch argues that people believe that the mares' tail hairs become inferior because they are wetted by their urine – which discomfort is, so the reader must add for himself, strange to stallions, since their tails, as opposed to those of mares, are not located in the vicinity of the urethra. Another example is found in *Q.N.* 6, where Plutarch examines why persons (probably hunters) who frequently walk through bushes wet with dew contract 'leprosy' on those body parts that come into contact with the brushwood. The first explanation is attributed to Laetus, who holds that the dewy moisture scrapes off the skin by means of its fineness. This implies that dew harms the human skin directly. The second explanation is more detailed than the first, since Plutarch there also includes the role of the plants in the phenomenon, as mentioned in the *quaestio*. He argues that some kind of dust is discharged from plants wet with dew, and that this dust causes harm to those who come into contact with it. As such, it is implied that the erosive effect of dew damages the skin indirectly, that is, via the damage

---

[277] Cf. also, e.g., *Quaest. Plat.* 1004D, where the second and final explanation is introduced by μᾶλλον οὐδὲν περιεργαστέον ἀλλ' ἁπλῶς ἀκουστέον ὅτι κτλ. (see J. Opsomer, 1996a, pp. 78–79, with n. 32 for the negative connotation of περιεργαστέον). See also E. Teixeira, 1992, p. 219 (regarding *Quaest. conv.* 722E): "On a l'impression, ainsi, que Plutarque adopte une démarche quelque peu surprenante, en ce sens qu'il présente une explication assez scientifique avant d'en proposer une autre, beaucoup plus simple et, pour ansi dire, terre à terre." For the idea of a "progressive structure" in Plutarch's writings (especially, but not exclusively, in his collections of *quaestiones*), see G. Roskam, 2011c, pp. 424–425.

it does to plants. Arguably, the custom of making such subtle aetiological distinctions ties in closely with the common rhetorical strategy in ancient scientific competition (ἀγών) to cap an opponent's arguments with one's personal ingenuities[278] (as is clear in Plutarch's criticism of Laetus here in *Q.N.* 6, 913E). This sensitivity to increasing aetiological specificity and detail also recurs in the sympotic discussions recorded in *Quaestiones convivales*. Excessive subtlety is avoided at all cost, though, because, as we saw previously, this suits sophistic disputations by its striving for rhetorical rivalry rather than philosophical dialogue[279] [see 3.1.3.].

A similar feature of aetiological subtlety can be found in the type of explanations that aim to refute or nuance the phenomenon at issue in the *quaestio* by explaining a different but seemingly similar phenomenon (this is sub-category 2B in the analysis of the different types of *causae* set out previously [see 4.3.3.1.]). This kind of explanation is usually found last in the aetiology. For instance, regarding the problem of why fishermen's nets decompose more in the winter than in the summer, whereas the opposite is true for other objects, Plutarch, in the second (and final) *causa* of *Q.N.* 13, 915BC, argues that these nets do not actually decompose, but undergo a process *similar* to decay (σήψει τι καὶ μυδήσει πάσχει παραπλήσιον κτλ.). The real process is that of weathering, which is due to the fact that the nets are desiccated by the cold and violently frayed by the waves. Likewise, in the final *causa* of *Q.N.* 32, regarding the problem of why the palm tree alone rises against a weight imposed upon it, Plutarch argues that the tree raises its twigs only in appearance. The supple and tender twigs cannot sustain the impetus of the weight, but when the weight comes to rest, the twigs gradually erect themselves and give the impression of rising up against it (*speciem quasi contra illud adsurgant praebent*). Another example can be found in the last *causa* of *Q.N.* 4, 913A, regarding the problem of why rainwater accompanied by thunder and lightning is more fertile for seeds. There, Plutarch argues that lightning and thunder are not the actual cause of increasing fertility in plants but rather an incidental aspect. The (spring)rains themselves are more nutritive because they come *before* the summer heat and, thus, protect the crops against it. Similar instances can be found in *Quaestiones convivales*, especially in those cases where Plutarch emphatically draws a distinction between cause

---

[278] See T. Barton, 1994a, pp. 13–14 (esp. p. 14, regarding "the emphasis on subtle distinctions, which seems highly unconvincing to us now"), and pp. 147–149 (on "the agonistic arena"). For professional rivalry among ancient physicians, see, e.g., G.E.R. Lloyd, 1970, pp. 142–143, 1979, pp. 97–98, 1983, pp. 208–209. For the aspect of debate in the context of Galen's anatomical show-case demonstrations (ἐπιδείξεις), see M. Gleason, 2009, pp. 88–89.

[279] For such σοφιστικοὶ ἀγῶνες cf., e.g., *De tuenda* 133E. Cf. also *Quaest. conv.* 614DE, 713F.

and effect in a given natural phenomenon. Plutarch uses this aetiological strategy, for instance, in dealing with the problem of the remora slowing down a ship (*Quaest. conv.* 641AE), and also regarding the (subsequent) problem of wolf-bitten horses being mettlesome (*Quaest. conv.* 641F–642B). He argues – twice at the end of the aetiology – that we should not mistake the effect for the cause in these problems[280] (σκοπῶμεν δ' εἶπον ὅτι πολλὰ συμπτώματος ἔχοντα φύσιν αἰτιῶν λαμβάνει δόξαν οὐκ ὀρθῶς). It is not so much the remora that slows down the ship, but rather the seaweed that sticks to the hull of the ship and that attracts this tiny fish. In the same way, the horses that are attacked by wolves do not at once become swifter, but they would not have escaped their attackers if they were not swift by nature in the first place. In each of these cases, Plutarch puts the popular belief as formulated in the *quaestio* somewhat in perspective, but he does not, as such, aim to reject it (nor to deny any aspect of divine providence to it). After all, he still takes the popular tradition (ἱστορία) as a credible starting-point for the discussion, by giving it the benefit of the doubt [see 4.1.1.2.]. Moreover, this kind of explanation, which sets out on denying the traditional view, is only a plausible one, among several other such explanations: eventually Plutarch postpones final judgement (ἐποχή).

Another subtle aetiological strategy is the use of argumentative rings, where a specific argument is wound up or completed only at the very end of the explanation or aetiology (e.g., *Q.N.* 1, 2, 9, 12, 18, 26). In some of these cases, we are dealing with a repetition of aformentioned elements in a conclusive fashion. In *Q.N.* 18, for instance, Plutarch wonders why the appearance of the calamary is a sign of a great impending storm. He opens his explanation by pointing out that all cephalopods by nature are sensitive to cold due to their bodily constitution. Their flesh is bare and naked, they are not covered with a shell, skin or scales, and their hard and bony parts are located inside, which is why they are called 'soft things' (μαλάκια). At the end of the explanation Plutarch adduces the specific case of the calamary, noting that it jumps out of the water in an attempt to escape the cold and the disturbance in the depths of the sea. He says that the calamary has the most fragile and delicate flesh of all of the cephalopods (καὶ γὰρ ἔχει μάλιστα τῶν μαλακίων εὔθρυπτον καὶ ἁπαλὸν τὸ σαρκῶδες), thus closing the argumentative ring by repeating an element that was adduced at the very beginning of the explanation. In other cases, however, the closing of the argumentative ring at the very end of the aetiology enables a clearer view on an explanation that was previously given. This is the case, for instance in *Q.N.* 1, regarding the

---

[280] The parallel argument in these two problem chapters was also marked by S.-T. Teodorsson, 1989, p. 262.

problem of why seawater cannot provide nourishment to trees. Here part of the first *causa* becomes clearer only after reading the last one. In the first *causa* Plutarch argues that it is not the case that seawater, for the reason that it is nutritious and potable for marine plants and fishes alike, feeds plants and trees on land just as well. The idea that marine plants feed on seawater is not further explained here, but recurs in the fourth and final *causa*, where Plutarch explains that such marine plants (growing in the Indian Ocean) actually receive their nourishment from rivers which deposit much silt in the sea. Plutarch, thus, suggests that the nourishment of marine plants does not essentially differ from that of terrestrial plants, since they are not fed by the seawater itself but by a specific nutritious constituent in it (c.q. silt deposits from rivers). As such, the reference to the nourishment of the marine plants (and perhaps also of fish) by seawater mentioned in the first *causa* is made more specific in the last one. A similar argumentative ring structures the aetiology in *Q.N.* 2, where an implicit connection is made between the first two explanations and the final one: the final *causa* suggests that the season of spring is probably the 'right season' (καθ' ὥραν) for the growth of plants in ponds, as mentioned, but not further clarified, in the second *causa* (see the commentary *ad loc.*). Such a careful, and at times very subtle, structuring gives the aetiology a sense of finalisation and completion, but, of course, the closing of the argumentative ring is not final or complete in the sense that it removes all possible doubt.

In conclusion, even if Plutarch's physical aetiologies are not – or better: cannot be – envisioned as fully logical demonstrations (ἀποδείξεις), the elementary rules and principles of logic and dialectics are still observed in them, albeit within the epistemological limits of physical aetiology. Several logical-rhetorical strategies regulate the concrete phrasing of Plutarch's problems, often in a very subtle and sophisticated way. A specific discursive feature that has remained unmentioned thus far but also contributes to the logical-rhetorical coherence of the problems is found in Plutarch's use of a more or less uniform set of scientific terms and concepts. Notably, most of the physical concepts and processes that Plutarch refers to recur among the most disparate natural phenomena. The theory of κρᾶσις, for instance, applies to the air in the atmosphere, to the composition of the soil and to the constitution of the body. In a similar way, πόροι are present in the earth, in the sea, in plants and animals. As such, these fixed terminological schemes hint at an aspect of conceptual coherence and aetiological unity in nature or at least indicate that the φυσικός is predisposed to find it there. It implies that there are certain conceptual rules and laws immanent and permanent in nature, at least within the limits of the contingent and unstable ontological status of the realm of natural phenomena to which they apply. These concepts deserve separate analysis in the following section, where I will zoom in on the terminological uniformity of the disourse in *Quaestiones naturales* and also on its level of technicality[281].

4.3.4. Uniformity and technicality of the scientific terminology

Even though Plutarch acknowledges that 'the ingenious organisation of nature's activities is beyond the range of words' (*Quaest. conv.* 699B [quoted 4.1.2.2.]), he is well aware of the fact that there is still a suitable scientific way to talk about the sensible world arond us. He also knows that in order to treat a scientific problem properly and, thus, to be regarded as a scientist, he must speak as one, that is, with the terminology that scientists most commonly employ. In *De def. or.* 436E, for instance, Plutarch (in the context of his dualistic view on causality [quoted 4.1.2.1.]) lists a number of concepts that the φυσικοί frequently use in formulating physical explanations: 'the younger generation of physicists [...] ascribes everything to bodies and their behaviour, to clashes, transmutations, and combinations' (οἱ δὲ νεώτεροι τούτων καὶ φυσικοὶ [...] ἀρχῆς ἐν σώμασι καὶ πάθεσι σωμάτων πληγαῖς τε καὶ μεταβολαῖς καὶ κράσεσι τίθενται τὸ σύμπαν). Plutarch is well acquainted with such scientific concepts, which he himself also recurrently employs in his natural problems. One may wonder, though, how technical they really are.

Compare, for instance, the unsystematic enumeration of physical concepts in the short excursion into magnetism in *Q.N.* 19, 916D. Plutarch there writes that 'people explain the phenomenon of attraction or jumping (in magnetism) by emanations, some assuming it to be due to entanglements, others to impacts, and still others to impulsions and circumventions' (καὶ γὰρ ἕλξεις ἢ ἐπιπηδήσεις ποιοῦσι ταῖς ἀπορροίαις, οἱ μὲν ἐμπλοκὰς αὐτῶν οἱ δὲ πληγὰς οἱ δ' ὤσεις τινὰς καὶ περιελάσεις ὑποτιθέμενοι). This unsystematic enumeration of technical terms contains references to the theories of Democritus/Epicurus (ἐμπλοκαί/περιπλοκαί: interlacing of atoms, πληγαί: clashing of atoms), Empedocles (ὤσεις: impulsions), and Plato (περιελεύσεις: circumventions). Plutarch does not explain these technical concepts any further, but it is rather unlikely that he is simply trying to impress the reader with his mere knowledge of such concepts. On the contrary, these terms are meant to illustrate Empedocles' theory of emanations just quoted (DK31B89: πάντων εἰσὶν ἀπορροαὶ ὅσσ' ἐγένοντο), as if to suggest that there is more to them, but that this is not the proper place to deal with them in any detail. Sandbach may well be right, therefore, that Plutarch means to summarise "some account of the

---

[281] For a separate treatment of this topic, see M. Meeusen, 2013b. See also L. Van der Stockt, 2013.

emanation theory that had no original connexion with the problem of the octopus" at issue in *Q.N.* 19, but this does not necessarily imply that the digression into magnetism "contains barely intelligible and certainly irrelevant detail"[282].

The section at hand will analyse and clarify the usage of such scientific terms in Plutarch's natural problems. Part of this analysis has already been provided earlier on (in light of the material principles and natural processes in *Quaestiones naturales* [see 4.3.1.]), and it will be further elaborated upon here in light of the collection's technicality. As we have already seen in chapter three, the genre of natural problems as treated in *Quaestiones naturales* is of an essentially 'technical' kind, being concerned with the type of knowledge that belongs to the τεχνῖται, but it is not therefore overly complex [see 3.1.4.]. What I will argue here, then, is that the aspect of technicality of Plutarch's scientific discourse never becomes a goal in itself but rather a means to communicate about natural phenomena in a clear and uniform way.

1. Let's talk science: the birth and use of technical vocabulary

A relevant passage for Plutarch's view on technical terminology can be found in *Cic.* 40, 2, which deals with Cicero's effort to come up with a Latin translation of several Greek terms of dialectics and natural philosophy:

αὐτῷ δ' ἔργον μὲν ἦν τότε τοὺς φιλοσόφους συντελεῖν διαλόγους καὶ μεταφράζειν, καὶ τῶν διαλεκτικῶν ἢ φυσικῶν ὀνομάτων ἕκαστον εἰς τὴν Ῥωμαϊκὴν μεταβάλλειν διάλεκτον· ἐκεῖνος γάρ ἐστιν ὥς φασιν ὁ καὶ τὴν φαντασίαν καὶ τὴν ἐποχὴν καὶ τὴν συγκατάθεσιν καὶ τὴν κατάληψιν, ἔτι δὲ τὴν ἄτομον, τὸ ἀμερές, τὸ κενὸν καὶ ἄλλα πολλὰ τῶν τοιούτων ἐξονομάσας πρῶτος ἢ μάλιστα Ῥωμαίοις, τὰ μὲν μεταφοραῖς, τὰ δ' οἰκειότησιν ἄλλαις γνώριμα καὶ προσήγορα μηχανησάμενος.

He made it his business also to compose and translate philosophical dialogues, and to render into Latin the several terms of dialectics and natural philosophy; for he it was, as they say, who first, or principally, provided Latin names for 'phantasia' (*visum*), 'synkatathesis' (*assensio*), 'epokhe' (*retentio*) and 'katalepsis' (*comprehensio*), as well as for 'atomon' (*individuum*), 'ameres', 'kenon' (*vacuum*), and many others like these, contriving partly by metaphors and partly by other proper senses to make them intelligible and familiar.

---

[282] *Pace* F.H. Sandbach, 1965, p. 137. Indeed, in a parallel passage in *Quaest. Plat.* 1005BD, Plutarch gives an interpretation of *Tim.* 79e–80c, where he provides a more extensive account of magnetism (the context is that of the effect of amber and the loadstone on other objects: see the commentary *ad loc.*).

What Plutarch is probably implying here is that Cicero's decision to translate these Greek terms into Latin is a practical one and originates from his personal experience that Latin, his native tongue, was somewhat inadequate for the task of philosophy, without the help of Greek philosophy[283]. It testifies to Cicero's intention to create a terminological system that is not only uniform in itself, but also uniform with the Greek philosophical tradition. These terms did not exist in Latin, or at least they did not have the extra, technical connotation that their Greek equivalents had before Cicero translated them from the Greek[284].

Notably, a significant amount of the terms and concepts that ancient Greek natural scientists employed were present in popular vocabulary long before they received their more specialised meanings in scientific discourses[285]. Other technical terms did not originate directly from popular vocabulary, but are rather the result of a linguistic process of neology (as is the case, for instance, with the Peripatetic concept of ἀντιπερίστασις, discussed earlier on [see 4.3.1.2.]). In many cases, popular vocabulary proved to be too imprecise to communicate a very specific scientific notion or idea, so that it often received a more specific and systematic 'pregnancy' in technical contexts[286]. Where in an unscientific context a concept like δύναμις simply denotes 'power', 'strength' or 'force', it receives a more

---

[283] On Cicero's translations from the Greek, see J.G.F. Powell, 1995. This passage is also discussed by L. Van der Stockt, 2013, pp. 440–441, who points at Plutarch's omission of the Latin equivalents for the Greek terms. He concludes (p. 441) that the Chaeronean "is eager to explain and to suggest the high level of Cicero's intellectual activity by a selection of technical terms to be translated. [...] Plutarch's audience was hardly eager to learn some translation theory, and this learned remark probably only testifies to Plutarch's eagerness to explain and illustrate, to his didacticism. And if any pride is involved here, it is not the pride of a Plutarch parading his knowledge of translation theory, but the Greek pride to have taught philosophy to the Romans." For the use of Greek terms in Latin technical literature more generally, see T. Fögen, 2009, pp. 92–105.

[284] Perrin's translation (B. Perrin, 1919, p. 185) of οἰκειότησιν ἄλλαις as "by new and fitting terms" is off the mark. Used of words and phrases, οἰκειότης means "proper sense, opp. μεταφορά" (LSJ, s.v. ii, with a reference to this passage). Cicero did not make 'new terms' (i.e. neologisms), but provided existing Latin words with 'other meanings' (i.e. proper senses). Cf. L. Van der Stockt, 2013, p. 441: "Cicero did not make new words: that would have been a *verborum fictio* acceptable only because of *necessitas*, sc. the *verborum inopia* of the Latin language. The μεταφορά Plutarch is hinting at, is probably the periphrastic metaphor (Lausberg #563 and 594, 3)."

[285] A considerable amount of the technical terminology of ancient scientific writings actually originates from the abstraction of everyday language (see B. Snell, 1960). E.g., for the close relationship between Hippocratic and popular vocabulary, see G.E.R. Lloyd, 1983, p. 204.

[286] E.g., for the development of anatomical terminology, see G.E.R. Lloyd, 1983, pp. 149–167.

technical connotation, for instance, in medical literature, where it often refers to the therapeutic properties of foodstuffs[287]. The concept of δύναμις also has its well-known technical meaning in Aristotelian philosophy, where it is typically translated as 'potentiality' and thought of as opposed to the concept of 'actuality' (ἐνέργεια)[288]. It is not always used in this strictly contrastive fashion by Aristotle, though. He often also uses it to speak more generally of an 'active characteristic' or 'property' of a specific substance, as does Plutarch[289]. According to Senzasono, Plutarch is using the concept of δύναμις in an essentially Aristotelian fashion in *Quaestiones naturales*[290], but this is not necessarily the case, because the concept was firmly entrenched in the ancient Greek scientific tradition more generally (it can also be found in this very meaning in earlier scientific authors, such as the Hippocratics or Plato)[291]. Moreover, the typically Aristotelian opposition between δύναμις and ἐνέργεια is absent in our collection and rare in Plutarch's writings more generally[292]. In *Q.N.* 29, 919A, Plutarch rather opposes the concept of δύναμις to the notion of στέρησις. He argues that some people believe that heat is a δύναμις (i.e. an active property), whereas cold is a στέρησις (i.e. a privation, absence or negation) of heat – an Aristotelian theory with which he disagrees here and elsewhere (*De prim. frig.* 945F–948A; see the commentary *ad loc.*). Plutarch's definition of cold as a δύναμις is more in line, then, with the term's connotation of "*elementary force*, such as heat, cold, etc."[293]. A στέρησις, by contrast, is in fact non-existent (*Q.N.* 29, 919B: ἐπεὶ πλειόνων αἴτιον ἐφαίνετο τὸ μὴ ὂν τοῦ ὄντος)[294].

---

[287] See, e.g., passim in the Hippocratic *De prisca medicina*. Galen wrote three books *De alimentorum facultatibus*.

[288] This Aristotelian opposition became popular also, e.g., in the writings of Imperial medical authors (cf. Gal., *Nat.Fac.* 2, 7, 2–3 Kühn).

[289] For a more systematic account of the concept of δύναμις in Aristotle, see A.L. Peck, 1953, pp. xlix–lv.

[290] L. Senzasono, 1999. Cf. also L. Senzasono, 2006, p. 36, n. 55 (with further references): "il termine in questione ha per noi senso essenzialmente aristotelico, o almeno è da considerarsi d'origine aristotelica: si tratta d'una proprietà materiale, che è potenza, e quindi causa materiale dei fenomeni".

[291] Cf. A.L. Peck, 1953, p. xlix. Cf. also G.E. Karamanolis, 2006, p. 113.

[292] It is mentioned only once in *Quaestiones convivales*, albeit not in a strictly contrastive fashion, viz. in *Quaest. conv.* 637D. Cf. also *Pars an fac.* 2, 16–17; 6, 15–17 and fr. 215f Sandbach.

[293] LSJ, s.v. ii, 2.

[294] According to F.H. Sandbach, 1965, p. 141, δύναμις sometimes has a specific material connotation for Plutarch, denoting a "substance of a distinctive character" (e.g., *Q.N.* 26, 918B, 29, 919A, 32) – a view that is much in line with Plutarch's custom to directly associate physical matter and its qualities (see n. 157). Cf., e.g., *De tuenda* 129F (ὕλην καὶ δύναμιν), *Quaest. conv.* 721F (οὐσία καὶ σῶμα καὶ δύναμις), *De prim. frig.* 945F (τοῦ

A similar case is found in the concept of περίττωμα, which is basically a nominalisation of the adjective περιττός (*Q.N.* 10, 914E). In a non-scientific context, περιττός means 'exceptional' or 'extraordinary', but in a natural scientific discourse it receives a metaphorical sense, implying 'superfluous'. The notion of περίττωμα, derived from it, denotes a residue of a substance that is secreted or excreted after a specific physical process has taken place (e.g., digestion of food). These residues often have various additional properties, among which Plutarch in *Quaestiones naturales* only mentions the generative one (the generative residue produces the offspring: see *Q.N.* 21, 917BC, 30, 919C). This concept occurs frequently in Aristotle's biological works, but it is not, therefore, of Aristotelian origin[295].

Due to the fact that Plutarch uses such scientific terms in a somewhat specialised sense, it is reasonable to assume that some familiarity with the scientific terminology and its use in the scientific tradition was required, or in any case helpful, for a proper understanding of the natural problems by the implied reader. It remains to be seen, however, whether we are really dealing with the kind of jargon that can only be understood by the Greek specialist (φυσικός) or not also by the generally educated layman (πεπαιδευμένος). Even if these terms received a specialised meaning at a certain point in history, their use had become relatively common by Plutarch's time[296]. Plutarch does use several rare Greek words and hapax

---

ψυχροῦ δύναμις ... πρώτη καὶ οὐσία). Sandbach also notes that it is sometimes used in a more abstract sense, thus simply referring to a "distinctive character" or quality in itself (e.g., *Q.N.* 10, 914D, 31, 919D). In regards to the use of *facultas* in *Q.N.* 32, cf. L. Senzasono, 2006, pp. 238–239, n. 186: "[i]l termine *facultas* quasi certamente corrisponde a δύναμις nel testo originale perduto". Senzasono adds that a more adequate Latin translation would have been *virtus* or *potentia*. Perhaps, the use of *facultas* can be explained, then, in light of Longolius' medical expertise and knowledge of Galen (see n. 287).

[295] On the constitution of these residues, including the generative one, and their importance in Aristotelian biology, see A.L. Peck, 1953, pp. lxv–lxvii and 1965, pp. lxxi–lxxii. The theory that sperm (and blood and marrow) is a περίττωμα τῆς τροφῆς is attributed to Pythagoras in Aët., *Plac.* 5, 3 = Ps.-Plut. 905A. Cf. also, e.g., Arist., *GA* 726a26–27. Notably, the concept of περίττωμα does not occur in the *corpus Hippocraticum*, and it may have been introduced in the field of nosology by Aristotle or one of his students: see A. Thivel, 1965 and P.J. van der Eijk, 1990, p. 53. For the distinction in meaning between περιττός and περίττωμα in Ps.-Arist., *Probl.* 953a10–955a40, cf. also B. Centrone, 2011b, p. 335.

[296] An analogy can be drawn with the modern English notion of 'landscape' (see OED, s.v.). This term derives from popular Dutch vocabulary, where the concept of 'landschap' initially denoted a 'region' or a 'tract of land'. The Dutch 'landschap' received a more technical meaning in the context of Dutch landscape painting around 1600 by taking on the artistic connotation of 'a picture depicting scenery on land'. The word was initially adopted

legomena in *Quaestiones naturales*, but in these cases we are dealing with derivations or compounds of already existing words rather than with fully neologistic technicisms[297]. For instance, the word πνευμονία ('disease of the lungs') in the quote from Mnesitheus in *Q.N.* 26, 918DE is rare, while περιπνευμονία or περιπλευμονία (also simply πλευμονία) are more common. One may wonder whether Plutarch is relying on an intermediary source in this case or is perhaps paraphrasing Mnesitheus' account in his own, non-specialist words[298].

2. Big words? High-tech vs. low-tech vocabulary

As Sandbach points out, "[i]t is an error to suppose that because a word needs explanation [for us non-native speakers] it is a 'technical' term". He argues that the scientific terms in *Quaestiones naturales* "are all common Greek words, and no ancient reader would have felt that they were used in an unusual or special way"[299]. The same and similar scientific terminology as found in *Quaestiones naturales* recurs in the natural problems in *Quaestiones convivales*. As seen before, sympotic decorum prescribes that the discussions be kept simple [see 3.1.3.]. This way, everyone can understand what learned people are saying, even those who have no erudition at all (*Quaest. conv.* 613E)[300]. In fact, the mood of the symposiasts would be affected too much if they bombarded each other with technicisms too abstruse for non-specialists (*Quaest. conv.* 614E). Plutarch even rebukes this as a practice of 'wranglers', as Democritus calls them, and of 'phrase-twisting' sophists (= DK68B150: 'ἐριδαντέων' δὲ κατὰ Δημόκριτον καὶ 'ἱμαντελικτέων' λόγους ἀφετέον). Such technicalities are banned from the dinner table, because they break convivial harmony: 'just as the wine must be common to all, so too the λόγος must be one in

---

in English with this technical (c.q. artistic) meaning, which can still be retraced today in our modern notion of 'landscape' as an 'inland natural scenery'. The technical sense is now largely lost, though, since the word is no longer used (only) in the context of painting. This example illustrates, then, how a popular concept can become technical at a certain point in time and afterwards bleed over again into popular vocabulary, with corresponding shifts in semantics. Similarly, the concept of 'panorama' came into existence as a neologism only near the end of the 18th century, again in the context of painting.

[297] These include the concepts of προδιαγωγή (*Q.N.* 5, 913C), ἀνοστότερον (*Q.N.* 15, 915E), ἀνθρηνιώδης (*Q.N.* 19, 916E), δυστίβευτος (*Q.N.* 23, 917E), ἰχνοσκοπία (*Q.N.* 24, 917F), πνευμονία (*Q.N.* 26, 918D) and περικαλινδήσεις (*Q.N.* 28, 919A). For a general study of Plutarch's use of rare words and hapax legomena, see S.-T. Teodorsson, 2005.

[298] See J. Bertier, 1972, p. 171 (with LSJ, s.vv.). Perhaps an emendation is needed (see the commentary *ad loc.*)?

[299] F.H. Sandbach, 1965, p. 141. For further discussion of the level of technicality in ancient (Latin) technical writings, see T. Fögen, 2009, pp. 19–25.

[300] Cf., e.g., P. Donini, 1992, p. 110.

which all can share' (δεῖ γὰρ ὡς τὸν οἶνον κοινὸν εἶναι καὶ τὸν λόγον, οὗ πάντες μεθέξουσιν). It is rather unlikely, then, that Plutarch – both as a symposiast and as an author – intended to overwhelm non-specialists with all too specialised concepts in his natural problems. He is not a sophist, after all, and does not intend to promote himself as one.

Of course, the discussion of natural problems was part of elite intellectual entertainment and education, but, even so, natural problems are essentially 'easy and persuasive' for Plutarch himself, as we saw (*De tuenda* 133E: ἐλαφρὰ καὶ πιθανά [quoted 3.1.3.]). This implies that he does not mean for this branch of inquiry to become too specialised. Consequently, when using technical scientific terms in explaining natural phenomena, Plutarch is not so much striving for verbal pretentiousness but rather for an efficient communication of scientific knowledge and ideas in a uniform way. Notably, most of the interlocutors in *Quaestiones convivales* have their own field of expertise, but they are still familiar – regardless of whether this may be a trick of Plutarch's literary fiction – with the basic features of the genre of natural problems. Grammarians, for instance, are also acquainted with the scientific literature, at least to such a degree that they are able to hold their ground when the discussion moves to their field of expertise. In *Quaest. conv.* 626E–627F, for instance, Theon, a grammarian and Homer specialist, contributes to the question of why fresh water washes clothes better than salt water by quoting Aristotle and not Homer (ᾤμην σε μᾶλλον Ὁμήρῳ τἀναντία λέγοντι πιστεύσειν). Notably, several symposiasts in *Quaestiones convivales* are philosophers and doctors, but they do not necessarily speak like specialists when discussing natural problems. If they had done so, this would probably not have been appreciated by their fellow symposiasts (including poets, grammarians, political figures, students etc.). As Van der Stockt rightly notes in regards to Plutarch's terminology in the scientific παρεκβάσεις in the *Vitae*, "that is in works not destined to experts in any τέχνη" [see 2.1.3.]:

> "it is inherent in such [technical] language that it is largely incomprehensible to those who are not familiar with the τέχνη; those laymen may take offence because of what they experience as obscurity in the communication [...] or as boasting with erudition on the part of the author/speaker, or even as a haughty or mindless insult of the ignorance of the audience."[301]

For people who are not at all acquainted with the scientific literature, such scientific terminology would probably sound too specialised. But then

---

[301] L. Van der Stockt, 2013, p. 440. Van der Stockt argues that in these scientific παρεκβάσεις, Plutarch is not addressing the specialist, but rather the layman, that is, the all-round πεπαιδευμένος.

again, Plutarch is not writing for the common plebs [see 3.1.]. Nevertheless, in some passages in his natural problems, he does emphatically employ the terminology of un(der)educated laymen, especially of farmers[302]. Farmers call water of rains that accompany thunder and lightning 'lightning water' (*Q.N.* 4, 912F: ἀστραπαῖα καλοῦσι), and they consider it 'more fertilising' than normal water (*Quaest. conv.* 664DE: εὐαλδῆ καλοῦσιν οἱ γεωργοὶ καὶ νομίζουσιν). They say that the sweet flavour that becomes mixed with the (unripe, sour) grape 'ripens' (*Q.N.* 27, 918E: πεπαίνεσθαι λέγεται), and that vines that do not fruit, but flourish luxuriously with branches and shoots, 'go goatish' (*Q.N.* 30, 919B: τραγᾶν λέγομεν)[303].

The sub-technical value of such popular colloquialisms can be inferred from the fact that Plutarch refers to them with phrases like 'this they call' (καλοῦσι), 'this is called' (λέγεται), or even '*we* call this' (λέγομεν). The last phrase is particularly intriguing as it hints at an aspect of intellectual community between author and reader by the use of the first person plural [see 3.1.5.]. It denotes something like 'we Greeks – including elite men and farmers alike – call this X', 'we call this X in Greek', 'Greek people is what we are and Greek is what we speak'. It is only likely, then, that in such cases, the use of popular vocabulary should not be taken as a narcissistic attempt to parade one's personal erudition and knowledge of sub-technical jargon, but more as a candid aspiration to strengthen the sense of a shared cultural heritage, as constructed, in this case, on linguistic grounds. It may seem strange, in any case, to measure a person's elitist παιδεία in terms of his knowledge of peasant lingo. By consequence, I am not inclined to evaluate the technicality of Plutarch's scientific terminology more generally in all too strict sociological categories, where the use of such terms is primarily understood as a means of promoting the user's own person rather than facilitating the efficient communication of his ideas [see 4.2.2.1.]. This does not imply, of course, that this lust for intentional obscurity was not observed in Antiquity[304], but it is rather unlikely, in my opinion, that it applies to Plutarch just as well[305].

---

[302] Also, e.g., fishermen, who call cephalopods 'soft things', 'molluscs' (*Q.N.* 18, 916A: κέκληται μαλάκια).

[303] For another, but similar, etymology, see, e.g., *Quaest. conv.* 692E: τρυγᾶν λέγομεν.

[304] Cf. Quint., *Inst. or.* 8, 2, 12–13.

[305] A similar *caveat* was made by L. Van der Stockt, 2013, p. 445: "It is indeed tempting to understand his [sc. Plutarch's] use of technical terminology and his fondness of quoting ancient authorities as a means to invest his 'cultural capital', in order to secure his own authority, and to obtain a distinguished social position. Yet concerning Plutarch's practice in these matters, I would rather read him as a spectator/investigator of human drama, than as a player or actor in the drama of his society, and I would regard any possible social gain from his practice rather as a non-pursued consequence than as his purpose. There is reason enough to make this stance at least acceptable." Van der Stockt concludes that Plutarch was more likely "driven by an honest desire to teach and explain".

3. Conclusion: Plutarch, Plato and Aristotle (again)

Most of the scientific terminology that Plutarch uses in his physical aetiologies has a specific Aristotelian imprint, as we saw, which is not so remarkable considering the Aristotelian origin of the tradition of natural problems. The same terminology is recurrent in the extant Ps.-Aristotelian *Problems*. This does not imply, however, that it is also necessarily of Aristotelian origin, or – more importantly – that Plutarch is using such terminology with the intention of being regarded as a Peripatetic natural philosopher[306].

As a way to conclude the first part of this study, I will here revisit the question of Plutarch's philosophical allegiance in his natural problems [see 1.1.2.]. Since Plutarch was first and foremost an enthusiastic adherent of Plato, one may wonder why he was so interested in this Aristotelian type of scientific research. Did he perhaps intend to inscribe himself in the scientific community of the Lyceum by positioning himself in the tradition of Peripatetic scholarship? Did he intend to ally Platonism with Aristotelianism, as other middle- and neo-Platonists did? Or is the answer more nuanced? The question is particularly intriguing, since it sheds a light on the influence of the Aristotelian and Peripatetic tradition – a tradition that was reinvigorated in the time of the early Roman Empire – on Plutarch's own philosophical thinking[307].

Regarding the natural problems treated in *Quaestiones convivales*, Kechagia has recently argued that while there is clearly a Peripatetic character to the general explanatory scheme employed by the symposiasts, Plutarch himself probably did not aspire any strict allegiance to this philosophical school. After all, the symposiasts that Plutarch stages in *Quaestiones convivales* often adhere to different, if not rivalling, philosophical schools, although they still share the very same interest and knowledge of the, in that case, more 'generic' Peripatetic tradition[308].

---

[306] See F.H. Sandbach, 1965, p. 134: "the terminology used in the proffered solutions largely coincides with that employed, but not necessarily invented, by the Peripatetics". See also M. Meeusen, 2016.

[307] That Plutarch was not a member of the Peripatetic tradition is emphasised by H. Flashar, 1962, p. 369, who notes that "[a]ußerhalb des peripatos läßt sich eine sachliche Nähe und productive Weiterbildung der arist. Probl. nur bei Plutarch beobachten". As M. Frede, 1987, p. 282 rightly observed, "someone who saw himself as basically a Platonist at this [sc. Imperial] time would be inclined to study Stoic or Peripatetic physics". Cf. also K. Oikonomopoulou, 2011, p. 129: "Peripatetic knowledge [is] the common property of a highly complex and diverse network of intellectual communities". On the reception of Aristotelian philosophy in the time of the Roman Empire, see most recently the contributions in Y. Lehmann, 2013.

[308] E. Kechagia, 2011a, p. 98: "It is characteristic that the speakers who appeal to this schema in the *Table Talk* often do so independently of sect-allegiance. What matters is not

In fact, the terminology that Plutarch employs is well-entrenched in the broader scientific tradition, and can, thus, be considered more 'universally' scientific[309].

Notably, Plutarch more often uses the terminology of philosophical schools other than Plato's, albeit always in the confines of an essentially Platonic framework[310]. The same is probably the case, then, in his natural problems. These should be interpreted in light of Plutarch's wider natural philosophical project, which is, as this chapter has shown, radically informed by Platonic dualism and generally inspired by Academic Scepticism. In the end, as we have learned from *Quaest. conv.* 699B, 'the ingenious organisation of nature's activities is beyond the range of words' for Plutarch [quoted 4.1.2.2.], that is, beyond a general physical style of discourse – either Peripatetic or more generally scientific –, since it does not enable the φυσικός to seize the divine, intelligible principles that lie behind the face of nature.

According to Teodorsson, however, Plutarch frequently draws on Aristotle's natural scientific writings and is also generally inspired by his critical scientific method, as would be attested by his sceptical attitude towards popular beliefs[311]. Karamanolis has objected to this view by pointing out that this alleged critical 'Aristotelian' attitude "is what one would expect from a Platonist of sceptical orientation like Plutarch anyway"[312]. I have tried to further nuance Teodorsson's view by showing that Plutarch actually gives these popular beliefs the benefit of the doubt before subjecting them to a thorough scrutiny. A specific anti-Aristotelian dynamic was detected in Plutarch's subordination of the actual reality of the natural phenomena (the ὅτι) to their causes (the διὰ τί) [see 4.1.1.3.].

---

the sect so much as the attempt at a plausible explanation." On philosophical allegiance in the Greco-Roman world more generally, see D. Sedley, 1997 (esp. pp. 117–118 on the role of Aristotelianism in Imperial philosophy).

[309] The concept of πόροι, for instance, is described in *Quaest. conv.* 689C as a popular theory (τοῖς πόροις τούτοις [...] ὧν ἔνιοι περιέχονται καὶ ἀγαπῶσι). It was given physical currency first by Empedocles, and it became very popular, indeed, later on in Peripatetic and medical writings. See n. 177.

[310] See D. Babut, 1969, pp. 533–534, esp. p. 533 (quoted also by R. Flacelière, J. Irigoin, J. Sirinelli and A. Philippon, 1987, p. clix): "Tandis que chez Plutarque, alors même que les *mots* sont les mêmes que dans les textes stoïciens, le *fond*, le soubassement d'idées et de croyances qu'ils traduisent se révèle inconciliable avec la vision stoïcienne du monde." See also J. Opsomer, 2014, p. 88. For Plutarch's use of Stoic and Epicurean terms also in *Quaestiones Platonicae*, see J. Opsomer, 1994a, p. 620. For the use of Stoic and Aristotelian terminology by Platonists more generally, see G.E. Karamanolis, 2006, pp. 24–25, esp. p. 25: "All this merely represents a modernization of the language of philosophy; it does not imply anything about the philosophical loyalties of the Platonists."

[311] S.-T. Teodorsson, 1999a, p. 674.
[312] G.E. Karamanolis, 2006, p. 90, n. 19.

Plutarch does this not just for methodological reasons but with a deeper philosophical and religious motive in mind, according to which wonder-inducing phenomena reveal not only the working of natural forces in the world around us but also hint at its providential ordering.

To set the record straight, then, *Quaestiones naturales* is not the work of a *Plutarchus Aristotelicus*, and Plutarch's science of natural problems more generally should not be seen as the product of his philosophical aspiration to be counted among the rangs of the Peripatetics. Even if there is much Aristotelian twinkling in Plutarch's physical aetiologies, it is a Platonic twilight that will eventually shatter darkness in his world.

COMMENTARY

0. Approach and structure

The second part of this study aims to provide further detail about the content and set-up of *Quaestiones naturales* in the form of a commentary. In discussing each problem chapter individually a tripartite structure will be followed: first a short synopsis of the chapter as a whole with a discussion of its topic and basic structure, second a paraphrase of the aetiology with further clarification of the line of reasoning in the explanations, and third a commentary for the lemmas in the text that require further clarification (this also includes a discussion of parallel passages, cross-references and possible sources). The structural and thematic unity – and also the disruption thereof – is studied not only as a feature of the problem chapters individually (as seen in the coherence and development of the *causae*) but also of the collection as a whole (viz. in the concatenative and variative principles between the problem chapters themselves [see 1.1.5.]). Each cluster of problems will be shortly introduced under a separate heading. The following scheme gives a tentative overview of the thematic clusters in the Greek problems (*Q.N.* 1–31).

Q.N.	Theme	Cluster
1–13	Salt and water	1
14–16	Wheat and barley	2
17–19	Sea animals and fishing	3
20–28	Land animals and hunting	4
30–31	Viniculture	5

Two further remarks are in place. First of all, this scheme might oversimplify the more complex concatenative processes at work in the collection: for instance, problems of agriculture (2) also occur in the category on hydrology (1), and viniculture (5) can also be considered part of agriculture (2). Furthermore, the division between sea animals (3) and land animals (4) may not be as strict as this scheme suggests. Alternative schemes are, therefore, possible[1].

Second, the original Greek text of a number of additional problems went lost in the manuscript tradition after the lacuna in *Q.N.* 31 [see the

---

[1] For instance, K. Oikonomopoulou, 2013a, p. 152 distinguishes the following thematic categories: "matters of nourishment (1–5); the sea (8-, or perhaps 7–12); plants and agriculture (14–16); animals and human activities connected with them (18–26)". She also notes that "[t]he reader is offered hints that these categories may expand to wine and drinking (10, 27, 30–31)".

prologue]. They – or at least part of them – did survive thanks to two side traditions of the text, albeit in adapted versions. *Q.N.* 32–39 are preserved in Longolius' 1542 Latin translation of a lost manuscript from Milan (6), and *Q.N.* 40–41 are found in Psellus' encyclopaedic *De omnifaria doctrina* (7). In general, these additional problems are related to the categories demarcated in the scheme above (with the exception perhaps of *Q.N.* 34; see the commentary *ad loc.* for further detail). Additionally, there is a small cluster of two problems concerning the behaviour of bees in *Q.N.* 35–36 (and perhaps another less obvious one about animal instincts in *Q.N.* 37–38).

The work of Sandbach (1965) has been a very useful source of inspiration for this commentary (I follow his text mostly). In several cases, though, my interpretation will be radically different from his. Other editions that have certainly proven to be useful are those of Hubert (1960) and Senzasono (2006). A new edition of the text is currently being prepared by Filippomaria Pontani and myself in the *Collection des Universités de France* (Budé).

1. *Salt and water* (Q.N. 1–13)

The first thirteen problem chapters (with the exception perhaps of *Q.N.* 6) form a close thematic cluster. The most prominent subject of this cluster concerns the physical properties of salt and water, especially based on a qualitative opposition between salty seawater and sweet drinking water. This is a theme that probably draws on Book 23 of Ps.-Aristotle's *Problems* (ὅσα περὶ τὸ ἁλμυρὸν ὕδωρ καὶ θάλατταν). Another potential source is Theophrastus' lost *De aquis* (who is quoted in *Q.N.* 7, 914A and 13, 915B; cf. also *Q.N.* 1, 911D and 5, 913CD).

Q.N. 1, 911CF

In *Q.N.* 1, Plutarch wonders **why seawater provides no nourishment to trees** (Διὰ τί τὸ θαλάττιον ὕδωρ οὐ τρέφει τὰ δένδρα;). He gives four explanations, each of which is concerned with the natural constitution and properties of the salty seawater. The first *causa* draws attention to its heavy, earthy constituent, the second to its drying property, the third to its oily character, and the last to its admixture with burnt earth. The final *causa* ties in closely with the first by alluding to the earthy constituent of seawater and by referring to plants growing in the sea (this results in an implicit argumentative ring [see 4.3.3.3.]).

In the **first** explanation, Plutarch takes into account the heavy, earthy constituent of seawater. He starts by drawing an analogy with the animal kingdom, arguing that the reason for the inability of seawater to nourish trees is the same as to why it does not nourish land animals either (Πότερον

COMMENTARY 369

δι' ἣν αἰτίαν οὐδὲ τῶν ζῴων τὰ χερσαῖα;). Plato, Anaxagoras and Democritus actually thought that a plant is an animal fixed in the earth (ζῷον γὰρ ἔγγαιον τὸ φυτὸν εἶναι οἱ περὶ Πλάτωνα καὶ Ἀναξαγόραν καὶ Δημόκριτον οἴονται). It is not the case, so Plutarch adds, that seawater, for the reason that it is nutritious and potable for marine plants and fishes alike, feeds plants and trees on land just as well (οὐ γὰρ διότι τοῖς ἐναλίοις φυτοῖς τρόφιμόν ἐστι καὶ πότιμον ὥσπερ τοῖς ἰχθύσιν, ἤδη καὶ τὰ ἐν τῇ χέρσῳ φυτά τε καὶ δένδρα τρέφει). After all, seawater cannot enter the roots nor rise in the plant, because of its thickness and heavy weight respectively (οὔτε γὰρ ἐνδύεται ταῖς ῥίζαις ὑπὸ πάχους οὔτ' ἀναφέρεται ὑπὸ βάρους). In a hyperbolical fashion, Plutarch then asserts that the heavy and earthy quality of seawater can be demonstrated from many other phenomena, as from the fact that it holds up and supports ships and swimmers better than sweet water does (ὅτι δ' ἐμβριθές ἐστι καὶ γεῶδες, ἄλλοις τε πολλοῖς ἀποδείκνυται καὶ τῷ μᾶλλον ἀνέχειν καὶ ὑπερείδειν τὰ πλοῖα καὶ τοὺς κολυμβῶντας).

The **second** explanation is based on the drying property of seawater and salt. Plutarch argues that trees are especially damaged by dryness, and that seawater has a drying property (ἢ μάλιστα μὲν βλάπτεται ξηρότητι τὰ δένδρα, ξηραντικὸν δὲ τὸ θαλάττιον;). This drying property explains why salt (which is present in seawater) is a safeguard against putrefaction, and why the bodies of people who have bathed in the sea instantly receive a dry and rough surface (ὅθεν πρός τε τὰς σήψεις οἱ ἅλες βοηθοῦσι, καὶ τὰ σώματα τῶν λουσαμένων ἐν θαλάττῃ ξηρὰν εὐθὺς ἴσχει καὶ τραχεῖαν τὴν ἐπιφάνειαν).

The oily and fatty character of seawater is central to the **third** explanation, where Plutarch argues that oil is hostile to plants and destroys those that are smeared with it (ἢ τὸ μὲν ἔλαιον τοῖς φυτοῖς πολέμιον καὶ φθείρει τὰ προσαλειφόμενα). He adds that the sea has a large fatty content, which is why it is combustible, and why we advise people not to throw seawater on flames (μετέχει δὲ πολλῆς ἡ θάλαττα λιπαρότητος· διὸ συνεξάπτει, καὶ παραινοῦμεν εἰς τὰς φλόγας μὴ ἐμβάλλειν θαλάσσιον ὕδωρ).

In the **fourth** and final explanation, Plutarch further refines the aspect of the earthy component in seawater (which was already referred to in the first explanation). On the authority of Aristotle, he argues that the (sea)water has become undrinkable and bitter by an admixture of burnt earth (ἢ γέγονεν ἄποτον καὶ πικρὸν τὸ ὕδωρ, ὡς Ἀριστοτέλης φησίν, ἀναμίξει κατακεκαυμένης γῆς;). He offers further illustrations regarding 1) the *burnt* earth and 2) the undrinkable character of (sea)water caused by *heat*. Plutarch explains 1) that lye (which is also undrinkable and bitter) is formed when fresh water is thrown onto ashes, and 2) that burning heat changes and ruins the useful and potable constituent, similar to how in our bodies fevers turn moisture into bile (καὶ γὰρ ἡ κονία γίνεται γλυκέος ὕδατος εἰς τέφραν ἐμπεσόντος, ἡ δὲ διάκαυσις ἐξίστησι καὶ φθείρει τὸ χρηστὸν καὶ πότιμον, ὡς ἐν ἡμῖν οἱ πυρετοὶ τὸ ὑγρὸν εἰς χολὴν τρέπουσιν). In order to connect this explanation more closely with the problem at hand, Plutarch adds that the bushes and plants that are reported to grow in the Indian Ocean do not bear any fruit and acquire their nourishment from rivers, which deposit a great deal of silt in the sea. Therefore, these plants do not grow far away from the shore but close to it (ἃ δ' ἱστοροῦσιν ἐν τῇ Ἐρυθρᾷ θαλάσσῃ βλαστάνειν ὑλήματα καὶ φυτά, καρπὸν μὲν οὐδένα φέρει τρέφεται δὲ τοῖς ποταμοῖς πολλὴν ἐμβάλλουσιν ἰλύν· ὅθεν οὐ πρόσω τῆς γῆς ἀλλὰ πλησίον ἔχει τὴν γένεσιν).

911C Διὰ τί τὸ θαλάττιον ὕδωρ οὐ τρέφει τὰ δένδρα;: The chapter as a whole is closely paralleled in the ninth problem of the first Book of *Quaestiones convivales* (esp. 627AD), albeit in the context of a different problem (viz. why clothes are washed with fresh water instead of seawater). The phenomenon itself is rejected by Theophrastus, who reports that saline waters feed land plants too (albeit worse than sweet drinking waters), and especially trees, because of their strength (*CP* 2, 5, 3).

911C ζῷον γὰρ ἔγγαιον τὸ φυτὸν εἶναι οἱ περὶ Πλάτωνα καὶ Ἀναξαγόραν καὶ Δημόκριτον οἴονται: The phrase 'οἱ περὶ X' can denote simply 'X', or more periphrastically 'X and his followers', that is, 'the school of X' or at least 'the school of thought of X' (for the ambiguity in phrases of this kind, see W.R. Roberts, 1910, p. 195, J. Dillon, 1977, p. 231, L. Torraca, 1998, pp. 3489–3494). The first meaning is doubtful here. In any case, these are not Plato's *ipsissima verba* but rather a paraphrase of what he writes on this topic – which is uncertain for the accounts of Anaxagoras (= DK59A116) and Democritus (not recorded among the DK fragments). The allusion is to *Tim.* 77ab, where Plato calls trees, plants and seeds ἕτερον ζῷον (cf. also Aët., *Plac.* 5, 26 = Ps.-Plut. 910B) with an ἀνθρωπίνης συγγενὲς φύσεως φύσις. Plato explains: πᾶν γὰρ οὖν ὅτιπερ ἂν μετάσχῃ τοῦ ζῆν, ζῷον μὲν ἂν ἐν δίκῃ λέγοιτο ὀρθότατα (see W.K.C. Guthrie, 1965, p. 316, with n. 2). Conversely, in *Tim.* 90a, man is called a φυτὸν οὐκ ἔγγειον ἀλλὰ οὐράνιον (which Plutarch repeats in *De Pyth. or.* 400B, *De genio Socr.* 591DE, *De exilio* 600F, *Amatorius* 757E; cf. F. Fuhrmann, 1964, pp. 120–121). See also Pl., *Rep.* 491d, 546a and *Epin.* 981d. For Anaxagoras' and Democritus' theory that plants have the power of thought, cf. Ps.-Arist., *De plant.* 815b16 (= DK31A70 and DK59A117), where Empedocles (Abrucalis) is also mentioned as an authority. On the difficulty in classifying some inanimate (vegetal) and animate lifeforms, cf. Arist., *HA* 588b16–17 and *PA* 681a12–15. Considering the different contexts, it seems unlikely that Plutarch was influenced by Aristotle here – *pace* L. Senzasono, 2006, p. 142, n. 1, who is probably right, however, that Plutarch preferred Plato's authority in the present context, albeit not so much to that of Aristotle, as to that of Anaxagoras and Democritus. In any case, the reference to Plato is not out of place at the very beginning of this collection of natural problems, perhaps as a subtle Platonic σφραγίς [see 4.2.1.1.].

911D οὐ γὰρ διότι τοῖς ἐναλίοις φυτοῖς τρόφιμόν ἐστι καὶ πότιμον ὥσπερ τοῖς ἰχθύσιν, ἤδη καὶ τὰ ἐν τῇ χέρσῳ φυτά τε καὶ δένδρα τρέφει: For the idea that seawater is undrinkable and bad for humans but provides nourishment for fish, cf. *De cap. ex inim.* 86E. Democritus may be the source of this theory: see H. Diels, 1905, pp. 314–315. Indeed, a similar belief, viz. that fish are not nourished by the salt water in the sea, but by the sweet water mixed with it, is recorded by Aelian on the authority of Democritus (DK68A155a) in combination with that of Aristotle and Theophrastus (*NA* 9, 64: Λέγει δὲ Ἀριστοτέλης, καὶ Δημόκριτος πρὸ ἐκείνου, Θεόφραστός τε ἐκ τρίτων καὶ αὐτός φησι, μὴ τῷ ἁλμυρῷ ὕδατι τρέφεσθαι τοὺς ἰχθῦς, ἀλλὰ τῷ παραμεμιγμένῳ τῇ θαλάττῃ γλυκεῖ ὕδατι). In this context, Aelian also refers to the Aristotelian account about the wax jar that filters seawater (which recurs in *Q.N.* 5, 913C: see the commentary *ad loc.*). If we are to follow Diels, it is not unlikely that Plutarch knew Democritus' account (and perhaps also that of Anaxagoras) indirectly via a lost Ps.-Aristotelian problem. This lost problem, in turn, was perhaps based on an intermediary doxographical passage in one of Theophrastus' smaller natural scientific treatises, presumably his lost *De aquis* (cf. D.L., 5, 45). A clear parallel is found in Theophr., *CP* 6, 10, 2 (ἐπεὶ καὶ τὰ ἐν τῇ θαλάττῃ φυόμενα γλυκύτητί τινι καὶ ἑτέροις χυμοῖς φύεται καὶ συνίσταται καθάπερ ἰχθῦς καὶ τἆλλα ζῷα τὰ ἐν αὐτῇ).

911D οὔτε γὰρ ἐνδύεται ταῖς ῥίζαις ὑπὸ πάχους οὔτ' ἀναφέρεται ὑπὸ βάρους: This implies that only sweet drinking water is fine enough to penetrate through the pores of the roots, due to their narrowness (cf. *Q.N.* 5, 913D: οἱ γὰρ πόροι διὰ λεπτότητα τὸ γεῶδες καὶ παχυμερὲς οὐ διηθοῦσιν). The concept of 'penetration into the roots' recurs in *Q.N.* 31, 919D (ἐνδύεται ταῖς ῥίζαις) and *Q.N.* 2, 911F (διαδύεται μᾶλλον εἰς τὴν ῥίζαν); cf. also, e.g., *Quaest. conv.* 664E (ἐνδύεσθαι τοῖς βλαστάνουσι). After ἀναφέρεται, Psellus in *De omn. doctr.* § 168, 3 Westerink adds ταχέως εἰς τὸ στέλεχος καὶ τοὺς ἀκρέμονας ('rise quickly into the stem and branches'). As F.H. Sandbach, 1965, p. 149, n. c notes, Psellus "may as well have invented this as found it in his text of Plutarch". Notably, the terms στέλεχος and ἀκρεμών in Psellus' addition recur in *Q.N.* 21, 917D and 30, 919B respectively. These problems, however, are not in Psellus (see M. Meeusen, 2012b, p. 113, n. 60). Moreover, the adverb ταχέως recurs in a similar context in *Q.N.* 2, 912B (τὸ ἐκ Διὸς ὕδωρ [...] ἀναπέμπεται ταχέως εἰς τὸ φυτόν). This is not, however, a strong argument for keeping the addition from Psellus' text, which is probably an interpolation.

911D ὅτι δ' ἐμβριθές ἐστι καὶ γεῶδες, ἄλλοις τε πολλοῖς ἀποδείκνυται καὶ τῷ μᾶλλον ἀνέχειν καὶ ὑπερείδειν τὰ πλοῖα καὶ τοὺς κολυμβῶντας: The idea that the admixture of earthy matter (γεῶδες) in seawater is responsible for its saltiness is ascribed to Aristotle (fr. 217 Rose) in *Quaest. conv.* 627AB, where it is also reported that this earthy constituent in seawater supports swimmers and makes heavy objects float better. Cf. also Arist., *Mete.* 359a7–21 (in the context of the experiment with the wax jar: cf. *Q.N.* 5, 913C), Ps.-Arist., *Probl.*, 933a9–13, Pliny, *NH* 2, 224. Cf. also M. Glycas, *Ann.* 1, 16 (p. 31, 1–4 Bekker): ἐν μὲν γὰρ τοῖς ποταμοῖς οὐ δύνανται πλοῖα βαρὺν ἔχοντα φόρτον πλεῖν, ἐν δὲ τῇ θαλάσσῃ ἐλαφρῶς ταῦτα βαστάζονται διὰ τὴν ὑποκειμένην παχύτητα.

911D ἢ μάλιστα μὲν βλάπτεται ξηρότητι τὰ δένδρα, ξηραντικὸν δὲ τὸ θαλάττιον;: For the drying property of salt, cf. *Q.N.* 5, 913E. Cf. also Ps.-Arist., *Probl.* 932a40–b8.

911D ὅθεν πρός τε τὰς σήψεις οἱ ἅλες βοηθοῦσι: For the preservative property of salt, cf. *Q.N.* 10, 914DE, 40 and *Quaest. conv.* 685BC. Because of this property, salt is also believed to have a divine character, which is not mentioned here (cf. also *Quaest. conv.* 697D and 684E–685F more generally, with the scholia on Hom., *Il.* 9, 214).

911D τὰ σώματα τῶν λουσαμένων ἐν θαλάττῃ ξηρὰν εὐθὺς ἴσχει καὶ τραχεῖαν τὴν ἐπιφάνειαν: Cf. *Quaest. conv.* 627D for the Aristotelian account (*Probl.* 932b25–28) that people who wash themselves in the sea, if they stand in the sun, dry faster than those who use fresh water – an idea that recalls the εὐθύς from our passage and seems to imply that seawater, as opposed to fresh water, contains natural heat (which was, indeed, commonly accepted). In *Quaest. conv.* 627EF, Plutarch further elaborates this view by making a subtle distinction between the constituents of seawater. He argues that the sun does not evaporate the seawater in its full extent but only its finest and lightest parts so that a salty and rough coating, that is, a briny scum, remains on the body (τὸ δ' ἁλμυρὸν αὐτὸ καὶ τραχὺ καταλειφθέν). This can be rinsed off with fresh drinking water. The distinction between fine and rough constituents is less relevant here in *Q.N.* 1, 911D, where the main emphasis is on the drying property of the seawater itself (ξηραντικὸν δὲ τὸ θαλάττιον). It is not unlikely, as F.H. Sandbach, 1965, p. 150, n. a points out, that Plutarch's allusion in *Q.N.* 1, 911D to a rough deposit on the surface of the skin is perhaps "due to an association of ideas caused by the writing of the other passage". Alternatively, it is not unlikely either, as Sandbach concludes, that the roughness may simply imply dryness here. However, in *Q.N.* 5, 913DE Plutarch again mentions the drying property of salt but says that it is οὐ τραχύν.

911E ἢ τὸ μὲν ἔλαιον τοῖς φυτοῖς πολέμιον καὶ φθείρει τὰ προσαλειφόμενα: In *Quaest. conv.* 640C, Plutarch ascribes the belief that oil is harmful (πολέμιον) to plants (and bees) both to learned people (σοφοί) and to farmers (γεωργικοί). Cf. also Pl., *Prot.* 334b, Theophr., *CP* 5, 15, 6, *HP* 4, 16, 5, Pliny, *NH* 17, 234; 18, 152.

911E μετέχει δὲ πολλῆς ἡ θάλαττα λιπαρότητος· διὸ συνεξάπτει, καὶ παραινοῦμεν εἰς τὰς φλόγας μὴ ἐμβάλλειν θαλάσσιον ὕδωρ: The belief that seawater contains a fatty (λιπαρός) content and is therefore combustible is ascribed to Aristotle in *Quaest. conv.* 627C (cf. Ps.-Arist., *Probl.* 932b4–6, 933a18–27, 935a5–8 and b18–20). Plutarch there reports that seawater, when sprinkled onto flames, flashes up with them. He also says that, compared to other types of water, seawater is particularly flammable (which explains his advice here). Notably, in the historical context of the battle of Actium, Dio Cass., *Hist. Rom.* 50, 34 reports that Antony's seamen were unable to extinguish the burning missiles that were fired from Octavian's fleet with salty seawater (ἡ γὰρ ἅλμη ἡ θαλαττία ἂν κατ᾽ ὀλίγον ἐπιχέηται φλογί, ἰσχυρῶς αὐτὴν ἐκκαίει). Plutarch also mentions these incendiary projectiles in *Ant.*, 66 (πυροβόλοι), albeit without reference to this peculiar natural phenomenon.

911E ἢ γέγονεν ἄποτον καὶ πικρὸν τὸ ὕδωρ, ὡς Ἀριστοτέλης φησίν, ἀναμίξει κατακεκαυμένης γῆς;: In *Q.N.* 5, 913D, the salty flavour is identified with the bitter (πικρόν), as seems to be the case here too. The quote from Aristotle may be a reference to *Mete.* 358a14–17, where it is reported that some people ascribe the saltiness of seawater to burnt earth (θάλατταν ἐκ κατακεκαυμένης φασὶ γενέσθαι γῆς). Aristotle himself, however, considers this ascription absurd (ἄτοπον). Even still, he concludes that the admixture of what he vaguely refers to as 'such earthy stuff' (ἐκ τοιαύτης <γῆς>) with water is undoubtedly what makes the sea salty. We may be dealing here with a simplifying paraphrase by Plutarch (cf. L. Senzasono, 2006, p. 144, n. 7). Yet, the possibility cannot be excluded that Plutarch is relying on a lost Ps.-Aristotelian problem, where that precise theory was supported (cf. F.H. Sandbach, 1982, p. 227). The second possibility is not unlikely, since in the Ps.-Aristotelian *Problems* there is a clear tendency to restore specific theories that Aristotle explicitly rejected (see H. Flashar, 1962, pp. 334–335 and M. Meeusen, forthcoming g; for instance, in Ps.-Arist., *Probl.* 934b34–36, an argument from the Heracliteans is restored that is considered ridiculous in Arist., *Mete.* 354b33.) In addition, F.H. Sandbach, 1965, p. 151, n. f is probably correct in pointing out that Plutarch assumes burnt earth to be essentially burning earth, in the sense that it retains a certain amount of heat, which spoils the useful and potable constituent in seawater. This is important for the connection between the quotation from Aristotle and the natural phenomena referred to in what follows in Plutarch's explanation (viz. the formation of lye from water and ashes, heat ruining drinking water and fever turning moisture into bile: see the following comments). Cf. Arist., *Mete.* 358b7–9: πάντα γὰρ ὅσα πεπύρωται, ἔχει δυνάμει θερμότητα ἐν αὑτοῖς. ὁρᾶν δ᾽ ἔξεστι καὶ τὴν κονίαν καὶ τὴν τέφραν κτλ.

911E ἡ κονία γίνεται γλυκέος ὕδατος εἰς τέφραν ἐμπεσόντος: What Plutarch is probably implying here is that the basic constitution of seawater is very similar to that of lye: both substances are basically a mixture of fresh water and burnt earth or ashes, which makes them both undrinkable and bitter. Another correspondence between lye (κονία) and seawater is the fact that lye is an alkaline fluid used for washing (cf. LSJ, s.v.), and that seawater, with its corrosive, earthy constituent, could also be used for this purpose, as is argued in *Quaest. conv.* 627BC on the authority of Aristotle (fr. 217 Rose). In *Quaest. conv.* 684C and 697A, Plutarch reports that lye produced from ashes of the wood of fig trees is most purgative. For the relation between lye and seawater, cf. also Gal., *SMT* 11, 630, 2–4 Kühn (more generally regarding lye, cf. Gal., *SMT* 12, 35, 3–7; 222, 15–223, 5 Kühn). Cf. also Pl., *Tim.* 60de and Arist., fr. 222 Rose.

911E ἡ δὲ διάκαυσις ἐξίστησι καὶ φθείρει τὸ χρηστὸν καὶ πότιμον: For the negative effects of (summer) heat on sweet drinking water, cf. *Q.N.* 9, 914C (with the commentary *ad loc.*).

911E ὡς ἐν ἡμῖν οἱ πυρετοὶ τὸ ὑγρὸν εἰς χολὴν τρέπουσιν: For diseases producing a change in the bodily κρᾶσις, cf. *Q.N.* 26, 918D. For Plato's comments on bile, see *Tim.* 83c.

911EF ἃ δ' ἱστοροῦσιν ἐν τῇ Ἐρυθρᾷ θαλάσσῃ βλαστάνειν ὑλήματα καὶ φυτά, καρπὸν μὲν οὐδένα φέρει τρέφεται δὲ τοῖς ποταμοῖς πολλὴν ἐμβάλλουσιν ἰλύν· ὅθεν οὐ πρόσω τῆς γῆς ἀλλὰ πλησίον ἔχει τὴν γένεσιν: The ancient Ἐρυθρὰ θάλασσα (literally 'red sea') can be identified with the Indian Ocean, which covers all known waters along the south coast of Asia, sometimes including the modern Red Sea itself, that is, the ancient Ἀράβιος κόλπος (cf. LSJ, s.v.). In *Quaest. conv.* 733B, Plutarch quotes from Agatharchides' work on the Ἐρυθρὰ θάλασσα (I found no parallel in the extant excerpts). As to the types of bushes and plants at issue (ὑλήματα καὶ φυτά), Plutarch in *De facie* 939D writes that plants of wondrous magnitude grow down in the deep of the (Indian) Ocean near Gedrosia and Ethiopia, some of which are called olives, some bay, and some Tresses of Isis. For the aquatic plants of the 'outer sea' (i.e. the Atlantic and Indian Ocean), see Theophr., *HP* 4, 7, esp. 2, where it is reported that so-called bay and olive grow there, but that the latter – in opposition to what Plutarch suggests in *Q.N.* 1, 911E – *does* carry fruit similar to genuine olives (cf. also Pliny, *NH* 2, 226; 13, 135; 139–142 and Eratosthenes *apud* Strabo, *Geogr.* 16, 3, 6 (c. 766)). With respect to plants growing in the Mediterranean sea, see also Theophr., *HP* 4, 6: some of these plants are reported to grow close to the shore (πρόσγεια, παράγειοι, πρὸς τῇ γῇ), but others do not, and some of them *do* bear fruit. If Plutarch's mention of the absence of fruits is his own invention (which is uncertain, though), he may be implying that these marine plants consume their entire nourishment for their own growth – *De facie* 939D, in any case, emphasises the amazing magnitude of these plants – so that no

COMMENTARY 375

generative residue remains to form fruits (cf. *Q.N.* 21, 917B, 30, 919C, *Quaest. conv.* 640F–641A, 724E). In addition, there seems to be a subtle opposition in this final *causa* between the non-nutritive (bitter) burnt earth (κατακεκαυμένη γῆ) in seawater and the nutritious silt (ἰλύς) rivers deposit in the sea. Plutarch, thus, seems to suggest that the nourishment of marine plants does not *essentially* differ from that of terrestrial plants, since they are *not* fed by the water of the sea but by the nutritive particles (c.q. silt) deposited by sweet river water in it. This specifies the reference in the first *causa* to the nourishment of the ἐνάλια φυτά (and perhaps also of fish?) by seawater, which are presumably nourished in the same way (the result then is a subtle argumentative ring [see 4.3.3.3.]). The belief that marine plants are not nourished by the salty seawater, but by the fresh water (and other flavours) present in it is paralleled, e.g., in Theophr., *CP* 6, 10, 2 (quoted in the comment on *Q.N.* 1, 911D above).

৵

Q.N. 2, 911F–912D

Q.N. 2 is closely linked to the theme of the previous problem by its hydrological and botanical interests. It again focuses on the nourishment that certain kinds of water provide to plants. Plutarch wonders **why trees and seeds naturally receive more nourishment from rainwater than from irrigation water** (Διὰ τί μᾶλλον ὑπὸ τῶν ὑετίων ἢ τῶν ἐπιρρύτων ὑδάτων τὰ δένδρα καὶ τὰ σπέρματα πέφυκε τρέφεσθαι;). Five explanations are given, which – with the exception of the first – deal with the material constitution and physical characteristics of rainwater (*vis-à-vis* irrigation water). The first explanation is of a mechanical kind and draws attention to the rainwater's impact on the earth, the second explanation draws attention to the rainwater's freshness, the third to its airy and breathlike composition, the fourth to the ease with which it changes, and the last to its sweetness. There is again a subtle argumentative ring in the aetiology by the implicit connection of the first two *causae* with the final one [see 4.3.3.3.]. The connective idea is that the raining/mating season (i.e. spring), as referred to in the final *causa*, is probably the 'right season' (καθ' ὥραν) for growth and procreation in plants and animals in ponds, as mentioned in the second *causa* (which, in turn, is closely connected to the first).

The **first** explanation focuses on the impact of rainwater on the earth. It is ascribed to Laetus, who said that raindrops make passages in the earth by separating it on impact, so that they better penetrate into the roots of plants (Πότερον, ὡς Λαῖτος ἔλεγε, τῇ πληγῇ τὰ ὄμβρια διιστάντα τὴν γῆν πόρους ποιεῖ καὶ διαδύεται μᾶλλον εἰς τὴν ῥίζαν;).

In the **second** explanation Plutarch criticises Laetus' theory. He argues that this theory is incorrect, Laetus failing to notice that plants that grow in ponds, such as reed mace, wool-tufted reed and rushes, also remain without growth or shoot if rain does not fall in the right season (ἢ τοῦτο μὲν οὐκ ἀληθές, ἀλλ' ἔλαθε τὸν Λαῖτον ὅτι καὶ τὰ λιμναῖα φυτά, τύφη καὶ φλέως καὶ θρύον, ἀναυξῆ καὶ ἀβλαστῆ μένει μὴ γενομένων ὄμβρων καθ' ὥραν). Plutarch then gives Aristotle's explanation, according to which rainwater is fresh and new as opposed to that of ponds, which is stale and old (τὸ δὲ τοῦ Ἀριστοτέλους ἀληθές, ὅτι πρόσφατόν ἐστι καὶ νέον ὕδωρ τὸ ὑόμενον ἕωλον δὲ καὶ παλαιὸν τὸ λιμναῖον;). He wonders, however, whether this theory is probable rather than true (ἢ καὶ τοῦτο πιθανὸν μᾶλλον ἢ ἀληθές ἐστι;). After all, the (running) waters of springs and rivers are (also) fresh and new-born, but are (still) less nourishing than rainwater (τὰ γὰρ πηγαῖα καὶ ποτάμια νάματα πρόσφατα μέν ἐστι καὶ νεογενῆ [...], τρέφει δὲ καὶ ταῦτα τῶν ὀμβρίων χεῖρον). The second point (about rainwater being more nourishing than water of springs and rivers) is explained no further here (but it is in the fourth *causa*; see also *Q.N.* 33), and the first point (about water of springs and rivers being fresh and new-born) is illustrated with a literal interpretation of Heraclitus' river statement: 'you could not step into the same rivers twice', as Heraclitus says, 'because other waters flow upon you' (ποταμοῖς γὰρ δὶς τοῖς αὐτοῖς οὐκ ἂν ἐμβαίης ὥς φησιν Ἡράκλειτος, ἕτερα γὰρ ἐπιρρεῖ ὕδατα).

The **third** explanation draws attention to the airy and breathlike composition of rainwater. Plutarch argues that the water from the heavens is light and airy and, being mixed with breath, is more quickly guided and transmitted into the plant by its tenuity (ἆρ' οὖν κοῦφόν ἐστι τὸ ἐκ Διὸς ὕδωρ καὶ ἀερῶδες, καὶ πνεύματι μεμιγμένον ὁδηγεῖται καὶ ἀναπέμπεται ταχέως εἰς τὸ φυτὸν ὑπὸ λεπτότητος). This is illustrated by the fact that rainwater makes bubbles by the admixture of air (δι' ὃ καὶ πομφόλυγας ποιεῖ τῇ ἀναμίξει τοῦ ἀέρος;).

The **fourth** explanation focuses on the ease with which rainwater changes. Plutarch's argument is relatively sophisticated and is based on two premises. He argues 1) that most nourishment is provided by what is mastered most by the thing fed (Plutarch clarifies that this is the process of concoction – non-concoction being the opposite, when the food is too strong to undergo that action), and more concretely 2) that light, simple and tasteless substances, like rainwater, are more subject to change (ἢ τρέφει μὲν μάλιστα τὸ μάλιστα κρατούμενον ὑπὸ τοῦ τρεφομένου (τοῦτο γάρ ἐστι πέψις· ἀπεψία δὲ τοὐναντίον, ὅταν ἰσχυρότερα τοῦ παθεῖν ᾖ), καὶ μεταβάλλει τὰ λεπτὰ καὶ ἁπλᾶ καὶ ἄχυμα μᾶλλον, οἷόν ἐστι τὸ ὄμβριον ὕδωρ;). In what follows, Plutarch then alludes to both of these premises in reversed order in an attempt to substantiate his theory 2) that rainwater is simpler in composition (i.e. unmixed) than irrigation water, and thus more liable to change, so that 1) it is more concocted, and thus provides more nourishment. 2) In regards to the simple composition of rainwater, he first explains that rain is formed in the air and in the wind (cf. the previous *causa*); it falls from the sky pure and unmixed, while springwaters, because of their assimilation both to the earth and to the places whence they emerge, become infected with many qualities, so that they change with less ease and convert more slowly by concoction into the thing nourished (γεννώμενον γὰρ ἐν ἀέρι καὶ πνεύματι καθαρὸν καὶ ἀμιγὲς κάτεισι· τὰ δὲ πηγαῖα καὶ τῇ γῇ καὶ τοῖς τόποις ὁμοιούμενα, δι' ὧν ἔξεισι, πολλῶν ἀναπίμπλαται ποιοτήτων, δι' ἃς ἧττόν ἐστιν εὔτρεπτα καὶ βράδιον αὐτὰ παρέχει τῇ πέψει μεταβάλλειν εἰς τὸ τρεφόμενον). The ease with which rainwater, on the other hand, changes is accounted for by its processes of putrefaction. Plutarch explains that it putrefies more easily than water from rivers and wells (τῶν δ' ὀμβρίων τὸ εὔτρεπτον αἱ σήψεις κατηγοροῦσιν· εὐσηπτότερα γάρ ἐστι τῶν ποταμίων καὶ φρεατιαίων). 1) In order to complete his argument, Plutarch then states that concoction appears to be a process of putrefaction. He borrows this point from Empedocles, who says that 'Wine is water from the bark, putrefied in the wood' (ἡ δὲ πέψις ἔοικεν εἶναι σῆψις, ὡς Ἐμπεδοκλῆς μαρτυρεῖ λέγων 'οἶνος ἀπὸ φλοιοῦ πέλεται σαπὲν ἐν ξύλῳ ὕδωρ').

The **fifth** and final explanation draws attention to the sweet constituent of rainwater and is introduced as being the most obvious and easiest of all the explanations (ἡ πάντων ἑτοιμότατόν ἐστι καὶ ῥᾷστον αἰτιάσασθαι). With an implicit allusion to the third *causa*, Plutarch argues that the sweet and useful part of rainwater is immediately lifted (into the plant) by the breath (τὸ γλυκὺ τῶν ὀμβρίων καὶ χρηστόν, εἰσπεμπόμενον εὐθὺς ὑπὸ τοῦ πνεύματος). In regards to the rain's sweet constituent, Plutarch explains that domestic animals also enjoy rainwater with more pleasure, and that frogs croak louder in joyful anticipation of the rain, looking forward to accepting it as a sweetening of the water of the pond and as a seed of their (sc. the ponds'?) sweetness (διὸ καὶ τὰ θρέμματα τούτων ἀπολαύει προθυμότερον, καὶ οἱ βάτραχοι προσδοκῶντες ὄμβρον ἐπιλαμπρύνουσι τὴν φωνὴν ὑπὸ χαρᾶς, ὥσπερ ἥδυσμα τοῦ λιμναίου τὸ ὑέτιον προσδεχόμενοι καὶ σπέρμα τῆς ἐκείνων γλυκύτητος). Plutarch illustrates this last point with a conclusive quotation from Aratus, who considers the croaking of frogs as a sign of coming rain. The poet says: 'straight from the pond, the tadpoles' fathers cry: truly wretched race, the victual of water snakes' (ἓν γὰρ καὶ τοῦτο ποιεῖται σημεῖον ὑετοῦ μέλλοντος Ἄρατος εἰπών 'ἦ μάλα δείλαιαι γενεαί, ὕδροισιν ὄνειαρ, / αὐτόθεν ἐκ λίμνης πατέρες βοόωσι γυρίνων').

911F Διὰ τί μᾶλλον ὑπὸ τῶν ὑετίων ἢ τῶν ἐπιρρύτων ὑδάτων τὰ δένδρα καὶ τὰ σπέρματα πέφυκε τρέφεσθαι;: A similar problem concerning the usefulness of rainwater for plants (and more specifically rain that accompanies thunder and lightning) is discussed in *Q.N.* 4. Notably, Theophrastus at several occasions denies that rainwater is better than irrigation water for nourishing several kinds of plants (cf. *CP* 3, 8, 3 (ὁμοίως); 2, 5, 5; *HP* 4, 7, 8; 8, 7, 3; fr. 159, 32–37 Wimmer = 214A, 26–30 FHSG), but he accepts it for others (cf. *HP* 7, 5, 2). F.H. Sandbach, 1965, p. 152 (with n. a) translates σπέρματα as 'seedlings', i.e. young plants shortly after the fase of germination (here and in *Q.N.* 4, 913A; he bases this translation on Theophr., *HP* 8, 8, 2, but the concept does not seem to be used in a different sense there). It is uncertain, however, that this is really what Plutarch means (cf. also L. Senzasono, 2006, p. 146, n. 11). Perhaps Plutarch is implying that the production of seeds, which is triggered by a residue of nourishment in the plant or tree (περίττωμα τῆς τροφῆς: cf. *Q.N.* 21, 917B and 30, 919C), increases when these seeds – and, by implication, the plant or tree itself from which they grow – receive more nourishment (c.q. by the rainwater). Alternatively, τὰ δένδρα καὶ τὰ σπέρματα may also refer to *all* kinds of flora in a metonymical fashion, thus including trees and seeds in specific but also plants, shrubs, bushes etc. more generally (i.e. the *genus* of τὰ φυτά).

911F Πότερον, ὡς Λαῖτος ἔλεγε, τῇ πληγῇ τὰ ὄμβρια διιστάντα τὴν γῆν πόρους ποιεῖ καὶ διαδύεται μᾶλλον εἰς τὴν ῥίζαν;: This Λαῖτος, also quoted in *Q.N.* 6, 913E, is probably to be identified with the Platonist Ofellius Laetus [see 4.2.1.1., n. 115]. It is implied in Laetus' explanation that irrigation water, as opposed to rainwater, causes no impact on the earth, because it flows over it. The concept of penetration into the roots recurs in *Q.N.* 1, 911D and 31, 919D (where Plutarch writes ἐνδύεται ταῖς ῥίζαις twice). Similarly, in *Quaest. conv.* 664B, we read that thunder also parts the earth (τὴν γῆν διίστασθαι) by using the air as a spike.

912A ἢ τοῦτο μὲν οὐκ ἀληθές, ἀλλ' ἔλαθε τὸν Λαῖτον ὅτι καὶ τὰ λιμναῖα φυτά [...] ἀναυξῆ καὶ ἀβλαστῆ μένει μὴ γενομένων ὄμβρων καθ' ὥραν: Plutarch's refutation of Laetus' thesis is clear (rain does not part the earth in the case of water plants), but Plutarch does not specify the phrase καθ' ὥραν any further. The meaning of this phrase will become clearer from the fifth *causa*, where we learn that the (raining/mating) season of spring is probably meant by the 'right season' here (see the commentary *ad loc.*).

912A τὸ δὲ τοῦ Ἀριστοτέλους ἀληθές, ὅτι πρόσφατόν ἐστι καὶ νέον ὕδωρ τὸ ὑόμενον ἕωλον δὲ καὶ παλαιὸν τὸ λιμναῖον;: There are no parallels for this quotation in Aristotle's surviving works (= Arist., fr. 215 Rose). We are probably dealing with a remnant of a Ps.-Aristotelian problem that is now lost (see F.H. Sandbach, 1982, pp. 224 and 227; according to L. Senzasono, 2006, pp. 148–149, n. 14, it is not impossible that the original Aristotelian text made reference to a physical process of change, by which fresh water, including rains, turns into the water of ponds, but this is conjecture). The usefulness of running water (including rainwater) as opposed to stagnant and slow water is also mentioned by Pliny, *NH* 31, 31, who invokes the authority of physicians. Considering the medical (i.e. non-agricultural) context there, Senzasono (*ibid.*) argues that Pliny possibly relies on Hipp., *Aer.* 7, where rain is characterised as ἀεὶ νέος. He states that Plutarch possibly had Hippocrates' passage on hand as well, but this is unlikely. The intermediation of a lost problem seems more plausible: after all, Plutarch is quoting Aristotle and the link between Ps.-Aristotle's *Problems* and the Hippocratic writings is well-known [see 4.2.1.1., nn. 100–101].

912A ἢ καὶ τοῦτο πιθανὸν μᾶλλον ἢ ἀληθές ἐστι;: As we saw earlier, this pattern of speech is relevant in light of Plutarch's search for plausible explanations in natural scientific matters and, thus, for his Platonic-Academic method more generally [see 4.3.2.2.]. It indicates that what is at first considered a correct explanation (τὸ δὲ τοῦ Ἀριστοτέλους ἀληθές;), after closer examination appears to have falsely aroused that appearance, so that it is degraded from being true to just being plausible.

912A ποταμοῖς γὰρ δὶς τοῖς αὐτοῖς οὐκ ἂν ἐμβαίης ὥς φησιν Ἡράκλειτος, ἕτερα γὰρ ἐπιρρεῖ ὕδατα: The same fragment is recorded in different forms in *De E* 392B and *De sera num.* 559C (= Heracl., DK22B12, 49a, 91; A6, 15). Therefore, the introductory ὥς φησιν Ἡράκλειτος does not guarantee that Plutarch is quoting Heraclitus κατὰ λέξιν or directly (cf. J.P. Hershbell, 1977, p. 190, n. 46). According to G.S. Kirk, 2010 (= 1954), pp. 366–380, Plutarch's quote in the passage at hand may be affected by Pl., *Crat.* 402a (perhaps via a sceptical source, viz. Aenesidemus, as M. Marcovich, 1978, p. 152 suggests): Λέγει που Ἡράκλειτος ὅτι πάντα χωρεῖ καὶ οὐδὲν μένει, καὶ ποταμοῦ ῥοῇ ἀπεικάζων τὰ ὄντα λέγει ὡς δὶς ἐς τὸν αὐτὸν ποταμὸν οὐκ ἂν ἐμβαίης (this may explain the same potential mood in Plutarch's version: οὐκ ἂν ἐμβαίης). Moreover, Kirk argues that the original saying was that of Arius Didymus *apud* Euseb., *PE* 15, 20: Ποταμοῖσι τοῖσιν αὐτοῖσιν ἐμβαίνουσιν ἕτερα καὶ ἕτερα ὕδατα ἐπιρρεῖ (hence perhaps the "awkward plural ποταμοῖς" in Plutarch's version, as F.H. Sandbach, 1965, p. 153, n. d notes). This theory, however, was countered by G. Vlastos, 1955, pp. 338–344, who argues that Plutarch's quotation in *Q.N.* 2 is closer to the original. According to W.K.C. Guthrie, 1962, pp. 488–492 Plato's is. I do not intend to get involved in this discussion. The least that can be said is that it is typical of Plutarch's method of citing that Heraclitus' saying is playfully lifted from its original context to receive a new meaning [see 4.2.1.1.]. It is interpreted here in a literal, physical fashion and not in its original cosmological sense (cf. L. Senzasono, 2006, p. 151, n. 16).

912A ἆρ' οὖν κοῦφόν ἐστι τὸ ἐκ Διὸς ὕδωρ καὶ ἀερῶδες, καὶ πνεύματι μεμιγμένον ὁδηγεῖται καὶ ἀναπέμπεται ταχέως εἰς τὸ φυτὸν ὑπὸ λεπτότητος: Plutarch again alludes to the airy and breathlike composition of rainwater in the fourth and fifth *causae* of *Q.N.* 2 and also in the first *causa* of *Q.N.* 4, 912F. For the lightness of rainwater, cf., e.g., Hipp., *Aer.* 8, Cels., *De med.* 2, 18, 12,

Pliny, *NH* 31, 31, Ps.-Arist./Alex. Aphr., *Suppl. probl.* 2, 22. The poetical formulation of the phrase τὸ ἐκ Διὸς ὕδωρ to denote rainwater falling from the sky may seem somewhat at odds with the general sub-literary style of the collection [see 1.2.3.]. However, the name of Zeus is a common metaphor for all celestial phenomena (cf. L. Senzasono, 2006, p. 13), including the διοσημία (cf. Aratus' quote in the fifth *causa*). Cf., e.g., *Alex.* 27, 2.

912B δι' ὃ καὶ πομφόλυγας ποιεῖ τῇ ἀναμίξει τοῦ ἀέρος;: The reference may be to the bubbles (filled with air) that are formed by the impact of raindrops on the surface of puddles, ponds and pools.

912B ἣ τρέφει μὲν μάλιστα τὸ μάλιστα κρατούμενον ὑπὸ τοῦ τρεφομένου (τοῦτο γάρ ἐστι πέψις· ἀπεψία δὲ τοὐναντίον, ὅταν ἰσχυρότερα τοῦ παθεῖν ᾖ), καὶ μεταβάλλει τὰ λεπτὰ καὶ ἁπλᾶ καὶ ἄχυμα μᾶλλον, οἷόν ἐστι τὸ ὄμβριον ὕδωρ;: For the easy concoction of simple foodstuffs, cf. *Quaest. conv.* 661BD (to the contrary, cf. *Quaest. conv.* 663B with Arist., *DA* 416a28–35; but cf. also Ps.-Arist., *Probl.* 861a6–9).

912B καὶ τῇ γῇ καὶ τοῖς τόποις ὁμοιούμενα, δι' ὧν ἔξεισι: It is not unlikely that 'the earth' (τῇ γῇ) and 'the places' (τοῖς τόποις) mentioned here are to be identified with one another (despite the double use of the copulative καί). In this sense, the phrase 'whence they emerge' (δι' ὧν ἔξεισι) implies that the springwaters emerge from (ἐξ) certain 'locations in the earth'. According to F.H. Sandbach, 1965, p. 155, however, Plutarch uses the term τόποι to imply that not only the earth, but also the air at the location where water exits a spring affects the quality of the water. A parallel passage for this is found in Sen., *NQ* 3, 21, 2: *locus atque aer aquas inficit similesque regionibus reddit per quas et ex quibus veniunt*. Cf. also *Q.N.* 33: *iniurias quas vel ab aeris mala qualitate vel a terra accipiunt digerere nequeant* (said of stagnant waters). Due to the fact, however, that Plutarch has just ascribed the *purity* of rainwater to the air (and breath) in which it is formed (γεννώμενον γὰρ ἐν ἀέρι καὶ πνεύματι καθαρόν, cf. also the third *causa*), it seems unlikely that the air at the localities of the spring renders the springwater impure, unless of course, this 'spring-air' is less pure (or breathlike) than that in the sky, but this seems far-fetched. In any case, Plutarch would probably have made this point more explicit if he really intended to draw a subtle distinction between both types of air (cf. also L. Senzasono, 2006, p. 156, n. 22). Sandbach also refers to M. Glycas, *Ann.* 1, 16 (p. 31, 12–13 Bekker), where the 'spring-air' is *not* mentioned, though: τὰ πάντα γὰρ τῆς ποιότητος ἐκεῖθεν μετέχουσιν ἀφ' ἧς γῆς διέρχονταί τε καὶ ἀνέρχονται. For the influence of the earth on the rivers that sprout from it, see also Pl., *Phd.* 112ac, where it is noted, moreover, that air and breath accompany the oscillating movements of the liquid in Tartarus.

According to L. Senzasono, p. 157, n. 22, Plutarch is actually relying on this passage in Plato so that the connection of rainwater with the air and breath just mentioned "difficilmente può essere casuale". This is very unlikely, though, because the contexts are completely different (and if we may assume that the rivers flowing out of the Tartaric liquid contain admixed air and breath themselves, this is contradictory to Plutarch's argument).

912B ἧττόν ἐστιν εὔτρεπτα: I take this to imply that springwater, as opposed to rainwater, is in fact already a changed substance by the fact that it is assimilated to the earthy locations whence it arises (see the previous comment), so that it has great difficulty in undergoing new change.

912BC τῶν δ' ὀμβρίων τὸ εὔτρεπτον αἱ σήψεις κατηγοροῦσιν· εὐσηπτότερα γάρ ἐστι τῶν ποταμίων καὶ φρεατιαίων: For the rapid putrefaction of rainwater, cf. Pliny, *NH* 31, 34. Also in Hipp., *Aer.* 8 reference is made to the putrefactive quality of rainwater, albeit on the basis that it is impure and of mixed origin. In *Quaest. conv.* 725CD, we read that mixing produces putrefaction and that standing waters in ponds are εὔσηπτα μᾶλλόν, because they are impure and mixed with earth (running waters, by contrast, shake off any admixed earth). On the putrefactive quality of standing waters, cf. also *Aqua an ignis* 957D and *De lat. viv.* 1129D (with F. Fuhrmann, 1964, p. 60).

912C ἡ δὲ πέψις ἔοικεν εἶναι σῆψις, ὡς Ἐμπεδοκλῆς μαρτυρεῖ λέγων 'οἶνος ἀπὸ φλοιοῦ πέλεται σαπὲν ἐν ξύλῳ ὕδωρ': Line also quoted in *Q.N.* 31, 919CD with syntactical adaptations (= Emp., DK31B81). There is discussion among scholars about the correct interpretation of this fragment: see H. Diels, 1901, p. 137, F.H. Sandbach, 1965, pp. 156–157 (n. a), M.R. Wright, 1981, pp. 225–226, B. Inwood, 2001, p. 131, L. Senzasono, 2006, p. 157, n. 24. I am not so sure of Sandbach's claim (approved of by Senzasono) that οἶνος implies grape juice here in Plutarch's account and not wine (being the resulting concoction from that grape juice). He refers to *Quaest. conv.* 676B, where we read that clay is hot, which is why it matures οἶνος (kept in clay vessels presumably). The formulation is, indeed, elliptic there, implying, so I take it, that heat concocts <grape juice into> wine. The distinction is clearer in *Q.N.* 27, 918EF: πέψις ἐστὶ τοῦ γλεύκους ἡ εἰς τὸ οἰνῶδες μεταβολή. In *Q.N.* 31, 919CD, the same line from Empedocles is quoted regarding the putrefactive nature of, what is abstractly called, 'the vinous' liquid, which is probably identical with wine (φύσει σηπτικὸν τὸ οἰνῶδές ἐστιν ὥς φησιν Ἐμπεδοκλῆς κτλ.). Sandbach adds that Plutarch undoubtedly wrote ἀπὸ φλοιοῦ ('from the bark') in the fragment at hand, but that it is possible that Empedocles originally wrote ὑπὸ φλοιοῦ ('under the bark', as already suggested by Xylander, but rejected by Diels). The term φλοιός itself may refer to the bark of the vine, but also to the skin of the grape (cf. LSJ, s.v. 1). The second option seems more plausible, albeit perhaps in a metonymical sense, as in Emp., DK31B80 (= *Quaest. conv.* 683D), where φλοιός probably refers not just to the skin but to the edible part of the apple surrounding the seeds (see Wright). As to ἐν ξύλῳ, Wright argues that it refers to the wooden casks or vats containing the pressed grapes, but this was already rejected by Diels ("noli cogitare de vino in dolio condito"). It is difficult to reconstruct the original sense of the fragment (especially because Plutarch may be twisting Empedocles' original wording). What matters most for Plutarch's argument, though, is the idea that wine – being a concoction from grape juice – is essentially putrid water, which illustrates the idea that the process of concoction resembles that of putrefaction (ἡ δὲ πέψις ἔοικεν εἶναι σῆψις). Notably, Arist., *Top.* 127a19 dismisses Empedocles' view that wine is putrefied water, but he does this in order to illustrate a common error in predication of the *genus* (ἁπλῶς γὰρ οὐκ ἔστιν ὕδωρ), rather than to reject the physical process as such.

912C τὸ γλυκὺ τῶν ὀμβρίων καὶ χρηστόν, εἰσπεμπόμενον εὐθὺς ὑπὸ τοῦ πνεύματος: We are probably dealing here with an allusion to the third *causa*, as can be illustrated by two points: 1) the phrase ὑπὸ τοῦ πνεύματος recalls πνεύματι, and 2) the participle εἰσπεμπόμενον implies a movement of the rainwater *into* the plant: as such, it recalls the phrase πνεύματι μεμιγμένον […] ἀναπέμπεται εἰς τὸ φυτόν (cf. also *Q.N.* 4, 912F: τὸ δὲ πνεῦμα τὴν ὑγρότητα […] ἀναπέμπει). However, εἰσπεμπόμενον is an emendation by F.H. Sandbach, 1965, p. 156, n. 1 of the manuscript reading ἐκπεμπόμενον (Hubert suggests ἐκπεμπομένων). If the manuscript reading is to be preferred, the prefix in ἐκπεμπόμενον probably explains the genitive τῶν ὀμβρίων (notion of separation). In this sense, the breath immediately exports the sweet and useful constituent *from* the rainwater (*into* the plant). Sandbach corrects the prefix ἐκ- in εἰσ- by comparing Hipp., *De flat.* 7 (in the context of breath entering the body together with the food). The correction is plausible from a paleographical perspective, and even if one would stick to the reading of the manuscripts, the connection with *causa* three remains clear.

912C διὸ καὶ τὰ θρέμματα τούτων ἀπολαύει προθυμότερον, καὶ οἱ βάτραχοι προσδοκῶντες ὄμβρον ἐπιλαμπρύνουσι τὴν φωνὴν ὑπὸ χαρᾶς, ὥσπερ ἥδυσμα τοῦ λιμναίου τὸ ὑέτιον προσδεχόμενοι καὶ σπέρμα τῆς ἐκείνων γλυκύτητος: F.H. Sandbach 1965, 156, n. 3 finds "[t]he text […] suspect" and suggests that "perhaps some words, to which ἐκείνων refers, are missing". He translates: "Frogs, when expecting rain, croak more loudly and clearly for joy, looking forward to the rain-water as a kind of sweetening for the water of the pond, and as a seed from which the freshness of the other waters will increase (?)." As Sandbach indicates, the "other waters" in his translation is unclear (he does not clarify it any further). There is no lacuna in the manuscripts, so the text is probably correct. L. Senzasono 2006, p. 160, n. 26 may be right, therefore, that ἐκείνων has anaphoric value, referring back to the water of the pond (τὸ λιμναῖον <ὕδωρ>). The variation between singular and plural forms may cause confusion, but, if Senzasono is correct, for Plutarch, the singular and the plural forms imply basically the same here. In other words, 'the water of the pond' (i.e. in *each* pond *specifically*) implies the same as 'the water of the ponds' (i.e. in ponds *in general*). Notably, Plutarch formulates the names of different types of water in the plural in 912C (ὀμβρίων, ποταμίων, φρεατιαίων), yet for rainwater he uses both plural and singular forms in the fifth *causa* (ὀμβρίων, ὄμβρον, ὑέτιον, ὑετοῦ). Thus, the same variation can be inferred for pond water(s). Furthermore, it is not improbable that Plutarch deliberately avoids using τούτων here (instead of ἐκείνων), because only a little bit earlier, this pronoun (in that precise form, as conjectured by Wyttenbach for ταύτης or τούτοις in the manuscripts) referred to rainwater (sc. τῶν ὀμβρίων). This, of course, cannot be the meaning of ἐκείνων here (seeing that the resulting meaning would make no sense: 'rain as a seed of the rain's sweetness'?). It is not unlikely, therefore, that in his attempt to avoid misunderstanding, Plutarch unintentionally obscures things a bit. I interpret the passage as follows: 'frogs croak more loudly and out of joy when they expect rain, looking forward to accepting it as a sweetening of the water of the pond and as a seed of their (sc. the ponds') sweetness' (see also M. Meeusen, 2015a).

Some further commentary is necessary, then, regarding the croaking of frogs. The scholiast on the Aratus passage Plutarch quotes (see the following comment) says that the croaking of frogs is a sign of storm (σημεῖον χειμῶνος). He explains that frogs become aware in advance (προαισθάνονται) that the rainwater from storms turns cold and that they croak very loudly from joy (χαίροντες), since they are fond of water (φίλυδροι). Especially the fact that the rainwater is sweeter than the water of the pond gives them joy and causes them to breed more, just like rainwater causes better flowering in plants (ὅτι τὸ ὄμβριον ὕδωρ γλυκύτερον ὂν τοῦ πηγαίου εὐφραίνει αὐτοὺς καὶ πλέον ζωογονεῖν ποιεῖ, ὡς καὶ τὰ φυτὰ μᾶλλον ὑπὸ τοῦ ὀμβρινοῦ θάλλουσι). This last point about rainwater (*vis-à-vis* pond water) improving the flowering of plants is remarkably close to the problem at hand in *Q.N.* 2. C. Hattink *apud* F.H. Sandbach, 1965, p. 157, n. c even argues that the Aratus scholiast may be relying on Plutarch's lost Αἰτίαι τῶν Ἀράτου Διοσημιῶν (see frs. 13–20 Sandbach; Lamprias catalogue nr. 119). However, the use of a communal source is also plausible (e.g., one of the many Aratus commentaries that circulated widely in Antiquity [see 4.1.2.2., n. 75]). Regarding the croaking of frogs itself, in *De soll. an.* 982E, a distinction is made between 1) the erotic/nuptial call of male frogs to attract female congeners (φωνὴν ἐρωτικὴν καὶ γαμήλιον: this is the so-called ὀλολυγών, which is not mentioned in *Q.N.* 2, 912C; cf. also Arist., *HA* 536a11), and 2) the shrill cry that they make when they expect rain (λαμπρύνουσι τὴν φωνήν, ὑετὸν προσδεχόμενοι· καὶ τοῦτο σημεῖον ἐν τοῖς βεβαιοτάτοις ἐστίν: Plutarch here does use the very same terminology as in *Q.N.* 2, 912C). This distinction is rendered explicit in *De soll. an.* 982E with the adverb ἄλλως (Helmbold: ἄλλοτε). Aelian makes exactly the same distinction in *NA* 9, 13 (as L. Senzasono, 2006, p. 159, n. 25 points out, it is uncertain whether Aelian relies on Plutarch directly or on the same source; for the difference in pitch of the 'erotic' ὀλολυγών, see also *NA* 6, 19; the scholiast on the Aratus passage adds: καὶ ἡ ὀλολυγὼν δὲ ὁμοίως ἐπὶ τούτῳ χαίρει καὶ κράζει ἅμα ἡμέρᾳ. ἔστι δὲ ζῷον λιμναῖον φιλόψυχρον, but this ὀλολυγών is presumably the tree-frog rather than the croaking itself produced by male frogs: see LSJ, s.v. ii). It is not unlikely, however, that these two distinct explanations for the croaking of frogs are to be considered complementary, because spring is *both* the mating and the raining season (for frogs as a token, σύμβολον, of springtime, cf. *De Pyth. or.* 400C; for the influence of the hot south wind on the mating of animals, including frogs, that arise *ex spermate*, cf. also Arist., fr. 245, 10 Rose). F.H. Sandbach, 1965, p. 157, n. c. even believes that "it is possible [...] that the word σπέρμα in 912C refers to the mating of frogs, which croak particularly in the breeding season" (cf. also L. Senzasono, 2006, p. 160, n. 26). This is, indeed, what is suggested by the Aratus scholiast (who writes: πλέον ζωογονεῖν ποιεῖ). Seeing, however, that no explicit reference is made to the 'erotic' ὀλολυγών in *Q.N.* 2, it remains uncertain as to whether σπέρμα really refers to the frog's mating here (the reference is implicit at most). The problem at hand in *Q.N.* 2 has a main hydrological and botanical interest, not a zoological one. It concerns the physical-meteorological circumstances connected to the rain's sweetness and its influence on plants rather than animals. In regards to the actual meaning of the concept of σπέρμα, then, L. Senzasono, 2006, pp. 12–13, n. 16 and p. 160, n. 26 is probably right that it has metaphorical value, denoting a cause, germ or origin of something (cf., e.g., fr. 136 Sandbach: τοιούτου πάθους σπέρμα μὴ παραδέχεσθαι μηδ' ἀρχήν, see LSJ, s.v. i, 2). The seed imagery may be more ingenious than that. It is from the rainwater-seed, then, that the sweet content of the ponds grows. In this sense, it is implied that pond water is initially unsweet (or, in any case, less sweet), and that it becomes sweet(er) from the rainwater-seeds that fall into it. There may be reason to connect the term σπέρμα with the introductory explanation of the problem by Laetus (who argues, as we saw, that the rain parts the earth by its impact and so better penetrates into the roots). The image of raindrops (ὄμβρια) being planted like seeds might not be that far-fetched in the final *causa* at hand: like seeds, raindrops fall on a certain surface and penetrate it, so that they provide fresh and sweet nourishment to plants. Moreover, in his criticism of Laetus' theory, Plutarch refers to plants growing in ponds when rain falls in the 'right season' (καθ' ὥραν), which is not further specified. One can infer from the final *causa* that the 'right season' probably is that of spring. The result, then, is a subtle argumentative ring composition, where Plutarch at the very end of the aetiology explains what remained unexplained initially [see 4.3.3.3.]. The (generalising) singular σπέρμα (cf. also ἥδυσμα) can be explained, then, from the fact that it refers to ὄμβρον/τὸ ὑέτιον <ὕδωρ> (which implies basically the same as τὰ ὄμβρια in Laetus' explanation). On the 'spermatic' faculty of rain falling from the heavens on earth (where the heavens resemble the father and the earth the mother), cf. also, e.g., Aët., *Plac.* 1, 6 = Ps.-Plut. 880B. In a similar vein, Ps.-Arist./Alex. Aphr., *Suppl. probl.* 1, 17, 32–35 compares sunbeams to seeds planted in the moon's surface.

912D ἓν γὰρ καὶ τοῦτο ποιεῖται σημεῖον ὑετοῦ μέλλοντος Ἄρατος εἰπών 'ἢ μάλα δειλαιαι γενεαί, ὕδροισιν ὄνειαρ, / αὐτόθεν ἐκ λίμνης πατέρες βοόωσι γυρίνων': = Arat., *Phaen.* 946–947 (= *Diosem.* 214–215). Theophr., *Sign.* 15 also includes the croaking of frogs among signs of rain; cf. also *Sign.* 40, Cic., *De div.* 1, 15 = *Progn.* 4, 1–3, and Pliny, *NH* 18, 361. See the previous comment for a discussion of the comments of the Aratus scholiast on this passage and their relevance for the problem at hand.

Q.N. 3, 912DF

Q.N. 3 deals with a dietetical problem concerning the nutritive effects of salt on animals (a theme that recurs in *Q.N.* 1, 911CD and 5, 913B). Plutarch examines **why herdsmen put salt down for their cattle** (Διὰ τί παραβάλλουσι τοῖς θρέμμασιν ἅλας οἱ νομεῖς;). He provides three explanations dealing with the physical constitution and properties of salt. The first one refers to the salt's pungency, the second to its ability to dissolve fat and the third to its generative properties. The first two explanations are 'contradictory' to each other [see 4.3.3.2.].

The **first** explanation focuses on the pungency of salt and is of a mechanical kind. It is a popular belief, so Plutarch writes, that salt produces a bulk of food and fattens the cattle (Πότερον, ὡς οἱ πολλοὶ νομίζουσι, πλήθους τροφῆς ἕνεκα καὶ τοῦ παχύνειν;). Salt produces this effect indirectly. The pungency of salt, so Plutarch explains, stimulates the appetite, and by opening up the pores, it ameliorates the passage of the food for distribution in the body (τήν τε γὰρ ὄρεξιν ἡ δριμύτης ἐκκαλεῖται καὶ τοὺς πόρους ἀναστομοῦσα μᾶλλον ὁδοποιεῖ τῇ τροφῇ πρὸς τὴν ἀνάδοσιν). That is why Apollonius, the follower of Herophilus, recommends that weak and ill-nourished persons should not be fed on syrup or porridge but on pickled and salty foods, the fineness of which, having become like a hair-sieve (?), adds the nourishment to the body through the pores (διὸ καὶ τοὺς ἰσχνοὺς καὶ τοὺς ἀτρόφους Ἀπολλώνιος ὁ Ἡροφίλειος ἐκέλευε μὴ γλυκεῖ μηδὲ χόνδρῳ τρέφειν ἀλλὰ τοῖς ταριχευτοῖς καὶ ὑφαλμυρίζουσιν, ὧν ἡ λεπτότης, ὥσπερ ἐντρίχωμα γενομένη, τὰ σιτία τοῖς σώμασι διὰ τῶν πόρων προστίθησιν).

The **second** explanation aims to refute the first. Plutarch argues that herdsmen accustom their cattle to licking salt, rather because it makes them healthy and *reduces* their bulk (ἢ μᾶλλον ὑγιείας ἕνεκα καὶ συγκοπῆς πλήθους τὸν ἅλα λείχειν ἐθίζουσι τὰ βοσκήματα;). He explains that excessive weight makes them ill, but that salt melts the fat away and dissolves it (νοσεῖ γὰρ ἄγαν πιαινόμενα, τὴν δὲ πιμελὴν τήκουσιν οἱ ἅλες καὶ διαχέουσιν). This is illustrated by the fact that herdsmen skin their animals easily and without difficulty after the slaughtering, because the fat that binds and fastens the skin becomes thin and weak due to the pungency of the salt (ὅθεν εὐμαρῶς καὶ ῥᾳδίως ἀποδέρουσιν αὐτὰ σφάξαντες· ἡ γὰρ κολλῶσα καὶ συνδέουσα τὸ δέρμα πιμελὴ λεπτὴ καὶ ἀσθενὴς γέγονεν ὑπὸ τῆς δριμύτητος). Plutarch adds that the blood of animals that lick salt also grows thin, and that there is no internal solidification when salt is admixed (λεπτύνεται δὲ καὶ τὸ αἷμα τῶν ἅλας λειχόντων οὐδὲ πήγνυται τὰ ἐντὸς ἁλῶν μιγέντων).

In the **third** explanation, Plutarch alludes to the generative and aphrodisiac property of salt. He urges the reader to consider that animals (by licking the salt) become more fertile and readier towards coition (σκόπει δὲ μὴ καὶ γονιμώτερα καὶ προθυμότερα πρὸς τὰς συνουσίας). After all, bitches also conceive quickly when they eat salted meat after mating, and ships that transport salt harbour a larger number of mice (or rats), because they frequently copulate (καὶ γὰρ αἱ κύνες κύουσι ταχέως τάριχος ἐπεσθίουσαι, καὶ τὰ ἁληγὰ τῶν πλοίων πλείους τρέφει μῦς διὰ τὸ πολλάκις συμπλέκεσθαι).

912D Διὰ τί παραβάλλουσι τοῖς θρέμμασιν ἅλας οἱ νομεῖς;: Plutarch is referring to salt stones that are licked by the animals (cf. 912E: λείχειν, λειχόντων). To this day, such salt stones are used to provide extra mineral nutrition to farm animals in order to foster their growth and overall health. According to G.W.M. Harrison, 2000b, p. 248, the problem at hand "presumably encapsulates Peripatetic views, although a precise source has yet to be identified". A potential source is Arist., *HA* 596a16–25, where it is reported that farmers put salt down for their animals so that they become thirsty. Aristotle explains that thirst fattens (πιαίνει) the animals thus improving their health (for further parallels with *HA*, see also the comments below on 912EF), It is uncertain whether Plutarch directly relies on this account or, rather, on a lost Ps.-Aristotelian problem based on it. A clear parallel is found in Ps.-Arist./Alex. Aphr., *Suppl. probl.* 2, 137, 2–4 (οἱ βόες χαίρουσι μᾶλλον τοῖς ἁλμυρωτέροις, διὸ καὶ εἰς τὴν τροφὴν παρεμβάλλονται αὐτοῖς ἅλες καὶ συμμίγνυται ὁ τοιοῦτος χυμός). Cf. also Theophr., *CP* 6, 4, 6, where it is said that salt functions as a ἥδυσμα (cf. *Q.N.* 5, 913B) tempering the food for assimilation both in humans and in some animals (οὐδὲ γὰρ δυνάμεθα κρατεῖν ἀκράτου καθάπερ οὐδὲ τῶν ἄλλων ἔνια ζώων δι' ὃ καὶ τούτοις παρέχομεν τοὺς ἅλας). For the vital nutritive value of salt for cattle, cf. also Ps.-Arist., *Mir. ausc.* 844b20–22 (ἁλίζουσι γὰρ αὐτὰ δὶς τοῦ ἐνιαυτοῦ. ἐὰν δὲ μὴ ποιήσωσι τοῦτο, συμβαίνει αὐτοῖς ἀπόλλυσθαι τὰ πλεῖστα τῶν βοσκημάτων).

912D τήν τε γὰρ ὄρεξιν ἡ δριμύτης ἐκκαλεῖται καὶ τοὺς πόρους ἀναστομοῦσα μᾶλλον ὁδοποιεῖ τῇ τροφῇ πρὸς τὴν ἀνάδοσιν: Plutarch here draws attention to the nutritive effect of salt, but he does not try to prove that salt is nutritive in and of itself for land-animals (this would be in contradiction with what he says in *Q.N.* 1, 911C). The effect is rather of a mechanical kind: the salt's pungency stimulates the appetite by opening up passages for the distribution of the food into the body. Cf. also *Q.N.* 5, 913B for the idea that salt acts as a relish for some animals by removing the satiety (τὸ πλήσμιον) caused by their food, and thus stimulating appetite.

912DE τοὺς ἰσχνοὺς καὶ τοὺς ἀτρόφους Ἀπολλώνιος ὁ Ἡροφίλειος ἐκέλευε μὴ γλυκεῖ μηδὲ χόνδρῳ τρέφειν ἀλλὰ τοῖς ταριχευτοῖς καὶ ὑφαλμυρίζουσιν, ὧν ἡ λεπτότης, ὥσπερ ἐντρίχωμα γενομένη, τὰ σιτία τοῖς σώμασι διὰ τῶν πόρων προστίθησιν: The manuscript reading χονδρῷ (perispomenon), as printed by Hubert, refers to the granular form of the food, which is not at issue here. The fragment at hand (= Apoll., fr. 33 von Staden [see 4.2.1.1., n. 110]) possibly originates from Apollonius' Εὐπόριστα (*On common remedies*). The precise meaning of ἐντρίχωμα is obscure and remains problematic. LSJ, s.v. give two possible meanings for it: 1) "edges of the eyelids, eyelashes" (with reference to Poll., 2, 69) and 2) "hair-sieve" (with reference to our passage). In regards to the second meaning ("hair-sieve"), LSJ give ἠθμός (i.e. a strainer or colander) as a synonym. The passage that comes closest to formulating this idea is Xen., *Mem.* 1, 4, 6, where we read that eye-lashes grow from the eye-lids in order to 'filter' the winds, so that they cannot injure the eye (ὡς δ' ἂν μηδὲ ἄνεμοι βλάπτωσιν, ἠθμὸν βλεφαρίδας ἐμφῦσαι). If this is the correct meaning here, it remains to be seen in what precise sense the fineness of salty foods becomes 'like a hair-sieve' in our passage. According to K. Oikonomopoulou (in personal correspondence), this fineness may refer to the fine grains of the salt. She argues that the 'hair-sieve' "would suggest that salt acts like some kind of filter of fine hairs, which perhaps attaches itself to the pores (? the role of the pores is not clear from the passage) and aids the digestion of food (note the verb προστίθημι: it 'adds the food to the body' [cf. also *Q.N.* 30, 919C and 31, 919E]), so it acts as some sort of medium of digestion; this would fit well with the meaning of 'filter'. If this is right, the correct translation would be: 'acting like a fine sieve'." As to the role of the pores (in the flesh presumably), Plutarch previously argued that they are opened by the salty constituent in the pickled foodstuffs and that the nourishment is transmitted through them to the rest of the body (τοὺς πόρους ἀναστομοῦσα μᾶλλον ὁδοποιεῖ). The 'sieve' thus created may, indeed, be a very fine one, with microscopic pores having the diameter of a single hair. I agree with Oikonomopoulou when she adds that "the theory was [perhaps] idiosyncratic to Apollonius" ("though it should be noted that we cannot be sure, on the basis of Plutarch's quotation, whether the explanation is actually Apollonian: maybe the Apollonius quote ends with ὑφαλμυρίζουσιν, the rest of the explanation supplied by Plutarch"). Moreover, the context in *Q.N.* 3 is very similar to that of *Quaest. conv.* 687D, where we read that salt recovers the appetite in ill people by its effect on the stomach (with reference to the theory of pores) – it is not unlikely that the word ἐντρίχωμα fell out in the lacuna there. For similar references to the working of a sieve in relation to the transmission of nourishment in the body, cf. also *Quaest. conv.* 689C (where Plutarch rejects the idea that dry and liquid food are separated by the pores in the flesh, as if through a strainer, ὥσπερ ἠθμοῖς) and 699AB (where Plutarch argues that the lung is created in the pattern of a sieve and that it contains many pores, ἠθμοειδὴς καὶ πολύπορος, for the transmission, διήησιν, of liquids and solids).

Other interpretations of the concept of ἐντρίχωμα have been suggested (sometimes involving textual emendations), but none as convincing as the one proposed. The translation by F.H. Sandbach, 1965, p. 159 (with n. b) as "some kind of hairy growth (?)" is accurate from an etymological perspective (cf. the first meaning in LSJ) but remains enigmatic (before him, Duebner and Bernardakis also thought of hairs). Sandbach notes that "[t]his passage awaits explanation. ἐντρίχωμα might be expected to mean a growth of hairs or filaments: its only known use is of that part of the eyelid from which the lashes grow, Pollux, ii. 69. Does Plutarch mean that the 'fine' parts of salty foods form fine threads, which pass through the passages and draw the rest of the food after them?" This is odd, but in any case there is a certain relationship between the residue (περίττωμα) of nourishment and the growth of hairs (cf. *Quaest. conv.* 651A, *De Is. et Os.* 352D; Ps.-Arist., *Probl.* 893a31, 893b7; Ps.-Arist./Alex. Aphr., *Suppl. probl.* 2, 59, 61, 64). There is no clear indication, however, that this is what Plutarch is hinting at here. Alternately, Xylander and Kaltwasser have connected the concept of ἐντρίχωμα with an aspect of friction, and Koechly conjectures ἔντριμμα ('ointment'). L. Senzasono, 2006, pp. 161–162, n. 29 connects ἐντρίχωμα with ταρσός ('mat of reeds') on the basis of Poll., 2, 69 (καλοῦνται ἔλυτρα καὶ ἐντριχώματα καὶ ὀρχοὶ καὶ ταρσοί), but this is uncertain in the present context (as are the other synonyms in Pollux). Doehner, by contrast, emends ἐντρίχωμα in θρίγκωμα ('coping', 'cornice') on the basis of *Quaest. conv.* 685B, where it is argued that salt adapts the food to our appetite. Metaphorically speaking, it becomes a θρίγκωμα to the food for the body, implying a 'completion', or 'finishing touch' (Hoffleit), or a 'topping'. If this emendation is correct – F. Fuhrmann, 1978, p. 182, n. 1

(to p. 87) speaks of ἐντρίχωμα as "une étrange paronymie" –, it perhaps contains a reference to salty desserts, which were believed to produce beneficial effects (cf. *Quaest. conv.* 669B: salty food improves digestion). This remains doubtful, but Doehner's emendation is not unsound from a paleographical perspective (as can be inferred from Flav. Jos., *Antiq. Jud.* 15, 11, 3, where θριγκώμασι was also conjectured for τριχώμασιν, cf. LSJ, s.v.). On the contrary, S.-T. Teodorsson, 1990a, pp. 228–229 (cf. also 1990b, pp. 64–66), would emend θρίγκωμα as (ἐν)τρίχωμα in *Quaest. conv.* 685B, arguing that we are dealing with a strange anatomical practice by which strong and pliable hairs (of horses or humans) were used as a kind of a 'probe' to examine the fine bloodvessels. It is difficult to either prove or disprove this theory. Another possible theory may involve an emendation based on *De Is. et Os.* 352F, where salt is considered to 'sharpen' the appetite (ἐπιθήγοντας τὴν ὄρεξιν). The imagery of salt as an ἐπίθημα (the head of a spear) is not implausible in the present context, but this remains conjecture. Another possibility can perhaps be adduced from Erasistratus' metaphor of water serving as an ὄχημα τῆς τροφῆς, a vehicle transporting (προστίθημι) the food through the pores into the body (cf. *Quaest. conv.* 690A, 698D; probably also 687E). The context is similar, but the imagery is not mentioned anywhere in regard to salt itself, only to seawater (in a different meaning, cf. *De cap. ex inim.* 86E) and honey (cf. Gal., *MM* 10, 300, 11).

912E τὴν δὲ πιμελὴν τήκουσιν οἱ ἅλες καὶ διαχέουσιν: On the melting property of salt, cf. also *Q.N.* 5, 913D (ἀποτήκειν), 10, 914D (ἀποτήκοντες), 40 (ἐκτηκομένης).

912E ὅθεν εὐμαρῶς καὶ ῥᾳδίως ἀποδέρουσιν αὐτὰ σφάξαντες· ἡ γὰρ κολλῶσα καὶ συνδέουσα τὸ δέρμα πιμελὴ λεπτὴ καὶ ἀσθενὴς γέγονεν ὑπὸ τῆς δριμύτητος: Cf. *Quaest. conv.* 697B, where salt is considered a powerful natural solvent owing to its heat, which counteracts the so-called interlocking and binding together of particles (καὶ πρὸς τοῦτο συνεργοῦσιν οἱ ἅλες, θερμοὶ γάρ εἰσι, πρὸς δὲ τὴν λεγομένην περιπλοκὴν καὶ σύνδεσιν ἀντιπράττουσι, διαλύειν γὰρ μάλιστα πεφύκασι). The property of heat in salt may also be implicitly present in our passage in the concept of melting (τήκουσιν); see the previous comment.

912E λεπτύνεται δὲ καὶ τὸ αἷμα τῶν ἅλας λειχόντων οὐδὲ πήγνυται τὰ ἐντὸς ἁλῶν μιγέντων: Similarly, in *Q.N.* 20, 917AB, Plutarch connects the salinity in the tears of deer (as opposed to the sweet ones of boars) with the thinness (λεπτόν) of their blood, from which the tears are secreted. The reference in τὰ ἐντός is not simply back to τὸ αἷμα (or to blood drops or the like), but to other 'internal parts', perhaps the animals' entrails. This meaning is found in *Quaest. conv.* 684A, where in the context of slaughtering, the 'outher pieces' (τὰ ἔξω, i.e. the limbs etc.) of sacrificial animals are contrasted to the 'inner pieces' (τὰ ἐντός). The same context of slaughtering is present in our passage (cf. Plutarch's previous mention of the skinning of animals). This may suggest that the salt not only ameliorates the skinning process but also the drainage of the blood and the removal of the organs. In this sense, there may be an implicit opposition between the 'external' skin on the one hand and the organs 'inside' of the body on the other. Alternatively, τὰ ἐντός perhaps refers to the *contents* of the bowels, the main point of interest being the laxative property of salt. This is, indeed, relevant in the context of the reduction of 'bulk' in cattle as mentioned in the second *causa* (but the formulation is obscure; cf. F.H. Sandbach, 1965, p. 160, n. a). In *Quaest. conv.* 690A, concerning the effect of bathing on nourishment, τὰ ἐντός is used for the contents of the stomach: ἡ ὑγρότης εὐχυμότερα ποιεῖ καὶ τροφιμώτερα τῷ ἐγχαλᾶσθαι τὰ ἐντός (cf. also R.M. Aguilar, 1994, p. 42). For the laxative property of salty foods, cf. also, e.g., Hipp., *De victu* 45, 7–10 (white chick-peas) and 56, 1ff. (salty meat). Cf. also Ps.-Arist./Alex. Aphr., *Suppl. probl.* 2, 115–116. In Hipp., *Aer.* 7, 68–72, the popular theory that salt water is a laxative (διαχωρητικός) is rejected: τὴν κοιλίην ὑπ' αὐτέων [sc. τῶν ἁλμυρῶν ὑδάτων] στύφεσθαι μᾶλλον ἢ τήκεσθαι. It seems far-fetched to assume that there is a subtle polemic against Hippocrates here (*pace* L. Senzasono, 2006, p. 163, n. 30; cf. also V. Andò, 2004, pp. 177–178, n. 34 vs. R.M. Aguilar, 1994, p. 42) [see 4.2.1.1., n. 100].

912EF σκόπει δὲ μὴ καὶ γονιμώτερα καὶ προθυμότερα πρὸς τὰς συνουσίας· καὶ γὰρ αἱ κύνες κύουσι ταχέως τάριχος ἐπεσθίουσαι, καὶ τὰ ἁληγὰ τῶν πλοίων πλείους τρέφει μῦς διὰ τὸ πολλάκις συμπλέκεσθαι: The imperative σκόπει δὲ μή may signal Plutarch's preference for this explanation, or that we are even dealing with Plutarch's personal contribution to the problem [see 1.1.4., n. 111]. In any case, it emphatically draws the reader's attention to what Plutarch is writing here. However, the same popular belief about bitches and mice is reported not by Plutarch but by Philinus (in the final explanation) in *Quaest. conv.* 685DE, in relation to the divine character of generation (earlier on in *Quaest. conv.* 685A Plutarch refers to the aphrodisiac properties of salt). Philinus argues that mice do not become pregnant simply by licking the salt, as some say, but that the salt probably stimulates copulation (as is argued here in *Q.N.* 3). The same popular belief is also recorded in Arist., *HA* 580b31 and Pliny, *NH* 10, 185. Regarding the relation between salt and fertility in cattle, cf. also Arist., *HA* 574a8.

ò∾

Q.N. 4, 912F–913A

Q.N. 4 concerns a meteorological phenomenon that, especially by its focus on the effect of rainwater on seeds, ties in closely with the theme of *Q.N.*

2. Plutarch wonders **why rain that accompanies thunder and lightning is more fertilising for seeds**. He adds that this kind of rain is called 'lightning water' (Διὰ τί τῶν ὀμβρίων ὑδάτων εὐαλδέστερα τοῖς σπέρμασι τὰ μετὰ βροντῶν καὶ ἀστραπῶν, ἃ δὴ καὶ ἀστραπαῖα καλοῦσι;). Three explanations are given for this problem. The first one focuses on the mixture of rainwater with breath, the second on its concocted nature, and the third on its cooling effect. The influence of lightning and thunder on the increase in fertility is – especially in the final *causa* – considered an incidental aspect rather than the main cause of this phenomenon. As such, Plutarch does not aim to reject the popular belief about lightning water, but certainly puts it into a broader physical perspective in his aetiology [see 4.3.3.3.].

In the **first** explanation, Plutarch alludes to the breathlike constituent of lightning water. He argues that these rains become breathlike by the disturbance and admixture of air, and that the breath better transmits and distributes the water in the plant by imparting movement to it (Πότερον ὅτι πνευματώδη διὰ τὴν τοῦ ἀέρος ταραχὴν καὶ ἀνάμιξιν, τὸ δὲ πνεῦμα τὴν ὑγρότητα κινοῦν μᾶλλον ἀναπέμπει καὶ ἀναδίδωσιν;).

The process of concoction is central to the **second** *causa*, where Plutarch first explains how thunder and lightning are produced. This is due to the conflict of atmospheric heat and cold in the air (ἢ βροντὰς μὲν καὶ ἀστραπὰς ποιεῖ τὸ θερμὸν ἐν τῷ ἀέρι πρὸς τὸ ψυχρὸν μαχόμενον). Therefore, thunder occurs least in winter and most in spring and autumn, owing to the irregular temperature in those seasons (διὸ χειμῶνος ἥκιστα βροντᾷ μάλιστα δ' ἔαρος καὶ φθινοπώρου διὰ τὴν ἀνωμαλίαν τῆς κράσεως). Plutarch concludes that it is in fact the heat (of the lightning flashes) that concocts the moisture (c.q. rain) and makes it agreeable and useful for growing things (ἡ δὲ θερμότης πέττουσα τὸ ὑγρὸν προσφιλὲς ποιεῖ τοῖς βλαστάνουσι καὶ ὠφέλιμον;).

In the **third** and final explanation, Plutarch makes a subtle distinction by arguing that it is true that thunder and lightning especially occur in spring for the reason given (see the previous *causa*), but that spring rains are more essential to the seeds, because they come *before* the summerheat (ἢ μάλιστα μὲν ἔαρος βροντᾷ καὶ ἀστράπτει διὰ τὴν εἰρημένην αἰτίαν, τὰ δ' ἐαρινὰ τῶν ὑδάτων ἀναγκαιότερα τοῖς σπέρμασι πρὸ τοῦ θέρους). This is illustrated by the fact that the land that receives most rain in spring, like that in Sicily, yields crops that are abundant in quantity and good in quality (ὅθεν ἡ πλεῖστον ὑομένη τοῦ ἔαρος χώρα καθάπερ ἡ ἐν Σικελίᾳ πολλοὺς καὶ ἀγαθοὺς καρποὺς ἀναδίδωσιν;).

912F Διὰ τί τῶν ὀμβρίων ὑδάτων εὐαλδέστερα τοῖς σπέρμασι τὰ μετὰ βροντῶν καὶ ἀστραπῶν, ἃ δὴ καὶ ἀστραπαῖα καλοῦσι;: Cf. *Quaest. conv.* 664DE, where the generative property of lightning water is mentioned among other διοσημία not unworthy of belief. We learn from that passage that the term εὐαλδής ('fertilising') is a farmer's word. Plutarch's explanation is very similar in the second *causa* here: he points out that the heat that is produced by the fire of lightning and that is mixed with the rain makes it fertile and enriches the soil (the idea that lightning, therefore, has a divine character is not repeated, though [see 3.1.4.]).

912F Πότερον ὅτι πνευματώδη διὰ τὴν τοῦ ἀέρος ταραχὴν καὶ ἀνάμιξιν, τὸ δὲ πνεῦμα τὴν ὑγρότητα κινοῦν μᾶλλον ἀναπέμπει καὶ ἀναδίδωσιν;: Presumably, it is implied that lightning and thunder are accompanied by disturbance of the air in the sky, which in turn causes the rain's admixture with breath (for the breathlike and airy constituent of rainwater, cf. also *Q.N.* 2, 912AB).

913A ἢ βροντὰς μὲν καὶ ἀστραπὰς ποιεῖ τὸ θερμὸν ἐν τῷ ἀέρι πρὸς τὸ ψυχρὸν μαχόμενον: A similar, but more detailed, explanation of the origin of thunder and lightning, based on the opposite temperatures of the exhalations in the sky, is found in Arist., *Mete.* 369a12–29.

913A διὸ χειμῶνος ἥκιστα βροντᾷ μάλιστα δ' ἔαρος καὶ φθινοπώρου: For the idea that thunderbolts occur most in spring and in autumn (for the reason previously given), cf. Lucr., *De rer. nat.* 6, 357–378. Cf. also Pliny, *NH* 2, 135–136 and Arrian, *Frag. Phys.* 3, 190, 7–12 Roos (= Stob., *Flor.* 1, 29, 2). By contrast, according to Arist., fr. 245, 8, 15–19 Rose, thunder occurs most often in the summer and in the winter. Cf. also Sen., *NQ* 2, 57, 2: lightning occurs more frequently in the summer. The reader may wonder why at this point in Plutarch's aetiology only the season of summer remains unmentioned. Presumably, there is not much conflict between atmospheric heat and cold in the hot summer to trigger lightning flashes (as is also the case in winter, which Plutarch does mention). The summer season will play an important role in the third *causa*, which may explain why it is not mentioned here.

913A διὰ τὴν ἀνωμαλίαν τῆς κράσεως: Presumably, the irregularity of temperature refers back to the atmospheric conflict between hot and cold air. The seasons of spring and autumn are some kind of 'intermediary seasons', situated between those of winter (cold) and summer (hot), which explains their irregular temperature (cold and hot).

913A ἡ δὲ θερμότης πέττουσα τὸ ὑγρὸν προσφιλὲς ποιεῖ τοῖς βλαστάνουσι καὶ ὠφέλιμον;: Plutarch is probably referring to the heat in the storm clouds, so caused by the lightning blazes (cf. *Quaest. conv.* 664DE). For the agreeableness of slightly warm moisture for the young crops, cf. fr. 68 Sandbach. Cf. also *Quaest. conv.* 676B (regarding hot soil).

913A ἢ μάλιστα μὲν ἔαρος βροντᾷ καὶ ἀστράπτει διὰ τὴν εἰρημένην αἰτίαν, τὰ δ' ἐαρινὰ τῶν ὑδάτων ἀναγκαιότερα τοῖς σπέρμασι πρὸ τοῦ θέρους: According to F.H. Sandbach, 1965, p. 162, n. b, the comparative ἀναγκαιότερα may indicate "that spring rain is more essential than winter rain" (he finds the Greek "awkward"; cf. also L. Senzasono, 2006, pp. 167–168, n. 38, who supposes an ellipse of ἢ τὰ χειμερινά or τῶν χειμερινῶν). His suggestion to emend it to mean "more essential than those which come just before the time of the harvest", however, is not very clear, nor plausible. The argument is probably more subtle. First of all, note that the adverb μάλιστα ('especially') in the second *causa* (said of the occurrence of thunder in specific seasons) concerns both spring and autumn, whereas in the third *causa*, it only concerns spring. Indeed, autumn, does not come *before* summer (πρὸ τοῦ θέρους) and, therefore, remains unmentioned here. This temporal specification is essential for a proper understanding of the *causa* at hand. I take the μέν [...] δέ construction to imply a disjunction between two (opposed) facts here, viz. thunder and lightning on the one hand and spring rain on the other. What Plutarch is probably implying, then, is that (μέν) thunder and lightning particularly occur in spring (for the reason given in *causa* two), but (δέ) that they have no actual effect on the growth of seeds, because the spring rains themselves are *more essential* (ἀναγκαιότερα) in that regard, that is, more essential than the effects of thunder and lightning, the reason being that they come before the summer. This last point, putting the emphasis on the temporal specification of spring rains falling before the summer (πρὸ τοῦ θέρους) is paralleled in *Q.N.* 14, 915D. There, Plutarch argues that spring rain is important for wheat and barley grains because it falls before the summer, soaking the earth, so that it protects the ears of the grain against the hot southerly winds (ὑόμενος δὲ πρὸ τοῦ θέρους ὁ σῖτος βοηθεῖται πρὸς τὰ θερμὰ καὶ νότια πνεύματα). It is unlikely, therefore, that Plutarch is referring to "some generative property of spring rains" here in *Q.N.* 3 (*pace* P.A. Clement and H.B. Hoffleit, 1969, p. 321, n. a). The reference is rather to the rain's cooling effect. With this final explanation, then, Plutarch does not, as such, aim to reject the popular belief about lightning water (as mentioned in the *quaestio*), but he certainly puts it in a broader physical perspective (compare the initial 'scepticism' of the symposiasts towards the belief that thunder generates truffles in *Quaest. conv.* 664B and Plutarch's subsequent reaction in 664DE [see 3.1.4.]).

913A ὅθεν ἡ πλεῖστον ὑομένη τοῦ ἔαρος χώρα καθάπερ ἡ ἐν Σικελίᾳ πολλοὺς καὶ ἀγαθοὺς καρποὺς ἀναδίδωσιν;: Regarding the useful character of spring rains for plants, Plutarch in his commentary on Hes., *Op.* 486–489 (= fr. 68 Sandbach) ascribes the growth of good crops (εὐκαρπεῖν) in Sicily to the copious quantity of spring rains that the land receives (ἀπὸ τοῦ τὴν Σικελίαν πολλοὺς δεχομένην ἐαρινοὺς ὄμβρους, for the three-month variety of wheat mentioned in this passage, cf. also *Q.N.* 15, 915DE). Cf. also Theophr., *HP* 8, 6, 6, for the idea that spring rains are most seasonable (καιριώτατα [...]

τὰ ἠρινά), and that Sicily is rich in corn (πολύσιτος), owing to the abundance of soft rain in spring and the lack of it in winter (cf. also *CP* 4, 9, 4–5 and *HP* 8, 4, 5). Notably, none of these parallel passages pay attention to the thunder and lightning that accompany the spring rains.

ₑ🙰

Q.N. 5, 913AE

In *Q.N.* 5, Plutarch deals with a problem concerning the generation of flavours (χυμοί), with specific attention to the salty. He asks **why we observe that only one of the eight generic flavours, namely the salty, is *not* produced by any fruit, whereas the others actually are** (Διὰ τί τῶν χυμῶν, ὀκτὼ τῷ γένει ὄντων, ἕνα μόνον, τὸν ἁλμυρόν, ἀπ᾽ οὐδενὸς καρποῦ γεννώμενον ὁρῶμεν; καίτοι κτλ.). Plutarch's aetiology comprises four explanations in total. The first one holds that salt is not generated at all, the second that it is destroyed by the fruit's heat, the third that it cannot enter the plant's narrow pores, and the fourth that it is a kind of bitterness. Still in the *quaestio* Plutarch gives an account of the different types of generic flavours and their generation in several kinds of fruits.

Plutarch writes that at first, the olive produces the *bitter* flavour while the grape produces the *acid*. Afterwards, they change and become *oily* and *vinous* respectively (καὶ τὸν πικρὸν ἡ ἐλαία φέρει πρῶτον καὶ τὸν ὀξὺν ὁ βότρυς, εἶτα μεταβάλλων ὁ μὲν γίνεται λιπαρὸς ὁ δ᾽ οἰνώδης). The *astringent* flavour in dates and the *sour* in pomegranates become *sweet*, although some pomegranates and apples only produce the acid; and, the *pungent* flavour is prominent in roots and seeds (μεταβάλλει δὲ καὶ ὁ στρυφνὸς ἐν ταῖς φοινικοβαλάνοις καὶ ὁ αὐστηρὸς ἐν ταῖς ῥόαις εἰς τὸν γλυκύν· ἔνιαι δὲ ῥόαι καὶ μῆλα τὸν ὀξὺν ἁπλῶς φέρουσιν, ὁ δὲ δριμὺς ἐν ταῖς ῥίζαις καὶ σπέρμασι πολύς ἐστι).

In the **first** explanation, Plutarch draws attention to the opposition between the processes of generation and corruption. He explains that the salty flavour is not generated by any fruit; in fact, it is not generated at all. On the contrary, it is a corruption of the other flavours (Πότερον οὖν οὐκ ἔστιν ἁλμυροῦ γένεσις ἀλλὰ φθορὰ τῶν ἄλλων τὸ ἁλμυρόν;). Therefore, the salty flavour is not nutritious for any animal that feeds on plants and seeds. For some, though, it acts as a relish by removing the satiety caused by their foods (διὸ καὶ πᾶσιν ἄτροφον τοῖς ἀπὸ φυτῶν καὶ σπερμάτων τρεφομένοις, ἥδυσμα δ᾽ ἐνίοις γίνεται τῷ τὸ πλήσμιον ἀφαιρεῖν τῶν τρεφόντων).

In the **second** explanation, Plutarch takes a step backwards by arguing that the salty flavour – initially present one way or another in fruits (as opposed to what was said in the *quaestio*) – loses its natural character by the action of heat, just like people remove the salty and pungent constituent from seawater by boiling it (ἤ, καθάπερ τῆς θαλάττης ἕψοντες ἀφαιροῦσι τὸ ἁλυκὸν καὶ δηκτικόν, ἐν τοῖς καρποῖς ὑπὸ θερμότητος ἐξαμαυροῦται τὸ ἁλμυρόν;).

In the **third** explanation, Plutarch quotes Plato and argues that a flavour is basically water that has been strained through a plant (ἢ χυμὸς μέν ἐστιν, ὡς Πλάτων εἶπεν, ὕδωρ ἠθημένον διὰ φυτοῦ). He adds that when seawater is filtered, it also loses its saltiness, because this is earthy and has large particles (διηθουμένη δὲ καὶ θάλαττα τὸ ἁλμυρὸν ἀποβάλλει· γεῶδες γὰρ καὶ παχυμερές ἐστιν). Three examples follow to illustrate the filtration of seawater. 1) Those who dig near the seashore find small drinkable springs (ὅθεν ὀρύττοντες παρὰ τὸν αἰγιαλὸν ἐντυγχάνουσι ποτίμοις λιβαδίοις). 2) People also frequently draw up sweet filtered water from the sea in jars made of wax, because the salty and earthy constituent is separated from it (πολλοὶ δὲ καὶ κηρίνοις ἀγγείοις ἀναλαμβάνουσιν ἐκ τῆς θαλάττης ὕδωρ γλυκὺ διηθούμενον, ἀποκρινομένου τοῦ ἁλυκοῦ καὶ γεώδους). 3) The previous passing and filtering through white clay also renders the seawater completely drinkable, because the clay retains the earthy constituent and does not let it through (ἡ δὲ δι' ἀργίλου προδιαγωγὴ παντάπασι τὴν θάλατταν διηθουμένην πότιμον ἀποδίδωσι τῷ κατέχειν ἐν ἑαυτῇ καὶ μὴ διιέναι τὸ γεῶδες). Since this is so, it is likely, according to Plutarch, that plants neither take up anything salty from their surroundings nor excrete any salty product into their fruit, should it be generated internally, because the pores, owing to their narrowness, do not strain through the earthy and large particles (οὕτως δὲ τούτων ἐχόντων, εἰκός ἐστι τὰ φυτὰ μήτ' ἔξωθεν ἀναλαμβάνειν ἁλμυρίδα μήτ', ἂν ἐν αὐτοῖς λάβῃ γένεσιν, ἐκκρίνειν εἰς τὸν καρπόν· οἱ γὰρ πόροι διὰ λεπτότητα τὸ γεῶδες καὶ παχυμερὲς οὐ διηθοῦσιν).

According to the **fourth** and final explanation, it could be posited that the salty flavour is a kind of bitterness, which Plutarch illustrates with the following Homeric lines: 'From his mouth he spat the bitter brine that ran gushing from his head.' (ἢ τῆς πικρότητος εἶδος τὴν ἁλμυρότητα θετέον, ὡς Ὅμηρος· στόματος δ' ἐξέπτυσεν ἅλμην / πικρήν, ἥ τοι πολλὸν ἀπὸ κρατὸς κελάρυζεν). In what follows, Plutarch combines the authority of Homer with that of Plato, who also connects both flavours with each other. Plato says that they both cleanse and dissolve, though the salty does this to a lesser degree and is not rough (καὶ ὁ Πλάτων φησὶν ἀμφοτέρους ῥύπτειν καὶ ἀποτήκειν τοὺς χυμούς, ἧττον δὲ ταῦτα ποιεῖν τὸν ἁλυκὸν καὶ οὐ τραχὺν εἶναι). In order to illustrate the last point (i.e. to specify the slight difference that Plato notes between the bitter and the salty), Plutarch adds that the bitter has a greater level of dryness, because the salty also has some drying property (δόξει δὲ τὸ πικρὸν τοῦ ἁλυκοῦ ξηρότητος ὑπερβολῇ διαφέρειν, ἐπεὶ ξηραντικόν τι καὶ τὸ ἁλυκόν).

913AB Διὰ τί τῶν χυμῶν, ὀκτὼ τῷ γένει ὄντων, ἕνα μόνον, τὸν ἁλμυρόν, ἀπ' οὐδενὸς καρποῦ γεννώμενον ὁρῶμεν;: The same problem was raised and solved (see the following comments) by Theophrastus (*CP* 6, 10, 1–2: διὰ τί ποθ᾽ οἱ μὲν ἄλλοι πάντες ἐν τοῖς φυτοῖς καὶ καρποῖς γίνονται [...] ὁ δ᾽ ἁλμυρὸς οὐκέτι· οὐδὲν γὰρ τῶν φυομένων ἁλυκὸν ὥστε καὶ ἐν ἑαυτῷ τοιοῦτον ἔχειν τὸν χυλὸν κτλ.). Theophrastus' account was probably known to Plutarch via a lost Ps.-Aristotelian problem (see already H. Diels, 1905, pp. 315–316).

913B καὶ τὸν πικρὸν ἡ ἐλαία φέρει πρῶτον καὶ τὸν ὀξὺν ὁ βότρυς, εἶτα μεταβάλλων ὁ μὲν γίνεται λιπαρὸς ὁ δ᾽ οἰνώδης· μεταβάλλει δὲ καὶ ὁ στρυφνὸς ἐν ταῖς φοινικοβαλάνοις καὶ ὁ αὐστηρὸς ἐν ταῖς ῥόαις εἰς τὸν γλυκύν· ἔνιαι δὲ ῥόαι καὶ μῆλα τὸν ὀξὺν ἁπλῶς φέρουσιν, ὁ δὲ δριμὺς ἐν ταῖς ῥίζαις καὶ σπέρμασι πολύς ἐστι: While Plutarch maintains that there are eight different flavours, he sums up nine in total (also noticed by F.H. Sandbach, 1965, p. 163, n. d). The confusion may originate from Theophrastus, who says that there are seven kinds of flavours (ἰδέαι τῶν χυμῶν), on the condition that saltiness and bitterness are not considered different (οὐχ ἕτερον, cf. the final *causa* here in *Q.N.* 5); otherwise saltiness becomes the eighth flavour (see *CP* 6, 4, 1; cf. also *CP* 6, 1, 2). After summing up the different kinds of flavours, Theophrastus concludes that some people would add the vinous – as does Plutarch here –, and rightly so, because this flavour has its own particular nature (ἰδία τις ἡ φύσις) and therefore stands apart by itself. As such, the total number of flavours amounts to nine, not eight (or seven). According to L. Senzasono, 2006, p. 169, n. 41 (who believes that Plutarch makes no mistake at all), Plutarch is following Theophrastus in implicitly setting the vinous flavour apart from the others. If this is true – which I doubt –, one may wonder why Plutarch did not explicitly say so, as Theophrastus did. I take it that the fact that the vinous flavour is not mentioned separately from the other eight by Plutarch, precisely suggests that he considers it to be a generic flavour (τῷ γένει) just as well. On the other hand, if there is one flavour that is clearly the odd one out, it is the salty one, because it is not generated in plants. To speak of a genuine miscalculation on Plutarch's side goes too far, since he really sees the salty as being distinct from the eight generic flavours. Plutarch's inaccuracy can perhaps be explained, then, by his untended writing (due to hypomnematic negligence [see 2.3.2.]). Alternatively, he perhaps wrote the intermediate part of the *quaestio* in view of his argument in the fourth *causa*, where the salty flavour is identified as a kind of bitterness. Whatever may be the case, this inaccuracy (if that is what it is) has no repercussions for the rest of Plutarch's argument. Theophrastus also concludes that the correct number of flavours would make no difference for the proper understanding of the rest of the subject; for him the number seven is 'most appropriate and natural' (*CP* 6, 4, 2: Ὁ δὲ ἀριθμὸς ὁ τῶν ἑπτὰ καιριώτατος καὶ φυσικώτατος – this expression may be based on Pythagorean numerology; cf. Alex. Aphr., *Comm. in Ar. Met.* 38, 16–20).

Scholars have argued that Theophrastus' source for the number of flavours was probably Democritus (see already H. Diels, 1905, pp. 312–316). In Theophrastus' doxography of Democritus' theory of the shapes of flavours, seven flavours are enumerated, excluding αὐστηρός (*CP* 6, 1, 6). In a parallel passage (*Sens*. 65–67), he lists only six flavours, excluding λιπαρός and αὐστηρός. On the assumption of a lacuna there, Diels (*ibid*., p. 314; cf. also 1879, p. 518, 18) argues that Democritus' theory also originally proposed eight flavours, and he adds λιπαρός and αὐστηρός (from *CP* 6, 4, 1), yet in his remark to Dem., DK68A135 (= Theophr., *Sens*., 67, 27), he notes that "[i]n den Hss. [...] nichts ausgefallen [ist]". There is also much speculation in the remark of L. Senzasono, 2006, pp. 169–170, n. 41 that the Democritean origin of the eight flavours can be accepted for *CP* 6, 4, 1 on the assumption that Theophrastus – for whatever reason – left out these two flavours (λιπαρός and αὐστηρός) in *Sens*. 65–67. Senzasono does not exclude the possibility, however, that Democritus' theory did embrace six flavours, but that Theophrastus added two more in *CP* 6, 4, 1.

There are also interesting accounts of the nature and number of the flavours in Aristotle and Plato (the latter is quoted by Plutarch in the third and fourth *causae*). Aristotle writes that there are seven flavours, excluding the vinous and identifying the salty and the bitter; he describes these flavours in terms of contraries and draws a link with the number of colours (see *DA* 422b10–14 and *De sensu* 442a17–21). Plato lists seven flavours in *Tim*. 65b–66c, excluding the oily and the vinous; he believes that taste is brought about by certain contractions and dilatations of the surface of the tongue and that the flavours involve roughness and smoothness (cf. also Theophr., *CP* 6, 1, 4). Additionally, Pliny, *NH* 15, 106 ff. records 13 *genera saporum*. For yet another account of the number of flavours, see Gal., *SMT* 11, 450, 14 ff. Kühn.

913B Πότερον οὖν οὐκ ἔστιν ἁλμυροῦ γένεσις ἀλλὰ φθορὰ τῶν ἄλλων τὸ ἁλμυρόν;: According to Theophrastus (*CP* 6, 10, 1–2), salt is neither nourishing nor procreative (ἄτροφον καὶ ὥσπερ ἀγέννητον τὸ ἁλμυρόν). A probable explanation for this is found in the fact that virtually no plant grows on salty land, because the salt would consume and take away its powers (δυνάμεις) and so prevent its growth. It is reasonable, so Theophrastus specifies, that what prevents other things from generating will do no generating itself (Ὁ δὲ καὶ τοῖς ἄλλοις τούτου αἴτιον εὔλογον μηδὲ καθ' αὑτὸ γεννᾶν). He then refers to plants growing in the sea: they are not nourished by the salt, but by some sweet constituent and by other flavours in the seawater, as are fish and other sea animals (this is close to Plutarch's argument in *Q.N.* 1, 911D: see the commentary *ad loc.*). Importantly, unlike Plutarch, Theophrastus does *not* so much claim that salt is 'not generated' or 'unoriginated' (passive), but that it 'does not generate', 'is not productive' (active). The adjective ἀγέννητον can have both meanings, but Theophrastus clearly uses it in the second sense (cf. LSJ, s.v. iii: "not productive", with a reference to this particular passage in Theophrastus). If we may assume that Theophrastus is Plutarch's source, the adaptation to the new context, where the γένεσις – φθορά opposition is central, may be intentional. If no adaptation had been made, that is, if Plutarch had followed Theophrastus in arguing that salt has no productive or generative property, this would be in direct contradiction with his arguments in *Q.N.* 3, 912EF and *Quaest. conv.* 685DE (which both concern the generative and aphrodisiac properties of salt). Moreover, by making this semantic shift, Plutarch is able to more closely connect the *causa* with the *quaestio* at hand (ἀπ' οὐδενὸς καρποῦ γεννώμενον – οὐκ ἔστιν ἁλμυροῦ γένεσις: salt is not generated in any fruit – *a fortiori* it is not generated at all). As to the idea that salt is a corruption of 'other things' (φθορὰ τῶν ἄλλων), I take it that Plutarch means the other flavours – these are generated from the fruit and turn salty when corrupted. For the idea that generation of a thing can be caused by the corruption of something else (and vice versa), cf. Arist., *GC* 318b33–35 (cf. L. Senzasono, 2006, p. 171, n. 43).

913B διὸ καὶ πᾶσιν ἄτροφον τοῖς ἀπὸ φυτῶν καὶ σπερμάτων τρεφομένοις, ἥδυσμα δ' ἐνίοις γίνεται τῷ τὸ πλήσμιον ἀφαιρεῖν τῶν τρεφόντων: The idea that the salty flavour is not nutritious for any animal that feeds on plants and seeds echoes the first *causa* of *Q.N.* 1, 911CD (where Plutarch argues that seawater is not nutritious for land animals). The second idea, that salt acts as a relish for some animals by removing the satiety caused by their foods, is presumably added to avoid contradiction (and thus to reinforce *concatenatio*) with the first *causa* of *Q.N.* 3, 912D (where Plutarch argues that salt stimulates the appetite of animals by its pungency). Theophrastus (*CP* 6, 4, 6) also says that salt functions as a ἥδυσμα, tempering the food for assimilation both in humans and in some animals (quoted in the commentary to *Q.N.* 3, 912D above).

913C τῆς θαλάττης ἕψοντες ἀφαιροῦσι τὸ ἁλυκὸν καὶ δηκτικόν: The idea that seawater becomes sweet by boiling (or more generally by heating) is paralleled in Arist., *Mete.* 358b16–18 (in the context of distillation). Cf. also Ps.-Arist., *Probl.* 933b11–16, Cass. Iatrosoph., *Probl.* 65 and *Geopon.* 2, 47, 11 (the account of Hipp., *Aer.* 8, 6 is more general: the heat of the sun produces coction and, thus, sweetens the rainwater, and in addition all other boiled liquids always become sweet as well). According to F.H. Sandbach, 1965, p. 164, n. a, this extraordinary idea "[p]erhaps [...] arose from a misunderstanding of some account of distillation". He refers to M. Glycas, *Ann.* 1, 9 (p. 19, 5–9 Bekker): καὶ αὐτὸ τὸ τῆς θαλάττης ὕδωρ ἴδοι τις ἂν ὑπὸ τῶν ναυτιλλομένων ἑψόμενον κἀντεῦθεν τὴν χρείαν μετρίως παραμυθούμενον· τηνικαῦτα γὰρ σπόγγοις ὑποδεχόμενοι τοὺς ἀναγομένους ἐκεῖθεν ἀτμοὺς ἴαμα δίψης ἐν ταῖς ἀνάγκαις εὑρίσκουσι. The context of distillation is also clear from

Arist., *Mete.* 358b18–20, where we read that wine can be evaporated and afterwards condensed into water (on Aristotle's experiments in this passage, see G.E.R. Lloyd, 1991, pp. 90–91 and L. Taub, 2003, pp. 102–103).

913C ἐν τοῖς καρποῖς: The manuscripts read ἐν τοῖς θερμοῖς, while Hubert suggests θερείοις (which implies basically the same), and Sandbach conjectures καρποῖς (cf. also V. Ramón Palerm, 2005, pp. 398–399). L. Senzasono, 2006, pp. 171–172, n. 44 follows the reading of the manuscripts against Sandbach's conjecture, but his stylistic argument is not convincing (I am doubtful about the alleged elegance of the phrase ἐν τοῖς θερμοῖς ὑπὸ θερμότητος). Even if we follow the manuscript reading, θερμοῖς would still imply καρποῖς, because the ripening process of fruit (which involves heat; see the following comment) is probably the issue here.

913C ὑπὸ θερμότητος: Heat is generated in the fruit when it ripens (ripening being a process of concoction), so that any salty constituent that may be a part of the fruit's nourishment is destroyed by it (cf. F.H. Sandbach, 1965, p. 164, n. a). A parallel process is described in *Q.N.* 20, 917A, where Plutarch argues that the salty tears of wild boars become sweet by the heat of their fiery temper. For the idea that the sweetness of a plant (c.q. the fig tree) is concentrated in the fruit, leaving the rest of the plant bitter and unmixed (without reference to a ripening process, though), cf. *Quaest. conv.* 684C (ὅσον ἂν ἐνῇ τῷ φυτῷ γλυκύτητος, ἅπαν τοῦτο συνθλιβόμενον εἰς τὸν καρπὸν εἰκότως δριμὺ ποιεῖν καὶ ἄκρατον τὸ λειπόμενον).

913C ἢ χυμὸς μέν ἐστιν, ὡς Πλάτων εἶπεν, ὕδωρ ἠθημένον διὰ φυτοῦ: Pl., *Tim.* 59e. For the relation between the flavour of a plant and the moisture that permeates its shoots, cf. *Quaest. conv.* 664E.

913C διηθουμένη δὲ καὶ θάλαττα τὸ ἁλμυρὸν ἀποβάλλει; γεῶδες γὰρ καὶ παχυμερές ἐστιν: For the idea that filtered seawater becomes drinkable, cf. Arist., *Mete.* 354b18 (διηθούμενον γὰρ γίγνεσθαι τὸ ἁλμυρὸν πότιμον). For the earthy matter in seawater, cf., e.g., *Q.N.* 1, 911D, 8, 914B and *Quaest. conv.* 627AC (Arist., fr. 217 Rose). The phrase γεῶδες καὶ παχυμερές recurs at the end of the explanation.

913CD ὅθεν ὀρύττοντες παρὰ τὸν αἰγιαλὸν ἐντυγχάνουσι ποτίμοις λιβαδίοις, πολλοὶ δὲ καὶ κηρίνοις ἀγγείοις ἀναλαμβάνουσιν ἐκ τῆς θαλάττης ὕδωρ γλυκὺ διηθούμενον, ἀποκρινομένου τοῦ ἁλυκοῦ καὶ γεώδους· ἡ δὲ δι' ἀργίλου προδιαγωγὴ παντάπασι τὴν θάλατταν διηθουμένην πότιμον ἀποδίδωσι τῷ κατέχειν ἐν ἑαυτῇ καὶ μὴ διιέναι τὸ γεῶδες: Numerous parallels can be traced for these three experiments. For the first (about digging near the seashore), cf. Ps.-Arist., *Probl.* 935b3–17 and 933b33–41 (also b17–20), Ps.-Arist./Alex.

Aphr., *Suppl. probl.* 2, 34 (with S. Kapetanaki and R.W. Sharples, 2006, p. 137, n. 255), Ps.-Alex. Aphr., *Probl.* 1, 55 (J.L. Ideler, 1841, p. 19, 22–31), Pliny, *NH* 2, 224, Sen., *NQ* 3, 5, Lucr., *De rer. nat.* 2, 474; 5, 269 ff. and 6, 635 ff. For the second (about the wax jar [see 4.3.2.3., n. 246]), cf. Arist., *Mete.* 358b34–359a5 (with Olympiod., *Comm in Ar. Mete.* 158, 27ff.), *HA* 590a22–27, *GA* 743a8–11, Ael., *NA* 9, 64, Pliny, *NH* 31, 70. For the third (about white clay), cf. Pliny, *NH* 31, 70 (and 48). The noun προδιαγωγή is a hapax (cf. LSJ, s.v.: "previous passing through" with a reference to the passage at hand).

As to the experiment with the wax jar, F.H. Sandbach, 1982, p. 227 argues that Plutarch possibly relies on Arist., *Mete.* 358b34–359a5, but that the intermediation of a lost Ps.-Aristotelian problem cannot be excluded. Arguably, in the latter case, Plutarch perhaps also relies on this intermediating problem for the other two experiments (these are not mentioned in Aristotle's *Meteorology*). Cf. H. Diels, 1905, pp. 310–316, esp. pp. 315–316, who argues that all three experiments were given currency by Democritus, but that Democritus was hardly read in the time of the Early Empire, so that Plutarch probably became acquainted with them via a lost collection of Ps.-Aristotelian *Problems*. This collection, so Diels argues, contained a large amount of *Democritea* (Democritus was, indeed, the first to compose a collection of problems, as we saw [1.1.3., n. 58]) and *Theophrastea* (viz. material from Theophrastus' smaller doxographical natural scientific treatises, such as the lost *De aquis*: cf. D.L., 5, 45). Interestingly, in regards to the belief that fish are nourished by the sweet water mixed with salt water in the sea (see the commentary to *Q.N.* 1, 911D), Aelian quotes Democritus, Theophrastus and Aristotle together; yet, for the experiment with the wax jar he only calls in the authority of the son of Nicomachus, that is, Aristotle (*NA* 9, 64).

913D οὕτως δὲ τούτων ἐχόντων, εἰκός ἐστι τὰ φυτὰ μήτ' ἔξωθεν ἀναλαμβάνειν ἁλμυρίδα μήτ', ἂν ἐν αὐτοῖς λάβῃ γένεσιν, ἐκκρίνειν εἰς τὸν καρπόν· οἱ γὰρ πόροι διὰ λεπτότητα τὸ γεῶδες καὶ παχυμερὲς οὐ διηθοῦσιν: The interjection of the phrase ἂν ἐν αὐτοῖς λάβῃ γένεσιν is contradictory to the first *causa*, where Plutarch argues that there is no generation of salt (οὐκ ἔστιν ἁλμυροῦ γένεσις). Clearly then, the element of εἰκός is what really matters here, suggesting that all possibilities remain open from an epistemological perspective [see 4.3.3.1.]. For the idea that the earthy and heavy quality of salt makes it impossible for seawater to enter into the roots of the plant, cf. also *Q.N.* 1, 911D.

913D ἢ τῆς πικρότητος εἶδος τὴν ἁλμυρότητα θετέον ὡς Ὅμηρος· στόματος δ' ἐξέπτυσεν ἅλμην πικρήν, ἥ τοι πολλὸν ἀπὸ κρατὸς κελάρυζεν: Hom., *Od.* 5, 322–323. For the identification of the salty and the bitter, cf. also, e.g., Arist., *De sensu* 442a17 and Theophr., *CP* 6, 4, 1.

913D καὶ ὁ Πλάτων φησὶν ἀμφοτέρους ῥύπτειν καὶ ἀποτήκειν τοὺς χυμούς, ἧττον δὲ ταῦτα ποιεῖν τὸν ἁλυκὸν καὶ οὐ τραχὺν εἶναι: See Pl., *Tim.* 65de. The idea that salt has a dissolving property (ἀποτήκειν) is paralleled in *Q.N.* 3, 912E (τήκουσιν), 10, 914D (ἀποτήκοντες) and 40 (ἐκτηκομένης). The idea that salt is not rough (οὐ τραχύν) is contradicted in *Q.N.* 1, 911D (τραχεῖαν), albeit in a different context.

913DE δόξει δὲ τὸ πικρὸν τοῦ ἁλυκοῦ ξηρότητος ὑπερβολῇ διαφέρειν, ἐπεὶ ξηραντικόν τι καὶ τὸ ἁλυκόν: Plutarch already dealt with the drying property of salt (in seawater) earlier in *Q.N.* 1, 911D, and this point does not seem to require further detail here. Indeed, Plutarch does not further specify the greater level of dryness in the bitter flavour *vis-à-vis* the salty, perhaps because it is only added in order to illustrate the difference that Plato mentions (viz. that the salty flavour cleanses and dissolves less than the bitter). Thus, it supports the claim that salt is a species (εἶδος) of bitterness, showing that these flavours are not simply identical.

ஓ

Q.N. 6, 913EF

Q.N. 6 does not fit in as well with the previous and the following problems that constitute the first thematic cluster of questions about salt and seawater (thus, we are dealing here with a case of thematic *variatio*). The focus returns to the central theme in the following problems of *Q.N.* 7–13. In *Q.N.* 6, by contrast, no mention is made of salt or seawater, but the problem is still generally concerned with hydrology (c.q. water in the form of dew), and Plutarch agains quotes Laetus here, in the first *causa*, as he did in *Q.N.* 2, 911F. Plutarch wonders **why persons (probably hunters) who frequently walk through bushes wet with dew contract 'leprosy' on those body parts that come into contact with the brushwood** (Διὰ τί τοῖς συνεχῶς διὰ τῶν δεδροσισμένων δένδρων βαδίζουσι λέπραν ἴσχει τὰ ψαύοντα τῆς ὕλης μόρια τοῦ σώματος;). The theme of dew will return in the cluster of problems on hunting in *Q.N.* 23–25. A similar context can be presumed for the problem at hand, then. Two explanations are given for the detrimental effects of dew on the skin. The first focuses on the fineness of dew, the second on its erosive property. As opposed to the first *causa*, the second includes the role of the plants in the explanation, suggesting that dew damages the skin indirectly.

The **first** explanation remains very concise and is attributed to Laetus, who said that the dewy moisture scrapes off the skin by means of its fineness (Πότερον, ὡς Λαῖτος ἔλεγε, τῇ λεπτότητι τὸ δροσῶδες ὑγρὸν ἀποξύει τοῦ χρωτός). This is not further explained but implies that the dew harms the skin directly.

In the **second** explanation, Plutarch further examines the harmful effects of dew, viz. its erosive property. As opposed to Laetus, Plutarch does not, however, connect this erosive property directly to the damage done to the skin, but examines more closely the role of plants in the phenomenon (as mentioned in the *quaestio*). By analogy with the rust that forms in wet seeds, Plutarch argues that some kind of dust is discharged from objects (c.q. from plants) that are tender and soft on the surface, but become rough and dissolved by the dew. This dust causes (contagious) harm (ἤ, καθάπερ ἐρυσίβη τοῖς ὑγραινομένοις ἐγγίνεται σπέρμασιν, οὕτως ὑπὸ τῆς δρόσου τῶν ἐπιπολῆς χλωρῶν καὶ ἁπαλῶν ἀναχαρασσομένων καὶ ἀποτηκομένων ἄχνη τις ἀπιοῦσα τοῦ σίνοντος ἀναπίμπλησι). When the dust settles on those parts of the flesh that are the most bloodless, such as the lower legs and the feet (of hunters), it scratches and eats away at the surface of the skin (προσχεομένη τοῖς ἀναιμοτάτοις μέρεσι τῆς σαρκός, οἷα κνῆμαι καὶ πόδες, ἀμύσσει καὶ δάκνει τὴν ἐπιφάνειαν). Testimony for this naturally erosive ('biting') property of dew is found in the fact that it makes people who imbibe it thinner. In any case, so Plutarch says, fat women imagine that they can dissolve their excess fat by soaking up dew on cloths or soft flocks of wool (ὅτι γὰρ φύσει τι δηκτικὸν ἔνεστι τῇ δρόσῳ, μαρτυρεῖ τὸ τοὺς πίνοντας ἰσχνοτέρους ποιεῖν· αἱ γοῦν πίονες γυναῖκες ἱματίοις ἢ ἐρίοις ἁπαλοῖς ἀναλαμβάνουσαι τῆς δρόσου δοκοῦσι συντήκειν τὴν πολυσαρκίαν).

913E Διὰ τί τοῖς συνεχῶς διὰ τῶν δεδροσισμένων δένδρων βαδίζουσι λέπραν ἴσχει τὰ ψαύοντα τῆς ὕλης μόρια τοῦ σώματος;: The topic of this problem falls under the general theme of 'sympathy' (in the medical sense of 'contagions'), as treated in Book 7 of Ps.-Aristotle's *Problems*: ὅσα ἐκ συμπαθείας (see R. Mayhew, 2011a, pp. 228–229). On infectious diseases, cf. esp. Ps-Arist., *Probl.* 886b4–9 and 887a22–40. According to G.W.M. Harrison, 2000b, p. 244, "[t]he picture is of people walking through low plants shedding dew and the vocabulary fairly frolics with bouncy light syllables and assonance, particularly nouns formed in composition with ἀνα- [three times], which is quite a feat given Plutarch's well known aversion to hiatus". There is great deal of sensitivity in Harrison's phonetic analysis, but there is only one (minor) case of hiatus in the aetiology: viz. in ἢ ἐρίοις.

913E λέπραν: The λέπρα in question is not the same as modern-day leprosy, but is, rather, some kind of scabby skin disease. Cf. F.H. Sandbach, 1965, p. 167: "scabbiness"; P.A. Clement and H.B. Hoffleit, 1969, p. 359, n. a: "any scaly condition, cf. psoriasis". According to L. Senzasono, 2006, p. 176, n. 53, the term is used in its Hippocratic meaning ("la scabbia, la psoriasi e l'eczema"). Senzasono identifies modern leprosy with ἐλεφαντίασις (cf. *Quaest. conv.* 731A, with S.-T. Teodorsson, 1996, pp. 259–260), but λεύκη is more plausible (cf., e.g., Hipp., *Prorrh.* 2, 43).

In *De Is. et Os.* 353F, λέπρα is bracketed together with ψωρικαί τραχύτητες (itchy scabies), and in *Quaest. conv.* 670F with ἐπὶ χρωτὸς λεῦκαι (white scale) and ψωρικά ἐξανθήματα (scaly eruptions). For the belief that dew causes *scabies*, cf. also Pliny, *NH* 17, 225; 31, 33 and Seneca, *NQ* 3, 25, 11.

913E ὡς Λαῖτος ἔλεγε: This Λαῖτος, also quoted in *Q.N.* 2, 911F, is probably to be identified with the Platonist Ofellius Laetus [see 4.2.1.1., n. 115].

913E καθάπερ ἐρυσίβη τοῖς ὑγραινομένοις ἐγγίνεται σπέρμασιν: For the harmful effect of rust (ἐρυσίβη) produced by moisture on seeds and plants, cf. Theophr., *CP* 3, 22, 1–2 and 4, 14, 3 (where it is treated as some kind of decomposition). F.H. Sandbach, 1965, p. 167 (with n. e) translates ἐρυσίβη as mildew (with reference to Pliny, *NH* 18, 91 and 275, who mentions *robigo*). Mildew is "a growth (typically a whitish and fluffy coating) of fungal mycelium and fructifications on the surface of a plant; plant disease characterized by this type of growth" (OED, s.v.). He notes that mildew implies honey-dew, that is, "a sweet sticky substance found on the leaves and stems of trees and plants [...] formerly imagined to be in origin akin to dew" (OED, s.v.). In addition, there is note of a rare disease called mildew-gangrene or mildew-mortification produced by diseased grain, related perhaps to Plutarch's λέπρα (OED, s.vv.: considered obsolete).

913E οὕτως ὑπὸ τῆς δρόσου τῶν ἐπιπολῆς χλωρῶν καὶ ἁπαλῶν ἀναχαρασσομένων καὶ ἀποτηκομένων ἄχνη τις ἀπιοῦσα τοῦ σίνοντος ἀναπίμπλησι: Conversely, for the beneficial effect of copper-ore dust (ἄχνη) on the eyes of miners, cf. *Quaest. conv.* 659C.

913E τοῖς ἀναιμοτάτοις μέρεσι τῆς σαρκός, οἷα κνῆμαι καὶ πόδες: The verb βαδίζουσι in the *quaestio* probably explains why Plutarch only refers to the lower legs and the feet. One may wonder, therefore, whether the dust is only harmful for these 'most bloodless' body parts, and not also, for instance, for the hands, arms, chest and face (perhaps these remain unmentioned because they are not 'most bloodless'; but what the bloodlessness of body parts has to do with the argument remains unspecified, though).

913F ὅτι γὰρ φύσει τι δηκτικὸν ἔνεστι τῇ δρόσῳ, μαρτυρεῖ τὸ τοὺς πίνοντας ἰσχνοτέρους ποιεῖν· αἱ γοῦν πίονες γυναῖκες ἱματίοις ἢ ἐρίοις ἁπαλοῖς ἀναλαμβάνουσαι τῆς δρόσου δοκοῦσι συντήκειν τὴν πολυσαρκίαν: Caelius Aurelianus reports, in the context of losing weight (πολυσαρκία), that physicians recommend imbibing the air moistened by nocturnal dew before sunrise (*Tard. pass.* 5, 139: *nocturni roris auram ante solis ortum bibendam*). Many thanks are due to C. Laes for drawing my attention to this passage. I could find no clear source for the popular belief about overweight women soaking up dew on their garments in order to dissolve excess fat. Plutarch may be reliant, therefore, on popular hearsay or (indirectly) on a gynaecological treatise (perhaps from the kind of Soranus' *Gynaecia* or the Hippocratic gynaecological writings). The belief that pregnant women eat both stones and dirt (as recorded in *Q.N.* 26, 918D: see the commentary *ad loc.*) may originate from the same source.

ᐧᕁ

Q.N. 7, 913F–914B

In *Q.N.* 7, Plutarch returns to the thematic line of the first section of questions concerning salt and seawater. He examines **why boats travel more slowly on rivers during the winter, while this is not even nearly true of travel on the sea** (Διὰ τί τὰ πλοῖα χειμῶνος ἐν τοῖς ποταμοῖς πλεῖ βράδιον, ἐν δὲ τῇ θαλάττῃ οὐ παραπλησίως;). Plutarch gives two explanations for this paradox, first attributing the cause to the heavy river air, and afterwards to the cold river water itself.

According to the **first** explanation the river air is an obstacle to shippers. It is always slow-moving and heavy, so Plutarch explains, and is further thickened in the winter by the frosty cold (Πότερον ὁ ποτάμιος ἀήρ, ἀεὶ δυσκίνητος ὢν καὶ βαρὺς ἐν δὲ χειμῶνι μᾶλλον παχυνόμενος διὰ τὴν περίψυξιν, ἐμποδών ἐστι τοῖς πλέουσιν;).

In the **second** explanation, the thickening of the river water itself, rather than the river air, is taken into consideration (ἢ τοῦτο μᾶλλον τοῦ ἀέρος πάσχουσιν οἱ ποταμοί;). Plutarch explains that the winter cold compresses the water and makes it heavy and solid (συνελαύνουσα γὰρ ἡ ψυχρότης τὸ ὕδωρ ποιεῖ βαρὺ καὶ σωματῶδες). He illustrates this with two examples. The same phenomenon can be observed in the clepsydrae, which draw up water more slowly in the winter than in the summer (ὡς ἔστιν ἐν ταῖς κλεψύδραις καταμαθεῖν, βράδιον γὰρ ἕλκουσι χειμῶνος ἢ θέρους). Additionally, Theophrastus records that there is a spring on Mt. Pangaeum in Thrace from which the same amount of water in a full vessel, when put on the scales, weighs twice as much in the winter as in the summer (ἐν δὲ Θράκῃ περὶ τὸ Πάγγαιον ἱστορεῖ Θεόφραστος εἶναι κρήνην, ἀφ᾿ ἧς ταὐτὸ γέμον ἀγγεῖον ὕδατος ἱστάμενον χειμῶνος ἕλκειν διπλάσιον σταθμὸν ἢ θέρους). After these two examples, Plutarch returns to the main argument and notes that the density of the water causes the ship's slow passage, which is obvious from the fact that river-ships can carry more cargo in the winter, as the water becomes denser and firmer and has more buoyancy (ὅτι δ᾿ ἡ πυκνότης τοῦ ὕδατος τὴν βραδυτῆτα ποιεῖ τοῦ πλοῦ, δῆλόν ἐστι τῷ πλείονα γόμον φέρειν τὰ ποτάμια πλοῖα τοῦ χειμῶνος· τὸ γὰρ ὕδωρ μᾶλλον ἀντερείδει πυκνότερον καὶ βαρύτερον γινόμενον). The sea, by contrast, is prevented from growing solid by its heat, and this explains why it does not freeze, since stiffening seems to be a process of 'condensation' (τὴν δὲ θάλατταν ἡ θερμότης κωλύει πυκνοῦσθαι, δι' ἣν οὐδὲ πήγνυται, μάλκη γὰρ ἔοικεν εἶναι πύκνωσις).

913F ὁ ποτάμιος ἀήρ, ἀεὶ δυσκίνητος ὢν καὶ βαρύς: Plutarch does not further specify why river air is always slow-moving and heavy. Perhaps it is implicitly contrasted with the windy weather on the open sea, for instance, during storms (these do not occur on rivers).

914A ὡς ἔστιν ἐν ταῖς κλεψύδραις καταμαθεῖν, βράδιον γὰρ ἕλκουσι χειμῶνος ἢ θέρους: Considering the notion of time in βράδιον ('more slowly'), it is perhaps more likely the water clock rather than the 'pipette' (ὑδράρπαξ) that is meant with κλεψύδρα here (cf. the following comment; pace F.H. Sandbach, 1965, pp. 168–169, n. a). The water clock was used in law-courts to measure the litigants' speaking time. It was also used for making other measurements: e.g., military (cf. Aen. Tact., 22, 24) or astronomical (cf. Procl., *Hyp.* 4, 74). For images, see M. Lewis, 2000, pp. 361–369.

914A ἐν δὲ Θράκῃ περὶ τὸ Πάγγαιον ἱστορεῖ Θεόφραστος εἶναι κρήνην, ἀφ' ἧς ταὐτὸ γέμον ἀγγεῖον ὕδατος ἱστάμενον χειμῶνος ἕλκειν διπλάσιον σταθμὸν ἢ θέρους: We learn from Athenaeus (*Deipn.* 2, 42b = Theophr., fr. 159, 15–21 Wimmer = 214A, 13–17 FHSG) that this account (= Theophr., fr. 161 Wimmer = 214C FHSG) originates from Theophrastus' lost *De aquis* (cf. D.L., 5, 45), and that Plutarch simplifies the weight ratio, which is recorded in a more precise way by Athenaeus (viz. 96 to 46). Interestingly, the same fragment also mentions the γνώμων in the – exceptional – meaning of a water clock, that is, the κλεψύδρα (see LSJ, s.v. with a reference to Theophrastus' fragment; cf. the previous comment). It is reported that the water that flows in these water clocks does not measure the hours correctly in the winter, but runs too long, which is attributed to the slower outflow of the water due to its thickness (βραδυτέρας οὔσης τῆς ἐκροῆς διὰ τὸ πάχος). It is not unlikely that Plutarch here (and presumably also in the previous account about the κλεψύδραι) relies on one or more lost Ps.-Aristotelian problems based on the passage in Theophrastus' *De aquis*.

914A ὅτι δ' ἡ πυκνότης τοῦ ὕδατος τὴν βραδυτῆτα ποιεῖ τοῦ πλοῦ, δῆλόν ἐστι τῷ πλείονα γόμον φέρειν τὰ ποτάμια πλοῖα τοῦ χειμῶνος· τὸ γὰρ ὕδωρ μᾶλλον ἀντερείδει πυκνότερον καὶ βαρύτερον γινόμενον: This is not necessarily incompatible with the idea from *Q.N.* 1, 911D that seawater supports ships better (μᾶλλον) than fresh water does, owing to its earthy constituent.

914AB τὴν δὲ θάλατταν ἡ θερμότης κωλύει πυκνοῦσθαι, δι' ἣν οὐδὲ πήγνυται, μάλκη γὰρ ἔοικεν εἶναι πύκνωσις: The idea here is probably that the natural heat of seawater (present in the salt) causes liquefaction, so that no solidification or 'condensation' can occur in it during the winter by freezing cold (cf. *Q.N.* 8, 914B: μὴ πήγνυσθαι). For similar effects of winter cold on sea and riverwater, cf. Pliny, *NH* 2, 234 (seawater freezes more slowly and boils more quickly) and 31, 56 (riverwater becomes heavier after the winter solstice). F.H. Sandbach, 1965, p. 170, n. a is probably correct that the term μάλκη implies an analogy between the body growing numb with cold and the freezing of water (cf. LSJ, s.v., with a reference to this passage; *pace* L. Senzasono, 2006, p. 177, n. 59).

Q.N. 8, 914B

From a thematic perspective *Q.N.* 8 is closely connected with the previous problem, especially by its focus on the inability of seawater to freeze (this theory recurs at the end of the aetiology). Plutarch wonders **why we see that the sea grows warmer when it is agitated, whereas all other liquids become colder when moved and disturbed** (Διὰ τί, τῶν ἄλλων ὑγρῶν ἐν τῷ κινεῖσθαι καὶ στρέφεσθαι ψυχομένων, τὴν θάλατταν ὁρῶμεν ἐν τῷ κυματοῦσθαι θερμοτέραν γιγνομένην;). He gives one explanation based on the seawater's innate heat.

Plutarch argues that motion dispels and dissipates the heat from other liquids, where it is an incidental and alien intrusion. But in seawater, to which it is innate, heat it is fanned and fed by the winds more and more (Ἡ τῶν μὲν ἄλλων ὑγρῶν ἐπεισόδιον οὖσαν καὶ ἀλλοτρίαν ἐξίστησιν ἡ κίνησις τὴν θερμότητα καὶ διαφορεῖ, τὴν δὲ τῆς θαλάττης σύμφυτον οὖσαν ἐκριπίζουσι μᾶλλον οἱ ἄνεμοι καὶ τρέφουσι;). Evidence for the seawater's natural heat is adduced from the facts that seawater is transparent and that it does not freeze, although it is earthy and heavy (μαρτύρια δὲ τῆς θερμότητος ἡ διαύγεια καὶ τὸ μὴ πήγνυσθαι, καίπερ οὖσαν γεώδη καὶ βαρεῖαν).

914B Διὰ τί, τῶν ἄλλων ὑγρῶν ἐν τῷ κινεῖσθαι καὶ στρέφεσθαι ψυχομένων, τὴν θάλατταν ὁρῶμεν ἐν τῷ κυματοῦσθαι θερμοτέραν γιγνομένην;: Cicero also notes that the sea becomes warm when violently stirred by the wind (*De nat. deor.* 2, 26). In *Q.N.* 18, 916B, however, Plutarch says that the calamary leaps out of the sea in an attempt to escape the *cold and the disturbance* in the depths (τὸ ψῦχος καὶ τὴν ἐν βάθει ταραχὴν τῆς θαλάττης). This implies that movement and cold can still occur together in the sea.

914B Ἡ τῶν μὲν ἄλλων ὑγρῶν ἐπεισόδιον οὖσαν καὶ ἀλλοτρίαν ἐξίστησιν ἡ κίνησις τὴν θερμότητα καὶ διαφορεῖ, τὴν δὲ τῆς θαλάττης σύμφυτον οὖσαν ἐκριπίζουσι μᾶλλον οἱ ἄνεμοι καὶ τρέφουσι;: Plutarch does not explain why this movement exactly increases the heat of the seawater (cf. also L. Senzasono, 2006, p. 29). One may wonder if this is perhaps due to the friction of the earthy particles in it (γεώδη). For the (Stoic?) opposition between the attributes of σύμφυτον and ἐπεισόδιον, cf. also, e.g., *De virt. mor.* 451C and *De comm. not.* 1070C (cf. also *De Stoic. rep.* 1041E).

914B μαρτύρια δὲ τῆς θερμότητος ἡ διαύγεια καὶ τὸ μὴ πήγνυσθαι, καίπερ οὖσαν γεώδη καὶ βαρεῖαν: Plutarch does not further specify 1) that seawater is transparent and 2) that it does not freeze, nor 3) how this relates to it being earthy and heavy. The following points, however, can be added. 1) Seawater is considered to be transparent, because it contains light and, by implication, fire and heat (cf. *Q.N.* 12, 915A and 39). According to Ps.-Arist., *Probl.* 932b8–16 and 935b17–27, seawater is even more transparent than fresh water (cf. F.H. Sandbach, 1965, p. 171, n. c). 2) The inability of seawater to freeze (οὐδὲ πήγνυται) is also mentioned at the end of *Q.N.* 7, 914B, in relation to its inability to solidify (ἡ θερμότης κωλύει πυκνοῦσθαι). This process of solidification (or 'condensation': πύκνωσις) is more likely to occur in liquids that contain a large amount of solid matter (cf. F.H. Sandbach, 1965, p. 171, n. d). 3) The fact that sea water is earthy and heavy (γεώδη καὶ βαρεῖαν, cf. also *Q.N.* 1, 911D) but at the same time transparent and unable to freeze may seem strange in this logic, but not for Plutarch. For him, it *a fortiori* underpins the underlying idea that the effects of the innate heat outdo those of the earthy and heavy constituents in seawater. Just like fire, the innate heat gives light (c.q. transparency) and heat (c.q. inability to freeze) to seawater.

Q.N. 9, 914BD

In *Q.N.* 9, we find yet another problem concerned with seawater and how it is perceived by our senses, viz. **why the sea becomes less bitter to the taste during the winter** (Διὰ τί τοῦ χειμῶνος ἧττον πικρὰ γίνεται γευομένοις ἡ θάλαττα;). Plutarch notes that people say that this phenomenon is also recorded by Dionysius, the designer of aqueducts (τοῦτο γάρ φασι καὶ Διονύσιον ἱστορεῖν τὸν ὑδραγωγόν). He goes on to provide one extensive explanation, in which he relates the taste of seawater to seasonal temperatures.

Plutarch argues that the sea's bitterness is not entirely destitute or devoid of sweetness, because so many rivers empty into it (Ἡ ὅτι παντελῶς μὲν ἔρημος οὐκ ἔστι γλυκύτητος οὐδ' ἄμοιρος ἡ πικρότης, ἅτε δὴ ποταμοὺς τοσούτους ὑποδεχομένης τῆς θαλάττης). The sun evaporates the sweet drinkable constituent, which lies on the surface because it is light (τοῦ δ' ἡλίου τὸ γλυκὺ καὶ πότιμον ἐξαιροῦντος ὑπὸ κουφότητος ἐπιπολάζον). This evaporation occurs more in the summer, because the winter sun has a gentler effect, due to the weakness of its heat. As such, a large amount of sweetness is left behind (in winter) that dilutes the purely bitter and poisonous constituent (καὶ μᾶλλον ἐν τῷ θέρει τοῦτο ποιοῦντος, ἐν δὲ τῷ χειμῶνι μαλακώτερον ἁπτομένου δι᾽ ἀσθένειαν θερμότητος, ὑπολειπομένη μοῖρα πολλὴ γλυκύτητος ἀνίησι τὸ ἀκράτως πικρὸν καὶ φαρμακῶδες). Plutarch adds that the same process occurs in a mild way in drinkable waters as well. In the summer they become less good, since the heat dissipates the lightest and sweetest part, while fresh new water flows in during the winter (τοῦτο δ᾽ ἡσυχῇ καὶ τοῖς ποτίμοις συμβέβηκε· θέρους γὰρ πονηρότερα γίνεται, τὸ κουφότατον καὶ γλυκύτατον τοῦ θερμοῦ διαφοροῦντος, ἐν δὲ χειμῶνι νέον ἐπιρρεῖ καὶ πρόσφατον). At the end of the explanation, Plutarch points out that seawater necessarily receives a share of this fresh new water as well, since rain falls upon it and rivers empty into it (οὗ μετέχειν ἀνάγκη καὶ τὴν θάλατταν, ὑομένην ἅμα καὶ τῶν ποταμῶν ἐπιδιδόντων). With this last point the argumentative ring is complete [see 4.3.3.3.].

914B Διὰ τί τοῦ χειμῶνος ἧττον πικρὰ γίνεται γευομένοις ἡ θάλαττα;: According to Pliny, *NH* 31, 52 *all* water is sweeter in winter and less so in summer. For the relation between the salt and the bitter flavours, cf. *Q.N.* 5, 913D (with the commentary *ad loc.*).

914C τοῦτο γάρ φασι καὶ Διονύσιον ἱστορεῖν τὸν ὑδραγωγόν: Nothing is known about this Dionysius, the designer of aqueducts [see 4.2.1.1, n. 114].

914C Ἢ ὅτι παντελῶς μὲν ἔρημος οὐκ ἔστι γλυκύτητος οὐδ᾽ ἄμοιρος ἡ πικρότης, ἅτε δὴ ποταμοὺς τοσούτους ὑποδεχομένης τῆς θαλάττης: For rivers flowing into the sea, thus rendering it more drinkable, cf. also, e.g., Pliny, *NH* 4, 79 and 6, 51. In *Q.N.* 1, 911E, Plutarch formulates a similar idea (viz. that rivers deposit silt into the sea).

914C τοῦ δ᾽ ἡλίου τὸ γλυκὺ καὶ πότιμον ἐξαιροῦντος ὑπὸ κουφότητος ἐπιπολάζον: The idea that the sun evaporates the sweet drinkable constituent in seawater recurs in *Q.N.* 40. Cf. also Hipp., *Aer.* 8, Arist., *Mete.* 355a32ff., Ps.-Arist., *Probl.* 934b27ff., Ps.-Alex. Aphr., *Probl.* 1, 55 (J.L. Ideler, 1841, p. 19, 23–25), Pliny, *NH* 2, 222.

914C τοῦτο δ᾽ ἡσυχῇ καὶ τοῖς ποτίμοις συμβέβηκε· θέρους γὰρ πονηρότερα γίνεται, τὸ κουφότατον καὶ γλυκύτατον τοῦ θερμοῦ διαφοροῦντος, ἐν δὲ χειμῶνι νέον ἐπιρρεῖ καὶ πρόσφατον: For the idea that burning heat spoils good drinking water, cf. also *Q.N.* 1, 911E (with the commentary *ad loc.*). Plutarch does not specify where fresh new water (νέον καὶ πρόσφατον) comes from in winter (perhaps from the mountains or the Northern regions?).

Q.N. 10, 914DE

In *Q.N.* 10, Plutarch deals with yet another problem related to the topic of seawater, this time in combination with that of wine (cf. also *Q.N.* 27, 30–31). He wonders **why people pour seawater into wine, and why those who live far away from the sea put baked gypsum from Zacynthus into it** (Διὰ τί τῷ οἴνῳ θάλασσαν παραχέουσι [...] οἱ δὲ πόρρω θαλάττης ἐμβάλλουσι γύψον Ζακυνθίαν ὀπτήσαντες;). In parenthesis, Plutarch refers to a story told by the people of Halieis (or by fishermen?) who say that they received an oracle instructing them to dip Dionysus into the sea (χρησμόν τινα λέγουσιν Ἁλιεῖς (ἁλιεῖς?) κομισθῆναι προστάττοντα βαπτίζειν τὸν Διόνυσον πρὸς τὴν θάλατταν). Two explanations are given: the first focuses on the heat of seawater, the second on its earthy constituent.

The **first** explanation is based upon an opposition between heating and cooling. Plutarch argues that heat (which is natural to seawater) is an aid against cooling, which, by itself, does more than anything to change the wine by extinguishing and destroying its (hot) faculty (Πότερον ἡ θερμότης βοηθεῖ πρὸς τὴν περίψυξιν, ἢ δι' αὑτῆς ἐξίστησι μάλιστα τὸν οἶνον ἀποσβεννύουσα καὶ φθείρουσα τὴν δύναμιν;).

In the **second** explanation, the elementary composition of wine *vis-a-vis* that of seawater is central. Plutarch argues that earthy substances, which naturally bring forth fixation and reduction, fasten the watery and breathlike parts of the wine, which are most inclined to suffer change (ἢ τὸ ὑδατῶδες καὶ πνευματῶδες τοῦ οἴνου πρὸς μεταβολὴν ἐπισφαλέστατ' ἔχον ἵστησι τὰ γεώδη πεφυκότα στύφειν καὶ κατισχναίνειν). By refining and dissolving foreign and superfluous parts (c.q. in the wine), the salty (c.q. earthy) crystals in the seawater prevent the development of unpleasant odours or putrefaction (οἱ δ' ἅλες μετὰ τῆς θαλάττης λεπτύνοντες καὶ ἀποτήκοντες τὸ ἀλλότριον καὶ περιττὸν οὐκ ἐῶσι δυσωδίαν οὐδὲ σῆψιν ἐγγίνεσθαι). At the end, Plutarch adds that all that is thick and earthy (c.q. in wine) is entangled and dragged down along with the heavier parts (c.q. in the seawater) to form a sediment, the lees, leaving the wine itself clear (πρὸς δὲ τούτοις, ὅσον ἐστὶ παχὺ καὶ γεῶδες, ἐμπλεκόμενον τοῖς βαρυτέροις καὶ συγκατασπώμενον ὑποστάθμην ποιεῖ καὶ τρύγα τὸν δ' οἶνον ἀπολείπει καθαρόν).

914D Διὰ τί τῷ οἴνῳ θάλασσαν παραχέουσι [...] οἱ δὲ πόρρω θαλάττης ἐμβάλλουσι γύψον Ζακυνθίαν ὀπτήσαντες;: The custom of adding seawater or gypsum to wine, common in Antiquity, is widely attested in the literature (see, e.g., Theophr., *Lap.* 67, Ath., *Deipn.* 1, 26b; 31f.; 32de; 33b, Pliny, *NH* 14, 73–75; 78; 120; 126, Pallad., *Op. agr.* 11, 14; 17; 21). The admixture of salt or seawater was considered typical of Greek wines, but some Romans also adopted this custom. Cf. Cato, *De agr.* 24; 104–106 (with a recipe on making seawater), Col., *De re rust.* 12, 21–22; 12, 37, Plautus, *Rudens* 588.

914D χρησμόν τινα λέγουσιν Ἁλιεῖς (ἁλιεῖς?) κομισθῆναι προστάττοντα βαπτίζειν τὸν Διόνυσον πρὸς τὴν θάλατταν: This report reminds the reader of the type of cultural-antiquarian inquiry in Plutarch's *Quaestiones Graecae* (cf. also *Q.N.* 14, 915C and 23, 917F [see 2.4.2.]). We may be dealing with a reference to the ritual submerging of the statue of Dionysus into the sea (see L. Senzasono, 2006, p. 180, n. 66). In Hom., *Il.* 6, 136, there is the story of Dionysus finding his refuge in the bosom of the sea nymph Thetis. In Eustathius' commentary to this passage, this is interpreted allegorically as a riddle about the usefulness of seawater for preserving wine (629, 63–64; cf. also 871, 36–38, Ath., *Deipn.* 1, 26b, Ps.-Heracl., *Quaest. Hom.* 35; for a similar riddle, cf. *Quaest. conv.* 716F–717A). As to the addressees of the χρησμός mentioned in Plutarch's *quaestio*, F.H. Sandbach, 1965, pp. 172–173 (nn. 6 and d) corrected the reading of the manuscript ἁλιεῖς into Ἁλαιεῖς ('inhabitants of Halae', a deme located on the north-east coast of Attica), but perhaps Ἁλιεῖς is more plausible ('inhabitants of Halieis', a city on the coast of Argolis, where Dionysus received the epiclesis Ἁλίευς; see O. Jessen, 1912). Sandbach relies on von Wilamowitz' corrections of the T-scholium to the Homeric passage at hand (*Il.* 6, 136), which is ascribed to Philochorus (FGrHist 328, 191) and may very well be Plutarch's source: χρησμὸς ἐδόθη Ἁλαιεῦσιν (von Wilamowitz, vs. Lobeck's emendation Ἁλιεῦσιν) ἐν πόντῳ (Tümpel's emendation for τόπῳ) Διόνυσον Ἁλαιέα (von Wilamowitz, vs. Tümpel's emendation Ἁλιέα) βαπτίζοιτε, ὡς Φιλόχορος. See C.A. Lobeck, 1829, p. 1088, von Wilamowitz *apud* E. Maaß, 1887, p. 210 and K. Tümpel, 1889. Tümpel (with Lobeck) prefers the reading of Ἁλιεῦσιν and Ἁλιέα. Considering the manuscipt reading of Plutarch's *quaestio* at hand (ἁλιεῖς), I am inclined to follow this reading. In that case, Plutarch refers to the 'inhabitants of Halieis', rather than 'the people of Halae' (paleographically, Ἁλιεῖς is closer to the manuscript reading than Ἁλαιεῖς). In any case, the reference is probably not just to unspecified 'fishermen', since the city or its inhabitants to which an oracle is given is commonly mentioned by name in ancient Greek literature (cf. L. Senzasono, 2006, p. 180, n. 66). Moreover, the Ἁλιεῖς were often considered fishermen from their name, which may explain the confusion here. Cf. Strabo, *Geogr.* 8, 6, 12 (Ἁλιεῖς λεγόμενοι θαλαττουργοί τινες ἄνδρες), Ephorus, FGrHist 70, 56 (ἐλέγοντο [sc. Ἁλιεῖς] δ' οὕτως διὰ τὸ πολλοὺς τῶν Ἑρμιονέων ἁλιευομένους κατὰ τοῦτο τὸ μέρος οἰκεῖν τῆς χώρας).

914D Πότερον ἡ θερμότης βοηθεῖ πρὸς τὴν περίψυξιν, ἢ δι' αὐτῆς ἐξίστησι μάλιστα τὸν οἶνον ἀποσβεννύουσα καὶ φθείρουσα τὴν δύναμιν;: The δύναμις of wine is in its fiery heat (cf., e.g., *Q.N.* 31, 919CD and *Quaest. conv.* 701F), which is destroyed by cooling. In *Quaest. conv.* 652B–653B, by contrast,

Plutarch argues *ex tempore* that the δύναμις of wine is cold. According to Aristotle and Theophrastus, some wines are hot by nature while others are colder (fr. 221 Rose). They also hold that the innate heat in wine is due to putrefaction (fr. 222 Rose): this heat is, so to say, 'killed' by vinegar, because the vinous parts in wine become cold when the wine turns into vinegar, while the watery residue putrefies and receives an additional amount of heat, like all putrefied things (in this sense, vinegar is constituted by parts that are opposed by their properties, viz. from cooled and heated substances, as is also the case with all ashes of burned wood). In regards to the excessive coldness of vinegar, which is even said to be able to extinguish fire, cf. also *Quaest. conv.* 652F.

914D ἢ τὸ ὑδατῶδες καὶ πνευματῶδες τοῦ οἴνου πρὸς μεταβολὴν ἐπισφαλέστατ' ἔχον ἵστησι τὰ γεώδη πεφυκότα στύφειν καὶ κατισχναίνειν: F.H. Sandbach, 1965, p. 175, n. d is probably correct in pointing out that "[t]he vaporous [c.q. breathlike] elements tend to evaporate, the watery ones to putrefy". An allusion to these two processes (of evaporation and putrefaction) can be found in the second part of the sentence that follows, viz. in the concepts of δυσωδία and σῆψις respectively. Note, moreover, that the verb κατισχναίνειν is related to the reduction and weakening of odours (see LSJ, s.v.). For the tenuity (λεπτότης) of πνεῦμα and the putrefactive quality (σήψεις) of (rain)water, cf. *Q.N.* 2, 912B (cf. also *Q.N.* 33 for the bad quality, *qualitas aliquis mala*, of stagnant waters).

914D οἱ δ' ἅλες μετὰ τῆς θαλάττης λεπτύνοντες καὶ ἀποτήκοντες τὸ ἀλλότριον καὶ περιττὸν οὐκ ἐῶσι δυσωδίαν οὐδὲ σῆψιν ἐγγίνεσθαι: The idea that salt is a safeguard against putrefaction is paralleled in *Q.N.* 1, 911D and 40; cf. also *Quaest. conv.* 685BC. For the melting (τήκειν) and thinning (λεπτύνειν) property of salt, cf. *Q.N.* 3, 912E (with the commentary *ad loc.*).

֍

Q.N. 11, 914EF

Q.N. 11 deals with the pathological condition of **seasickness caused by navigation, which, so Plutarch writes, occurs on the sea rather than on rivers, even when the weather is calm** (Διὰ τί μᾶλλον ναυτιῶσι τὴν θάλατταν πλέοντες ἢ τοὺς ποταμούς, κἂν ἐν γαλήνῃ πλέωσι;). One explanation is given, which is based on the opposition between strange and familiar smells. The aetiology draws on a close relation between human physiology and psychology showing how our senses (c.q. of smell) stir our soul into an erroneous emotion (c.q. of fear of things to come), and how the body reacts on this erroneous impulse, thereby leading to sickness.

Plutarch explains that of all the sensations, that of smell, and of all emotions, that of fear, is the most conducive to seasickness (ἡ μάλιστα ναυτίαν κινεῖ τῶν αἰσθήσεων ἡ ὄσφρησις, τῶν δὲ παθῶν ὁ φόβος;). After all, when people imagine some danger, they tremble and shiver, and their bowels turn to water (καὶ γὰρ τρέμουσι καὶ φρίττουσι καὶ κοιλίας ἐξυγραίνονται φαντασίαν κινδύνου λαβόντες). Neither of these causes (viz. smell or fear), though, worries those who travel by river, because everyone's smell is accustomed to the drinkable and fresh river water, and the passage is riskless (τούτων δ' οὐδέτερον ἐνοχλεῖ τοῖς διὰ ποταμοῦ πλέουσιν· ἡ γὰρ ὄσφρησις παντὶ ποτίμῳ καὶ γλυκεῖ συνήθης ἐστὶν ὁ δὲ πλοῦς ἀκίνδυνος). At sea, however, people feel uncomfortable by the strange smell, and they are afraid of what may become, since they distrust the present situation (ἐν δὲ τῇ θαλάττῃ τήν τ' ὀσμὴν ἀηθείᾳ δυσχεραίνουσι καὶ φοβοῦνται, μὴ πιστεύοντες τῷ παρόντι, περὶ τοῦ μέλλοντος). Thus the calm in their surroundings does not offer any help. On the contrary, the disturbance and wavering of the soul stirs up the body and infects it with turmoil (οὐδὲν οὖν ὄφελος τῆς ἔξω γαλήνης, ἀλλὰ ἡ ψυχὴ σάλον ἔχουσα καὶ θορυβουμένη συγκινεῖ καὶ ἀναπίμπλησι τὸ σῶμα τῆς ταραχῆς).

914E Διὰ τί μᾶλλον ναυτιῶσι τὴν θάλατταν πλέοντες ἢ τοὺς ποταμούς, κἂν ἐν γαλήνῃ πλέωσι;: The theme of this chapter ties in closely with that of Book 27 of Ps.-Aristotle's *Problems*, where the emotions of fear and courage are treated in terms of physiological processes in the body (ὅσα περὶ φόβον καὶ ἀνδρείαν). For the etymological link between ναυτία (seasickness) and the world of navigation, see *Quaest. conv.* 694B (ὡς δὲ ναυτιᾶν ὠνομάσθη μὲν ἐπὶ τῶν ἐν νηὶ καὶ κατὰ πλοῦν τὸν στόμαχον ἐκλυομένων, ἔθει δ' ἴσχυκεν ἤδη καὶ κατὰ τῶν ὁπωσοῦν τοῦτο πασχόντων ὄνομα τοῦ πάθους εἶναι, κτλ.). Cf. also Pl., *Leg.* 639b.

914E καὶ γὰρ τρέμουσι καὶ φρίττουσι καὶ κοιλίας ἐξυγραίνονται φαντασίαν κινδύνου λαβόντες: Similarly, in *Arat.* 29, 6, the phenomenon of people's bowels turning to water (τὴν κοιλίαν ἐξυγραίνεσθαι) in the presence of seeming peril, is ascribed to cowardice or some faulty temperament and chilliness in the body (this last point recalls the phrase τρέμουσι καὶ φρίττουσι here) [quoted 3.1.1.]. For the connection between seasickness and fear, cf. also Ps.-Arist./Alex. Aphr., *Suppl. probl.* 2, 106 (S. Kapetanaki and R.W. Sharples, 2006, p. 2 report a lack of parallels in the medical literature for this problem, but *Q.N.* 11 may provide an exception).

914F ἐν δὲ τῇ θαλάττῃ τήν τ' ὀσμὴν ἀηθείᾳ δυσχεραίνουσι καὶ φοβοῦνται: By contrast, in *De prim. frig.* 954C, Plutarch argues that living by the sea in the winter provides a welcome refuge from living on land, because we can wrap ourselves in the comfortable and warm salty sea air there.

914F οὐδὲν οὖν ὄφελος τῆς ἔξω γαλήνης, ἀλλὰ ἡ ψυχὴ σάλον ἔχουσα καὶ θορυβουμένη συγκινεῖ καὶ ἀναπίμπλησι τὸ σῶμα τῆς ταραχῆς: The same imagery, where the upset in the human soul is compared with that of the sea, recurs in *De tranq. an.* 475EF, *Demetr.* 38, 4 and *Per.* 33, 5 (see F. Fuhrmann, 1964, p. 50). For the physiological implications of psychological θόρυβος and its metaphorical value in light of the decorum of the symposium, see M. Vamvouri Ruffy, 2011, pp. 135–139 (with a reference to the problem at hand).

ಜು

Q.N. 12, 914F–915B

In *Q.N.* 12, Plutarch examines **why oil that is sprinkled on seawater causes clearness and calm** (Διὰ τί τῆς θαλάττης ἐλαίῳ καταρραινομένης γίνεται καταφάνεια καὶ γαλήνη;). The problem ties in closely with the general theme of seawater in the previous problems; and on a more minute level, the opposition γαλήνη – σάλον mentioned in the conclusion to the previous problem (*Q.N.* 11, 914F) will recur in the first *causa* here. The aetiology at hand contains three explanations that are closely interconnected with each other. The first two explanations alternately deal with the exterior and interior effects of oil on the sea (the first one deals with the aspect of γαλήνη, the second with that of καταφάνεια), and the final explanation provides an answer for both of these effects together (in doing so, it combines elements from the preceding two explanations).

The **first** explanation, which Plutarch borrows from Aristotle, is kept fairly brief: he says that the wind, slipping off the smoothness (so caused by the oil), makes no impact and raises no surge (Πότερον, ὡς Ἀριστοτέλης φησί, τὸ πνεῦμα τῆς λειότητος ἀπολισθαῖνον οὐ ποιεῖ πληγὴν οὐδὲ σάλον;).

In the **second** explanation, Plutarch highlights the incompleteness of Aristotle's theory, which he considers to be plausible, but only so regarding the external aspect of the phenomenon (ἢ τοῦτο μὲν πιθανῶς εἴρηται πρὸς τὰ ἐκτός). He then draws attention to the internal, submarine aspect of the phenomenon, by referring to the popular account that divers take oil into their mouth and blow it out in the depths, so that they get light and transparency in the water (ἐπεὶ δέ φασι καὶ τοὺς κατακολυμβῶντας, ὅταν ἔλαιον εἰς τὸ στόμα λαβόντες ἐκφυσήσωσιν ἐν τῷ βυθῷ φέγγος ἴσχειν καὶ δίοψιν). As Plutarch notes, it is of course impossible to allege slipping of the wind as the cause there (οὐκ ἔστιν ἐκεῖ πνεύματος ὄλισθον αἰτιάσασθαι). He explains that because of its density, the oil (in its movement out of the divers' mouth) pushes and forces aside the sea, which is earthy and irregular. When the sea flows back to itself and draws together afterward, intermediate passages are left, which provide transparency and clearness to the eyes (σκόπει δὴ μὴ τὴν θάλατταν γεώδη καὶ ἀνώμαλον οὖσαν ἐξωθεῖ καὶ διαστέλλει τῇ πυκνότητι τὸ ἔλαιον, εἶτ' ἀνατρεχούσης εἰς αὐτὴν καὶ συστελλομένης ἀπολείπονται πόροι μεταξὺ ταῖς ὄψεσι διαύγειαν καὶ καταφάνειαν διδόντες).

By combining the internal and the external aspect of the problem in the **third** and final *causa*, Plutarch tries to explain both the effects of καταφάνεια and γαλήνη as mentioned in the *quaestio*. He starts by pointing out that the air that is mixed with the seawater (internally, but probably close to the surface of the sea, where the sunbeams can reach it) is naturally full of light because of its heat (which explains the sea's transparency), but that it becomes irregular (ἀνώμαλος, cf. the previous *causa*) and dark when it is disturbed (ἢ φύσει μέν ἐστι φωτεινὸς ὑπὸ θερμότητος ὁ τῇ θαλάττῃ καταμεμιγμένος ἀήρ, γίνεται δὲ ταραχθεὶς ἀνώμαλος καὶ σκιώδης). Therefore, when the oil with its density (πυκνότητι, cf. the previous *causa*) smoothes out the irregularity of the surface (externally, cf. *causa* one), the air regains its evenness and transparency (ὅταν οὖν τὴν ἀνωμαλίαν ἐπιλεάνῃ πυκνότητι τὸ ἔλαιον, ἀπολαμβάνει τὴν ὁμαλότητα καὶ τὴν διαύγειαν;).

914F Διὰ τί τῆς θαλάττης ἐλαίῳ καταρραινομένης γίνεται καταφάνεια καὶ γαλήνη;: Theophylactus Simocatta raises the same problem (*Quaest. phys.* 7: J.L. Ideler, 1841, p. 175, 3–24), and the solution he provides is very close to the one Plutarch attributes to Aristotle in the first *causa* here (for Theophylactus' sources, see M. Marcovich, 1954). The problem also recurs in *De prim. frig.* 950BC, albeit in an adapted form and somewhat in passing (Aristotle's account is again quoted and criticised). Plutarch there argues that, among other liquids, oil is the most transparent (διαφανές), because it contains the most air (and *not* the sea, as here in the third *causa*; contrary also to *Quaest. conv.* 696B and 702BC). This is evidenced by the oil's lightness (κουφότης), which causes it to remain on the surface (ἐπιπολάζει) of all things (c.q. liquids), being carried up by the air. Plutarch notes that oil calms the sea when it is sprinkled on the waves (ποιεῖ δὲ καὶ τὴν γαλήνην ἐν τῇ θαλάττῃ τοῖς κύμασιν ἐπιρραινόμενον). He criticises Aristotle's explanation (cf. *causa* one), according to which the winds slip off the surface of the sea because of the oil's smoothness. In fact, so Plutarch argues, the waves are dissipated when they are hit by *any* liquid (οὐ διὰ τὴν λειότητα τῶν ἀνέμων ἀπολισθαινόντων, ὡς Ἀριστοτέλης ἔλεγεν, ἀλλὰ παντὶ μὲν ὑγρῷ τὸ κῦμα διαχεῖται πληττόμενον). This last point is not repeated in *Q.N.* 12, but the focus on both the internal and the external aspect of the phenomenon remains. It is a typical feature of oil, so Plutarch continues, that it provides light and transparency (αὐγὴν καὶ καταφάνειαν) at the bottom of the sea, because the liquids there are dispersed by the air contained by the oil (διαστελλομένων τῷ ἀέρι). F.H. Sandbach, 1965, p. 178, n. a is probably correct that Plutarch is implying that "the oil contains much air (hence its lightness), which provides the transparency". Plutarch believes that it *literally* gives light, not only on the surface for those who pass the night at sea, but also below the surface for spongedivers when they blow it out of their mouth (κάτω τοῖς σπογγοθήραις διαφυσώμενον ἐκ τοῦ στόματος ἐν τῇ θαλάττῃ, cf. the second *causa* in *Q.N.* 12). There are clear parallels with the aetiology in *Q.N.* 12, but also some slight differences. Importantly, the problem of oil poured on the surface of the sea is not central in *De prim. frig.* 950BC. What is central there is the (Empedoclean and anti-Stoic) theory that water, rather than air, is the principle of cold (οὐ μᾶλλον οὖν τῷ ἀέρι τοῦ μέλανος ἢ τῷ ὕδατι μέτεστιν, ἧττον δὲ τοῦ ψυχροῦ). It is in this context that the phenomenon is mentioned. The fact that oil is most transparent and gives light to divers supports the idea that it contains much air. Water, by contrast, contains darkness, and is, therefore, cold (darkness being related to coldness: cf. *De prim. frig.* 948E). In *Q.N.* 12, however, the natural phenomenon is treated separately and on its own terms, so that the aetiology is more elaborated and systematic (the same conclusion can be drawn for the parallel between *De prim. frig.* 950AB and *Q.N.* 39 about the darkness of deep water: see the commentary *ad loc.*).

914F Πότερον, ὡς Ἀριστοτέλης φησί, τὸ πνεῦμα τῆς λειότητος ἀπολισθαῖνον οὐ ποιεῖ πληγὴν οὐδὲ σάλον;; No clear parallel can be found for this quotation in Aristotle's works (nor for the similar one in *De prim. frig.* 950B). Neither is it included among Aristotle's fragments, but *Q.N.* 12 and several of its parallel passages are mentioned in V. Rose, 1863, p. 219. As we will see below, there is, indeed, a large amount of parallel material, in 1) Ps.-Aristotle's *Problems* and in 2) the *Supplementary problems* ascribed to Ps.-Aristotle/Alexander of Aphrodisias. Therefore, we may be dealing in *Q.N.* 12 with the remains from a lost Ps.-Aristotelian problem (cf. F.H. Sandbach, 1965, p. 177, n. a and 1982, p. 227). It is not unlikely either, though, that Plutarch is simply reorganising some of this material in *Q.N.* 12, rather than that he is inaccurate in reproducing it (*pace* L. Senzasono, 2006, pp. 185–186, n. 74). 1) A passage similar to the opening problem is found in *Probl.* 935b17–27 (concerning the greater transparency of seawater *vis-à-vis* fresh water; cf. also *Probl.* 932b8–24). It is argued there, but not further explained, that oil after being poured on the water renders it more transparent (τὸ δὲ ἔλαιον ἐπιχυθὲν ποιεῖ μᾶλλον εὐδίοπτον). Another parallel is found in *Probl.* 961a18–23 (regarding oil being poured in the ear to clear water from it), where we read that oil lies on the surface of the water (ἐπιπολῆς τοῦ ὕδατος), that the water is held to it due to is stickiness (διὰ γλισχρότητα), and that it is smooth so that it causes slippage (λεῖον ὂν ποιεῖ ὀλισθαίνειν). Similarly, in *Probl.* 961a24–30 (concerning the ears of divers breaking less easily when oil has been poured in them), it is argued, in very similar wording as in Plutarch, that oil causes the seawater in the ears to slip off (ἀπολισθαίνειν ποιεῖ), so that it cannot produce a choc in the ears (πληγήν οὐ ποιεῖ). In the Aristotelian *causa* in *Q.N.* 12, 914F, however, the context is different and the subject of ἀπολισθαῖνον οὐ ποιεῖ πληγὴν is not the seawater but the wind (according to Senzasono, Plutarch may be rendering this account in an inexact way due to the inaccuracy of his memory or personal notes, but perhaps he is just rephrasing it). 2) The topic of the sea's calmness being produced by the oil recurs in Ps.-Arist./Alex. Aphr., *Suppl. probl.* 3, 29 and 47 (Διὰ τί τὸ ἔλαιον τῇ θαλάσσῃ ἐπιχεόμενον / ἐπιρραινόμενον γαλήνιᾳ / γαλήνην ποιεῖ;). According to V. Rose, 1863, p. 219, *Suppl. probl.* 3, 47 derives from 3, 29 (he says that the latter is 'used' for the former), but this is questionable (cf. S. Kapetanaki and R.W. Sharples, 2006, p. 277, n. 563). In any case, two different answers are given: in *Suppl. probl.* 3, 29, it is argued that the oil, being smooth, causes the *wind* to slip off the surface of the sea such that it does not cause the watter to ripple (ἢ ὅτι λεῖόν ἐστιν; ἀπολισθαίνει τοίνυν τὸ πνεῦμα, καὶ οὐ ποιεῖ φρίκην), whereas in *Suppl. probl.* 3, 47 it is argued that the oil, which is sticky and moist, causes the *waves* to slip over each other and lose their rapid motion and impetus (ὅτι γλίσχρον ὂν καὶ ὑγρὸν διολισθαίνειν ποιεῖ τὰ κύματα πρὸς ἑαυτὰ καὶ ἐνδιδόναι τῆς οἰκείας ὁρμῆς καὶ φορᾶς). The first explanation is closest to the Aristotelian *causa* in *Q.N.* 12 and the second to *Probl.* 961a24–30 mentioned above. According to H. Flashar, 1962, p. 360, n. 1 and p. 740 (cf. also R. Mayhew, 2011a, p. 351, n. 14), *Q.N.* 12, 915A (and, I presume, especially the reference to divers in *causa* two) probably refers to Ps-Arist., *Probl.* 961a24–30 rather than to Ps.-Arist./Alex. Aphr., *Suppl. probl.* 3, 29 and 47 (he concludes that "in die Probl. ined. im einzelnen Material eingegangen [ist], das Plutarch [...] in einem dem Ar. zugeschrieben Corpus von Probl. gefunden [hat]"). However, S. Kapetanaki and R.W. Sharples, 2006, p. 12 (cf. also R.W. Sharples, 2006, p. 25, n. 17) argue, on the contrary, that *Q.N.* 12 (and, I presume, especially the *quaestio* and *causa* one) probably refers to *Suppl. probl.* 3, 47. They argue 1) that "the question in Plutarch, and the point for which Aristotle is cited, is the general one raised in 3.47 rather than the specific point about divers in the Bekker *Problem* [= *Probl.* 961a24–30]", and 2) that "Plutarch goes on to attack the claim of those (plural) who say that the same explanation applies to divers". 1) In regards to the first point, however, it seems that *Suppl. probl.* 3, 29 is closer (than 3, 47) to Plutarch's formulation of the *quaestio* and the first *causa* of *Q.N.* 12. 2) In regards to the second point, it should be noted that Plutarch does not argue that 'those people' – Kapetanaki and Sharples are clearly referring to the subject in φασι – claim that "the same [c.q. Aristotelian, i.e. external] explanation applies to divers"; rather, he adduces this popular account in order to add a new point, which introduces another (c.q. internal) perspective. This is not meant as an attack on 'those people' but is part of Plutarch's criticism of Aristotle (who only provides an external explanation). They also add 3) that the point that Plutarch makes about the oil causing the wind to slip over the waves in *De prim. frig.* 950B probably refers to *Suppl. probl.* 3, 29, but the same is actually true, as we just saw, for the *quaestio* and the first *causa* of *Q.N.* 12.

The most problematic aspect of this kind of *Quellenforschung*, however, is that it downplays Plutarch's exploratory attitude to received knowledge and his own innovative contributions to the problem. It is not unlikely that Plutarch joins together material that he found in, or generally remembers from, a collection of problems that is quite different from the collections we have today [see 1.1.3., n. 87]. Moreover, if we compare the argumentative strategies in his other problems, it can only be expected that Plutarch not only rearranged this material afresh, but also added his own original insights and comments (as may be marked by the imperative σκόπει δὴ μή in the second *causa* [see 1.1.4., n. 111]).

915A ἢ τοῦτο μὲν πιθανῶς εἴρηται πρὸς τὰ ἐκτός, ἐπεὶ δέ φασι καὶ τοὺς κατακολυμβῶντας, ὅταν ἔλαιον εἰς τὸ στόμα λαβόντες ἐκφυσήσωσιν ἐν τῷ βυθῷ φέγγος ἴσχειν καὶ δίοψιν, οὐκ ἔστιν ἐκεῖ πνεύματος ὄλισθον αἰτιάσασθαι;: For the remark about divers, cf. *De prim. frig.* 950B. Cf. also Pliny, *NH* 2, 234 and Opp., *Hal.* 5, 638–648.

915A σκόπει δὴ μὴ τὴν θάλατταν γεώδη καὶ ἀνώμαλον οὖσαν ἐξωθεῖ καὶ διαστέλλει τῇ πυκνότητι τὸ ἔλαιον, εἶτ' ἀνατρεχούσης εἰς αὐτὴν καὶ συστελλομένης ἀπολείπονται πόροι μεταξὺ ταῖς ὄψεσι διαύγειαν καὶ καταφάνειαν διδόντες: The earthy and irregular character of seawater (γεώδη καὶ ἀνώμαλον, cf. *Q.N.* 1, 911D) hinders clear vision into it. The density (πυκνότης) of the oil, on the other hand, probably refers to its lack of pores (cf. *Quaest. conv.* 696AB, 702BC) so that it cannot mingle with water but pushes it aside (cf. F.H. Sandbach, 1965, p. 178, n. b). In so doing, it creates pores in the seawater, providing transparency and clearness to our sight. Sandbach is right that "[t]his explanation applies particularly to the submarine phenomenon". The reader may object, therefore, that this second *causa* is equally deficient as the first one. It only explains the aspect of καταφάνεια and not also that of γαλήνη, as mentioned in the *quaestio*. Plutarch will make an attempt to explain both of these aspects at the same time in the third *causa*.

915A ἢ φύσει μέν ἐστι φωτεινὸς ὑπὸ θερμότητος ὁ τῇ θαλάττῃ καταμεμιγμένος ἀήρ: Cf. *Q.N.* 39, for the idea that surface water, as opposed to water in the depths, takes on the clearness of the sunbeams, which renders it white in colour.

915AB ὅταν οὖν τὴν ἀνωμαλίαν ἐπιλεάνῃ πυκνότητι τὸ ἔλαιον, ἀπολαμβάνει τὴν ὁμαλότητα καὶ τὴν διαύγειαν;: As F.H. Sandbach, 1965, p. 179, n. c notes, we are primarily dealing with the effects of oil *on* the surface of the sea in this third *causa* (cf. ἐπιλεάνῃ). However, Plutarch also clearly refers to the air mixed *with* the seawater (cf. καταμεμιγμένος, see the previous comment). This is not necessarily incompatible, if we may assume that

Plutarch is referring to the top-layer of the seawater including the sea surface itself (he thus combines both the external and the internal aspects of the phenomenon at the same time). What Plutarch is basically arguing, then, is that, on account of its density, oil raises no swelling of the waves (cf. *causa* one), but smoothes out the sea and, thus, the air contained in it, leading it to become calm and, by implication, transparent. The words τὴν ὁμαλότητα καὶ τὴν διαύγειαν at the end of the explanation imply basically the same thing as γαλήνη and καταφάνεια respectively, as mentioned in the *quaestio* (as such, the aetiology is 'complete' at this point, resulting in an argumentaive ring [see 4.3.3.3.]). Psellus in *De omn. doctr.* §169, 7–10 Westerink rewrites this sentence completely. He agrees, however, that oil on the surface is under scrutiny (externally) as well as the seawater itself and the air it contains (internally): ἢ ὅτι τὸ ἔλαιον ὁμαλώτατον καὶ λιπαρώτατον ὄν, ἐπιχεόμενον τῇ θαλάσσῃ διασκίδνησι τὸν ἐν αὐτῇ ζοφερὸν ἀέρα καὶ λαμπρότατον ἀπεργάζεται, γαλήνην δὲ ἐμποιεῖ ἐπιπλέον ἄνωθεν καὶ οὐκ ἐῶν κάτωθεν αὐτὴν ἀναβράττεσθαι. See also M. Meeusen, 2012b, p. 112, n. 54.

Q.N. 13, 915BC

In *Q.N.* 13, Plutarch deals with a last problem concerning seawater. He wonders **why fishermen's nets decompose more in the winter than in the summer, whereas the opposite is true of other objects** (διὰ τί χειμῶνος μᾶλλον ἢ θέρους τὰ τῶν ἁλιέων σήπεται δίκτυα, καίτοι τά γ' ἄλλα μᾶλλον ἐν τῷ θέρει τοῦτο πάσχει;). Two explanations are given: the first draws attention to the process of ἀντιπερίστασις, the second to a process that is said to be similar to, but not to be mistaken for, σῆψις (this certainly puts the σήπεται from the *quaestio* into perspective [see 4.3.3.3.]).

The **first** explanation is based on the idea that the process of putrefaction is triggered by heat being concentrated in the sea. It opens with a quotation from Theophrastus, who believes that heat, withdrawing under the cold, is concentrated and makes the water in the depths of the sea warmer, as is also true of the (interior of the) earth (Πότερον, ὡς Θεόφραστος οἴεται, τῷ ψυχρῷ τὸ θερμὸν ὑποχωροῦν ἀντιπεριίσταται καὶ θερμότερα ποιεῖ τὰ ἐν βάθει τῆς θαλάττης, ὥσπερ τῆς γῆς;). Regarding the analogy between the interior part of the sea and that of the earth, Plutarch explains that springwaters are also warmer in the winter. He adds that ponds and rivers emit more vapour, because the heat is enclosed in the depths, as it is mastered by the cold (διὸ καὶ τὰ πηγαῖα τῶν ὑδάτων χλιαρώτερα τοῦ χειμῶνός εἰσι καὶ μᾶλλον ἀτμίζουσιν αἱ λίμναι καὶ οἱ ποταμοί· κατακλείεται γὰρ εἰς βάθος ἡ θερμότης ὑπὸ τοῦ ψυχροῦ κρατήσαντος).

In the **second** explanation, Plutarch makes a subtle distinction by reinterpreting the process of σῆψις mentioned in the *quaestio*. He argues that the nets do not actually rot but undergo some kind of a process that is very similar to rotting and putrefaction when they become rough and frozen, as they are desiccated by the cold and violently frayed by the waves (ἡ σῆψις μὲν οὐκ ἔστι τῶν δικτύων, ὅταν δὲ φρίξῃ καὶ παγῇ διὰ τὸ ψῦχος ἀναξηραινόμενα, καὶ θρυπτόμενα μᾶλλον ὑπὸ τοῦ κλύδωνος, σήψει τι καὶ μυδήσει πάσχει παραπλήσιον;). Plutarch adds that the nets also suffer more damage by the cold, as they are pulled to pieces like overstrained chords. This is due to the fact that the sea is repeatedly agitated by the wintery weather (καὶ γὰρ πονεῖ μᾶλλον ἐν κρύει, καθάπερ τὰ νεῦρα συντεινόμενα σπαράττεται, πλεονάκις ἐκταραττομένης διὰ τὸν χειμῶνα τῆς θαλάττης). This is also why fishermen, out of fear that their nets will come loose, treat them with dyes and make them more solid (διὸ καὶ στύφουσιν αὐτὰ ταῖς βαφαῖς καὶ πυκνοῦσι, φοβούμενοι τὰς ἀναλύσεις). An additional reason is that fish would not notice them (and, by implication, not be attracted to them) if they were not dyed or tinted, since the normal colour of the net is like that of air and is hardly visible in the sea (ἐπεὶ μὴ βαφέντα μηδὲ χρισθέντα μᾶλλον ἂν ἐλάνθανε τοὺς ἰχθῦς· ἐνάερον γὰρ τὸ τοῦ λίνου χρῶμα καὶ ἀπατηλὸν ἐν θαλάττῃ).

915B διὰ τί χειμῶνος μᾶλλον ἢ θέρους τὰ τῶν ἁλιέων σήπεται δίκτυα, καίτοι τά γ' ἄλλα μᾶλλον ἐν τῷ θέρει τοῦτο πάσχει;: The paradox is that while (summer) heat normally triggers putrefaction, the reverse seems to be the case for nets in the sea, whereas this is not true in the case of other objects (τά γ' ἄλλα). No further mention is made of the other objects in the aetiology, but presumably they are not located in the sea but on land, so that they are directly exposed to atmospheric temperatures.

915B Πότερον, ὡς Θεόφραστος οἴεται, τῷ ψυχρῷ τὸ θερμὸν ὑποχωροῦν ἀντιπεριίσταται καὶ θερμότερα ποιεῖ τὰ ἐν βάθει τῆς θαλάττης, ὥσπερ τῆς γῆς;: See Theophr., fr. 163 Wimmer (= 173 FHSG). The ensuing argument (διὸ καὶ κτλ.: see the following comment) is interpreted by Wimmer and FHSG as part of the quotation from Theophrastus and correctly so if we may assume that Plutarch did not add the analogy with the earth himself (ὥσπερ τῆς γῆς). This remains unclear, though. The quote may very well originate from Theophrastus' lost *De aquis* (just like the one in *Q.N.* 7, 914A, which also concerns the influence of the seasons on water; see the commentary *ad loc.*), but it may be known to Plutarch indirectly via a lost Aristotelian problem. Pliny, *NH* 2, 234 also says that the sea is warmer in winter.

915B διὸ καὶ τὰ πηγαῖα τῶν ὑδάτων χλιαρώτερα τοῦ χειμῶνός εἰσι καὶ μᾶλλον ἀτμίζουσιν αἱ λίμναι καὶ οἱ ποταμοί· κατακλείεται γὰρ εἰς βάθος ἡ θερμότης ὑπὸ τοῦ ψυχροῦ κρατήσαντος: For the working of this process (viz. of ἀντιπερίστασις) in similar natural contexts, see Arist., *Mete.* 348b2–5,

Theophr., *Ign.* 16, Ps.-Alex. Aphr., *Probl.* 1, 56 (J.L. Ideler, 1841, pp. 19, 32–20, 4), Oenopides of Chios, DK41A11 (drawn from Ps.-Aristotle's *De Nilo*, fr. 248 Rose), Cic., *De nat. deor.* 2, 25, Sen., *NQ* 6, 13, 2–3 (= Strato, fr. 89 Wehrli), Lucr., *De rer. nat.* 6, 840–847, Pliny, *NH* 31, 50. For vapours rising from rivers, cf. also *De prim. frig.* 951BC and fr. 75 Sandbach.

915C ἐπεὶ μὴ βαφέντα μηδὲ χρισθέντα μᾶλλον ἂν ἐλάνθανε τοὺς ἰχθῦς· ἐνάερον γὰρ τὸ τοῦ λίνου χρῶμα καὶ ἀπατηλὸν ἐν θαλάττῃ: This point is contradicted in *De soll. an.* 976EF, where Phaedimus, regarding the construction of fishing lines, points out that fishermen take care that the hairs forming the leader be as white as possible (λευκά, being the colour of seawater: cf. *Q.N.* 39). In this way, they are *less* noticeable in the sea due to their similar colour (μᾶλλον γὰρ οὕτως ἐν τῇ θαλάττῃ δι᾽ ὁμοιότητα τῆς χρόας λανθάνουσι). Notably, in what follows in *De soll. an.* 976F, there is a clear parallel to *Q.N.* 17 (on the construction of fishing lines from the hairs of horses; see the commentary *ad loc.*). Indeed, *Q.N.* 13 and 17 are closely connected to each other by their shared interest in fishing utensils. Considering the shared link with *De soll. an.* 976EF, they perhaps even rely on the same hypomnematic material. Arguably, the problems that follow in *Q.N.* 18–19 (on sea animals) may also pertain to this hypomnematic cluster (see the commentary *ad loc.*). The *concatenatio* of the problems is complicated at this point in the collection, however, by the incorporation of a cluster of three problems (*Q.N.* 14–16) regarding the natural properties of wheat and barley.

2. Wheat and barley (*Q.N. 14–16*)

The cluster of problems in *Q.N.* 14–16 deals with three agricultural questions and is concerned, more particularly, with problems related to the constitution, growth, and nourishment of wheat and barley, the Greeks' two main cereal crops. The theme of this cluster ties in with the focus on the physical differences, and especially contrarieties, between wheat and barley in Book 21 of Ps.-Aristotle's *Problems* (ὅσα περὶ ἄλφιτα καὶ μᾶζαν καὶ τὰ ὅμοια). A significant number of parallels can also be drawn with Theophrastus' botanical works, who himself on a number of occasions reflects on the physical differences between wheat and barley (e.g., *HP* 8, 4, 1–6). In *CP* 3, 21, 5, Theophrastus says that the growth of grain (including wheat and barley: cf. *HP* 8, 1, 1) depends on three universal variables: the nature of the land, the strength or weakness of the seeds and the temperature of the air. These three variables recur in the aetiologies of *Q.N.* 14–16. This does not necessarily imply, however, that Theophrastus is Plutarch's direct source, since the intermediation of a set of lost problems cannot be excluded.

Q.N. 14, 915CD

In *Q.N.* 14, Plutarch wonders **why the people of Doris pray for a bad harvest of hay** (Διὰ τί Δωριεῖς εὔχονται κακὴν χόρτου συγκομιδήν;). He provides one explanation, which boils down to the idea that they actually pray for rainfall before the summer heat.

Plutarch argues that hay is 'badly' harvested if it gets rained on (Ἡ κακῶς μὲν συγκομίζεται χόρτος ὑόμενος;). Hay is cut when it is green, not when it is dry, so it quickly putrefies if it gets soaked (κόπτεται γὰρ οὐ ξηρὸς ἀλλὰ χλωρός, ὥστε σήπεται ταχὺ διάβροχος γενόμενος). Rainfall before the summer is unfavourable for hay. Grain, however, can use it as protection against the hot southerly (summer)winds (ὑόμενος δὲ πρὸ τοῦ θέρους ὁ σῖτος βοηθεῖται πρὸς τὰ θερμὰ καὶ νότια πνεύματα). In fact, these winds do not allow the grain to grow firm as it forms in the ear, but they inhibit or reverse the hardening process by their heat, unless the earth is soaked so that there is lasting moisture to keep the ears cool and damp (ταῦτα γὰρ οὐκ ἐᾷ πυκνωθῆναι συνιστάμενον ἐν τῷ στάχυι τὸν καρπόν, ἀλλ' ἐξίστησι καὶ διαχεῖ τῇ θερμότητι τὴν πῆξιν, ἂν μὴ βεβρεγμένης τῆς γῆς ὑγρότης παραμένῃ ψύχουσα καὶ νοτίζουσα τὸν στάχυν).

915C Διὰ τί Δωριεῖς εὔχονται κακὴν χόρτου συγκομιδήν;: What this paradoxical prayer implies, so we learn from the aetiology, is that a bad harvest of hay is connected with a good harvest of grain (cf. L. Senzasono, 2006, p. 190, n. 82). This problem reminds the reader of the type of cultural-antiquarian inquiry in Plutarch's *Quaestiones Graecae* (cf. also *Q.N.* 10, 914D and 23, 917F [see 2.4.2.]). Presumably, the proper name Δωριεῖς does not refer to all the Doric speaking Greeks, but rather to the inhabitants of Doris, a small state located on the border between Thessaly and Boeotia (see F.H. Sandbach, 1965, p. 181 n. d). Doris can also refer to the region in Asia Minor, consisting of several Doric settlements and islands.

915D ὑόμενος δὲ πρὸ τοῦ θέρους ὁ σῖτος βοηθεῖται πρὸς τὰ θερμὰ καὶ νότια πνεύματα· ταῦτα γὰρ οὐκ ἐᾷ πυκνωθῆναι συνιστάμενον ἐν τῷ στάχυι τὸν καρπόν, ἀλλ' ἐξίστησι καὶ διαχεῖ τῇ θερμότητι τὴν πῆξιν, ἂν μὴ βεβρεγμένης τῆς γῆς ὑγρότης παραμένῃ ψύχουσα καὶ νοτίζουσα τὸν στάχυν: The importance of cooling rains falling *before* the summer season (πρὸ τοῦ θέρους) to foster the growth of grain is echoed in *Q.N.* 4, 913A (see the commentary *ad loc.*). For 'wind-blown' (ἐξανεμοῦσθαι) wheat and barley, cf. Theophr., *HP* 8, 10, 2–3. Theophrastus says, however, that the winds evaporate the moisture in the grains so that they dry and wither, whereas according to Plutarch, the southerly winds reverse the hardening process by their heat. Cf. also *CP* 4, 13, 4.

Q.N. 15, 915DE

In *Q.N.* 15, Plutarch wonders **why a rich, deep soil bears wheat, while thin soils are better for barley** (Διὰ τί πυροφόρος ἡ πίων καὶ βαθεῖα χώρα, κριθοφόρος δὲ μᾶλλον ἡ λεπτόγεως;). He provides one explanation, based on the distinct physical properties of wheat and barley seeds (esp. their nourishment and strength).

Plutarch argues that strong seeds need more food (which is provided by the soil in which they are planted), whereas weak ones require thin and light nourishment. He also states that barley is weaker and more open in texture, so that it will not bear much or heavy food (Ἤ ὅτι τῶν σπερμάτων τὰ ἰσχυρὰ πλείονος τροφῆς δεῖται τὰ δ' ἀσθενῆ λεπτῆς καὶ ἐλαφρᾶς, ἀσθενέστερον δ' ἡ κριθὴ καὶ μανότερον· ὅθεν οὐ φέρει τὴν πολλὴν τροφὴν καὶ βαρεῖαν;). Testimony is found in the fact that the three-month variety of wheat, which gives a lower yield and requires less nourishment, grows better in dry places, and therefore also matures sooner (μαρτυρεῖ δὲ τῷ λόγῳ τούτῳ τοῦ πυροῦ τὸν τρίμηνον ἐν τοῖς ὑποξήροις φύεσθαι βέλτιον, ἀνοστότερον ὄντα καὶ τροφῆς ἐλάττονος δεόμενον· διὸ καὶ συντελεῖται τάχιον).

915D Διὰ τί πυροφόρος ἡ πίων καὶ βαθεῖα χώρα, κριθοφόρος δὲ μᾶλλον ἡ λεπτόγεως;: For the idea that rich soil bears wheat best, whereas thin soil barley, cf. Theophr., *CP* 3, 21, 2 (Ὡς δ' ἁπλῶς εἰπεῖν ἡ μὲν λεπτὴ γῆ κριθοφόρος ἀμείνων, ἡ δὲ πίειρα πυροφόρος) and *HP* 8, 9, 1 (ὁ [πυρὸς] μὲν ἀγαθὴν ζητεῖ χώραν ἡ δὲ κριθὴ δύναται καὶ ἐν ταῖς ψαφαρωτέραις ἐκφέρειν). Cf. also Varro, *De re rust.* 1, 23–24.

915D Ἤ ὅτι τῶν σπερμάτων τὰ ἰσχυρὰ πλείονος τροφῆς δεῖται τὰ δ' ἀσθενῆ λεπτῆς καὶ ἐλαφρᾶς, ἀσθενέστερον δ' ἡ κριθὴ καὶ μανότερον· ὅθεν οὐ φέρει τὴν πολλὴν τροφὴν καὶ βαρεῖαν;: On the nourishment of both wheat and barley in relation to their strength, cf. Theophr., *CP* 4, 13, 4–5: [ἡ κριθὴ] τὸ ὅλον ἀσθενής […] [τὸ πυρὸν] πυκνότερον καὶ ἰσχυρότερον ὥστε τὴν μὲν ὀλίγης δεῖσθαι τροφῆς τὸν δὲ πλείονος […] ἡ δὲ [κριθὴ] πλείω τε ἕλκει μανὴ τὴν φύσιν οὖσα.

915DE μαρτυρεῖ δὲ τῷ λόγῳ τούτῳ τοῦ πυροῦ τὸν τρίμηνον ἐν τοῖς ὑποξήροις φύεσθαι βέλτιον, ἀνοστότερον ὄντα καὶ τροφῆς ἐλάττονος δεόμενον· διὸ καὶ συντελεῖται τάχιον: The three-month variety of wheat is sown in spring and is named after the fact that it takes only three months to grow (cf. Theophr., *HP* 8, 1, 4). In his commentary on Hes., *Op.* 486–489 (= fr. 68 Sandbach), Plutarch reports that this kind of grain owes its growth to the spring rains (cf. *Q.N.* 4, 913A and 14, 915D). For the lower nutritive demands of the three-month variety of wheat, cf. Theophr., *CP* 3, 21, 2: Τῶν δὲ πυρῶν ὁ μὲν τρίμηνος ἐν τοῖς λεπτογείοις καλλίων· σύμμετρος γὰρ ἡ τροφὴ κούφη κούφοις. Cf. also Cato, *De agr.* 35. The comparative ἀνοστότερον ('giving a lower yield') is only found here in *Q.N.* 15, 915E and in Theophr., *CP* 4, 13, 2 (cf. also 3, 21, 1: ἀνοστιμώτατα). C. Hubert, 1960, p. 14 suggests correcting ἀνοστότερον to ἀνοτιστότερον ('drier'), which is not implausible. It is perhaps safer, though, to stay with the reading of most manuscripts (cf. also L. Senzasono, 2006, pp. 190–191, n. 85), especially because it is closer also to Theophrastus' word use (ἄνοστος and ἀνόστιμος are synonymous: see LSJ, s.vv., ii). Kaltwasser's emendation (ἀνοτώτερον, 'without the south wind') is not very useful.

ം

Q.N. 16, 915EF

Q.N. 16 is closely related to the previous problem by the topic of planting wheat and barley (and more explicitly by the phrase ὡς εἰρήκαμεν in the first *causa*). Plutarch seeks to explain **the saying 'plant wheat in mud, but barley in dust'** (Διὰ τί λέγεται 'σῖτον ἐν πηλῷ φύτευε, τὴν δὲ κριθὴν ἐν κόνει'). He provides four explanations in a rapid succession. The first one focuses on the amount of food mastered by both types of grain, the second on their composition, the third on the composition of the soil, and the fourth on the damage done to wheat by ants.

In the **first** explanation, Plutarch cross-refers to what was said in the previous chapter (*Q.N.* 15, 915D). He argues that wheat can master more food, whereas barley cannot stand a great and overwhelming amount (Πότερον, ὡς εἰρήκαμεν, ὁ μὲν δύναται πλείονος τροφῆς κατακρατεῖν ἡ δ' οὐ φέρει τὸ πολὺ καὶ κατακλύζον).

According to the **second** explanation, wheat, being solid and woody, grows better in damp soil, where it is softened and moistened, whereas a drier soil at the start (i.e. when the seeds are planted?) is suitable for barley because of its open structure (ἢ πυκνὸς ὢν ὁ πυρὸς καὶ ξυλώδης φύεται βέλτιον ἐν ὑγρῷ μαλαττόμενος καὶ χυλούμενος, τῇ δὲ κριθῇ διὰ μανότητα σύμφορον ἐν ἀρχῇ τὸ ξηρότερον;).

In the **third** explanation, Plutarch argues that the composition (of the muddy soil) is suitably proportioned and harmless (to the wheat) because of the (wheat's) heat, while barley is colder (ἢ διὰ θερμότητα σύμμετρος καὶ ἀβλαβὴς ἡ κρᾶσις ψυχρότερον δ' ἡ κριθή;).

In the **fourth** and final explanation, Plutarch draws attention to farmers' fear of planting wheat in dry soil because of the damage done to it by ants, which immediately attack it (ἢ φοβοῦνται τὸν πυρὸν ἐν ξηρῷ τρίβειν διὰ τοὺς μύρμηκας, εὐθὺς γὰρ ἐπιτίθενται). On the other hand, ants are less likely to plunder barley, since their grains are difficult to lift and carry off due to their size (τὰς δὲ κριθὰς ἧττον φέρονται, δυσβάστακτοι γάρ εἰσι καὶ δυσπαρακόμιστοι διὰ μέγεθος;).

915E Διὰ τί λέγεται 'σῖτον ἐν πηλῷ φύτευε, τὴν δὲ κριθὴν ἐν κόνει': Notably, the proverb reads σῖτον, not πυρόν, but Plutarch interprets it as wheat in his aetiology (i.e. barley's counterpart). E. Diehl, 1925, p. 197, 16 included this verse in his collection of *carmina popularia*. According to L. Senzasono, 2006, p. 191, n. 86, it is an agricultural proverb (rather than a riddle) with a clear instructional meaning. It is perhaps a bit strange, however, to literally plant barley, or anything else, in dust or ashes (κόνις). The same is true for planting wheat in the mud. A similar saying is explained in Ps.-Arist., *Probl.* 923a9–10 regarding mint: Διὰ τί λέγεται μίνθην ἐν πολέμῳ μήτ' ἔσθιε μήτε φύτευε; (this saying is listed in M. Apostolius, *Coll. paroem.* 11, 61; see E.L. Leutsch, 1965, pp. 530–531).

915E Πότερον, ὡς εἰρήκαμεν, ὁ μὲν δύναται πλείονος τροφῆς κατακρατεῖν ἡ δ' οὐ φέρει τὸ πολὺ καὶ κατακλύζον: For the idea that wheat can master food better, cf. Theophr., *CP* 3, 21, 4 (κατακρατεῖ γὰρ τῆς τροφῆς [...] ὁ πυρὸς μᾶλλον) and 4, 13, 5 ([Πυρὸς] ἔτι δὲ ἰσχυρότερος ὢν καὶ κατακρατεῖ καὶ συμπέττει μᾶλλον).

915E ἢ πυκνὸς ὢν ὁ πυρὸς καὶ ξυλώδης φύεται βέλτιον ἐν ὑγρῷ μαλαττόμενος καὶ χυλούμενος, τῇ δὲ κριθῇ διὰ μανότητα σύμφορον ἐν ἀρχῇ τὸ ξηρότερον;: According to Theophr., *CP* 3, 21, 4, wheat grows better in rainy regions and is more resistant to rains than barley. Regarding the open texture of barley (also mentioned in *Q.N.* 15, 915D), cf. *CP* 4, 13, 5: ἡ κριθὴ [...] μανὴ τὴν φύσιν οὖσα.

915E ἢ διὰ θερμότητα σύμμετρος καὶ ἀβλαβὴς ἡ κρᾶσις ψυχρότερον δ' ἡ κριθή;: The formulation of this *causa* is concise. I take διὰ θερμότητα to refer to the natural heat of the wheat and not of the muddy soil in which it is planted, although this is not impossible from a syntactical perspective (cf. F.H. Sandbach, 1965, pp. 184–185 with n. a and L. Senzasono, 2006, p. 192, n. 88). After all, muddy soil is more likely to be cold as it is drained with cool water (cf. *Q.N.* 14, 915D), as opposed to dusty soil, which is drier and, by implication, hotter. If this is correct, Plutarch implies that a hot plant (c.q. wheat) requires a cold soil (c.q. mud) and a cold plant (c.q. barley) a hot one (c.q. dust). A relevant parallel for this is found in *Quaest. conv.* 648CD, where Ammonius refers to Harpalus' failure to plant ivy in Babylonian soil: this is ascribed to the natural heat of the plant (cf. Theophr., *CP* 4, 4, 1; cf. also *Alex.* 35, 15). Ivy, being naturally hot itself, cannot become acclimated to the hot Babylonian soil and withers because of the excessive heat: 'what is cold loves heat, and what is hot cold' (φιλόθερμόν ἐστι τὸ ψυχρὸν καὶ φιλόψυχρον τὸ θερμόν). What is probably meant in *Q.N.* 16, then, is that the cold, muddy soil is commensurate (σύμμετρος) to the wheat, which is hot (διὰ θερμότητα), and – vice versa – that barley, being colder (ψυχρότερον), requires a warmer, drier, dusty soil.

For the required commensuration (συμμετρία) of the composition (κρᾶσις) of the soil to the plant growing in it, cf. also Theophr., *CP* 2, 9, 7: εἴ τις ἑτέρα χώρα τοιαύτην ἔχει τὴν κρᾶσιν ὥστε σύμμετρον ἐκδιδόναι τὴν τροφήν. For the coldness of barley, cf. Hipp., *De victu*. 2, 40, and for the heat of wheat, cf. Theophr., *CP* 3, 21, 4. In *Quaest. conv.* 697B, the natural heat of wheat is illustrated by the fact that wine is quickly evaporated when wine jars are placed in the wheat pits.

915EF ἢ φοβοῦνται τὸν πυρὸν ἐν ξηρῷ τρίβειν διὰ τοὺς μύρμηκας, εὐθὺς γὰρ ἐπιτίθενται· τὰς δὲ κριθὰς ἧττον φέρονται, δυσβάστακτοι γάρ εἰσι καὶ δυσπαρακόμιστοι διὰ μέγεθος;: This seems to imply that ants cannot cross humid soil (c.q. mud); otherwhise they would be able to steal the wheat grains. According to this logic though, one may wonder why barley is not planted in humid soil just as well – the ants cannot carry its grains anyway, so it would not really matter –, but this point was already explained in the previous *causae*. On the cleverness (and virtuousness) of ants more generally, see *De soll. an*. 967D–968B (where allusion is again made to ants' reluctance toward humidity in the quotation from Arat., *Phaen*. 956). Michael Glycas also refers to ants' reluctance toward barley and their preference for wheat (c.q. σῖτον). This is not explained in light of the magnitude of the grains of barley but of the odour given off by their ears (*Ann*. 1, 62 = p. 118, 17–19 Bekker: Ὁ μύρμηξ ἐπὶ τὸν στάχυν ἀνερχόμενος ὀσφραίνεται, καὶ ἐὰν κριθή, φεύγει ἀπ' αὐτοῦ καὶ ἐπὶ τὸν τοῦ σίτου στάχυν ἔρχεται). For ants' perception of smells, cf. Arist., *De sensu* 444b12. Theophr., *HP* 8, 10, 4 is concerned with the damage caused by grubs (σκώληκες) to the roots and ears of wheat. According to L. Senzasono, 2006, pp. 192–193, n. 89 the term σκώληκες in that passage may indicate 'larvae of insects', and thus also of ants. This, however, is speculation and Plutarch does not speak of grubs or larvae at all.

∂♥

3. Sea animals and fishing (Q.N. 17–19)

The three problems that follow in *Q.N.* 17–19 deal with the topic of sea animals and fishing. From a thematic perspective *Q.N.* 13 links up closely with *Q.N.* 17 (both problems deal with fishermen utensils). Based on several parallel accounts in a relatively short section of *De sollertia animalium* (viz. 976E–977A, 978EF, 979B; see the commentary below), there may be reason to assume that *Q.N.* 13 and 17–19 are modelled on the same or similar hypomnematic material (it is unlikely, therefore, that there is a chronological rupture in composition between *Q.N.* 18 and 19 [see the prologue, n. 24]). No clear source can be appointed for these chapters, but it can be presumed that Plutarch, at least in part, relies on

Peripatetic zoology (see, e.g., the quote from Theophrastus' *De animalibus colorem mutantibus* in *Q.N.* 19, 916B). As always, the intermediation of lost Ps.-Aristotelian problems cannot be excluded.

Q.N. 17, 915F–916A

In *Q.N.* 17, Plutarch wonders **why people use the hairs of stallions rather than those of mares to manufacture fishing lines** (Διὰ τί τῶν ἀρρένων ἵππων μᾶλλον ἢ τῶν θηλειῶν τὰς τρίχας εἰς τὴν ὁρμιὰν λαμβάνουσι;). He provides two short explanations. According to the first, the properties of the horse's body as a whole are also present in its smaller body parts, even in its smallest ones (c.q. its hairs). In the second, Plutarch makes a more concrete distinction between the male and the female body, where attention (implicitly) goes to their different anatomy.

In the **first** explanation, Plutarch argues, by analogy with the relative strength in other body parts, that the stallion has stronger hair than the mare (Πότερον, ὡς τοῖς ἄλλοις τὸ ἄρρεν τοῦ θήλεος μέρεσι, καὶ ταῖς θριξὶν εὐτονώτερόν ἐστιν;).

In the **second** explanation, he suggests that people rather believe that the mares' hairs become inferior because they are wetted by their urine (ἢ μᾶλλον διὰ τὸ οὖρον οἴονται τὰς τρίχας τῶν θηλειῶν βρεχομένας γίνεσθαι χείρονας;).

915F Διὰ τί τῶν ἀρρένων ἵππων μᾶλλον ἢ τῶν θηλειῶν τὰς τρίχας εἰς τὴν ὁρμιὰν λαμβάνουσι;: As appears from the second explanation, "[t]he hairs of the tail are in question" (F.H. Sandbach, 1965, p. 185, n. c).

916A ἢ μᾶλλον διὰ τὸ οὖρον οἴονται τὰς τρίχας τῶν θηλειῶν βρεχομένας γίνεσθαι χείρονας;: The argument is based on the anatomical dissimilarity between the male and the female body. The reported discomfort is strange to stallions, since their tails, as opposed to those of mares, are not located in the vicinity of the urethra. The same explanation is given in a parallel passage in *De soll. an.* 977A: ἱππείαις γὰρ θριξὶ χρῶνται, τὰς τῶν ἀρρένων λαμβάνοντες· αἱ γὰρ θήλειαι τῷ οὔρῳ τὴν τρίχα βεβρεγμένην ἀδρανῆ ποιοῦσιν. This passage probably originates from the scholia on Hom., *Il.* 24, 80–82 (quoted in *De soll. an.* 976F, with reference also to Archil., fr. 57 Edmonds). Cf. also Eustathius' commentary *ad loc.*: οἱ δὲ ὕστερον καὶ ἱππείαις θριξὶ χρῶνται. Cf. H. Cherniss and W.C. Helmbold, 1957, p. 423, n. f.

ଛ

Q.N. 18, 916AB

The following problem in *Q.N.* 18 originates from popular weather lore (cf. also *Q.N.* 2 and 4). Plutarch wonders **why the appearance of the**

calamary is a sign of a big storm (Διὰ τί τευθὶς φαινομένη σημεῖόν ἐστι μεγάλου χειμῶνος;). He gives only one explanation, which draws specific attention to the etymology of the word μαλάκια (cephalopods; literally 'soft things', 'molluscs').

Plutarch argues that all cephalopods by nature are sensitive to cold due to their bodily construction. Their flesh is bare and naked, they are not covered with a shell, skin or scales, and their hard and bony parts are located inside, which is why they are called 'soft things' (Ἡ πάντα φύσει τὰ μαλάκια δύσριγα διὰ γυμνότητα τῆς σαρκὸς καὶ ψιλότητα, μήτ' ὀστράκῳ μήτε δέρματι μήτε λεπίδι σκεπομένης ἀλλ' ἐντὸς ἐχούσης τὸ σκληρὸν καὶ ὀστεῶδες, διὸ καὶ κέκληται μαλάκια;). By reason of their sensitivity, cephalopods quickly become aware of the impending storm (ταχὺ δὴ προαισθάνεται δι' εὐπάθειαν τοῦ χειμῶνος). That is why the octopus makes a retreat to land, and when it grasps small rocks, it is a sign of imminent winds, whereas the calamary jumps out of the water in an attempt to escape the cold and the disturbance in the depths of the sea (ὅθεν ὁ μὲν πολύπους εἰς γῆν ἀνατρέχει καὶ τῶν πετριδίων ἀντιλαμβανόμενος σημεῖόν ἐστι πνεύματος ὅσον οὔπω παρόντος, ἡ δὲ τευθὶς ἐξάλλεται, φεύγουσα τὸ ψῦχος καὶ τὴν ἐν βάθει ταραχὴν τῆς θαλάττης). At the end of the explanation, Plutarch adds that the calamary has the most fragile and delicate flesh of all of the cephalopods (this results in an implicit argumentative ring [see 4.3.3.3.]) (καὶ γὰρ ἔχει μάλιστα τῶν μαλακίων εὔθρυπτον καὶ ἁπαλὸν τὸ σαρκῶδες).

916A τευθίς: See LSJ, s.v.: "calamary or squid, *Loligo vulgaris*". See also A. Dalby, 2003, p. 311, s.v. "squid": "When used specifically, Greek *teuthis* is the name for the small *Loligo vulgaris* and related species. Greek *teuthos* is the name for full-grown specimens of the much larger species *Todarodes sagittatus* and its relatives. This is the squid that can fly, though it does not fly far." According to Varro's Latin etymology (*De ling. Lat.* 5, 79), the *lolligo* was originally called *volligo* because it flies up. It is probably better to speak of 'leaping', though, as Plutarch does (cf. ἐξάλλεται), rather than 'flying'. Moreover, L. Senzasono, 2006, p. 195, n. 94 notes that the *loligo vulgaris* (the calamary) does not fly out of the water, but that the *loligo volitans* does (on which, cf. Pliny, *NH* 18, 361 and 32, 149: probably to be identified with the flying fish, i.e. the family of exocoetidae). He points out that both animals were probably mistaken for one another in Antiquity. For further references on squids 'flying' in the manner of flying fish, see D'A.W. Thompson, 1947, p. 260.

916A Ἡ πάντα φύσει τὰ μαλάκια δύσριγα διὰ γυμνότητα τῆς σαρκὸς καὶ ψιλότητα, μήτ' ὀστράκῳ μήτε δέρματι μήτε λεπίδι σκεπομένης ἀλλ' ἐντὸς ἐχούσης τὸ σκληρὸν καὶ ὀστεῶδες, διὸ καὶ κέκληται μαλάκια;: Regarding the octopus' intolerance of cold (and sweet) water, cf. also Theophylactus Simocatta, *Quaest. phys.* 9 (J.L. Ideler, 1841, p. 176, 12–34).

916AB ὅθεν ὁ μὲν πολύπους εἰς γῆν ἀνατρέχει καὶ τῶν πετριδίων ἀντιλαμβανόμενος σημεῖόν ἐστι πνεύματος ὅσον οὔπω παρόντος, ἡ δὲ τευθὶς ἐξάλλεται, φεύγουσα τὸ ψῦχος καὶ τὴν ἐν βάθει ταραχὴν τῆς θαλάττης: As to the image of the octopus clinging to small stones in order to remain balanced, a similar remark can be found in *De soll. an.* 979B, where Plutarch reports that the sea-hedgehog has the same habit: it clings to small rocks (λιθιδίοις) in order not to capsize or be swept away by the swell when it senses a storm coming (ὅταν αἴσθωνται μέλλοντα χειμῶνα καὶ σάλον). Similar reports about the octopus, calamary, cuttlefish and the like clinging to (small) rocks in order to remain steady are found, e.g., in Arist., *PA* 685a30–b2, *HA* 523b32–33, Ael., *NA* 5, 41, Ath., *Deipn.* 7, 323d (= Arist., fr. 338 Rose), Pliny, *NH* 9, 83; 18, 361. Cf. also *Sept. Sap. conv.* 163D (a giant octopus carries along a stone) and *De soll. an.* 967AB (bees ballast themselves with little stones). For the calamary's peculiar ability to leap out (ἐξάλλεται) of the sea and 'fly' in the air, cf. Opp., *Hal.* 1, 429–432 (in order to escape its attackers); 3, 166–167, Ael., *NA* 9, 52 (out of fear), Cic., *De div.* 2, 145, 12–14, Pliny, *NH* 9, 84; 18, 361; 32, 15 and 149 (*loligo volitans*). According to A.F. Scholfield, 1958, p. xxii "there can be little doubt that Aelian has paraphrased Oppian". If this is correct, Plutarch and (Aelian via) Oppian possibly rely (either directly or indirectly) on the same or similar source, presumably of Peripatetic origin, perhaps Theophrastus' *De signis* (cf. *Sign.* 40: Κολοιοὶ ἐκ τοῦ νότου πετόμενοι καὶ τευθίδες χειμέριαι). There is no reason to assume, however, that Plutarch relies on personal observation for the differences in explanation with Aelian (*pace* L. Senzasono, 2006, p. 194, n. 94). Moreover, the idea that the calamary jumps out of the water in an attempt to escape the cold and the disturbance in the depths of the sea is contradicted in *Q.N.* 8, 914B, where Plutarch reports that the sea becomes warmer when it grows rough, not colder [see 4.3.3.1.].

916B καὶ γὰρ ἔχει μάλιστα τῶν μαλακίων εὔθρυπτον καὶ ἁπαλὸν τὸ σαρκῶδες: The idea that the calamary has the most fragile and delicate flesh of all cephalopods implies that it is also the most sensitive. Cf. Arist., *PA* 678b32: τὸ σῶμα πᾶν ἐκ μαλακωτέρας συνεστάναι σαρκός. In the margin, F.H. Sandbach, 1965, p. 186, n. d remarks that the calamary's "bony processes are smaller than those of the cuttlefish but not than those of the octopus". The main focus here is, of course, on the calamary and less on the octopus (to which Plutarch turns in the next problem).

Q.N. 19, 916BF

Q.N. 19 is the most extensive problem in the collection and concerns **the octopus' change of colour** (Διὰ τί τὴν χρόαν ὁ πολύπους ἐξαλλάττει;). Regarding *concatenatio*, the πολύπους has made its appearance already in *Q.N.* 18, 916A, and the image of it sitting on a rock recurs in this problem. There is a relatively large cluster of parallel passages for this phenomenon throughout Plutarch's oeuvre (cf. *Alc.* 23, 4–5, *De ad. et am.* 51D–53D, *De am. mult.* 96F, *De soll. an.* 978EF), which I have presented schematically earlier on [see 2.1.2.]. Plutarch provides three explanations for the phenomenon here, which are closely connected to each other. The first one deals with the octopus' change of colour by a physiological process in the animal's body involving its breath (πνεῦμα), the second draws attention to the fact that the octopus not only changes, but also adapts its colour to its surroundings and the third tries to explain the natural mechanism that lies behind this adaptation (therefore, it is also the most comprehensive of the three, explaining both the change and the adaptation of the octopus' colour).

In the **first** *causa*, Plutarch gives Theophrastus' opinion, according to which the octopus is by nature a cowardly animal. Thus, when it is agitated, it undergoes a change by its breath and simultaneously alters its colour, much like in human beings (Πότερον, ὡς Θεόφραστος ᾤετο, δειλόν ἐστι φύσει ζῷον· ὅταν οὖν ταραχθῇ τρεπόμενον τῷ πνεύματι, συμμεταβάλλει τὸ χρῶμα καθάπερ ἄνθρωπος). The reference to human physiology is illustrated with the Homeric saying that 'the coward's complexion alters' (διὸ καὶ λέλεκται 'τοῦ μὲν γάρ τε κακοῦ τρέπεται χρώς').

In the **second** *causa*, Theophrastus' argument is considered insufficient. Even though it is a plausible explanation for the change of the octopus' colour, it does not account for its assimilation to its surroundings (ἢ τοῦτο πρὸς τὴν μεταβολὴν πιθανῶς λέλεκται πρὸς δὲ τὴν ἐξομοίωσιν οὐχ ἱκανῶς;). Plutarch further explains that the octopus changes in such a way that it assimilates its colour to the rocks that are closeby (μεταβάλλει γὰρ οὕτως, ὥστε τὴν χρόαν αἷς ἂν πλησιάζῃ πέτραις ὁμοιοῦν). Plutarch's critical attitude becomes more pronounced when he quotes two poets in order to illustrate the aspect of assimilation. First he gives an excerpt from Pindar, who wrote: 'matching most in mind the sea beast's complexion, take in every town your place' (πρὸς ὃ καὶ Πίνδαρος ἐποίησε 'ποντίου θηρὸς χρωτὶ μάλιστα νόον / προσφέρων πάσαις πολίεσσιν ὁμίλει'). The ability to adapt and assimilate to one's surroundings (ὁμιλεῖ) is echoed in the distich by Theognis that follows: 'acquire the mind of the many-coloured octopus, that looks to the eye like the rock on which it settles' (καὶ Θέογνις 'πουλύποδος νόον ἴσχε πολυχρόου, ὃς ποτὶ πέτρῃ, / τῇ προσομιλήσῃ, τοῖος ἰδεῖν ἐφάνη'). Plutarch continues that they (viz. the poets) say that people who are very clever and cunning also have this habit; they always imitate the octopus in order to remain unseen and unnoticed by their neighbours (τοῦτο δὴ καὶ τοὺς πανουργίᾳ καὶ δεινότητι ὑπερφέροντας ἔχειν τὸ ἐπιτήδευμα λέγουσιν, ὡς ὑπὲρ τοῦ λαθεῖν καὶ διαφυγεῖν τοῖς πλησίον ἑαυτοὺς ἀεὶ ἀπεικάζειν πολύποδι). At the end of the *causa*, Plutarch ridicules Pindar and Theognis, wondering ironically whether they believe that the octopus treats its colour like a garment that can be easily changed whenever the animal wishes (ἢ καθάπερ ἐσθῆτι τῇ χρόᾳ νομίζουσι χρῆσθαι, ῥᾳδίως οὕτως ᾗ βούλεται μετενδυόμενον;).

At the beginning of the **third** *causa*, Plutarch suggests that the octopus itself initiates the effect by feeling fright, but that the determining factor of the explanation lies elsewhere (ἆρ' οὖν τὴν μὲν ἀρχὴν αὐτὸς ἐνδίδωσι τοῦ πάθους δείσας, τὰ δὲ κύρια τῆς αἰτίας ἐν ἄλλοις ἐστί;). He emphatically incites the reader to consider Empedocles' theory of emanations, according to which 'there are emanations from all things that ever were' (σκόπει δή, κατ' Ἐμπεδοκλέα γνοὺς ὅτι πάντων εἰσὶν ἀπορροαὶ ὅσσ' ἐγένοντο). Plutarch explains that many streams of particles continuously emanate not only from animals, plants, the earth and the sea, but also from stones, bronze and iron (οὐ γὰρ ζῴων μόνον οὐδὲ φυτῶν οὐδὲ γῆς καὶ θαλάττης, ἀλλὰ καὶ λίθων ἄπεισιν ἐνδελεχῶς πολλὰ ῥεύματα καὶ χαλκοῦ καὶ σιδήρου). The emanation theory is further illustrated in two somewhat digressive ways (καὶ γάρ). Plutarch first notes that everything that decays or gives off a smell does so because some part is continuously streaming away and departing from it (καὶ γὰρ φθείρεται πάντα καὶ ὄδωδε τῷ ῥεῖν ἀεί τι καὶ φέρεσθαι συνεχῶς). Then a short digression into magnetism follows. Plutarch notes that people explain the phenomenon of attraction or jumping (in magnetism) by emanations, some assuming it to be due to entanglements, others to impacts, and still others to impulsions and circumventions (καὶ γὰρ ἕλξεις ἢ ἐπιπηδήσεις ποιοῦσι ταῖς ἀπορροίαις, οἱ μὲν ἐμπλοκὰς αὐτῶν οἱ δὲ πληγὰς οἱ δ' ὤσεις τινὰς καὶ περιελεύσεις ὑποτιθέμενοι). After these digressive illustrations, the discourse returns to the main line of explanation. Plutarch argues that it is likely that many fine particles are constantly detached from rocks lying by the seashore as they are sprayed and battered by the sea. These particles do not adhere to the body of any animal (except the octopus), and they cannot be seen because they either slip off the skin of animals that have narrower pores or pass through those that have more open ones (μάλιστα δὲ τῶν παράλων πετρῶν ἐπιρραινομένων καὶ ψηχομένων ὑπὸ τῆς θαλάττης ἀπιέναι μέρη καὶ θραύσματα πολλὰ καὶ λεπτὰ <εἰκὸς> συνεχῶς, ἃ τ<οῖς μὲν ἄλ>λοις οὐ προσί<σχεται> σώμα<σιν> ἀλλὰ λανθάνει περιολισθάνοντα τῶν πυκνοτέρους ἐχόντων πόρους ἢ διεκθέοντα τῶν μανοτέρους). The octopus, by contrast, has flesh that is obviously honeycombed in appearance and full of pores and (thus) receptive of emanations (ὁ δὲ πολύπους τήν τε σάρκα προσιδεῖν αὐτόθεν ἀνθρηνιώδης καὶ πολύπορος καὶ δεκτικὸς ἀπορροιῶν ἐστιν). When the animal is frightened, it undergoes and effects a change by its breath (cf. *causa* one), by compressing, as it were, and contracting its body, so that it receives and shelters the emanations from nearby objects on its surface (ὅταν τε δείσῃ, τῷ πνεύματι τρεπόμενος καὶ τρέπων οἷον ἔσφιγξε τὸ σῶμα καὶ συνήγαγεν, ὥστε προσδέχεσθαι καὶ στέγειν ἐπιπολῆς τὰς τῶν ἐγγὺς ἀπορροίας). This is further explained by the fact that (the skin's) roughness, in combination with (its) softness, provides places of lodgement for the particles that settle on it and do not disperse but collect and remain in place. As such, the roughness causes the surface (of the skin) to take on the same colour as the objects that are most nearby (καὶ γὰρ ἡ τραχύτης μετὰ τῆς μαλακότητος ἕδρας παρέχουσα τοῖς ἐπιφερομένοις μέρεσι, μὴ σκεδαννυμένοις ἀλλ᾽ ἀθροιζομένοις καὶ προσμένουσι, σύγχρου<ν ἀπεργάζεται> τὴν ἐπιφάνειαν <τοῖς ἐγγύ>τατα). As an important piece of evidence, at the end of the explanation, Plutarch refers to the fact that the octopus does not take on a likeness to all of the objects nearby, nor does the chameleon take on a likeness to pale objects. In fact, both animals only take on a likeness to those objects that have emanations to which their pores are commensurate (τεκμήριον δὲ τῆς αἰτίας μέγα τὸ μήτε τοῦτον πᾶσιν ἐξομοιοῦσθαι τοῖς πλησίον μήτε τὸν χαμαιλέοντα τοῖς λευκοῖς χρώμασιν, ἀλλὰ μόνοις ἑκάτερον, ὧν ταῖς ἀπορροίαις πόρους συμμέτρους ἔχουσιν).

916B Διὰ τί τὴν χρόαν ὁ πολύπους ἐξαλλάττει;: The verb ἐξαλλάττει means to "change utterly" (LSJ, s.v.) and does not imply an element of adaptation. Nevertheless, this aspect of adaptation will play an important role in Plutarch's aetiology (viz. in the second and third *causae*).

916B Πότερον, ὡς Θεόφραστος ᾤετο, δειλόν ἐστι φύσει ζῷον· ὅταν οὖν ταραχθῇ τρεπόμενον τῷ πνεύματι, συμμεταβάλλει τὸ χρῶμα καθάπερ ἄνθρωπος: The basic idea here (= fr. 188 Wimmer = 365C FHSG) is that the octopus' passive sensation of fear triggers an active movement and change in the body (c.q. metachrosis) by the intermediation of the πνεῦμα (cf. F.H. Sandbach, 1965, p. 187, n. f). This is reminiscent of Aristotle's πνεῦμα theory, where πνεῦμα is famously described as an ὄργανον of the soul, by which movement is imparted to the body (*DA* 433b18, *GA* 789b8–9; cf. also esp. *MA* 703a3–b1). As we learn from Athenaeus, Theophrastus' account originates from his lost *De animalibus colorem mutantibus* (D.L. 5, 44), where the octopus' metachrosis is attributed to its fear and need for protection (Theophr., fr. 173 Wimmer = 365B FHSG = Ath., *Deipn.* 7, 317 f.: τὸν πολύποδα φησὶ τοῖς πετρώδεσι μάλιστα τόποις συνεξομοιοῦσθαι τοῦτο ποιοῦνται φόβῳ καὶ φυλακῆς χάριν). Cf. also, e.g., Pliny, *NH* 9, 87: *colorem mutat ad similitudinem loci, et maxime in metu*. Theophrastus is probably following Aristotle here (see *PA* 679a13: ὁ πολύπους τὰς πλεκτάνας ἔχει χρησίμους καὶ τὴν τοῦ χρώματος μεταβολήν, ἣ συμβαίνει αὐτῷ [...] διὰ δειλίαν, cf. also *HA* 622a15). Several parallels for this phenomenon have been collected by V. Rose, 1863, pp. 362–365 (to which Clem. Al., *Paed.* 3, 11, 80, 2 can be added).

916B διὸ καὶ λέλεκται 'τοῦ μὲν γάρ τε κακοῦ τρέπεται χρώς': Line taken from the conversation between Idomeneus and Meriones on heroes and cowards (*Il.* 13, 279). In *De virt. mor.* 452A Plutarch quotes from the same Homeric passage (*Il.* 13, 284–285a: τοῦ δ' ἀγαθοῦ οὔτ' ἄρ τρέπεται χρὼς οὔτε τι λίην ταρβεῖ). In *Q.N.* 19, however, the focus is on the physical rather than the ethical side of cowardice (which is reminiscent of the general line of argumentation in Book 27 of Ps.-Aristotle's *Problems*: ὅσα περὶ φόβον καὶ ἀνδρείαν [see 1.2.5, n. 226]). Pliny, *NH* 11, 224–225 ascribes the change in colour to the variation of the blood supply.

916B ἢ τοῦτο πρὸς τὴν μεταβολὴν πιθανῶς λέλεκται πρὸς δὲ τὴν ἐξομοίωσιν οὐχ ἱκανῶς;: Notably, Plutarch did not mention the issue of assimilation in the *quaestio*, where he only refers to the octopus' change of colour (cf. ἐξαλλάττει: see the commentary *ad loc.*). For the aspect of assimilation, cf. also, e.g., Ael., *VH* I, 1 and Antig. Car., *Hist. mir.* 25a, 1–2; 50, 1–2.

916BC πρὸς ὃ καὶ Πίνδαρος ἐποίησε 'ποντίου θηρὸς χρωτὶ μάλιστα νόον / προσφέρων πάσαις πολίεσσιν ὁμίλει,' καὶ Θέογνις 'πουλύποδος νόον ἴσχε πολυχρόου, ὃς ποτὶ πέτρῃ, / τῇ προσομιλήσῃ, τοῖος ἰδεῖν ἐφάνη': The original context of Pindar's quote (= fr. 43 Snell) is Amphiaraus' advice to his son Amphilochus, that he has to adapt to the circumstances in which he finds himself. Theognis' lines (215–216 West) are part of the poet's advice to Cyrnus to adapt his ἦθος to each one of his friends' characters (alternatively, the poet may also be addressing his own θυμός: cf. 213 West, see B.A. Van Groningen, 1966, p. 82). The two quotations may have been taken from one and the same intermediate source where they were already combined. They both recur in *De soll. an.* 978E and are also clustered in Ath., *Deipn.* 12, 513cd (see also 7, 317a). The lines from Theognis are also quoted separately in *De am. mult.* 96F. Plutarch may be relying on a Peripatetic tradition, more precisely on an intermediary derivation from Theophrastus, e.g., via a lost problem (see A. Peretti, 1953, p. 96, n. 1 and F.R. Adrados, 1958, pp. 3–4; see also F.H. Sandbach, 1965, p. 136 and L. Senzasono, 2006, pp. 197–198, nn. 99–100).

916C τοῦτο δὴ καὶ τοὺς πανουργίᾳ καὶ δεινότητι ὑπερφέροντας ἔχειν τὸ ἐπιτήδευμα λέγουσιν, ὡς ὑπὲρ τοῦ λαθεῖν καὶ διαφυγεῖν τοῖς πλησίον ἑαυτοὺς ἀεὶ ἀπεικάζειν πολύποδι: F.H. Sandbach, 1965, p. 189 translates λέγουσιν as "men say", thus interpreting it impersonally (λέγεται). However, it seems more natural that Pindar and Theognis are the subjects of λέγουσιν and also of νομίζουσι further on (cf., e.g., *De esu* 997E, where Pythagoras and Empedocles are the subjects of φήσουσι). Indeed, *what* 'they say' according to Plutarch's interpretation is in agreement with the original context of the quotes just recorded (see the previous comment). Sandbach marks a lacuna between ἀπεικάζειν and πολύποδι, and adds: "<But what do they

suppose to be the mechanism of change> in the octopus?" (cf. *De soll. an.* 978E: μηχανῇ χρώμενος, and *Alc.* 23, 4: μηχανή). Pohlenz proposes to insert ὁμοίως. There is no lacuna in the manuscripts, though, but, even so, Sandbach's interpolation does nicely capture what Plutarch seems to be implying (no modifications should be made to the text to make this more explicit, though). In what follows it is clearly Plutarch's intention to criticise the poets and their moralising interpretation of the natural phenomenon at hand. What Plutarch is probably implying, then, is that the octopus' colour change is not the effect of a deliberate choice: it is triggered not on purpose but by fear (τὴν μὲν ἀρχὴν αὐτὸς ἐνδίδωσι τοῦ πάθους δείσας κτλ.). According to L. Senzasono, 2006, p. 198, n. 101, an element of fear rather than friendship is already present in Plutarch's interpretation of the poets, but this seems unlikely (the context of friendship is clear from the parallels in *De ad. et am.* 51D and *De am. mult.* 96F–97A).

916C ἢ καθάπερ ἐσθῆτι τῇ χρόᾳ νομίζουσι χρῆσθαι, ῥᾳδίως οὕτως ᾗ βούλεται μετενδυόμενον;: The irony in Plutarch's literary criticism of the two poets is clear here (the register of which is exceptional to the overall denotative style of the collection [see 1.2.4., n. 204]). Cf. also *Amatorius* 767A: ὥσπερ ἱματίων. At the same time, this phrase raises the question of whether the octopus deliberately changes its colour or not, that is, whether the adaptation is based on a rational decision or on the emotion of fear. In *De soll. an.* 978EF, Phaedimus argues that the octopus' metachrosis (as opposed to the chameleon's; see further) is not due to an emotional reaction (c.q. fear), but is, instead, a deliberate change (μεταβάλλει γὰρ ἐκ προνοίας, μηχανῇ χρώμενος τοῦ λανθάνειν ἃ δέδιε [*contradictio in terminis*!] καὶ λαμβάνειν οἷς τρέφεται). Plutarch will disagree with this theory in the following *causa* in *Q.N.* 19, where he argues that the octopus itself initiates the effect by feeling fright (τὴν μὲν ἀρχὴν αὐτὸς ἐνδίδωσι τοῦ πάθους δείσας κτλ.). Cf. also F.H. Sandbach, 1965, pp. 136–137. Additionally, the quotations from Pindar and Theognis (and also from Homer in the first *causa*) oblige Plutarch to incorporate a certain degree of ethical 'depth' in his explanation (cf. *De am. mult.* 96F [see 1.2.4.]). See L. Senzasono, 2006, p. 198, n. 101. He complies with this only for the purposes of a refutation, that is, a proof of default thereof. Plutarch knows, after all, that the octopus' change of colour cannot be adequately explained in terms of purely mental or ethical categories, but that it must have a genuine natural cause and a physical and corporeal specificity (τὰ δὲ κύρια τῆς αἰτίας ἐν ἄλλοις ἐστί). Therefore, in the third and final *causa*, Plutarch – presumably in his personal contribution to the problem – provides an explanation based on Empedocles' theory of emanations.

916D σκόπει δή, κατ' Ἐμπεδοκλέα γνοὺς ὅτι πάντων εἰσὶν ἀπορροαὶ ὅσσ' ἐγένοντο: The introductory phrase γνοὺς ὅτι is probably no part of Empedocles' fragment (= DK31B89) (see F.H. Sandbach, 1965, p. 188 and L. Senzasono, 2006, p. 199, n. 102), although this is not impossible from a metrical viewpoint (see C. Hubert, 1960, p. 16). J.P. Hershbell, 1971, p. 181 considers this quote from Empedocles to be an example of Plutarch's polemical and eclectic attitude towards the Peripatetic school in *Quaestiones naturales*. The least that can be said is that Plutarch is trying to complete – rather than to reject – Theophrastus' theory with it, as formulated in the first *causa*. Moreover, we have no account of the octopus' change and assimilation of colour by Empedocles – if he actually had an opinion about it. In fact, according to F.H. Sandbach, 1965, p. 137, "Plutarch would seem to be summarizing some account of the emanation theory that had no original connexion with the problem of the octopus". There may be reason to assume, therefore, that we are dealing in this third *causa* with Plutarch's personal contribution to the problem (see M. Meeusen, 2012a). This may be indicated by the introduction of the explanation with the phrase σκόπει δή [see 1.1.4., n. 111], and by, what Sandbach calls, a return to a "full style" (he additionally marks the presence of several coupled synonyms: μέρη καὶ θραύσματα, ἀνθρηνιώδης καὶ πολύπορος, ἔσφινγξε καὶ συνήγαγεν, ἀθροιζομένοις καὶ προσμένουσι). At least, the use of this imperative and the circumstantiality of discourse may hint at Plutarch's enthusiasm for expounding what he believes is the 'determining factor of the explanation' (τὰ δὲ κύρια τῆς αἰτίας).

916D καὶ γὰρ φθείρεται πάντα καὶ ὄδωδε τῷ ῥεῖν ἀεί τι καὶ φέρεσθαι συνεχῶς: For Empedocles' theory of smells and our perception thereof, see the (critical) doxography of Theophr., *Sens.* 20 (= DK31A86): εἰ ἡ φθίσις διὰ τὴν ἀπορροήν [...] συμβαίνει δὲ καὶ τὰς ὀσμὰς ἀπορροῇ γίνεσθαι, τὰ πλείστην ἔχοντα ὀσμὴν τάχιστ' ἐχρῆν φθείρεσθαι. νῦν δὲ σχεδὸν ἐναντίως ἔχει κτλ.

916D καὶ γὰρ ἕλξεις ἢ ἐπιπηδήσεις ποιοῦσι ταῖς ἀπορροίαις, οἱ μὲν ἐμπλοκὰς αὐτῶν οἱ δὲ πληγὰς οἱ δ' ὤσεις τινὰς καὶ περιελεύσεις ὑποτιθέμενοι: According to F.H. Sandbach, 1965, p. 137, this digression into magnetism "contains barely intelligible and certainly irrelevant detail". On closer inspection, however, the passage is relatively clear and is not that irrelevant for the aetiology either. The same is true for the previous account about decay and smell (see the previous comment). By incorporating these two accounts Plutarch does, indeed, postpone the actual point that he aims to make with the quote from Empedocles, but they do still come in handy to illustrate the emanation theory. In fact, the octopus' metachrosis is presented as a third instance of this theory (besides that of smell and magnetism). The enumeration of technical terms does, however, seem to be unsystematic: the passage contains references to Democritus/Epicurus (ἐμπλοκαί/περιπλοκαί: interlacing of atoms, πληγαί: clashing of atoms), Empedocles (ὤσεις: impulsions), and Plato (περιελεύσεις:

circumventions; see below). An important doxographical source for our understanding of ancient theories on magnetism is Alex. Aphr., *Quaest*. 2, 23 (= Psellus, *De lapid*. 26). This passage summarises the magnetic theories of Empedocles (DK31A89), Democritus (DK68A165) and Diogenes of Apollonia (DK64A33). For Epicurus' view on magnetism, see fr. 293 Usener (= Gal., *Nat.Fac*. I, 14 = 2, 44–56 Kühn). According to L. Senzasono, 2006, p. 200, n. 104, the quotation from Empedocles in *Q.N.* 19 originates from the same text on which Alex. Aphr., *Quaest*. 2, 23 relies, but this is uncertain. Perhaps more important is the parallel between the passage at issue in *Q.N.* 19, 916D and *Quaest. Plat*. 1005BD, where Plutarch provides an exegesis of Pl., *Tim*. 79e–80c (esp. 80c on magnetism). Plutarch there takes a Platonic position in the debate on magnetism, by rejecting the mechanisms of ἕλξις and ἐπιπήδησις, as mentioned here (*Quaest. Plat*. 1005B: Τὸ δ' ἤλεκτρον οὐδὲν ἕλκει τῶν παρακειμένων ὥσπερ οὐδ' ἡ σιδηρῖτις λίθος, οὐδὲ προσπηδᾷ τι τούτοις ἀφ' αὑτοῦ τῶν πλησίον). For further commentary, see A.E. Taylor, 1928, pp. 578 ff. and J. Opsomer, 1994a, pp. 336–337. The correspondence in thought and words between *Quaest. Plat*. 1005D and *Q.N.* 19, 916D is striking, so that it is not unlikely that Plutarch is relying on the same hypomnematic material in both passages. Most notably, the same Platonic position that Plutarch supports in the Platonic question is echoed in the technical terms enumerated here in *Q.N.* 19 (ἐν κύκλῳ περιιών = περιέλευσις). Again, the Empedoclean theory of emanations is very prominent there. Both passages also formulate the same idea of commensuration (συμμετρία, see further). They also contain more subtle verbal reminiscences, such as the fact that particles do not 'slip off' (ἀπολισθαίνειν) of a 'rough surface' (τραχύτητας), or the description of the holes in this surface as 'lodgements' (ἕδρας: notably, F.H. Sandbach, 1965, p. 190, n. 7 corrects ἕλικας in ἕδρας in *Q.N.* 19, 916D precisely on the basis of *Quaest. Plat*. 1005D: μὴ ἀπολισθαίνειν ἀλλ' ἕδραις τισὶν ἐνισχόμενον, cf. also *Tim*. 80c: ἕδραν). For further accounts of magnetism in Plutarch, see *De Is. et Os*. 376B and *Quaest. conv*. 641C (with J. Opsomer, 1994a, pp. 351–361). As to Plutarch's source, it is not unlikely that we should think of a commentary on Plato's *Timaeus* (see A.E. Taylor, 1928, p. 579: "Plutarch does not mention the source of any of his explanations, but as his various essays show him to have been directly or indirectly acquainted with the explanations of or commentaries on the *Timaeus* by Xenocrates, Crantor, and Posidonius, probably some of these are responsible for what he says.").

916E ὁ δὲ πολύπους τήν τε σάρκα προσιδεῖν αὐτόθεν ἀνθρηνιώδης καὶ πολύπορος καὶ δεκτικὸς ἀπορροιῶν ἐστιν: The detail of the honeycombed appearance (ἀνθρηνιώδης: literally, 'like a wasp's nest') of the octopus' skin, which is 'full of pores' (πολύπορος) is echoed in *Quaest. conv*. 721EF, where Plutarch comments on the structure of iron: ὁ σίδηρος ἔχων τι σαθρὸν καὶ πολύκενον καὶ τενθρηνῶδες. The adjective ἀνθρηνιώδης is a hapax (as is τενθρηνῶδες, perhaps to be corrected in τενθρηνιῶδες, which is more, albeit not very, common).

916E ὅταν τε δείσῃ, τῷ πνεύματι τρεπόμενος καὶ τρέπων οἷον ἔσφιγξε τὸ σῶμα καὶ συνήγαγεν, ὥστε προσδέχεσθαι καὶ στέγειν ἐπιπολῆς τὰς τῶν ἐγγὺς ἀπορροίας: With the phrase ὅταν τε δείσῃ, τῷ πνεύματι τρεπόμενος καὶ τρέπων Plutarch implies that the octopus, when afraid, simultaneously undergoes and effects a bodily change by its breath. The shift in verbal diatheses clearly echoes 916B: ὅταν οὖν ταραχθῇ τρεπόμενον τῷ πνεύματι, συμμεταβάλλει τὸ χρῶμα. The same shift recurs in *De sera num.* 565C (where the squid's ejection of ink is mentioned): ἐκεῖ γὰρ ἥ τε κακία τῆς ψυχῆς τρεπομένης ὑπὸ τῶν παθῶν καὶ τρεπούσης τὸ σῶμα τὰς χρόας ἀναδίδωσι. On the basis of the parallel passage in *De am. mult.* 96F (τοῦ πολύποδος αἱ μεταβολαὶ βάθος οὐκ ἔχουσιν, ἀλλὰ περὶ αὐτὴν γίγνονται τὴν ἐπιφάνειαν), F.H. Sandbach, 1965, p. 191 (with n. d) adds the following in his translation: "without allowing them [sc. the emanations] to penetrate it [sc. the surface of the skin]". However, this interpolation is not based on the reading of the manuscripts and seems unnecessary.

916E καὶ γὰρ ἡ τραχύτης μετὰ τῆς μαλακότητος ἕδρας παρέχουσα τοῖς ἐπιφερομένοις μέρεσι, μὴ σκεδαννυμένοις ἀλλ' ἀθροιζομένοις καὶ προσμένουσι, σύγχρου<ν ἀπεργάζεται> τὴν ἐπιφάνειαν <τοῖς ἐγγύ>τατα: For the softness of the octopus' skin and that of the μαλάκια more generally, cf. *Q.N.* 18, 916AB.

916F τεκμήριον δὲ τῆς αἰτίας μέγα τὸ μήτε τοῦτον πᾶσιν ἐξομοιοῦσθαι τοῖς πλησίον μήτε τὸν χαμαιλέοντα τοῖς λευκοῖς χρώμασιν, ἀλλὰ μόνοις ἑκάτερον, ὧν ταῖς ἀπορροίαις πόρους συμμέτρους ἔχουσιν: For the chameleon's metachrosis, see also Antig. Car., *Hist. mir.* 25b (and the passages collected by V. Rose, 1863, pp. 362–365 more generally). More specifically, for the chameleon's inability to adapt to pale or red colours, see Theophr., fr. 172, 1 Wimmer (= 365A FHSG: Μεταβάλλει δ' ὁ χαμαιλέων εἰς πάντα τὰ χρώματα, πλὴν τὴν εἰς τὸ λευκὸν καὶ τὸ ἐρυθρὸν οὐ δέχεται μεταβολήν). See also Pliny, *NH* 8, 122; 11, 225. Notably, in *Q.N.* 19, Plutarch only mentions the chameleon's inability to adapt to pale colour, thus leaving out the red one, as he also does in the parallel passages in *De ad. et am.* 53D and *Alc.* 23, 5. Therefore, Plutarch is presumably relying on an intermediary source that was equally incomplete (cf. L. Senzasono, 2006, p. 202, n. 107: "È probabile che Plutarco attinga a una fonte meno completa per la varietà dei colori"). Senzasono's suggestion to translate τοῖς λευκοῖς χρώμασιν as 'bright colours' rather than 'pale colours' is not convincing (cf. *Q.N.* 39: *Cur aqua in summa parte alba, in fundo vero nigra spectatur?*). In yet another parallel passage, viz. in *De soll. an.* 978EF, Plutarch again relies on Theophrastus (= fr. 189 Wimmer = 365D FHSG) and writes that the chameleon's metachrosis is triggered by fear and is due to the fact that the animal is full of breath since it has very large lungs – this may well be an echo of Theophrastus' πνεῦμα theory in the first *causa* in *Q.N.* 19 (cf. also Theophr., fr. 172, 4 Wimmer = 365A FHSG: Ὁ δὲ χαμαιλέων δοκεῖ τῷ πνεύματι ποιεῖν τὰς μεταβολὰς, πνευματικὸν γὰρ φύσει. Σημεῖον δὲ τὸ τοῦ πνεύμονος μέγεθος· σχεδὸν γὰρ δι' ὅλου τοῦ σώματος τέταται· ἅμα δὲ καὶ αὐτὸς ἐξαιρόμενος καὶ φυσώμενος). Without a doubt, this idea also originates from Theophrastus' *De animalibus colorem mutantibus* (see above).

4. Land animals and hunting (Q.N. 20–28)

The following cluster of problems in *Q.N.* 20–28 centres on the topic of land animals (*Q.N.* 20–22, 26, 28) and hunting (*Q.N.* 23–25) – with the exception of *Q.N.* 27, which deals with the production of wine from must and, thus, adheres more naturally to the following cluster of problems on viniculture (*Q.N.* 30–31). No clear source can be appointed for these chapters, but it can be presumed that Plutarch, at least in part, draws on Peripatetic zoology (esp. Aristotle, quoted in *Q.N.* 21, 917CD). Notably, the problems on hunting (*Q.N.* 23–25) have a very high concentration of parallels in Xenophon, *Cyn.* 5, 1–5 (presented schematically in [4.2.1.2.]).

Q.N. 20, 916F–917B

In *Q.N.* 20, Plutarch inquires into the physiology of tears. He examines **how it is that the tears of wild boars are sweet, whereas those of deer are salty and 'ordinary'** (Διὰ τίν' αἰτίαν τὸ τῶν ἀγρίων συῶν δάκρυον ἡδὺ τὸ δὲ τῶν ἐλάφων ἁλμυρόν ἐστι καὶ φαῦλον;). Two explanations are given, both of which are – remarkably enough – formulated dogmatically rather than interrogatively for unclear reasons. The first one focuses on the animals' bodily temperature, the second on the constitution of their blood.

According to the **first** explanation, heat and cold are the main cause of this difference (Αἰτία δὲ θερμότης καὶ ψυχρότης τούτων). The deer has a cold nature but the boar a very hot and fiery one (καὶ ψυχρὸν μὲν ὁ ἔλαφος περίθερμον δὲ καὶ πυρῶδες ὁ σῦς). That is why the former flees and the latter defends itself when chased, and it is then that the boar especially sheds its tears in rage (ὅθεν τὸ μὲν φεύγει τὸ δ' ἀμύνεται τοὺς ἐπιόντας, ὅτε καὶ μάλιστα διὰ τὸν θυμὸν ἐκβάλλει τὸ δάκρυον). A great amount of heat passes into its eyes, so that the (salty tears) melt and become sweet (πολλῆς γὰρ ἐπὶ τὰ ὄμματα θερμότητος φερομένης […] γλυκὺ γίνεται τὸ ἀποτηκόμενον). In order to illustrate this, Plutarch quotes a line from Homer: 'Setting up his bristly manes, flashing fire with his eyes' (ὡς εἴρηται 'φρίξας εὖ λοφιήν, πῦρ ὀφθαλμοῖσι δεδορκώς').

The **second** explanation connects the quality of tears with that of blood. Plutarch reports that according to some people, like Empedocles, tears are discharged from the blood when it is stirred, just as whey (i.e. milk serum) is from milk (ἔνιοι δέ φασιν, ὥσπερ γάλακτος ὀρρὸν τοῦ αἵματος ταραχθέντος ἐκκρούεσθαι τὸ δάκρυον, ὡς Ἐμπεδοκλῆς). Since, accordingly, the blood of boars is rough and black because of their heat (cf. the previouse *causa*), while that of deer is thin and watery, it is reasonable that the secretion, in both states of rage and of fear (respectively), is like (the quality of) the blood of either animal (ἐπεὶ τοίνυν τραχὺ καὶ μέλαν τὸ τῶν κάπρων αἷμα διὰ θερμότητα λεπτὸν δὲ καὶ ὑδαρὲς τὸ τῶν ἐλάφων, εἰκότως καὶ τὸ ἀποκρινόμενον ἐν τοῖς θυμοῖς καὶ τοῖς φόβοις ἑκατέρου τοιοῦτον).

916F Διὰ τίν' αἰτίαν τὸ τῶν ἀγρίων συῶν δάκρυον ἡδὺ τὸ δὲ τῶν ἐλάφων ἁλμυρόν ἐστι καὶ φαῦλον;: The tears of deer are 'ordinary' (φαῦλον), presumably because they are salty (as opposed to those of wild boars) and are, thus, comparable to 'normal', human tears. Therefore, the sweet taste of those of boars is, indeed, 'extraordinary'. The same problem is mentioned in passing in *Quaest. conv.* 700F, where it remains unsolved. There may be a connection in this problem with Book 27 of Ps.-Aristotle's *Problems* (ὅσα περὶ φόβον καὶ ἀνδρείαν), where the emotions of fear and courage are also treated in physiological terms (c.q. bodily cold and heat) [see 1.2.5, n. 226].

917A καὶ ψυχρὸν μὲν ὁ ἔλαφος περίθερμον δὲ καὶ πυρῶδες ὁ σῦς· ὅθεν τὸ μὲν φεύγει τὸ δ' ἀμύνεται τοὺς ἐπιόντας, ὅτε καὶ μάλιστα διὰ τὸν θυμὸν ἐκβάλλει τὸ δάκρυον: For the rage and fiery character of boars, cf. fr. 106 Sandbach. Cf. also Arist., *PA* 651a2–3, Xen., *Cyn.* 10, 8; 15–17, Pliny, *NH* 8, 207.

917A πολλῆς γὰρ ἐπὶ τὰ ὄμματα θερμότητος φερομένης ὡς εἴρηται 'φρίξας εὖ λοφιήν, πῦρ ὀφθαλμοῖσι δεδορκώς' γλυκὺ γίνεται τὸ ἀποτηκόμενον: = Hom., *Od.* 19, 446. The idea that heat destroys salty flavours is paralleled in *Q.N.* 5, 913C in the context of ripening fruit. No heating takes place in deer, which explains why their tears remain salty and 'ordinary'.

917A ἔνιοι δέ φασιν, ὥσπερ γάλακτος ὀρρὸν τοῦ αἵματος ταραχθέντος ἐκκρούεσθαι τὸ δάκρυον, ὡς Ἐμπεδοκλῆς: = Emp., DK31A78. Cf. Aët., *Plac.* 5, 22, 1 = Ps.-Plut. 909C.

917AB ἐπεὶ τοίνυν τραχὺ καὶ μέλαν τὸ τῶν κάπρων αἷμα διὰ θερμότητα λεπτὸν δὲ καὶ ὑδαρὲς τὸ τῶν ἐλάφων, εἰκότως καὶ τὸ ἀποκρινόμενον ἐν τοῖς θυμοῖς καὶ τοῖς φόβοις ἑκατέρου τοιοῦτον: It remains unexplained precisely why rough, black blood makes for sweet tears and thin, watery blood for salty ones. Nevertheless, there seems to be a connection with *Q.N.* 3, 912E, where Plutarch argues that the blood of animals (c.q. cattle) that lick salt becomes thin (λεπτύνεται). This may explain the thinness of deer blood, and, by implication, the secretion of salty tears from it. According to Aristotle, the blood of deer is watery and cold (*Mete.* 384a26–28) and that of boars fibrous (*PA* 651a3: ἰνωδέστατον). Cf. also *PA* 648a2–5 (thick and warm blood makes for strength, while thin and cold blood is conducive to sensation and intelligence) and 650b28 (animals that have very watery blood are somewhat timorous).

ം

Q.N. 21, 917BD

Q.N. 21 is closely related to the previous problem by its zoological focus (and especially by its attention to wild boars in the final *causa*). Plutarch wonders **why domesticated sows farrow more than once (perhaps implying twice; see the final comment), and at various moments, while wild sows farrow only once and almost during the same period of days** (Διὰ τί τῶν ὑῶν αἱ μὲν ἥμεροι πλεονάκις τίκτουσι καὶ κατ' ἄλλον ἄλλαι χρόνον, αἱ δ' ἄγριαι καὶ ἅπαξ καὶ περὶ τὰς αὐτὰς ἁπάσας σχεδὸν ἡμέρας;). Plutarch adds that this is at the beginning of the summer, when the rainy months are over. This is illustrated by the proverb: 'It no longer rains the night that the wild sow farrows' (αὗται δ' εἰσὶν ἀρχομένου θέρους· διὸ καὶ λέλεκται 'μηκέτι νυκτὸς ὕειν, ᾗ κεν τέκῃ ἀγροτέρη σῦς'). Four explanations are provided: the first is based on the theory of generative residues, the second focuses on the effects of leisure and action on the animal's body, the third draws attention to the visual stimulus for animal impulses (ὁρμαί) and the fourth deals with the fertility of wild boars (as implicitly linked to that of sows). Owing to the close structural connections between the separate explanations (by means of several connective phrases: cf. ἅμα συνημμένον in *causa* two and ἢ καί in *causa* three and four [see 4.3.3.1., n. 262]), there is a strong sense of argumentative coherence in the aetiology.

In the **first** *causa*, Plutarch explains that domestic sows have a great amount of food to produce the generative residue. He supports this with a Euripidean fragment: 'Love is in satiety' (Ἡ διὰ πλῆθος τροφῆς, ὡς ὄντως 'ἐν πλησμονῇ Κύπρις'). The generative residue is produced by an abundance of food both in plants and in animals (ἀφθονία γὰρ τροφῆς τὸ γόνιμον περίττωμα ποιεῖ καὶ φυτοῖς καὶ ζῴοις). While wild sows have to search for their own food in the wilderness and in fear, domesticated sows constantly have it at hand, either growing naturally or prepared for them (αἱ μὲν οὖν ἄγριαι δι' αὑτῶν καὶ μετὰ φόβου τὴν τροφὴν ζητοῦσι, ταῖς δ' ἡμέροις ὑπάρχει διὰ παντὸς ἡ μὲν αὐτοφυὴς ἡ δ' ἐκ παρασκευῆς).

The **second** explanation focuses on the effects of leisure and action on the animal's body. It is directly connected with the previous one, with which it can be combined (ἢ τὸ τῆς σχολῆς καὶ ἀσχολίας ἅμα συνημμένον;). Plutarch argues that while domesticated sows are lazy and unwilling to wander far away from their swineherds, wild ones traverse mountains and run around. Thus, they dissipate their entire nourishment and use it up for maintaining their bodies (αἱ μὲν γὰρ ἀργοῦσι, μὴ βουλόμεναι πόρρω πλανᾶσθαι τῶν συφορβῶν, αἱ δ' ὀρειβατοῦσαι καὶ περιθέουσαι τὴν τροφὴν διαφοροῦσι καὶ καταναλίσκουσιν εἰς τὸ σῶμα πᾶσαν). As a consequence of this permanent exertion, no (generative) residue is formed (ὥστε διὰ τὸ ἀεὶ συντείνειν μὴ γίνεσθαι περίττωμα).

In the **third** explanation, Plutarch argues that the fact that domesticated sows are fed and herded together with the males reminds them of sex and excites their desire (ἢ καὶ τὸ συντρέφεσθαι καὶ συναγελάζεσθαι τὰ θήλεα τοῖς ἄρρεσιν ἀνάμνησιν ποιεῖ τῶν ἀφροδισίων καὶ συνεκκαλεῖται τὴν ὄρεξιν). A line from Empedocles, originating from the context of human passions, is quoted to illustrate this: 'Desire comes upon him, being reminded by his sight' (ὡς ἐπ' ἀνθρώπων Ἐμπεδοκλῆς ἐποίησε 'τῷ δ' ἐπὶ καὶ πόθος εἶσι δι' ὄψιος ἀμμιμνήσκων'). In wild sows, by contrast, which are reared separately from each other, their lack of natural affection and unsociability blunts and quenches their impulses (ἐν δὲ τοῖς ἀγρίοις, ἀποτρόφοις οὖσιν ἀλλήλων, τὸ ἄστοργον καὶ δυσεπίμικτον ἀμβλύνει καὶ ἀνασβέννυσι τὰς ὁρμάς).

The **fourth** and final explanation takes into account the fertility of wild boars. Plutarch wonders whether perhaps Aristotle's account is true (ἢ καὶ τὸ λεγόμενον ὑπ' Ἀριστοτέλους ἀληθές ἐστιν). Aristotle says that Homer gives the name χλούνης ('castrated boar') to the boar that has only one testicle, and that the testicles of most boars get crushed by their rubbing themselves against tree-stumps (ὅτι 'χλούνην' Ὅμηρος ὠνόμασε σῦν τὸν μόνορχιν; τῶν γὰρ πλείστων φησὶ προσκνωμένων τοῖς στελέχεσι θρύπτεσθαι τοὺς ὄρχεις).

917B Διὰ τί τῶν ὑῶν αἱ μὲν ἥμεροι πλεονάκις τίκτουσι καὶ κατ' ἄλλον ἄλλαι χρόνον, αἱ δ' ἄγριαι καὶ ἅπαξ καὶ περὶ τὰς αὐτὰς ἅπασαι σχεδὸν ἡμέρας; αὗται δ' εἰσὶν ἀρχομένου θέρους: As F.H. Sandbach, 1965, p. 195, n. d observes, "Plutarch does nothing to answer the second half of his question, namely why all the wild sows farrow at the same time" (περὶ τὰς αὐτὰς ἅπασαι σχεδὸν ἡμέρας). Neither does he explain that domesticated sows farrow 'at various moments' (κατ' ἄλλον ἄλλαι χρόνον). One may wonder whether this temporal specification perhaps originally belonged to an introductory πότερον *causa* that was later rewritten and integrated in the *quaestio* (note that the first *causa* opens with Ἦ). This remains unclear. Alternatively, L. Senzasono, 2006, pp. 209–210, n. 125 speaks of Plutarch's "scrupulo di completezza". In addition, although the formulation of the *quaestio* is not very clear for these points, it seems unlikely that Plutarch implies 1) that wild sows farrow only once in their entire life rather than once a year (cf. F.H. Sandbach, 1965, pp. 194–195, n. c and L. Senzasono, 2006, p. 205, n. 117), and 2) that the domesticated sows farrow on several *fixed* times a year. 1) The idea that wild sows farrow once a year is confirmed by Ps.-Arist., *Probl.* 896a20–29, Ps.-Arist./Alex. Aphr., *Suppl. probl.* 2, 155 (= *Probl. ined.* 2, 152), Pliny, *NH* 8, 212 and M. Glycas, *Ann.* 1, 62 (p. 119, 22–120, 2 Bekker). As to the season in which they litter, Arist., *HA* 578a26 writes spring, instead of the beginning of summer – on the basis of which H. Flashar, 1962, p. 526 prefers to read ἦρος here in *Q.N.* 21 with manuscript B. 2) Regarding the second point, F.H. Sandbach, 1965, pp. 194–195, n. c notes that "[i]t is unlikely that Plutarch believed that domesticated sows farrow several times a year". However, Plutarch clearly writes that they do this 'at various moments' (κατ' ἄλλον ἄλλαι χρόνον). Therefore, I take it that Sandbach means that it is unlikely that domesticated sows farrow several *fixed* times a year, which may very well be implied, indeed (cf. Ps.-Arist., *Probl.* 896a21, regarding the mating of domesticated pigs: ὅτε ἔτυχεν). Notably, domesticated sows bear twice per year according to Ps.-Arist./Alex. Aphr., *Suppl. probl.* 2, 155 (= *Probl. ined.* 2, 152); cf. also Pliny, *NH* 8, 205 and M. Glycas, *Ann.* 1, 62 (p. 119, 22–120, 2 Bekker). In Ps.-Arist./Alex. Aphr., *Suppl. probl.* 2, 144 (= *Probl. ined.* 2, 141), we read that domesticated sows πλεονάκις τίκτουσι (the same wording as in Plutarch's *quaestio*, where it is clearly contrasted with ἅπαξ). Cf. also Arist., *HA* 542a27–30, 572a5–8 and Ps.-Arist., *Probl.* 896a23. Sandbach notes in the margin that "it is only by very early weaning that modern pig-keepers obtain three [litters]". As regards the number of piglets yielded at each farrowing, according to *Suppl. probl.* 2, 144 (= *Probl. ined.* 2, 141), wild pigs bear seven young at the most, whereas domesticated sows bear thirteen.

917B διὸ καὶ λέλεκται 'μηκέτι νυκτὸς ὕειν, ἥ κεν τέκῃ ἀγροτέρη σῦς': The author and origin of this hexameter are unknown. According to R. Strömberg, 1954, p. 91, we are dealing with a proverb that may originate from a poem on weather signs or from a comic pastiche of such a weather prognosis (cf. Ar., *Pax* 1083, 1086). As such, it may have an ironic connotation, implying that "circumstances must not be unfavourable when a villain makes a coup". Strömberg argues, moreover, that the particle κεν does not necessarily imply that the verse has an Aeolic or Tessalian origin, and that the infinitive (ὕειν) may depend on a δεῖ (it is perhaps more plausible that Plutarch simply adapted the verse to the new syntactical context, depending on λέλεκται). Erasmus includes the verse in *Adag. Chil.* 2, 5, 43 (= 1443): *non iam nocte pluet, sus qua enitetur agrestis*, "The wild sow's farrow will bring a fine morrow" (R.A.B. Mynors, 1991, pp. 260–261). He explains that it has proverbial value: *per iocum usurpari poterit si dicamus res fore tranquilliores, posteaquam morosus ac rixosus quispiam animo suo morem gerat* (Mynors: "It will be possible to use it in jest, suppose we wish to say that things will be quieter, once some quarrelsome person who is difficult to please has relieved his feelings."). According to L. Senzasono, 2006, pp. 205–206, n. 118, however, the proverb should be interpreted literally, like most other agricultural or meteorological proverbs (as a potential source for this proverb, Senzasono thinks of a work similar to Aratus' *Phaenomena*). Plutarch does, indeed, interpret the verse literally here, but then again, he has a well-known custom of lifting verses from their original context and providing new meanings to them [see 4.2.1.1.]. According to F.H. Sandbach, 1965, p. 195, n. d, the verse quoted in *De prim. frig.* 949B – i.e. Call., fr. 787 Schneider (rejected by Pfeiffer): εἰ δὲ νότος βορέην προκαλέσσεται, αὐτίκα νίψει – may derive from the same source, but this is uncertain. In any case, the same metrum, introductory phrase (διὸ καὶ λέλεκται), and meteorological context do not lend any absolute certainty (cf. also Senzasono).

917B Ἡ διὰ πλῆθος τροφῆς, ὡς ὄντως 'ἐν πλησμονῇ Κύπρις': The same fragment (= Eur., TGF 895; cf. also *De tuenda* 126C) is cited by Ps.-Arist., *Probl.* 896a22–24 regarding the fertility of wild *vis-à-vis* domesticated animals (c.q. boars). The full verse runs as follows: ἐν πλησμονῇ τοι Κύπρις, ἐν πεινῶντι δ' οὔ. For the relation between the fertility of sows and their nourishment, cf. also Arist., *HA* 542a27–28, Ps.-Arist./Alex. Aphr., *Suppl. probl.* 2, 144 (= *Probl. ined.* 2, 141); 2, 155 (= *Probl. ined.* 2, 152), M. Glycas, *Ann.* 1, 62 (p. 119, 22–120, 2 Bekker).

917B ἀφθονία γὰρ τροφῆς τὸ γόνιμον περίττωμα ποιεῖ καὶ φυτοῖς καὶ ζῴοις: For the idea that the generative residue is produced by an abundance of food, cf. also *Q.N.* 30, 919C [see 4.3.4.1., n. 295].

917B αἱ μὲν οὖν ἄγριαι δι' αὑτῶν καὶ μετὰ φόβου τὴν τροφὴν ζητοῦσι, ταῖς δ' ἡμέροις ὑπάρχει διὰ παντὸς ἡ μὲν αὐτοφυὴς ἡ δ' ἐκ παρασκευῆς: Seeing that wild sows do not consume much nourishment (since they have to search for their own food in the wilderness and in fear), they can produce only a small amount of generative residue, rendering them less fertile than domesticated sows. The addition of μετὰ φόβου (related to wild sows) should perhaps be interpreted in opposition to Κύπρις (related to domesticated sows). Where there is fear, there can be no love and, by implication, no satiety or generative residue.

917C ἢ τὸ τῆς σχολῆς καὶ ἀσχολίας ἅμα συνημμένον;: I follow F.H. Sandbach, 1965, p. 194, n. 3, who conjectures συνημμένον for συμμένον in the manuscripts (preferred by L. Senzasono, 2006, p. 207, n. 121; cf. also V. Ramón Palerm, 2005, p. 401). The former (συνημμένον) metaphorically denotes a "combination in thought" (LSJ, s.v. συνάπτω, i, 2), the latter "hold together" (LSJ, s.v. συμμένω). One late manuscript reads συμβαῖνον (also Hubert). Wyttenbach suggests συναίτιον, and Bernardakis συμβαῖνον <αἴτιον>. On the assumption that Sandbach's conjecture is correct, note that the technical, logical meaning of τὸ συνημμένον (ἀξίωμα) as a "hypothetical proposition as premiss in a syllogism" is not at issue here (LSJ, s.v. συνάπτω, iii, 3; cf., e.g., *De aud.* 43C, *De E* 387A).

917C αἱ μὲν γὰρ ἀργοῦσι, μὴ βουλόμεναι πόρρω πλανᾶσθαι τῶν συφορβῶν, αἱ δ' ὀρειβατοῦσαι καὶ περιθέουσαι τὴν τροφὴν διαφοροῦσι καὶ καταναλίσκουσιν εἰς τὸ σῶμα πᾶσαν: According to Arist., *HA* 578a26–28, wild sows leave for inaccessible and shady regions with precipices and ravines in order to litter. The aspect of fear (μετὰ φόβου) from the first *causa* may be implicit here.

917C ἢ καὶ τὸ συντρέφεσθαι καὶ συναγελάζεσθαι τὰ θήλεα τοῖς ἄρρεσιν ἀνάμνησιν ποιεῖ τῶν ἀφροδισίων καὶ συνεκκαλεῖται τὴν ὄρεξιν: It is unlikely that the first καί should be read in conjunction with the καί that follows, because Plutarch does not repeat the article τό. The first καί (in conjunction with ἤ) may be used adverbially, expressing "emphatic assertion or assent" (see LSJ, s.v.); alternatively, it may connect the *causa* at hand with the previous one [see 4.3.3.1., n. 262].

917C ὡς ἐπ' ἀνθρώπων Ἐμπεδοκλῆς ἐποίησε 'τῷ δ' ἐπὶ καὶ πόθος εἶσι δι' ὄψιος ἀμμιμνήσκων': This fragment is only known from this passage in Plutarch (= Emp., DK31B64; the soundness of the text is debated). For text critical remarks, see C. Hubert, 1960, p. 19, F.H. Sandbach, 1965, p. 197, n. a, M.R. Wright, 1981, p. 218, B. Inwood, 2001, p. 283, V. Ramón Palerm, 2005, pp. 401–402, L. Senzasono, 2006, pp. 207–208, n. 122. Wright's position is the most critical one: "The fragment is hopelessly corrupt, and as with other lines having Plutarch as the only source [...], it may be that his memory failed him." However, it is not out of the question that Plutarch deliberately adapted the verse to the new context, as he more often does [see 4.2.1.1.]. The sensory aspect of sight (δι' ὄψιος, being Wyttenbach's conjecture for διὰ πέψεως in the manuscripts) is central to Plutarch's argument. For sight as an impulse to love, cf. also *Quaest. conv.* 681A.

917D ἢ καὶ τὸ λεγόμενον ὑπ' Ἀριστοτέλους ἀληθές ἐστιν ὅτι 'χλούνην' Ὅμηρος ὠνόμασε σῦν τὸν μόνορχιν; τῶν γὰρ πλείστων φησὶ προσκνωμένων τοῖς στελέχεσι θρύπτεσθαι τοὺς ὄρχεις: In *HA* 578a32–b5, Aristotle quotes the same Homeric passage (*Il.* 9, 539) and writes that young boars catch a disease that causes itching of the testicles. To stop the irritation, they scratch their testicles against trees (πρὸς τὰ δένδρα) and damage them, which results in *full* castration (τομίας). The same phenomenon is recorded with a similar (but not the same) Homeric reference in Ps.-Arist./Alex. Aphr., *Suppl. probl.* 2, 145 (= *Probl. ined.* 2, 142; see S. Kapetanaki and R.W. Sharples, 2006, p. 225, n. 457), where it illustrates another problem, viz. why there are many wild boars but few wild swine (Διὰ τί ἄρρενες μὲν ὗες ἄγριοι πολλοί εἰσι, κάπροι δ' ὀλίγοι;). This is ascribed to the fact that young swine catch an itching of the testicles, so that they scratch and destroy them. Xenophon reports that small boars use stones to rub themselves against (*Mem.* 1, 2, 30). The author of *Suppl. probl.* 2, 145 may rely on Arist., *HA* 578a32–33, which specifically concerns the relation between castration, savagery and magnitude in wild boars: see F.H. Sandbach, 1965, p. 197, n. b (for the opposite idea that boars become gentle by castration, see however fr. 106 Sandbach). Sandbach also believes that Plutarch perhaps found the passage at hand in Aristotle's Περὶ Ὁμήρου (F.H. Sandbach, 1982, pp. 225–226). But Plutarch probably only knew this work indirectly (see J. Bouffartigue, 2012, p. 108, n. 285). There may still be reason to assume that Plutarch in *Q.N.* 21 generally draws on one or more lost Aristotelian problems, which, in their turn, probably originate from Aristotle's original text. Notably, the theme of *Q.N.* 21 ties in closely with that of Book 10 of the Ps.-Aristotelian *Problems* (the Ἐπιτομὴ Φυσικῶν [see 1.1.1., n. 4]), which draws heavily on Aristotle's zoological and biological writings, and often specifically deals with copulation, generation and with the number and nature of offspring in animals. According to U.C. Bussemaker, 1857, p. xviii, the reference in the last *causa* in *Q.N.* 21, 917CD is to *Suppl. probl.* 2, 145, but H. Flashar, 1962, p. 360 believes that Plutarch is rather referring to Arist., *HA* 578b1, or perhaps even to a lost problem combining that passage with *Probl.* 896a20–29 (concerning the number and time of copulation in wild and domesticated animals *vis-à-vis* humans; with a reference to Eur., TGF 895). Flashar adds (p. 307) that *Probl.* 896a20–29 is probably an abridged version of that lost problem, which in itself must have looked more or less like Plutarch's *Q.N.* 21 as a whole. Flashar does not, however, pay attention to the adaptation of the Aristotelian material to the new context in *Q.N.* 21 ("Das echt-arist. Problem muß dann etwa so ausgesehen haben, wie der ganze Abschnitt bei Plut."). Cf. also *ibid.*, pp. 526–527 (with further parallels between *Q.N.* 21 and *Probl.* 896a20–29). Three further remarks should be made, though. 1) First of all, there is no parallel for Plutarch's third *causa* in Aristotle (regarding sociability in swine, and the lack of it), unless this is implied – though this is unlikely – in the concept of ἀλέα in *Probl.* 896a23–24 (ἢ διὰ τὴν τροφὴν καὶ ἀλέαν καὶ πόνον;). This concept is translated as 'heat' by Flashar (and probably correctly so: see J. Jouanna, 1982 and R. Mayhew, 2011a, p. 323), rather than as 'escape' or 'shelter' (which is the meaning of the homonymous ἀλέα, cf. LSJ, s.v. A). No reference is made to this 'heat' in the third *causa* (unless this is implied in the sow's sexual desire). 2) Second, what is supplemented in *Probl.* 896a24–29 (viz. the reference to sheep in Magnesia and Lybia bearing young twice, and the reason being the long period of gestation) is nowhere to be found in Plutarch. 3) Third and most importantly, Flashar pays no attention to the adaptation of the

Aristotelian material to the new context in *Q.N.* 21 – in fact, Plutarch is simply reduced to Aristotle, or even worse, to a lost Aristotelian problem. Most notably, Plutarch changes τομίας into μόνορχις ('with one testicle only'). It would, of course, be absurd to claim that all boars become fully castrated during their youth, because in this way, the species of boars would soon be extinct (Plutarch writes πλεῖστοι rather than πάντες, whereas Aristotle simply has νέοι). The claim of L. Senzasono, 2006, p. 209, n. 123 that Plutarch knew that the term τομίας in Aristotle does not necessarily indicate the full castration and privation of both testicles but only one is not convincing, since it does not explain why Plutarch did not simply copy τομίας, then (Senzasono also says that the root τεμ- indicates a cut and thus implies a mutilation, but in Arist., *HA* 575b1 it clearly refers to full castration). Bearing in mind the number of litters vaguely referred to in the *quaestio* – viz. πλεονάκις for the domesticated sow and ἅπαξ for the wild sow – Plutarch's adaptation may perhaps imply that half the number of testicles diminishes the fertility of wild boars and thereby that of wild sows – perhaps, indeed, by a factor of two. The fact that domesticated sows have two litters, while wild sows have only one is confirmed by several ancient authors (see the comments to the *quaestio*). It seems plausible, then, that we are not dealing with a redundant *fait divers* in this final explanation, because it not only concerns the fertility of wild boars, but indirectly also that of wild sows. Thus, it links up closely with the *quaestio* at hand. A digression into total infertility and castration (τομίας) would be of no use here, and thus an adaptation of Aristotle's text was mandatory for Plutarch (μόνορχις). For the idea that impregnation is possible with only one testicle, see Arist., *GA* 765a23–31. For the opposite idea that horses with only one testicle are infertile or beget such offspring, see *Hipp. Berol.* 14, 1, 11–13 (= *Corp. Hipp. Graec.* 1, p. 78, 15–17 Oder – Hoppe): τοὺς μονόρχεις δὲ οὐ δεῖ παραλαμβάνειν, ἀγόνους ὄντας κατὰ τὸ πλεῖστον ἢ ὅμοιον γεννῶντας.

917D 'χλούνην': 'Castrated boar' (τομίας <ὗς>) is one possible meaning of the word χλούνης (cf. LSJ, s.v. and the scholia to the Homeric passage at hand). In his *Lexicon Homericum*, Apollonius the Sophist analyses the etymology of χλούνης on the basis of a contraction of χλόη and εὐνή (viz. χλοεύνης, ὁ ἐν τῇ χλόῃ εὐναζόμενος, "couching in the grass or greenwood"; cf. also *Etym. magn.* 812, 44–51, s.v. χλόη and LSJ, s.v. ii, 4). According to P. Louis, 1968, p. 165, n. 2, this is the meaning of χλούνης in Arist., *HA* 578b1: he bases his argument on the phrase πρὸς τὰ δένδρα, but it is clear from the broader context that the primary connotation there is rather that of castration (cf. τομίας).

Q.N. 22, 917D

In *Q.N.* 22 Plutarch wonders **why people say that the she-bear's 'hands' (c.q. fore-paws) have the sweetest flesh and why they are the most delicious to eat** (Διὰ τί τῆς ἄρκτου φασὶ τὴν χεῖρα γλυκυτάτην ἔχειν σάρκα καὶ φαγεῖν ἡδίστην;). The treatment of the topic of sweetness in a zoological context recalls the problem in *Q.N.* 20 (regarding the sweet tears of wild boars). The explanation that is given here may again contain an allusion to the relation between sweetness and heat. Plutarch provides only one explanation drawing attention to the processes of concoction and 'transpiration'.

Plutarch argues that those body parts that most concoct the food provide the most delicious meat (Ἢ ὅτι τὰ πέττοντα τὴν τροφὴν μάλιστα τοῦ σώματος παρέχει τὸ κρέας ἥδιστον;). Subsequently, he asserts that the best concoction is by what 'transpires', being most in motion and most exercised (πέττει δὲ κάλλιστα τὸ διαπνέον, κινούμενον μάλιστα καὶ συγγυμναζόμενον). Plutarch then applies the case of the she-bear to these two premises (ὥσπερ ἡ ἄρκτος): the bear makes the most movement with its fore-paws, which it uses as feet when walking or running and as hands when grasping (τῷ μέρει τούτῳ πλεῖστα κινεῖται· καὶ γὰρ ὡς ποσὶ τοῖς ἐμπροσθίοις βαδίζουσα χρῆται καὶ τρέχουσα καὶ ὡς χερσὶν ἀντιλαμβανομένη).

917D Διὰ τί τῆς ἄρκτου φασὶ τὴν χεῖρα γλυκυτάτην ἔχειν σάρκα καὶ φαγεῖν ἡδίστην;: *Pace* O. Keller, 1887, p. 122 and 1980, p. 180, who believes that the delicacy of the bear's paws (and frog's legs) was a medieval invention unknown in Antiquity. On boiled bear meat, cf. Pliny, *NH* 8, 128. In Petr., *Sat.* 66, 5–6 bear meat is considered disgusting and is compared with that of wild boars. In the Finnish epic *Kalevala*, the bear has several stock euphemisms, which refer to the sweetness of its paws (as reported by F.H. Sandbach, 1965, p. 199, n. a), but the references are mostly to the sweet taste of honey rather than of the flesh itself: *mesikämmen* (honey palm), *mesikäpälä* (honey paw), *mesiloappa* (honey greedy). Thanks are due to Kristiina Näyhö from the Finnish Literature Society (SKS) for providing useful information on this matter.

917D Ἢ ὅτι τὰ πέττοντα τὴν τροφὴν μάλιστα τοῦ σώματος παρέχει τὸ κρέας ἥδιστον; πέττει δὲ κάλλιστα τὸ διαπνέον, κινούμενον μάλιστα καὶ συγγυμναζόμενον: Without further explanation, Plutarch connects the process of 'transpiration' (τὸ διαπνέον) with that of concoction, leaving unspecified what is their precise relation and what is actually meant with the former. On the basis (presumably) of the phrase τὰ πέττοντα τὴν τροφήν, C. Hubert, 1960, p. 20 suggests that τὸ διαπνέον implies τὴν τροφήν ('what transpires the food'), but he remains uncertain, and rightly so in my belief (in any case, concoction is not simply the same as 'transpiration'). According to L. Senzasono, 2006, pp. 211–212, n. 128, it is not the food that is 'transpired', but the residues that it leaves behind (he refers to Ps.-Arist., *Probl.* 966a17–25, where no mention is made of 'transpiration' but only of a 'well ventilated body', σῶμα τὸ εὔπνουν). This does not help us much in clarifying the meaning of τὸ διαπνέον here. F.H. Sandbach, 1965, p. 199 (with n. b) translates τὸ διαπνέον as "what transpires", but he adds that "[t]he meaning is dubious and the text not above suspicion". The text is not suspect, but still I am inclined to side with Sandbach's abstract translation. In any case, διαπνεῖν, in the sense of 'to transpire', does not simply mean 'to sweat', but has to do with a more general process of evaporation (see LSJ, s.v. iii; the concept of sweat is not found in *Q.N.* 22, and in Ps.-Arist., *Probl.* 967a3, the 'transpiration' of the body by air is, in fact, explicitly opposed to sweating, c.q. by an enclosure of heat). It involves an active passage of air through an object or body (cf. LSJ, s.v. i: "blow through"; see, e.g., *Quaest. conv.* 702C and Ps.-Arist., *Probl.* 943b23 (with Hom., *Od.* 4, 567)). In the passage at hand, the 'transpiring' object (τὸ διαπνέον) is clearly the bear's hand (τὴν χεῖρα): it is this body part that "admits air" most by moving or exercising (see LSJ, s.v. i, 2 for the intransitive use). Since the bear's hand 'transpires' most, it is this body part that concocts the food most, so that it is sweetest. As to the connection between transpiration and concoction, a link can be drawn with *Q.N.* 27, 918E, where the verb διαπνεῖν is directly related to τὸ θερμόν, which in turn is related to τὴν γλυκύτητα (cold does not allow 'transpiration', but by shutting in the heat preserves the sweetness of the must). As such, both the processes of 'transpiration' and concoction imply an element of heat (for the connection between 'transpiration' and heat, cf. also, e.g., Ps.-Arist., *Probl.* 927b12, 936b18, 939b37). This heat may be implied in the passage at hand by the frequent motion and exertion of the bear's paws (κινούμενον μάλιστα, cf. esp. *De tuenda* 123A; cf. also *Q.N.* 8, 914B for the idea that motion fans the heat of the sea). Moreover, in *Q.N.* 20, 917A, heat also causes sweetness (by melting the salty tears of boars). Cf. also esp. *Quaest. conv.* 642C, for the belief that sheep that are bitten by wolves have the sweetest flesh, because the wolf's breath (πνεῦμα), which is very hot and fiery, makes it tender.

917D τῷ μέρει τούτῳ πλεῖστα κινεῖται· καὶ γὰρ ὡς ποσὶ τοῖς ἐμπροσθίοις βαδίζουσα χρῆται καὶ τρέχουσα καὶ ὡς χερσὶν ἀντιλαμβανομένη: The bear also uses its hands for tearing apart the nets of hunters: cf. *Q.N.* 28, 919A.

ֆֆ

The problems treated in *Q.N.* 23–25, 917E–918B are closely related to each other through the topic of hunting and, more specifically, by the influence of meteorological conditions on animal tracks and trails (c.q. spring, winter, dew and full moons). There is a dense cluster of parallels with Xen., *Cyn.* 5, 1–5 (presented schematically in [4.2.1.2.]). As noted earlier on, *Q.N.* 6 links up with this cluster of problems by its focus on the physical properties of dew probably in the broader context of hunting.

Q.N. 23, 917EF

In *Q.N.* 23 Plutarch wonders **why the season of spring is unfavourable for tracking scents** (Διὰ τί δυστίβευτος ἡ τοῦ ἔαρος ὥρα;). One explanation is given, where Plutarch refers to Empedocles' theory of emanations. At the end he incorporates a mythological coda.

In the first part, Plutarch refers to the theory of emanations. He argues that hounds, 'with nostrils tracking fragments of the limbs of wild beasts' as Empedocles says, pick up the emanations left behind by animals in the brushwood, but that these are obscured and confused in the spring by the profusion of scents from plants and shrubs (Πότερον αἱ κύνες, ὥς φησιν Ἐμπεδοκλῆς 'κέρματα θηρείων μελέων μυκτῆρσιν ἐρευνῶσαι', τὰς ἀπορροίας ἀναλαμβάνουσιν, ἃς ἐναπολείπει τὰ θηρία τῇ ὕλῃ, ταύτας δὲ τοῦ ἔαρος ἐξαμαυροῦσι καὶ συγχέουσιν αἱ πλεῖσται τῶν φυτῶν καὶ τῶν ὑλημάτων ὀσμαί). These blended odours that overflow from blooming flowers distract and deceive the hounds so that they cannot pick up the scent of animals (καὶ ὑπὲρ τὴν ἄνθησιν ὑπερχεόμεναι καὶ κεραννύμεναι περισπῶσι καὶ διαπλανῶσι τὰς κύνας τῆς τῶν θηρίων ὀσμῆς ἐπιλαβέσθαι). This is further illustrated in a mythological coda: as people say, nobody hunts around Mt. Etna in Sicily, because a great amount of mountain violets grows and flourishes in its meadows throughout the year. The sweet fragrance of this flower always occupies the place and captures the exhalations from the animals (διὸ περὶ τὴν Αἴτνην ἐν Σικελίᾳ φασὶ μηδένα κυνηγεῖν· πολὺ γὰρ ἀναφύεσθαι καὶ τεθηλέναι δι' ἔτους ἴον ὀρεινὸν ἐν τοῖς λειμῶσι, καὶ τὸν τόπον εὐωδίαν ἀεὶ κατέχουσαν ἁρπάζειν τὰς τῶν θηρίων ἀναπνοάς). A myth is told that Korè was abducted by Pluto when she was picking flowers in that region (λέγεται δὲ μῦθος, ὡς τὴν Κόρην ἐκεῖθεν ἀνθολογοῦσαν ὁ Πλούτων ἀφαρπάσειε). People therefore honour and worship the place as a sanctuary and do not attack the animals that graze there (καὶ διὰ τοῦτο τιμῶντες καὶ σεβόμενοι τὸ χωρίον ὡς ἄσυλον οὐκ ἐπιτίθενται τοῖς ἐκεῖ νεμομένοις).

917E Διὰ τί δυστίβευτος ἡ τοῦ ἔαρος ὥρα;: The adjective δυστίβευτος (LSJ, s.v.: "bad for scent") is very rare and recurs in *Q.N.* 25, 918A (cf. S.-T. Teodorsson, 2005, p. 409; perhaps to be corrected in δυσθήρευτος, 'bad for hunting': cf. Pl., *Soph.* 218d and 261a). For the concept of στιβεία/στιβεύω in Plutarch, cf. *Q.N.*, 25, 918B, *De Pyth. or.* 399A and *De soll. an.* 966D.

917E Πότερον αἱ κύνες, ὥς φησιν Ἐμπεδοκλῆς 'κέρματα θηρείων μελέων μυκτῆρσιν ἐρευνῶσαι', τὰς ἀπορροίας ἀναλαμβάνουσιν, ἃς ἐναπολείπει τὰ θηρία τῇ ὕλῃ, ταύτας δὲ τοῦ ἔαρος ἐξαμαυροῦσι καὶ συγχέουσιν αἱ πλεῖσται τῶν φυτῶν καὶ τῶν ὑλημάτων ὀσμαί: This fragment (= Emp., DK31B101) is also quoted in *De cur.* 520F. It may originate from the same Plutarchan hypomnema (see H. Martin, 1969, p. 70). For more on the fragrance of flowers obscuring the smell of tracks in spring, cf. Xen., *Cyn.* 5, 5 and Theophr., *CP* 6, 20, 4. See also Book 12 of the Ps.-Aristotelian *Problems* more generally (ὅσα περὶ τὰ εὐώδη), where the same emanation theory of odour is present throughout.

917EF διὸ περὶ τὴν Αἴτνην ἐν Σικελίᾳ φασὶ μηδένα κυνηγεῖν· πολὺ γὰρ ἀναφύεσθαι καὶ τεθηλέναι δι' ἔτους ἴον ὀρεινὸν ἐν τοῖς λειμῶσι, καὶ τὸν τόπον εὐωδίαν ἀεὶ κατέχουσαν ἁρπάζειν τὰς τῶν θηρίων ἀναπνοάς: According to F.H. Sandbach, 1965, p. 200, n. a, "[t]his version is perhaps somewhat forced. An emendation may be made, to give the meaning 'the nostrils of the hunters seize on the fragrance ...'". The emendation is based on a similar account in *Quaest. conv.* 647E (ὀσμαὶ ἁρπαζόμεναι ταῖς ὀσφρήσεσι), but it is unnecessary in the passage at hand (cf. also L. Senzasono, 2006, p. 214, n. 133).

917F λέγεται δὲ μῦθος, ὡς τὴν Κόρην ἐκεῖθεν ἀνθολογοῦσαν ὁ Πλούτων ἀφαρπάσειε: According to the Homeric *Hymn to Demeter* 6–8, Persephone gathers several kinds of flowers, including violets (albeit at the plain of Nysa: 17). For further commentary on 'the flower catalogue' and the actual location of Persephone's abduction (with a list of other locations), see N.J. Richardson, 1974, pp. 140–144 and 148–150. For Etna as the place of Persephone's abduction, see already Carcinus, TGF p. 799, fr. 5, 6 (cf. also Moschus, *Epit. Bionis* 121 and Hyg., *Fab.* 146). According to Ps.-Arist., *Mir. ausc.* 836b13ff. and Diod. Sic. 5, 3, 2, the abduction took place in the vicinity of Enna. For further references, see F. Bräuninger, 1937, cols. 952 and 966.

917F καὶ διὰ τοῦτο τιμῶντες καὶ σεβόμενοι τὸ χωρίον ὡς ἄσυλον οὐκ ἐπιτίθενται τοῖς ἐκεῖ νεμομένοις: This type of aetiology reminds the reader of the cultural-antiquarian inquiry in Plutarch's *Quaestiones Graecae* (cf. also *Q.N.* 10, 914D and 14, 915C [see 2.4.2.]). From the opening of the aetiology in the next problem (*Q.N.* 24, 917F: Ἦ διὰ τὴν εἰρημένην αἰτίαν;: see the commentary *ad loc.*), it can be inferred that the ending of this problem is lacunary (see F.H. Sandbach, 1965, p. 201, n. c). Perhaps, Plutarch intervened in the text to obtain a mythological finale here [see 4.1.2.2., n. 82].

Q.N. 24, 917F–918A

In *Q.N.* 24, we find yet another problem from the world of hunting. There is an explicit reference in the aetiology to a preceding explanation that cannot be clearly traced in *Q.N.* 23, and may therefore be missing (Ἡ διὰ τὴν εἰρημένην αἰτίαν;). Plutarch wonders **why hunters are least successful in tracking during full moons** (Διὰ τί περὶ τὰς πανσελήνους ἥκιστα ταῖς ἰχνοσκοπίαις ἐπιτυγχάνουσιν;). He gives one explanation, which draws attention to the precipation of dew by the moon. Again, the explanation contains a mythographical part.

As noted, the *causa* opens with a reference to a preceding explanation that cannot be traced (Ἡ διὰ τὴν εἰρημένην αἰτίαν;). Plutarch explains that full moons precipitate dew, which he illustrates with a quotation from Alcman, who allegorically calls Dew the daughter of Zeus and Moon: 'Dew, the daughter of Zeus and Selene, provides food' (δροσοβόλοι γὰρ αἱ πανσέληνοι· διὸ καὶ τὴν δρόσον ὁ Ἀλκμὰν Διὸς θυγατέρα καὶ Σελήνης προσεῖπε ποιήσας 'Διὸς θυγάτηρ Ἔρσα τρέφει καὶ Σελάνας δίας'). Dew is a weak and impotent kind of rain, and the heat of the moon is also weak. Therefore, the moon draws it up from the earth like the sun, but being unable to lift it to a height and to raise it up, drops it again (ἡ γὰρ δρόσος ἀσθενής ἐστι καὶ ἀδρανὴς ὄμβρος, ἀσθενὲς δὲ καὶ τὸ τῆς σελήνης θερμόν· ὅθεν ἕλκει μὲν ἀπὸ γῆς ὥσπερ ὁ ἥλιος, ἄγειν δ' εἰς ὕψος μὴ δυναμένη μηδ' ἀναλαμβάνειν μεθίησιν).

917F Διὰ τί περὶ τὰς πανσελήνους ἥκιστα ταῖς ἰχνοσκοπίαις ἐπιτυγχάνουσιν;: The noun ἰχνοσκοπία is a hapax (see LSJ, s.v.: "looking at the tracks"), but the verb ἰχνοσκοπῶ occurs more often.

917F Ἡ διὰ τὴν εἰρημένην αἰτίαν;: According to J. Opsomer (in personal communication), "the back reference is merely to the idea that some physical influences (the flow of water, for instance) may wipe away traces of scents". This is not implausible, but it seems too general (compare the back reference, ὡς εἰρήκαμεν, in *Q.N.* 16, 915E to *Q.N.* 15, 915DE, which is much more concrete). By contrast, F.H. Sandbach, 1965, p. 201, n. c believes that the explanation to which Plutarch refers probably drew attention to the fact that "dew is frequent in the spring and spoils the scent" of tracks (for the view that heavy dew obliterates the scent of tracks by carrying it downwards, cf. Xen., *Cyn.* 5, 3). At the end of *Q.N.* 23, "an alternative answer, anticipated by the introductory word πότερον, has been lost". It is, indeed, odd that no second *causa* follows after the πότερον *causa* in *Q.N.* 23 (for exceptions, see R. Kühner and B. Gehrt, 1966, §589, 10). We may be dealing with a case of hypomnematic negligence [see 2.3.2.] or intermittent composition [see the prologue, n. 23]. In either case, it is not impossible that *Q.N.* 24 is a separate reformulation and elaboration of what was originally the now lost second *causa* in *Q.N.* 23, which may have run as follows: 'Or is it because dew, which is frequent in springs, obscures the scent of the tracks? Full moons also precipitate dew, which is why hunters are least successful in following tracks during full moons.' Plutarch then felt that further explanation was necessary, thus why he devoted a new, separate chapter to it (c.q. *Q.N.* 24) with a ghost-reference to what was previously said. This remains hypothetical. One point that may support this theory is that Plutarch may have intended to maintain the mythological conclusion at the end of the (first) *causa* in *Q.N.* 23. In any case, Plutarch more often concludes his physical aetiologies with a mythological account [see 4.1.2.2.]. Moreover, if the myth had remained between the two original physical explanations, it would have seemed to be a redundant *fait divers*, which Plutarch does not, of course, hold it to be (cf. L. Senzasono, 2006, p. 215, n. 135: "Se è cosí, un motivo religioso era calettato entro due motivi naturalistici e quindi in un certo senso soverchiato da essi."). According to J. Schellens, 1864, pp. 19–20, however, there can be no doubt ("haud dubie") that *Q.N.* 25 originally preceded *Q.N.* 24. At the end of *Q.N.* 25, he would add: "Quare etiam circa plenilunia minime indagantur e vestigiis ferae. Nam plenilunia rorem spargunt qui et per se languidus est (οὐ κινεῖ τὴν ὄσφρησιν), et calore lunae parum valido nequit auferri." There is no problem with the ending of *Q.N.* 25, though, as is rather the case with that of *Q.N.* 23, as we saw.

917F–918A: δροσοβόλοι γὰρ αἱ πανσέληνοι: It was a common ancient belief that the moon generates dew. Cf., e.g., Macrob., *Sat.* 7, 16, 31 (with reference to Alcman) and Manilius, *Astron.* 4, 501–502 (with A.E. Housman, 1920, p. 61 for further references).

918A διὸ καὶ τὴν δρόσον ὁ Ἀλκμὰν Διὸς θυγατέρα καὶ Σελήνης προσεῖπε ποιήσας 'Διὸς θυγάτηρ Ἔρσα τρέφει καὶ Σελάνας δίας.' ἡ γὰρ δρόσος ἀσθενής ἐστι καὶ ἀδρανὴς ὄμβρος, ἀσθενὲς δὲ καὶ τὸ τῆς σελήνης θερμόν· ὅθεν ἕλκει μὲν ἀπὸ γῆς ὥσπερ ὁ ἥλιος, ἄγειν δ' εἰς ὕψος μὴ δυναμένη μηδ' ἀναλαμβάνειν μεθίησιν: For the idea that dew is a kind of rain, cf. also Arist., *Mete.* 347b17–22. For the weakness of the moon's heat, cf. *Quaest. conv.* 658B and *De facie* 929A. Xenophon reports that the moon obscures the scent directly by its heat (τῷ θερμῷ), especially when at full (*Cyn.* 5, 4). The idea that the moon draws up moisture from the earth ties in closely with Stoic exhalation theory (cf. SVF 2, p. 197, fr. 663 = *De Is. et Os.* 367E: οἱ δὲ Στωικοὶ τὸν μὲν ἥλιον ἐκ θαλάττης ἀνάπτεσθαι καὶ τρέφεσθαί φασι, τῇ δὲ σελήνῃ τὰ κρηναῖα καὶ λιμναῖα νάματα γλυκεῖαν ἀναπέμπειν καὶ μαλακὴν ἀναθυμίασιν). That Plutarch is probably drawing from a Stoic source is also supported by the allegorical explanation of Alcman's mythological account (= Alcm., 43 Diehl), this type of exegesis being particularly privileged by the Stoics (cf. P.R. Hardie, 1992, p. 4772, with n. 114). The same quote from Alcman and a similar allegorical interpretation recurs in the parallel passages in *Quaest. conv.* 659B and *De facie* 940A, where a slightly different explanation for the role of the moon in the production of dew is given. Attention is drawn there to a process of μεταβολή (the moon has a liquefying effect, and the air – Zeus in the quotation – is liquefied by the moon into dew), while in *Q.N.* 24 the explanation is based on the motive force of ὁλκή, which is of a mechanical kind (the aspect of lunar liquefaction is absent here). Cf. also F.H. Sandbach, 1965, p. 203, n. a. The attempt of L. Senzasono, 2006, p. 218, n. 141 to reconcile the differences between both explanations is at the risk of neglecting the different argumentative contexts in this cluster of parallels (cf. B. Van Meirvenne, 2001, pp. 292–293, n. 27) [see 2.1.2.].

ತಿ

Q.N. 25, 918AB

The topic of hunting continues in *Q.N.* 25 (cf. L. Senzasono, p. 219, n. 142). Plutarch wonders **why a domain that has become dewy due to coldness is unfavourable for following traces** (Διὰ τί τὸ δρόσιμον γενόμενον διὰ τοῦ ψύχους δυστίβευτον;). He gives two explanations: the first one refers to the quantity and (implicitly) to the visual aspect of the traces (ἴχνη), and the second to their smell (ὄσφρησις). Both features of vision and smell are closely related to each other by the phrase ἢ δεῖ μὴ μόνον κτλ. in the second *causa*.

In the **first** explanation, Plutarch argues that wild animals hesitate to go far from their lairs because of the frost. Thus, they do not produce many traces (Πότερον ὅτι τὰ θηρία πόρρω τῶν κοιτῶν ὀκνοῦντα προϊέναι διὰ τὸ κρύος οὐ ποιεῖ πολλὰ σημεῖα;). People say that for the same reason they spare the food close at home (in other seasons), so that they are not worried to wander far off in winter, but always have some food close to home (διὸ καί φασιν αὐτὰ φείδεσθαι τῶν πλησίον, ὅπως μὴ κακοπαθῇ πλανώμενα μακρὰν τοῦ χειμῶνος ἀλλ' ἀεὶ ἐγγύθεν ἔχῃ νέμεσθαι).

At the beginning of the **second** explanation, Plutarch makes a subtle distinction. He argues that hunting grounds must not only be marked by (visible) tracks, but must also affect the sense of smell (ἢ δεῖ μὴ μόνον ἔχειν ἴχνη τὸν στιβευόμενον τόπον ἀλλὰ κινεῖν τὴν ὄσφρησιν). This is the case when scents are loosened and gently released by heat, whereas excessive cold freezes the odours and does not let them flow or affect sensation (κινεῖ δὲ λυόμενα καὶ χαλώμενα μαλακῶς ὑπὸ θερμότητος, ἡ δ' ἄγαν περίψυξις πηγνύουσα τὰς ὀσμὰς οὐκ ἐᾷ ῥεῖν οὐδὲ κινεῖν τὴν αἴσθησιν;). People say that perfume and wine give off less smell in cold weather and in the winter. This is the case because the frozen air arrests the scents and does not allow them to be given off (ὅθεν καὶ τὰ μύρα καὶ τὸν οἶνον ἧττον ὄζειν ψύχους καὶ χειμῶνος λέγουσιν· ὁ γὰρ ἀὴρ πηγνύμενος ἵστησι τὰς ὀσμὰς ἐν αὑτῷ καὶ οὐκ ἐᾷ ἀναδίδοσθαι).

918A Διὰ τί τὸ δρόσιμον γενόμενον διὰ τοῦ ψύχους δυστίβευτον;: Bernardakis corrects the adjective δρόσιμον (a hapax) into the more, but still not very, common δροσινόν (cf. LSJ, s.v.: "= δροσερός"). For more on the rarity of the adjective δυστίβευτος, see the comment to *Q.N.* 23, 917E above. F.H. Sandbach, 1965, p. 203, n. b doubts the soundness of the text here and notes that "dew has nothing to do with the answer". There is nothing suspicious in the manuscripts, though, and the answer that Plutarch gives is still very pertinent. According to some scholars, it is not unlikely that Plutarch is referring to frozen dew. Longolius summarised the problem as follows: *Ros hybernus, hoc est pruina, indagationem difficilem reddit*. J. Schellens, 1864, p. 20 also speaks of "pruina" (i.e. frost or snow). Cf. also L. Senzasono, 2006, pp. 218–219, n. 142 ("brina"). Yet, as J. Opsomer notes (in personal correspondence), Plutarch literally speaks of 'dew due to cold' (it is unlikely that διὰ τοῦ ψύχους relates to δυστίβευτον); thus, the reference is not necessarily to hoarfrost.

918A Πότερον ὅτι τὰ θηρία πόρρω τῶν κοιτῶν ὀκνοῦντα προϊέναι διὰ τὸ κρύος οὐ ποιεῖ πολλὰ σημεῖα;: For more on the cleverness of animals to leave no traces, cf. *De soll. an.* 971D. Cf. also Ael., *NA* 6, 3.

918B κινεῖ δὲ λυόμενα καὶ χαλώμενα μαλακῶς ὑπὸ θερμότητος, ἡ δ' ἄγαν περίψυξις πηγνύουσα τὰς ὀσμὰς οὐκ ἐᾷ ῥεῖν οὐδὲ κινεῖν τὴν αἴσθησιν;: The negative effect of wintery cold on smells and on the hounds' perception thereof is also mentioned by Xenophon (*Cyn.* 5, 1–2), who notes, in addition, that it is only when the sun loosens (διαλύσῃ) the tracks or as the day advances that the dogs are able to smell and that the scent revives. For the dulling effect of cold on flavours and odours (and on our sensation of them), cf. also Theophr., *CP* 6, 17, 5 and *Od.* 40 with. Ps.-Arist., *Probl.* 907a8–12 and Ps.-Arist./Alex. Aphr., *Suppl. probl.* 2, 101.

918B ὅθεν καὶ τὰ μύρα καὶ τὸν οἶνον ἧττον ὄζειν ψύχους καὶ χειμῶνος λέγουσιν· ὁ γὰρ ἀὴρ πηγνύμενος ἵστησι τὰς ὀσμὰς ἐν αὑτῷ καὶ οὐκ ἐᾷ ἀναδίδοσθαι: The analogy with wine ties in closely with the problems on viniculture in *Q.N.* 30–31 (see also *Q.N.* 10 and 27). F.H. Sandbach, 1965, p. 202, n. 3 corrects πηγνύμενος in πηγνυμένας on the basis of the phrase πηγνύουσα τὰς ὀσμάς above and with a reference to *De prim. frig.* 951A, where the inability of air to freeze is mentioned (αὐτὸν τὸν ἀέρα μηδαμοῦ πηγνύμενον ὁρῶντες), but this seems unnecessary (cf. also L. Senzasono, 2006, p. 220, n. 145).

ॐ

Q.N. 26, 918BE

Q.N. 26 concerns another zoological problem, viz. **why animals seek and pursue substances that have remedying properties when they are ill, and often restore themselves to health by using them** (Διὰ τί τὰ ζῷα τὰς βοηθούσας δυνάμεις, ὅταν ἐν πάθει γένηται, ζητεῖ καὶ διώκει καὶ χρώμενα πολλάκις ὠφελεῖται;). The problem connects with the previous ones by the reference to animal sense perception (αἴσθησις).

The *quaestio* is further illustrated in an intermediate section where Plutarch gives several examples drawn from natural history. Bitches eat grass in order to vomit bile, sows capture and eat river crabs to relieve headaches, tortoises feed on marjoram as an antidote for the viper flesh they have eaten, and people say that the she-bear recovers from nausea by licking up and swallowing ants (καθάπερ αἱ κύνες ἐσθίουσι πόαν, ἵνα τὴν χολὴν ἐξεμῶσιν· αἱ δ' ὕες ἐπὶ τοὺς ποταμίους καρκίνους φέρονται, βοηθοῦνται γὰρ ἐσθίουσαι πρὸς κεφαλαλγίαν· ἡ δὲ χελώνη φαγοῦσα τὴν σάρκα τοῦ ἔχεως ὀρίγανον ἐπεσθίει· τὴν δ' ἄρκτον λέγουσιν ἀσωμένην τοὺς μύρμηκας ἀναλαμβάνειν τῇ γλώττῃ καὶ καταπίνουσαν ἀπαλλάττεσθαι). The actual problem is formulated at the very end of the *quaestio*, where Plutarch points out that these animals have no previous experience or have never tried these remedies before (τούτων δ' οὔτε πεῖρα οὔτε περίπτωσις γέγονεν αὐτοῖς). This means that they do not act on the basis of knowledge or insight, and that there must be another, more physical, reason for it.

Plutarch gives two explanations that are closely connected to each other. According to the first *causa*, it is by the attractive qualities of odours that animals find the proper cure for their disease. However, this does not explain why animals are attracted by these odours only when they are ill. Plutarch explains this point in the second *causa*, where he argues that the animal's bodily constitution (κρᾶσις) follows the disease, and as such influences the appetite. The two explanations are complementary to each other and must be read together for a proper understanding of the problem.

In the **first** explanation, Plutarch adds two new cases: he argues that in the same way as the odour of honeycombs excites and attracts bees from far off and that of carrion vultures, so do crabs attract sows, marjoram the tortoise and ants the she-bear (Πότερον οὖν, ὥσπερ τὰ κηρία τὴν μέλιτταν τῇ ὀσμῇ καὶ τὰ κενέβρεια τὸν γῦπα κινεῖ καὶ προσάγεται πόρρωθεν, οὕτως καὶ σῦς οἱ καρκίνοι καὶ τὴν χελώνην ἡ ὀρίγανος, αἱ δὲ μυρμηκιαὶ τὴν ἄρκτον). The attraction comes about by the odours and streams that are conducive and suitable to the animal's well-being, under the governance of the animal's sensation without any calculation of profit (ὀσμαῖς καὶ ῥεύμασι προσφερέσι καὶ οἰκείοις ἕλκουσιν, οὐ λογισμῷ τοῦ συμφέροντος ἀγούσης τῆς αἰσθήσεως).

COMMENTARY   455

The **second** explanation centres on the animal's κρᾶσις (i.e. its bodily constitution), as related to its ὄρεξις (appetite) and νόσος (disease). First, Plutarch generally argues that the animals' appetites are produced by their bodily constitutions, which are brought about by their diseases. These diseases create various pungencies, sweetnesses, or other strange and abnormal qualities in the body by the changes in its fluids (ἢ τὰς ὀρέξεις ἐπιφέρουσι τοῖς ζῴοις αἱ τῶν σωμάτων κράσεις, ἃς αἱ νόσοι ποιοῦσι, διαφόρους δριμύτητας ἢ γλυκύτητας ἤ τινας ἄλλας ἐντίκτουσαι ποιότητας ἀήθεις καὶ ἀτόπους, τῶν ὑγρῶν τρεπομένων;). This point is further illustrated by two cases of abnormal appetites. Plutarch first refers to the fact that pregnant women eat both stones and dirt (ὡς δῆλόν ἐστιν ἐπὶ τῶν γυναικῶν, ὅταν κύωσι, καὶ λίθους καὶ γῆν προσφερομένων). The second example is more circumstantial. Plutarch argues that accomplished physicians know in advance which patients are past recovery and which are capable of recovery on the basis of their appetites (διὸ καὶ τῶν νοσούντων ταῖς ὀρέξεσιν οἱ χαρίεντες ἰατροὶ προΐσασι τοὺς ἀσώτως ἢ σωτηρίως ἔχοντας). According to the physician Mnesitheus, at least, patients that are in the initial stage of 'pneumonia' recover when they have an appetite for onions but die when they long for figs, because the appetite follows the bodily constitution, which in turn, follows the disease (ἱστορεῖ γοῦν Μνησίθεος ἰατρὸς ἐν ἀρχῇ πνευμονίας τὸν ἐπιθυμήσαντα κρομμύων σῴζεσθαι τὸν δὲ σύκων ἀπόλλυσθαι, διὰ τὸ ταῖς κράσεσι τὰς ὀρέξεις τὰς δὲ κράσεις τοῖς πάθεσιν ἕπεσθαι). Plutarch concludes (with an argumentative ring [see 4.3.3.3.]) that it is plausible that animals also (i.e. just like human beings) that catch not entirely lethal and destructive diseases acquire precisely that bodily condition and constitution that leads and guides each one of them via their appetites to the things that save them (πιθανὸν οὖν ἐστι καὶ τῶν θηρίων τὰ μὴ παντελῶς ὀλεθρίοις μηδ' ἀναιρετικοῖς περιπίπτοντα νοσήμασι ταύτην τὴν διάθεσιν καὶ κρᾶσιν ἴσχειν, ὑφ' ἧς ἐπὶ τὰ σῴζοντα φέρεται καὶ ἄγεται ταῖς ὀρέξεσιν ἕκαστον αὐτῶν).

918B Διὰ τί τὰ ζῷα τὰς βοηθούσας δυνάμεις, ὅταν ἐν πάθει γένηται, ζητεῖ καὶ διώκει καὶ χρώμενα πολλάκις ὠφελεῖται;: This chapter is also recorded in the sourcebook of G.L. Irby-Massie and P.T. Keyser, 2002, p. 277 on *Greek Science of the Hellenistic Era*.

918CD καθάπερ αἱ κύνες ἐσθίουσι πόαν, ἵνα τὴν χολὴν ἐξεμῶσιν· αἱ δ' ὗες ἐπὶ τοὺς ποταμίους καρκίνους φέρονται, βοηθοῦνται γὰρ ἐσθίουσαι πρὸς κεφαλαλγίαν· ἡ δὲ χελώνη φαγοῦσα τὴν σάρκα τοῦ ἔχεως ὀρίγανον ἐπεσθίει· τὴν δ' ἄρκτον λέγουσιν ἀσωμένην τοὺς μύρμηκας ἀναλαμβάνειν τῇ γλώττῃ καὶ καταπίνουσαν ἀπαλλάττεσθαι: This intermediate section is meant to illustrate that such auto-remediation is a relatively common phenomenon in the animal kingdom. There are numerous parallels for these paradoxical accounts, both in the *corpus Plutarcheum* (for tortoises, dogs, and the she-bear, cf. *De soll. an.* 974B; for pigs and tortoises, cf. *Gryllus* 991EF), and in the ancient scientific tradition more generally, sometimes with minor variations. For bitches, cf. Arist., *HA* 612a5–6, 594a28–29, Ael., *NA* 5, 46; 8, 9, Gal., *Ven.Sect.Er.* 11, 168, 2–3 Kühn, Sext. Emp., *HP* 1, 71, Pliny, *NH* 25, 91, Cic., *De nat. deor.* 2, 126; for sows, cf. Pliny, *NH* 8, 98 (boars eat sea-crabs); for tortoises, cf. Arist., *HA* 612a24–25, Ps.-Arist., *Mir. ausc.* 831a27–28, Antig. Car., *Hist. mir.* 34, Ael., *NA* 3, 5; 6, 12 (they eat marjoram *before* attacking vipers), Pliny, *NH* 8, 98 and 20, 169 (they eat *cunila bubula*); for she-bears, cf. Arist, *HA* 594b9 (they eat crabs and ants), Ael., *NA* 6, 3, Sext. Emp., *HP* 1, 57, Pliny, *NH* 8, 101 (as an antidote against mandragora) and 29, 133 (they eat ants' eggs).

918C τούτων δ' οὔτε πεῖρα οὔτε περίπτωσις γέγονεν αὐτοῖς: F.H. Sandbach, 1965, p. 204, n. 3 adds διδασκαλία ποθὲν οὔτε ('they did not receive any instruction') between δ' οὔτε and πεῖρα on the basis of both *Gryllus* 991E (τίς δὲ τὰς χελώνας ἐδίδαξε κτλ.;) and Pliny, *NH* 27, 7 (*feris ratio et usus inter se tradi non possit*). This seems unnecessary, though (cf. also L. Senzasono, 2006, p. 223, n. 148).

918C Πότερον οὖν, ὥσπερ τὰ κηρία τὴν μέλιτταν τῇ ὀσμῇ καὶ τὰ κενέβρεια τὸν γῦπα κινεῖ καὶ προσάγεται πόρρωθεν, οὕτως καὶ σῦς οἱ καρκίνοι καὶ τὴν χελώνην ἡ ὀρίγανος, αἱ δὲ μυρμηκιαὶ τὴν ἄρκτον: For the effect of honeycombs on bees and of carrion on vultures, cf. Lucr., *De rer. nat.* 4, 678–680. For bees, cf. also Col., *De re rust.* 9, 15, 10. Plutarch will deal with the effect of strong smells on bees in *Q.N.* 35–36 in more detail.

918D ἢ τὰς ὀρέξεις ἐπιφέρουσι τοῖς ζῴοις αἱ τῶν σωμάτων κράσεις, ἃς αἱ νόσοι ποιοῦσι, διαφόρους δριμύτητας ἢ γλυκύτητας ἤ τινας ἄλλας ἐντίκτουσαι ποιότητας ἀήθεις καὶ ἀτόπους, τῶν ὑγρῶν τρεπομένων;: For the relation between ὄρεξις and κρᾶσις, cf. *Quaest. conv.* 687DE and 688A. By contrast, in *Quaest. conv.* 733D, Plutarch writes that the νόσοι follow the κρᾶσις of the body and not the other way around. Regarding the change of bodily fluids, cf., e.g., *Q.N.* 1, 911E, where Plutarch indicates that fevers turn moisture into bile.

918D ὡς δῆλόν ἐστιν ἐπὶ τῶν γυναικῶν, ὅταν κύωσι, καὶ λίθους καὶ γῆν προσφερομένων: Pregnancy is presented as some kind of a 'disease', which brings about a change in the bodily constitution of women and, by implication, in their appetite. The belief that pregnant women long for stones is repeated in *Praec. ger. reip.* 801A. Plutarch's source is uncertain: he may rely on hearsay or (indirectly) on a medical treatise, perhaps from the kind of Soranus' *Gynaecia* or the Hippocratic gynaecological writings. He may have found it in the same work from which he draws Mnesitheus' account (in what follows). Notably, the Lamprias catalogue mentions a lost work Περὶ γεωφάγων (nr. 191). Cf. also Hipp., *De superfetat.* 18, Arist., *HA*

584a19, *EN* 1148b24–29, Pliny, *NH* 28, 247. On geophagy more generally (which is considered rare among Greeks and Romans), see B. Laufer, 1930 (p. 164).

918D διὸ καὶ τῶν νοσούντων ταῖς ὀρέξεσιν οἱ χαρίεντες ἰατροὶ προΐσασι τοὺς ἀσώτως ἢ σωτηρίως ἔχοντας: For the formula χαρίεντες ἰατροί, cf. Arist., *Div. som.* 463a5 and *EN* 1102a22.

918DE ἱστορεῖ γοῦν Μνησίθεος ἰατρὸς ἐν ἀρχῇ πνευμονίας τὸν ἐπιθυμήσαντα κρομμύων σῴζεσθαι τὸν δὲ σύκων ἀπόλλυσθαι, διὰ τὸ ταῖς κράσεσι τὰς ὀρέξεις τὰς δὲ κράσεις τοῖς πάθεσιν ἕπεσθαι: = Mnesith., fr. 16 Bertier [see 4.2.1.1., n. 111]. The word πνευμονία is rare while περιπλευμονία (also simply πλευμονία) or περιπνευμονία are more common (see LSJ, s.vv. and J. Bertier, 1972, p. 171). Is Plutarch perhaps relying on an intermediate source or paraphrasing Mnesitheus' account in his own (less technical) words? Or should we, rather, correct the reading of the manuscripts (manuscript ψ has περὶ πνευμονίας)?

᎒

Q.N. 27, 918EF

Q.N. 27 concerns a vinicultural problem, and, as such, links up more naturally with *Q.N.* 30–31 (cf. also *Q.N.* 10) than with the problems concerning land animals and hunting in *Q.N.* 20–28. Plutarch wonders **why must (i.e. freshly pressed grape juice, in which the sugar has not yet changed into alcohol) remains sweet for a long time if the vessel is kept in cold surroundings** (Διὰ τί τὸ γλεῦκος, ἂν ὑπὸ ψύχους περιέχηται τὸ ἀγγεῖον, γλυκὺ διαμένει πολὺν χρόνον;). He provides two solutions, which are explicitly opposed to each other (ἢ τοὐναντίον;). The first explanation is based on the idea that heat destroys the sweetness of the must whereas cold preserves it. According to the second, cold shuts heat in and thus preserves the sweetness of the must.

The **first** explanation maintains that the change of must into the vinous liquid is a concoction. Cold, on the other hand, hinders concoction, because this process is triggered by heat (Πότερον ὅτι πέψις ἐστὶ τοῦ γλεύκους ἡ εἰς τὸ οἰνῶδες μεταβολὴ κωλύει δὲ τὴν πέψιν ἡ ψυχρότης, ὑπὸ θερμοῦ γὰρ ἡ πέψις;).

The **second** explanation is explicitly opposed to the first one by the phrase ἢ τοὐναντίον; ('Or is the opposite the case?'). Plutarch argues that the sweet flavour is proper to the (ripe) grape. Hence it is said that the sweet flavour that becomes mixed (with the unripe, sour grape) 'ripens' (οἰκεῖός ἐστι τῆς σταφυλῆς χυμὸς ὁ γλυκύς, διὸ καὶ πεπαίνεσθαι λέγεται τὸ γλυκὺ κιρνώμενον). Plutarch continues that cold does not allow 'transpiration', but shuts in the heat and thus conserves the sweetness of the must (ἡ δὲ ψυχρότης οὐκ ἐῶσα διαπνεῖν, ἀλλὰ συνέχουσα τὸ θερμὸν τὴν γλυκύτητα διατηρεῖ τοῦ γλεύκους). That is the same reason as to why the must of grapes that are gathered in the rain ferments less. After all, fermentation is caused by heat and the heat is confined and contracted by the cold (αὕτη δ' ἐστὶν αἰτία καὶ τῶν τρυγωμένων ὄμβρῳ τὸ γλεῦκος ἧττον ἀναζεῖν· ἡ γὰρ ζέσις ὑπὸ θερμότητος, τὴν δὲ θερμότητα κατέχει καὶ συστέλλει τὸ ψυχρόν).

918E Διὰ τί τὸ γλεῦκος, ἂν ὑπὸ ψύχους περιέχηται τὸ ἀγγεῖον, γλυκὺ διαμένει πολὺν χρόνον;: Plutarch may be implying that the container with must is submerged into fresh water in order to keep it sweet: cf. Pliny, *NH* 14, 83 and Col., *De re rust.* 12, 29.

918E οἰκεῖός ἐστι τῆς σταφυλῆς χυμὸς ὁ γλυκύς, διὸ καὶ πεπαίνεσθαι λέγεται τὸ γλυκὺ κιρνώμενον: F.H. Sandbach, 1965, pp. 208–209 (with nn. 1 and a for further remarks) marks a lacuna between λέγεται and τὸ γλυκὺ κιρνώμενον (he translates "<when the warmth leaves the must there is also released>"), but there is no lacuna in the manuscripts. Even so the text is rather concise at this point. If we interpret the term σταφυλή as a 'ripe grape', as (implicitly) opposed to an 'unripe grape' (ὄμφαξ), the meaning becomes clearer (for the opposition between the ripe and the unripe grape, see LSJ, s.v. σταφυλή). Plutarch's argument then amounts to the idea that the admixture of sweetness matures the unripe grape and ripens it (sweetness being proper to a ripe grape). Cf. Gal., *SMT* 11, 657, 2–4 (Kühn): ἡ ὄμφαξ μὲν ὀξεῖα, γλυκεῖα δὲ ἡ στραφυλὴ καὶ τὸ πεπαίνεσθαι τοῖς καρποῖς ἅπασι παρὰ τῆς ἡλιακῆς ἐγγίνεται θερμότητος. For the ripening of grapes, cf. also *Quaest. conv.* 641D (ὡς εἴ τις οἴοιτο τῇ ἀνθήσει τοῦ ἄγνου πεπαίνεσθαι τὸν τῆς ἀμπέλου καρπόν κτλ.) and 658C (μέλας γὰρ αὐταῖς οὐ πεπαίνεται βότρυς). Presumably, an aspect of heat is implicit in the verb πεπαίνεσθαι, since ripening is a process that involves increasing heat (cf. *Q.N.* 5, 913C and LSJ, s.v.; on an etymological basis, F.H. Sandbach, 1965, pp. 208–209 n. a even links the concept of πεπαίνεσθαι with that of πέψις). According to Ps.-Arist., *Probl.* 930b23–25, fruit contains a great deal of fire and moisture, so that because of the fire, the juice causes something like boiling (ζέσις: the same concept recurs at the end of *Q.N.* 27). The association between the grape's heat and its sweetness is central to Plutarch's argument. The point seems to be that the sweetness of the ripe grape somehow *contains* heat owing to the process of maturisation that is triggered by its admixture. This heat is also present in the must itself that is made from the ripe, sweet grapes (as is clear from what follows in the argument). This is, indeed, opposite to the first *causa* (cf. ἢ τοὐναντίον;), where Plutarch argues, to the contrary, that heat (of concoction) destroys sweetness in the must, turning it into a vinous liquid. For the association between heat and sweetness, cf. also *Q.N.* 20, 917A (and 5, 913C).

918EF ἡ δὲ ψυχρότης οὐκ ἐῶσα διαπνεῖν, ἀλλὰ συνέχουσα τὸ θερμὸν τὴν γλυκύτητα διατηρεῖ τοῦ γλεύκους: The notion of 'transpiration' is also related to sweetness and heat in *Q.N.* 22, 917D (τὸ διαπνέον: see the commentary *ad loc.*). The idea that coldness shuts the heat in is probably an allusion to the process of ἀντιπερίστασις. Cf. also Theophr., *CP* 2, 8, 2–3.

918F αὕτη δ' ἐστὶν αἰτία καὶ τῶν τρυγωμένων ὄμβρῳ τὸ γλεῦκος ἧττον ἀναζεῖν· ἡ γὰρ ζέσις ὑπὸ θερμότητος, τὴν δὲ θερμότητα κατέχει καὶ συστέλλει τὸ ψυχρόν: For the opposite idea, viz. that wine made from grapes collected in the rain is sour rather than sweet, see Gal., *SMT* 11, 656, 11–14 (Kühn). According to Arist., *Mete.* 380b31–381a1, the production of wine from must is due to a process of boiling (ἔψησις, cf. also Pliny, *NH* 14, 83: *fervere*). This concept of 'boiling' is present in the terms ἀναζεῖν and ζέσις here (cf. LSJ, s.vv.).

ಞ

Q.N. 28, 918F–919A

In *Q.N.* 28, Plutarch returns to the theme of land animals and hunting. He wonders **why she-bears, least of all animals, gnaw through nets, although both wolves and foxes do so** (Διὰ τί τῶν θηρίων ἡ ἄρκτος ἥκιστα διεσθίει τὰ δίκτυα, καίτοι καὶ λύκοι καὶ ἀλώπεκες διεσθίουσι;). Plutarch provides three explanations that follow each other in a rapid succession (the conclusion of the problem is lacunary). The first explanation concerns the anatomy of the bear's mouth, the second, the strength in its fore-paws, and the third refers to both the bear's mouth and its fore-paws, but adds a further specification. The second and third *causae*, which mention the bear's 'hands', will especially remind the reader of *Q.N.* 22, where Plutarch discusses the sweet taste of the bear's fore-paws in relation to their frequent movement.

In the **first** explanation, Plutarch argues that the bear's teeth are set at the very back of its open mouth, so that it (sc. the mouth) is least able to reach the cords (of the net). After all, the bear's lips meet them first due to their thickness and volume (Πότερον ἐνδοτάτω τοὺς ὀδόντας ἔχουσα τοῦ χάσματος ἥκιστα πρὸς τὰ λίνα ἐξικνεῖται, προεμπίπτει γὰρ τὰ χείλη διὰ πάχος καὶ μέγεθος;).

The **second** explanation draws attention to the bear's fore-paws, which, so Plutarch argues, have greater power to shred and tear apart the mesh of the net (ἢ μᾶλλον ἰσχύουσα ταῖς χερσὶ ῥήγνυσι καὶ διασπᾷ τὸν βρόχον;).

The **third** explanation combines elements from the preceding two, but the text is lacunary at the end. Plutarch argues that the bear uses both its paws and its mouth at the same time: the former to tear the net apart and the latter to defend itself against hunters (ἢ καὶ ταῖς χερσὶν ἅμα χρῆται καὶ τῷ στόματι, ταῖς μὲν διασπῶσα τὸ λίνον τῷ δ' ἀμυνομένη τοὺς διώκοντας;). A final consideration involves the idea that nothing helps the bear more than rolling around. Rather than trying to tear the cords (of the net) apart, it is in this way that the bear often tumbles out of the nets and escapes, so that it no longer needs (the help of) its teeth (οὐδενὸς δ' ἧττον αὐτῇ βοηθοῦσιν αἱ περικαλινδήσεις· διὸ μᾶλλον ἢ διασπᾶν τὰ λίνα πραγματευομένη πολλάκις ἐκκυβιστᾷ καὶ σῴζεται, ὥστ' ἂν μηκέτι δέοιτο τῶν ὀδόντων).

919A οὐδενὸς δ' ἧττον αὐτῇ βοηθοῦσιν αἱ περικαλινδήσεις· διὸ μᾶλλον ἢ διασπᾶν τὰ λίνα πραγματευομένη πολλάκις ἐκκυβιστᾷ καὶ σῴζεται, ὥστ' ἂν μηκέτι δέοιτο τῶν ὀδόντων: The manuscripts read .... ἀμὴ καὶ δέοι ἡ τῶν ὀδόντων. I follow the reading of C. Hubert, 1960, p. 24 ('so that the bear no longer needs (the help of) its teeth'), but the alternative reading of F.H. Sandbach, 1965, p. 210, n. 1 is also appealing, as it draws attention to a more subtle and specific detail: προνοοῦσα μὴ καὶ δεθῇ ("avoiding the possibility of being entangled by its teeth"). *Alii alia*. The noun περικαλινδήσεις is a hapax (the verb περικαλινδέομαι occurs only once in Greek literature: *Martyr. Sebast.*, 6, 236). It should perhaps be corrected in the more common περιδινήσεις (cf., e.g., *De facie* 923C, *Flam.* 10, 6, *Lys.* 12, 6).

Q.N. 29, 919AB

In *Q.N.* 29, we find a problem that stands somewhat on its own in the collection. It is relatively atypical not only from a formal perspective, but its content is also much more reflective and rhetorical than is the case in Plutarch's other natural problems [see 1.2.3.]. Initially, the reader might expect Plutarch to simply treat yet another natural problem here, but the tone of the discourse rapidly changes. The problem is **why we marvel at hot springs, but not at cold ones** (Τίς ἡ αἰτία, δι' ἣν τὰ ψυχρὰ τῶν ὑδάτων οὐ θαυμάζομεν ἀλλὰ τὰ θερμά;). Plutarch points out that there is, in fact, not much reason to marvel at this phenomenon, because it is obvious that heat is the reason for the former and cold for the latter (καίτοι δῆλον ὅτι θερμότης αἰτία τούτων ὡς ψυχρότης ἐκείνων). This seems quite right, but it remains unclear where this heat or cold exactly comes from. In fact, we will learn from the aetiology that it is not so much the natural phenomena of hot or cold springs themselves, but people's short-sighted marvelling at them that is the issue here. Unfortunately, the ending of the chapter is lacunary: it breaks off abruptly, but the original argument can be restored from several parallel accounts (see the final comment).

At the beginning of the explanation, Plutarch makes an abstract and sophisticated remark about the essence of cold, pointing out that it is

COMMENTARY 461

not true, as some believe, that heat is an active property, whereas cold is a privation of heat. After all, in this way, the non-existent would appear to be responsible for more phenomena than the existent (Οὐ γάρ, ὡς ἔνιοι νομίζουσιν, ἡ μὲν θερμότης δύναμίς ἐστιν ἡ δὲ ψυχρότης στέρησις θερμότητος, ἐπεὶ πλειόνων αἴτιον φαίνεται τὸ μὴ ὂν τοῦ ὄντος). By contrast, it appears that nature attributes marvellousness to rarity and stimulates the research of how a phenomenon comes to be only if it occurs infrequently (ἀλλ' ἔοικε τῷ σπανίῳ τὸ θαυμάσιον ἡ φύσις νέμουσα πῶς γίνεται ζητεῖν τὸ μὴ πολλάκις γινόμενον). In what follows, Plutarch describes his own personal marvel for the cosmic spectacles that nature puts on display. He first quotes the following lines from Euripides: 'You see this infinite heaven up high / surrounding earth in a damp embrace' (ὁρᾷς τὸν ὑψοῦ τόνδ' ἄπειρον αἰθέρα / καὶ γῆν πέριξ ἔχονθ' ὑγραῖς ἐν ἀγκάλαις). He then hymnically calls out: 'What a multitude of spectacles does it bring at night, how great is the beauty it exhibits by day!' (ὅσα μὲν ἔρχεται φέρων θεάματα νυκτός, ὅσον δὲ μεθ' ἡμέραν κάλλος ἀναδείκνυσιν;). The discourse receives a more biting tone when Plutarch starts to target the common people, whom he accuses of not feeling any wonder for the nature of these phenomena (οὐ μέντοι θαυμάζουσιν οἱ πολλοὶ τὴν τούτων φύσιν). Their attention only goes to rare phenomena such as rainbows, the variety of clouds by day, meteors bursting like bubbles, and comets ... – and then the text breaks off (ἴριδες δὲ καὶ ποικίλματα νεφῶν ἡμέρας καὶ σέλα ῥηγνύμενα πομφόλυγος δίκην καὶ κομῆται ****).

919A Τίς ἡ αἰτία, δι' ἣν τὰ ψυχρὰ τῶν ὑδάτων οὐ θαυμάζομεν ἀλλὰ τὰ θερμά;: Notably, the formulation of the *quaestio* with the phrase τίς ἡ αἰτία δι' ἣν is rather exceptional (but cf. also *Q.N.* 40), and the aetiology is not based on the typical structure of πότερον [...] ἢ [...] ἢ [...]. The topic of this chapter falls under the general theme of Book 24 of Ps.-Aristotle's *Problems* (ὅσα περὶ τὰ θέρμα ὕδατα). According to Plutarch's personal theory put forward in *De prim. frig.* 952C–955C, it is not water, but earth that is the principle of cold. From this perspective, the problem of *Q.N.* 29 seems legitimate: if all springs rise from the earth, why are some hot and others cold? As G. Sarton, 1965, p. 388, n. 28 notes, "[h]ot and mineral springs were highly appreciated and exploited by the Romans, as they had been before them by the Greeks, Etruscans, Carthaginians, and Gauls. Balneology began in prehistoric times." Plutarch in *Ca. Ma.* 21, 5 reports, for instance, that Cato the Elder bought ὕδατα θερμά. On the generation and disappearance of νάματα θερμά, see *De def. or.* 433F (cf. also *Mar.* 19, 2). On the generation of springs in general, see *Aem. Paul.* 14.

919AB Οὐ γάρ, ὡς ἔνιοι νομίζουσιν, ἡ μὲν θερμότης δύναμίς ἐστιν ἡ δὲ ψυχρότης στέρησις θερμότητος, ἐπεὶ πλειόνων αἴτιον φαίνεται τὸ μὴ ὂν τοῦ ὄντος: This point is not explained any further. What Plutarch probably implies is that cold springs occur more frequently than hot springs, making them less 'wonderful', as was put forward in the *quaestio* (οὐ θαυμάζομεν). The same theory, according to which cold is a δύναμις in itself rather than a στέρησις of heat, is elaborated upon in the first part of *De primo frigido* (945F–948A). The idea that cold is a δύναμις in its own right is paralleled in Plato (cf. *Tim*. 33a: θερμὰ καὶ ψυχρὰ καὶ πάνθ' ὅσα δυνάμεις ἰσχυρὰς ἔχει περιιστάμενα ἔξωθεν). The theory that cold is a στέρησις of heat is Aristotelian (cf. *Met*. 1070b9–13, *De caelo* 286a25–26, *GC* 318b14–17; but by contrast, cf. *PA* 649a18–19; see also O. Longo, 1992). If Plutarch with ἔνιοι implicitly refers to the Peripatetics here, it appears that these are criticised without their name being explicitly mentioned. This procedure is not uncommon in the *Moralia* (as marked by D. Babut, 1994, p. 574, with n. 137 and 1969, p. 95 with nn. 5 and 6). Aristotle is anonymously criticised, also in *De def. or*. 426D. Let it be clear, moreover, that these ἔνιοι should *not* be identified with οἱ πολλοί later on.

919B ἀλλ' ἔοικε τῷ σπανίῳ τὸ θαυμάσιον ἡ φύσις νέμουσα πῶς γίνεται ζητεῖν τὸ μὴ πολλάκις γινόμενον: The idea that 'less marvellous' phenomena (c.q. cold springs) deserve attention just as much as 'more marvellous' ones (c.q. hot springs) is relatively common (cf. Arist., *PA* 645a16–17, Sen., *NQ* 7, 1–4, Cic., *De nat. deor*. 2, 96, Ps.-Cic., *Rhet. ad Her*. 3, 36, Lucr., *De rer. nat*. 2, 1030–1039, Pliny, *Ep*. 8, 20, esp. 1–2 – reference by J.J. Hartman, 1916, p. 556). Most scholars translate φύσις as 'human nature' here, which may well be correct (cf. F.H. Sandbach, 1965, p. 211; L. Senzasono, 2006, p. 121). But perhaps a more denotative and referential interpretation of the term is worth considering, in line with the general stylistics of the collection (otherwise, one could expect a more periphrastic wording, cf., e.g., *Quaest. conv*. 734D: αἱ φιλόσοφοι φύσεις). C.F. Schnitzer, 1860, p. 2732 combines both options in his translation (my italics: "Aber es scheint in der *Natur* zu liegen daß *man* dem Seltenen den Charakter des Wunderbaren beilegt und bei Allem was nicht oft vorkommt nach der Entstehung fragt."). The meaning of φύσις may, indeed, be zeugmatic, in that human nature is strongly related to and dependent on nature itself (cf., e.g., *De E* 386F: θεωρίαν καὶ κρίσιν ἀνθρώπῳ μόνῳ παραδέδωκεν ἡ φύσις). In that case, it is perhaps implied that nature itself attributes marvellousness to rarity and stimulates the research of how a phenomenon comes to be only if it occurs infrequently. It sounds more natural, then, to translate the infinitive ζητεῖν as a *causativum* with the ellipse of an object ('nature *incites* <people> to inquire'). The aspect of wonder (τὸ θαυμάσιον) is not so much considered a human πάθος, then, but an inherent attribute of φύσις itself. In this sense, nature (φύσις), in presenting its 'wonders' to us, strongly appeals to a proper understanding by researching how these phenomena come to be (πῶς γίνεται ζητεῖν).

COMMENTARY 463

919B ὁρᾷς τὸν ὑψοῦ τόνδ' ἄπειρον αἰθέρα / καὶ γῆν πέριξ ἔχονθ' ὑγραῖς ἐν ἀγκάλαις: The same Euripidean lines (= TGF 941) are also quoted in *De exilio* 601A and *Ad princ. iner.* 780D. In the third verse, which is not quoted here, the poet identifies the αἰθήρ with Zeus.

919B ὅσα μὲν ἔρχεται φέρων θεάματα νυκτός, ὅσον δὲ μεθ' ἡμέραν κάλλος ἀναδείκνυσιν;: For similar rhetorical questions and exclamations, cf. Sen., *Ben.* 4, 23 and *Ad Helv.* 8, 4. The expressive couple of rhetorical questions further underlines the awesome sights that the περιέχον puts on display. In combination with the evocative fashion in which Plutarch quotes Euripides' lines (see the previous comment), they lift the discourse to a more rhetorical level (which is exceptional to the collection's general register [see 1.2.3.]).

919B οὐ μέντοι θαυμάζουσιν οἱ πολλοὶ τὴν τούτων φύσιν: The text is corrupt at this point. F.H. Sandbach, 1965, p. 210, n. 5 adds οὐ μέντοι θαυμάζουσιν (after Wyttenbach). Cf. also C. Hubert, 1960, p. 25: οἱ <δὲ> πολλοὶ τὴν τούτων φύσιν <οὐ θαυμάζουσιν>. The phrase οἱ πολλοί traditionally refers to the un(der)educated plebs (cf., e.g., *Q.N.* 3, 912D and *Quaest. conv.* 664BC). I take this to imply that the wonder of the common people for rare natural phenomena remains superficial and does not lead on to actual natural philosophical inquiry. After all, they do not look into the φύσις, i.e. the natural causes (cf. πῶς γίνεται ζητεῖν), of natural phenomena. For φύσις denoting natural causes, cf. G.E.R. Lloyd, 1979, p. 31.

919B ἴριδες δὲ καὶ ποικίλματα νεφῶν ἡμέρας καὶ σέλα ῥηγνύμενα πομφόλυγος δίκην καὶ κομῆται ****: F.H. Sandbach, 1965, p. 211, n. e is probably right that the ποικίλματα νεφῶν refer to the "coloration rather than shapes or patterns" of the clouds. As to σέλας, we read in Ps.-Arist., *De mund.* 395b3-4 that the noun refers to the lighting of a column of fire in the air, either flashing or fixed (Σέλας δέ ἐστι πυρὸς ἀθρόου ἔξαψις ἐν ἀέρι. Τῶν δὲ σελάων ἃ μὲν ἀκοντίζεται, ἃ δὲ στηρίζεται). Seneca translates σέλα as *fulgores* (*NQ* I, 15, 1-3, cf. also *De fort. Rom.* 323C: σέλας ἀστραπῇ παραπλήσιον). F. Fuhrmann, 1964, p. 77 marks only one case of literary imagery for *Quaestiones naturales*, viz. "[l]es météores éclatent comme des bulles". For the bursting of fiery bubbles, cf. also *De sera num.* 563F-564A and Sen., *NQ* I, 1, 3. For a similar polysyndetic enumeration of 'wonderful' meteorological spectacles, see *De Pyth. or.* 409CD, where children's amazement for celestial phenomena is particularly reprimanded (καὶ γὰρ οἱ παῖδες ἴριδας μᾶλλον καὶ ἅλως καὶ κομήτας ἢ σελήνην καὶ ἥλιον ὁρῶντες γεγήθασι καὶ ἀγαπῶσι κτλ.). A similar passage is found in *Amatorius* 766A (ὥσπερ οἱ παῖδες προθυμούμενοι τὴν ἶριν ἑλεῖν τοῖν χεροῖν, ἑλκόμενοι πρὸς τὸ φαινόμενον). One can expect from these parallels that Plutarch originally concluded his invective against the common people in *Q.N.* 29 with the same topic, namely that the 'childish' astonishment of the ignorant plebs for such wonderful phenomena – and hence also for similar phenomena, such as the hot springs – is motivated on irrational grounds, presumably superstition (cf. *Per.* 6, 1 [quoted 3.2.2.], see also, e.g., *Alex.* 75, 1–2, Sen., *NQ* 7, 2 and Critias, fr. DK88B25, esp. 27–36). I have elaborated this point elsewhere: see M. Meeusen, 2015b. Ironically enough, though, with all of this, Plutarch does not really provide a detailed solution to the *quaestio* at hand about hot and cold springs (did it perhaps follow in the lost part?). He does, of course, explain in the *quaestio* that it is clear that heat is the reason for the hot springs and cold for the cold ones, but this requires further elaboration to be satisfactory. Cf. also C.F. Schnitzer, 1860, p. 2732, n. 2: "Die Erklärung der Ursache fehlt, wenn sie nicht darin liegt daß das Seltene als das Wunderbare gelte."

5. *Viniculture* (Q.N. 30–31)

The two chapters that follow in *Q.N.* 30–31 are closely connected to each other through the topic of vines and viniculture, which ties in with the subject of wine drinking and drunkenness in Book 3 of the *Problems* (ὅσα περὶ οἰνοποσίαν καὶ μέθην). The same theme is discussed in *Q.N.* 10 and 27. The main focus in *Q.N.* 30–31 is of a mostly botanical kind, though. One may, therefore, presume a Theophrastan source, possibly through the intermediation of lost problems (see esp. *Q.N.* 30). Unfortunately, the original Greek text breaks off abruptly at the end of *Q.N.* 31.

Q.N. 30, 919BC

In *Q.N.* 30, Plutarch examines **the meaning and etymology of the verb τραγᾶν ('to go goatish'), which is said of vines that do not bear fruit, but flourish with branches and shoots** (Διὰ τί τῶν ἀμπέλων τὰς ἀκάρπους, τοῖς δ' ἀκρέμοσι καὶ ἔρνεσιν εὐτροφούσας τραγᾶν λέγομεν;). Plutarch offers one explanation, which draws attention to the theory of generative residues.

By analogy with the vines, Plutarch argues that exceedingly fat male goats (τράγοι) are also less fertile and have difficulty copulating due to their fat (Ἢ ὅτι καὶ τῶν τράγων οἱ σφόδρα πίονες ἧττόν εἰσι γόνιμοι καὶ μόλις ὑπὸ πιμελῆς ὀχεύουσι;). He explains that seed is a residue of the food that is added to the body (τὸ γὰρ σπέρμα περίττωμα τῆς τροφῆς ἐστι τῆς τῷ σώματι προστιθεμένης). Thus, when either an animal or a tree is in good condition and increases in volume, it is a sign that the food is consumed in the body and either does not produce a residue or produces only a small and modest amount (ὅταν οὖν ἢ ζῷον ἢ δένδρον εὐεκτῇ καὶ παχύνηται, τοῦτο σημεῖόν ἐστι τοῦ τὴν τροφὴν ἐν αὐτῷ καταναλισκομένην μηθὲν ἢ μικρόν τι καὶ ἀγεννὲς περίττωμα ποιεῖν).

COMMENTARY 465

919B Διὰ τί τῶν ἀμπέλων τὰς ἀκάρπους, τοῖς δ' ἀκρέμοσι καὶ ἔρνεσιν εὐτροφούσας τραγᾶν λέγομεν;: The text of the *quaestio* is very lacunary. I follow Wyttenbach's reading. The same phenomenon is recorded by Aristotle and Theophrastus, who provide basically the same explanation for it, at times also mentioning the decreased fertility of fat goats (see Arist., *HA* 546a1-4, *GA* 725b32-726a6, Theophr., *CP* 5, 9, 10; cf. also *HP* 2, 7, 6 and 4, 14, 6). There is reason to assume that Plutarch's source was Peripatetic and that he perhaps relies on a lost problem. The concept of τραγᾶν is also used with respect to boys' voices cracking when they reach puberty. Cf. Ps.-Alex. Aphr., *Probl.* 1, 125 (J.L. Ideler, 1841, pp. 42, 36-43, 32): ἐκ μεταφορᾶς τῶν τράγων οὕτως κραζόντων. Cf. also Gal., *Sem.* 4, 633, 12 Kühn and *UP* 4, 172, 13 Kühn.

919C τὸ γὰρ σπέρμα περίττωμα τῆς τροφῆς ἐστι τῆς τῷ σώματι προστιθεμένης: For the idea that seed is formed from a residue of food, cf. *Q.N.* 21, 917B [see 4.3.4.1., n. 295].

919C ὅταν οὖν ἢ ζῷον ἢ δένδρον εὐεκτῇ καὶ παχύνηται, τοῦτο σημεῖόν ἐστι τοῦ τὴν τροφὴν ἐν αὐτῷ καταναλισκομένην μηθὲν ἢ μικρόν τι καὶ ἀγεννὲς περίττωμα ποιεῖν: For the same reason, no residue is left in strong palm trees and athletes (cf. *Quaest. conv.* 724E with *Q.N.* 32), or in other trees (like fir and cypress) and overweight people (cf. *Quaest. conv.* 640F–641A).

Q.N. 31, 919CE

Q.N. 31 is closely connected with the previous chapter through the topic of vines (cf. also the repetition of the verb προστίθημι, which refers to the addition of food to the body). Plutarch wonders **why a vine wilts if it is sprinkled with wine, especially with wine made from its own grapes** (Διὰ τί ἄμπελος οἴνῳ ῥαινομένη, μάλιστα τῷ ἐξ αὐτῆς, ἀναξηραίνεται;). Four explanations are provided. The first three focus on the first part of the *quaestio* (viz. that a vine wilts if it is sprinkled with wine). The first one refers to the wine's heat, the second to its putrefactive character and the third to its astringent nature. The fourth explanation, by contrast, focuses on the specification of 'wine made from its own grapes' in the *quaestio*, more precisely by referring to the vine's inability to receive a substance that has left it (the ending of the problem is lacunary).

In the **first** explanation, Plutarch draws an analogy between the wilting of vines and baldness that occurs in hard drinkers (ὥσπερ ἐν τοῖς πολυπόταις γίνεται φαλάκρωσις). Plants and hairs alike grow only if their roots receive moisture, but the wine evaporates the moisture due to its heat (ὑπὸ θερμότητος τοῦ οἴνου τὸ ὑγρὸν ἐξατμίζοντος).

In the **second** explanation, Plutarch draws attention to the putrefactive nature of vinous liquid by quoting Empedocles, who asserts that 'wine is water from the bark, putrefied in the wood' (ἢ φύσει σηπτικὸν τὸ οἰνῶδές ἐστιν, ὥς φησιν Ἐμπεδοκλῆς οἶνον ἀπὸ φλοιοῦ πέλεσθαι σαπὲν ἐν ξύλῳ ὕδωρ). Thus, when the vine is externally moistened by wine, fire is generated for the vine, and the mixture/'temperature' changes the specific property of the nutrient moisture (ὅταν οὖν ἔξωθεν οἴνῳ βρέχηται, γίνεται πῦρ ἀμπέλῳ καὶ τοῦ τρέφοντος <ὑγροῦ τὴν οἰκείαν> δύναμιν ἐξίστησιν ἡ κρᾶσις).

The **third** explanation draws attention to the clogging effect of wine. Plutarch argues that unmixed wine with its astringent nature enters the roots and there narrows and contracts the pores. In this way, it does not allow water into the plant by which it would naturally flourish and grow (ἢ στυπτικὴν φύσιν ἔχων ὁ ἄκρατος ἐνδύεται ταῖς ῥίζαις, καὶ τοὺς πόρους συναγαγὼν καὶ πυκνώσας οὐ διίησι τὸ ὕδωρ εἰς τὸ φυτόν, ᾧ εὐθαλεῖν καὶ βλαστάνειν πέφυκεν;).

In the **fourth** and final explanation, Plutarch argues that it is even more unnatural for the vine to receive a substance that leaves and returns back to it (ἢ καὶ τοῦτο μᾶλλον εἶναι τῇ ἀμπέλῳ παρὰ φύσιν, τὸ ἐξ αὐτῆς ἀπιὸν εἰς αὐτὴν ἐπανιὸν πάλιν δέχεσθαι;). He explains this by pointing out that the part of the moisture in plants that cannot feed them, or be added, or be part of them is strained out <into the fruit> (τῆς γὰρ ἐν τοῖς φυτοῖς ὑγρότητος ἠθεῖται τὸ μὴ τρέφειν μηδὲ προστίθεσθαι μηδὲ μέρος εἶναι τοῦ φυτοῦ δυνάμενον ***).

919C Διὰ τί ἄμπελος οἴνῳ ῥαινομένη, μάλιστα τῷ ἐξ αὐτῆς, ἀναξηραίνεται;: For the related, though contrary, view that wine positively effects seeds, see Theophr., *CP* 3, 24, 4.

919C ὑπὸ θερμότητος τοῦ οἴνου τὸ ὑγρὸν ἐξατμίζοντος: Aristotle also compares the loss of foliage in a plant (and of feathers in a bird) with baldness in men, explaining both phenomena on account of a lack of warm moisture (*GA* 783b18: ἔνδεια ὑγρότητος θερμῆς, cf. Ps.-Arist., *Probl.* 880a34–b3). Cf. also *Quaest. conv.* 649CD, where Plutarch argues that loss of foliage is not a sign of coldness (οὐδὲ γὰρ ψυχρότητος τὸ φυλλορροεῖν), since some cold plants, like myrtle and maidenhair, are evergreens. For the idea that dryness causes baldness, cf. also Hipp., *De sem., de nat. pu., de morb.* 4, 20, 30–33 and Ps.-Alex. Aphr., *Probl.* 1, 2 (J.L. Ideler, 1841, p. 6, 6–29). Galen also believes that baldness is due to a lack of moisture (*Comp.Med.Loc.* 12, 381, 16–17 Kühn: δι' ἔνδειαν μὲν ὑγρῶν ἡ φαλάκρωσις γίγνεται). In *Quaest. conv.* 652F, we read that drunkards resemble old men, because many get bald or grey at an early age. Plutarch explains this, however, from a deficiency of heat (θερμότητος ἐνδείᾳ). Cf. also Arist., fr. 235 Rose (εἴτε γὰρ αὔανσις τριχὸς ἡ πολιὰ εἴτε ἔνδεια θερμοῦ). For the idea that dryness causes greyness (whereas moisture renders young men's hair black), cf. *De Is.*

et Os. 364B and Arist., *GA* 780b6 (greyness is a weakness, viz. a non-concoction, of the moisture in the brain). For the place of such theories in ancient popular medicine, see I. Rodríguez Alfageme, 1999a, p. 421.

919CD ἡ φύσει σηπτικὸν τὸ οἰνῶδές ἐστιν, ὥς φησιν Ἐμπεδοκλῆς οἶνον ἀπὸ φλοιοῦ πέλεσθαι σαπὲν ἐν ξύλῳ ὕδωρ: The same fragment (= DK31B81) is quoted without syntactical adaptations in *Q.N.* 2, 912C, where Plutarch argues that the process of concoction resembles that of putrefaction (ἡ δὲ πέψις ἔοικεν εἶναι σῆψις). Especially the hot property of wine is important for Plutarch's argument at hand. As Empedocles' quote illustrates, this heat is due to a process of putrefaction, more precisely the putrefaction of water.

919D ὅταν οὖν ἔξωθεν οἴνῳ βρέχηται, γίνεται πῦρ ἀμπέλῳ καὶ τοῦ τρέφοντος <ὑγροῦ τὴν οἰκείαν> δύναμιν ἐξίστησιν ἡ κρᾶσις: This passage is corrupt, but the basic meaning of the second *causa* is clear: wine is putrefied water, which is harmful for the vine due to its heat and fiery constituent. Some further consideration is required regarding the concepts of 1) πῦρ and 2) κρᾶσις. 1) Regarding the phrase γίνεται πῦρ ἀμπέλῳ, Sandbach follows Wyttenbach in adding ἐπὶ πῦρ τῇ between πῦρ and ἀμπέλῳ, thus referring to the theory of adding fire to fire (πῦρ ἐπὶ πῦρ). According to this theory, a smaller amount of fire or heat (c.q. the innate heat of the vine) is extinguished by a larger one (c.q. of the wine). Alternatively, this theory can also imply that a smaller amount of fire or heat merges into a larger one and increases its volume. In this sense, the heat becomes excessive for the vine, by it being moistened with hot wine. For more detail on this theory, see H. Flashar, 1962, p. 328. It is frequently seen in the work of Aristotle (e.g., *GC* 323b8–10) and also in the *Problems* (866a26–28 (repeated in a36–b1), 874b6–7, 937a26–27, 961b31–32). Plutarch is well acquainted with this theory (cf. also Pl., *Leg.* 666a): it has proverbial value in *De ad. et am.* 61A, *De tuenda* 123E, *Coni. praec.* 143F, *Cons. ad ux.* 610C and *Art.* 28, 1 (see J.A. Fernández Delgado, 1991, pp. 202–203). However, even if it is not impossible that Plutarch had this process in mind in the present passage, there is no lacuna in the manuscripts between πῦρ and ἀμπέλῳ, so that there is no need for an editorial intervention. The meaning of the text remains clear with or – preferably – without it (see also V. Ramón Palerm, 2005, p. 403 and L. Senzasono, 2006, p. 52 and pp. 234–235, n. 176: "In realtà il testo tràdito è chiaro"; the same theory remains implicit also, e.g., in *Quaest. conv.* 648D: αἱ γὰρ ὑπερβολαὶ φθείρουσι τὰς δυνάμεις κτλ.). Therefore, I take it that Plutarch is simply using the term πῦρ in reference to the fiery constituent of the wine, the heat of which was indeed commonly approved (see the previous *causa*; cf. also, e.g., Ps.-Arist., *Probl.* 871a2 and Arist., fr. 222 Rose). Alternatively, according to V. Ramón Palerm, 2005, p. 403, wine transforms into fire for the vine ("mediante el riego con vino, éste se convierte en fuego para una vid (γίνεται πῦρ ἀμπέλῳ)"), but this is not what Plutarch actually writes (πῦρ is the subject, not the predicate, of γίνεται: 'fire is generated for the vine'). The *causa* at this point amounts to the idea that the fiery constituent of the wine (πῦρ) dries out the vine. 2) In what follows, the concept of κρᾶσις remains vague (τοῦ τρέφοντος <ὑγροῦ τὴν οἰκείαν> δύναμιν ἐξίστησιν ἡ κρᾶσις). There is a lacuna of 12 to 16 letters in the manuscripts between τρέφοντος and δύναμιν (neglected by L. Senzasono, 2006, pp. 235–236, nn. 177–179 (cf. also p. 52), who does not, moreover, clarify the meaning of κρᾶσις). I follow Sandbach in supplementing ὑγροῦ τὴν οἰκείαν, which implies that the κρᾶσις changes the specific property of the nutrient moisture. As to the concept of κρᾶσις, then, this may not just be a reference to the fiery 'temperament' of the vine itself, but to the mixed composition of the wine, which is a 'blend' of cold, watery and hot, vinous constituents. In this sense, the wine's 'temperature' is too hot – viz. like πῦρ – for the vine [see 4.3.1.2., n. 171]. Thus, it destroys the property of the nutrient substance (c.q. the water) in the wine, so that the vine sprinkled with it withers. Alternatively, the nutrient moisture to which Plutarch refers is already present in the vine itself, and the 'admixture' (κρᾶσις) of the wine changes its nutrient property by its fiery constituent.

919D ἢ στυπτικὴν φύσιν ἔχων ὁ ἄκρατος ἐνδύεται ταῖς ῥίζαις, καὶ τοὺς πόρους συναγαγὼν καὶ πυκνώσας οὐ διίησι τὸ ὕδωρ εἰς τὸ φυτόν, ᾧ εὐθαλεῖν καὶ βλαστάνειν πέφυκεν;: For the (Empedoclean) theory that the narrowness of pores is related to leaf fall in plants, cf. *Quaest. conv.* 649D (with Emp., DK31B77–78). For the phrase εὐθαλεῖν καὶ βλαστάνειν, cf. also, e.g., *Quaest. conv.* 745A. The phrase ἐνδύεται ταῖς ῥίζαις, recurs in *Q.N.* 1, 911D; cf. also *Q.N.* 2, 911F (διαδύεται μᾶλλον εἰς τὴν ῥίζαν) and *Quaest. conv.* 664E (ἐνδύεσθαι τοῖς βλαστάνουσι).

919D ἢ καὶ τοῦτο μᾶλλον εἶναι τῇ ἀμπέλῳ παρὰ φύσιν, τὸ ἐξ αὐτῆς ἀπιὸν εἰς αὐτὴν ἐπανιὸν πάλιν δέχεσθαι;: The idea is probably that it is unnatural for the vine to receive wine, as is explained in the previous *causae*, but that it is even more unnatural (μᾶλλον παρὰ φύσιν) to receive wine made *from its own grapes* (τὸ ἐξ αὐτῆς), as is implied in the *quaestio* (μάλιστα τῷ ἐξ αὐτῆς). See the following comment.

919DE τῆς γὰρ ἐν τοῖς φυτοῖς ὑγρότητος ἠθεῖται τὸ μὴ τρέφειν μηδὲ προστίθεσθαι μηδὲ μέρος εἶναι τοῦ φυτοῦ δυνάμενον ***: According to L. Senzasono, 2006, pp. 237–238, n. 184, there is a climax in the tricolon of the verbs τρέφειν, προστίθεσθαι and μέρος εἶναι, but it is not unlikely either that the latter two verbs simply specify the first. The verb προστίθημι recurs in *Q.N.* 3, 912E and 30, 919C, again regarding the assimilation of nourishment. What Plutarch is probably implying, then, is that the grapes contain the residual (excremental) part of the vine's nourishment, which is useless and presumably even detrimental for the vine. There seems to be a link with the generative residue, which, as Plutarch argues in *Q.N.* 30, 919C, is a περίττωμα τῆς τροφῆς τῆς τῷ σώματι προστιθεμένης (cf. the μηδὲ προστίθεσθαι here). Arguably, by discharging the residue via the grapes, the vine purifies itself. Therefore, Sandbach's addition <into the fruit> seems essential (cf. *Q.N.* 5, 913D: ἐκκρίνειν εἰς τὸν καρπόν). It elegantly explains the second part of the *quaestio* – viz. why wine is harmful for the vine, especially when it is produced ἐξ αὐτῆς, that is, from its own grapes –, which would otherwise have remained unsolved. It is not unlikely, therefore, that the aetiology also originally ended with this fourth explanation. Notably, Psellus completely rewrites the last explanation. In doing so, he actually contradicts Plutarch: ῥᾷον δὲ ἡ ἄμπελος δέχεται τὸ ἐξ αὐτῆς ἀπιὸν εἰς αὐτὴν ἐπανιόν (*De. omn. doctr.* §187 Westerink). It is not unlikely that Psellus found Plutarch's argument to be obscure. Perhaps the text was already illegible in the manuscript that he used (see M. Meeusen, 2012b, p. 115, n. 68).

ཤ

6. Longolius (Q.N. 32–39)

The eight chapters that follow in *Q.N.* 32–39 were first published by Longolius in his 1542 Latin translation of *Quaestiones naturales*. Longolius notes in a marginal note that they are extracted from a Milanese manuscript, but this manuscript has been lost ever since, and the Greek text is still missing today. Considering the numerous parallels in Plutarch's other works and the same general style and method of explanation, there can be no doubt that these problems are authentic (for further detail on Longolius' translation, see A. Morales Ortiz, 1999 and M. Meeusen, forthcoming a).

Q.N. 32

It is unclear whether *Q.N.* 32 followed immediately after *Q.N.* 31 in Plutarch's original Greek text. In any case, there is a faint thematic link between *Q.N.* 32 and *Q.N.* 30–31, in that they each concern botanical problems (viz. plants and their specific properties). The link with the aetiology in *Q.N.* 30 becomes more concrete in the shared parallel passage in *Quaest. conv.* 724EF (both concerning generative residues). In the problem at hand in *Q.N.* 32, Plutarch wonders **why the palm tree alone among all trees rises against a weight imposed on it** (*Cur inter omnes arbores sola palma contra impositum onus adsurgit?*). Three explanations are given: the first one focuses on the palm tree's fiery and breathlike property, the second on the compressed air in its branches and the third on the slow rising of its twigs, which gives them the impression (*speciem*) of rising up against the weight. With this final explanation, Plutarch tries to show that the tree raises its twigs only in appearance. He does not, as such, aim to reject the popular belief (as formulated in the *quaestio*), but he certainly puts it in a broader physical perspective [see 4.3.3.3.].

In the **first** explanation, Plutarch argues that the palm tree's fiery and breathlike property, which gives it its great strength, exerts itself when tested and vexed, and so raises (the palm) more and more (*Utrum quod ignea et spirabilis facultas, qua maxime pollet, cum tentatur et irritatur, sese exercens magis et magis erigit?*).

According to the **second** explanation, the sudden impetus of the weight on the branches pushes back all the air compressed within them (*An quoniam pondus ramos subito urgens aerem omnem qui in his est oppressum cedere retro cogat*). Afterwards, when the air slightly recovers its strength, it pushes against the weight with greater force (*qui deinde resumptis paulo viribus adversum onus acrius rursus instat?*).

In the **third** and final explanation, Plutarch argues that the supple and tender twigs cannot sustain the impetus of the weight. When the weight comes to rest, though, they gradually erect themselves and give the impression of rising up against it (*An molles et tenerae virgae impetum non sustinentes, cum onus quiescit, paulatim se erigunt et speciem quasi contra illud adsurgant praebent?*).

***Cur inter omnes arbores sola palma contra impositum onus adsurgit?*:** Scholars have found the formulation of Plutarch's problem rather obscure. There is especially discussion as to whether Plutarch is referring to a living palm tree, or to the beams or logs made from its timber. On the basis of several parallel accounts (collected below), F.H. Sandbach, 1965, p. 215, n. b believes that Plutarch is probably implying that "a log laid horizontally hunches itself in an upward curve against a superimposed weight". By contrast, L. Senzasono, 2006, p. 238, n. 185 argues that the problem concerns a growing and living palm tree. He refers to the terms *ramos* and *tenerae virgae* in the second and third explanation respectively. Sandbach notes that "[i]t is possible that Longolius misunderstood the Greek text here, and wrongly supposing it to refer to a growing tree, introduced the words *ramos* and *virgae*, without warrant". Then again, since Longolius uses these two *different* terms for the palm's 'branches', it seems only logical that these render two different Greek terms. Indeed, *ramos* is perhaps a poor translation of 'logs' or 'beams', but this is implausible for *virgae molles et tenerae*. Therefore, Senzasono is probably correct that "non c'è ragione di congetturare la possibilità che il traduttore abbia frainteso il testo qui". The popular belief has a rich tradition and seems to originate from the world of architecture and building. Xenophon reports that Cyrus used palm trees for the construction of the foundations of watchtowers (*Cyr.* 7, 5, 11–12). He notes that this kind of wood is the right material for this purpose, because of the well-known fact that palm trees bend upward under heavy pressure like the backs of pack-asses (οἱ φοίνικες ὑπὸ βάρους ἄνω κυρτοῦνται, ὥσπερ οἱ ὄνοι οἱ κανθήλιοι). Similarly, Theophrastus writes that among other kinds of wood, that of palm trees is strong, because it bends the opposite way to other woods. Whereas other kinds of wood bend downwards, palm wood bends upwards (*HP* 5, 6, 1: ἰσχυρὸν δὲ καὶ ὁ φοῖνιξ· ἀνάπαλιν γὰρ ἡ κάμψις ἢ τοῖς ἄλλοις γίνεται· τὰ μὲν γὰρ εἰς τὰ κάτω κάμπτεται, ὁ δὲ φοῖνιξ εἰς τὰ ἄνω). Theophrastus notes that fir and silver fir also have an upward thrust, and that they are strong when set slant-wise (πλάγιαι τιθέμεναι); he adds that wood of the chestnut tree is used for roofing (χρῶνται πρὸς τὴν ἔρεψιν). Theophrastus is referring to struts and supports for walls and roofs here (cf. the *impositum onus* in Plutarch's *quaestio*). Notably, the term πλάγιαι in Theophrastus' text is an emendation based on Pliny, *NH* 16, 222: *in traversum*. Theophrastus was probably Pliny's source. For the palm tree's special bending ability, see *NH* 16, 223 (*et palmae arbor valida; in diversum enim curvatur,* [*et populus*] *cetera omnia in inferiora pandantur, palma ex contrario fornicatim*). See L&S, s.v. *fornicatim*: "in the form of an arch, archwise". See also S. Amigues, 1993, p. 84, n. 1. A similar account of the use of palm trees in construction works is found in Strabo (parallel only marked by S. Amigues, 1993, p. 84, n. 3). Strabo writes that the palm tree, when aged, does not give way downwards, but curves upwards because of the weight and gives better support to the roof (*Geogr.* 15, 3, 10: τὴν φοινικίνην δοκόν· στερεὰν γὰρ οὖσαν, παλαιουμένην οὐκ εἰς τὸ κάτω τὴν ἔνδοσιν λαμβάνειν, ἀλλ' εἰς τὸ ἄνω μέρος κυρτοῦσθαι τῷ βάρει καὶ βέλτιον ἀνέχειν τὴν ὀροφήν). Notably, the palm tree's natural resilience remained a popular topic well beyond Antiquity in the form of a moral 'emblem' (see L. Holford-Strevens, 2005). Similar value is ascribed to it in *Quaest. conv.* 724EF, where the symposiasts discuss the award of the palm frond as a symbol of victory. At the end of the discussion, the palm tree is compared with an athlete: the *tertium comparationis* is their largeness, shapeliness, sterility, and strength. It is argued that all of the nourishment is used to build up the body, so as not to form any generative residue (cf. *Q.N.* 21, 917B, 30, 919C and *Quaest. conv.* 640F–641A; see F. Fuhrmann, 1964, p. 81, with n. 3). The athlete's bodily and mental strength (acquired through heavy training) is compared with the unique character of palm wood (ἴδιον δὲ παρὰ ταῦτα πάντα καὶ μηδενὶ συμβεβηκὸς ἑτέρῳ). It does not bend down and give way, but curves up in the opposite direction when a weight is imposed upon it, as though resisting the person who would force it (φοίνικος γὰρ ξύλον ἂν ἄνωθεν ἐπιθεὶς βάρη πιέζῃς, οὐ κάτω θλιβόμενον ἐνδίδωσιν, ἀλλὰ κυρτοῦται πρὸς τοὐναντίον ὥσπερ ἀνθιστάμενον τῷ βιαζομένῳ). Interestingly, Gellius paraphrases this passage with an explicit reference to the eighth Book of Plutarch's *Symposiacs* and also to the seventh of Aristotle's *Problems* (*NA* 3, 6 = Arist., fr. 229 Rose: *Per hercle rem mirandam Aristoteles in septimo problematorum et Plutarchus in octauo symposiacorum dicit* etc.; see F. Klotz and K. Oikonomopoulou, 2011, p. 235). The problem at hand cannot be traced in the extant Ps.-Aristotelian *Problems*, though, meaning that it is probably lost.

Utrum quod ignea et spirabilis facultas, qua maxime pollet, cum tentatur et irritatur, sese exercens magis et magis erigit?: F.H. Sandbach, 1965, p. 215 translates the first *magis* as an adverb modifying *sese exercens* and the second modifying *erigit* ("more than before"). The reduplication of *magis*, though, can also denote an aspect of the gradual rise in *erigit* only, which seems preferable here ("more and more": L&S, s.v. b, 2, b; cf. LSJ, s.v. μᾶλλον: "denoting a constant increase, *more and more*, sts. doubled"). Cf. also L. Senzasono, 2006, p. 239, n. 187.

An quoniam pondus ramos subito urgens aerem omnem qui in his est oppressum cedere retro cogat, qui deinde resumptis paulo viribus adversum onus acrius rursus instat?: The argument is of a mechanical kind and can perhaps be placed in the larger frame of an anti-κενόν theory. To clarify Plutarch's explanation, one may think of the compression of air in a pump, syringe, or bellows. When one closes off the opening and presses the handle, the compressed air will not allow the handle to be pushed down completely. When the handle is released again, the pressure of the air pushes it backwards. A similar effect occurs in the palm tree's branches when the air that is present in them is compressed by the imposed weight and released again. The fact that the branches raise themselves is because the air in them naturally rises. F.H. Sandbach, 1965, p. 215 translates *paulo* as "slowly", but it is not synonymous with *paulatim* (see the third *causa*); thus, 'slightly' seems more apposite here. Cf. also L. Senzasono, 2006, pp. 239–240, n. 188.

An molles et tenerae virgae impetum non sustinentes, cum onus quiescit, paulatim se erigunt et speciem quasi contra illud adsurgant praebent?: Plutarch does not further explain why and how the twigs gradually erect themselves. What is more important for Plutarch's argument, though, is that they only give the impression (*speciem*) of rising up against the imposed weight, meaning that the reported belief should be nuanced [see 4.3.3.3.].

Q.N. 33

Q.N. 33 connects with the chapters on salt and water in *Q.N.* 1–13 (esp. *Q.N.* 1, 2 and 4, which also deal with the nutritive properties of water). Plutarch examines **why water that is drawn from wells is less nutritious than water that flows from a spring or falls from the sky** (*Quare aqua de puteis hausta minus alit quam quae de fonte aut caelo manat?*). Three explanations are given in a rapid succession: the first takes into account the coldness of the water in the well and its small amount of air, the second its salty constituent and the third its immobility.

Plutarch **first** argues that water drawn from wells is colder and also contains (only) a small amount of air (*An quia frigidior magis sit et parum quoque aeris habeat?*).

In the **second** explanation, he argues that well water contains a large amount of salt from the earth that is mixed with it, and that salt especially causes thinness (*An quod salis multum immixta sibi de terra habeat; sal autem maciem, si quid aliud, facit?*).

In the **third** explanation, Plutarch argues that well water, because it is inactive and immobile, acquires a bad quality, which is hostile to plants and animals. This explains why it is not well concocted and cannot nurture anything (*An quod pigra nec cursu exercitata qualitatem aliquam malam adquirat, quae stirpibus et animantibus contraria in causa est quod nec bene concoquatur nec nutrire quicquam possit?*). Hence, stagnant waters are considered less good, since they cannot disperse the damage they receive from bad qualities that originate either in the air or the earth (*Hinc et stagnantes aquae minus probae censentur, quod iniurias quas vel ab aeris mala qualitate vel a terra accipiunt digerere nequeant*).

An quia frigidior magis sit et parum quoque aeris habeat?: The comparative *frigidior magis* seems odd. The opposition between *magis* and *parum* was probably clearer in the original Greek text. Plutarch does not clarify this argument any further. It may be implied that coldness slows down concoction of food – concoction being brought about by heat. The greater coldness of water drawn from wells can perhaps be explained by the fact that it does not come into contact with sunlight, as opposed to water from rivers or water that falls from the sky. Moreover, water from rivers stand more in contact with air than that from wells. Rainwater also contains more air because it falls from the sky. They are both more nutritious, probably because airy water is more quickly guided and transmitted into the body (e.g., into a plant: cf. *Q.N.* 2, 912A and 4, 912F).

An quod salis multum immixta sibi de terra habeat; sal autem maciem, si quid aliud, facit?: The idea that salt causes thinness is paralleled in

Q.N. 3, 912E (λεπτύνεται). It could be objected here that water flowing from springs is also affected by the earth through and over which it flows (cf. *Q.N.* 2, 912B).

Hinc et stagnantes aquae minus probae censentur, quod iniurias quas vel ab aeris mala qualitate vel a terra accipiunt digerere nequeant: For the idea that pond water is old and stale, while spring and river water is fresh and new-born and therefore more nutritive (albeit still less nutritive than rainwater), see *Q.N.* 2, 912A. For the idea that the admixture of earth makes stagnant waters putrid, while running waters avoid mixture or throw off any earth that enters their course, see *Quaest. conv.* 725D (cf. also *Aqua an ignis* 957D and *De lat. viv.* 1129D with F. Fuhrmann, 1964, p. 60). For the disease inducing properties of stagnant waters, cf. Hipp., *Aer.* 7 and Ps.-Arist., *Probl.* 884a32–34. According to L. Senzasono, 2006, pp. 240–241, n. 192, the phrase *stagnantes aquae* is a translation of στάσιμα ὕδατα, or perhaps λιμναῖα ὕδατα, but the second seems unlikely, because Longolius in *Q.N.* 2, 912A translates τὸ λιμναῖον <ὕδωρ> as *aqua palustris*. Additionally, the *aqua de puteis* (water drawn from wells) in the *quaestio* is probably a translation of τὸ φρεατιαῖον ὕδωρ (cf. *Quaest. conv.* 690B). Senzasono also argues (p. 241, n. 193) that *digerere* is a translation of διακρίνω in the medical sense of 'dissolve', 'dissociate', 'dissipate', 'destroy' (cf., e.g., Cels., *De med.* 2, 17, 1; see L&S, s.v. i, b, 2: "In medic. lang., *to dissolve, dissipate* morbid matter"). Yet, a less technical term is not impossible either (e.g., διακρούεται: *Quaest. conv.* 725D).

Q.N. 34

In dealing with winds, the chapter at hand stands relatively isolated in the collection. Plutarch examines **why the west wind is commonly considered the swiftest of all, as Homer writes: 'we too could run as fast as the west wind's blast'** (*Cur Zephyrus ventorum omnium celerrimus vulgo fertur, et Homerus 'nos quoque Zephyri curramus flatibus una'*). Two explanations are given that are closely connected to each other. Whereas the first explanation only deals with the atmospheric circumstances when the west wind blows (viz. the cleanness of the sky), the second accounts for the actual cause of the wind and its motive force (viz. the heat of the sun).

The **first** explanation draws attention to the atmospheric circumstances. Plutach argues that the west wind normally blows when the air is entirely clear and the least clouded. This is because the density and impurity of the air is no mean obstacle for the rapid passage of the winds (*An quod* *aere perpurgato et minime nebuloso flare soleat? Aeris enim densitas et impuritas ventorum cursum non mediocriter impedit*).

Central to the **second** explanation is the opposition between heat and cold. Plutarch argues that the sunlight is responsible for this phenomenon, because it strikes the cold wind so that it moves faster (*An quod sol radiis suis flatum frigidum perstringens, quo velocius feratur, auctor est?*). He finds it credible that anything cold that is contracted by the force of the winds is forced by the heat into prolonged and accelerated flight, as if overcome by an enemy (*Quicquid enim frigidi ventorum vi contrahitur, id a calore veluti hoste superatum longius et citius propelli credendum est*).

Cur Zephyrus ventorum omnium celerrimus vulgo fertur: The topic of winds falls under the general theme of Book 26 of Ps.-Aristotle's *Problems* (ὅσα περὶ τοὺς ἀνέμους). Notably, the exact opposite problem, viz. why the west wind is the calmest and most pleasant, is treated in Ps.-Arist., *Probl.* 943b21–23 (with a reference to Hom., *Od.* 4, 567) and 946b21–22. According to Theophr., *Vent.* 38 (with a reference to Hom., *Il.* 23, 200), this depends on the season.

et Homerus 'nos quoque Zephyri curramus flatibus una': The original Homeric lines (= *Il.* 19, 415–416), spoken by the horse Xanthus to Achilles, run as follows: νῶϊ δὲ καί κεν ἅμα πνοιῇ Ζεφύροιο θέοιμεν, / ἥν περ ἐλαφροτάτην φάσ' ἔμμεναι.

An quod sol radiis suis flatum frigidum perstringens, quo velocius feratur, auctor est? Quicquid enim frigidi ventorum vi contrahitur, id a calore veluti hoste superatum longius et citius propelli credendum est: It seems reasonable to assume that the sun would also strike other winds by its heat, and not only the west wind. Therefore, this second *causa* should probably be read in combination with the first, concerning the atmospheric circumstances when the west wind blows. Note, moreover, that since the cold is not concentrated by the opposite heat of the sun but by the force of the winds (*contrahitur ventorum vi*), this is probably not an instance of the process of ἀντιπερίστασις (*pace* L. Senzasono, 2006, pp. 241–242, n. 196). The beating of the sunlight (*perstringens*) can be linked with the rising and setting of the sun, described as the cause of the winds in Ps.-Arist., *Probl.* 944a11–12 (where we read, more in specific, that the west wind blows towards the late afternoon but not in the morning). According to F.H. Sandbach, 1965, p. 218, n. a, "[o]ne would think the rising sun as well able to set the east wind in motion as the setting sun the west. But it may be assumed that the west wind is colder […] and therefore flees faster." Its coldness is, indeed, commented on in Ps.-Arist., *Probl.* 946a17–32, albeit in connection with its mildness (moreover, in the same passage the north wind is said to be even colder than the west wind: ἧττον μὲν οὖν ψυχρὸς τοῦ βορέου). Cf. also Theophr., *Vent.* 40. The imagery in *veluti hoste superatum* is exceptional to the general stylistics of the collection. By its obvious military register, it possibly alludes to the Homeric line in the *quaestio*. Cf. also *Coni. praec.* 139D, where Plutarch refers to a fable attributed to Aesop (nos. 306–307), according to which the sun earned a victory (ἐνίκησεν) over the north wind.

The two chapters of *Q.N.* 35–36 are closely linked to each other through the focus on the natural and instinctive behaviour of bees towards certain bad odours. The bees' strong perception of smells was already alluded to in *Q.N.* 26, 918C, where Plutarch notes that honeycombs excite bees by their scent and attract them from a distance (cf. *Q.N.* 36: *olfactus sensu valet plurimum*). Interest in bees is not unusual among the ancients, because it was the only domesticated insect useful to humans (for further reading on bees and apiculture in Antiquity, see O. Keller, 1980, pp. 421–431 (esp. p. 424)). Flies, by contrast, were considered ungovernable creatures (cf. *Quaest. conv.* 728A). Despite the frequent references to bees in the *Moralia*, there is no reason to assume that Plutarch was a bee-keeper himself (cf. also L. Van der Stockt, 2005, p. 13; *pace* G. Siefert, 1908, p. 19, n. 1; on bee imagery in Plutarch, see F. Fuhrmann, 1964, pp. 58, 94 and E.K. Borthwick, 1991).

Q.N. 35

In *Q.N.* 35, Plutarch examines **why bees cannot bear smoke** (*Cur apes fumum ferre nequeunt?*). He gives two explanations: the first one focuses on the narrow passages of the bees (it is formulated in an assertoric rather than interrogative fashion for unclear reasons), the second on their aversion towards pungency and bitterness.

According to the **first** explanation, bees have very narrow passages for their vital breath (*Quod meatus spiritus vitalis sane quam angustos habeant*). The breath is cut off and clogged by the smoke and suffocates the bees, nearly killing them (*At is fumo interceptus et conclusus angit et propemodum ad mortem apes adigit*).

In the **second** explanation, Plutarch argues that the cause can be found in the pungent and bitter constituent of smoke (*An acredo amaritudoque fumi in causa est?*). He explains that bees find delight in sweet things and have no other form of nourishment (*Gaudent enim dulcibus apes neque alio nutrimento aluntur*). Thus, they hate smoke as a contrary (to sweetness) and harmful thing for it being bitter (*itaque ut contrariam et noxiam rem propter amaritudinem fumum detestantur*). This is illustrated by the common practice of bee-keepers to make smoke by burning bitter plants like hemlock or centaury in order to drive the bees away (*Qua de causa mellarii cum fumum abigendis apibus faciunt, amaras herbas, ut cicutam et centaurium, incendere solent*).

Cur apes fumum ferre nequeunt?: For the idea that bees cannot bear smoke, cf. *De aud.* 42C (ἄν τις ὥσπερ καπνῷ σμῆνος λόγῳ δριμεῖ τὴν διάνοιαν ἀχλύος πολλῆς καὶ ἀμβλύτητος γέμουσαν ἐκκαθήρῃ) and *Praec. ger. reip.* 821B (ταύτας [sc. τὰς μελίττας] μὲν καπνῷ κολάζουσιν). Cf. also, e.g., Arist., *HA* 623b20–21, Pliny, *NH* 11, 45, Col., *De re rust.* 9, 14, 7, Verg., *Georg.* 4, 230; 241–242.

Quod meatus spiritus vitalis sane quam angustos habeant: Longolius probably translates πνεῦμα as *spiritus vitalis* here (cf. F.H. Sandbach, 1965, p. 219; L. Senzasono, 2006, pp. 242–243, n. 199). It is unclear why the *causa* opens with *quod* rather than *an quod* and why it is formulated in an assertoric rather than interrogative fashion. F.H. Sandbach, 1965, p. 218, n. 1 is correct that "[e]ither Longolius's Greek text had the mistake of omitting ἤ (cf. 914C, E, 917D) or *an* is omitted from the Latin text".

Gaudent enim dulcibus apes neque alio nutrimento aluntur: For the bee's love of sweet things, cf. *Quaest. conv.* 673E: ἡ μέλιττα τῷ φιλόγλυκυς. Cf. also Arist., *HA* 535a2.

Qua de causa mellarii cum fumum abigendis apibus faciunt, amaras herbas, ut cicutam et centaurium, incendere solent: For the use of smoke in driving off bees, cf., e.g., Pliny, *NH* 11, 45: *apes abigi fumo utilissimum* (see also the comment on the *quaestio*). There is no reason to assume, *pace* L. Senzasono, 2006, p. 243, n. 201, that "l'umanista [Longolius] avesse presente il testo di Plinio nel tradurre questo passo". It is not impossible, however, that Plutarch and Pliny rely on the same or a similar (Peripatetic?) source.

ॐ

Q.N. 36

Q.N. 36 is closely related to the previous problem by its focus on bees. Plutarch examines **why bees are quicker to sting people who have just committed adultery** (*Cur apes citius pungunt qui stuprum dudum fecerunt?*). One explanation is given, which draws attention to the bee's perception of smells and their devotion to cleanliness. The explanation closes with two mythological references.

Plutarch argues that the bee is extremely devoted to neatness and hygiene, and, furthermore, that it has a very powerful sense of smell (*An quod animal est munditiae et elegantiae perquam studiosum; praeterea olfactus sensu valet plurimum?*). He explains that irregular coition is usually more impure (than regular coition) through unchasteness and unrestrained lust. Bees discover it more quickly and dislike it more violently (*Quum itaque impuri congressus propter impudicitiam et immoderatam libidinem soleant esse immundiores, et citius ab apibus deprehenduntur et odium vehementius adversus illos concipiunt*). This is further illustrated in a mythological coda. Plutarch writes that Theocritus, in veiled ironic terms, describes how the (dying) herdsman Daphnis sends off Aphrodite to Anchises in order to be stung by bees because of their adultery (*Unde apud Theocritum iocose Venus ad Anchisen a pastore ablegatur, uti apum aculeis propter adulterium commissum pungatur*): 'So go to Ida, go to Anchises, where oak and galingale grows, and the mellifluous home buzzes loudly by the humming of bees' ('*Te confer ad Idam, / confer ad Anchisen, ubi quercus atque cypirus crescit, / apum strepitatque domus melliflua bombis*'). Pindar, too, associates the bee-sting with perfidy, when he addresses the bee as follows: 'Tiny builder of honeycombs, who pierced Rhoecus with your sting, taming his perfidy' (*et Pindarus: 'Parvula favorum fabricatrix quae Rhoecum pupugisti aculeo domans illius perfidiam'*).

Cur apes citius pungunt qui stuprum dudum fecerunt?: It is an ancient commonplace that the bee is an honourable and virtuous creature, a true model of diligence and purity, and famous for its chastity, neatness, and abstinence. The priestesses of Delphi, Demeter, Artemis, and Cybele, for instance, were called *Melissai* (cf. LSJ, s.v. μέλισσα ii, 2). Likewise, it was a popular ancient belief that bees procreate a-sexually (cf. Pliny, *NH* 11, 16). The *locus classicus* is provided by Semonides of Amorgos' comparison of the perfect wife to a bee (fr. 7 West, 83–93): the bee-wife manages a thriving household and does not like to sit with her female fellows to gossip about sex. Nevertheless, she makes her husband happy by producing an illustrious and handsome offspring. For the queen bee as a model of oeconomia, cf. also Xen., *Oec.* 7, 32–35 (with further commentary by S.B. Pomeroy, 1995, pp. 276–280).

An quod animal est munditiae et elegantiae perquam studiosum; praeterea olfactus sensu valet plurimum? Quum itaque impuri congressus propter impudicitiam et immoderatam libidinem soleant esse immundiores, et citius ab apibus deprehenduntur et odium vehementius adversus illos concipiunt: The aetiology in *Q.N.* 36 runs parallel to that in *Q.N.* 35 in two ways. First of all, the strong emotional category of hate returns in both problems (*detestantur – odium vehementius*). Notably, this aggressive aversity is not morally but physically motivated, since the bee's fundamental dislike and potentially lethal irritation is provoked by natural causes. Second, the idea that something contrary is also harmful to the bee (*Q.N.* 35: *contrariam et noxiam*), is paralleled in *Q.N.* 36 in the causal opposition between cleanliness and dirtiness (the negative prefix *in-* recurs frequently in *Q.N.* 36: *impuri, impudicitiam, immoderatam, immundiores*). The 'dirtiness' of the *impuri congressus* bears on the material aspect of the impure mixture of body odours and perfumes (for further parallels, see below), meaning that it is not based on the ethical depravity of the act of adultery itself (*impudicitiam et immoderatam libidinem*). As such, adultery is punished by the bee for physical reasons, without direct ethical motives. Cf. also L. Senzasono, 2006, p. 244, n. 202: "Plutarco non attribuisce alle api un discernimento moralistico, ma si limita a rilevare che il coito impuro è piú sudicio e che quindi l'olfatto delle api ne è piú rapidamente sollecitato per la loro fine sensibilità (*elegantia*) e per il loro forte senso della pulizia." In *Coni. praec.* 144D, by contrast, Plutarch incorporates the same natural phenomenon in the moral dynamics of the marital advice that he gives to the groom [see 1.2.4.] (I have treated this parallel elsewhere: see M. Meeusen, 2013c). A large amount of other parallels can be found in Plutarch and other authors. On the unpleasant body odour of those who have or are able to have sexual intercourse (as opposed to children), see Ps.-Arist., *Probl.* 879a23–26. For the evil smell of adulteresses, cf. also, e.g., Cat., *Carm.* 42 (*moecha putida*). On the relationship between odours and mating in animals, see, e.g., *Gryllus* 990BC. For the idea that the emanations from the female body defile that of the male, cf. fr. 97 Sandbach (ἀπόρροιαί τινες ἐκ τῶν γυναικείων σωμάτων καὶ περιττωμάτων χωροῦσιν, ὧν ἀναπίμπλασθαι τοὺς ἄνδρας μολυσματῶδές ἐστι) and *Quaest. conv.* 651E (ἀναπίμπλανται γὰρ αὐτοὶ τοῦ χρίσματος ἐν τῷ συγκαθεύδειν, κἂν μὴ θίγωσι μηδὲ προσάψωνται τῶν γυναικῶν, διὰ θερμότητα καὶ μανότητα τοῦ σώματος ἕλκοντος). For the hostility of bees towards irritating smells, see Arist., *HA* 626a26–28, where we read that bees are very tidy creatures (καθαριώτατον ζῷον) and are annoyed by unpleasant odours or perfumes; in fact, they sting those who wear them (cf. also Ps.-Arist., *Mir. ausc.* 832a3–4: myrrh). L. Senzasono, 2006, p. 244, n. 202 believes that Aristotle's reference (ὥσπερ εἴρηται) back to *HA* 623b20 (on bees eating honey in order to build up a reserve when they are confused by smoke) also explains the narrow connection between *Q.N.* 35–36, but this is very uncertain (the contexts are different). On the assumption that Aristotle was, indeed, Plutarch's source, the intermediation of a lost Ps.-Aristotelian problem cannot be excluded (however, F.H. Sandbach, 1982, pp. 207–232, esp. 230 believes that Plutarch was probably acquainted with Aristotle's *Historia animalium* directly). For further parallels, see also Theophr., *CP* 6, 5, 1 (bees are extremely hostile to persons wearing perfume), Ael., *NA* 1, 58 (bees equally dislike bad smells and perfumes); 5, 11 (bees chase off men wearing perfume and they also recognise and pursue adulterers, which is considered a sign of their σωφροσύνη), Pliny, *NH* 11, 61 (bees hate bad odours and artificial perfumes and attack persons wearing perfume); 11, 44 (they hate menstruating women); 7, 64 (beehives wither from the smell of menstrual fluids); 28, 79 (if menstruating women touch a beehive, the bees flee from it), Col., *De re rust.* 9, 14, 3 (beekeepers are advised not to enjoy veneric pleasures the day before they handle the beehives, nor to be drunk or unwashed; they should also abstain from food with a strong aroma), Varro, *De re rust.* 3, 16, 6 (bees follow everything that is pure and avoid places that are befouled or have an evil odour; they also cannot bear perfumes or people wearing them), *Geop.* 15, 2, 19 (bees attack persons more severely if they reek of wine and of perfume, especially women that are of an amorous complexion; according to L. Senzasono, 2006, pp. 243–244, n. 202 this has nothing to do with the problem at hand, but I do not see how); 15, 3, 4 (the bee is an extraordinarily clean animal, settling on nothing that has a disagreeable smell or that is impure), Pallad., *Op. agr.* 1, 37, 4–5 and 4, 15, 4 (the bee-keeper has to be pure and chaste).

Unde apud Theocritum iocose Venus ad Anchisen a pastore ablegatur, uti apum aculeis propter adulterium commissum pungatur: The actual pun (*iocose*) is probably in the fact that Daphnis, in his description of the *locus amoenus* setting (see the following comment), makes specific mention of bees, which will not, of course, give a very hearty welcome to the goddess-adulteress if she is to follow Daphnis' 'good advice' (cf. R. Hunter, 1999, p. 97). According to F.H. Sandbach, 1965, p. 220, the Latin *pungatur* can imply that he (Anchises) or she (Venus) was stung, but the second option seems more likely. L. Senzasono, 2006, p. 244, n. 203 doubts Plutarch's understanding of Theocritus' text, but I see no reason why ("Riguardo all'averbio *iocose*, se pur decodifica fedelmente il messagio del testo originale, non è chiaro se Plutarco, intento a documentare il suo parere sull'istinto delle api, abbia veramente colto il tono del testo teocriteo, tenuto conto che spesso, quando cita, non se ne cura, preoccupato com'è solo d'illustrare con esempi autorevoli la sua tesi o di conferire un certo tono al suo contesto."). Plutarch does, indeed, lift these verses from their original context, but this custom is common in his writings and testifies to his ability to use the available sources in an original way [see 4.2.1.1.].

Te confer ad Idam, / confer ad Anchisen, ubi quercus atque cypirus / crescit, apum strepitatque domus melliflua bombis: These verses (= Theocr., *Id.* 1, 105–107) are spoken by Daphnis as a taunt of Aphrodite. The original Greek text runs as follows: ἕρπε ποτ' Ἴδαν, / ἕρπε ποτ' Ἀγχίσαν· τηνεὶ δρύες ἠδὲ κύπειρος, / αἱ δὲ καλὸν βομβεῦντι ποτὶ σμάνεσσι μέλισσαι. In the original story, Anchises was struck by lightning for revealing his secret affair with Aphrodite (see A.S.F. Gow, 1950, pp. 23–24). U. von

Wilamowitz, 1906, p. 234 refers to an otherwise unattested myth according to which Anchises was blinded by bees (cf. the myth of Rhoecus: see the following comment).

Parvula favorum fabricatrix quae Rhoecum pupugisti aculeo domans illius perfidiam: This fragment (= Pind., fr. 252 (= 165) Snell) is only known from Longolius' Latin translation. According to E.K. Borthwick, 1991, p. 562, the Greek text read τεχνῖτις or the like for *fabricatrix* (cf., e.g., *De soll. an.* 982F: τῷ τεχνικῷ παραβάλλειν μελίττας). The myth of Rhoecus can be reconstructed from several sources: viz. the version of Charon of Lampsacus (in *Etym. magn.* 75, 26–44, s.v. Ἀμαδρυάδες = FGrHist 262, 12), the scholia on Apollonius Rhodius (2, 476–477) and those on Theocritus (*Id.* 3, 13c). The story goes that a Cnidian named Rhoecus, who lived in Nineveh (?), instructed his servants to support a tree that was falling with age. The tree nymph – an 'hamadryad', whose life was connected with that of the tree in which she lived – thanked her saviour by granting him a wish. He chose to lie with her and she agreed. She promised that a bee would summon him to her but also demanded that he avoided intercourse with other women. The bee eventually came to Rhoecus when he was playing draughts, but he spoke impatiently to it. Therefore, the insulted nymph punished him by cursing him with some bodily handicap, perhaps blindness (πηρωθῆναι). F.H. Sandbach, 1965, p. 221, n. b may be right that "Pindar's story must have been different, Rhoecus suffering for infidelity, as is hinted in the inconsequent ban on other intercourse in Charon's version, and not for lack of tact". This is also what Plutarch is implying. Yet, the variation and confusion of motives is a relatively common feature of ancient mythography in general. See also H.W. Prescott, 1913, p. 180: "It is probable, therefore, that Rhoecus, like Daphnis, suffered because of his faithlessness toward the nymph." According to L. Senzasono, 2006, p. 245, n. 204, Rhoecus' aggressive reaction towards the bee was perhaps considered an act of *perfidia* in itself by the nymph, meaning that he did not necessarily commit adultery (Plutarch is clearly thinking otherwise, though). Whether this implies, moreover, as Senzasono adds, that the punishment was reduced to a simple sting of the bee, and that precisely this would be in line with the jesting tone (*iocose*), seems unlikely. From the opposition between the gentile character of the *parvula favorum fabricatrix* and the fact that it punishes *perfidia* with stings, Senzasono suggests that the quotation from Pindar allows an even more ironic interpretation than that from Theocritus. Strictly speaking, though, the adverb *iocose* depends on *ablegatur*, so that the humoristic aspect is restricted to the myth told by Theocritus, where a clear ironic effect is, indeed, present (see the commentary *ad loc.*).

The following two chapters in *Q.N.* 37–38 are closely connected to the problems on land animals and hunting in *Q.N.* 20–28, especially through their focus on animal instincts (cf. *Q.N.* 26).

Q.N. 37

Q.N. 37 deals with the intelligence of dogs (or rather, their lack of it). Plutarch wonders **why dogs chase and bite a stone, ignoring the person who flung it** (*Quare canes, relicto homine qui iecit, lapidem morsu insectantur?*). For a good understanding of the *quaestio*, one should note that the stone is thrown *at* the animal in order to injure it, and not simply thrown away so that the dog should fetch it. Plutarch provides three explanations: the first draws attention to the dog's intellectual inferiority to man, the second to its instinct and false opinion and the third to its emotion of hate. The aspect of hate (*odit*) in the last *causa* recalls the strong aversity of bees towards certain odours in *Q.N.* 35–36 (*detestantur – odium vehementius*).

In the **first** explanation, Plutarch argues that the dog cannot understand anything by means of its intellect and that it has no memory, which are virtues proper to human beings (*An quia neque cogitatione comprehendere quicquam nec reminisci (quibus solus homo virtutibus valet) potest?*). Due to the fact that the dog cannot distinguish in its mind where the injury comes from, it believes that the object alone is its enemy, considering the threatening movement that it makes before its eyes. Therefore, it takes out its revenge on the stone (*Itaque quum mente non discernat a quo iniuria fuerit illata, id tantum quod ob oculos minaciter versatur inimicum esse existimat deque eo vindictam sumere parat*).

In the **second** explanation, Plutarch argues that the dog imagines that the stone is some kind of animal, as it rolls over the ground, and that it instinctively tries to catch it at first. Once it realises that it is mistaken by its imagination, though, it returns to attack the man (*An lapidem, dum per terram mittitur, feram aliquam esse autumnans, pro ingenio hanc prius capere conatur, deinde cum viderit se opinione sua frustrari, hominem rursus invadit?*).

According to the **third** explanation, the dog hates the projectile just as much as it hates the man who threw it, but that it pursues the closer of the two (*An quod et id quod missum fuerit et hominem ipsum aequaliter odit, et id quod proximius est insectatur?*).

Quare canes, relicto homine qui iecit, lapidem morsu insectantur?: The same phenomenon is recorded by Plato (*Rep.* 469e). It also serves as an example in *De gar.* 514D, where Plutarch explains the nickname of the Stoic Antipater – 'pen-valiant' (καλαμοβόας) – in light of the fact that the man wrote entire books against Carneades' attacks on the Stoa instead of addressing him personally (SVF 3, p. 244, fr. 5; this Antipater is quoted in the following chapter in *Q.N.* 38). The link with the phenomenon at hand is clear (cf. F. Fuhrmann, 1964, pp. 24 and 148), but only in *Q.N.* 37 does Plutarch provide an aetiology for it. Cf. also *Sept. sap. conv.* 147C and *De tranq. an.* 467C for the story of a man who threw a stone at his dog, missed, and hit his stepmother (whereupon he exclaimed: 'Not so bad after all!').

An quia neque cogitatione comprehendere quicquam nec reminisci (quibus solus homo virtutibus valet) potest?: By contrast, in *Q.N.* 21, 917C, Plutarch ascribes a certain aspect of ἀνάμνησις to animals (c.q. pigs), which implies that they have some rational capability. If we bear in mind Plutarch's writings on animal psychology, in which he defends the rational abilities of animals (cf., e.g., *Gryllus* 992A), it seems rather unlikely that Plutarch would personally ascribe much credibility to the explanation at hand (cf. also L. Senzasono, 2006, pp. 245–246, n. 206). As to the potentially Stoic background to this argument, cf. Sext. Emp., *HP* 1, 64–72.

Itaque quum mente non discernat a quo iniuria fuerit illata, id tantum quod ob oculos minaciter versatur inimicum esse existimat deque eo vindictam sumere parat: For the idea that rage often dims the sight (cf. *ob oculos*) of dogs or even blinds them in their struggle with their prey, cf. *Quaest. conv.* 681DE (θυμοὶ κυνῶν ἐν ταῖς πρὸς τὰ θηρία γινομέναις ἁμίλλαις ἀποσβεννύουσι τὰς ὁράσεις πολλάκις καὶ τυφλοῦσι).

Q.N. 38

In *Q.N.* 38 Plutarch examines **why she-wolves give birth to their young at a fixed time of the year within twelve days** (*Cur lupae certo anni tempore omnes intra xii dies pariunt?*). The problem is much in line with *Q.N.* 21 (on the number and time that sows litter). The aetiology contains two explanations: first, a physical one (with an implicit allusion to the Stoic theory of natural sympathy) and then a mythological one. Both are formulated in an assertoric rather than interrogative fashion for unclear reasons. They are also commensurable with each other in that the first explanation tackles the first part of the *quaestio* (*certo anni tempore*) and the second explanation the second part (*intra xii dies*).

The **first** explanation is borrowed from Antipater, who, in his book *On animals*, asserts that she-wolves litter at the same time when trees that bear nuts (or acorns) shed their flowers (*Antipater in libro de animalibus partum lupas proiicere adserit, cum glandiferae arbores florem abiiciunt*). By eating these flowers, their wombs are opened, but when these flowers are not available, the offspring perishes in the mother's very body and cannot see the light of day (*quo gustato uteri illarum reserantur: cum eius copia non est, partum in ipso corpore emori nec in lucem venire posse*). Plutarch adds that those regions that are not fertile with chestnut-trees or oak-trees are not inhabited by wolves (*praeterea regiones illas a lupis non vastari, quae glandium quercuumque feraces non sunt*).

In the **second** explanation, Plutarch paraphrases the story about Leto to which certain people refer (*Quidam ad fabulam Latonae referunt*). When Leto became pregnant from Zeus, she could not find a safe haven from Hera anywhere. Thus, Zeus transformed her into a wolf for a period of twelve days, during which she travelled to Delos (*quae cum uterum ferret nec uspiam tuta prae Iunone esse posset, duodecim diebus, quibus in Delum proficiscebatur, in lupum a Iove mutata*). In this way, she procured that all wolves should be able to litter in that same period from then on (*ut deinceps omnes lupae eo ipso tempore parere possint impetravit*).

Cur lupae certo anni tempore omnes intra xii dies pariunt?: The belief that she-wolves (and dogs) mate and litter within 12 days is also found in Aristotle, who remains sceptical about it (*HA* 580a11–22: οὐδέν πω συνῶπται μέχρι γε τοῦ νῦν, ἀλλ' ἢ ὅτι λέγεται μόνον). The fact that she-wolves give birth to their young at a fixed time of the year probably implies that they litter once a year but presumably more than once in their lifetime, as is confirmed by Aristotle (*HA* 580a21–22: Οὐκ ἀληθὲς δὲ φαίνεται ὂν οὐδὲ τὸ λεγόμενον ὡς ἅπαξ ἐν τῷ βίῳ τίκτουσιν οἱ λύκοι). Cf. also Ael., *NA* 4, 4, Antig. Car., *Hist. mir.* 56, Pliny, *NH* 8, 83, Isid., *Et.* 12, 2, 24, Solinus, *Mem.* 2, 36.

Antipater in libro de animalibus partum lupas proiicere adserit, cum glandiferae arbores florem abiiciunt: Presumably a quote from the Stoic of Tarsus (= SVF 3, p. 251, fr. 48) [see 4.2.1.1., n. 113].

Quidam ad fabulam Latonae referunt, quae cum uterum ferret nec uspiam tuta prae Iunone esse posset, duodecim diebus, quibus in Delum proficiscebatur, in lupum a Iove mutata, ut deinceps omnes lupae eo ipso tempore parere possint impetravit: The story goes that after Leto travelled from the land of the Hyperboreans to Delos in twelve days, she gave birth to Apollo and Artemis. This perhaps accounts for Apollo's title Lykeios (see F.H. Sandbach, 1965, p. 224, n. a). Among the *quidam* mentioned at the beginning of the explanation, Aristotle comes first, since he records the same mythological account in *HA* 580a14–19 (cf. also Ael., *NA* 4, 4 and Antig. Car., *Hist. mir.* 56). According to L. Senzasono, 2006, pp. 246–247,

n. 207, Plutarch relies directly on Aristotle's account (F.H. Sandbach, 1982, pp. 207–232, esp. 230 argues that Plutarch was probably acquainted with Aristotle's *Historia animalium*), but the intermediation of a lost problem cannot be excluded. Especially Book 10 of Ps.-Aristotle's *Problems* comes to mind (i.e. the Ἐπιτομὴ Φυσικῶν [see 1.1.1., n. 4]), which draws heavily on Aristotle's biological and zoological writings and is often concerned with animal copulation and generation (see R. Mayhew, 2011a, pp. 279–281).

ఎ

Q.N. 39

From a thematic perspective, *Q.N.* 39 is closely related to the hydrological problems in *Q.N.* 1–13. In the chapter at hand, Plutarch examines **why water is seen as white in its upper layer, but black at the bottom** (*Cur aqua in summa parte alba, in fundo vero nigra spectatur?*). He provides three explanations: the first focuses on the clarifying effect of sunlight on the water, the second on the water's colour as a reflection of light and the third on the earthy constituent in the water itself.

In the **first** explanation, Plutarch metaphorically calls the depth 'the mother of blackness', because it dulls and weakens the sunbeams before they can descend to it (*An quod profunditas nigredinis mater est, ut quae solis radeos prius quam ad eam descendant obtundat et labefactet?*). Due to the fact that the surface, by contrast, is continuously affected by the sun, it must take on the clearness of its light (*Superficies autem, quoniam continuo a sole adficitur, candorem luminis recipiat oportet*). Empedocles is quoted to support this: 'And the black colour at the bottom of the river arises from shadow and is also seen in cavernous hollows' (*Quod ipsum et Empedocles approbat 'Et niger in fundo fluvii color exstat ab umbra / atque cavernosis itidem spectatur in antris'*).

In the **second** explanation, Plutarch argues that the depths of rivers and of the sea are bursting with mud, and that they (viz. the depths) from themselves produce the same (viz. dark) colour by reflection of the sun, as characterises the mud (*An limo plerumque oppletus fluminum marisque fundus talem de se colorem per solis reflexum parit, quali utique is praeditus est?*).

In the **third** explanation, Plutarch argues that it is more plausible that the water of rivers and of the sea is the least pure and clean but that it is stained by an earthy quality, since it continuously carries along an amount of the earth over which the river runs or the sea tosses (*An probabilius est aquam minime quae illis est puram et sinceram esse, sed terrea qualitate (utpote quae continuo, qua currit vel agitur, aliquid ex ea advehat) imbutam*). When the earth sinks to the bottom, the water becomes more turbid and less transparent there (*cum ad fundum residet, turbidiorem et minus perspicuam effici*).

Cur aqua in summa parte alba, in fundo vero nigra spectatur?: The problem is that if water is essentially black, why is it seen as white in its upper layer and black only at the bottom? The idea that water is black was common in Antiquity (cf., e.g., Arist., *GA* 779b30–33 and 735a32 with b35). On the basis of a few parallels in Homer, where allusion is made to the blackness of water (*Il.* 2, 825; 9, 14; *Od.* 4, 359; 20, 158), L. Senzasono, 2006, p. 248, n. 210 argues that Plutarch is probably interpreting a Homeric passage here but does not quote the poet explicitly as he presupposes the reader to know the lines by heart. This is unlikely, because Plutarch does not refrain from quoting Homer elsewhere (running up to five quotations in total in *Quaestiones naturales*). A parallel account for the darkness of deep water is found in *De prim. frig.* 950AB (cf. also, e.g., *De Is. et Os.* 364B), but Plutarch does not give a very detailed or systematic explanation of it there (since it is only part of the sub-argument, the main argument being the Empedoclean, anti-Stoic theory that water, rather than air, is the principle of cold). Plutarch there ascribes the darkness to the mass density of the water (ὑπὸ πλήθους), whereas the water that comes in contact with air is illuminated and looks bright (αὐτοῦ μὲν οὖν τοῦ ὕδατος σκοτεινότατον ὑπὸ πλήθους φαίνεται τὸ βαθύτατον, οἷς δ' ἀὴρ πλησιάζει, ταῦτα περιλάμπεται καὶ διαγελᾷ). The element of πλῆθος is not repeated here in *Q.N.* 39; the relation between air and light, on the other hand, is relatively common: cf. also *Q.N.* 12, 915AB, *De prim. frig.* 952F (in the sea, lakes and rivers) and *De facie* 922DE (in the hollows of the earth).

An quod profunditas nigredinis mater est, ut quae solis radeos prius quam ad eam descendant obtundat et labefactet?: The imagery in *profunditas nigredinis mater* is rather exceptional to the general register and style of the collection [see 1.2.3.]. One may wonder if this is perhaps Longolius' invention.

Superficies autem, quoniam continuo a sole adficitur, candorem luminis recipiat oportet: F.H. Sandbach, 1965, p. 225 translates *continuo* as "immediately" here, but a little bit further on as "perpetually", which suits both cases better (as is also seen, e.g., in Quint., *Inst. or.* 2, 20, 3 and 9, 1, 11; see L&S, s.v. b).

Quod ipsum et Empedocles approbat 'Et niger in fundo fluvii color exstat ab umbra / atque cavernosis itidem spectatur in antris': The fragment is only preserved in Longolius' Latin translation (= Emp., DK31B94). H. Diels, 1901, p. 141, 94 tried to restore the original Greek text as follows: καὶ πέλει ἐν βένθει ποταμοῦ μέλαν ἐκ σκιόεντος / καὶ σπηλαιώδεσσιν ὁμῶς ἐνορᾶται ἐν ἄντροις. According to Empedocles, blackness belongs to water and whiteness to fire: cf. DK31A69a (Ἐμπεδοκλῆς δὲ καὶ περὶ τῶν χρωμάτων καὶ ὅτι τὸ μὲν λευκὸν τοῦ πυρός, τὸ δὲ μέλαν τοῦ ὕδατος) and DK31A91 (καθάπερ Ἐμπεδοκλῆς φησι, τὰ δὲ μέλανα [sc. ὄμματα] πλεῖον ὕδατος ἔχειν ἢ πυρός κτλ.).

An limo plerumque oppletus fluminum marisque fundus talem de se colorem per solis reflexum parit, quali utique is praeditus est?: For the idea that reflection of light is only possible with respect to solid bodies, cf. *De facie* 931B (δεῖ τὸ ποιῆσον ἀντιτυπίαν τινὰ καὶ κλάσιν ἐμβριθὲς εἶναι καὶ πυκνόν, ἵνα πρὸς αὐτὸ πληγὴ καὶ ἀπ' αὐτοῦ φορὰ γένηται). In his translation, F.H. Sandbach, 1965, p. 225 adds between brackets that the (dark) colour that reflects from the mud "may not be that of the water" itself, which is, indeed, what Plutarch seems to be implying here.

An probabilius est aquam minime quae illis est puram et sinceram esse, sed terrea qualitate (utpote quae continuo, qua currit vel agitur, aliquid ex ea advehat) imbutam: For the earthy quality of seawater, cf. *Q.N.* 1, 911D and 8, 914B, and for that of spring waters, cf. *Q.N.* 2, 912B. By contrast, in *Quaest. conv.* 725D, Plutarch writes that running water, as opposed to stagnant water, avoids mixture or shakes off any earth that enters its course.

ತಿ

7. *Psellus (Q.N. 40–41)*

The additional problems in *Q.N.* 40–41 are part of the 11th century side tradition of *Quaestiones naturales* in Michael Psellus' *De omnifaria doctrina*, where they are numbers §§ 170 and 188 Westerink respectively (= §§ 134 and 152 Migne; the latter was copied by M. Glycas, *Ann.* 1, 13 (= pp. 25, 21–26, 5 Bekker)). The least that can be said is that there is a Plutarchan core to these two problems, so they deserve to be studied here, if only for the parallel material and related physical topics in Plutarch. In spite of the controversy about the authenticity of these two problems, there may be reason to assume – not only on the basis of the parallel material in Plutarch's writings but also of the order of Psellus' sources in the first redaction of *De omnifaria doctrina* (which only covers natural scientific topics) – that they contain the remains of two lost *Quaestiones naturales*. If this is true, these chapters are not the result of a mere *bric-à-brac* by Psellus (though his adaptation of Plutarch's text is obvious). For further discussion, see L.G. Westerink, 1948, p. 3, F.H. Sandbach, 1965, p. 143 and M. Meeusen, 2012b, pp. 107–110. L. Senzasono, 2006, pp. 50–51 refuses to include both problems in his edition on grounds of the uncertain attribution in comparison to Longolius' problems.

Q.N. 40 (= Psellus, *De omn. doctr.* 170 Westerink)

Q.N. 40 links up with the problems concerning salt and water in *Q.N.* 1–13. It examines **why brine bursts forth from the sea when a thunderbolt falls in it** (Τίς ἡ αἰτία δι' ἥν, ὅταν εἰς τὴν θάλασσαν ἐμπέσῃ κεραυνός, ἅλες ἐξανθοῦσιν;). This phenomenon is explained in one lengthy explanation, where the very first word takes central position, viz. the process of solidification.

The argument goes that the solidification of seawater produces salt, and that it is solidified when a thunderbolt falls in it and draws off the sweet and drinkable water (πηγνύμενον τὸ θαλάσσιον ὕδωρ τοὺς ἅλας ποιεῖ, πήγνυται δὲ τοῦ κεραυνοῦ ἐμπεσόντος ἐν τῇ θαλάσσῃ καὶ τὸ γλυκὺ καὶ πότιμον ὕδωρ ἐξάγοντος). Light and drinkable water, by contrast, is not solidified when scorched by the sun or by a thunderbolt, but both things, and especially the thunderbolt, have this effect on seawater (ὅθεν τὸ μὲν λεπτὸν καὶ πότιμον ὕδωρ οὔθ' ὑπὸ ἡλίου καιόμενον πήγνυται οὔθ' ὑπὸ κεραυνοῦ, τὸ δ' ἁλμυρὸν ὑπ' ἀμφοτέρων τοῦτο πάσχει, καὶ μάλιστα ὑπὸ κεραυνοῦ). This is explained by the fact that when the lightning fire, which is sulphureous, falls into the sea, it evaporates and dries up the drinkable substance and solidifies the earthy and salty constituents within it (θειῶδες γὰρ ὂν τὸ κεραύνιον πῦρ, ὅταν εἰς τὴν θάλασσαν ἐμπέσῃ, ἐξατμίζει μὲν καὶ ἀναξηραίνει τὸ πότιμον, πήγνυσι δὲ τὸ γεῶδες καὶ ἁλμυρόν). An illustration is found in the fact that lightning protects corpses against decay and that salt also conserves them undecayed, as they both melt out the moisture (ὅθεν ἄσηπτα μὲν οἱ κεραυνοὶ τὰ σώματα ποιοῦσιν, ἄσηπτα δὲ οἱ ἅλες διαφυλάττουσιν, ἐκτηκομένης ὑπ' αὐτῶν τῆς ὑγρότητος). At the end, we read that Aristotle, along with the better scientists, approves of this explanation (ταύτην τὴν αἰτίαν καὶ Ἀριστοτέλης ὁ φιλόσοφος ἀποδέχεται καὶ οἱ κρείττους τῶν φυσικῶν).

Τίς ἡ αἰτία δι' ἥν, ὅταν εἰς τὴν θάλασσαν ἐμπέσῃ κεραυνός, ἅλες ἐξανθοῦσιν;: For a similar formulation of the *quaestio* with the introductory phrase τίς ἡ αἰτία δι' ἥν, cf. *Q.N.* 29, 919A.

θειῶδες γὰρ ὂν τὸ κεραύνιον πῦρ, ὅταν εἰς τὴν θάλασσαν ἐμπέσῃ, ἐξατμίζει μὲν καὶ ἀναξηραίνει τὸ πότιμον, πήγνυσι δὲ τὸ γεῶδες καὶ ἁλμυρόν: For the idea that heat evaporates the drinkable constituent in seawater, cf. *Q.N.* 9, 914C (with the commentary *ad loc.* for further parallels). For heat (of the sun) evaporating the finest and lightest part of the seawater on the skin of bathers and leaving behind a briny scum (ἁλώδης ἐπίπαγος), cf. also *Quaest. conv.* 627EF (with a reference to Ps.-Arist., *Probl.* 932b25). See also *Q.N.* 1, 911D (with the commentary *ad loc.*). In *Quaest. conv.* 665CD, the orator Dorotheüs ascribes a divine character to sulphur (θεῖον) on the basis of a dubious etymology. See S.-T. Teodorsson, 1990a, p. 57. Ps.-Arist., *Probl.*

937b29 also calls sulphur and lightning sacred (the problem is why hot baths are considered holy: cf. *Q.N.* 29).

ὅθεν ἄσηπτα μὲν οἱ κεραυνοὶ τὰ σώματα ποιοῦσιν, ἄσηπτα δὲ οἱ ἅλες διαφυλάττουσιν, ἐκτηκομένης ὑπ' αὐτῶν τῆς ὑγρότητος: For the preservative quality of salt, cf. *Q.N.* 1, 911D, 10, 914DE and *Quaest. conv.* 685BC. For that of lightning, cf. *Quaest. conv.* 665C and 685C (where it illustrates the divine character of lightning). For the contrary belief that bodies struck by lightning decay in a few days, cf. Sen., *NQ* 2, 31, 2 (*fulmine icta intra paucos dies verminant*). According to S.-T. Teodorsson, 1990a, p. 231, the connection between the preservative qualities of lightning and salt was presumably "first made in a Peripatetic work". This may be true, because at the end of *Q.N.* 40, the authority of Aristotle is adduced (see the following comment).

ταύτην τὴν αἰτίαν καὶ Ἀριστοτέλης ὁ φιλόσοφος ἀποδέχεται καὶ οἱ κρείττους τῶν φυσικῶν: It is unclear whether this conclusive remark (= Arist., fr. 218 Rose) is original or part of the reformulation of the chapter by Psellus. In any case, the Stagirite's natural scientific acumen is also explicitly praised in *Quaest. conv.* 656C, where Plutarch's father says that Aristotle is very sharp in solving natural problems (ὀξύτατος ὢν ἐν τοῖς τοιούτοις [sc. φυσικοῖς] ζητήμασι). Plutarch more often identifies the Stagirite as a φιλόσοφος, perhaps to distinguish him from the historian Aristotle of Chalcis (cf. *Amatorius* 761A). Therefore, it is probably "neither an *epitheton ornans* nor a *cognomen ex virtute*" (G. Roskam, 2011b, p. 42). Cf. *Thes.* 3, 2, *Lyc.* 1, 1, *Sol.* 32, 4, *Cam.* 22, 3, *Comp. Arist. et Ca. Ma.* 2, 4, *Comp. Alc. et Cor.* 3, 2, *Mul. virt.* 254EF, *Quaest. Rom.* 265B (cf. also fr. 122 Sandbach, *De gar.* 503A, *Alex.* 17, 9).

ॐ

Q.N. 41 (= Psellus, *De omn. doctr.* 188 Westerink)

Q.N. 41 concerns a botanical problem, viz. **why roses flower better when certain ill-smelling plants have been planted beside them** (Διατί τὰ ῥόδα μᾶλλον ἀνθεῖ δυσόδμων τινῶν παραπεφυτευμένων αὐτοῖς;). One lengthy explanation is given, which is based on the theory of like attracts like, as effected, more specifically, by the processes of attraction and motion of material effluences (ὁλκή and φορά). The (Stoic) concept of natural sympathy and antipathy is not far away (for the antipathetic property of the fig tree, which is mentioned as a parallel case in the explanation, cf. *Quaest. conv.* 664C: ἀλλ' ἔχων δύναμιν ἀντιπαθῆ, καθάπερ ἡ συκῆ).

First, the phenomenon at hand is generalised as being commonly observed in the botanical world. Not only roses, but also lilies, violets and all flowers that have a sweet efflux become even sweeter-smelling when (malodorous) garlic and onions are planted beside them (οὐ τὰ ῥόδα μόνον, ἀλλὰ καὶ τὰ κρίνα καὶ τὰ ἴα καὶ πάντα ὅσα ἔχει ἡδεῖαν ἀποφοράν, ὅταν σκόροδα καὶ κρόμμυα τούτοις παραφυτεύηται, εὐωδέστερα γίνεται). The reason for this is that if there is anything pungent and ill-smelling in them, it naturally emanates to the more pungent plants. What remains becomes very sweet-smelling and fragrant (διότι πᾶν εἴ τι δριμὺ καὶ δύσοδμον ἐν τούτοις ᾖ, ἐν τοῖς δριμυτέροις τῶν σπερμάτων φυσικῶς ἀπορρεῖ, καὶ γίνεται τὸ καταλιμπανόμενον εὐωδέστατον καὶ ὀσφραντικώτατον). This is further illustrated by two examples, both regarding the bad smell of fig trees. Rue also becomes more pungent than normal, when planted under a fig tree, because what is malodorous in the fig tree is transposed to the (rue) plant (καὶ τὸ πήγανον δὲ ὑπὸ τῇ συκῇ φυτευόμενον δριμύτερον ἑαυτοῦ γίνεται. μετατίθεται γὰρ εἰς τὸ φυτὸν τὸ ἐν τῇ συκῇ βαρύοσμον). In addition, figs improve when wild figs are planted beside them (καὶ ταῖς συκαῖς δὲ ἀγρίων παραπεφυτευμένων συκῶν βελτίω τὰ σῦκα γίνεται). This is due to the fact that in each of these plants, there is an attraction and motion towards things congeneric and alike. Thus, all that is pungent in the sweet fig tree passes over to the wild one, preserving the sweetness of its figs undiluted (ὁλκῆς γὰρ ἑκάστῳ καὶ φορᾶς πρὸς τὰ σύμφυλα καὶ ὅμοια γινομένης, ὅσον ἐστὶν ἐν τῇ γλυκείᾳ συκῇ δριμὺ εἰς τὴν ἀγρίαν μεταβαίνει συκῆν καὶ ἄμικτον τὴν τοῦ σύκου φυλάττει γλυκύτητα).

οὐ τὰ ῥόδα μόνον, ἀλλὰ καὶ τὰ κρίνα καὶ τὰ ἴα καὶ πάντα ὅσα ἔχει ἡδεῖαν ἀποφοράν, ὅταν σκόροδα καὶ κρόμμυα τούτοις παραφυτεύηται, εὐωδέστερα γίνεται, διότι πᾶν εἴ τι δριμὺ καὶ δύσοδμον ἐν τούτοις ᾖ, ἐν τοῖς δριμυτέροις τῶν σπερμάτων φυσικῶς ἀπορρεῖ, καὶ γίνεται τὸ καταλιμπανόμενον εὐωδέστατον καὶ ὀσφραντικώτατον: The belief that roses and violets are improved by planting garlic and onions beside them is paralleled in *De cap. ex inim.* 92B, where the same explanation is given (Plutarch ascribes the belief to accomplished farmers there: οἱ χαρίεντες γεωργοί).

καὶ τὸ πήγανον δὲ ὑπὸ τῇ συκῇ φυτευόμενον δριμύτερον ἑαυτοῦ γίνεται. μετατίθεται γὰρ εἰς τὸ φυτὸν τὸ ἐν τῇ συκῇ βαρύοσμον: By contrast, in *Quaest. conv.* 684D, Plutarch's grandfather Lamprias reports (invoking the authority of gardeners) that rue actually becomes sweeter and milder instead of more pungent (ἥδιον εἶναι καὶ τῷ χυμῷ μαλακώτερον). Cf. also Dioscor., *De mat. med.* 3, 45, 1, Pallad., *Op. agr.* 4, 9, 14, Pliny, *NH* 19, 156. Similarly, Ps.-Arist., *Probl.* 924b35–925a5 examines why rue grows best and most if it is grafted onto a fig tree (cf. Theophr., *CP* 5, 6, 10). The contradiction may be due to Psellus' rewriting of the chapter, but an adaptation by Plutarch himself cannot be excluded (either way, to bring this example more in line with the one that follows). For the pungency of fig trees and the intense vapour that they produce, cf. *Quaest. conv.* 696EF.

καὶ ταῖς συκαῖς δὲ ἀγρίων παραπεφυτευμένων συκῶν βελτίω τὰ σῦκα γίνεται: The same phenomenon is mentioned in passing in *Quaest. conv.* 700F. In *Amatorius* 753A, the artificial pollination of dates and figs is compared to joining a young man to an older woman. Cf. also Arist., *HA* 557b29, Theophr., *CP* 2, 9, 5, Pliny, *NH* 15, 79–81.

Synopsis

By way of conclusion, I here provide a synopsis of the main arguments as elaborated in the four introductory essays that form the first part of this book. I hope to have shown that Plutarch's *Quaestiones naturales* demonstrate that, among many other intellectual and philosophical predilections, the Chaeronean had numerous particular – and at times rather peculiar – questions about the natural world on his mind and took them to heart. By providing a systematic study and commentary of this generally neglected work in light of Plutarch's natural scientific programme more generally the volume at hand is meant to usefully contribute to our understanding of Plutarch's world view and, thus, to our knowledge of ancient natural science in the Imperial Era more generally.

The **first chapter** provides a general outline of the Aristotelian genre of natural problems and the place of Plutarch in the wider tradition of the Ps.-Aristotelian *Problems*. A seminal point that is raised here and serves as a conceptual framework for the study as a whole is that Plutarch's natural problems have an obvious Aristotelian, or more generally Peripatetic, character, which is problematic in light of his philosophical allegiance to Plato and the Academy. A good understanding of Plutarch's natural problems proves to be indispensable for contemporary scholarship not only because it provides precious insight into the reception of Ps.-Aristotle's *Problems* in the Imperial Era, but also because it sheds an important light on the Stagirite's influence on Plutarch's philosophy, a problem that is settled only at the end of chapter four (see below). Against this backdrop, the first chapter examines the 'problematic' organisation of *Quaestiones naturales* both on a micro- and on a macrostructural level. As indicated by its original Greek title, the aspect of physical aetiology is central to the collection's scientific set-up, which explains the sub-literary style of discourse and the general avoidance of moralising dynamics. This type of discourse is characteristic of the Aristotelian genre of natural problems more generally, which served as Plutarch's model. It is not strictly representative of the author's scientific intentions, since Plutarch's concept of natural science is by no means reducible to these features. This raises questions about the position of *Quaestiones naturales* in relation to the *corpus Plutarcheum* more generally.

The **second chapter** further elaborates on this topic. Special attention there goes to the incorporation of the same and similar *Quaestiones naturales* material in Plutarch's other treatises, especially *Quaestiones convivales*. We have seen that Plutarch's collections of problems should not be mistaken for his personal notes (ὑπομνήματα), as traditional scholarship has often done. By contrast, *Quaestiones naturales* provides an independent aetiological framework for Plutarch to collect his thoughts on particular natural questions and to deal with them in an autonomous fashion (i.e. to a large degree on their own terms and free from any other preoccupations, such as stylistic embellishment or moralising dynamics). Eventually, the possiblity of the collection's publication by Plutarch himself is considered, where the usability of this kind of literature in a philosophical school context is emphasised.

This last point is further elaborated upon in **chapter three**, which addresses the intended reading and educational value of *Quaestiones naturales*. I here show that natural problems were a popular subject for discussion in Plutarch's philosophical school and also during convivial gatherings of his intellectual milieu. In a seminal passage from *De tuenda* 133E, natural problems are described as being ἐλαφρὰ καὶ πιθανά ('easy and persuasive'), a phrase that highlights the low level of complexity of the genre and its general utility as exercises in natural scientific debate. Thus, I elaborate on the idea of intellectual gymnastics promoted by Plutarch's natural problems, while also stressing that the solutions to these problems are not simply meant as forms of sophistic playfulness. On the contrary, the search for physical causes in explaining wonder-inducing natural phenomena can be seen as an intellectual exercise aimed at the eradication of irrational, superstitious beliefs about God and his influence in the natural world around us – an idea that ties in closely with Plutarch's broader philosophical-religious project.

The place of Plutarch's *Quaestiones naturales* in this broader philosophical-religious project is further elaborated upon in **chapter four**. Here, I first focus on the collection's aetiological design and its link with the ancient genre of paradoxography and *mirabilia* literature. Plutarch was not so much concerned – for underlying philosophical and religious motives – with the veracity of the natural phenomena but with their physical causes. An explanation is found in Plutarch's Platonic-Academic outlook on the world and his dualistic view on causality, wherein it is accepted that natural phenomena are based on physical causes but also have a higher, divine motivation. In order to support this, I provide an analysis and interpretation of the mythological material that Plutarch incorporates in his physical aetiologies, arguing that these may hint at a higher type of causality and at a 'mystification' of the aetiological discourse. An analysis of the material Plutarch borrows from the poets and from authors of scientific prose then follows. We see that in his attempt to formulate plausible explanations to the problems, Plutarch often relies on received knowledge by 'problematising', that is, reframing in the problem format, a wide array of ancient Greek scientific learning. At the same time, he tries to balance this approach with his own innovative contributions to the problems, thus demonstrating his own argumentative creativity. In the remainder of the chapter, I provide a general outline of Plutarch's scientific methodology, focusing successively on the material principles and natural processes mentioned in the physical aetiologies, Plutarch's generally sceptical and anti-empiricist approach to natural phenomena, the logical-rhetorical organisation of the collection and the use of a more or less uniform set of technical terms. At the end of chapter four, I revisit the question of Aristotle's influence on Plutarch's Platonism, arguing that *Quaestiones naturales* is not the product of his aspirations to be regarded as an Aristotelian scientist. In the end, Plutarch's science of natural problems is, by its inquisitive method and philosophical purpose, framed in a wider Platonic view of the world.

Bibliography

A

Abramowiczówna, Z., *Komentarz Krytyczny i egzegetyczny do Plutarcha quaestiones convivales Ks. I i II*, Toruń, 1960.

—, "Plutarch's 'Tischgespräche'", *Altertum*, 8 (1962), pp. 80–88.

Adam, H., *Plutarchs Schrift* Non posse suaviter vivi secundum Epicurum. *Eine Interpretation*, Amsterdam, 1974.

Adrados, F.R., "El poema del pulpo y los orígenes de la colección teognídea", *Emerita*, 26 (1958), pp. 1–10.

Aguilar, R.M., "Hipócrates en Plutarco", *Cuadernos de Filologia Clásica*, 4 (1994), pp. 35–45.

—, "Plutarco y los médicos helenísticos", in Casanova, A., 2005, pp. 417–434.

Amboglio, D., "Fra hypomnemata e storiografia", *Athenaeum*, 78 (1990), pp. 503–508.

Amigues, S., *Théophraste. Recherches sur les plantes*, vol. 3, Paris, 1993.

Anderson, G., *The Second Sophistic: A Cultural Phenomenon in the Roman Empire*, London, 1993.

Andò, V., "La ricezione ippocratica in Plutarco", in Gallo, I., 2004, pp. 159–183.

Anton, J.P. (ed.), *Science and the Sciences in Plato*, New York, 1980.

Armstrong, A.H., *The Cambridge history of later Greek and early Medieval Philosophy*, Cambridge, 1967.

Arnould, D., *Le rire et les larmes dans la littérature grecque: d'Homère à Platon*, Paris, 1990.

Ashbaugh, A.F., *Plato's Theory of Explanation. A Study of the Cosmological Account in the Timaeus*, New York, 1988.

Asmis, E., *Epicurus' Scientific Method*, New York, 1984.

Asper, M., *Kallimachos. Werke. Griechisch und Deutsch*, 2004, Darmstadt.

— (ed.), *Writing Science. Mathematical and Medical Authorship in Ancient Greece*, Berlin, 2013.

B

Babbitt, F.C., *Plutarch's Moralia in Sixteen Volumes*, vol. 2, Cambridge, Mass. – London, 1928.

—, *Plutarch's Moralia in Sixteen Volumes*, vol. 4, Cambridge, Mass. – London, 1936 [= 1936a].

—, *Plutarch's Moralia in Sixteen Volumes*, vol. 5, Cambridge, Mass. – London, 1936 [= 1936b].

Babilas, W., *Tradition und Interpretation. Gedanken zur philologischen Methode*, München, 1961.

Babut, D., *Plutarque et le stoïcisme*, Paris, 1969.

—, "Sur l'unité de la pensée d'Empédocle", *Philologus*, 120 (1976), pp. 139–164.

—, "Du scepticisme au dépassement de la raison: philosophie et foi religieuse chez Plutarque", in Babut, D. (ed.), *Parerga: choix d'articles de Daniel Babut (1974–1994)*, Lyon, 1994, pp. 549–581.

—, "Plutarque, Aristote, et l'Aristotélisme", in Van der Stockt, L., 1996a, pp. 1–28.

—, "L'unité de l'Académie selon Plutarque. Notes en marge d'un débat ancien et toujours actuel", in Bonazzi, M., Lévy, C. and Steel, C. (eds.), *A Platonic Pythagoras: Platonism and Pythagoreanism in the Imperial Age*, Turnhout, 2007, pp. 63–98.

Bakker, F., *Three Studies in Epicurean Cosmology*, Utrecht, 2010.

Baldassari, M. "Osservazioni sulla struttura del periodo e sulla costruzione ritmica del discorso nei *Moralia* di Plutarco", in Van der Stockt, L., 2000a, pp. 1–13.

Barigazzi, A., "Implicanze morali nella polemica plutarchea sulla psicologia degli animali", in Gallo, I., 1992, pp. 297–315.

Barrow, R.H., *Plutarch and his times*, London, 1967.

Barthelmess, J., "Recent Work on the *Moralia*", in Brenk, F.E. and Gallo, I., 1986, pp. 61–81.

Barton, T., *Power and knowledge: astrology, physiognomics, and medicine under the Roman Empire*, Michigan, 1994 [= 1994a].

—, *Ancient Astrology*, London, 1994 [= 1994b].

Battegazzore, M., "L'atteggiamento di Plutarco verso le scienze", in Gallo, I., 1992, pp. 19–59.

Beagon, M., "The curious eye of the Elder Pliny", in Gibson, R.K. and Morello, R., 2011, pp. 71–88.

Becchi, F., "Aristotelismo ed antistoicismo nel *De virtute morali* di Plutarco", *Prometheus*, 1 (1975), pp. 160–180.

—, "Aristotelismo funzionale nel *De virtute morali* di Plutarco", *Prometheus*, 4 (1978), pp. 261–275.

—, "Plutarco tra platonismo ed aristotelismo: la filosofia come παιδεία dell'anima", in Pérez Jiménez, A., Garciá López, J. and Aguilar, R.M., 1999, pp. 25–43.

—, "Lignes directrices de la doctrine zoopsychologique de Plutarque", *Myrtia*, 17 (2002), 159–174.

—, "Le traduzioni latine dei Moralia di Plutarco tra XIII e XVI secolo", in Volpe Cacciatore, P. (ed.), *Plutarco nelle traduzioni latine di età umanistica: Seminario di studi, Fisciano, 12–13 luglio 2007*, Napoli, 2009, pp. 9–52.

—, "Plutarch, Aristotle, and the Peripatetics", in Beck, M., 2014, pp. 73–87.

Beck, M., "Plutarch's *Hypomnemata*, Standard *Topoi* and Idiosyncratic Composition in the *Moralia*", in Horster, M. and Reitz, C., 2010a, pp. 349–367.

— (ed.), *A Companion to Plutarch*, Oxford, 2014.

Bernard, W., *Spätantike Dichtungstheorien: Untersuchungen zu Proklos, Herakleitos und Plutarch*, Stuttgart, 1990.

Bertier, J., *Mnésithée et Dieuchès*, Leiden, 1972.

—, "A propos de quelques resurgences des *Épidémies* dans les *Problemata* du Corpus aristotelicien", in Baader, G. and Winau, R. (eds.), *Die hippokratischen Epidemien: Theorie, Praxis, Tradition*, Stuttgart, 1989, pp. 261–269.

Beta, S., "Riddling at table: trivial ainigmata vs. philosophical problemata", in Ferreira, J.R., Leão, D., Tröster, M. and Barata Dias, P., 2009, pp. 97–102.

Bétolaud, V., *Oeuvres complètes de Plutarque. Oeuvres morales et oeuvres diverses traduites en français*, vol. 4, Paris, 1870.

Blair, A., "The *Problemata* as a Natural Philosophical Genre", in Grafton, A. and Siraisi, N.G. (eds.), *Natural Particulars. Nature and the Disciplines in Renaissance Europe*, Cambridge, 1999, pp. 171–204.

Bolkestein, H., *Adversaria critica et exegetica ad Plutarchi quaestionum convivalium librum primum et secundum*, Amsterdam, 1946.

Bömer, F., "Der Commentarius. Zur Vorgeschichte und literarischen Form des Schriften des Caesars", *Hermes*, 81 (1953), pp. 210–250.

Bonazzi, M., "Contro la rappresentazione sensibile: Plutarco tra l'Academia e il platonismo", *Elenchos*, 25 (2004), pp. 41–71.

—, "Plutarch and the Skeptics", in Beck, M., 2014, pp. 121–134.

Bonazzi, M. and Opsomer, J. (eds.), *The Origins of the Platonic System. Platonisms of the Early Empire and their Philosophical Context*, Louvain – Paris – Namur – Walpole, Mass., 2009.

Bonitz, H., *Index Aristotelicus*, Berlin, 1870.

Borthwick, E.K., "Bee Imagery in Plutarch", *CQ*, 41 (1991), pp. 560–562.

Bouffartigue, J., *Plutarque. L'intelligence des animaux*, vol. 14, 1, Paris, 2012.

Boulogne, J., "Les 'Questions Romaines' de Plutarque", *ANRW* 2, 33, 6 (1992), pp. 4682–4708.

—, *Plutarque. Un aristocrate grec sous l'occupation romaine*, Lille, 1994.

—, "Plutarque et la médecine", *ANRW* 2, 37, 3 (1996), pp. 2762–2792.

—, "Les 'Étiologies romaines': une herméneutique des moeurs à Rome", in Payen, P., 1998a, pp. 31–38.

—, *Plutarque. Conduites méritoires de femmes, Étiologies romaines, Étiologies grecques, Parallèles mineurs*, vol. 4, Paris, 2002.

—, *Plutarque dans le miroir d'Épicure. Analyse d'une critique systématique de l'épicurisme*, Villeneuve d'Asq, 2003.

— (ed.), *Les Grecs de l'antiquité et les animaux. Le cas remarquable de Plutarque*, Lille, 2005 [= 2005a].

—, "Le culte égyptien des animaux vu par Plutarque. Une étiologie égyptienne (*Isis et Osiris*, 71–76, 379D–382C)", in Boulogne, J., 2005a, pp. 197–205 [= 2005b].

—, "Plutarque lecteur de Théophraste", in Casanova, A., 2005, pp. 287–300 [= 2005c].

—, "Le paradigme de la crase dans la pensée de Plutarque," *Ploutarchos*, 4 (2006/7), pp. 3–17.

—, "Les digressions scientifiques dans les *Vies* de Plutarque", in Nikolaidis, A.G., 2008, pp. 733–749.

Bowersock, G.W.M., "Plutarch and the Sublime Hymn of Ofellius Laetus", *GRBS*, 23 (1982), pp. 275–279.

—, "Plutarch", in Easterling, P.E. and Knox, B.M.W. (eds.), *The Cambridge History of Classical Literature, 1: Greek Literature*, 1985, pp. 665–669.

Bowie, E., "Plutarch's Habits of Citation", in Nikolaidis, A.G., 2008, pp. 143–157.

Boyancé, P., "Platon et le Vin", *BAGB Lettres d'Hum.*, 10 (1951), pp. 3–19.

Boys-Stones, G., "Plutarch on the Probable Principle of Cold: Epistemology and the *De Primo Frigido*", *CQ*, 47 (1997), pp. 227–238 [= 1997a].

—, "Thyrsus-Bearer of the Academy or Enthusiast for Plato? Plutarch's *de Stoicorum repugnantiis*", in Mossman, J., 1997a, pp. 41–58 [= 1997b].

Bräuninger, F., "Persephone", *RE*, 19, 1 (1937), cols. 944–972.

Bréchet, C., "Vers une Philosophie de la Citation Poétique: Écrit, Oral et Mémoire chez Plutarque", in Castelnérac, B., 2007, pp. 101–134.

Brenk, F.E., *In Mist Apparelled: religious themes in Plutarch's Moralia and lives*, Leiden, 1977.

—, "Plutarch's Middle-Platonic God: About to Enter (or Remake) the Academy", in Hirsch-Luipold, R. (ed.), *Gott und die Götter bei Plutarch. Götterbilder – Gottesbilder – Weldbilder*, Berlin – New York, 2005, pp. 27–49.

Brenk, F.E. and Gallo, I. (eds.), *Miscellanea Plutarchea. Atti del I convegno di studi su Plutarco (Roma, 23 novembre 1985)*, Ferrara, 1986.

Brisson, L. (transl. C. Tihanyi), *How Philosophers Saved Myths: Allegorical Interpretation and Classical Mythology*, Chicago, 2004.

—, "Why Is the *Timaeus* Called an *Eikôs Muthos* and an *Eikôs Logos*?", in Collobert, C., Destrée, P., and Gonzalez, F.J. (eds.), *Plato and Myth. Studies on the Use and Status of Platonic Myths*, Leiden – Boston, 2012, pp. 369–391.

Broadie, S., *Nature and Divinity in Plato's Timaeus*, Cambridge, 2012.

Bucher-Isler, B., *Norm und Individualität in den Biographien Plutarchs. Untersuchungen zu seiner Charakterdarstellung*, Stuttgart, 1972.

Buckler, J., "Plutarch and Autopsy", in *ANRW* 2, 33, 6 (1992), pp. 4788–4830.

Bühler, W., "Die Philologie der Griechen und ihre Methoden", *Jahrbuch der Akademie der Wissenschaften in Göttingen* (1977), pp. 44–62.

Burnyeat, M.F., "Eikōs muthos", in Partenie, C. (ed.), *Plato's myths*, Cambridge, 2009, pp. 167–186.

Bussemaker, U.C., *Aristotelis opera omnia graece et latine*, vol. 4, Paris, 1857.

Butterfield, H., *The Whig Interpretation of History*, London, 1931.

C

Caballero Sanchez, R., "Excursus geografici nella *Vita Alexandri* di Plutarco", in Gallo, I., 1992, pp. 91–97.

Cameron, A., *Callimachus and His Critics*, Princeton, 1995.

Candau Morón, J., González Ponce, F. and Chávez Reino, A. (eds.), *Plutarco transmisor. Actas del X simposio internacional de la sociedad española de Plutarquistas (Sevilla, 12–14 de noviembre de 2009)*, Sevilla, 2011.

Capelle, W., "Auf Spuren alter Φυσικοί", *Hermes*, 45 (1910), pp. 321–336.

Carrano, A., *Questioni greche*, Napoli, 2007.

Casanova, A. (ed.), *Plutarco e l'età ellenistica*, Firenze, 2005.

Castelli, L.M., "Manifestazioni somatiche e fisiologia delle "affezioni dell'anima" nei *Problemata* aristotelici", in Centrone, B., 2011a, pp. 239–274.

Castelnérac, B. (ed.), *Philosophia and Philologia: Plutarch on Oral and Written Language*, Hermathena, 182 (2007).

Centrone, B. (ed.), *Studi sui* Problemata Physica *aristotelici*, *Elenchos*, 58 (2011) [= 2011a].

—, "Μελαγχολικός in Aristotele e il *Problema* XXX 1", in Centrone, B., 2011a, pp. 309–339 [= 2011b].

von Christ, W., *Geschichte der griechischen Literatur*, vol. 2, 1, München, 1959.

Chassignet, M. (ed.), *L'étiologie dans la pensée antique*, Turnhout, 2008.

Cherniss, H., *Plutarch's Moralia in Sixteen Volumes*, vol. 13, 1, Cambridge, Mass. – London, 1976.

Cherniss, H. and Helmbold, W.C., *Plutarch's Moralia in Sixteen Volumes*, vol. 12, Cambridge, Mass. – London, 1957.

Claes, P., *Concatenatio Catulliana: A New Reading of the* Carmina, Amsterdam, 2002.

Clagett, M., *Greek Science in Antiquity*, New York, 1955.

Clement, P.A. and Hoffleit, H.B., *Plutarch's Moralia in Sixteen Volumes*, vol. 8, Cambridge, Mass. – London, 1969.

Collobert, C., "Aristotle's Review of the Presocratics: Is Aristotle Finally a Historian of Philosophy?", *Journal of the History of Philosophy*, 40 (2002), pp. 281–295.

Conte, G.B., "Empirical and Theoretical Approaches to Literary Genre", in Galinsky, K. (ed.), *The Interpretation of Roman Poetry: Empiricism or Hermeneutics?*, Frankfurt am Main, 1992, pp. 104–123.

Corcoran, T.H., *Seneca, Naturales Quaestiones*, vol. 1, London – Cambridge, Mass., 1971.

Cornford, F.M., *Plato's Cosmology. The Timaeus translated with a running commentary*, London, 1937.

Cribiore, R., *Gymnastics of the Mind. Greek Education in Hellenistic and Roman Egypt*, Princeton, 2001.

Croiset, A. and M., *Histoire de la littérature grecque, V: Période alexandrine; période romaine*, Paris, 1899.

Culham, P., "Plutarch on the Roman Siege of Syracuse: the Primacy of Science over Technology", in Gallo, I., 1992, pp. 179–197.

Cunningham, A., "Getting the Game Right: Some Plain Words on the Identity and Invention of Science", *Studies in History and Philosophy of Science*, 19 (1988), pp. 365–389.

D

Dalby, A., *Food in the ancient world from A–Z*, London, 2003.

Darbo-Peschanski, C., "Pourquoi chercher des causes aux coutumes?", in Payen, P., 1998a, pp. 21–30.

D'Arms, J.H., "The Roman *Convivium* and the Idea of Equality", in Murray, O., 1990a, pp. 308–320.

—, "Heavy Drinking and Drunkenness in the Roman World: Four Questions for Historians", in Murray, O. and Tecuşan, M. (eds.), *In vino veritas*, London, 1995, pp. 304–317.

De Lacy, P.H., "Plutarch and the Academic Sceptics", *The Classical Journal*, 49 (1953), pp. 79–85.

De Leemans, P. and Goyens, M. (eds.), *Aristotle's Problemata in Different Times and Tongues*, Leuven, 2006.

Del Re, R., "Il pensiero metafisico di Plutarco: Dio, la natura, il male", *Studi Italiani di Filologia Classica*, 24 (1950), pp. 33–64.

Denniston, J.D., *The Greek Particles*, Oxford, 1966.

Démarais, L., "L'animal, les *mirabilia* et l'étiologie dans les *Propos de table*. L'exemple du problème sur le rémora (II 7)", in Boulogne, J., 2005a, pp. 157–171.

De Rosalia, A., "Il Latino di Plutarco", in D'Ippolito, G. and Gallo, I., 1991, pp. 445–459.

Desideri, P., "Scienza nelle *Vite* di Plutarco", in Gallo, I., 1992, pp. 73–89.

Diehl, E., *Anthologia lyrica graeca*, vol. 2, Leipzig, 1925.

Diels, H., *Doxographi Graeci*, Berlin, 1879.

—, *Poetarum Philosophorum Fragmenta*, Berlin, 1901.

—, "Aristotelica", *Hermes*, 40 (1905), pp. 301–316.

Diller, H., *Wanderarzt und Aitiologe: Studien zur hippokratischen Schrift* περὶ ἀέρων ὑδάτων τόπων, *Philologus Suppl.*, 26, Leipzig, 1934.

Dillon, J.M., *The Middle Platonists: a study of Platonism 80 B.C. to A.D. 220*, London, 1977.

—, ""Orthodoxy" and "Eclecticism": Middle Platonists and Neo-Pythagoreans", in Dillon, J.M. and Long, A.A., 1988, pp. 103–125.

—, "Plutarch and God: Theodicy and Cosmogony in the Thought of Plutarch", in Frede, D. and Laks, A. (eds.), *Traditions of Theology. Studies in Hellenistic Theology, its Background and Aftermath*, Leiden, 2002, pp. 223–237.

—, "Plutarch and Platonism", in Beck, M., 2014, pp. 61–72.

Dillon, J.M. and Long, A.A. (eds.), *The Question of "Eclecticism": Studies in Later Greek Philosophy*, Berkeley, 1988.

D'Ippolito, G. and Gallo, I. (eds.), *Strutture formali dei "Moralia" di Plutarco*, Napoli, 1991.

D'Ippolito, G. and Nuzzo, G., *Plutarco. L'origine del freddo – Se sia più utile l'acqua o il fuoco. Introduzione, testo critico, traduzione e commento*, Napoli, 2012.

Doehner, T., *Quaestiones Plutarcheae*, vol. 2, Misniae, 1858.

—, *Vindiciarum Plutarchearum liber*, Zwiccaviae, 1864.

Donini, P., "Problemi del pensiero scientifico a Roma: il primo e il secondo secolo d.C.", in G. Giannantoni, G. and Vegetti, M. (eds.), *La scienza ellenistica : atti delle tre giornate di studio tenutesi a Pavia dal 14 al 16 aprile 1982*, Napoli, 1984, pp. 353–374.

—, "Lo scetticismo academico. Aristotele e l'unità della tradizione platonica secondo Plutarco", in Cambiano, G. (ed.), *Storiografia e dossografia nella filosofia antica*, Torino, 1986, pp. 203–226 [= 1986a].

—, "Plutarco, Ammonio e l'Academia", in Brenk, F.E. and Gallo, I., 1986, pp. 97–110 [= 1986b].

—, "Science and Metaphysics: Platonism, Aristotelianism, and Stoicism in Plutarch's *On the Face in the Moon*", in Dillon, J.M. and Long, A.A., 1988, pp. 126–144.

—, "I fondamenti della fisica e la teoria delle cause in Plutarco", in Gallo, I., 1992, pp. 99–120.

—, "Plutarco e la rinascita del platonismo", in Cambiano, G., Canfora, L. and Lanza, D. (eds.), *Lo spazio letterario della Grecia antica. 1: La produzione e la circolazione del testo. Vol. 3: I Greci e Roma*, Roma, 1994, pp. 35–60 [= 1994a].

—, "Testi e commenti, manuali e insegnamento: la forma sistematica e i metodi della filosofia in età postellenistica", *ANRW* 2, 36, 7 (1994), pp. 5027–5100 [= 1994b].

—, "Platone e Aristotele nella tradizione pitagorica secondo Plutarco", in Pérez Jiménez, A., Garciá López, J. and Aguilar, R.M., 1999, pp. 9–24.

—, "L'eredità academica e i fondamenti del platonismo in Plutarco", in Barbanti, M., Giardina, G.R. and Manganaro, P. (eds.), *Henosis kai philia. Unione e amicizia. Omaggio a F. Romano*, Catania, 2002, pp. 247–273.

—, "Il silenzio di Epaminonda, i demoni e il mito: il platonismo di Plutarco nel *De genio Socratis*", in Bonazzi, M. and Opsomer, J., 2009, pp. 187–214.

—, *Plutarco. Il volto della luna. Introduzione, testo critico, traduzione e commento*, Napoli, 2011.

Dorandi, T., "Den Autoren über die Schulter geschaut: Arbeitsweise und Autographie bei den antiken Schriftstellern", *ZPE*, 87 (1991), pp. 11–33.

—, *Le stylet et la tablette: dans le secret des auteurs antiques*, Paris, 2000.

Dörrie, H., *Porphyrios' "Symmikta Zetemata". Ihre Stellung in System und Geschichte des Neuplatonismus nebst einem Kommentar zu den Fragmenten*, München, 1959.

Dronkers, A.I., *De comparationibus et metaphoris apud Plutarchum*, Utrecht, 1892.

Düring, I., "Aristotle's method in biology. A note on *De Part. An.* I i, 639b30–640a2", in Mansion, S., 1961, pp. 213–221.

Duff, T., *Plutarch's Lives. Exploring Virtue and Vice*, Oxford, 1999 [= 1999a].

—, "Plutarch, Plato and 'Great Natures'", in Pérez Jiménez, A., Garciá López, J. and Aguilar, R.M., 1999, pp. 313–332 [= 1999b].

—, "The Prologues", in Beck, M., 2014, pp. 333–349.

Dyroff, A., "Zur stoischen Tierpsychologie", *Bayerische Blätter für das Gymnasial-Schulwesen*, 33 (1897), pp. 399–404.

E

van der Eijk, P.J., "Aristoteles über die Melancholie", *Mnemosyne*, 43 (1990), pp. 33–72.

—, "Towards a Rhetoric of Ancient Scientific Discourse. Some Formal Characteristics of Greek Medical and Philosophical Texts (Hippocratic Corpus, Aristotle)", in Bakker, E.J. (ed.), *Grammar as Interpretation. Greek Literature in its Linguistic Contexts*, Leiden, 1997, pp. 77–129.

—, "Quelques observations sur la réception d'Aristote dans la médecine gréco-romaine de l'époque impériale", in Lehmann, Y., 2013, pp. 183–193.

Emerson, R.W., *The Complete Works of Ralph Waldo Emerson, vol. 10: Lectures and Biographical Sketches*, Boston, 1891.

F

Fairbanks, A., "On Plutarch's Quotations from the Early Greek Philosophers", *TAPA*, 28 (1897), pp. 75–87.

Farrington, B., *Greek Science. Its Meaning For Us*, Harmondsworth, 1961.

Fernández Delgado, J.A., "Los proverbios en los "Moralia" de Plutarco", in D'Ippolito, G. and Gallo, I., 1991, pp. 195–212.

Ferrari, F., *Dio, idee e materia. La struttura del cosmo in Plutarco di Cheronea*, Napoli, 1995.

—, "Plutarco e lo scetticismo ellenistico", in Casanova, A., 2005, pp. 369–384.

Ferrari, F. and Baldi, L., *Plutarco. La generazione dell'anima nel* Timeo. *Introduzione, testo critico, traduzione e commento*, Napoli, 2002.

Ferreira, A., "The Power of Nature and Its Influence on Statesmen in the Work of Plutarch", in Meeusen, M. and Van der Stockt, L., 2015, pp. 155–165.

Ferreira, J.R., Leão, D., Tröster, M. and Barata Dias, P. (eds.), *Symposium and Philantropia in Plutarch*, Coimbra, 2009.

Ferreira, J.R., Leão, D. and Martins de Jesus, C.A. (eds.), *Nomos, Kosmos & Dike in Plutarch*, Coimbra, 2012.

Filius, L., *The* Problemata physica *attributed to Aristotle: the Arabic version of Hunain ibn Isḥāq and the Hebrew version of Moses ibn Tibbon*, Leiden, 1999.

Fischer, K.-D., "Beiträge zu den pseudosoranischen *Quaestiones medicinales*", in Fischer, K.-D., Nickel, D. and Potter, P. (eds.), *Text and Tradition. Studies in Ancient Medicine and its Transmission presented to Jutta Kollesch*, Leiden, 1998, pp. 1–54.

Fitzgerald, W., *Martial: The World of the Epigram*, Chicago, 2007.

Flacelière, R., "Plutarque et les éclipses de lune", *Revue des Etudes Anciennes*, 53 (1951), pp. 203–221.

—, "La lune selon Plutarque", in Ducrey, P., Bérard, C., Dunant, C. and Paschoud, F. (eds.), *Mélanges d'histoire ancienne et d'archéologie offerts à Paul Collart*, Lausanne, 1976, pp. 193–195.

Flacelière, R., Irigoin, J., Sirinelli, J. and Philippon, A., *Plutarque. De l'éducation des enfants; Comment lire les poètes*, vol. I, 1, Paris, 1987.

Flashar, H., *Problemata physica*, Berlin, 1962.

Fögen, T., *Wissen, Kommunikation und Selbstdarstellung. Zur Struktur und Charakteristik römischer Fachtexte der frühen Kaiserzeit*, München, 2009.

Forster, E.S., "The Pseudo-Aristotelian *Problems*: Their Nature and Composition", *CQ*, 22 (1928), pp. 163–165.

Franke, W.A. and Mircea, M., "Plutarch's Report on the Blue Patina of Bronze Statues at Delphi: A Scientific Explanation", *Journal of the American Institute for Conservation*, 44 (2005), pp. 103–116.

Frazier, F., "Théorie et pratique de la παιδιά symposiaque dans les *Propos de table* de Plutarque", in Trédé, M. and Hoffmann, P. (eds.), *Le rire des anciens*, 1998, Paris, pp. 281–292.

—, "Philosophie et religion dans la pensée de Plutarque. Quelques réflexions autour des emplois du mot πίστις", *Études platoniciennes*, 5 (2008), pp. 41–61.

—, "Quand Plutarque actualise le mythe d'Er. Delphes, la Justice et la Providence dans le mythe de Thespésios (*De sera* 22, 563 B–33, 568 A)", in Van der Stockt, L., Titchener, F.,

Ingenkamp, H.G. and Pérez Jiménez, A. (eds.), *Gods, Daimones, Rituals, Myths and History of Religions in Plutarch's Works. Studies Devoted to Professor Frederick E. Brenk by the I.P.S.*, Málaga – Logan, 2010, pp. 193–210.

—, "Ordre et désordre dans la pensée de Plutarque. Réseaux lexicaux et problématiques philosophiques autour de δίκη, κόσμος, νόμος", in Ferreira, J.R., Leão, D. and Martins de Jesus, C.A., 2012, pp. 215–242.

—, "The Perils of Ambition", in Beck, M., 2014, pp. 488–502.

Frazier, F. and Leão, D.F. (eds.), *Tychè et pronoia. La marche du monde selon Plutarque*, Coimbra, 2010.

Frazier, F. and Sirinelli, J., *Plutarque. Propos de table*, vol. 9, 3, Paris, 1996.

French, R., *Ancient Natural History*, London, 1994.

Frede, M., *Essays in Ancient Philosophy*, Oxford, 1987.

Froidefond, C., *Plutarque. Isis et Osiris*, vol. 5, 2, Paris, 1988.

Frost, F.J., *Plutarch's* Themistocles. *A Historical Commentary*, Princeton, 1980.

Fuhrmann, F., *Les images de Plutarque*, Paris, 1964.

—, *Plutarque. Propos de table*, vol. 9, 1, Paris, 1972.

—, *Plutarque. Propos de table*, vol. 9, 2, Paris, 1978.

—, *Plutarque. Apophthegmes de rois et de généraux; Apophtegmes laconiens*, vol. 3, Paris, 1988.

G

Gallo, I. (ed.), *Plutarco e le scienze. Atti del IV Convegno plutarcheo, Genova-Bocca di magra, 22–25 aprile 1991*, Genova, 1992.

—, "Forma letteraria nei 'Moralia' di Plutarco: Aspetti e problemi", in *ANRW* 2, 34, 4 (1998), pp. 3511–3540 [reprinted in *id.* (ed.), *Parerga Plutarchea*, Napoli, 1999, pp. 39–86].

— (ed.), *La biblioteca di Plutarco. Atti del IX Convegno plutarcheo, Pavia, 13–15 giugno 2002*, Napoli, 2004.

Gallo, I. and Moreschini, C. (eds.), *I generi letterari in Plutarco. Convegno plutarcheo. Pisa, 2–4 Juin 1999*, Napoli, 2000.

García López, J., "La Naturaleza en las comparaciones de Plutarco", in García López, J. and Calderón Dorda, E. (eds.), *Estudios sobre Plutarco: paisaje y naturaleza*, Madrid, 1991, pp. 203–220.

Garzya, A. and Masullo, R., *I problemi di Cassio Iatrosofista*, Napoli, 2004.

Georgiadou, A., "The Corruption of Geometry and the Problem of Two Mean Proportionals", in Gallo, I., 1992, pp. 147–164.

Gianakaris, C.J., *Plutarch*, New York, 1970.

Gibson, R.K. and Morello, R. (eds.), *Pliny the Elder: Themes and Contexts*, Leiden – Boston, 2011.

Giesen, K., "Plutarchs *Quaestiones graecae* und Aristoteles' *Politien*", *Philologus*, 60 (1901), pp. 446–471.

Gleason, M., *Making Men: Sophists and Self-presentation in Ancient Rome*, Princeton, 1995.

—, "Shock and awe: the performance dimension of Galen's anatomy demonstrations", in Gill, C., Whitmarsh, T. and Wilkins, J. (eds.), *Galen and the World of Knowledge*, Cambridge, 2009, pp. 85–114.

Glucker, J., *Antiochus and the Late Academy*, Göttingen, 1978.

Goldhill, S., "The Anecdote: Exploring the Boundaries between Oral and Literate Performance in the Second Sophistic", in Johnson, W.A. and Parker, H.N. (eds.), *Ancient Literacies: The Culture of Reading in Greece and Rome*, Oxford, 2009, pp. 96–113.

Göldi, O., *Plutarchs sprachliche Interessen*, Diss. Zürich, 1920.

Goodwin, W.W., *Plutarch's Morals*, vol. 3, Boston, 1878.

Görgemanns, H., *Das Mondgesicht*, Zürich, 1968.

—, *Untersuchungen zu Plutarchs Dialog De facie in orbe lunae*, Heidelberg, 1970.

—, "Biologie bei Platon", in Wöhrle, G. (ed.), *Geschichte der Mathematik und der Naturwissenschaften in der Antike*, vol. 1, *Biologie.*, Stuttgart, 1999, pp. 74–88.

Gow, A.S.F., *Theocritus*, vol. 2, Cambridge 1950.

Graf, E., "Plutarchisches. Entstehungsweise der *Symposiaca*. Keine Excerpte. Chronologische Reihenfolge. Ἑταῖρος. *Amatorius*", in *Commentationes philologae quibus Ottoni Ribbeckio praeceptori inlustri sexagensimum* [sic] *aetatis magisterii Lipsiensis decimum annum exactum congratulantur discipuli Lipsienses*, Leipzig, 1888, pp. 57–70.

Grandjean, T., "Le recours à l'étiologie chez Dion de Pruse et chez Plutarque de Chéronée", in Chassignet, M., 2008, pp. 147–164.

Grant, E., *A History of Natural Philosophy. From the Ancient World to the Nineteenth Century*, Cambridge, 2007.

Gregory, A., *Plato's Philosophy of Science*, London, 2000.

Griffith, M., "Public and Private in Early Greek Institutions of Education", in Too, Y.L. (ed.), *Education in Greek and Roman Antiquity*, Leiden, 2001, pp. 23–84.

Grimaudo, S., "La medicina ellenistica in Plutarco", in Gallo, I., 2004, pp. 417–437.

Guardasole, A., "Les *Problemata hippocratiques*: un exemple original de catéchisme et commentaire dans la tradition médicale et religieuse", *REG*, 120 (2007), pp. 142–160.

Gudeman, A., "λύσεις", *RE*, 13, 2 (1927), cols. 2511–2529.

Guidorizzi, G., *Il mondo letterario greco: storia, civiltà, testi. 3: Dall'età ellenistica all'età cristiana*, Milano, 2000.

Gunderson, E., *Nox Philologiae: Aulus Gellius and the Fantasy of the Roman Library*, Wisconsin, 2009.

Guthrie, W.K.C., *A History of Greek Philosophy*, Cambridge, 1962 (= vol. 1), 1965 (= vol. 2).

H

Hadot, P., *What is ancient philosophy?*, Cambridge, 2002.

Halliday, W.R., *The Greek questions of Plutarch*, Oxford, 1928.

Halliwell, S., "*On Poets* and *Homeric Problems*", in Kennedy, G. (ed.), *The Cambridge History of Literary Criticism*, vol. 1, Cambridge, 1989, pp. 149–151.

Hamilton, J.R., *Plutarch: Alexander, A Commentary*, Oxford, 1969.

Hani, J., *La religion égyptienne dans la pensée de Plutarque*, Paris, 1976.

Hankinson, R.J., "Saying the Phenomena", *Phron.*, 35 (1990), pp. 194–215.

Harder, A., *Callimachus. Aetia. Vol. 1: Introduction, Text, and Translation, Vol. 2: Commentary*, Oxford, 2012.

Hardie, P.R., "Plutarch and the interpretation of myth", *ANRW* 2, 33, 6 (1992), pp. 4743–4787.

Harrison, G.W.M., "Problems with the Genre of *Problems*: Plutarch's Literary Innovations", *Class. Phil.*, 95 (2000), pp. 193–199 [= 2000a].

—, "Tipping his Hand: Plutarch's Preferences in the *Quaestiones Naturales*", in Van der Stockt, L., 2000a, pp. 237–249 [= 2000b].

Hartman, J.J., *De Plutarcho scriptore et philosopho*, Lugduni Batavorum, 1916.

Healy, J.F., *Pliny the Elder on Science and Technology*, Oxford, 1999.

Hein, G., *Quaestiones Plutarcheae. Quo ordine Plutarchus nonnulla scripta moralia composuerit, agitur*, Diss. Berlin, 1916.

Helmbold, W.C., and O'Neil, E.N., *Plutarch's Quotations*, Baltimore, 1959.

Hershbell, J.P., "Plutarch as a Source for Empedocles Re-Examined", *AJPh*, 92 (1971), pp. 156–184.

—, "Empedoclean Influences on the *Timaeus*", *Phoenix*, 28 (1974), pp. 145–166.

—, "Plutarch and Heraclitus", *Hermes*, 105 (1977), pp. 179–201.

—, "Plutarch and Anaxagoras", *ICS*, 7 (1982), pp. 141–158 [= 1982a].

—, "Plutarch and Democritus", *QUCC*, 10 (1982), pp. 81–111 [= 1982b].

Hett, W.S., *Aristotle. Problems*, vol. 1, London – Cambridge, Mass., 1936.

Hillyard, B.P., *Plutarch: De audiendo: a text and commentary*, New York, 1981.

Hine, H.M., *An edition with commentary of Seneca, Natural questions, book two*, Salem (New Hampshire), 1984.

—, "Subjectivity and Objectivity in Latin Scientific and Technical Literature", in Taub, L. and Doody, A., 2009, pp. 13–30.

Hirsch-Luipold, R., *Plutarchs Denken in Bildern: Studien zur literarischen, philosophischen und religiösen Funktion des Bildhaften*, Tübingen, 2002.

—, "Religion and Myth", in Beck, M., 2014, pp. 163–176.

Hirzel, R., *Der Dialog. Ein literarhistorischer Versuch*, 2 vols., Leipzig, 1895.

—, *Plutarch*, Leipzig, 1912.

Holford-Strevens, L., "Favorinus: The Man of Paradoxes", in Barnes, J. and Griffin, M. (eds.), *Philosophia Togata II. Plato and Aristotle at Rome*, Oxford, 1997, pp. 188–217.

—, *Aulus Gellius: An Antonine Scholar and his Achievement*, Oxford, 2003.

—, "*Recht as een Palmen-Bohm* and Other Facets of Gellius' Medieval and Humanistic Reception", in Holford-Strevens, L. and Vardi, A. (eds.), *The Worlds of Aulus Gellius*, Oxford, 2005, pp. 249–281.

Holzberg, N., *Martial und das antike Epigramm*, Darmstadt, 2002.

Horster, M. and Reitz, C. (eds.), *Condensing texts – condensed texts*, Stuttgart, 2010 [= 2010a].

—, "'Condensation' of literature and the pragmatics of literary production", in Horster, M. and Reitz, C., 2010a, pp. 3–14 [= 2010b].

Housman, A.E., *M. Manilii, Astronomicon, liber quartus*, London, 1920.

Hubert, C., "Zur Entstehung der Tischgespräche Plutarchs", in Χάριτες, *Friedrich Leo zum sechzigsten Geburtstag dargebracht*, Berlin, 1911, pp. 170–187.

—, *Plutarchi Moralia*, vol. 6, 1, Lipsiae, 1959.

— (Pohlenz, M. and Drexler, H.), *Plutarchi Moralia*, vol. 5, 3, Lipsiae, 1960 [Hubert edited *Quaestiones Naturales*].

Huit, C., *La philosophie de la nature chez les anciens*, Paris, 1901.

Hunter, R., *Theocritus. A Selection*, Cambridge, 1999.

Huxley, G.L., "Historical Criticism in Aristotle's "Homeric Questions"", *Proceedings of the Royal Irish Academy*, 79 (1979), pp. 73–81.

I

Ideler, J.L., *Physici et medici Graeci minores*, vol. 1, Berlin, 1841.

Ieraci Bio, A.M., "L'ἐρωταπόκρισις nella letteratura medica", in Moreschini, C. (ed.), *Esegesi, parafrasi e compilazioni in età tardoantica. Atti del terzo convegno dell'associazione di studi tardoantichi*, Napoli, 1995, pp. 187–207.

Ierodiakonou, K., "Alexander of Aphrodisias on medicine as a stochastic art", in van der Eijk, P.J., Horstmanshoff, H.F.J. and Schrijvers, P.H. (eds.), *Ancient Medicine in its Social-Cultural Context*, vol. 2, Amsterdam, 1995, pp. 473–485.

Ingenkamp, H.G., "Plutarch by D.A. Russell", *Gnomon*, 48 (1976), pp. 546–551.

—, "Οὐ ψέγεται τὸ πίνειν. Wie Plutarch den übermäßigen Weingenuß beurteilte", in Montes Cala, J.G., Sanchez Ortiz de Landaluce, M. and Gallé Cejudo, R., 1999, pp. 277–290.

Inglese, L., *Plutarco. La curiosità. Introduzione, testo critico, traduzione e commento*, Napoli, 1996.

Inwood, B., *The Poem of Empedocles*, *Phoenix Suppl.*, 29, London, 2001.

Ioppolo, A.M., "La posizione di Plutarco nei confronti dello scetticismo", in Gallo, I., 2004, pp. 289–310.

Irby-Massie, G.L. and Keyser, P.T., *Greek Science of the Hellenistic Era: A Sourcebook*, London – New York, 2002.

Irigoin, J., "Le Catalogue de Lamprias: tradition manuscrite et éditions imprimées", *REG*, 99 (1986), pp. 318–331.

Isnardi Parente, M., "Plutarco et la matematica Platonica", in Gallo, I., 1992, pp. 121–145.

J

Jacob, C., "De l'art de compiler à la fabrication du merveilleux. Sur la paradoxographie grecque", *Lalies*, 2 (1983), pp. 121–140.

—, "Questions sur les questions: archéologie d'une pratique intellectuelle et d'une forme discursive", in Volgers, A. and Zamagni, C. (eds.), *Erotapokriseis. Early Christian Question-and-Answer Literature in Context. Proceedings of the Utrecht Colloquium, 13–14 October 2003*, Leuven, 2004, pp. 25–54.

Janssen, G., *Moralia 9: Biologie En Natuurkunde*, Leeuwarden, 2004.

Johansen, T.K., *Plato's Natural Philosophy*, Cambridge, 2004.

Jones, C.P., "Towards a Chronology of Plutarch's Works", *JRS*, 56 (1966), pp. 61–74 [= 1966a].

—, "The Teacher of Plutarch", *HSCP*, 71 (1966), pp. 205–213 [= 1966b].

Jones, R.M., *The Platonism of Plutarch and selected Papers*, New York – London, 1980.

Jones, W.H.S., *Hippocrates*, vol. 1, Cambridge, Mass. – London, 1923.

—, *Hippocrates*, vol. 4, Cambridge, Mass. – London, 1931.

Jouanna, J., "Plutarque et la patine des statues à Delphes (*Sur les oracles de la Pythie*, 395B–396C)", *Rev. Philol.*, 49 (1975), pp. 67–71.

—, "Sens et étymologie de ἀλέα (i et ii) et de ἀλκή", *REG*, 95 (1982), pp. 15–36.

—, "Hippocrate et les *Problemata* d'Aristote: essai de comparaison entre *Airs, eaux, lieux*, c. 10; *Aphorismes* iii, 11–14 et *Problemata* i 8–12 et 19–20", in Wittern, R. and Pellegrin, P. (eds.), *Hippokratische Medizin und antike Philosophie*, Hildesheim, 1996, pp. 273–293.

Jeanneret, M. (Whiteley, J. and Hughes, E. transls.), *A feast of Words. Banquets and Table Talk in the Renaissance*, Cambridge, 1991.

Jessen, O., "Halieus", *RE*, 7, 2 (1912), col. 2252.

Jürss, F., "Wissenschaft und Erklärungspluralismus im Epikureismus", *Philologus*, 138 (1994), pp. 235–251.

K

Kahle, C., *De Plutarchi ratione dialogorum componendorum*, Diss. Gottingae, 1912.

Kapetanaki, S. and Sharples, R.W., *Pseudo-Aristoteles (Pseudo-Alexander). Supplementa problematorum*, Berlin, 2006.

Karamanolis, G.E., *Plato and Aristotle in Agreement? Platonists on Aristotle from Antiochus to Porphyry*, Oxford, 2006.

—, "Plutarch", in Zalta, E.N. (ed.), *Stanford Encyclopedia of Philosophy*, 2010, http://plato.stanford.edu/archives/fall2014/entries/plutarch/ (accessed 13/10/2014).

Kechagia, E., "Philosophy in Plutarch's *Table Talk*. In Jest or in Earnest?", in Klotz, F. and Oikonomopoulou, K., 2011, pp. 77–104 [= 2011a].

—, *Plutarch Against Colotes: A Lesson in History of Philosophy*, Oxford, 2011 [= 2011b].

Keller, O., *Thiere des classischen Alterthums in culturgeschichtlicher Beziehung*, Innsbruck, 1887.

—, *Die antike Tierwelt*, Hildesheim, 1980 (Leipzig, 1909–1913).

Keyser, P.T. and Irby-Massie, G.L. (eds.) *The Encyclopedia of Ancient Natural Scientists. The Greek Tradition and its Many Heirs*, New York, 2008 [= 2008a].

—, "Asklēpiodotos (of Nikaia?) (40 BCE–30 CE)", in Keyser, P.T. and Irby-Massie, G.L., 2008, p. 172 [= 2008b].

Kirk, G.S., *Heraclitus. The Cosmic Fragments*, Cambridge, 2010 (= 1954).

Klotz, F., "Portraits of the Philosopher: Plutarch's Self-Presentation in the *Quaestiones Convivales*", *CQ*, 57 (2007), pp. 650–667.

—, "The Sympotic Works", in Beck, M., 2014, pp. 207–222.

Klotz, F. and Oikonomopoulou, K. (eds.), *The Philosopher's Banquet. Plutarch's* Table Talk *in the Intellectual Culture of the Roman Empire*, Oxford, 2011.

König, J., "Fragmentation and coherence in Plutarch's *Sympotic Questions*", in König, J. and Whitmarsh, T., 2007, pp. 43–68.

—, "Sympotic dialogue in the first to fifth centuries CE", Goldhill, S. (ed.), *The End of Dialogue in Antiquity*, Cambridge, 2008, pp. 85–113.

—, "Conversational and Citational Brevity in Plutarch's *Sympotic Questions*", in Horster, M. and Reitz, C., 2010a, pp. 321–348.

—, "Self-Promotion and Self-Effacement in Plutarch's *Table Talk*", in Klotz, F. and Oikonomopoulou, K., 2011, pp. 179–203.

König, J. and Whitmarsh, T. (eds.), *Ordering Knowledge in the Roman Empire*, Cambridge, 2007.

Kowalski, J., *De Plutarchi scriptorum iuvenilium colore rhetorico*, Cracoviae, 1918.

Krafft, F., "ΧΕΡΝΙΚΑ ΠΡΟΒΛΗΜΑΤΑ. Vermutungen zum Titel einer Schrift Demokrits", in Manegold, K.-H. (ed.), *Wissenschaft, Wirtschaft und Technik. Studien zur Geschichte, Wilhelm Treue zum 60. Geburtstag*, München, 1969, pp. 448–453.

Krauss, F., *Die rhetorischen Schriften Plutarchs und ihre Stellung im Plutarchischen Schriftenkorpus*, Diss. Nürnberg, 1912.

Kroll, W., "P. Fabianus", *RE*, 18, 3 (1949), cols. 1056–1059.

Kühner, R. and Gerth, B., *Ausführliche Grammatik der griechischen Sprache*, Hannover, 1966.

L

Labhardt, A., "*Curiositas*. Notes sur l'histoire d'un mot et d'une notion", *MH*, 17 (1960), pp. 206–224.

Lachenaud, G., *Scholies à Apollonios de Rhodes*, Paris, 2010.

Laes, C., *Children in the Roman Empire: Outsiders Within*, Cambridge, 2011.

Lakmann, M.L., *Der Platoniker Tauros in der Darstellung des Aulus Gellius*, Leiden, 1995.

Lao, E., *Restoring the Treasury of Mind. The practical Knowledge of the* Natural History, Diss. Princeton, 2008.

Laufer, B., *Geophagy*, Chicago, 1930.

Lawn, B., *The Salernitan Questions: an Introduction to the History of Medieval and Renaissance Problem Literature*, Oxford, 1963.

Leão, D., "Plutarch on Solon's Simplicity Concerning Natural Philosophy: *Sol.* 3,6–7 and Frs. 9 and 12 West", in Meeusen, M. and Van der Stockt, L., 2015, pp. 227–238.

Lee, H.D.P., *Aristotle. Meteorologica*, London – Cambridge, Mass., 1952.

Lehmann, Y. (ed.), *Aristoteles Romanus. La réception de la science aristotélicienne dans l'Empire gréco-romain*, Turnhout, 2013.

Lehoux, D., "Tropes, Facts and Empiricism", *Perspectives on Science*, 11 (2003), pp. 326–345 [= 2012, pp. 133–154].

—, "Observers, Objects, and the Embedded Eye; or, Seeing and Knowing in Ptolemy and Galen", *Isis*, 98 (2007), pp. 447–467 [= 2012, pp. 106–132].

—, *What Did the Romans Know?: an Inquiry into Science and Worldmaking*, Chicago – London, 2012.

Leith, D., "Question-Types in Medical Catechisms on Papyrus", in Taub, L. and Doody, A., 2009, pp. 107–123.

Lelli, E., "Plutarco", in Radici Colace, P., Medaglia, S.M., Rossetti, L. and Sconocchia, S. (eds.), *Dizionario delle scienze e delle techniche di Grecia e Roma*, vol. 2, Pisa – Roma, 2010, pp. 848–849.

Lennox, J.G., *Aristotle's Philosophy of Biology: Studies in the Origins of Life Science*, Cambridge, 2001.

—, "Aristotle's *Posterior Analytics* and the Aristotelian *Problemata*", in Mayhew, R., 2015c, pp. 36–60.

Leo, F., *De Plutarchi Quaestionum Romanarum auctoribus*, Diss. Halis Saxonum, 1864.

Lesage Gárriga, L., "The Light of the Moon: An Active Participant on the Battlefield in Plutarch's *Parallel Lives*", in Meeusen, M. and Van der Stockt, L., 2015, pp. 145–153.

Leutsch, E.L., *Corpus paroemiographorum Graecorum*, vol. 2, Hildesheim, 1965.

Levi, P., *The Pelican History of Greek Literature*, Harmondsworth, 1985.

Lewis, M. "Theoretical Hydraulics, Automata, and Water Clocks", in Wikander, O. (ed.), *Handbook of Ancient Water Technology*, Leiden – Boston – Köln, 2000, pp. 343–369.

Liedmeier, C., *Plutarchus' Biographie van Aemilius Paullus: historische commentaar*, Utrecht – Nijmegen, 1935.

Lindberg, D.C., *The Beginnings of Western Science. The European Scientific Tradition in Philosophical, Religious and Institutional Context, 600 B.C. to A.D. 1450*, Chicago, 1992.

Lloyd, G.E.R., "The Hot and the Cold, the Dry and the Wet in Greek Philosophy", *JHS*, 84 (1964), pp. 92–106.

—, *Polarity and Analogy: Two Types of Argumentation in Early Greek Thought*, Cambridge, 1966.

—, "Plato as a Natural Scientist", *JHS*, 88 (1968), pp. 78–92.

—, *Early Greek Science: Thales to Aristotle*, London, 1970.

—, *Greek Science after Aristotle*, New York, 1973.

—, *Magic, Reason and Experience: Studies in the Origin and Development of Greek Science*, Cambridge, 1979.

—, *Science, Folklore and Ideology: Studies in the Life Sciences in Ancient Greece*, Cambridge, 1983.

—, *Science and Morality in Greco-Roman Antiquity*, Cambridge, 1985 [= 1991, pp. 352–371].

—, *The Revolutions of Wisdom: Studies in the Claims and Practice of Ancient Greek Science*, Berkeley, 1987.

—, *Methods and Problems in Greek Science*, Cambridge, 1991.

—, *Adversaries and Authorities: Investigations into Ancient Greek and Chinese Science*, Cambridge, 1996.

—, "Science in Antiquity. The Greek and Chinese Cases and Their Relevance to the Problems of Culture and Cognition", in Biagioli, M. (ed.), *The Science Studies Reader*, London, 1999, pp. 302–316.

—, *Ancient Worlds, Modern Reflections. Philosophical Perspectives on Greek and Chinese Science and Culture*, Oxford, 2004.

Lobeck, C.A., *Aglaophamus sive de theologiae mysticae graecorum causis, libri tres*, Regimontii Prussorum, 1829.

Longo, O., "La teoria Plutarchea del *primum frigidum*", in Gallo, I., 1992, pp. 225–230.

Lopes, R., "The Omnipresence of Philosophy in Plutarch's *Quaestiones Convivales*", in Ferreira, J.R., Leão, D., Tröster, M. and Barata Dias, P., 2009, pp. 415–424.

Louis, P., *Aristote, Histoire des animaux*, Paris, 1968.

—, *Aristote, Problèmes*, vol. 1, Paris, 1991.

M

Maaß, E., *Scholia Graeca in Homeri Iliadem Townleyana*, Oxford, 1887.

Magnelli, E., "Poeti ellenistici in Plutarco: tipologie e preferenze", in Casanova, A., 2005, pp. 215–242.

Mansfeld, J., "Sources", in Algra, K., Barnes, J., Mansfeld, J. and Schofield, M. (eds.), *The Cambridge History of Hellenistic Philosophy*, Cambridge, 1999, pp. 3–30.

—, "*Physikai doxai* and *problêmata physika* in Philosophy and Rhetoric: From Aristotle to Aëtius (and Beyond)", in Mansfeld, J. and Runia, D.T. (eds.), *Aëtiana. The Method and Intellectual Context of a Doxographer, vol. 3: Studies in the Doxographical Traditions of Ancient Philosophy*, Leiden – Boston, 2010, pp. 33–97.

Mansfeld, J. and Runia, D.T., *Aëtiana. The Method and Intellectual Context of a Doxographer, vol. 1: the Sources*, Leiden – New York – Köln, 1997.

Mansion, S. (ed.), *Aristote et les problèmes de méthode*, Louvain – Paris, 1961.

Marcovich (Markovic), M., "On the sources of Theophylactus Simocatta's *Quaestiones physicae*", *ZAnt.*, 4 (1954), pp. 120–135 [in Croatian with English summary].

—, *Eraclito, Frammenti*, Firenze, 1978.

Marganne, M.-H., *La chirurgie dans l'Égypte gréco-romaine d'après les papyrus littéraires grecs*, Leiden – Boston – Köln, 1998.

Marrou, H.-I., *Histoire de l'éducation dans l'Antiquité*, Paris, 1948,

Martin, H., "Plutarch's Citation of Empedocles at *Amatorius* 756D", *GRBS*, 10 (1969), pp. 57–70.

Martin, J., *Symposion: die Geschichte einer literarischen Form*, Paderborn, 1931.

Mayhew, R., *Aristotle, Problems*, Cambridge, Mass. – London, 2011 [= 2011a].

—, "On *Problemata* XXIX 13: Peripatetic Legal Justice and the Case of Jury Ties", in Centrone, B., 2011a, pp. 275–307 [= 2011b].

— (ed.), *The Aristotelian* Problemata Physica: *Philosophical and Scientific Investigations*, Leiden – Boston, 2015 [= 2015a].

—, "Aristotle on Fever in *Problemata* I", *Apeiron*, 48 (2015), pp. 176–194 [= 2015b].

—, "Aristotle's Biology and his Lost *Homeric Puzzles*", *QC*, 65 (2015), pp. 109–133 [= 2015c].

Meeusen, M., "From Reference to Reverence: Five Quotations of Aristotle in Plutarch's *Quaestiones Naturales*", in Candau Morón, J., González Ponce, F. and Chávez Reino, A., 2011, pp. 347–363.

—, "Matching in Mind the Sea Beast's Complexion. On the Pragmatics of Plutarch's Hypomnemata and Scientific Innovation: the Case of *Q.N.* 19 (916BF)", *Philologus*, 156 (2012), pp. 234–259 [= 2012a].

—, "Salt in the Holy Water: Plutarch's *Quaestiones Naturales* in Michael Psellus' *De omnifaria doctrina*", in Roig Lanzillotta, L. and Muñoz Gallarte, I., 2012, pp. 101–121 [= 2012b].

—, "Opening up the Heavens over Athens: Plutarch and Laetus Discussing Physical Causes", in Casanova, A. (ed.), *Figure d'Atene nelle opere di Plutarco*, Firenze, 2013, pp. 249–262 [= 2013a].

—, "Natural Philosophy, Technè and Technicality in Plutarch", in Santana Henríquez, G., 2013, pp. 157–167 [= 2013b].

—, "How to Treat a Bee-Sting? On the Higher Cause in Plutarch's *Causes of Natural Phenomena*: the Case of *Q.N.* 35–36", *QUCC*, 105 (2013), pp. 131–157 [= 2013c].

—, "Plutarch and the Wonder of Nature. Preliminaries to Plutarch's Science of Physical Problems", *Apeiron*, 47 (2014), pp. 310–341.

—, "A Note on Croaking Frogs: Plu. *Q.N.* 2.912C", *Mnemosyne*, 68 (2015), pp. 115–120 [= 2015a].

—, "Plutarch Solving Natural Problems: for What Cause? (The Case of *Quaest. nat.* 29,919AB)", in Meeusen, M. and Van der Stockt, L., 2015, pp. 129–142 [= 2015b].

—, "Aristotle's Authority in the Tradition of Natural Problems. The Case of Plutarch of Chaeronea", in Boodts, S., Leemans, J. and Meijns, B. (eds.), *Shaping Authority. How Did a Person Become an Authority in Antiquity, the Middle Ages and the Renaissance?*, eds. Turnhout, 2016, pp. 47–85.

—, "Natural Problems Lost and Found: Gisbert Longolius Translating Plutarch's *Quaestiones Naturales*", forthcoming in *Humanistica Lovaniensia* [= forthcoming a].

—, "The Shifting Realities of Plutarch's Natural Problems. A Note on the Reception of *Quaestiones Naturales*", forthcoming in Guerrier, O. and Frazier, F., *Plutarque: éditions, traductions, paratextes* [= forthcoming b].

—, "Pagan Garlands and Christian Roses: Plutarch's *Quaestiones Convivales* in Michael Psellus' *De Omnifaria Doctrina*", forthcoming in Van Deun, P., Van Pee, S. and Demulder, B. (eds.), *Building the Kosmos. Greek Patristic and Byzantine Question and Answer Literature* [= forthcoming c].

—, "Egyptian Knowledge at Plutarch's Table: Out of the Question?", forthcoming in

Georgiadou, A. and Oikonomopoulou, K. (eds.), *Space, Time and Language in Plutarch* [= forthcoming d].

—, "'Why are Dionysian Artists Mostly Worthless People?' Aristotle's Προβλήματα Ἐγκύκλια in Context", forthcoming in *CQ* [= forthcoming e].

—, "An interpretation of Ps.-Alexander of Aphrodisias, *Medical Puzzles* I, *Praef.* in light of medical-philosophical school practice", forthcoming in Bouras-Vallianatos, P. and Xenophontos, S., *Greek Medical Literature and its Readers: From Hippocrates to Islam and Byzantium* [= forthcoming f].

—, "Aristotle's Second Breath: Pneumatic Processes in the *Natural Problems* (on Sexual Intercourse)", forthcoming in Coughlin, S., Leith, D. and Lewis, O. (eds.), *The Concept of* Pneuma *After Aristotle* [= forthcoming g].

Meeusen, M. and Van der Stockt, L. (eds.), *Natural Spectaculars. Aspects of Plutarch's Philosophy of Nature*, Leuven, 2015.

Menn, S., "Democritus, Aristotle, and the *Problemata*", in R. Mayhew, 2015a, pp. 10–35.

Messeri Savorelli, G. and Pintaudi, R., "Frammenti di rotoli letterari Laurenziani", *ZPE*, 115 (1997), pp. 171–177.

Michaëlis, C.T., *De ordine vitarum parallelarum Plutarchi*, Berlin, 1875.

Middleton, W.E.K., *A history of the thermometer and its use in meteorology*, Baltimore, 1966.

Minar, E.L., Sandach, F.H. and Helmbold, W.C., *Plutarch's Moralia in Sixteen Volumes*, vol. 9, Cambridge, Mass. – London, 1961.

Mittelhaus, K., *De Plutarchi Praeceptis gerendae reipublicae*, Diss. Berlin, 1911.

Mohr, R.D. and Sattler, B. (eds.), *One Book, the Whole Universe: Plato's* Timaeus *Today*, Las Vegas, 2010.

Montes Cala, J.G., Sanchez Ortiz de Landaluce, M. and Gallé Cejudo, R. (eds.), *Plutarco, Dioniso y el vino. Actas del VI Simposio espanol sobre Plutarco, Cadiz, 14–16 de mayo de 1998 Sociedad espanola de Plutarquistas*, Madrid, 1999.

Morales Ortiz, A., "Pedro Juan Núñez, traductor de Plutarco", in Rodríguez Adrados, F. and Martínez Díez, A. (eds.), *Literatura griega, Actas del IX Congreso español de Estudios Clásicos: Madrid, 27 al 30 de septiembre de 1995*, Vol. 4, Madrid, 1998, pp. 253–257.

—, "Observaciones a la traducción latina de G. Longueil de *Aetia Physica* de Plutarco", *Myrtia*, 14 (1999), pp. 143–151.

—, *Plutarco en España: Traducciones de* Moralia *en el siglo XVI*, Murcia, 2000.

Moraux, P., *Les listes anciennes des ouvrages d'Aristote*, Louvain, 1951.

Morel, P.-M., *Démocrite et la recherche des causes*, Paris, 1996.

Morel, W., "Griechisch σιγοποιός", *Glotta*, 47 (1969), p. 219.

Morgan, T., *Literate Education in the Hellenistic and Roman Worlds*, Cambridge, 1998.

—, "The Miscellany and Plutarch", in Klotz, F. and Oikonomopoulou, K., 2011, pp. 49–73

Mossman, J. (ed.), *Plutarch and his Intellectual World*, London, 1997 [= 1997a].

—, "Plutarch's Dinner of the Seven Wise Men and its place in symposion literature", in Mossman, J., 1997a, pp. 119–140 [= 1997b].

—, "Travel Writing, History, and Biography", in McGing, B. and Mossman, J. (eds.), *The Limits of Ancient Biography*, Swansea, 2006, pp. 281–303.

Mossman, J. and Titchener, F., "Bitch is Not a Four-Letter Word. Animal Reason and Human Passion in Plutarch", in Roskam, G. and Van der Stockt, L., 2011, pp. 273–296.

Müller, K.K., "Asklepiodotos", *RE*, 2, 2 (1896), cols. 1637–1641.

Murphy, T., *Pliny the Elder's Natural History. The Empire in the Encyclopedia*, Oxford, 2004.

Murray, O. (ed.), *Sympotica: a Symposium on the Symposion*, Oxford, 1990 [= 1990a].

—, "The Affair of the Mysteries: Democracy and the Drinking Group", in Murray, O., 1990a, pp. 149–161 [= 1990b].

Mynors, R.A.B., *Collected works of Erasmus. Adages II i 1 to II vi 100*, Toronto, 1991.

N

Naas, V., "Imperialism, *mirabilia* and knowledge: some paradoxes in the *Naturalis Historia*", in Gibson, R.K. and Morello, R., 2011, pp. 57–70.

Nachstädt, W., Sieveking, W., Titchener, J.B., *Plutarchi Moralia*, vol. 2, Lipsiae, 1935.

Negri, M., "Plutarco lettore (e commentatore) di Arato", in Gallo, I., 2004, pp. 275–288.

Newmyer, S.T., "Plutarch on Justice Toward Animals: Ancient Insights on a Modern Debate", *Scholia*, 1 (1992), pp. 38–54.

—, *Animals, Rights and Reason in Plutarch and Modern Ethics*, New-York, 2006.

—, "Animal *Philanthropia* in the *Convivium Septem Sapientium*", in Ferreira, J.R., Leão, D., Tröster, M. and Barata Dias, P., 2009, pp. 496–504.

—, "Animals in Plutarch", in Beck, M., 2014, pp. 223–234.

Nikolaidis, A.G., "Plutarch's Contradictions", *Classica et Mediaevalia*, 42 (1991), pp. 153–186.

—, "Plutarch's Attitude to Wine", in Montes Cala, J.G., Sanchez Ortiz de Landaluce, M. and Gallé Cejudo, R., 1999, pp. 337–348.

— (ed.), *The unity of Plutarch's work: "Moralia" themes in the "Lives", features of the "Lives" in the "Moralia"*, Berlin, 2008.

—, "Plutarch's Views on Art and specially on Painting and Sculpture", in Santana Henríquez, G., 2013, pp. 169–181.

Nutton, V., "Galen's Authorial Voice: a Preliminary Enquiry", in Taub, L. and Doody, A., 2009, pp. 53–62.

Nuzzo, G., ""La natura del freddo": struttura e valore nel corpus dei "Moralia"", in D'Ippolito, G. and Gallo, I., 1991, pp. 409–417.

O

O'Grady, P. and Silvermintz, D., "The *Anonymus Iamblichi* and the *Double Arguments*", in O'Grady, P. (ed.), *The Sophists: an Introduction*, London, 2008, pp. 138–151.

Oikonomopoulou, K., "Peripatetic Knowledge in Plutarch's *Table Talk*", in Klotz, F. and Oikonomopoulou, K., 2011, pp. 105–130.

—, "Plutarch's corpus of *quaestiones* in the tradition of imperial Greek encyclopaedism", in König, J. and Woolf, G. (eds.), *Encyclopaedism from Antiquity to the Renaissance*, Cambridge, 2013, pp. 129–153 [= 2013a].

—, "Ancient question-and-answer literature and its role in the tradition of dialogue", in Föllinger, S. and Müller, G.M. (eds.), *Der Dialog in der Antike. Formen und Funktionen einer literarischen Gattung zwischen Philosophie, Wissensvermittlung und dramatischer Inszenierung*, Berlin, 2013, pp. 37–64 [= 2013b].

—, "The *Problemata*'s Medical Books: Structural and Methodological Aspects", in Mayhew, R., 2015a, pp. 61–78.

Oltramare, P., *Sénèque. Questions naturelles*, vol. 1, Paris, 1961.

O'Neil, E.N., *Plutarch, Moralia*, vol. 16, Cambridge, Mass. – London, 2004.

Opsomer, J., *Geschiedenis van het Platonisme en Plato-exegese in Plutarchus' Quaestiones Platonicae*, Diss. Leuven, 1994 [= 1994a].

—, "L'âme du monde et l'âme de l'homme chez Plutarque," in Valdés, M.G. (ed.), *Estudios sobre Plutarco: ideas religiosas*, Madrid, 1994, pp. 33–49 [= 1994b].

—, "Ζητήματα: Structure et argumentation dans les *Quaestiones Platonicae*", in Fernández Delgado, J.A. and Pordomingo Pardo, F. (eds.), *Estudios sobre Plutarco: aspectos formales. Actas del IV simposio español sobre Plutarco, Salamanca, 26 a 28 de Mayo de 1994*, Salamanca, 1996, pp. 71–83 [= 1996a].

—, "Divination and Academic 'Scepticism' according to Plutarch", in Van der Stockt, L., 1996a, pp. 165–194 [= 1996b].

—, "Favorinus versus Epictetus on the Philosophical Heritage of Plutarch. A Debate on Epistemology," in Mossman, J., 1997a, pp. 17–40.

—, *In Search of the Truth, Academic Tendencies in Middle Platonism*, Brussel, 1998.

—, "Antiperistasis: A Platonic Theory", in Pérez Jiménez, A., Garciá López, J. and Aguilar, R.M., 1999, pp. 417–429.

—, "*Eirôneia* in the Corpus Plutarcheum (with an Appendix on Plutarch's Irony)", in Van der Stockt, L., 2000a, pp. 309–329.

—, "Plutarch's Platonism Revisited", in Bonazzi, M. and Celluprica, V. (eds.), *L'eredità platonica: Studi sul platonismo da Arcesilao a Proclo*, Napoli, 2005, pp. 163–200.

—, "Ofellius Laetus (*ca* 50–95 CE)", in Keyser, P.T. and Irby-Massie, G.L., 2008a, pp. 586–587.

—, "M. Annius Ammonius. A philosophical profile", in Bonazzi, M. and Opsomer, J., 2009, pp. 123–186.

—, "Arguments non-linéaires et pensée en cercles. Forme et argumentation dans les *Questions Platoniciennes* de Plutarque", in Brouillette, X. and Giavatto, A. (eds.), *Les dialogues platoniciens chez Plutarque: stratégies et méthodes exégétiques*, Leuven, 2010, pp. 93–116.

—, "Plutarch and the Stoics", in Beck, M., 2014, pp. 88–103.

—, "Plutarch on the Geometry of the Elements", in Meeusen, M. and Van der Stockt, L., 2015, pp. 29–55.

O'Sullivan, T.M., *Walking in Roman Culture*, Cambridge, 2011.

Owen, G.E.L., "Τιθέναι τὰ φαινόμενα", in Mansion, S., 1961, pp. 83–103.

P

Pailler, J.-M., "Les *Questions* dans les plus anciennes *Vies* romaines. Art du récit et rhétorique de la fondation", in Payen, P., 1998a, pp. 77–94.

Parke, H.W. and Wormell, D.E.W., *The Delphic Oracle*, vol. 1, Oxford, 1956.

Parroni, P., *Seneca, Ricerche sulla natura*, Milano, 2002.

Payen, P. (ed.), *Plutarque: Grecs et Romains en Questions*, Saint-Bertrand-de-Comminges, 1998 [= 1998a].

—, "Rhétorique et géographie dans les *Questions romaines* et *Questions grecques* de Plutarque", in Payen, P., 1998a, pp. 39–73 [= 1998b].

—, "Les recueils de *Questions* et la tradition "antiquaire" dans le corpus de Plutarque", Pallas, 90 (2013), pp. 217–233.

—, "Plutarch the Antiquarian", in Beck, M., 2014, pp. 235–248.

Pease, A.S., "Things without honour", *Class. Phil.*, 21 (1926), pp. 27–42.

Pecere, O. and Stramaglia, A. (eds.), *La Letteratura di Consumo nel Mondo Greco-Latino*, Cassino, 1996.

Peck, A.L., *Aristotle, Generation of Animals*, Cambridge, Mass. – London, 1953.

—, *Aristotle, Historia Animalium*, vol. 1, Cambridge, Mass. – London, 1965.

Pelling, C., "Plutarch's Method of Work in the Roman Lives", *JHS*, 99 (1979), pp. 74–96 [reprinted in Scardigli, B., 1995, pp. 265–318 (with a postscript) and Pelling, C., 2002, pp. 1–44].

—, "Plutarch's Adaptation of His Source-Material", *JHS*, 100 (1980), pp. 127–140 [reprinted in Scardigli, B., 1995, pp. 125–154 and Pelling, C., 2002, pp. 91–115].

—, *Plutarch and History: Eighteen Studies*, Swansea – London, 2002.

—, "Plutarch", in de Jong, I., Nünlist, R. and Bowie, A. (eds.), *Narrators, Narratees, and Narratives in Ancient Greek Literature: Studies in Ancient Greek Narrative*, vol. 1, Boston, 2004, pp. 403–422.

—, "Plutarch's Socrates", *Hermathena*, 175 (2005), pp. 105-139.

—, "Putting the -viv- into 'Convivial': The *Table Talk* and the *Lives*", in Klotz, F. and Oikonomopoulou, K., 2011, pp. 207-231.

Peretti, A., *Teognide nella tradizione gnomologica*, Pisa, 1953.

Pérez Jiménez, A., "Alle frontiere della scienza: Plutarco e l'astronomia", in Gallo, I., 1992, pp. 271-286.

—, "Δεισιδαιμονία: el miedo a los dioses en Plutarco", in Van der Stockt, L., 1996a, pp. 195-225.

Pérez Jiménez, A., Garciá López, J. and Aguilar, R.M. (eds.), *Plutarco, Platón y Aristóteles. Actas del V Congreso Internacional de la I.P.S. (Madrid – Cuenca, 4-7 de Mayo de 1999)*, Madrid, 1999.

Pérez Jiménez, A. and Titchener, F. (eds.), *Valori letterari delle Opere di Plutarco. Studi offerti al Professore Italo Gallo dall'International Plutarch Society*, Málaga – Logan, 2005.

Perrin, B., *Plutarch's Lives in eleven volumes*, vol. 7: *Demosthenes and Cicero, Alexander and Caesar*, Cambridge, Mass. – London, 1919.

Pomeroy, S.B., *Xenophon* Oeconomicus. *A Social and Historical Commentary*, Oxford 1995.

Pordomingo Pardo, F., "El banquete de Plutarco: ¿Ficción literaria o realidad histórica?", in Montes Cala, J.G., Sanchez Ortiz de Landaluce, M. and Gallé Cejudo, R., 1999, pp. 379-392.

Poschenrieder, F., *Die naturwissenschaftlichen Schriften des Aristoteles in ihrem Verhältnis zu den Büchern der hippokratischen Sammlung*, Bamberg, 1887.

Powell, J.G.F., "Cicero's Translations from Greek", in Powell, J.G.F. (ed.), *Cicero the Philosopher*, Oxford, 1995, pp. 273-300.

Prantl, C., *Ueber die Probleme des Aristoteles*, Abh. Bayr. Akad. d. Wiss. 6, München, 1852, pp. 339-377.

von Premerstein, A., "commenarii", *RE*, 4, 1 (1900), cols. 726-759.

Prescott, H.W., "EBA POON (Theocritus, *Id.* I. 139, 140)", *CQ*, 7 (1913), pp. 176-187.

Preston, R., "Roman questions, Greek answers: Plutarch and the construction of identity", in Goldhill, S. (ed.), *Being Greek Under Rome*, Cambridge, 2001, pp. 86-119.

Puech, B., "Prosopographie des amis de Plutarque", *ANRW* 2, 33, 6 (1992), pp. 4831-4893.

R

Raingeard, P., *Le περὶ τοῦ προσώπου de Plutarque*, Paris, 1935.

Ramón Palerm, V., "*Cuestiones sobre la naturaleza*: Notas críticas", in Pérez Jiménez, A. and Titchener, F., 2005, pp. 397-404.

—, "Plutarco y Juan de Pineda", in Candau Morón, J., González Ponce, F. and Chávez Reino, A., 2011, pp. 621-632.

Ramón Palerm, V. and Bergua Cavero, J., *Obras Morales y de Costumbres, ix. Sobre la malevolencia de Heródoto; Cuestiones sobre la naturaleza; Sobre la cara visible de la luna; Sobre el principio del frío; Sobre si es más útil el agua o el fuego; Sobre la inteligencia de los animales; 'Los animales son racionales' o 'Grilo'; Sobre comer carne*, Madrid, 2002.

Regenbogen, O., "Die Naturwissenschaft der Peripatetiker", *Scientia*, 50 (1931), pp. 345–354.

Reinhardt, K., *Kosmos und Sympathie: neue Untersuchungen über Poseidonios*, München, 1926.

Relihan, J.C., "Rethinking the History of the Literary Symposium", *ICS*, 17 (1992), pp. 213–244.

Rescigno, A., *Plutarco. L'eclissi degli oracoli. Introduzione, testo critico, traduzione e commento*, Napoli, 1995.

Ricard, D., *Oeuvres de Plutarque, traduites du grec*, vol. 4, Paris, 1844.

Richardson, N.J., *The Homeric Hymn to Demeter*, Oxford, 1974.

—, "Aristotle and Hellenistic Scholarship", in Montanari, F. (ed.), *La philologie grecque à l'époque hellénistique et romaine, Entretiens Fondation Hardt*, 40 (1994), pp. 7–38.

Rihll, T.E., *Greek Science*, Oxford, 1999.

Risselada, R., *Imperatives and Other Directive Expressions in Latin: A Study in the Pragmatics of a Dead Language*, Amsterdam, 1993.

Roberts, W.R., *Dionysius of Halicarnasus*, On Literary Composition, London, 1910.

Rodríguez Alfageme, I., "Medicina popular en Plutarco", in Montes Cala, J.G., Sanchez Ortiz de Landaluce, M. and Gallé Cejudo, R., 1999, pp. 411–422 [= 1999a].

—, "Fisiología en Plutarco: Antecedentes Aristotélicos", in Pérez Jiménez, A., Garciá López, J. and Aguilar, R.M., 1999, pp. 613–628 [= 1999b].

—, "Aspectos de la medicina helenística en Plutarco", in Casanova, A., 2005, pp. 435–465

Roig Lanzillotta, L., "Plutarch's Anthropology and Its Influence on His Cosmological Framework", in Meeusen, M. and Van der Stockt, L., 2015, pp. 179–195.

Roig Lanzillotta, L. and Muñoz Gallarte, I. (eds.), *Plutarch in the Religious and Philosophical Discourse of Late Antiquity*, Leiden – Boston, 2012.

Rolleston, J.D., "Alcoholism in Classical Antiquity", *British Journal of Inebriety*, 24 (1927), pp. 101–120.

Rose, H.J., *The Roman Questions of Plutarch*, Oxford, 1924.

Rose, V., *Aristoteles Pseudepigraphus*, Leipzig, 1863.

—, *Anecdota Graeca et Graecolatina*, vol. 2, Berlin, 1870.

—, *Aristoteles qui ferebantur librorum fragmenta*, Leipzig, 1886.

Roskam, G., "From Stick to Reasoning. Plutarch on the Communication between Teacher and Pupil", *Wiener Studien*, 2004, pp. 93–114.

—, *On the Path to Virtue. The Stoic Doctrine of Moral Progress and its Reception in (Middle-)Platonism*, Leuven, 2005.

—, *A commentary to Plutarch's* De latenter vivendo, Leuven, 2007.

—, "Plutarch on Aristotle as the First Peripatetic", *Ploutarchos*, 6 (2008/9), pp. 25–44.

—, "Educating the young ... over wine? Plutarch, Calvenus Taurus, and Favorinus as convivial teachers", in Ferreira, J.R., Leão, D., Tröster, M. and Barata Dias, P., 2009, pp. 369–383.

—, "Plutarch's 'Socratic Symposia'. The *Symposia* of Plato and Xenophon as Literary Models in the *Quaestiones convivales*", *Athenaeum*, 98 (2010), pp. 45–70.

—, "Plutarch against Epicurus on Affection for Offspring. A reading of *De amore prolis*", in Roskam, G. and Van der Stockt, L., 2011, pp. 175–201 [= 2011a].

—, "Aristotle in Middle Platonism. The Case of Plutarch of Chaeronea", in Bénatouïl, T., Maffi, E. and Trabattoni, F. (eds.), *Plato, Aristotle, or Both? Dialogues between Platonism and Aristotelianism in Antiquity*, Hildesheim – Zürich – New York, 2011, pp. 35–61 [= 2011b].

—, "Two *Quaestiones Socraticae* in Plutarch", in Candau Morón, J., González Ponce, F. and Chávez Reino, A., 2011, pp. 419–431 [= 2011c].

Roskam, G. and Van der Stockt, L. (eds.), *Virtues for the People. Aspects of Plutarchan Ethics*, Leuven, 2011.

Roskam, G. and Verdegem, S., ""This Topic Belongs to Another Kind of Writing". The Digressions in Plutarch's *Life of Coriolanus*", in Van der Stockt, L. and Stadter, P.A. (eds.), *Weaving Text and Thought. On Composition in Plutarch*, forthcoming, Leuven.

Russell, D.A., *Plutarch*, New York, 1973.

—, "Plutarch, *Quaestiones Convivales* IV–VI", *CR*, 30 (1980), pp. 12–14.

—, "Self-Disclosure in Plutarch and in Horace", in Most, G.W., Petersmann, H. and Ritter, A.M. (eds.), *Philanthropia kai Eusebeia: Festschrift für Albrecht Dihle zum 70. Geburtstag*, Göttingen, 1993, pp. 426–437.

—, "The Rhetoric of the *Homeric Problems*", in Boys-Stones, G. (ed.), *Metaphor, Allegory, and the Classical Tradition: Ancient Thought and Modern Revisions*, Oxford, 2003, pp. 217–234.

Russell, D.A. and Konstan, D., *Heraclitus:* Homeric problems, Atlanta, 2005.

S

Saïd, S., Trédé, M. and Le Boulluec, A., *Histoire de la littérature grecque*, Paris, 1997.

Sambursky, S., *The physical world of the Greeks*, London, 1963.

Sandbach, F.H., "Rhythm and Authenticity in Plutarch's *Moralia*", *CQ*, 33 (1939), pp. 194–203.

— (and Pearson, L.), *Plutarch's Moralia in Sixteen Volumes*, vol. 11, Cambridge, Mass. – London, 1965 [Sandbach edited *Quaestiones Naturales*].

—, *Plutarch's Moralia in Sixteen Volumes*, vol. 15, Cambridge, Mass. – London, 1969.

—, "Plutarch and Aristotle", *ICS*, 7 (1982), pp. 207–232.

Sansone, D., "Plutarch, Alexander, and the Discovery of Naphtha", *GRBS*, 21 (1980), pp. 63–74.

Santana Henríquez, G. (ed.), *Plutarco y las artes*. XI Simposio Internacional de la Sociedad Española de Plutarquistas. Las Palmas de Gran Canaria, 8–10 November 2012, Madrid, 2013.

Santaniello, C., "Plutarco e i presocratici", in Gallo, I., 2004, pp. 107–133.

Sarton, G., *A History of Science: Ancient Science through the Golden Age of Greece*, New York, 1970.

—, *A History of Science: Hellenistic Science and Culture in the Last Three Centuries B.C.*, New York, 1965.

Sassi, M., "Mirabilia", in Cambiano, G., Canfora, L. and Lanza D. (eds.), *Lo spazio letterario della Grecia antica. 1: La produzione e la circolazione del testo. Vol. 2: L'ellenismo*, Roma, 1993, pp. 449–468.

Scarcella, A.M., *Conversazioni a tavola. Libro primo. Introduzione, testo critico, traduzione e commento*, Napoli, 1998.

Scardigli, B. (ed.), *Essays on Plutarch's Lives*, Oxford, 1995.

Scheffel, W., *Aspekte der Platonischen Kosmologie. Untersuchungen zum Dialog "Timaios"*, Leiden, 1976.

Scheid, J., *À Rome sur les pas de Plutarque*, Paris, 2012.

Schellens, J., *De hiatu in Plutarchi Moralibus*, Bonnae, 1864.

Schepens, G. and Delcroix, K., "Ancient Paradoxography: Origin, Evolution, Production, and Reception", in Pecere, O. and Stramaglia, A., 1996, pp. 375–460.

Schmidt, J., "Paradoxa", *RE*, 18, 3 (1949), cols. 1134–1137.

Schmidt, T.S., *Plutarque et les barbares*, Leuven, 1999.

—, "Les *Questions barbares* de Plutarque: un essai de reconstitution", in Chassignet, M., 2008, pp. 165–183.

Schmitt-Pantel, P., *La cité au banquet. Histoire des repas publics dans les cités grecques*, Rome, 1992.

Schmitz, T., *Bildung und Macht. Zur sozialen und politischen Funktion der zweiten Sophistik in der griechischen Welt der Kaiserzeit*, München, 1997.

—, "Plutarch and the Second Sophistic", in Beck, M., 2014, pp. 32–42.

Schnitzer, C.F., *Plutarch's Werke. Moralische Schriften*, vol. 22, Stuttgart, 1860.

Scholfield, A.F., *Aelian. On the Characteristics of Animals*, vol. 1, Cambridge, Mass. – London, 1958.

Schopenhauer, A., *Parerga und Paralipomena*, vol. 5, Darmstadt, 1976 (1851).

Schroeter, J., *Plutarchs Stellung zur Skepsis*, Diss. Königsberg, 1911.

Schuster, M., *Untersuchungen zu Plutarchs Dialog* De sollertia animalium *mit besonderer Berücksichtigung der Lehrtätigkeit Plutarchs*, Diss. Augsburg, 1917.

Sedley, D., "Philosophical Allegiance in the Greco-Roman World", in Griffin, M. and Barnes, J. (eds.), *Philosophia Togata I. Essays on Philosophy and Roman Society*, Oxford, 1989, pp. 97–119.

Seide, R., *Die mathematische Stellen bei Plutarch*, Diss. Regensburg, 1981.

Senzasono, L., *Plutarco. Precetti igienici. Introduzione, testo critico, traduzione e commento*, Napoli, 1992.

—, "Il concetto di 'potenza' nelle *Nat. Quaest.* di Plutarco", in Pérez Jiménez, A., Garciá López, J. and Aguilar, R.M., 1999, pp. 657–664.

—, *Plutarco. Cause dei fenomeni naturali. Introduzione, testo critico, traduzione e commento*, Napoli, 2006.

Setaioli, A., "Truffles and thunderbolts (Plu., *Quaest. conv.* 4.2, 1–2)", in Ferreira, J.R., Leão, D., Tröster, M. and Barata Dias, P., 2009, pp. 439–446.

Sharples, R.W., "Alexander of Aphrodisias: Scholasticism and Innovation", in *ANRW* 2, 36, 2 (1987), pp. 1176–1243.

—, "Science, Philosophy and Human Life in the Ancient World", in Wolff, J. and Stone, M.W.F. (eds.), *The Proper Ambition of Science*, London, 2000, pp. 7–27.

—, "Introduction: Philosophy and the Sciences in Antiquity", in Sharples, R.W. (ed.), *Philosophy and the Sciences in Antiquity*, Aldershot, 2005, pp. 1–7.

—, "Pseudo-Alexander or Pseudo-Aristotle, *Medical Puzzles and Physical Problems*", in De Leemans, P. and Goyens, M., 2006, pp. 21–31.

Siefert, G., *Plutarchs Schrift* Περὶ εὐθυμίας, Naumburg, 1908.

Simonetti, E.G., "Il caso della lampada di Ammone come *eikôn* dell'universo", forthcoming in Volpe Cacciatore, P. (ed.), *Immagini letterarie e iconografia nelle opere di Plutarco*, Salerno.

Sirinelli, J., *Plutarque de Chéronée: Un philosophe dans le siècle*, Paris, 2000.

Slater, W.J., "Aristophanes of Byzantium and Problem-Solving in the Museum", *CQ*, 32 (1982), pp. 336–349.

Slotty, F., "Die Stellung des Griechischen und anderer idg. Sprachen zu dem soziativen und affektischen Gebrauch des Plural der ersten Person", *Indogermanische Forschungen*, 45 (1928), pp. 348–363.

Small, J.P., *Wax Tablets of the Mind: Cognitive Studies of Memory and Literacy in the Ancient World*, London, 1997.

Smits, J.P.H.M., *Plutarchus en de Griekse muziek: de mentaliteit van de intellectueel in de tweede eeuw na Christus*, Bilthoven, 1970.

Snell, B., "The Forging of a Language for Science in Ancient Greece", *CJ*, 56 (1960), pp. 50–60.

Snyder, H.G., *Teachers and Texts in the Ancient World. Philosophers, Jews and Christians*, London, 2000.

Soury, G., "Les "Questions de table" et la philosophie religieuse de Plutarque", *REG*, 62 (1949), pp. 320–327.

von Staden, H., *Herophilus: the art of medicine in early Alexandria*, Cambridge, 1989.

—, "Author and Authority. Celsus on the Construction of a Scientific Self", in Vázquez Buján, M.E. (ed.), *Tradición e Innovación de la Medicina Latina de la Antigüedad y de la Alta Edad Media*, Santiago de Compostela, 1994, pp. 103–117.

Stadter, P.A., "The Proems of Plutarch's *Lives*", *ICS*, 13 (1988), pp. 275–295.

—, *A Commentary on Plutarch's Pericles*, Chapel Hill, 1989.

—, "Pericles Among the Intellectuals", *ICS*, 16 (1991), pp. 111–124.

—, "Drinking, *Table Talk*, and Plutarch's Contemporaries", in Montes Cala, J.G., Sanchez Ortiz de Landaluce, M. and Gallé Cejudo, R., 1999, pp. 481–490.

—, "Notes and Anecdotes: Observations on Cross-Genre *Apophthegmata*", in A.G. Nikolaidis, 2008, pp. 53–66.

Stahl, G., "Die "Naturales Quaestiones" Senecas: Ein Beitrag zum Spiritualisierungsprozeß der römischen Stoa", *Hermes*, 92 (1964), pp. 425–454.

Stahl, W.H., *Roman Science*, Madison, 1962.

Starr, R., "The Circulation of Literary Texts in the Roman World", *CQ*, 37 (1987), pp. 231–223.

Stoltz, C., *Zur relativen Chronologie der Parallelbiographien Plutarchs*, Lund, 1929.

Stoyles, B.J., "Material and Teleological Explanations in *Problemata* 10", in Mayhew, R, 2015a, pp. 124–150.

Strömberg, R., *Greek Proverbs: A Collection of Proverbs and Proverbial Phrases which are not Listed by the Ancient and Byzantine Paroemiographers*, Göteborg, 1954.

Stückelberger, A., "Meerwasserentsalzung Nach Aristoteles (?). Ein Nachtrag", *Hermes*, 124 (1996), pp. 378–380.

Swain, S., "Plutarch: Chance, Providence, and History", *AJPh*, 110 (1989), pp. 272–302.

T

Tanga, F., "Some Notes on Plutarch's *Quaestiones naturales*", in Meeusen, M. and Van der Stockt, L., 2015, pp. 113–128.

Tappe, G., *De Philonis libro qui inscribitur Ἀλέξανδρος ἢ περὶ τοῦ λόγον ἔχειν τὰ ἄλογα ζῷα quaestiones selectae*, Diss. Gottingae, 1912.

Taub, L., *Ancient Meteorology*, London, 2003.

—, *Aetna and the Moon: Explaining Nature in Ancient Greece and Rome*, Oregon, 2008.

—, "Explaining a Volcano Naturally: *Aetna* and the Choice of Poetry", in Taub, L. and Doody, A., 2009, pp. 125–141.

Taub, L. and Doody, A. (eds.), *Authorial Voices in Greco-Roman Technical Writing*, Trier, 2009.

Taylor, A.E., *A commentary on Plato's Timaeus*, Oxford, 1928.

Tecuşan, M., "*Logos Sympotikos*: Patterns of the Irrational in Philosophical Drinking: Plato Outside the *Symposium*", in Murray, O., 1990a, pp. 238–260.

Teixeira, E., "Remarques sur l'esprit scientifique de Plutarque d'après quelques passages des *Propos de table*", in Gallo, I., 1992, pp. 211–223.

Teodorsson, S.-T., *A commentary on Plutarch's Table Talks*, Göteborg, 1989 (= vol. 1), 1990 (= vol. 2 [= 1990a]), 1996 (= vol. 3).

—, "A Forgotten Anatomical Method Traced?", *Eranos*, 88 (1990), pp. 64–66 [= 1990b].

—, "Plutarch and Peripatetic Science", in Pérez Jiménez, A., Garciá López, J. and Aguilar, R.M., 1999, pp. 665–674 [= 1999a].

—, "Dionysus Moderated and Calmed: Plutarch on the Convivial Wine", in Montes Cala, J.G., Sanchez Ortiz de Landaluce, M. and Gallé Cejudo, R., 1999, pp. 57–69 [= 1999b].

—, "Plutarch's Use of Synonyms: A Typical Feature of his Style", in Van der Stockt, L., 2000a, pp. 511–518.

—, "Plutarco, innovatore del vocabolario greco", in Pérez Jiménez, A. and Titchener, F., 2005, pp. 405–418.

—, "The place of Plutarch in the literary genre of *Symposium*", in Ferreira, J.R., Leão, D., Tröster, M. and Barata Dias, P., 2009, pp. 3–16.

—, "Plutarch, a main source for the Presocratics and the Sophists", in Candau Morón, J., González Ponce, F. and Chávez Reino, A., 2011, pp. 433–446.

Thévenaz, P., *L'Âme du monde. Le devenir et la matière chez Plutarque. Avec une traduction du traité De la Genèse de l'Âme dans le Timée*, Neuchâtel, 1938.

Thivel, A., "La doctrine des περισσώματα et ses parallèles hippocratiques", *Revue de philologie*, 39 (1965), pp. 266–282.

Thomas, O., "Creating *Problemata* with the Hippocratic Corpus", in Mayhew, R., 2015a, pp. 79–99.

Thomas, R., *Herodotus in Context. Ethnography, Science and the Art of Persuasion*, Cambridge, 2000.

Thompson, D'A.W., *A Glossary of Greek Fishes*, London, 1947.

Titchener, F., "The role of reality in Plutarch's *Quaestiones Convivales*", in Ferreira, J.R., Leão, D., Tröster, M. and Barata Dias, P., 2009, pp. 395–401.

—, "Plutarch's *Table Talk*: Sampling a Rich Blend. A Survey of Scholarly Appraisal", in Klotz, F. and Oikonomopoulou, K., 2011, pp. 35–48.

—, "Fate and Fortune", in Beck, M., 2014, pp. 479–487.

Titchener, J.B., *The Manuscript-Tradition of Plutarch's* Aetia Graeca *and* Aetia Romana, Illinois, 1924.

Torraca, L., "L'astronomia lunare in Plutarco", in Gallo, I., 1992, pp. 231–261.

—, "Problemi di lingua e stile nei 'Moralia' di Plutarco", *ANRW* 2, 34, 4 (1998), pp. 3487–3510.

Tsekourakis, D., "Orphic and Pythagorean Views on Vegetarianism in Plutarch", in Brenk, F.E. and Gallo, I., 1986, pp. 127–138.

Tümpel, K., "Διόνυσος Ἁλιεύς", *Philologus*, 48 (1889), pp. 681–696.

U

Ulacco, A., "Malattia e alterazione del calore naturale: medicina ippocratica e fisiologia aristotelica negli *hosa iatrika* e in altri *Problemata* pseudo-aristotelici", in Centrone, B., 2011a, pp. 59–88.

Usener, H., *Alexandri Aphrodisiensis quae feruntur problematorum libri 3 et 4*, Berlin, 1859.

V

Valgiglio, E., *Plutarco. Gli oracoli della Pizia. Introduzione, testo critico, traduzione e commento*, Napoli, 1992.

Vallance, J.T., "A Second Look: Marshall Clagett's 'Greek Science in Antiquity': Thirty-Five Years Later", *Isis*, 81 (1990), pp. 713–721.

Vamvouri Ruffy, M., "Symposium, Physical and Social Health in Plutarch's *Table Talk*", in Klotz, F. and Oikonomopoulou, K., 2011, pp. 131–157.

—, *Les Vertus thérapeutiques du banquet: Médecine et idéologie dans les* Propos de Table *de Plutarque*, Paris, 2012.

Van der Stockt, L., "Plutarch's Use of Literature. Sources and Citations in the *Quaestiones Romanae*", *AncSoc*, 18 (1987), pp. 281–292.

—, *Twinkling and Twighlight: Plutarch's Reflections on Literature*, Brussel, 1992 [= 1992a].

—, "Plutarch on τέχνη", in Gallo, I., 1992, pp. 287–295 [= 1992b].

— (ed.), *Plutarchea Lovaniensia. A Miscellany of Essays on Plutarch*, Lovanii, 1996 [= 1996a].

—, "Some remarks on two Plutarchean introductions", in Van der Stockt, L., 1996a, pp. 265–272 [= 1996b].

—, "A Plutarchan Hypomnema on Self-Love", *The American Journal of Philology*, 120 (1999), pp. 575–599 [= 1999a].

—, "Three Aristotle's Equal but one Plato. On a Cluster of Quotations in Plutarch", in Pérez Jiménez, A., Garciá López, J. and Aguilar, R.M., 1999, pp. 127–140 [= 1999b].

— (ed.), *Rhetorical Theory and Praxis in Plutarch. Acta of the IVth International congress of the International Plutarch Society, Leuven, July 3–6, 1996*, Louvain, 2000 [= 2000a].

—, "Aspects of the Ethics and Poetics of the Dialogue in the *Corpus Plutarcheum*", in Gallo, I. and Moreschini, C., 2000, pp. 93–116 [= 2000b].

—, "Plutarch in Plutarch: the problem of the hypomnemata", in Gallo, I., 2004, pp. 331–340.

—, "Plutarch and dolphins: love is all You need", in Boulogne, J., 2005a, pp. 13–21.

—, "Some aspects of Plutarch's view of the physical world. Interpreting *Causes of Natural Phenomena*", in Candau Morón, J., González Ponce, F. and Chávez Reino, A., 2011, pp. 447–455.

—, "*Eunomia* in heaven and on earth. Plutarch's *nomos* between rhetoric and science", in Ferreira, J.R., Leão, D. and Martins de Jesus, C.A., 2012, pp. 203–213.

—, "Technical terminology in Plutarch's *Lives*: addressing the layman", in Pace, G. and Volpe Cacciatore, P. (eds.), *Gli scritti di Plutarco: tradizione, traduzione, ricezione, commento. Atti del IX Convegno Internazionale della International Plutarch Society Ravello – Auditorium Oscar Niemeyer 29 settembre–1 ottobre 2011*, Napoli, 2013, pp. 439–445.

Van Groningen, B.A., "General Literary Tendencies in the Second Century A.D.", *Mnemosyne*, 18 (1965), pp. 41–56.

—, *Théognis: le premier livre*, Amsterdam, 1966.

Van Hoof, L., *Plutarch's Practical Ethics. The Social Dynamics of Philosophy*, Oxford, 2010.

Van Kooten, G., "A Non-Fideistic Interpretation of πίστις in Plutarch's Writings: The Harmony between πίστις and Knowledge", in Roig Lanzillotta, L. and Muñoz Gallarte, I., 2012, pp. 215–233.

Van Meirvenne, B., "'Puzzling over Plutarch': Traces of a Plutarchean Plato-study Concerning *Lg.* 729a–c in *Adulat.* 32 (*Mor.* 71B), *Coniug. praec.* 46–47 (*Mor.* 144F) and *Aet. Rom* 33 (*Mor.* 272C)", in Montes Cala, J.G., Sanchez Ortiz de Landaluce, M. and Gallé Cejudo, R., 1999, pp. 527–540.

—, "'Earth and Ambrosia" (*De facie* §§ 24–25): Plutarch on the Habitability of the Moon", in Pérez Jiménez, A. and Casadesús Bordoy, F. (eds.), *Estudios sobre Plutarco: misticismo y religiones mistéricas en la obra de Plutarco, Actas del VII Simposio Español sobre Plutarco (Palma de Mallorca, 2–4 de nov. de 2000)*, Madrid – Malaga, 2001, pp. 283–296.

Van Nuffelen, P., "Words of Truth: Mystical Silence as a Philosophical and Rhetorical Tool in Plutarch", in Castelnérac, B., 2007, pp. 9–39.

Vegetti, M., *Il coltello e lo stilo: animali, schiavi, barbari e donne alle origine della razionalità scientifica*, Milano, 1979.

Verbeke, G., *L'évolution de la doctrine du pneuma: du stoïcisme à S. Augustin*, Paris – Louvain, 1945.

—, "Plutarch and the development of Aristotle", in Düring, I. and Owen, G.E.L. (eds.), *Aristotle and Plato in the mid-fourth century*, Göteborg, 1960, pp. 236–247.

Verdenius, W.J., "Der Ursprung der Philologie", *Studium Generale*, 19 (1966), pp. 103–114.

Vernière, Y., *Symboles et mythes dans la pensée de Plutarque. Essai d'interprétation philosophique et religieuse des Moralia*, Paris, 1977.

Vetta, M., "Plutarco e il 'Genere Simposio'", in Gallo, I. and Moreschini, C., 2000, pp. 217–229.

Veyne, P., *Les Grecs ont-ils cru à leurs mythes? Essai sur l'imagination constituante*, Paris, 1983.

Vlastos, G., "On Heraclitus", *AJP*, 76 (1955), pp. 337–368.

—, "The Socratic Elenchus", *Oxford Studies in Ancient Philosophy*, 1 (1983), pp. 27–58.

Volkmann, R., *Leben, Schriften und Philosophie des Plutarch von Chaeronea*, Berlin, 1869.

W

Weidlich, T., *Die Sympathie in der antiken Literatur*, Stuttgart, 1894.

Weiss, D., *De nonnullis Plutarchi Moralium locis ab Herwerdeno tractatis*, Biponti, 1888.

Weissenberger, B., *Die Sprache Plutarchs von Chaeronea und die pseudoplutarchischen Schriften*, Diss. Würzburg, 1895.

Wellmann, M., "Antipatros, 32", *RE*, I, 2 (1894), col. 2517.

—, "Bolos, 3", *RE*, 3, 1 (1897), cols. 676–677.

—, *Die Φυσικά des Bolos Demokritos und der Magier Anaxilaos aus Larissa*, Berlin, 1928.

Wendel, C., "Späne III", *Hermes*, 77 (1942), pp. 216–218.

Westerink, L.G., *Michael Psellus, De Omnifaria Doctrina*, Nijmegen, 1948.

Whitmarsh, T., *Greek Literature and the Roman Empire: The Politics of Imitation*, Oxford, 2001.

—, "Alexander's Hellenism and Plutarch's Textualism", *CQ*, 52 (2002), pp. 174–192.

—, *Ancient Greek Literature*, Cambridge, 2004.

—, *The Second Sophistic*, Oxford, 2005.

Williams, G.D., *The cosmic viewpoint: a study of Seneca's Natural questions*, Oxford, 2012.

von Wilamowitz, U., *Griechisches Lesebuch*, vol. 2, Berlin, 1902.

—, *Die Textgeschichte der griechischen Bukoliker*, Berlin, 1906.

Wright, M.R., *Empedocles: the Extant Fragments*, London, 1981.

X

Xenophontos, S.A., "Plutarch's Compositional Technique in the *An Seni Respublica Gerenda Sit*: Clusters vs. Patterns", *The American Journal of Philology*, 133 (2012), pp. 61–91.

Y

Yaginuma, S., "Plutarch's Language and Style", *ANRW* 2, 33, 6 (1992), pp. 4726–4742.

Z

Zeyl, D.J., *Plato. Timaeus*, Indianapolis, 2000.

Ziegler, K., "Plutarchos von Chaironeia", *RE*, 21, 1 (1951), cols. 636–962.

Zucker, A., "Papirius Fabianus (*ca* 35 BCE–*ca* 30 CE)", in Keyser, P.T. and Irby-Massie, G.L., 2008a, pp. 610–611.

INDEX LOCORUM

Index Locorum

Aelian
NA
1, 58	479
3, 5	456
4, 4	484
5, 11	122, 479
5, 41	428
5, 46	456
6, 3	453, 456
6, 12	456
6, 19	384
8, 9	456
9, 13	384
9, 52	428
9, 64	371, 400
Epil. 43–46	99

VH
1, 1	432

Aeneas Tacticus
Pol.
22, 24	405

Aesop
Fab.
306–307	476

Alcman
Fr. Diehl
43	140, 277, 451

Alexander (and Ps.–Alexander) Aphrodisiensis
In Ar. Met.
38, 16–20	396

In Ar. Top.
62, 30–63, 19	84

Probl.
1, 2	466
1, 55	400, 408
1, 56	420
1, 125	465
2, 46	83

Quaest.
2, 23	435

Anaxagoras
Fr. DK
59A35	18
59A116	281, 370
59A117	370

Anaximenes
Fr. DK
13B1	186

Antigonus Carystus
Hist. mir.
25a, 1–2	432
25b	436
34	456
50, 1–2	432
56	484

Apollonius Mys
Fr. von Staden
33	281, 387

Apollonius Rhodius
Arg.
2, 88–89a	285
2, 476–477	481

Apostolius, Michael
Coll. paroem.
11, 61	424

Aratus
Phaen.
946–947	269, 277, 385
956	269, 425

Archilochus
  Fr. Edmonds
    57   426

Aristophanes
  Nub.
    174   18
  Pax
    1083   441
    1086   441

Aristotle (and *corpus Aristotelicum*)
  APr.
    24a16–17   78
  APo.
    72a8   78
    89b24–35   79
    89b29–30   250
  Top.
    100a25–101a24   339
    101b28–37   78
    104b1–28   78
    104b3–5   78
    104b13–18   78
    105b13–19   78
    105b19–21   78
    127a3–13   301
    127a19   382
    149b25   339
    162a17   326
  De caelo
    286a25–26   462
  GC
    318b14–17   462
    318b33–35   398
    323b8–10   467
  Mete.
    347b17–22   451
    348b2   309
    348b2–5   419
    354b18   399
    354b33   373
    355a32 ff.   408
    358a   281
    358a14–17   373
    358b6–9   302
    358b7–9   372
    358b16–18   398
    358b18–20   399
    358b34–359a5   400
    359a7–21   372
    363a24   180
    369a12–29   392
    379a11–b9   307
    379b10–381b23 307
    380b31–381a1   459
    381b13   180
    384a26–28   439
  De mund.
    395b3–4   463
  DA
    415b2–3   266
    416a28–35   380
    420b29   147
    422b10–14   397
    433b18   302, 431
  De sensu
    442a17   400
    442a17–21   397
    444b12   425
  Somn. vig.
    456a29   180
  Div. som.
    463a5   457
  Juv.
    470a18   180
  De respir.
    472b6–473a14   309
    472b16   309
  HA
    523b32–33   428
    535a2   477
    536a11   384
    542a27–28   442
    542a27–30   441
    546a1–4   465
    557b29   491
    572a5–8   441
    574a8   390
    575b1   445
    578a   281
    578a26   441
    578a26–28   443
    578a32–b5   443
    578a32–33   444
    578b1   444, 445

580a11–22	484	783b18	466
580a14–19	251, 484	789b8–9	302, 431
580a20–22	251	*De plant.*	
580a21–22	484	815b16	370
580b31	390	*Mir. ausc.*	
584a19	457	831a27–28	456
588b16–17	370	832a3–4	479
590a22	333	836b13ff.	449
590a22–27	400	844b20–22	387
594a28–29	456	*Probl.*	
594b9	456	861a6–9	380
596a16–25	386	866a26–28	467
610b29	136	866a36–b1	467
612a5–6	456	871a2	305, 467
612a24–25	456	871a8–16	294
622a15	431	872b25–32	295
623b20	479	872b32–873a4	294
623b20–21	477	874b6–7	467
626a26–28	479	874b13–21	295
PA		875a29–40	294
640a14–15	249	875a34–35	82
645a16–17	462	879a23–26	479
646a15	303	880a34–b3	466
648a2–5	439	884a32–34	474
649a18–19	462	886b4–9	402
650b28	439	887a11	65
651a2–3	438	887a22–40	402
651a3	439	887b38–888a23	149, 295
676a18	180	893a31	388
678b32	428	893b7	388
679a13	431	896a20–29	441, 444
681a12–15	370	896a21	441
685a30–b2	428	896a22–24	442
MA		896a23	441
703a3–b1	302, 431	896a23–24	444
GA		896a24–29	444
725b32–726a6	465	907a8–12	453
726a26–27	358	923a9–10	424
735a32	486	924b35–925a5	490
735b35	486	927b12	447
736b37–737a1	301	930b23–25	458
743a8–11	400	932a40–b8	372
747b5	180	932b8	337
765a23–31	445	932b4–6	373
772b11	180	932b8–16	407
775b37	180	932b8–24	415
779b30–33	486	932b25	488
780b6	467	932b25–28	372

Probl. (*cont.*)

933a9–13	372	70	275
933a18–27	373	112	84
933b11–16	398	209–245	83
933b17–20	399	217	399
933b33–41	399	218	489
934b27ff.	408	215	281, 378
934b34–36	373	217	371, 374
935a5–8	373	218	282
935b	281	220	295
935b3–17	399	221	411
935b17–27	407, 415	222	374, 411, 467
935b18–20	373	223	81
936b18	447	229	472
937a26–27	467	233	197
937b29	489	235	466
939b37	447	243	190
943b21–23	475	245, 8, 15–19	392
943b23	447	245, 10	384
944a11–12	475	248	420
946a17–32	475	338	428

Vita Marciana

427, 8 R³	83

Index Hesychii

116	104
168	83

946b21–22	475		
948b35–949a8	189		
953a10–955a40	143, 358		
935b17	337		
961a18–23	415		

Ps.–Aristotle/Alexander Aphrodisiensis
Suppl. probl.

961a24–30	415, 416	1, 17, 32–35	385
961b31–32	467	2, 22	380
966a17–25	447	2, 34	400
967a3	447	2, 51	65

Met.

		2, 59	388
980a26–27	265	2, 61	388
981aff.	317	2, 64	388
982b11–15	243	2, 101	453
995b2–4	78	2, 106	412
1055b32ff.	85	2, 115–116	390
1070b9–13	462	2, 137, 2–4	386

EN

		2, 144	83, 441, 442
1102a22	457	2, 145	83, 443, 444
1148b24–29	457	2, 155	83, 441, 442

Rh.

		2, 156	181
1355a6–8	339	3, 17	83, 133
1355b26–27	339	3, 29	83, 416

Poet.

		3, 47	83, 416
1460b6ff.	84		
1447b18	275		

Probl. ined.

3, 51	83

Fr. Rose

62	220

INDEX LOCORUM

Arrian
Frag. Phys. (Roos)
3, 190, 7–12	392

Athenaeus
Deipn.
1, 26b	409, 410
1, 31 f.	409
1, 32de	409
2, 42b	311, 405
7, 317a	432
7, 317 f.	431
7, 323d	428
11, 33b	409
12, 513cd	432

Aurelianus, Caelius
Tard. pass.
5, 139	280, 403

Callimachus
Fr. Pfeiffer
1, 1–5	95
43, 12–17	152
178	152
465	95
787 Schneider	442

Carcinus
TGF
5, 6	449

Cassius Dio
Hist. Rom.
50, 34	373

Cassius Iatrosophista
Probl.
65	398

Cato
De agr.
24	410
35	422
104–106	410

Catullus
Carm.
42	479

Celsus
De med.
2, 17, 1	474
2, 18, 12	379

Charon of Lampsacus
FGrHist
262, 12	481

Cicero
De div.
1, 8, 13	269
1, 15	385
2, 21, 47	269
2, 33	285
2, 126	456
2, 145, 12–14	428

De nat. deor.
2, 25	420
2, 26	406
2, 96	462

De or.
2, 5, 21	191
2, 86–87	265

Ep. ad Attic.
7, 1, 1	196

Progn.
4, 1–3	385

Rhet. ad Her.
3, 36	462

Clemens Alexandrinus
Paed.
3, 11, 80, 2	431

Columella
De re rust.
9, 14, 3	480
9, 14, 7	477
9, 15, 10	456
12, 21–22	410
12, 29	458
12, 37	410

Critias
Fr. DK
88B25	464

Democritus
Fr. DK
68A17a	244
68A33	77
68A99a	245
68A135	397
68A155a	371
68A165	309, 435
68B118	104
68B150	359
68B299h	77
68B300	77

Diodorus Siculus
Bibl. hist.
5, 3, 2	449

Diogenes of Apollonia
Fr. DK
64A33	435

Diogenes Laertius
Vit. phil.
1, 10	62
5, 23	84
5, 26	104
5, 44	431
5, 45	79, 371, 400, 405
5, 46	61, 62
5, 48	61, 79, 84
5, 49	79, 84
7, 132–133	109
9, 21	61
9, 47	104
9, 49	77
10, 27	62
10, 26–28	104
10, 104	348

Dioscorides
De mat. med.
3, 45, 1	490

Elias (David)
Comm. in Ar. Cat.
114, 2	172
114, 8	172
114, 12–13	83
114, 13–14	173

Empedocles
Fr. DK
31A69a	487
31A70	370
31A78	277, 438
31A91	487
31A92	309
31B17	301
31B64	277, 443
31B74	155
31B77	309
31B77–78	468
31B80	115, 382
31B81	277, 278, 381, 467
31B86	434
31B89	277, 310, 354, 434, 435
31B94	278, 486
31B101	277, 449

Ephorus
FGrHist
70, 56	410

Epicurus
Ep. ad Her.
35, 2–5	62

Fr. Usener
18–21	104
57–65	104
293	435
524	122

Erasmus
Adag. Chil.
2, 5, 43 (= 1443)	441

Etymologicum magnum
75, 26–44	481
812, 44–51	445

Euripides
TGF
895	277, 442, 444
941	118, 277, 463

Eusebius
PE
1, 7, 16	154
15, 20	379

Eustathius
Comm. ad Hom. Il.
629, 63–64	410
871, 36–38	410

Favorinus
Fr. Barigazzi
1–2	236

Flavius Josephus
Antiq. Jud.
15, 11, 3	389

Galen (Kühn)
Temp.
1, 509, 1–4	308

Nat. Fac.
2, 7, 2–3	357
2, 44–56	435

UP
4, 172, 13	465

Sem.
4, 633, 12	465

MM
10, 300, 11	389

SMT
11, 450, 14ff.	397
11, 471, 13–14	325
11, 474	81
11, 630, 2–4	374
11, 656, 11–14	459
11, 657, 2–4	458
12, 35, 3–7	374
12, 222, 15–223, 5	374

Ven. Sect. Er.
11, 168, 2–3	456

Comp. Med. Loc.
12, 381, 16–17	466

Gellius, Aulus
NA
Praef. 2	99
Praef. 2–4	171
Praef. 7	154
Praef. 9	96
Praef. 25	97
1, 26	193
3, 5	171
3, 6	171, 472
4, 11, 13	171
7, 13	200, 222
17, 11	171
19, 4	22, 189, 199
19, 6	190
20, 4	190

Geminus
El. astr.
17, 48	269

Geoponica
2, 47, 11	398
15, 2, 19	480
15, 3, 4	480

Glycas, Michael
Ann. (Bekker)
1, 9 (p. 19, 5–9)	398
1, 13 (pp. 25, 21–26, 5)	487
1, 16 (p. 31, 1–4)	372
1, 16 (p. 31, 12–13)	380
1, 62 (p. 118, 17–19)	425
1, 62 (p. 119, 22–120, 2)	441, 442

Heraclitus
Fr. DK
22A6, 15	379
22B12	281, 379
22B49a	379
22B86	253
22B91	379

Ps.–Heraclitus
Quaest. Hom.
35	410

538　　INDEX LOCORUM

Herodotus
Hist.
2, 22　　　　　　　　325

Herophilus
Fr. von Staden
50a (and b)　　　　249

Hesiod
Op.
486–489　　　　　　393, 422

Hippiatrica Berolinensia
14, 1, 11–13　　　　445

Hippocrates (and *corpus Hippocraticum***)**
Aer.
7　　　　　　　　　379, 474
7, 68–72　　　　　　390
7–8　　　　　　　　280
8　　　　　　　　　379, 381, 408
8, 6　　　　　　　　398
De diaet. in morb. ac.
7 (3, 5–14 Littré)　　77
De flat.
7　　　　　　　　　382
De prisc. med.
18–19　　　　　　　307
De sem., de nat. pu., de morb.
4, 20, 30–33　　　　466
De superfetat.
18　　　　　　　　　280, 456
De victu
2, 40　　　　　　　425
45, 7–10　　　　　　390
56, 1ff.　　　　　　 390
Epid.
6, 2, 5 (5, 278–280 Littré)
　　　　　　　　　　77
Prorrh.
2, 43　　　　　　　402

Hippolytus
Haer.
5, 20, 5　　　　　　274

Homer
Il.
2, 825　　　　　　　486
6, 136　　　　　　　410
9, 14　　　　　　　 486
9, 214　　　　　　　154, 372
9, 539　　　　　　　277, 443
13, 279　　　　　　 119, 277, 432
13, 284–285a　　　　432
19, 415　　　　　　 278
19, 415–416　　　　 275, 475
23, 200　　　　　　 475
24, 80–82　　　　　 426
Od.
4, 359　　　　　　　486
4, 567　　　　　　　447, 475
5, 322–323　　　　　277, 400
19, 446　　　　　　 277, 438
20, 158　　　　　　 486
Hymn. Dem.
6–8　　　　　　　　449
17　　　　　　　　　449

Hyginus Mythographus
Fab.
146　　　　　　　　449

Inscriptions
I.Eph. VII, 2, *3901*　286
I.G. II², 3816　　　286

Isidorus Hispalensis
Et.
12, 2, 24　　　　　484

Isocrates
Panath.
135　　　　　　　　236
Paneg.
188　　　　　　　　236
Hel.
12　　　　　　　　　236

Lucretius
De rer. nat.
2, 474　　　　　　　400
2, 1030–1039　　　　462
4, 678–680　　　　　456

5, 269 ff.	400	II, 14	409
5, 526–532	347	II, 17	409
6, 357–378	392	II, 21	409
6, 635 ff.	400		
6, 840–847	420		

Macrobius
Sat.
7, 3, 23   200

Manetho
FGrHist
609, 19–22   62

Martialis
Spect.
22   331

Mnesitheus of Athens
Fr. Bertier
16   282, 457

Moschus
Epit. Bionis
121   449

Oenopides of Chios
Fr. DK
41A11   420

Olympiodorus
Comm. in Ar. Mete.
158, 27 ff.   400

Oppian
Hal.
1, 429–432   428
5, 638–648   417

Orpheus
Fr. Kern
168   263

Palladius
Op. agr.
1, 37, 4–5   480
4, 9, 14   490
4, 15, 4   480

Papyri
P. Antinoopolis 85   181
P. Antinoopolis 213   181
P. Berol. Inv. 9764, 17–18
                           181
PL III 543A   181
P. Oxy. 2688–2689   181
P. Oxy. 2744   181
PSI inv. 2055   181

Pausanias
Graec. descr.
I, 32, 1   331
3, 20, 4   331
7, 18, 12   331
8, 23, 9   331

Petronius
Sat.
66, 5–6   446

Philochorus
FGrHist
328, 191   410

Philolaus
Fr. DK
44A7a   312

Photius
Bibl.
Cod. 175, 119b   99

Pindar
Fr. Snell
43   120, 277, 432
252   277, 481

Plato (and *corpus Platonicum*)
Apo.
19b   18
26d   18
27a   90

Charm.
164e–165a 320
Crat.
402a 379
Epinom.
981c 301
981d 370
984a 229
Leg.
639b 412
666a 467
Men.
76c 309
Phd.
74a 90
95e–99d 18
96a 5, 128
97b–99d 262
112ac 380
Phdr.
229e 320
230d 18
249c 317
250d 265
270a 113
Philebus
55e 311
Pol.
285d 77
Prot.
334b 372
Rep.
401cd 277
469e 482
491b–492a 143
491d 281, 370
495b 143
527b 317
530b 77
531c 77
532bc 317
546a 281, 370
603ab 277
607be 277
Soph.
218d 448
245b 77
261a 77, 448

Symp.
177b 236
Tht.
155d 243
162e 19, 90
180c 77
Tim.
27d–28a 313
28a 313
28c 262, 267
29a 262
29d 313
29e 262
30a 262
30c 256
33a 462
46ce 262
47ab 265
47e 262
48a 136, 262
52a 313
53b 262
55cd 170
59e 281, 304, 399
60de 374
65de 281, 401
67b 147
68e 227, 262
70c 313
76d 262
77ab 370
79a–80c 309
79e–80c 183, 355, 435
80c 256, 435
83c 374
86b 129
90a 281, 370
91a 313
92c 229, 313

Plautus
Rudens
588 410

Plinius Maior
NH
2, 135–136 392
2, 222 408

2, 224	372, 400	31, 31	378, 380
2, 226	374	31, 33	403
2, 234	406, 417, 419	31, 34	381
4, 79	408	31, 48	400
6, 51	408	31, 50	420
7, 64	480	31, 52	408
8, 83	484	31, 56	406
8, 98	456	31, 70	400
8, 101	456	32, 15	428
8, 122	436	32, 149	427
8, 128	446	36, 125	109
8, 130	331		
8, 205	441	**Plinius Minor**	
8, 207	438	*Ep.*	
8, 212	441	3, 5, 17	179
9, 83	428	8, 20, 1–2	462
9, 84	428		
9, 87	431	**Plutarch (and** *corpus Plutarcheum*)	
10, 185	390		
11, 16	478	*MORALIA*	
11, 44	480	*De aud. poet.*	
11, 45	477	16A	276
11, 61	480	16C	274, 277, 326
11, 224–225	432	17DE	277
13, 135	374	17DF	268
13, 139–142	374	26B	71
14, 73–75	409	28A	293
14, 78	409	28E	75
14, 83	458, 459	31E	206, 293
14, 120	409	*De aud.*	
14, 126	409	42A	191
15, 79–81	491	42C	111, 477
15, 106ff.	397	42E–44A	193
16, 222	471	42F	193
16, 223	471	43A	198, 206
17, 234	372	43C	194, 312, 443
17, 225	403	44BC	243
18, 91	403	44F	236
18, 152	372	47B	206
18, 275	403	48BC	194, 298
18, 361	385, 427, 428	48C	296
19, 156	490	*De ad. et am.*	
20, 169	456	49AB	319
25, 91	456	51D	433
27, 7	456	51D–53D	138, 429
28, 79	480	53D	436
28, 247	457	61A	467
29, 133	456	65E	319

De ad. et am. (*cont.*)
		142F–143A	307
66CD	253	143F	467
70E	29	144D	122, 127, 138, 479
71A	292	*Sept. sap. conv.*	
72A	320	146E	195

De prof. in virt.
		147C	483
75F	313	147EF	222
78E	112, 113	149CE	260

De cap. ex inim.
		153E–154A	76
86E	371, 389	154DF	202
89A	319	156D	204
90C	71	163D	238, 319, 428
92B	121, 138, 490	163EF	126, 268

De am. mult.
		164B	319
96F	120, 121, 139, 429, 432, 433, 436	*De sup.*	
		164E	253
96F–97A	138, 433	165C	253

De fortuna
		169AB	143
98BC	147, 328	169EF	258
		169F	253

Cons. ad Apoll.
		171AB	253
116D	319	171F	254
119D	128	*Reg. et imp. apophth.*	
121E	168	172BE	167

De tuenda
		Mul. virt.	
122F	196	243A	167
123A	447	253E	100
123E	467	254EF	489
126C	442	*Quaest. Rom.*	
129F	357	263E	183
130A	196, 221	264B	183
130BD	197	265B	489
130D	199	267E	343
130F	197	268CD	183
130F–131A	197	269CD	183
131A	199	269F	148
133B	220	270B	343
133BF	197	271F–272B	148
133C	198, 199, 200	271D	148
133CE	339	273E	183, 343
133DF	198	275C	343
133E	106, 119, 199, 200, 206, 209, 211, 232, 322, 351, 360, 493	275E	343
		276E	26
		277D	343
		279DE	169
133F–134A	197	279E	343

Coni. praec.
		279F	343
138C	168	280E	343
139D	476		

281F	169, 343	385D	319
282CD	183	386B	134
282EF	183	386E	31
283A	343	386F	462
283C	293	387A	339, 443
285BC	148	387B	346
286B	343	387F	300, 319
288B	183, 343	389F	71
288C	183	389F–390A	301
289C	183	391E	28
290AB	183	392A	28, 319
291A	343	392AE	28
Quaest. Graec.		392B	379
292CD	183	392E	318
292E	183	393A	318
293A	183	393D	317
Parall. Graec. et Rom.		394C	319
315CD	76	*De Pyth. or.*	
De fort. Rom.		394E	134
323C	463	395A–396C	50, 133, 134
326A	167	395F	83, 91
De Al. Magn. fort.		396AC	134
328A	167, 196	396E	343
Bellone an pace		397DE	134
350DE	112	399A	448
De Is. et Os.		400B	370
351C	268	400C	384
351CD	266	400DE	134
352D	388	400F–401A	134
352F	149, 389	401E	134
353F	403	402E	240
355D	253	402BC	134
364B	337, 467, 486	404B–405A	126
367E	451	406E	111
372F	258	408E	319
375C	71	409CD	226, 227, 463
376B	435	*De def. or.*	
378A	253	410B–411D	133, 135
379E	253	411D	238
381B	268	411EF	67
382AB	66, 125, 268	412D	196
382D	71	414A	293
De E		420C	238
384E	177, 179	421E	286
385A	191	421E–431A	170
385AB	193	424B	103
385B	28	424C	300, 325
385C	243	424D	309

De def. or. (*cont.*)
426D	462
426DE	123, 255
430A–431A	301
430E	258
430E–431A	267
431A	319
433F	461
435F	103
435F–436A	261, 263
436D	103
436DE	262, 264
436E	354
437C–438D	134, 302
438D	315
438CD	302

De virt. mor.
440E	193, 292
451C	407
452A	432

De coh. ira
457D	168
457DE	168
458F–459A	82
461F–462A	279

De tranq. an.
464E	166, 178
464E–465A	165, 166, 172
464F	105, 166, 167, 168, 169, 170, 178
467C	483
472D	283
475EF	413
475F–476A	138
477CD	228, 231, 266

De am. prol.
493A–495A	124
493B	124, 125

De gar.
503A	489
511B	319
514D	138, 482

De cur.
515C	274
517CE	230
517D	264, 268
518D	285
520F	449

De vit. pud.
534F	112

De se ipsum laud.
544A	291

De sera num.
548AB	195
550DE	265
556F–557F	272
558E	136
559C	379
563B–568A	271
563F–564A	463
565C	436

De fato
568F	106

De genio Socr.
588E	147
591C	270
591DE	370

De exilio
600F	370
601A	463

Cons. ad ux.
610C	467

Quaest. conv.
612D	204
612DE	152, 162, 165
612E	29, 100, 156, 160, 162, 172, 200, 203
613C	191
613D	202
613E	199, 201, 202, 359
613F	191
614A	152, 201, 204, 209, 267
614C	201
614CD	339
614D	200, 202, 206
614DE	351
614E	200, 201, 206, 359
614F	201
615B	201, 206
615C	160
615CD	203
616F	202
619BF	164
620A	164
620A–622B	193

620E	307	643AB	202
621A	202	643C	202
621B	202	644D	202
621C	204	645C	160
621DE	195	645D–646A	192
623D	97	645D–649F	28
625AC	164	646A	28, 221
625C	293	646B	203
626D	307	647E	449
626EF	285	648CD	144, 424
626E–627F	360	648D	307, 467
626F	285, 326	649A	28
627AB	371	649CD	466
627AC	399	649D	308, 309, 468
627AD	25, 139, 370	649EF	144
627B	325	650A	82
627BC	374	651A	388
627C	373	651C	326
627D	372	651E	479
627E	275	651F	296
627EF	89, 372, 488	652B	305
628BD	244	652B–653B	410
628D	221	652F	411, 466
629A	324	653B	296
629C	222	654D	265
629D	98, 100, 160, 162, 165	654E	64
		655D–656B	294
629E	156	656A	293
629E–634F	164	656B	326
629F	195	656C	72, 489
634F	204	656D	293
635C	302	657B	202
635CD	89	657BE	331
637D	357	657E	271
640C	372	657F–659D	139
640F–641A	375, 465, 471	658B	116, 290, 451
641AE	352	658C	458
641B	256	659B	140, 452
641C	41, 247, 313, 314, 335, 435	659C	403
		660AB	204
641CE	208	660C	202
641D	458	660D	156, 160, 196
641F–642B	352	661BC	139
642A	293	661BD	380
642AB	208	662D	326
642C	447	663B	380
642BC	149	663F	139
643A	202	664A	173

Quaest. conv. (*cont.*)

664A–665A	206, 228, 257
664A–666D	241
664B	378, 393
664B–665A	208, 209
664BC	463
664C	241, 247, 489
664CD	257
664D	25, 222, 324
664DE	361, 391, 392, 393
664D–665A	206, 208
664D–665C	139
664E	371, 399, 468
665A	207, 209, 267
665C	158, 159, 241, 489
665CD	488
665E	224
666A	139
667C–669E	344
668D	222
669B	389
670B	206
670F	403
671BC	271
672C	331
672D–673A	200
673A	203
673D–674C	89
673E	477
674E–675D	89
675B	177
675DC	291
675EF	134
676B	381, 392
677E–678B	89
679DE	271
680CD	242, 340
680D	237, 245, 249
680F–681A	310
681A	443
681C	144
681DE	483
682A	222
682B	293
682BC	293
683D	382
683E	115, 274
684A	389
684C	374, 399
684D	490
684E	139, 158, 173
684E–685F	156, 157, 267, 372
684F	275
684F–685A	158
685A	390
685AB	158
685B	388, 389
685BC	372, 411, 489
685BD	139
685C	489
685D	25, 214
685DE	91, 154, 205, 390, 398
685DF	154
685EF	156, 271
686D	152, 162, 165, 195
686E	160
687BC	309
687D	388
687DE	325, 456
687E	389
688A	456
689AB	309
689B	104, 310
689BC	309
689C	363, 388
689E–690B	83
690A	389, 390
690B	474
690DE	89
690F	82, 247
690F–691C	89
691D–692A	89
692E	361
693E–695E	141, 149
694B	222, 412
694D	295, 326
694DE	308
695E	139
696AB	417
696B	414
696EF	491
697A	374
697B	389, 425
697D	201, 372
697DE	202

697E	156	718E	312
697F	170	718EF	260, 317
697F–700B	171, 314	719EF	222
698A	276	720B	262
698B	262	720C	29, 271
698D	389	720C–722F	28
698E	242	720E	227, 300, 301
698EF	275	721EF	147, 435
699A	275	721F	357
699AB	388	722E	350
699B	267, 274, 354, 363	723A	173, 203
699D	313, 314	723E–724D	134
700B	71, 74, 314, 328	724E	135, 138, 375, 465
700BC	173	724EF	171, 174, 469, 471
700C	85, 170	725C	313
700D	136, 175, 245, 247, 248	725CD	381
		725D	474, 487
700DE	175	725F	173
700E	174, 175, 293	727A	173
700EF	330	728A	476
700F	174, 438, 491	728E	213
701A	175, 242, 247	728F	268
701F	305, 326, 410	730B	171
702A	191	731A	206, 402
702BC	414, 417	733B	374
702C	447	733D	456
702D	220	734CD	79, 160, 190, 196, 219, 250
702D ff.	169		
704C–706E	331	734D	292, 295, 462
705B	191	734F	69
705E	171	734F–735A	296
706E	173, 224	735C	296
710C–711A	331	736C	76, 156, 160, 174
712A	196	736D	29
713C	191	737D	76
713F	351	738C	97
714C	271	739D	271
716BC	271	739E	170
716D	203	740F	173
716F	202	741B	331
716F–717A	410	741C	324
717A	160, 173	741D	97
717E	213	743BC	271
718AB	271	744D	300
718B	170, 211	744F	286
718C	213, 287, 296	745A	468
718D	265	747A	271
718DE	229, 300	747B	28

INDEX LOCORUM

Quaest. conv. (cont.)		17, 915F–916A	426
748D	76, 162	18, 916AB	426–428
Amatorius		19, 916BF	429–437
753A	491	20, 916F–917B	437–439
756B	240	21, 917BD	439–445
757E	370	22, 917D	446–447
761A	489	23, 917EF	448–449
761B	76, 286	24, 917F–918A	450–452
766A	226, 227, 463	25, 918AB	452–453
767A	433	26, 918BE	454–457
Maxime cum principibus		27, 918EF	457–459
776F–777A	135	28, 918F–919A	459–460
Ad princ. iner.		29, 919AB	460–464
780D	463	30, 919BC	464–465
781A	268	31, 919CE	465–469
Praec. ger. reip.		32	469–472
801A	138, 456	33	473–474
816DE	217	34	474–476
821B	477	35	476–477
De Her. mal.		36	477–481
855EF	237	37	482–483
Plac.		38	483–485
874D–911C	61	39	485–487
880B	385	40	488–489
900BD	19	41	489–491
905A	358	*De facie*	
909C	438	920B–945E	61
910B	370	922DE	486
Q.N.		922E–923A	216
(the list only includes separate		922F–928D	66
discussions of the problem chapters		923A	264
in the commentary)		923C	460
1, 911CF	368–375	927CD	256
2, 911F–912D	375–385	928AC	143
3, 912DF	385–390	929A	290, 451
4, 912F–913A	390–394	929B	191
5, 913AE	394–401	930C	324
6, 913EF	401–404	930F	305
7, 913F–914B	404–406	931B	487
8, 914B	406–407	932BC	83
9, 914BD	407–408	932D	340
10, 914DE	409–411	933A	313
11, 914EF	411–413	933CD	324
12, 914F–915B	413–418	935DE	300
13, 915BC	418–420	937CD	195
14, 915CD	421	938C	210, 246
15, 915DE	422–423	939D	374
16, 915EF	423–425	940A	139, 140, 452

INDEX LOCORUM

940F–945D	64, 271	960B	195
942C	191	961A	68
De prim. frig.		962F	285
945F	66, 267, 303, 357	964D	191
945F–948A	69, 357, 462	965C	286
946A–948A	141	965DE	193
946C	64	966B	125
946DE	307, 308	966D	448
946EF	266	967AB	428
946F	318	967D–968B	425
947E	301, 302	967E	285
947F	286, 306, 341	967F	269
948B	317	971D	453
948BC	205, 229, 316, 331	974B	455
948C	283, 317	974BD	139
948CD	68, 69	975AC	268
948D	64, 318	975B	66, 125
948E	415	975DE	313
949B	309, 442	976C	268
949F	69	976E–977A	27, 425
950A	337	976EF	420
950AB	415, 486	976F	420, 426
950B	70, 139, 415, 416, 417	977A	426
		977D	331
950BC	329, 414, 415	978E	432, 433
951A	453	978EF	27, 138, 139, 425, 429, 433, 436
951BC	420		
951D	301, 306	979B	27, 425, 428
951F	324	982E	383, 384
952A	318	982F	481
952B	325	*Gryllus*	
952C	91	989Cff.	124
952C–955C	293, 461	990BC	479
952D	319	991E	139, 456
952F	486	991EF	455
954C	412	992A	483
955A	301	*De esu*	
955C	69, 91, 315, 318, 320, 327, 328, 344, 348	997E	432
		Quaest. Plat.	
		1000B	216
Aqua an ignis		1000E	194
957A	301	1000E–1001C	178
957B	267	1001D	328
957D	381, 474	1002E–1003B	178
958E	64, 265	1003A	177, 213
De soll. an.		1003B–1004C	178
959B	290	1004D	309, 350
959C	23	1004D–1006B	178

INDEX LOCORUM

Quaest. Plat. (*cont.*)
1004DE	309
1004D–1006B	178, 183, 309
1004E	82, 309
1005B	435
1005BD	183, 256, 355, 435
1005D	309, 435
1005F	309
1006B–1007E	178
1006D	71
1006F	105

De an. procr.
1012B	178, 193

De Stoic. rep.
1033B	293
1035D	110
1035F–1036A	344
1040A–1041B	71
1041A	71
1041E	407
1045F	68
1047CD	314

De comm. not.
1060Bff.	192
1070C	407
1070F	293
1071D	197
1072F	293

Non posse
1086D	191, 195, 213
1086E	71
1091D	214
1095CD	203
1096BC	203
1096C	290

Adv. Col.
1108Bff.	192
1108BC	283
1114C	328
1115B	68, 167, 218
1115C	246
1115E	317
1117D	320
1118C	319
1122F	328
1124A	344
1124B	320
1124C	71, 283

De lat. viv.
1129B	122, 123, 124
1129D	381, 474

Pars an fac.
2, 16–17	357
6, 15–17	357

Fr. Sandbach
13–20	84, 106, 183, 269, 383
24	274
68	392, 393, 422
75	420
97	479
106	438, 444
122	489
122–127	75, 184
127	184, 275
136	384
156	270
157	272
179	154
215f	357
216g	68

Lampr. cat.
24	269
40	269
42	275
43	274
45	344
51	62
62	154
63	320
99	62
119	84, 269, 383
125	85
126	84
132	315
136	85
138	84, 102
139	84
149	84
160	84
161	84
166	84, 102
167	84
170	85

177	319	6, 1	226, 464
183	61, 62	6, 2–4	259
191	456	6, 4	64
193	62, 84	8, 1	71
196	62	8, 1–2	113
200	62	13, 7–8	253
200a	61	16, 7	219, 312
212	62	33, 5	139, 413
218	61, 85, 102	35, 2	143, 189
		39, 2–3	142

VITAE
  Thes.

		16, 6	245
1, 3	273	146	449
3, 2	489		

Rom.

Nic.

		1, 5	132, 168
3, 1	89	23	143
15, 7	26, 102, 103, 148, 179	23, 2–4	260
		23, 3	18
21, 8	76		

Comp. Nic. et Crass.

Lyc.

		5, 3	254
1, 1	489		

Alc.

Sol.

		1, 5	214
11, 2	167	6, 2	214
32, 4	489	23, 4	433
		23, 4–5	138, 139, 429

Them.

2	114	23, 5	436
32, 6	29		

Cor.

Cam.

		15, 4	214
6, 4	254	38	252
6, 5–6	252, 253	38, 2	260
6, 6	319		

Comp. Alc. et Cor.

19, 8	26, 148, 179	3, 2	489
22, 3	489		

Lys.

33, 7	167	2, 3	143

Arist.

		12	145, 149, 227
6, 2–3	123	12, 6	460
19, 7	260	12, 7	145, 148, 180

Ca. Ma.

		18, 1	133
21, 5	461		

Sull.

25, 4	204	5, 5	167

Comp. Arist. et Ca. Ma.

Pomp.

2, 4	76, 489	25, 6–7	147

Per.

Marc.

4–5	113	14	260
4, 6	19	17	260
4, 6–5, 1	43		

Dion

5, 1	113, 231	2, 4–7	142
6	258	21, 9	132

Brut.
25, 4–6 — 141
25, 6 — 149
Timol.
15, 11 — 142
Aem. Paul.
14 — 141, 145, 461
14, 11 — 146
15, 9–11 — 312
17, 7–13 — 143
Dem.
2, 2 — 177
2, 2–4 — 279
Cic.
40, 2 — 355
Comp. Dem. et Cic.
3, 1 — 167
Alex.
1 — 245
1, 2 — 145
7, 2 — 71, 218
7, 4 — 196
17, 9 — 489
27, 2 — 380
35 — 144
35, 10–12 — 273
35, 15 — 424
35, 16 — 132, 145
75, 1–2 — 464
Caes.
63, 2 — 147
Phoc.
2, 3 — 199
2, 6–9 — 136
3, 1–3 — 136
Demetr.
38, 4 — 139, 413
Ant.
15, 5 — 167
66 — 373
Mar.
19, 2 — 461
Flam.
10, 6 — 146, 460
Arat.
29, 6 — 139, 143, 188, 412
Art.
28, 1 — 467

Galba
2, 5 — 245
20, 1 — 279

Pollux
On.
2, 69 — 387, 388

Porphyry
Quaest. Hom. ad Il.
9, 682 — 76

Proclus
Hyp.
4, 74 — 405

Psellus, Michael
De omn. doctr.
168, 3 — 371
169, 7–10 — 418
170 — 15, 487, 488
187 — 469
188 — 15, 487, 489
De lap.
26 — 435

Quint.
Inst. or.
2, 20, 3 — 486
8, 2, 12–13 — 361
9, 1, 11 — 486

Semonides Amorgensis
Fr. West
7, 83–93 — 478

Seneca
Ad Helv.
8, 4 — 463
Ben.
4, 23 — 463
7, 1, 5 — 109
7, 1, 7 — 109
Brev. vit.
13, 1–3 — 110
De prov.
1, 2–4 — 255

Dial.
  5, 30, 1   331
Ep.
  88, 36   110
  118, 14   19
NQ
  1, 1, 3   463
  1, 15, 1–3   463
  2, 31, 2   108, 489
  2, 57, 2   108, 392
  2, 58   109
  3, *Praef.* 18   110
  3, 21, 2   380
  3, 25, 11   403
  3, 27, 3   109
  3, 5   108, 400
  3, 11   109
  3, 16   109
  3, 20   109
  3, 25, 11   108
  6, 13, 2   108
  6, 17, 3   109
  6, 32, 1   110
  7, 1–4   108, 462
  7, 2   464

Sextus Empiricus
HP
  1, 31 ff.   320
  1, 57   456
  1, 64–72   483
  1, 71   456
  1, 192–193   320
  1, 200–201   320
  1, 202–206   344

Solinus
Mem.
  2, 36   484

Sophocles
TGF
  787   231

Spartianus Aelius
Hadr.
  20, 2   76

Stesimbrotus
FGrHist
  107, 1   114

Stobaeus
Flor.
  1, 29, 2   392

Strabo
Geogr.
  2, 3, 8   109, 257
  8, 6, 12   410
  15, 3, 10   471
  16, 3, 6   374

Strato
Fr. Wehrli
  89   420

SVF
  1, pp. 100–101, frs. 451–453
                                 160
  1, p. 116, fr. 515   285
  1, p. 116, fr. 516   159
  2, p. 197, fr. 663   451
  2, p. 206, frs. 722–723
                                 159
  2, pp. 232–233, fr. 863
                                 265
  2, p. 233, fr. 866   265
  2, p. 333, 1154   159
  3, p. 17, fr. 68   110
  3, p. 146, fr. 546   285
  3, p. 205, lx   110
  3, p. 244, fr. 5   483
  3, p. 251, fr. 47   285
  3, p. 251, fr. 48   281, 285, 484
  3, p. 255, fr. 63, 15   307

Themistius
Or.
  26, 329c   113

Theocritus
Id.
  1, 105–107   278, 480
  3, 13c   481

Theognis
El.
213	432
215–216	120, 277, 432

Theophylactus Simocatta
Quaest. phys.
7	414
9	428

Theophrastus
CP
2, 3, 3	145
2, 5, 3	370
2, 5, 5	377
2, 7, 3	145
2, 8, 2–3	459
2, 9, 5	491
2, 9, 7	425
3, 8, 3	377
3, 21, 1	423
3, 21, 2	422
3, 21, 4	424, 425
3, 21, 5	420
3, 22, 1–2	403
3, 24, 4	466
4, 4, 1	424
4, 9, 4–5	394
4, 13, 2	423
4, 13, 4	421
4, 13, 4–5	422
4, 13, 5	424
4, 14, 3	403
5, 6, 10	490
5, 9, 10	465
5, 15, 6	372
6, 1, 2	396
6, 1, 4	397
6, 1, 6	397
6, 4, 1	396, 397, 400
6, 4, 2	396
6, 4, 6	386, 398
6, 5, 1	479
6, 10, 1–2	334, 396, 397
6, 10, 2	371, 375
6, 17, 5	453
6, 20, 4	449

HP
2, 7, 6	465
4, 4, 1	145
4, 6	374
4, 7, 3	239
4, 7, 2	374
4, 7, 8	377
4, 14, 6	465
4, 16, 5	372
5, 6, 1	471
7, 5, 2	378
8, 1, 1	420
8, 1, 4	422
8, 4, 1–6	420
8, 4, 5	394
8, 6, 6	393
8, 7, 3	377
8, 8, 2	378
8, 9, 1	422
8, 10, 2–3	421
8, 10, 4	425

Ign.
16	420

Lap.
67	409

Mete.
192 FHSG	183

Od.
40	453

Sign.
15	385
40	385, 428

Sens.
20	434
65–67	397
67, 27	397

Vent.
38	475
40	476

Fr. FHSG
26a	84
137	62
137, 26	79
173	281, 308, 419
214a, 13–17	311, 405
214a, 26–30	377
214c	281, 311, 405
263	238

		Xenophon	
362a	248	*Cyn.*	
365a	436, 437	5, 1–2	453
365b	431	5, 1–5	289, 437, 448
365c	281, 302, 431	5, 3	450
365d	437	5, 4	451
		5, 5	449
Varro		5, 33	203, 290
De ling. Lat.		10, 8	438
5, 79	427	10, 15–17	438
De re rust.		11, 1	331
1, 23–24	422	*Cyr.*	
3, 16, 6	480	7, 5, 11–12	471
		Mem.	
Vergilius		1, 2, 30	444
Georg.		1, 4, 6	387
4, 230	477	*Oec.*	
4, 241–242	477	7, 32–35	478

www.ingramcontent.com/pod-product-compliance
Ingram Content Group UK Ltd.
Pitfield, Milton Keynes, MK11 3LW, UK
UKHW021833210426
5322IPUK00004B/160